ELECTRONIC CIRCUITS

Devices, Models, Functions, Analysis, and Design

Electronic Circuits

Devices, Models, Functions, Analysis, and Design

Mohammed S. Ghausi
New York University

D. VAN NOSTRAND COMPANY
New York Cincinnati Toronto London Melbourne

D. Van Nostrand Company Regional Offices:
  New York, Cincinnati, Millbrae

D. Van Nostrand Company International Offices:
  London, Toronto, Melbourne

Copyright © 1971 by Litton Educational Publishing, Inc.

Library of Congress Catalog Card No. 70-154319
ISBN: 0-442-22653-5

All rights reserved. No part of this work covered by the
copyrights hereon may be reproduced or used in any form or by
any means — graphic, electronic or mechanical, including
photocopying, recording, taping, or information storage and
retrieval systems — without written permission of the publisher.
Manufactured in the United States of America.

Published by D. Van Nostrand Company
450 West 33rd Street, New York, N.Y. 10001

Published simultaneously in Canada by
Van Nostrand Reinhold Ltd.

10   9   8   7   6   5   4

To Marilyn, Nadjya, and Simine

Preface

This book is intended as a text for junior/senior level electrical engineering courses in electronic circuits. The prerequisite to the use of this text is a one-semester course in elementary circuit theory. Previous or concurrent exposure to a one-semester course on semiconductor physical electronics is helpful but not necessary.

The primary aim of this book is to introduce and familiarize undergraduate students with the analysis and design of discrete-component and integrated electronic circuits. Both linear and nonlinear circuits are covered. Field-effect transistors, of both the junction and insulated-gate types, as well as bipolar transistors are discussed concurrently.

Approximate, simple, paper-and-pencil design techniques are covered in some detail in order to understand the basic features and gain insight into and familiarity with the circuit functions. Once the proper role of the circuits and devices is understood, validity and improvement in the accuracy of circuit performance are treated by using more accurate models and computer-aided design. No particular program is specified where computer-aided analysis and design are mentioned, in order to maintain flexibility in the use of this text. It is desirable for the student to solve some of the more complex problems with a computer, especially using some of the packaged programs—if available. Access to a digital computer, however, is not essential to the use of this text.

The book is designed in the following manner for a two-semester course. The first-semester course may cover Chapters 1 through 5 and, perhaps, parts of Chapter 9 along with Appendices A and B. The first two chapters are in the form of a review, to refresh the reader and to introduce the notations used in the text. If the background of the students is adequate, most of the first two chapters may be skipped. Sections 2-5 to 2-8 may then be covered at a later time, along with amplifier response characteristics and design. (The author has found that the inclusion of these topics is very useful and desirable.) Thus, the subject matter of the text really begins in Chapter 3. Even this chapter can be covered rather quickly if the students have some familiarity with *p–n* junctions. Parts of Chapter 9, specifically piecewise-linear analysis using junction diodes and large-signal models (Sections 9-1 to 9-5 and Section 10-17), may be covered along with Chapter 3, if desired.

The second-semester course might begin with Chapter 6 and cover the rest of the text. Again, there are some topics, such as signal-flow graphs, crystal filters, active-distributed *RC* filters (or even the entire chapter on active *RC* filters), blocking oscillators, etc., which may be omitted if sufficient time is not available to cover the other topics adequately.

A large degree of flexibility exists in the choice and sequence of topics for a one-year course using this text. More material is purposely included in the text than can be covered in a one year course.

The devices emphasized in the text are bipolar transistors and field-effect transistors with vacuum-tube devices used only occasionally. A unified approach is stressed throughout the text so that FET devices are not covered in just one chapter and then abandoned. Also, the physics of the devices is kept to a minimum in order to concentrate on electronic circuitry. Many illustrative problems are included in the text and practical parameter values are used to give the student a feeling of the real values encountered in practice.

The contents of the book are arranged as follows.

Chapter 1 covers in review form the fundamentals of circuit theory. In particular, methods of circuit analysis are covered together with some important network theorems.

Chapter 2 deals with two-port networks and related topics. Two-port descriptions, power-gain calculation, Bode plot, and frequency and transient response determination are discussed. Transmission criteria such as maximally flat magnitude and linear phase responses are also included.

Chapter 3 introduces various devices with their *V-I* characteristics, models, and biasing. The devices include junction diodes, tunnel diodes, breakdown (Zener) diodes, bipolar transistors, field-effect transistors of both the junction (JFET) and insulated-gate (IGFET or MOSFET) types, unijunction transistors (UJT's), semiconductor controlled rectifiers (SCR's), and vacuum-tube devices, such as triodes and pentodes. In order not to confuse the student with many transistor models, only the hybrid π incremental model for the common-emitter transistor circuit is used throughout the text. Biasing techniques and stabilization are also discussed for discrete-component and integrated circuits.

Chapter 4 is concerned with the analysis and design of single-stage amplifiers. Low- and high-frequency responses, step response, and gain-bandwidth calculations are discussed together with some simple broadbanding techniques.

Chapter 5 considers the analysis and design of multistage amplifiers. Calculation of the effect of cascading interacting and noninteracting stages on the frequency and transient response of the amplifier by simple approximate methods and more accurate computer-aided analysis is also discussed. DC and differential amplifiers, which are often used in integrated circuits, are also included. Input and output stages in a multistage amplifier, which often require different considerations, are discussed in some detail. For example, noise calculations are included with the input stage, and power amplifier design is included with the output stage discussions.

Chapter 6 deals with elementary feedback theory and design-oriented analysis of feedback amplifiers. Operational amplifiers are discussed, together with their applications. The root-locus technique and stability criteria for linear systems are also included. This chapter also covers briefly *RC*, *LC*, and crystal oscillators.

Chapter 7 covers tuned amplifiers. The causes of instability and their cures are discussed. Analysis and design with mismatch, using practical examples, are also included. Crystal bandpass filters are discussed briefly. Low-frequency tuned amplifiers using active *RC* circuits are introduced, along with sensitivity considerations.

Chapter 8 introduces the subject of active *RC* filters. Various techniques used in the design of integrated-circuit filtering functions are discussed at an elementary level with practical examples. The use of distributed *RC* networks in active filters is also discussed briefly.

Chapter 9 covers elementary analysis of nonlinear electronic circuits. Piecewise-linear analysis, graphical methods, such as isocline and phase-plane analyses for the first- and second-order nonlinear differential equations, and numerical solutions of higher-order nonlinear differential equations are introduced with illustrative examples.

Chapter 10 considers regenerative, switching, and waveshaping circuits. The switching speed of the transistor from off to on, and vice versa, is discussed. The use of transistors in multivibrators, blocking oscillators, and time-base generators is discussed with pertinent waveforms. Waveform shaping for digital circuits is also discussed.

Chapter 11 is concerned with logic circuits and digital integrated-circuit functional blocks. The binary number system, Boolean relations, and the various logic gates used in computer circuitry are discussed briefly. The use of gates to realize flip-flops and the application of flip-flops in shift registers and counters are also included.

The book includes four appendices. Appendix A covers *p–n* junction properties in a deliberately brief manner, providing just the minimum semiconductor background necessary to understand the *V-I* characteristics and operation of various *p–n* junction devices discussed in

Chapter 3. Appendix B deals with fabrication of integrated circuits (IC's). The various IC components and design guidelines are discussed to familiarize the student with the different aspects of discrete-component and integrated-circuit design. Appendix C introduces the signal-flow graph, which may be covered together with Chapter 6. Some typical semiconductor device characteristic sheets and IC functional-block data sheets are provided in Appendix D. Other important topics, such as modulation and detection, are not covered in this text because these are usually covered in some detail in other undergraduate courses.

Each chapter is complemented by a large number of problems which are carefully made up to reinforce the understanding of the student. Some of the problems are taken from the professional literature and often simplified; however, the references for these are intentionally not given in order to avoid unprofitable search for the solutions. The author wishes to acknowledge the contributions to these problems of the various authors. Problems in which only numbers are to be plugged into the formulas are kept to a minimum, since they do not add much to the understanding of the subject matter discussed in the body of the text. Answers to selected problems are also included at the end of the book. Because of the design-oriented approach, the book should be useful to practicing engineers for self-study.

Part of this book was written in the École Polytechnique Fédérale de Lausanne in Switzerland, where the author was on sabbatical leave for one term.

I am indebted to many of my friends and colleagues for their help and advice. These include Professors L. Chua, A. Lo, J. Narud, Drs. J. Golembski and D. Hodges, Professors J. Kelly and R. Williams also offered comments and suggestions for improving the presentation of the material while teaching the course for which this book is written. Professor L. Huelsman reviewed the entire manuscript and I express my appreciation for his numerous helpful suggestions and comments. Thanks also are due to Mr. K. Laker and Dr. K. Sakuda, my graduate student assistants in the course, who provided much help in various forms.

The secretarial staff of the NYU Electrical Engineering Department is, of course, to be thanked: Mrs. Onna Shaw, Mrs. Nora Roback, Mrs. Pearl Granoff, Mrs. Rose Slotkin, and Mrs. Emily Colavito did a tremendous job in typing the final version of the manuscript. Lastly (but not least!), I should acknowledge my wife's continuous encouragement and support, which were instrumental in the completion of the project.

ENGLEWOOD, NEW JERSEY *M. S. Ghausi*
FEBRUARY, 1971

4-4	Low-Frequency Response Characteristics Due to Bias and Coupling Circuitry in BJT's	194
4-5	Low-Frequency Response of FET's and Vacuum Tubes Due to Bias and Coupling Circuitry	200
4-6	Step Response of a Typical High-Pass Circuit and Sag Calculation	203
4-7	High-Frequency Response Characteristics of BJT's	206
4-8	The Miller Effect and the Unilateral Approximation	209
4-9	Gain and Bandwidth Calculation for BJT's Using the Unilateral Model	213
4-10	The Dominant Pole-Zero Concept and Approximation	216
4-11	High-Frequency Response of FET's and Vacuum Tubes	220
4-12	Gain-Bandwidth Relations in BJT's	221
4-13	Simple Broadbanding Techniques	224
4-14	Step Response of an Amplifier Stage	235
4-15	Summary	237
	References and Suggested Reading	241
	Problems	241

CHAPTER 5 ANALYSIS AND DESIGN OF MULTISTAGE AMPLIFIERS   249

5-1	Introduction	249
5-2	Gain and Bandwidth Calculations for an *RC*-Coupled FET Multistage Amplifier	252
5-3	The Effect of a Number of Dominant Poles on the Bandwidth, Frequency Response, and Step Response of an Amplifier	258
5-4	Gain and Frequency Response Calculations for *RC*-Coupled Cascaded *CE* Transistor Stages	263
5-5	Midband Gain and Impedance Calculation in Multistage Circuits	269
5-6	Design of *RC*-Coupled *CE* Transistor Amplifiers for Specified Gain and Bandwidth	273
5-7	Multistage Broadband Transistor Amplifiers	276
5-8	Stagger-Tuning and Synchronous-Tuning in Amplifier Design	282
5-9	DC and Differential Amplifiers	285
5-10	Input Stage Design Considerations: Noise	292
5-11	Noise Models	294
5-12	Noise Figure	297

5-13	Output Stage Design Considerations: Power Amplifiers	302
5-14	Class A Power Amplifiers	303
5-15	Distortion Characterization and Calculation	309
5-16	Class B Push-Pull Power Amplifiers	312
5-17	Summary	317
	References and Suggested Reading	318
	Problems	318

CHAPTER 6 FEEDBACK AMPLIFIERS AND OSCILLATORS 327

6-1	Basic Feedback Concepts and Definitions	328
6-2	Feedback Configurations and Classifications	330
6-3	Advantages and Disadvantages of Negative Feedback	344
6-4	Design-Oriented Analysis of Some Feedback Amplifier Circuits	348
6-5	The Root-Locus Technique	361
6-6	Stability Considerations in Linear Feedback Systems	370
6-7	Operational Amplifiers and Applications	376
6-8	Practical Considerations in Operational Amplifiers	382
6-9	Linear Analysis of Sinusoidal Oscillators	384
6-10	RC Oscillators	385
6-11	LC Oscillators	388
6-12	Crystal Oscillators	389
	References and Suggested Reading	390
	Problems	390

CHAPTER 7 ANALYSIS AND DESIGN EXAMPLES OF BANDPASS AMPLIFIERS 399

7-1	Introduction	399
7-2	Single-Tuned Interstage	401
7-3	Impedance Transformation and Transformer Coupling	408
7-4	Transistor Single-Tuned Amplifier with Tuned Circuit at the Output	409
7-5	The Narrow-Band Approximation	412
7-6	Synchronous-Tuning and Stagger-Tuning	414
7-7	Double-Tuned Interstage	416
7-8	Oscillation Possibilities in Tuned Amplifiers	420

7-9	Maximum Frequency of Oscillation of a Bipolar Transistor	421
7-10	The Cascode Tuned Amplifier	423
7-11	Neutralization in Narrow-Band Tuned Amplifiers	427
7-12	Mismatch in Tuned Amplifiers	429
7-13	A Design Example of a Nonunilateral Tuned Amplifier	431
7-14	Crystal Bandpass Filters	437
7-15	Tuned Amplifiers Using Active-*RC* Circuits	444
7-16	Sensitivity in Active-*RC* Circuits	448
	References and Suggested Reading	453
	Problems	453

CHAPTER 8 ACTIVE RC FILTERS   461

8-1	*RC* Driving-Point Functions	463
8-2	Controlled-Source Realizations (*RC*: −*RC* Decomposition)	468
8-3	Realization by Cascade of Simple Second-Order Networks	475
8-4	The Negative-Impedance Converter as a Network Element	478
8-5	Transfer Function Synthesis Using Negative-Impedance Converter	482
8-6	The Gyrator as a Network Element	485
8-7	Operational Amplifier Realizations	487
8-8	Low-Sensitivity High-Q Transfer-Function Realizations	490
8-9	Distributed *RC* Circuit as a Network Element	494
8-10	Applications of Distributed *RC* Networks	498
8-11	Active-Distributed *RC* Low-Pass Filters	499
8-12	Summary	511
	References and Suggested Reading	513
	Problems	514

CHAPTER 9 ELEMENTARY ANALYSIS OF NONLINEAR ELECTRONIC CIRCUITS   520

9-1	Introduction	520
9-2	Modeling of Nonlinear Resistive Circuits by Piecewise-Linear Approximation	521

9-3	Synthesis of Piecewise-Linear Functions	524
9-4	Piecewise-Linear Models of Some Electronic Devices	529
9-5	Piecewise-Linear Analysis	531
9-6	Negative-Resistance Characteristics	537
9-7	Relaxation Oscillators	539
9-8	Nonlinear Analysis Techniques	540
9-9	First-Order System, Isocline Method	541
9-10	Second-Order System, Phase-Plane Diagram	543
9-11	Lienard's Method	547
9-12	Example of a Nonlinear Oscillator: Van der Pols' Equation	551
9-13	Numerical Method of Solution for the nth Order Nonlinear System	553
9-14	Nonlinear Reactance	557
	References and Suggested Reading	558
	Problems	559

CHAPTER 10 REGENERATIVE, SWITCHING, AND WAVESHAPING CIRCUITS   565

10-1	Introduction	565
10-2	The Transistor Switch: On–Off Time Interval Calculations	566
10-3	Circuitry to Improve the Switching Time of a Transistor	574
10-4	Classification of Regenerative Switching Circuits	577
10-5	Bistable Multivibrators	579
10-6	Monostable Multivibrators	584
10-7	Astable Multivibrators	586
10-8	Application of Multivibrators	588
10-9	Blocking Oscillators	588
10-10	Monostable Blocking Oscillator	589
10-11	Astable Blocking Oscillator	592
10-12	Time-Base Generators (Sweep Circuits)	594
10-13	Free-Running Time-Base Generators	595
10-14	Triggered Voltage Time-Base Generators	597
10-15	Triggered Current Time-Base Generators	599
10-16	RC Waveform Shaping	600
10-17	Clipping and Clamping Circuits	602
	References and Suggested Reading	605
	Problems	606

CHAPTER 11 LOGIC CIRCUITS AND DIGITAL INTEGRATED-CIRCUIT FUNCTIONAL BLOCKS   612

11-1	Introduction	612
11-2	Binary System	613
11-3	Boolean Relations	616
11-4	Basic Building Blocks	616
11-5	Circuit Implementation of the Building Blocks	620
11-6	Diode Logic (DL)	621
11-7	Resistor-Transistor Logic (RTL)	622
11-8	Direct-Coupled Transistor Logic (DCTL)	625
11-9	Diode-Transistor Logic (DTL)	626
11-10	Transistor-Transistor Logic (TTL)	627
11-11	Emitter-Coupled Logic (ECL or CML)	629
11-12	Comparison of Various Logic Gates	631
11-13	Figure of Merit of a Gate	632
11-14	Memory Circuits: The Flip-Flop (FF)	634
11-15	Register Circuits and Shift Registers	637
11-16	Counters	639
11-17	Summary	642
	References and Suggested Reading	642
	Problems	643

APPENDIX A SUMMARY OF SEMICONDUCTOR AND p–n JUNCTION PROPERTIES   648

A-1	Semiconductors: Intrinsic and Extrinsic	648
A-2	Drift and Diffusion Currents	651
A-3	The p-n Junction and its Properties	654
A-4	V-I Characteristic of p-n Junctions	659
	References and Suggested Reading	662

APPENDIX B INTEGRATED CIRCUITS   663

B-1	IC Transistor Fabrication	665
B-2	IC Diodes	669
B-3	IC Capacitors	669
B-4	IC Resistors	670
B-5	IC Inductors	671
B-6	IC Design Guidelines	671
B-7	Large-Scale Integration (LSI)	673
	References and Suggested Reading	674

APPENDIX C INTRODUCTION TO SIGNAL FLOW GRAPHS — 675

- C-1 Simplification of a Signal-Flow Graph — 677
- C-2 Mason's Gain Formula — 680
- C-3 Examples — 681
- References and Suggested Reading — 685

APPENDIX D DISCRETE DEVICE AND INTEGRATED CIRCUIT FUNCTIONAL BLOCK DATA SHEETS — 686

- D-1 Discrete Bipolar Transistor Data Sheets
  - (a) Texas Instruments (*pnp*) 2N3250, 2N3251 — 687
  - (b) Fairchild 2N3011 (*npn*) — 689
- D-2 Discrete FET Data Sheet
  - (a) Texas Instruments n-channel FET 2N3823 — 692
  - (b) RCA-MOSFET 40461 — 695
- D-3 Power Transistor Data Sheet
  - (a) RCA types 40546, 40547 — 697
- D-4 Operational Amplifier Data Sheet
  - (a) Fairchild μA702C — 699
  - (b) Motorola MC1520 — 702
- D-5 IC Logic Gates
  - (a) Transistor ECL Gate, Motorola MC300, 306, 307 — 706
  - (b) MOS Gate, Fairchild 3102 — 711

ANSWERS TO SELECTED PROBLEMS — 715

INDEX — 721

ELECTRONIC CIRCUITS

Devices, Models, Functions, Analysis, and Design

1
Review of Selected Topics from Circuit Theory

1-1 Introduction

Electronic circuits cover a broad range of signal processing systems. A signal is almost always a function of time. It may be completely known (such as a sinusoid) or largely unknown (such as random noise). Physically, a signal is a measurable quantity, such as voltage, current, velocity, or pressure. Usually a nonelectric signal is converted into an electric signal in the form of a time-varying voltage or current. The purpose of an electronic system is to transform and process information-carrying signals. Computers, television, communication satellites, radar, radio are examples of such systems.

The individual systems, which may be analog, digital, or a combination of both, may be divided into subsystems, such as amplifiers, filters, oscillators, switches, memories, logic gates, time-base generators, and counters. Each of these subsystems may consist, in some form or other, of components such as resistors, capacitors, inductors, transformers, transistors, and diodes.

This book will discuss the analysis and design of these subsystems. In particular, the various electronic circuit functions to be considered are *amplification, filtering, switching, logic-function implementation,* and *waveform generation*. To analyze and design such systems, we must be thoroughly familiar with the electrical properties of the components, their voltage-current relationships, and the interactions among various components. Before we embark upon the study of electronic circuits, therefore, we shall review some of the basic topics from circuit theory.

The discussion that follows is, of necessity, brief; it is assumed that the reader has already been exposed to most of these topics. This chapter is intended to serve as a refresher and to provide an opportunity to introduce the terminology and various definitions which will be used throughout this text. The mathematical defining equations for the idealized components are, therefore, given first, followed by elementary network topology and various general methods of circuit analysis. Some useful network theorems are discussed subsequently, since a knowledge of these, in addition to giving insight, will help to simplify the analysis. Network functions and natural frequencies for linear time-invariant circuits are also reviewed. The chapter concludes by briefly discussing electronic circuit analysis by computers.

1-2 Lumped Circuit Elements

The various idealized circuit elements and their voltage-current relationships are as follows:

Resistor: A resistor is a two-terminal element, shown symbolically in fig 1-1. The instantaneous voltage-current relationship of a resistance is given by

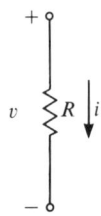

$$v(t) = Ri(t) \quad \text{or} \quad i(t) = Gv(t) \tag{1-1}$$

Fig 1-1 Graphic symbol for a resistor with voltage polarity and current reference directions

where R is the *resistance* in ohms and G is the *conductance* in mhos; $G = 1/R$. For a linear time-invariant circuit, R and G are constants, independent of i, v, and time. For a linear time-varying network, R and G are independent of i and v, but not independent of time. For a nonlinear circuit, R and G are functions of i or v.

1-2 Lumped Circuit Elements

Capacitor: A capacitor is a two-terminal element, shown symbolically in fig 1-2. The defining equation for a capacitance is

$$q(t) = Cv(t) \tag{1-2}$$

where C is the *capacitance* in farads and q is the *instantaneous charge* in coulombs.

For a linear time-invariant circuit, C is a constant, and the instantaneous voltage-current relationship of a capacitance from (1-2) is given by

$$i = \frac{dq(t)}{dt} = C\frac{dv(t)}{dt} \tag{1-3a}$$

or

$$v(t) = v(0) + \frac{1}{C}\int_0^t i(\tau)\,d\tau \tag{1-3b}$$

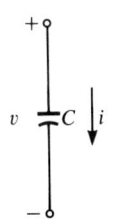

Fig 1-2 Graphic symbol for a capacitor with voltage polarity and current reference directions

Occasionally we may write $\varepsilon = 1/C$, which is usually referred to as the *elastance*. Note that to specify the v-i relationship completely, the initial voltage across the capacitor must be known. For a linear time-varying circuit, C may be a function of time, and the v-i relationship, from (1-2), is given by

$$i = \frac{dq}{dt} = C(t)\frac{dv}{dt} + v(t)\frac{dC(t)}{dt} \tag{1-4}$$

For a nonlinear capacitance, charge and voltage are related, not by a constant, but by a function, and the q-v relationship is nonlinear, i.e., $q = f[v(t)]$. For a nonlinear, time-varying capacitance, $q = f[v(t), t]$.

Inductor: An inductor is a two-terminal element, shown symbolically in fig 1-3. The defining equation for an inductance is

$$\phi(t) = Li(t) \tag{1-5}$$

where L is the *inductance* in henries and $\phi(t)$ is the *instantaneous flux linkage* in webers.

For a linear time-invariant circuit, L is a constant, and the instantaneous voltage-current relationship of an inductance, from (1-5), is

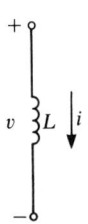

Fig 1-3 Graphic symbol for an inductor with voltage polarity and current reference directions

given by

$$v = \frac{d\phi}{dt} = L\frac{di(t)}{dt} \tag{1-6a}$$

or

$$i(t) = i(0) + \frac{1}{L}\int_0^t v(\tau)\,d\tau \tag{1-6b}$$

Occasionally, we may write $\Gamma = 1/L$ for convenience. Note that, to completely specify the v-i relationship for an inductor, the initial current through the inductor must be known. For a linear time-varying circuit, L may be a function of time, and the v-i relationship, from (1-5), is given by

$$v = \frac{d\phi(t)}{dt} = L(t)\frac{di(t)}{dt} + i(t)\frac{dL(t)}{dt} \tag{1-7}$$

For a nonlinear inductance, the ϕ-i relationship is nonlinear, i.e., $\phi = f[i(t)]$. For a nonlinear, time-varying inductance, $\phi = f[i(t), t]$.

Ideal Transformer: An ideal transformer is a four-terminal device, shown symbolically in fig 1-4a. The defining voltage-current relationships for the ideal transformer are given by

$$\frac{v_1}{v_2} = \frac{n_1}{n_2} \tag{1-8a}$$

$$\frac{i_1}{i_2} = -\frac{n_2}{n_1} \tag{1-8b}$$

where n_1 and n_2 are the number of turns in the primary and secondary windings (coils), respectively. Since n_1 and n_2 are constants, the ideal transformer is a linear time-invariant circuit element. An actual transformer (if losses are ignored) may be represented as in fig 1-4b, where L_m is the magnetizing inductance and $n = n_2/n_1$.

Coupled Inductors: Two coupled inductors are shown symbolically in fig 1-5. For linear time-invariant inductors, the instantaneous voltage-current relations are given by

$$v_1(t) = L_{11}\frac{di_1(t)}{dt} \pm L_{12}\frac{di_2}{dt} \tag{1-9a}$$

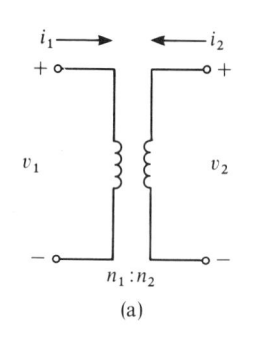

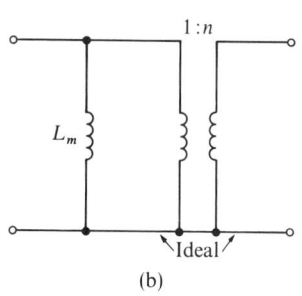

Fig 1-4 (a) Ideal transformer; (b) Actual transformer (losses are ignored)

1-3 Independent and Controlled Sources

$$v_2(t) = \pm L_{21}\frac{di_1(t)}{dt} + L_{22}\frac{di_2(t)}{dt} \quad (1\text{-}9b)$$

where the plus sign of M ($M = L_{12} = L_{21}$) is used for fig 1-5a, and the minus sign of M for fig 1-5b. The magnitude of M in a physical system is constrained by the relation

$$|M| = k\sqrt{L_{11}L_{22}} \quad (1\text{-}10)$$

where $0 < k < 1$ and k is called the coefficient of coupling.

Any circuit that consists of resistors, inductors, capacitors, and coupled inductors (or transformers) is called an *RLCM* circuit.

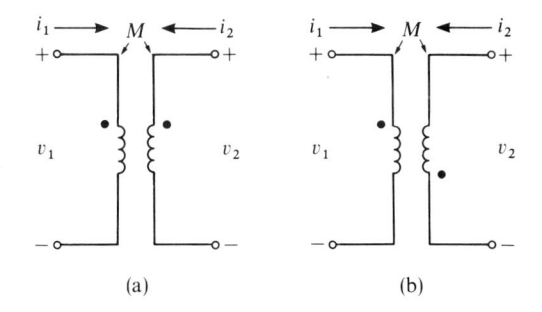

Fig 1-5 Coupled coils

Ideal Gyrator: An ideal gyrator is shown symbolically in fig 1-6. The defining voltage-current relationships of an ideal gyrator are given by

$$v_1(t) = \alpha i_2(t) \quad (1\text{-}11a)$$

$$v_2(t) = -\alpha i_1(t) \quad (1\text{-}11b)$$

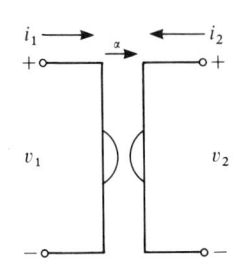

Fig 1-6 Symbol for a gyrator

The constant α is called the gyration resistance and has the dimension of ohms. Since α is a constant, the gyrator is a linear time-invariant circuit element. Unlike other passive circuit elements (RLCM), an ideal gyrator is a nonreciprocal passive two-port element (see Chapter 2).

Again, note that these models are idealizations which often are satisfactory approximations to the real physical devices, but in certain cases the idealizations may not be acceptable, and a more complicated model must be used. The more complicated models, however, still consist of these basic elements and others which will be discussed in the next section.

1-3 Independent and Controlled Sources

There are two basic types of sources: independent sources and controlled (or dependent) sources. The graphic symbols and references for independent voltage and current sources are shown in figs 1-7a and 1-7b, respectively. The independent voltage source in fig 1-7a is sometimes written in double subscript notation, without plus and minus signs in the figure but labeled as v_{jk}. In this text, we shall use the

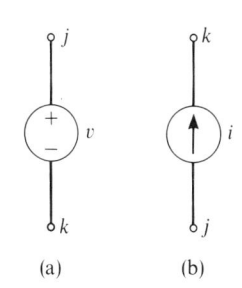

Fig 1-7 (a) Independent voltage source (v_{jk}); (b) Independent current source (i_{jk})

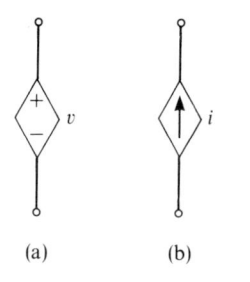

Fig 1-8 (a) Dependent voltage source; (b) Dependent current source

polarity signs and sometimes we also shall use v_1 with an arrow, where the point of the arrowhead will correspond to the plus reference direction. The independent current source in fig 1-7b also may be represented by a double subscript notation with no arrow in the figure but labeled as i_{jk}. We shall use an arrow to indicate the direction of current flow of the independent source.

The graphic symbols used in this book for controlled (dependent) voltage and current sources are shown in figs 1-8a and 1-8b, respectively. This convention has been used in some recent texts (Refs 1, 2, 3); it emphasizes the difference between the dependent and independent sources.

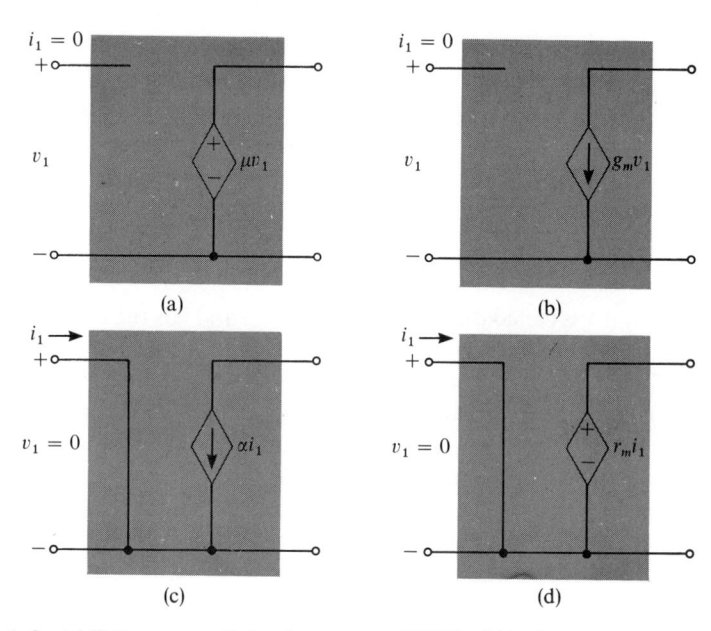

Fig 1-9 (a) Voltage-controlled voltage source (VCVS); (b) Voltage-controlled current source (VCCS); (c) Current-controlled current source (CCCS); (d) Current-controlled voltage source (CCVS)

The four basic idealized controlled-source representations for two-port circuits are shown in fig 1-9. Note that these are three-terminal circuits. For the voltage-controlled voltage source (abbreviated VCVS), the input is a voltage source, which may be dependent or independent, and the output is a dependent voltage source. The other circuits are self-explanatory in the same manner. The controlled-source elements are very important in electronic circuits and are basic constituents of electronic-device models. Although these models are

1-3 Independent and Controlled Sources

idealizations, in conjunction with appropriate RLC elements (discussed in Chapter 3), they can adequately represent actual electronic devices in the linear operating mode. Controlled sources can be used to represent active as well as passive elements. For example, the ideal transformer and the ideal gyrator, both of which are passive elements, may be represented by the controlled-source models shown in figs 1-10 and 1-11, respectively.

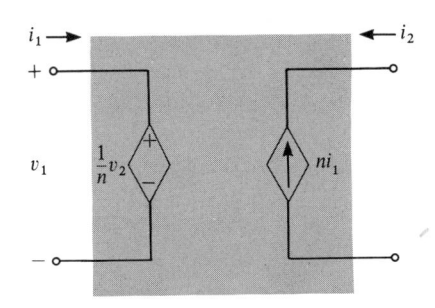

Fig 1-10 Controlled-source representation of an ideal transformer

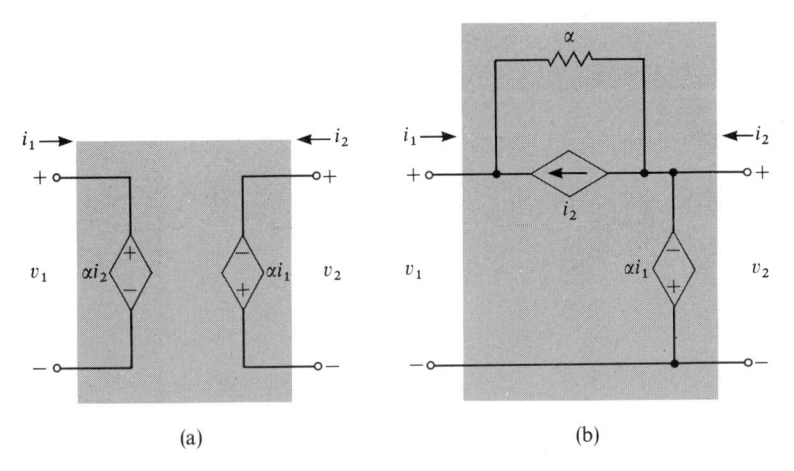

Fig 1-11 Circuits representing an ideal gyrator

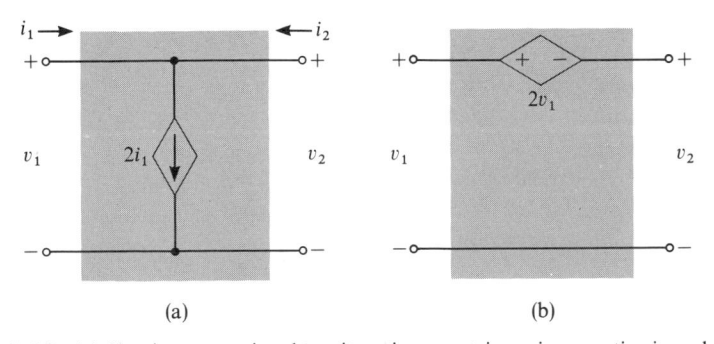

Fig 1-12 (a) Circuit representing the unity-gain current-inversion negative-impedance converter (INIC); (b) Circuit representing the unity-gain voltage-inversion negative-impedance converter (VNIC)

A useful active element employed in active filters (Chapter 8) is the unity gain *negative impedance converter* (abbreviated NIC). There are two types of NIC's: The current inversion, or INIC, and the voltage inversion, or **VNIC**. The defining equations for the unity gain INIC are

$$v_1 = v_2 \quad \text{and} \quad i_1 = i_2 \tag{1-12a}$$

The controlled-source representation for (1-12a) is shown in fig 1-12a. The defining equations for the unity gain VNIC are

$$v_1 = -v_2 \quad \text{and} \quad i_1 = -i_2 \tag{1-12b}$$

The controlled-source representation of (1-12b) is shown in fig 1-12b.

1-4 Linear and Time-Invariant Systems

Since, in this book, we shall consider both linear and nonlinear circuits, it is advisable at the outset to define a linear system.

A system is said to be linear if the *homogeneity* and *additivity* properties hold. The homogeneity property says that, for every constant α, $f(\alpha x) = \alpha f(x)$. The additivity property says that $f(x_1 + x_2) = f(x_1) + f(x_2)$. These two properties are summed up by the *superposition principle*, which states that

$$f(\alpha x_1 + \beta x_2) = \alpha f(x_1) + \beta f(x_2) \tag{1-13}$$

for every input x_1 and x_2 and every constant α and β. In other words, the superposition principle states that the response of any linear system to more than one independent excitation is given by the sum of the responses due to each excitation acting one at a time. For example, if input (excitation) x_1 yields an output (response) y_1, as shown schematically in fig 1-13, we may write symbolically $x_1(t) \to y_1(t)$. Also, if $x_2(t) \to y_2(t)$, then $k_1 x_1(t) + k_2 x_2(t) \to k_1 y_1(t) + k_2 y_2(t)$ for all constants k_1 and k_2. To determine whether a system is linear, we must establish (1-13).

A system is linear and time-invariant if input $x(t + \tau)$ gives an output $y(t + \tau)$ for all $x(t)$ and all $\tau > 0$. In other words, the specific time of applying the excitation has no effect on the waveform of the response. For a causal system, the response cannot precede the excitation; that is, if the excitation is applied at $t = t_0$, the response is zero for $t \leq t_0$. The equilibrium equations describing a linear time-invariant circuit are characterized by a set of linear integrodifferential equations with constant coefficients. For such systems, transform techniques can be used to reduce the set of integrodifferential equations to a set of linear algebraic equations.

We define the *zero-input response* of the system as the output y_0 when the input is identically zero. Such a response is not necessarily zero because there may be initial charges on the capacitors and/or

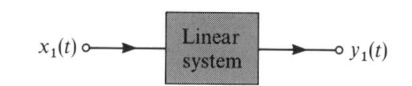

Fig 1-13 Schematic of a linear system with input and output variables

initial fluxes in the inductors. In system theory, the initial conditions also are referred to as the initial state of the system. (For a detailed discussion of the concept of state, the reader is encouraged to consult Ref 1.) We define the *zero-state response* as the output y due to an arbitrary input when all initial conditions are zero, i.e., when the initial state is zero. The initial charge on a capacitor can be represented by an independent constant voltage source in series with the capacitor having zero initial state. Similarly, the initial flux in an inductor can be represented by an independent constant-current source in shunt with the inductor having zero initial state. It then follows that, for a linear system, the *complete response* is equal to the sum of the zero-input response and the zero-state response. Superposition implies that the zero-state response caused by several independent sources is the sum of the zero-state responses caused by each independent source acting alone.

For a single-input, single-output, lumped linear time-invariant system, the input–output relationship can be expressed in general by an nth-order linear differential equation with constant coefficients:

$$A_n \frac{d^n y}{dt^n} + A_{n-1} \frac{d^{n-1} y}{dt^{n-1}} + \cdots + A_0 y$$
$$= B_m \frac{d^m x}{dt^m} + B_{n-1} \frac{d^{n-1} x}{dt^{n-1}} + \cdots + B_0 x \tag{1-14}$$

where x and y designate the input and output variables, respectively.

For a given input and the initial conditions

$$y(0), \quad \frac{dy(0)}{dt}, \ldots, \frac{d^{n-1} y(0)}{dt^{n-1}},$$

the complete solution $y(t)$ is uniquely determined. Note that the solution to the homogeneous equation, i.e., the solution of (1-14) when the right-hand side is set equal to zero, determines the zero-input response; whereas a particular solution, which depends upon x, when all the initial conditions are zero, determines the zero-state response.

1-5 The Laplace Transform

The Laplace transform is one of a family of integral transforms useful in the analysis and design of *linear time-invariant systems*. Since the first few chapters of this text cover mainly linear time-invariant circuits, the Laplace transform is briefly reviewed here.

The unilateral or one-sided Laplace transform (abbreviated $\mathscr{L}$) is defined by the following *definition integral*:[1]

$$F(s) = \mathscr{L}[f(t)] = \int_0^\infty f(t)e^{-st}\,dt \qquad (1\text{-}15)$$

This definition is meaningful if $F(s)$ converges (where s is the complex frequency, i.e., $s = \sigma + j\omega$) and $f(t) = 0$ for all $t < 0$. A function which is Laplace transformable has a unique transform, and vice versa.

The Laplace transform is a linear operation, i.e.,

$$\mathscr{L}[A_1 f_1(t) + A_2 f_2(t)] = A_1 F_1(s) + A_2 F_2(s) \qquad (1\text{-}16)$$

for all values of A_1 and A_2.

Some elementary and basic properties of the transform, which can be derived from (1-15), are given in table 1-1.

A function $F(s)$ is changed to the function $f(t)$ by the inverse Laplace transform using the inversion integral, or $F(s)$ is broken into simpler functions by partial fraction expansion and then a table of $\mathscr{L}$-transforms, such as table 1-1, is used to find $f(t)$.

Partial Fraction Expansion

Consider the following function of the complex frequency variable:

$$F(s) = \frac{N(s)}{D(s)} = \frac{a_n s^n + a_{n-1} s^{n-1} + a_{n-2} s^{n-2} + \cdots + a_0}{b_m s^m + b_{m-1} s^{m-1} + b_{m-2} s^{m-2} + \cdots + b_0} \qquad (1\text{-}17)$$

If the degree of the numerator is less than that of the denominator, i.e., $n < m$, then $F(s)$ is a proper rational function. If $F(s)$ is not a proper rational function, simply divide the numerator by the denominator to obtain a polynomial and a proper rational function. Then find the $\mathscr{L}$-transform of each part.

We now describe how to reduce a proper rational function into simpler functions.

The roots of the denominator, i.e., poles of the function, may all be distinct, real or complex, and/or some poles may be repeated.

---

[1] Some authors use 0^- as the lower limit of integration, to include the impulse in $f(t)$ at $t = 0$, if there is any. For example, if $f(t) = \delta(t)$, then

$$F(s) = \int_{0^-}^\infty \delta(t)e^{-st}\,dt = \int_{0^-}^\infty \delta(t)\,dt = 1.$$

For the 0^- definition, $f(t)$ need not be zero for $t < 0$.

TABLE 1-1 Laplace Transforms of Simple Functions and Operations

$f(t)$	$F(s) = \mathscr{L}[f(t)] = \int_0^\infty f(t)e^{-st}\,dt$
$\delta(t)$	1
$u(t)$	$1/s$
df/dt	$sF(s) - f(0^+)$
$\int_0^t f(t)\,dt$	$F(s)/s$
$-tf(t)$	$dF(s)/ds$
$(1/t)f(t)$	$\int_s^\infty F(z)\,dz$
$e^{-\lambda t}f(t)$	$F(s+\lambda)$
$f(t-a)u(t-a)$	$e^{-as}F(s)$
$f(t/a)$	$a[F(as)]$
t^n	$n!/s^{n+1}\qquad n = 1,2,3,\ldots$
e^{-at} (a real or complex)	$1/(s+a)$
$[1/(n-1)!]t^{n-1}e^{-at}$	$1/(s+a)^n\qquad n = 1,2,3,\ldots$
$1 - e^{-at}\sum_{k=0}^{n-1}[(at)^k/k!]$	$a^n/s(s+a)^n\qquad n = 1,2,3,\ldots$
$\sin \alpha t$	$\alpha/(s^2+\alpha^2)$
$\cos \alpha t$	$s/(s^2+\alpha^2)$
$e^{-bt}\sin \alpha t$	$\alpha/[(s+b)^2+\alpha^2]$
$e^{-at}\cos \beta t$	$(s+a)/[(s+a)^2+\beta^2]$
$2\|K\|e^{-at}\cos(bt + \arg K)$	$K/(s+a-jb) + \bar{K}/(s+a+jb)$ (bar indicates conjugate)
$f_1*f_2 = \int_0^t f_1(t-\tau)f_2(\tau)\,d\tau$	$F_1(s)F_2(s)$ (* denotes convolution)
$f(0^+)$	$\lim_{s\to\infty} sF(s)$
$f(\infty)$	$\lim_{s\to 0} sF(s)$

(1) For distinct poles p_i (real or complex), we may rewrite the proper rational function $F(s)$ as follows:

$$F(s) = \frac{N(s)}{D(s)} = \frac{K_1}{s+p_1} + \frac{K_2}{s+p_2} + \cdots + \frac{K_n}{s+p_n} + \frac{N_1(s)}{D_1(s)} \qquad (1\text{-}18)$$

From table 1-1 we obtain

$$f(t) = \frac{-21}{2}t^2 e^{-t} + 40te^{-t} - 47e^{-t} + 48e^{-2t} \qquad t \geq 0$$

$$= 0 \qquad \text{elsewhere}$$

As we have already mentioned, the Laplace transform is useful only for linear time-invariant circuits. If the network is nonlinear and/or time varying, we must use the analysis and design directly in the time domain. Only for a linear time-invariant network can we use either the frequency or time-domain analysis.

It also should be noted that, for a linear time-invariant circuit where we wish to obtain the time-domain response with a computer, the computer program usually solves the problem directly from the state-variable equations (Section 1-9). In other words, a recursive formula (1-58) is often used in the program to perform the computation and the inverse Laplace transform is not used.

We shall now briefly review circuit analysis. In particular, we shall consider node analysis, loop analysis, and the state-variable formulation. Before we discuss these methods, however, we shall consider elementary network topology, since familiarity with some of its basic ideas is of great aid in setting up the linear independent equations for the network.

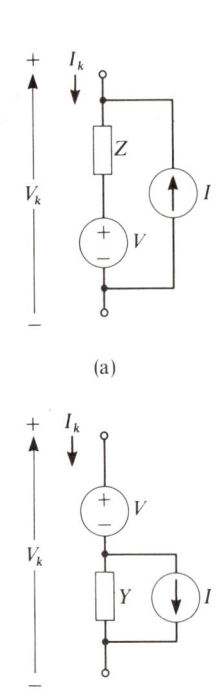

Fig 1-14 Two general forms of a branch

1-6 Elementary Network Topology and Terminology

A *network* constitutes an interconnection of circuit elements and devices. The terminals of an element are called *nodes*. A *branch* is a two-terminal network that contains a two-terminal element and/or voltage and current sources, as shown in fig 1-14. The sources may be controlled or independent. The rectangular box, represented by Z or Y, represents a circuit element that is not a source. Note that a junction between two or more branches constitutes a *node*. A *loop* is defined as a closed path formed by branches. A *network graph* is a simplified drawing of a network in which every branch is represented by a line segment. The graph represents the topological relationship of the elements and the nodes of the network without direct concern for the nature or kinds of elements. For example, the graph of the networks in figs 1-15a and 1-15b is shown in fig 1-15c. A graph may be connected or

1-6 Elementary Network Topology and Terminology

unconnected. The graph in fig 1-15c is connected while the one in fig 1-16b is unconnected. The latter graph is also referred to as having two separate parts.

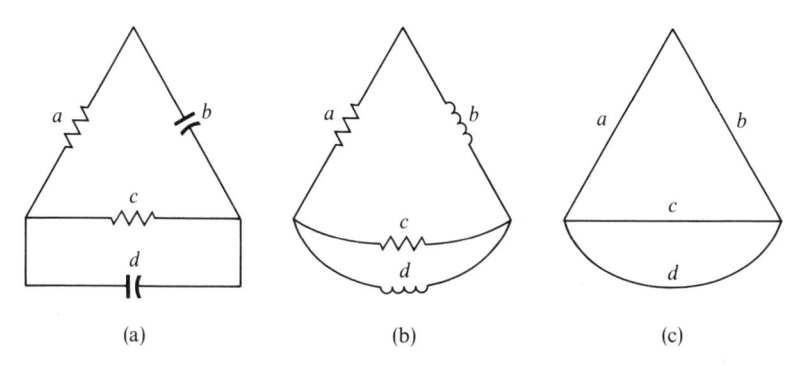

Fig 1-15 (a) and (b) Two arbitrary circuits; (c) Graph of (a) and (b)

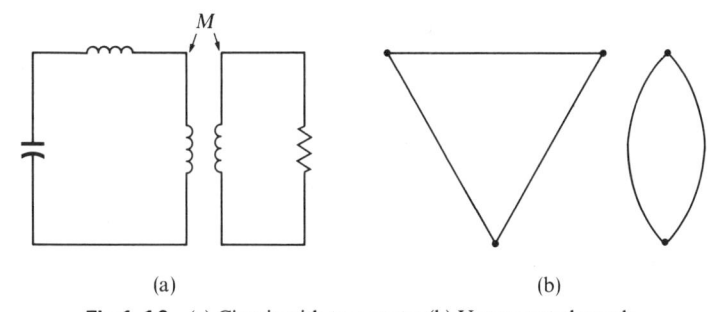

Fig 1-16 (a) Circuit with two parts; (b) Unconnected graph

A *tree* is a connected subgraph containing all the nodes of the graph but not forming any loop; i.e., there should exist a unique path through the tree between any pair of nodes in a connected graph. A *link* is a branch of the graph that is not part of the tree; e.g., for the network fig 1-17a, the graph and several trees and the associated links are shown in fig 1-18. The dashed lines are the links, while the solid lines are the branches of the tree. A *basic loop* in a graph is a loop which contains the branches of a tree and exactly one link.

Having introduced some of the definitions and terminologies of interest in the analysis of circuits, we shall now consider general methods of network analysis.

In the analysis of networks, two sets of *postulates* are used. One set is the voltage-current relationships of the branches in a network.

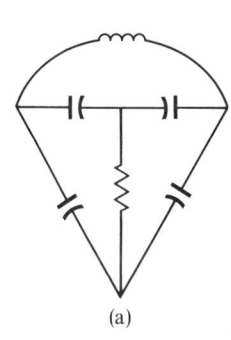

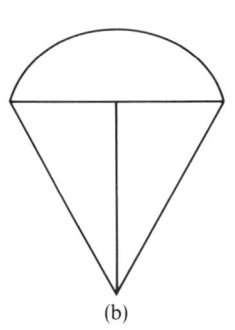

Fig 1-17 (a) Network; (b) Graph of (a)

The other set is Kirchhoff's current law and Kirchhoff's voltage law. Kirchoff laws are *fundamental laws* which govern the voltages and currents in a circuit. These laws must be satisfied at all times regardless of the composition of the circuit. These two sets of postulates are used in the analysis of linear and nonlinear time-invariant and time-varying

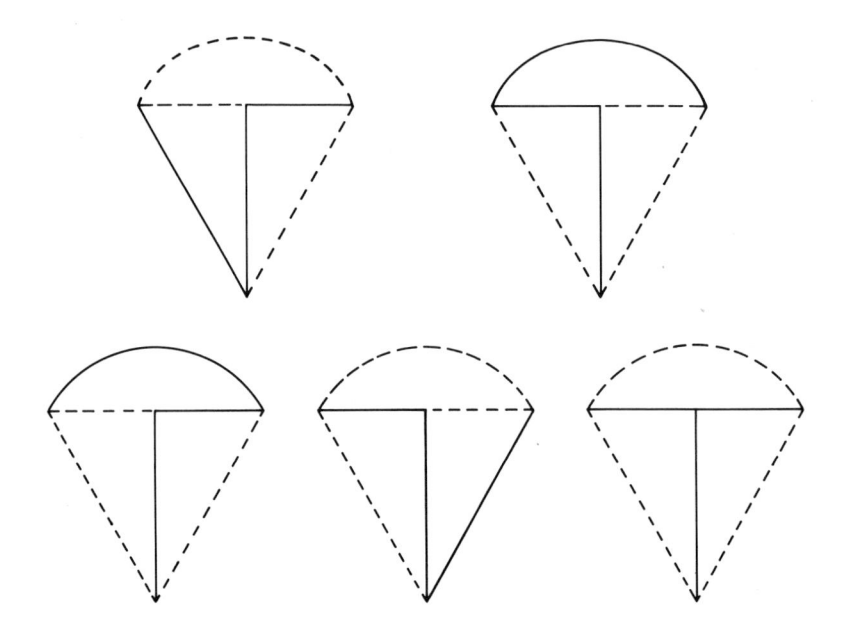

Fig 1-18 Various trees and links of the network graph in fig 1-17(a)

networks. For the nonlinear and/or time varying circuits, we can use only the time-domain analysis. For linear time-invariant networks, we may use either the time-domain or the frequency-domain (Laplace transform) analyses. In the following sections, we consider the analysis of linear time-invariant circuits.

1-7 Nodal Analysis

The equilibrium of the network on the node basis formulation is written by summing the total currents flowing into and out of the nodes, excluding the reference (datum) node. In a connected graph there are then $n - 1$ nodes. We apply Kirchhoff's current law (KCL) at $n - 1$

1-7 Nodal Analysis

nodes to obtain $n - 1$ linearly independent equations for the nodes to datum voltages. If the graph has s separate parts, *the number of independent node equations is exactly equal to $n - s$*, where n is the total number of nodes (including the reference node).

EXAMPLE

Consider the circuit shown in fig 1-19. We shall set up:

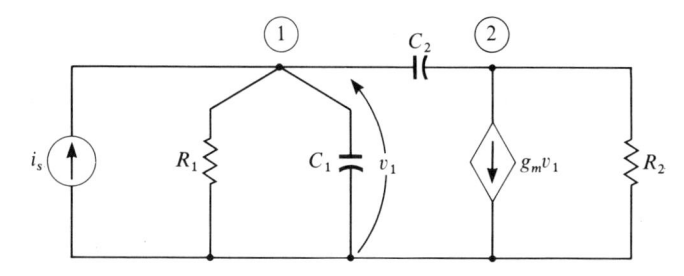

Fig 1-19 Example circuit

(a) the linearly independent integrodifferential nodal equations, and
(b) the linearly independent transform nodal equations, and analyze the circuit.

Assume the initial conditions to be zero, i.e., let all the capacitors have zero initial voltage.

We apply KCL and the voltage-current relationship of the associated branches to obtain the node equations. From application of KCL, which states that $\Sigma i_k = 0$, at nodes (1) and (2) we obtain:

$$i_s - v_1\left(\frac{1}{R_1}\right) - C_1\frac{dv_1}{dt} - C_2\frac{d}{dt}(v_1 - v_2) = 0 \qquad \text{(1-22a)}$$

$$C_2\frac{d}{dt}(v_1 - v_2) - g_m v_1 - v_2\left(\frac{1}{R_2}\right) = 0 \qquad \text{(1-22b)}$$

Using the operational notation $s = d/dt$ and $1/s = \int_0^t d\tau$, the above equations can be rearranged and solved.

In the transformed s domain, the corresponding rearranged equations are

$$V_1\left(\frac{1}{R_1} + sC_1 + sC_2\right) + V_2(-sC_2) = I_s \qquad \text{(1-23a)}$$

$$V_1(sC_2 - g_m) - V_2\left(sC_2 + \frac{1}{R_2}\right) = 0 \qquad \text{(1-23b)}$$

The above equations can be readily solved for V_1 and V_2. Note that capital letters are used for the voltages and currents in the transform domain and lower-case (small) letters are used for the time domain.

The transformed nodal equations for a linear time-invariant network are of the general form

$$\begin{aligned} \beta_{11}V_1 + \beta_{12}V_2 + \cdots + \beta_{1n}V_n &= I_1 \\ \beta_{21}V_1 + \beta_{22}V_2 + \cdots + \beta_{2n}V_n &= I_2 \\ &\cdots \\ \beta_{n1}V_1 + \beta_{n2}V_2 + \cdots + \beta_{nn}V_n &= I_n \end{aligned} \qquad (1\text{-}24)$$

where $\beta_{ij} = G_{ij} + sC_{ij} + (1/s)\Gamma_{ij}$ (in this case, Γ_{ij} denotes reciprocal inductance). Note that, in general, the currents I_i may include current sources due to the initial conditions. Further, in general, $\beta_{ij} \neq \beta_{ji}$; only for circuits comprising of R, L, C, transformers, and coupled coils is $\beta_{ij} = \beta_{ji}$.

Equation (1-24) may be written compactly in matrix form[2]

$$[\beta][V] = [I] \qquad (1\text{-}25)$$

where $[\beta]$, in general, is an $n \times n$ matrix and $[V]$ and $[I]$ are column vectors, i.e., $n \times 1$ matrices.

The solution of (1-24) may be obtained by a number of methods. One method, which is mathematically convenient, is by *Cramer's rule*, namely,

$$V_j = \sum_{i=1}^{n} \frac{\Delta_{ij}}{\Delta} I_i \qquad j = 1, 2, \ldots, n \qquad (1\text{-}26)$$

where

$$\Delta = \det[\beta] = \begin{vmatrix} \beta_{11} & \beta_{12} & \cdots & \beta_{1n} \\ \beta_{21} & \beta_{22} & \cdots & \beta_{2n} \\ \beta_{n1} & \beta_{n2} & \cdots & \beta_{nn} \end{vmatrix} \neq 0 \qquad (1\text{-}27)$$

and

$$\Delta_{ij} = (-1)^{i+j} M_{ij} \qquad (1\text{-}28)$$

is the cofactor. M_{ij} is the determinant of β with the ith row and jth column canceled out.

---

[2] For a good, brief discussion of the definitions and basic properties of matrices, see Appendix B, Ref 1. For a more detailed treatment, see Chapter 4, Ref 5.

For example, the solution of (1-23) is

$$V_1 = \frac{\begin{vmatrix} \beta_{22} I_s & -\beta_{12} \times 0 \end{vmatrix}}{\begin{vmatrix} \beta_{11} & \beta_{12} \\ \beta_{21} & \beta_{22} \end{vmatrix}}$$

$$= \frac{-(1/R_2 + sC_2)I_s}{[1/R_1 + s(C_1 + C_2)][-(1/R_2) - sC_2] + sC_2(sC_2 - g_m)} \quad (1\text{-}29)$$

$$V_2 = \frac{\begin{vmatrix} -\beta_{21} I_s + \beta_{11} \times 0 \end{vmatrix}}{\begin{vmatrix} \beta_{11} & \beta_{12} \\ \beta_{21} & \beta_{22} \end{vmatrix}}$$

$$= -\frac{(sC_2 - g_m)I_s}{[1/R_1 + s(C_1 + C_2)][-(1/R_2) - sC_2] + sC_2(sC_2 - g_m)} \quad (1\text{-}30)$$

For a given I_s and the circuit-element values, the node-to-datum voltages V_1 and V_2 (and hence all the branch voltages and currents of the circuits) can be determined.

For n unknowns, in the above method there are $n + 1$ determinants, each $n \times n$ in size, to be evaluated. If n is large, this is a task too large for even a high-speed computer. For that reason, this method of calculation, namely (1-26), is seldom used. The form of the solution, however, is useful for analytic purposes. A method of elimination, called Gaussian elimination (Ref 2), is an efficient, systematic, machine-oriented procedure for solving large systems of simultaneous equations.

1-8 Loop Analysis

Equilibrium equations of the network on a loop basis are written by using Kirchhoff's voltage law (KVL), $\Sigma v_k = 0$, and the voltage-current relationships of the associated branches. In order to write a linearly independent set of loop equations, we choose basic loops in applying KVL. Since a basic loop contains only one link, the number of links in a tree of a connected graph determines the number of independent loop equations. The number of links is readily determined as follows. For a connected graph with n nodes, there are $n - 1$ tree branches. If the total number of branches in the graph is b, the number of links l is given by $l = b - (n - 1)$. If the network has s separate parts, the corresponding number of links is given by $l = b - (n - s)$.

20　1/Review of Selected Topics from Circuit Theory

Reviewing: *the number of independent loop equations is equal to the number of links in the tree of the network graph.* The loop current variables are chosen such that each link has an independent current variable.

EXAMPLE

Consider the network shown in fig 1-20a. We write:
(a) the linearly independent integrodifferential loop equations, and
(b) the linearly independent transform loop equations.

Let the initial state of the circuit be zero, i.e., let all the capacitors and inductors have zero initial voltage and current, respectively. Note that if the initial state of a capacitor and/or an inductor is not zero, we may replace it by the equivalent circuit containing a battery or a constant current source, respectively, as shown in fig 1-21.

We redraw the circuit (fig 1-20b) and show the tree and the links (dotted). Applying KVL around the three basic loops yields

$$\text{Loop 1:} \quad v_1 - R_1(i_1 - i_3) - L\frac{d(i_1 - i_3)}{dt}$$

$$-\frac{1}{C_1}\int_0^t (i_1 - i_2)\, d\tau = 0 \tag{1-31a}$$

$$\text{Loop 2:} \quad \frac{1}{C_1}\int_0^t (i_1 - i_2)\, d\tau - R_2(i_2 - i_3) - R_4 i_2$$

$$-\frac{1}{C_2}\int_0^t i_2\, d\tau = 0 \tag{1-31b}$$

$$\text{Loop 3:} \quad r_m i_2 + R_2(i_2 - i_3) + L\frac{d}{dt}(i_1 - i_3)$$

$$+ R_1(i_1 - i_3) - R_3 i_3 = 0 \tag{1-31c}$$

Now consider the transform domain. Apply KVL around the same three loops to obtain:

$$(1) \quad V_1 - R_1(I_1 - I_3) - sL(I_1 - I_3) - \frac{1}{sC_1}(I_1 - I_2) = 0 \tag{1-32a}$$

$$(2) \quad \frac{1}{sC_1}(I_1 - I_2) - R_2(I_2 - I_3) - R_4 I_2 - \frac{1}{sC_2}I_2 = 0 \tag{1-32b}$$

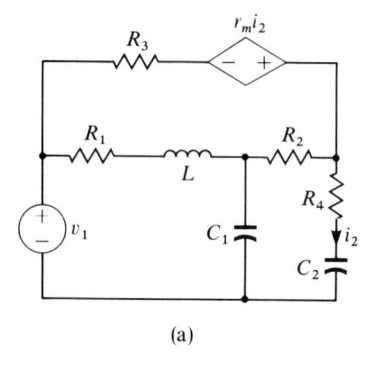

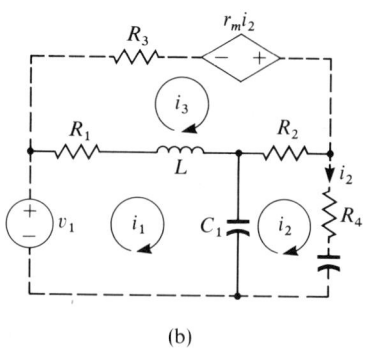

Fig 1-20 (a) Example circuit; (b) Circuit branches (links are dotted)

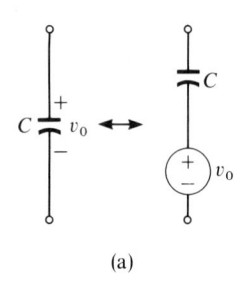

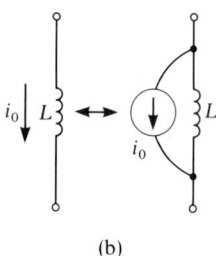

Fig 1-21 Representation of initial conditions

(3) $r_m I_2 + R_2(I_2 - I_3) + sL(I_1 - I_3)$
$$+ R_1(I_1 - I_3) - R_3 I_3 = 0 \qquad (1\text{-}32c)$$

These equations are rearranged as follows:

$$I_1\left(R_1 + sL_1 + \frac{1}{sC_1}\right) + I_2\left(-\frac{1}{sC_1}\right) + I_3(-R_1 - sL) = V_1$$

$$I_1\left(\frac{1}{sC_1}\right) + I_2\left(-\frac{1}{sC_1} - R_2 - R_4 - \frac{1}{sC_2}\right) + I_3 R_2 = 0 \qquad (1\text{-}33)$$

$$I_1(R_1 + sL) + I_2(R_2 + r_m) + I_3(-R_2 - R_1 - R_3 - sL) = 0$$

The above equations can now be solved for $I_1, I_2,$ and I_3 as discussed previously.

In the transform domain, the set of equations on the loop basis of formulation are of the general form

$$\begin{aligned}
\alpha_{11}I_1 + \alpha_{12}I_2 + \cdots + \alpha_{1n}I_n &= E_1 \\
\alpha_{21}I_1 + \alpha_{22}I_2 + \cdots + \alpha_{2n}I_n &= E_2 \\
&\cdots \\
\alpha_{n1}I_1 + \alpha_{n2}I_2 + \cdots + \alpha_{nn}I_n &= E_n
\end{aligned} \qquad (1\text{-}34)$$

where $\alpha_{ij} = R_{ij} + sL_{ij} + (1/s)\varepsilon_{ij}$ (here ε denotes reciprocal capacitance or elastance). The voltages $E_1, E_2, \ldots,$ include the voltages due to initial conditions. Further, in general, $\alpha_{ij} \neq \alpha_{ji}$; only for RLCM circuits is $\alpha_{ij} = \alpha_{ji}$. The solution for the unknown currents may be written by Cramer's rule. The set of solutions for (1-34) is

$$I_j = \sum_{i=1}^{n} \frac{\Delta'_{ij}}{\Delta'} E_i \qquad j = 1, 2, \ldots, n \qquad (1\text{-}35)$$

where

$$\Delta' = \det[\alpha] = \begin{vmatrix} \alpha_{11} & \alpha_{12} & \cdots & \alpha_{1n} \\ \alpha_{21} & \alpha_{22} & \cdots & \alpha_{2n} \\ \vdots & & & \vdots \\ \alpha_{n1} & \alpha_{n2} & & \alpha_{nn} \end{vmatrix} \qquad (1\text{-}36)$$

and Δ'_{ij} are the cofactors of Δ'.

1-9 The State-Variable Formulation

In nodal analysis, we chose the set of node-to-datum voltages as independent variables, and in the loop analysis we chose a set of basic loop currents as independent variables. We can also use a mixed set of capacitor voltages and inductor currents as independent variables. In state variable formulation it is convenient to define a special type of tree and then to write KCL and KVL for the nodes and loops of the tree respectively. The special tree, which is referred to as a *normal tree*, is constructed with the following hierarchy: (1) voltage sources; (2) capacitors; (3) resistors; (4) inductors; (5) current sources. For each capacitor of the normal tree the nodal equations and for each inductor, the basic loop equations are written and then the equilibrium equations are rearranged in the normal form [Eq. (1-42)]. This method, as we shall see, leads to a first-order matrix differential equation. The state-variable approach is particularly attractive when digital or analog computers are used in the solution of the equations and/or when we are dealing with a time-varying or nonlinear network. For nonlinear and/or time-varying circuits, it is preferable to use the capacitor charges and the inductor fluxes as the independent variables.

Since the capacitor voltages and the inductor currents determine the state of the network, i.e., all the voltages and currents of the network are determined by the state at t_0 and the input for all $t \geq t_0$, these quantities are called the *state variables*. We shall first describe the method by a simple example and then consider the general formulation of network analysis via the state-variable approach.

EXAMPLE

Consider the circuit shown in fig 1-22. A normal tree of the circuit is shown by the emphasized heavy lines. In this case, we use both KVL

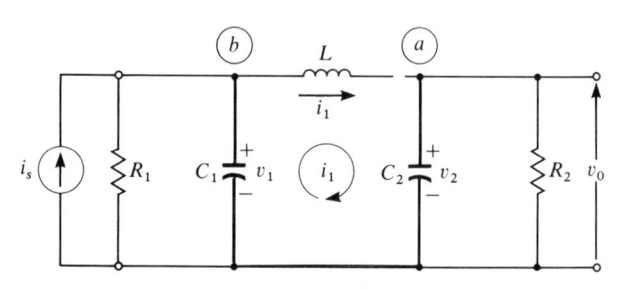

Fig 1-22 Example circuit

1-9 The State-Variable Formulation

and KCL. We write KCL at node a:

$$C_2 \frac{dv_2}{dt} = i_1 - \frac{v_2}{R_2} \tag{1-37}$$

KVL around the loop (indicated by an arrow) yields:

$$v_1 = L\frac{di_1}{dt} + v_2 \tag{1-38}$$

KCL at node b gives

$$i_s(t) = \frac{v_1}{R_1} + C_1 \frac{dv_1}{dt} + i_1 \tag{1-39}$$

Rearranging (1-37), (1-38), and (1-39), we have

$$\frac{dv_1}{dt} = -\frac{v_1}{R_1 C_1} + \frac{i_s}{C_1} - \frac{i_1}{C} \tag{1-40a}$$

$$\frac{dv_2}{dt} = -\frac{v_2}{R_2 C_2} + \frac{i_1}{C_2} \tag{1-40b}$$

$$\frac{di_1}{dt} = \frac{v_1}{L} - \frac{v_2}{L} \tag{1-40c}$$

Equation (1-40) may be written in matrix form as follows:

$$\frac{d}{dt}\begin{bmatrix} v_1 \\ v_2 \\ i_1 \end{bmatrix} = \begin{bmatrix} -1/(R_1 C_1) & 0 & -1/C_1 \\ 0 & -1/(R_2 C_2) & 1/C_2 \\ 1/L & & 0 \end{bmatrix}$$
$$\times \begin{bmatrix} v_1 \\ v_2 \\ i_1 \end{bmatrix} + i_s \begin{bmatrix} 1/C_1 \\ 0 \\ 0 \end{bmatrix} \tag{1-41}$$

The solution of (1-41) will be discussed shortly.

For a general linear time-invariant network, the state equations can be written in the following matrix form:[3]

$$[\dot{x}] = [A][x] + [B][\omega] \tag{1-42a}$$

---

[3] In some cases, the derivatives of the input may appear in the state equation (1-42a) and/or the output equations, but this can be rewritten in the forms of (1-42a) and (1-42b) by a change of variables.

$$[y] = [C][x] + [D][\omega] \qquad (1\text{-}42b)$$

where

$[x]$ is a column matrix (column vector) with n components for the state variables,
$[A]$ is a square matrix of order $n \times n$,
$[B]$ is a matrix of order $n \times m$,
$[C]$ is a matrix of order $r \times n$,
$[D]$ is a matrix of order $r \times m$,
$[y]$ is the output column matrix of order r,
$[\omega]$ is the input column matrix of order m.

For a single input-single output case, ω, y, and D are scalar quantities.
For example, in the circuit of fig (1-22), the input is $i_s(t)$ and the output is $v_0(t)$.

$$\omega = i_s(t), \qquad y = v_2(t) = v_0(t)$$

$$[x] = \begin{bmatrix} v_1 \\ v_2 \\ i_1 \end{bmatrix}, \quad [A] = \begin{bmatrix} -1/(R_1 C_1) & 0 & -1/C_1 \\ 0 & -1/(R_2 C_2) & 1/C_2 \\ 1/L & -1/L & 0 \end{bmatrix}$$

$$[B] = \begin{bmatrix} 1/C_1 \\ 0 \\ 0 \end{bmatrix}, \quad [C] = [0 \ 1 \ 0], \quad D = 0$$

It should be noted that in the state-variable formulation, the representation is not unique. Of course, the solution is unique.

The solution of (1-42a) can be written directly from the scalar first-order differential equation. For convenience, let us first define

$$[f(t)] = [B][\omega] \qquad (1\text{-}43)$$

so that (1-42a) may be written as

$$\frac{d}{dt}[x] = [A][x] + [f(t)] \qquad (1\text{-}44)$$

The solution of (1-44) is

$$[x(t)] = e^{[A](t-t_0)}[x(t_0)] + \int_{t_0}^{t} e^{[A](t-\tau)}[f(\tau)] \, d\tau \qquad (1\text{-}45)$$

The first term in (1-45) is the zero-input response and the second term is the zero-state response. The means for the evaluation of $e^{[A]t}$ will be discussed shortly. The matrix $e^{[A]t}$ is also called the fundamental or *state-transition* matrix of the matrix $[A]$. Differentiation and integration of a matrix mean differentiating and integrating each element of the matrix, respectively. For a time-invariant network, the matrix $[A]$ is independent of t.

To show that (1-45) is the solution of (1-44), differentiate (1-45) to get

$$\frac{d}{dt}[x] = [A]e^{[A](t-t_0)}[x(t_0)] + [A]e^{[A]t}\int_{t_0}^{t} e^{-[A]\tau}[f(\tau)]\,d\tau \\ + e^{[A]t}e^{-[A]t}[f(t)] \quad (1\text{-}46)$$

As we shall soon see, it can easily be shown that $e^{[A]t}e^{-[A]t} = [1]$. Equation (1-44) immediately follows from (1-45) and (1-46). Note that the initial conditions in (1-45) are also satisfied. Hence (1-45) is a solution of (1-44).

Now consider the meaning and evaluation of the state-transition matrix $e^{[A]t}$. The matrix $e^{[A]t}$ is defined by

$$e^{[A]t} = [1] + [A]t + [A]^2\frac{t^2}{2} + \cdots + [A]^k\frac{t^k}{k!} + \\ = \sum_{i=0}^{\infty} [A]^i \frac{t^i}{i!} \quad (1\text{-}47)$$

where $[1]$ is the identity matrix. From (1-47) we can see that $e^{[A]t}e^{-[A]t} = [1]$. To evaluate $e^{[A]t}$, we first determine the eigenvalues, which are also called the characteristic roots of the matrix.

For any square matrix $[A]$ of order $n \times n$, if some nonzero column matrix $[x]$ of order n can be found such that:

$$[A][x] = \lambda[x] \quad (1\text{-}48)$$

where λ is a scalar quantity, then the vector $[x]$ is called an *eigenvector* and λ is called an *eigenvalue* of the matrix $[A]$. To find the eigenvalues of $[A]$, we rewrite (1-48) as

$$\{[A] - \lambda[1]\}[x] = [0] \quad (1\text{-}49)$$

A nontrivial solution of $[x]$ in (1-49) exists if the determinant of the coefficient matrix is equal to zero, i.e.,

$$P(\lambda) = \det\{[A] - \lambda[1]\} = 0 \quad (1\text{-}50)$$

Equation (1-50) is called the characteristic equation of $[A]$. If $[A]$ is an $n \times n$ matrix, (1-50) is a polynomial of the nth degree and there are

n values of λ, namely, $\lambda_1, \lambda_2, \ldots, \lambda_n$. These values, which may or may not be distinct roots of the characteristic equation, are called the characteristic roots or eigenvalues.

EXAMPLE

Consider the following 2×2 matrix:

$$[A] = \begin{bmatrix} -6 & -3 \\ 1 & -2 \end{bmatrix}$$

$$P(\lambda) = \det\{[A] - \lambda[1]\} = \begin{vmatrix} -6-\lambda & -3 \\ 1 & -2-\lambda \end{vmatrix}$$

$$= \lambda^2 + 8\lambda + 15 = (\lambda + 5)(\lambda + 3) = 0$$

The eigenvalues are $\lambda_1 = -3$ and $\lambda_2 = -5$. The evaluation of $e^{[A]t}$ is as follows: we put

$$e^{[A]t} = C_0[1] + C_1[A] \tag{1-51}$$

If the matrix is of order $n \times n$, we put:

$$e^{[A]t} = C_0[1] + C_1[A] + \cdots + C_{n-1}[A]^{n-1} \tag{1-52}$$

This is possible since the higher powers of $[A]$ may be expressed in terms of the lower by the Cayley–Hamilton theorem, which states that every square matrix satisfies its own characteristic equation. From (1-51), after substitution of the eigenvalues, we get:

$$e^{-3t} = C_0 - 3C_1$$
$$e^{-5t} = C_0 - 5C_1$$

The solutions of these equations are

$$C_0 = \frac{5}{2}e^{-3t} - \frac{3}{2}e^{-5t}$$

$$C_1 = \frac{e^{-3t} - e^{-5t}}{2}$$

Therefore

$$e^{[A]t} = \frac{1}{2}\begin{bmatrix} -e^{-3t} + 3e^{-5t} & -3(e^{-3t} - e^{-5t}) \\ e^{-3t} - e^{-5t} & 3e^{-3t} - e^{-5t} \end{bmatrix}$$

If $P(\lambda)$ has repeated roots, of multiplicity r, differentiate $e^{[A]t}$ a total of $r-1$ times with $[A]$ replaced by λ.

EXAMPLE

Let

$$[A] = \begin{bmatrix} -2 & 0 \\ 1 & -2 \end{bmatrix}$$

$$P(\lambda) = \begin{vmatrix} -2-\lambda & 0 \\ 1 & -2-\lambda \end{vmatrix} = (\lambda+2)(\lambda+2) = 0$$

Hence $[A]$ has the eigenvalue $\lambda = -2$, with multiplicity 2. To find $e^{[A]t}$, we put

$$e^{[A]t} = C_0[1] + C_1[A]$$

and replace $[A]$ with λ.
Differentiate $e^{\lambda t}$ once with respect to λ to obtain

$$te^{\lambda t} = C_1$$

Thus

$$C_1 = te^{-2t}$$

and

$$C_0 = e^{-2t}(1+2t)$$

Therefore

$$e^{[A]t} = e^{-2t}\begin{bmatrix} 1 & 0 \\ t & 1 \end{bmatrix}$$

The number of independent state variables is equal to the total number of independent initial conditions required for the network. In other words, it is equal to the number of degrees of freedom for the network.

It should be noted, however, that not all the capacitor voltages or inductor currents are independent. If a loop consisting of only capacitors exists in the network, the voltage across one of the capacitors is not independent. This voltage, by KVL, is equal to the negative of the sum of the other capacitor voltages. The same is true for the inductors.[4] This is illustrated in fig 1-23. Also, if we have a node to which only

---

[4] It should be noted that in fig 1-23b, KVL around the loop relates the *derivatives* of the currents through the inductors; hence, one of the currents is determined by the other two, provided all three *arbitrary* initial conditions are given. The same is true of fig 1-23c: KCL at the node relates the *derivatives* of the voltages across the capacitors; hence, one of the voltages is determined by the other two, provided all three *arbitrary* initial conditions are known.

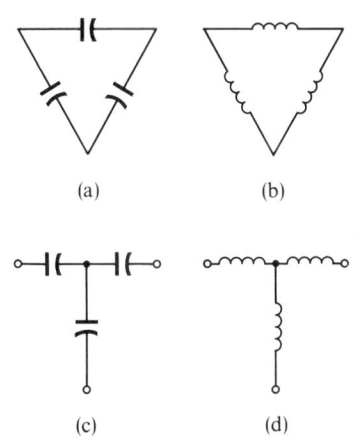

Fig 1-23 Circuits with one less degree of freedom than the number of energy-storage elements: (a) Capacitor only loop; (b) Inductor only loop; (c) Capacitors only connected to a node; (d) Inductors only connected to a node

inductors are connected, the current in one of the inductors is not independent of the current through the other inductors. This current is equal to the negative of the sum of all other inductor currents, assuming all the reference directions are into the node. The same is true for capacitors only connected to one node.[4] For example, in the network shown in fig 1-24, there are seven energy-storage elements, but the

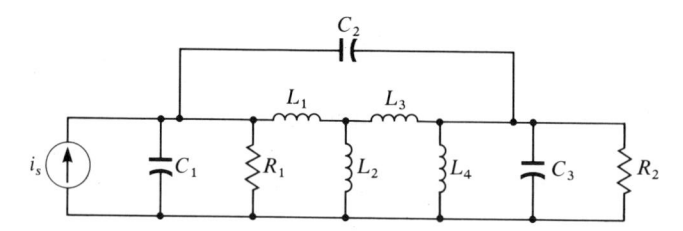

Fig 1-24 Network with four state variables

number of degrees of freedom for the network is only four. One dependence arises from the π of the capacitors and two from the π and T of the inductors. Hence, the number of state variables is four, i.e., four independent initial conditions are required for the network. It is left as an exercise (Problem 1-12) for the reader to set up the state equation for fig 1-24.

Approximate Method of Calculating $[x(t)]$

An approximate method for the solution of

$$\frac{d}{dt}[x] = [A][x] \tag{1-53}$$

where $[x(0)]$ is given, is as follows: We assume a sufficiently small time interval Δt, such that during this time interval $d[x(t)]/dt$ is approximately constant. Then, from the given initial state $[x(0)]$, we have

$$\frac{d[x(0)]}{dt} = [A][x(0)] \tag{1-54}$$

For a small interval $(0, \Delta t)$, we may express $[x(\Delta t)]$ approximately as

$$\frac{[x(\Delta t)] - [x(0)]}{\Delta t} \simeq \frac{d}{dt}[x(0)] \tag{1-55}$$

Thus

$$[x(\Delta t)] \approx [x(0)] + \frac{d[x(0)]}{dt}\Delta t = [x(0)] + [A][x(0)]\Delta t \quad (1\text{-}56)$$

We repeat the above step for the next interval to get

$$\frac{d[x(\Delta t)]}{dt} = [A][x(\Delta t)]$$

and

$$[x(2\,\Delta t)] \simeq [x(\Delta t)] + [A][x(\Delta t)]\Delta t \quad (1\text{-}57)$$

Continuing the successive approximation, we get

$$\{x[(n+1)\Delta t]\} \simeq [x(n\,\Delta t)] + [A][x(n\,\Delta t)]\Delta t \quad n = 0, 1, 2, \ldots, N$$
$$= \{[1] + \Delta t[A]\}\{[x(n\,\Delta t)]\} \quad (1\text{-}58)$$

The computation of (1-58) is usually done on a computer. The choice of the value of Δt depends upon the desired accuracy, the condition of the problem, and the number of significant figures used in the computation.

Note that the above method holds for nonlinear equations as well, since for a nonlinear system, the equation in the *normal form* may be written as

$$\frac{d[x(t)]}{dt} = \{f[x(t)]\} \quad (1\text{-}59)$$

The same approximation is utilized to obtain

$$[x(\Delta t)] \approx [x(0)] + \{f[x(0)]\}\Delta t$$
$$[x(2\,\Delta t)] \approx [x(\Delta t)] + \{f[x(\Delta t)]\}\Delta t \quad (1\text{-}60)$$
$$\{x[(n+1)\Delta t]\} \approx [x(n\,\Delta t)] + \{f[x(n\,\Delta t)]\}\Delta t \quad n = 0, 1, 2, \ldots, N$$

The computations in (1-60) can be readily performed on a computer. Nonlinear analysis is covered in Chapter 9 and will not be discussed further here.

1-10 Some Network Theorems

Familiarity with network theorems can be of great assistance in the analysis and design of circuits. In this section we discuss briefly, with examples, some of the useful theorems. For a detailed discussion, the reader is referred to the basic circuit theory texts listed in the references at the end of this chapter.

The Superposition Principle

The superposition principle is a direct consequence of the definition of linearity of a system. We have already encountered the definition of linearity in Section 1-4. The importance of this theorem cannot be overemphasized.

The superposition principle states that the response of any linear system to more than one independent excitation is given by the sum of the responses to each excitation acting alone.

EXAMPLE

Consider the voltage response across the 2 Ω resistor for the independent excitations shown in fig 1-25. When $i = 0$, $v = 3$ V, then $v_r = 2$ V. When $v = 0, i = 1$ A, then $v_r = 2/3$ V. Therefore (a) $v_r = 8/3$ V for $i = 1$ A and $v = 3$ V; and (b) if i and v are both multiplied by 6, then $v_r = 16$ V. However, if i is multiplied by 6 and v is multiplied by 3, then

$$v_r = 6(\tfrac{2}{3}) + 3(2) = 10 \text{ V}$$

In system theory the superposition theorem is usually worded as follows: For a linear network, the zero-state response caused by several independent sources is the sum of the zero-state responses caused by each independent source acting alone.

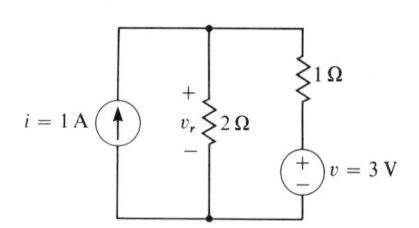

Fig 1-25 Example circuit

Thévenin and Norton Equivalent Theorems

The Thévenin and Norton equivalent theorems sometimes provide a significant simplification in calculating the response of complicated networks. By means of Thévenin's or Norton's theorem, a complicated network having a number of sources can be replaced by a simpler network with a series or shunt single-equivalent source.

1-10 Some Network Theorems

In Thévenin's theorem a linear network N connected to an arbitrary load (which can be nonlinear and/or time varying), as shown in fig 1-26a, is replaced by its equivalent representation, as shown in fig 1-26b. In fig 1-26b, v_e is the voltage across terminals 1–1' when the load is disconnected, i.e., when it is an open-circuit voltage.

Note that the interaction between N and N' is only through the voltage and current at terminals 1–1'. The linear network N can have independent and/or controlled sources. The voltage v_e in fig 1-26b is due to the independent sources of N and the initial conditions. The network N_e is obtained by setting all the initial conditions of N equal to zero, short-circuiting independent voltage sources, open-circuiting all the independent current sources, and not altering the controlled sources.

EXAMPLE

Consider the network shown in fig 1-27a. In this case the open-circuit voltage across 1–1' is $v_e = E/2$ and the impedance across 1–1' with E short circuited is $N_e = 2\,\Omega$, as shown in fig 1-27b.

In the Norton theorem, the linear network of fig 1-28a is replaced by its equivalent representation, as shown in fig 1-28b. In fig 1-28b, i_e is the current in the wire connecting terminals 1–1', i.e., it is a

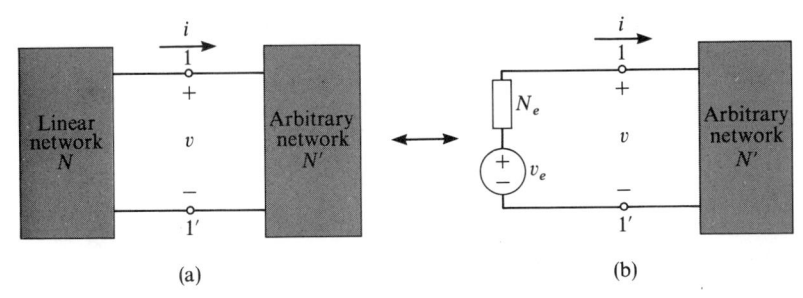

Fig 1-26 Thévenin equivalent representation

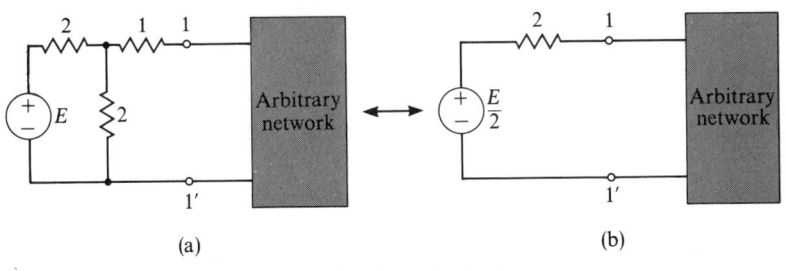

Fig 1-27 Example circuit

short-circuit current. N_e is obtained as for the Thévenin theorem. The Norton equivalent representation for the above example is shown in fig 1-29b.

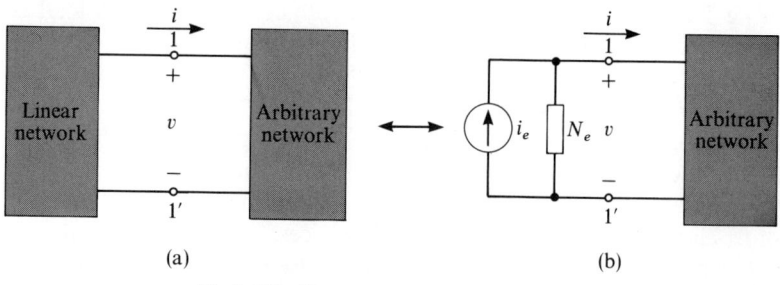

Fig 1-28 Norton equivalent representation

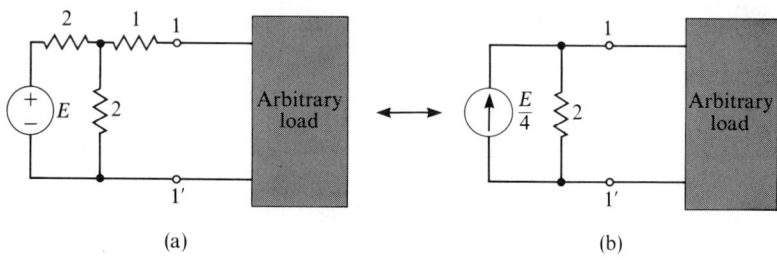

Fig 1-29 Example circuit

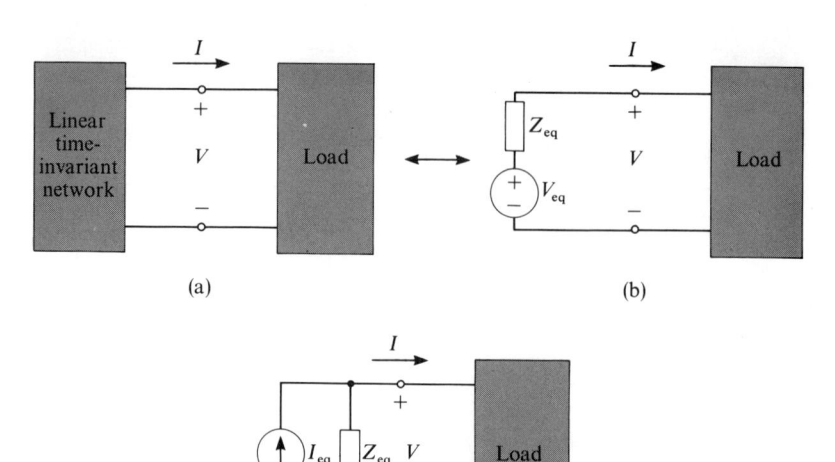

Fig 1-30 Thévenin and Norton equivalent representations of a general linear time-invariant network

1-10 Some Network Theorems

For a linear time-invariant network, it is sometimes more convenient to work with the transformed equations. In this case we have, from the Thévenin theorem, an equivalent impedance, Z_{eq}, in series with an equivalent voltage source, as shown in fig 1-30b. From the Norton theorem, we have the same impedance in shunt with an equivalent current source, as shown in fig 1-30c. In other words,

$$Z_{eq} = \frac{1}{Y_{eq}} = \frac{V_{eq}}{I_{eq}}$$

EXAMPLE

Consider the circuit shown in fig 1-31a. We wish to obtain the Thévenin and Norton equivalent representations. Redraw the circuit to

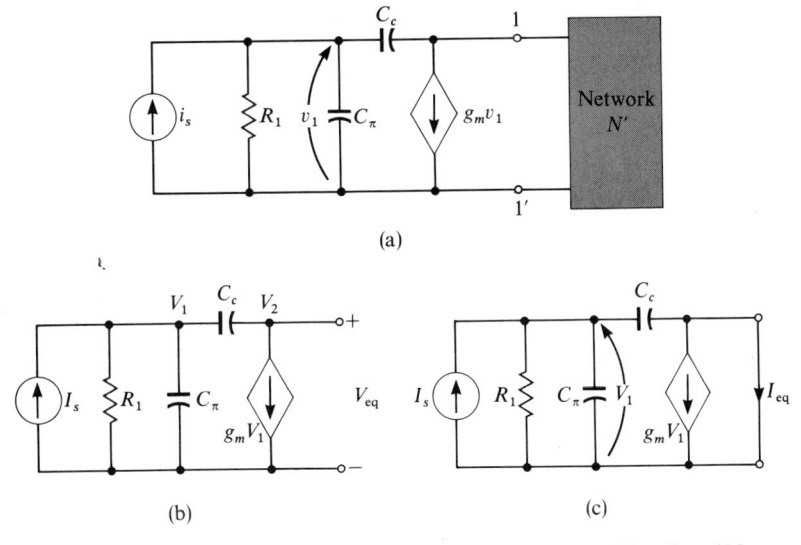

Fig 1-31 (a) Example circuit; (b) and (c) Circuits for obtaining the Thévenin and Norton equivalent networks

show the unknown quantities to be determined. From (b), by nodal analysis, we find the following transform equations:

$$\begin{bmatrix} G_1 + s(C_\pi + C_c) & -sC_c \\ g_m - sC_c & sC_c \end{bmatrix} \begin{bmatrix} V_1 \\ V_2 \end{bmatrix} = \begin{bmatrix} I_s \\ 0 \end{bmatrix}$$

$$V_{eq} = V_2 = \frac{\begin{vmatrix} G_1 + s(C_\pi + C_c) & I_s \\ g_m - sC_c & 0 \end{vmatrix}}{\begin{vmatrix} G_1 + s(C_\pi + C_c) & -sC_c \\ g_m - sC_c & sC_c \end{vmatrix}} = \frac{(sC_c - g_m)I_s}{s^2 C_c C_\pi + sC_c(g_m + G_1)}$$

To find I_{eq} from fig 1-31c, we have

$$I_{eq} = sC_c V_1 - g_m V_1$$

But

$$V_1 = \frac{I_s}{G_1 + s(C_\pi + C_c)}$$

Thus

$$I_{eq} = \frac{(sC_c - g_m)I_s}{G_1 + s(C_\pi + C_c)}$$

Hence

$$Z_{eq} = \frac{V_{eq}}{I_{eq}} = \frac{s(C_\pi + C_c) + G_1}{s^2 C_c C_\pi + sC_c(G_1 + g_m)}$$

Reciprocity Theorem

A network is said to be *reciprocal* when an ideal voltage source and a short circuit can be interchanged between any two pairs of terminals without changing the current in the short circuit, regardless of the topology of network. This is illustrated in fig 1-32. If the circuit is reciprocal, $i_1 = i_2$ for all t. Similarly, on a dual basis, a network is reciprocal when an ideal current source and an open circuit can be interchanged between any two terminals without changing the voltage across the open circuit, regardless of the topology of the network. This is illustrated in fig 1-33. If the circuit is reciprocal, $v_1 = v_2$ for all t. In the node and loop analysis, if the network consists of resistors, capacitors, inductors, coupled coils, and transformers, the elements $\beta_{ij} = \beta_{ji}$ and $\alpha_{ij} = \alpha_{ji}$.

It should be noted that all time-invariant RLCM networks are reciprocal regardless of the topology of the network. Passive networks are not reciprocal if they contain gyrators. In general, active networks are nonreciprocal; only in special cases can they be reciprocal. Two-terminal *negative-valued* elements, even though they are active, still produce reciprocal networks. Note that no sources, whether dependent or independent, are allowed in the reciprocity theorem. Thus reciprocity applies only for zero-state conditions.

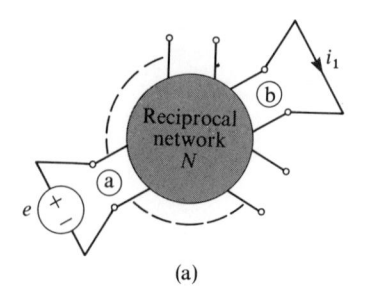

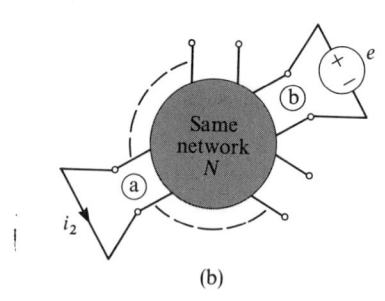

Fig 1-32 (a) and (b) Interchange of ideal voltage source and short-circuit to determine reciprocity

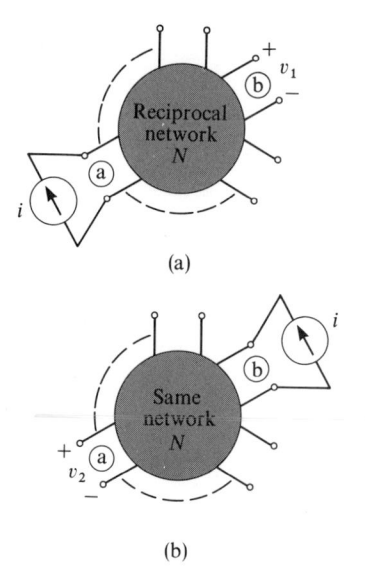

Fig 1-33 (a) and (b) Interchange of ideal current source and open-circuit to determine reciprocity

1-10 Some Network Theorems

EXAMPLE

Consider the linear time-invariant network in the transform domain shown in fig 1-34. If the input current at the terminals 1–1' is I_1, the output voltage is

$$V_0 = Z_4 \left(\frac{Z_2}{Z_2 + Z_3 + Z_4} \right) I_1$$

Now, if the same current is transferred at the terminals 2–2', the voltage across 1–1' is equal to V_0.

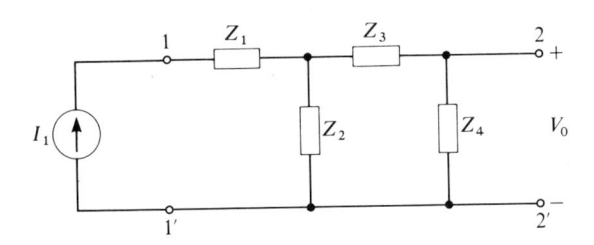

Fig 1-34 Example circuit

Source Transformations

Sometimes it is convenient to transform sources from one branch to another simply for the particular analysis, or when it is desired that an ideal source have a series or shunt elements such that the branches are of the form of fig 1-14. Two such methods of source transformations are shown in figs 1-35 and 1-36.

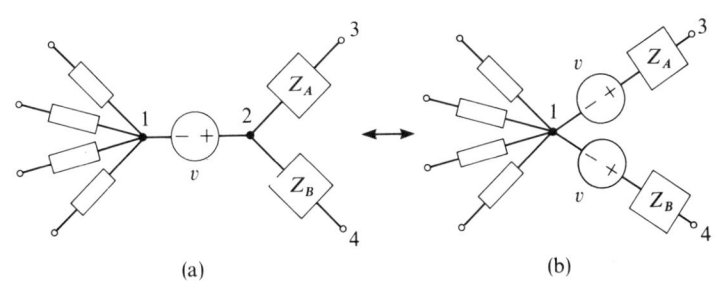

Fig 1-35 (a) and (b) Voltage-source transformation

The reader can readily verify that the above transformations do not affect the branch voltages and currents of the network. It should be noted that the above transformations apply for independent as well as for controlled sources.

EXAMPLE

Consider the circuit shown in fig 1-37. It is desired to find the voltage v_0 for a given input i_s. This problem can, of course, be solved by node analysis or some other method. We shall use it, however, to illustrate the use of the source transformation theorem.

According to fig 1-36, the circuit of fig 1-37 can be redrawn as shown

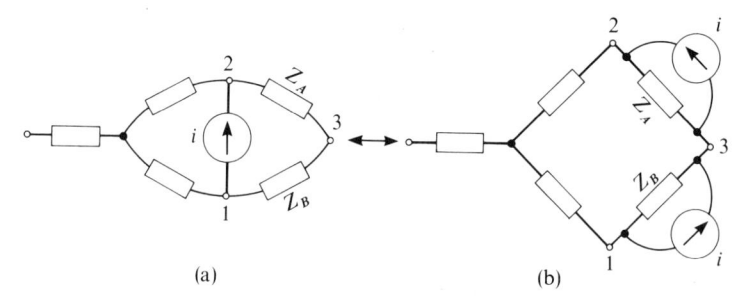

Fig 1-36 (a) and (b) Current-source transformation

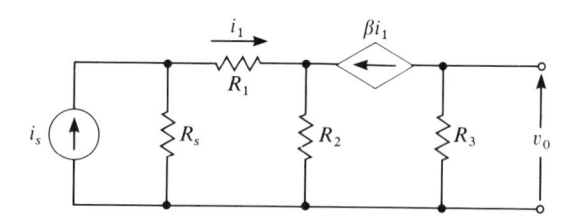

Fig 1-37 Example circuit

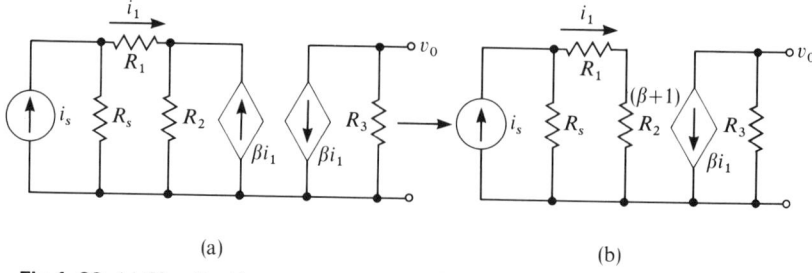

Fig 1-38 (a) Circuit with current-source transformation applied; (b) Simplified circuit

1-11 Network Functions, Poles, Zeros, and Natural Frequencies

in fig 1-38a. The circuit can, in turn, be simplified as in fig 1-38b. From fig 1-38b, we have

$$i_1 = \frac{R_s}{R_s + R_1 + (\beta + 1)R_2} i_s$$

$$v_0 = -(\beta i_1)R_3 = -\frac{\beta R_s R_3}{R_s + R_1 + (\beta + 1)R_2} i_s$$

1-11 Network Functions, Poles, Zeros, and Natural Frequencies

For the linear time-invariant circuit, the complex frequency-domain ($s = \sigma + j\omega$) approach is sometimes more convenient for analysis and design. For example, an amplifier design is readily made in the frequency domain. Consider a linear time-invariant network N having an excitation $x(t)$, which may be a current or a voltage, and a response $y(t)$, which may also be either a current or a voltage, as depicted in fig 1-39.

The network function is defined as the ratio of the transform of the response to the transform of the excitation, provided all the initial conditions of the network are zero. In other words, it relates the excitation and the response in the s domain as follows:

$$\mathscr{L}(\text{Response with zero initial state}) = (\text{Network function}) \times \mathscr{L}(\text{excitation}) \quad (1\text{-}61)$$

Fig 1-39 Linear time-invariant network

where $\mathscr{L}$ designates the Laplace transformation. We may rewrite (1-61) as follows:

$$\begin{pmatrix} \text{Network function} \\ \\ H(s) \end{pmatrix} = \begin{cases} \dfrac{\mathscr{L}(\text{Zero-state response})}{\mathscr{L}(\text{excitation})} \\ \\ \dfrac{Y(s)}{X(s)} \Big|_{\substack{\text{initial conditions of} \\ \text{the network set to zero}}} \end{cases} \quad (1\text{-}62)$$

Since in an electrical network the excitation (or input) and the response (or output) are each of two possible types, namely current or voltage, there are four possible combinations, as shown in table 1-2.

When the input and output quantities are defined at the *same* pair of terminals, these network functions are called *driving-point* impedance

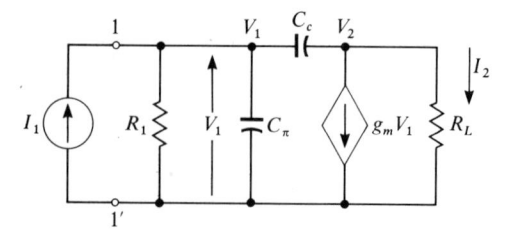

Fig 1-40 Example circuit

or *driving-point* admittance functions. For example,

$$Z(s) = \frac{V_i(s)}{I_i(s)} \qquad Y(s) = \frac{I_i(s)}{V_i(s)} \qquad i = 1, 2 \qquad (1\text{-}63)$$

EXAMPLE

For the circuit shown in fig 1-40, determine the transfer current ratio I_2/I_1 and the driving-point impedance function at port 1–1'. Using nodal analysis, we have

$$\begin{bmatrix} G_1 + s(C_\pi + C_c) & -sC_c \\ g_m - sC_c & sC_c + G_L \end{bmatrix} \begin{bmatrix} V_1 \\ V_2 \end{bmatrix} = \begin{bmatrix} I_1 \\ 0 \end{bmatrix}$$

$$V_2 = \frac{\begin{vmatrix} G_1 + s(C_\pi + C_c) & I_1 \\ g_m - sC_c & 0 \end{vmatrix}}{\begin{vmatrix} G_1 + s(C_\pi + C_c) & -sC_c \\ g_m - sC_c & sC_c + G_L \end{vmatrix}}$$

$$= \frac{(g_m - sC_c)I_1}{s^2 C_c C_\pi + s[C_c(G_1 + G_L + g_m) + C_\pi G_L] + G_1 G_L}$$

TABLE 1-2 *Transfer Functions*

Input	Output	Network function and name
$I_1(s)$	$I_2(s)$	$\dfrac{I_2(s)}{I_1(s)}$ = Transfer current ratio
$I_1(s)$	$V_2(s)$	$\dfrac{V_2(s)}{I_1(s)}$ = Transfer impedance
$V_1(s)$	$V_2(s)$	$\dfrac{V_2(s)}{V_1(s)}$ = Transfer voltage ratio
$V_1(s)$	$I_2(s)$	$\dfrac{I_2(s)}{V_1(s)}$ = Transfer admittance

and $I_2 = V_2 G_L$. Therefore, the transfer current ratio is given by

$$\frac{I_2}{I_1} = -\frac{G_L}{C_\pi} \frac{s - g_m/C_c}{s^2 + s[(1/C_\pi)(G_1 + G_L + g_m) + (1/C_c)G_L] + G_L G_1/C_c C_\pi}$$

1-11 Network Functions, Poles, Zeros, and Natural Frequencies

Similarly we find

$$V_1 = \frac{\begin{vmatrix} I_1 & -sC_c \\ 0 & sC_c + G_L \end{vmatrix}}{\Delta}$$

$$= \frac{(sC_c + G_L)I_1}{s^2 C_c C_\pi + s[C_c(G_1 + G_L + g_m) + C_\pi G_L] + G_L G_1}$$

and the driving-point impedance function between the terminals 1–1' is given by

$$\frac{V_1}{I_1} = \frac{1}{C_\pi} \frac{s + G_L/C_c}{s^2 + s[(1/C_\pi)(G_L + G_1 + g_m) + (1/C_c)G_L] + G_L G_1/C_c C_\pi}$$

It should be noted that for all linear lumped time-invariant networks, the network function can be written in the general form

$$H(s) = \frac{N(s)}{D(s)} = K \frac{s^m + a_{m-1}s^{m-1} + \cdots + a_1 s + a_0}{s^n + b_{n-1}s^{n-1} + \cdots + b_1 s + b_0} \quad (1\text{-}64)$$

in which the coefficients a_i and b_i are all real quantities. We may rewrite (1-64) in factored form as follows:

$$H(s) = \frac{N(s)}{D(s)} = K \frac{(s + z_1)(s + z_2) \cdots (s + z_m)}{(s + p_1)(s + p_2) \cdots (s + p_n)} \quad (1\text{-}65)$$

The roots of the numerator $N(s)$, namely z_i, are called the zeros of the network function, while the roots of the denominator $D(s)$, namely p_i, are called the poles of the network function.

As an example, consider fig (1-40) for the following circuit element values: $R_1 = R_L = 1\,\text{k}\Omega$; $C_\pi = 200\,\text{pF}$ ($1\,\text{pF} = 10^{-12}$ farad); $C_c = 10\,\text{pF}$; $g_m = 0.1$ mho. The transfer current ratio becomes

$$H(s) = \frac{I_2}{I_1} = -\frac{s}{s^2 + 13.1s + 12.1} = -\frac{s}{(s + 1.01)(s + 12.09)}$$

The pole-zero plot for this example is shown in fig 1-41, where ○ indicates the zero and × indicates the poles.

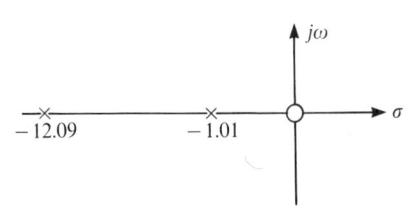

Fig 1-41 Pole-zero locations of the transfer function

Some Properties of Network Functions

The following properties of the network function can be readily established from (1-64). This is left as an exercise (Problem 1-25).
(1) $H(s)$ is real when s is real.
(2) Poles and zeros of $H(s)$ are either real or occur in complex conjugates.
(3) $H(-j\omega) = \overline{H(j\omega)}$, where the bar denotes conjugation.
(4) The real part of $H(j\omega)$ and the magnitude of $H(j\omega)$ are even functions of ω, i.e., $ReH(j\omega) = ReH(-j\omega)$, and so on. The imaginary part of $H(j\omega)$ and argument of $H(j\omega)$ are odd functions of ω, i.e., $ImH(j\omega) = -ImH(-j\omega)$, and so on.

Natural Frequencies

The knowledge of the location of *poles of a network function* gives some important information regarding the dynamic behavior of the network. They are, therefore, of particular interest and are called the *natural frequencies* of the network. In other words, each pole of a network function is called a natural frequency of the circuit. From (1-61), if the excitation is a current (i.e., open-circuit condition) the natural frequencies are referred to as the *open-circuit natural frequencies*. If the excitation is a voltage (i.e., short-circuit condition) the natural frequencies are referred to as the *short-circuit natural frequencies*. For (1-65), it is possible to cancel some natural frequencies with some zeros for some response functions. For those special locations, we say that the canceled natural frequencies are unobservable in certain portions of the network.

The natural frequencies of a network may be determined by any of the following methods:

(1) By setting the circuit determinant of the zero-state circuit from either node or loop analysis equal to zero, i.e., from (1-27) and (1-36):

$$\Delta = 0 \quad \text{or} \quad \Delta' = 0 \qquad (1\text{-}66)$$

The roots of (1-66) are the poles of the network functions, hence the natural frequencies of the circuit.

(2) By setting the loop impedance around any loop equal to zero: First set the initial conditions of the network equal to zero, short-circuit the independent voltage sources and open-circuit the independent current sources, and then determine the zeros of Z loop at any convenient loop.

(3) By setting the nodal admittance across any node pair equal to zero: First set the initial conditions of the network equal to zero,

1-11 Network Functions, Poles, Zeros, and Natural Frequencies

short-circuit the independent voltage sources, open-circuit the independent current sources, and then determine the zeros of Y node taken across any convenient port.

(4) By finding the eigenvalues of the A matrix associated with the zero-state system in the state variable formulation: Find the roots of

$$\det\{[A] - \lambda[1]\} = |[A] - \lambda[1]| = 0 \tag{1-67}$$

EXAMPLE

Consider the network shown in fig 1-42. We determine the natural frequencies of the network by the first two methods discussed above. (Problem 1-23 verifies the use of the other methods.) We first set the initial state of the circuit and the independent sources equal to zero, since the natural frequencies of the network are independent of these. Using the first two methods to find the natural frequencies, we have

(1) By node analysis: The circuit determinant of fig 1-42b on a nodal basis is

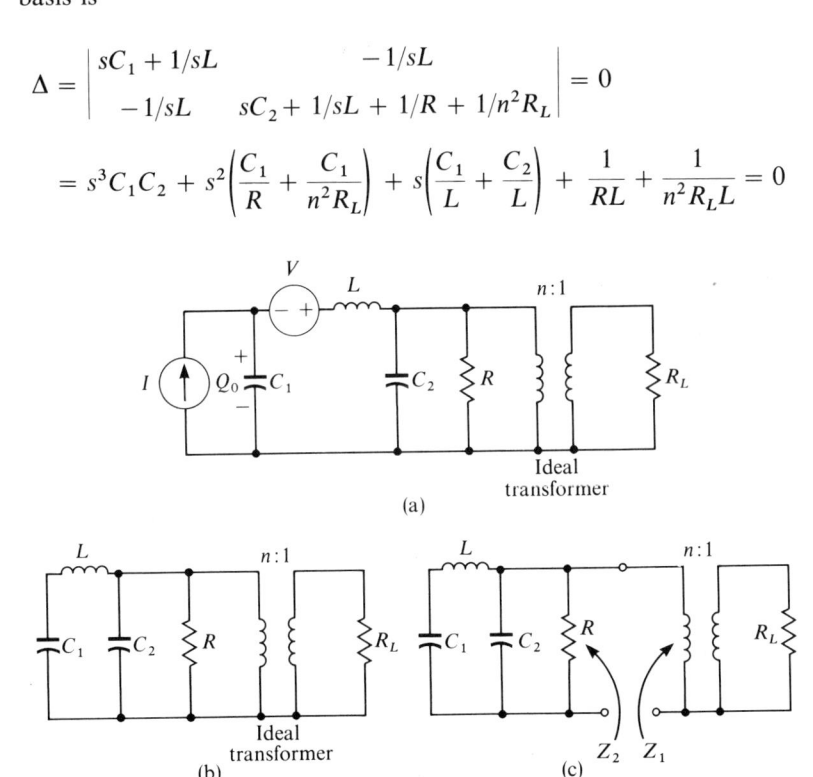

$$\Delta = \begin{vmatrix} sC_1 + 1/sL & -1/sL \\ -1/sL & sC_2 + 1/sL + 1/R + 1/n^2 R_L \end{vmatrix} = 0$$

$$= s^3 C_1 C_2 + s^2 \left(\frac{C_1}{R} + \frac{C_1}{n^2 R_L} \right) + s \left(\frac{C_1}{L} + \frac{C_2}{L} \right) + \frac{1}{RL} + \frac{1}{n^2 R_L L} = 0$$

Fig 1-42 (a) Example circuit; (b) Circuit with initial state and independent sources set equal to zero; (c) Loop-impedance calculation applied to circuit

The natural frequencies are the roots of $\Delta = 0$, namely, the roots of

$$s^3 + s^2\left(\frac{1}{RC_2} + \frac{1}{n^2 R_L C_2}\right) + s\left(\frac{1}{LC_2} + \frac{1}{LC_1}\right)$$
$$+ \frac{1}{RC_1 LC_2} + \frac{1}{n^2 R_L C_1 LC_2} = 0$$

If the circuit element values are given, the roots can be readily obtained (hence the natural frequencies).

(2) By the loop impedance method: Consider fig 1-42b as redrawn in fig 1-42c for clarity. We break open the loop as shown in the figure:

$$Z \text{ loop} = Z_1 + Z_2 = \underbrace{n^2 R_L}_{Z_1} + \underbrace{\left(sL + \frac{1}{sC_1}\right) \| \left(\frac{R}{1 + sRC_2}\right)}_{Z_2}$$

where $\|$ denotes parallel combination.

$$Z \text{ loop} = \frac{s^3 + s^2[(1/RC_2) + (1/n^2 R_L C_2)] + s[(1/LC_2) + (1/LC_1)] + (1/RC_1 LC_2) + (1/n^2 R_L C_1 LC_2)}{\text{Denominator of } Z \text{ loop}}$$

The zeros of Z loop $= 0$ are the natural frequencies, the same as above.

1-12 Circuit Normalization (Frequency and Magnitude Scaling)

In the analysis and design of circuits using paper and pencil, it is often desirable to use a suitable change of scale in amplitude and frequency. Under an appropriate scale change the tedium of computation with large numbers and powers of 10 is reduced to simple numerical operations. If a digital computer is used in the analysis, no advantage, in general, is accrued by normalization. A change in the frequency scale is referred to as frequency normalization and a change in the scale of amplitude is called resistance normalization. If the radian frequency is normalized with respect to Ω_0 and the resistance is normalized with respect to R_0, the normalized element values, designated by the sub-

1-12 Circuit Normalization (Frequency and Magnitude Scaling)

script n, are given by

$$s_n = \frac{s}{\Omega_0} \quad R_n = \frac{R}{R_0}$$

$$L_n = \frac{L\Omega_0}{R_0} \quad C_n = R_0\Omega_0 C \tag{1-68}$$

From (1-68), it is apparent that the normalized element values are dimensionless, hence dimensional analysis cannot be used as a check after normalization. Normalization should always be done with respect to known values of Ω_0 and R_0. The choice of R_0 and Ω_0 is usually obvious in a given problem.

EXAMPLE

Consider the simple circuit shown in fig 1-43. The locus of the natural frequencies as L is varied and the value of L which produces two identical poles is to be determined. The transform voltage transfer ratio of the circuit V_0/V_1 is given by

$$\frac{V_0}{V_1} = \frac{-g_m R}{s^2 LC + sRC + 1} = \frac{-g_m R}{LC} \frac{1}{s^2 + s(R/L) + 1/LC}$$

In this case, one appropriate set of values for R_0 and Ω_0 is $R_0 = 10\,\text{k}\Omega$ and $\Omega_0 = 10^7 (RC)^{-1}$. The normalized transfer function using (1-68) is

$$\left(\frac{V_0}{V_1}\right)_n = \frac{-(g_m R_0)(RR_0^{-1})}{(L_n R_0/\Omega_0)(C_n/R_0\Omega_0)}$$

$$\times \frac{1}{s^2 + s[(R_n R_0/L_n R_0)\Omega_0] + 1/[(L_n R_0/\Omega_0)(C_n/R_0\Omega_0)]}$$

$$\left(\frac{V_0}{V_1}\right)_n = \frac{-g_m R_0}{L_n C_n} \frac{1}{(s/\Omega_0)^2 + (s/\Omega_0)(R_n/L_n) + (1/L_n C_n)}$$

$$= \frac{-10}{L_n} \left(\frac{1}{s_n^2 + s_n(1/L_n) + (1/L_n)}\right)$$

The locus of the natural frequencies as L_n is varied is shown in fig 1-44. For $L_n = \frac{1}{4}$ we have identical poles and

$$A_v = \left(\frac{V_0}{V_1}\right)_n = -\frac{40}{(s_n + 2)^2} \tag{1-69}$$

The actual value of the inductance for this case is $L = L_n R_0/\Omega_0 = 0.25\,\text{mH}$.

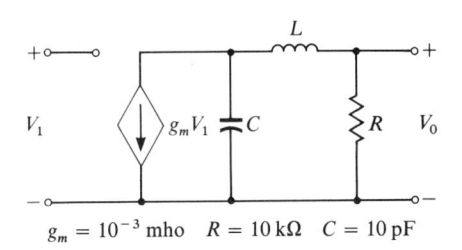

Fig 1-43 Example circuit

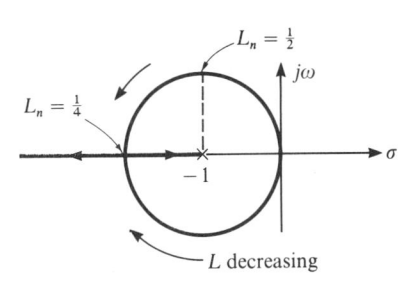

$g_m = 10^{-3}$ mho $\quad R = 10\,\text{k}\Omega \quad C = 10\,\text{pF}$

Fig 1-44 Normalized s-plane pole locus as L_n is varied

1-13 Electronic Circuit Analysis by Computers

Whenever the number of nodes and/or loops is large, the analysis is best performed by computer. One could write a program for the solution of such a problem; however, this is time-consuming and unnecessary, as there are now many computer programs available for performing such circuit analysis. Indeed, more versatile programs will be appearing on the scene in the near future.

Some of the present computer programs perform only ac or dc analysis for linear RLC circuits, while others are capable of analyzing nonlinear and active circuits. Efficient iterative analysis at low cost can be performed by most of the available computer programs. In addition to normal circuit analysis, these programs can simulate difficult laboratory problems. For example, the effect of individual circuit parameter variations on the response may be easily obtained by a computer analysis which otherwise is rather difficult to obtain experimentally. The effect of component failure on the circuit can easily be observed without destroying the component. Worst-case analysis, which is very difficult to achieve in the laboratory, can easily be made on a computer. Tolerance studies and iterative analysis can be efficiently handled by computers. The importance of these become quite evident when integrated circuits are considered, because the size of the problem necessitates the use of a computer and, also, because in these cases breadboard circuits are very expensive and time consuming.

Some of the general packaged computer programs now in use are the following[5]: **BIAS-3, CIRCAL, CIRCUS, CIRPAC, CORNAP, ECAP, LISA, NASAP, NET-1, SCEPTRE** and Time Sharing Library System. For illustrative purposes, brief descriptions of two of the user-oriented programs are given here.

---

[5] Detailed information about the various packaged programs mentioned above may be obtained from the originators of the programs listed below:

BIAS-3	University of California, Department of Electrical Engineering and Computer Sciences, Berkeley, California
CIRCAL	Massachusetts Institute of Technology, Department of Electrical Engineering, Cambridge, Massachusetts
CIRCUS	Boeing Company, Seattle, Washington
CIRPAC	Bell Telephone Laboratories, Inc., Murray Hill, New Jersey
CORNAP	Cornell University, Department of Electrical Engineering, Ithaca, New York
ECAP	IBM Corporation, White Plains, New York
LISA	IBM Corporation, San Jose, California
NASAP	NASA Electronics Research Center, Cambridge, Massachusetts
NET-1	Los Alamos Scientific Laboratory, Los Alamos, New Mexico
SCEPTRE	IBM Corporation, Owego, New York

ECAP

Electronic Circuit Analysis Program is written in FORTRAN language (the user, however, employs a modified simpler language) and is usable directly on large IBM computers (7090, 360 series), CDC 6600, Univac 1108, and any other machine that has a suitably large memory. The ECAP program performs ac, dc, and transient analysis. The program makes use of the topological, loop impedance, and nodal admittance matrices. Since elements of the ECAP matrices are numerical, poles and zeros cannot be obtained, but frequency and time responses are easily obtained. Sensitivity analysis by parameter variations and dc worst-case studies can also be made. The ECAP allows dependent and independent sources, switches, and RLCM elements. ECAP can handle electrical networks that contain as many as 20 nodes and 60 branches. If the network is nonlinear, the nonlinear elements must be described by a piecewise-linear approximation by means of switches and resistors. (This general topic is discussed in Chapter 9.) Although piecewise-linear approximation provides flexibility, it also has drawbacks. Exact analysis of nonlinear circuits cannot be performed. For accurate nonlinear analysis, CIRCUS and/or SCEPTRE may be used. ECAP is particularly suitable for the ac analysis of linear networks.

The topology of the network to be analyzed, the specific analysis desired, the circuit elements comprising the network, the nature of the excitations, and the desired output are furnished by the user in an ECAP program. The techniques involved in using ECAP can be readily learned by the reader since the input language of ECAP is in the terminology of electronic circuits. One needs no knowledge of the internal construction of the program and no previous computer experience is necessary. An example illustrates the input–output data of ECAP for dc and ac analyses. Consider the circuit shown in fig 1-45. In order to

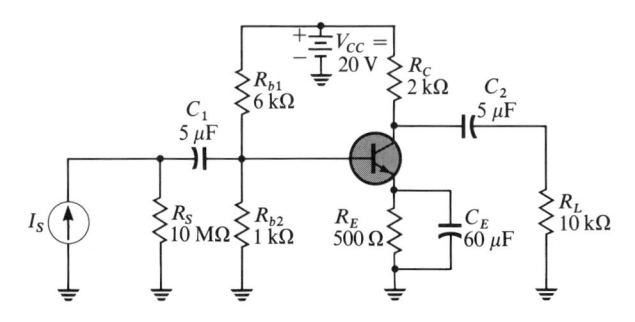

Fig 1-45 A single-stage common-emitter amplifier

analyze any electronic circuit with ECAP (dc, ac, or transient analysis), the devices must be replaced with equivalent circuits composed of circuit elements recognized by ECAP. These are RLCM elements, independent voltage source, independent current source, dependent current source, and ideal switch.

The dc performance including worst-case and sensitivity analysis can be analyzed with the aid of ECAP dc analysis. First, the transistor is replaced by its low-frequency model (which includes the emitter-base voltage V_{EB}). More elaborate models may be used if necessary. The dc circuit is shown in fig 1-46, where the transistor is assumed to have the following parameters:

$$(V_{EB} = 0.5 \text{ V}, h_{FE} = \beta_1 = 50, h_{ie} = 350 \text{ }\Omega, h_{oe}^{-1} = 11.1 \text{ k}\Omega)$$

The node numbers and branch numbers assigned with positive current directions are indicated by the arrowhead. The voltage source V is inserted in both branches B1 and B2 to eliminate the need for an additional node and branch. Note that the dependent current source in B6 is $h_{FE}I_B$, where I_B is the current in the controlling branch B4. The

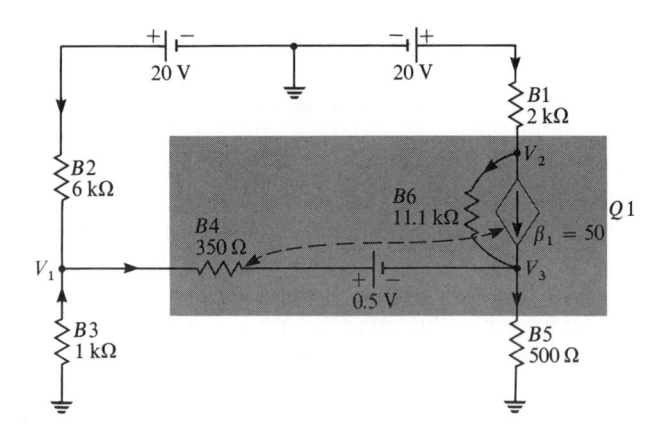

Fig 1-46 Dc equivalent circuit for ECAP

input data for the circuit and the dc nominal solution are shown in fig 1-47. (The examples are the same as those used in the ECAP user's manual). When worst-case analysis is desired, the tolerance data should be supplied. This is done by entering as input data the maximum and minimum values of each circuit parameter, as well as its nominal value, as shown in fig 1-48.

The ECAP ac analysis input-data preparation is quite similar to that for the dc analysis. For low-frequency analysis, the transistor model

1-13 Electronic Circuit Analysis by Computers

```
SINGLE-STAGE AMPLIFIER
C
      DC ANALYSIS
C
   B1 N(0,2),R=2000,E=20
   B2 N(0,1),R=6000,E=20
   B3 N(0,1),R=1000
   B4 N(1,3),R=350,E=-0.5
   B5 N(3,0),R=500
   B6 N(2,3),R=11.1E3
   T1 B(4,6),BETA=50
      PRINT,VOLTAGES,CURRENTS
      EXECUTE

NODE VOLTAGES

NODES            VOLTAGES

 1-   3        .27942207+01       .11072695+02       .22685272+01

ELEMENT CURRENTS

BRANCHES         CURRENTS

 1-   4        .44636525-02      .28676298-02     -.27942207-02     .73410293-04
 5-   6        .45370542-02      .44636829-02
```

Fig 1-47 Computer printout of input data and output results

```
SINGLE-STAGE AMPLIFIER
C
      DC ANALYSIS
C
   B1 N(0,2),R=2000(1860,2140),E=20(19,21)
   B2 N(0,1),R=6000(5580,6420),E=20(19,21)
   B3 N(0,1),R=1000(930,1070)
   B4 N(1,3),R=350(.10),E=-0.5
   B5 N(3,0),R=500(465,535)
   B6 N(2,3),R=11.1E3(.10)
   T1 B(4,6),BETA=50(45,55)
      WORST CASE
      PRINT,WORST CASE
      EXECUTE

WORST CASE SOLUTIONS FOR NODE VOLTAGES

  NODE    WCMIN             NOMINAL           WCMAX

    1    .23476473+01      .27942207+01      .33096024+01
    2    .65167675+01      .11072695+02      .14693142+02
    3    .18212073+01      .22685272+01      .27850159+01
```

Fig 1-48 Computer printout of input data and output worst-case results

is the same, except that V_{EB} is not needed (see fig 3-29b). The ac circuit is shown in fig 1-49 with the node numbers and branch numbers assigned, with positive current indicated by the arrowhead. The necessary data input for ECAP is shown in fig 1-50a. The command MODIFY statement is used to get ac parameter modification for the frequency-response calculation. It starts at 10 Hz and proceeds to 2.56 kHz with a multiplication constant of 2; for example, 10 Hz,

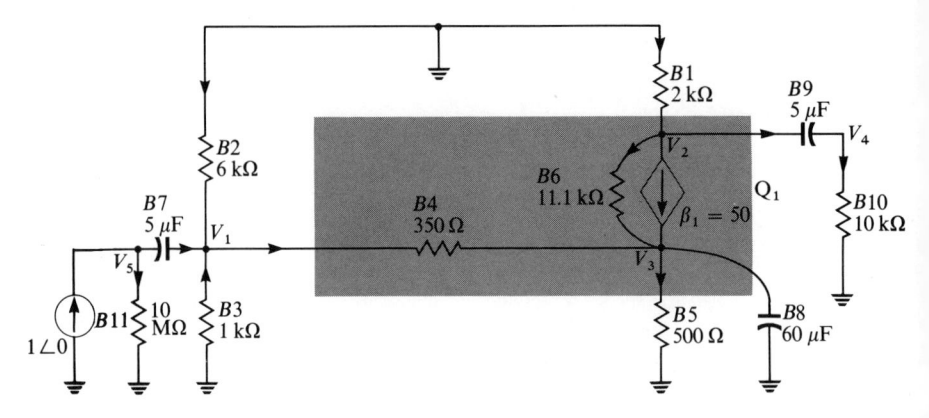

48 1/Review of Selected Topics from Circuit Theory

Fig 1-49 Ac equivalent circuit for ECAP

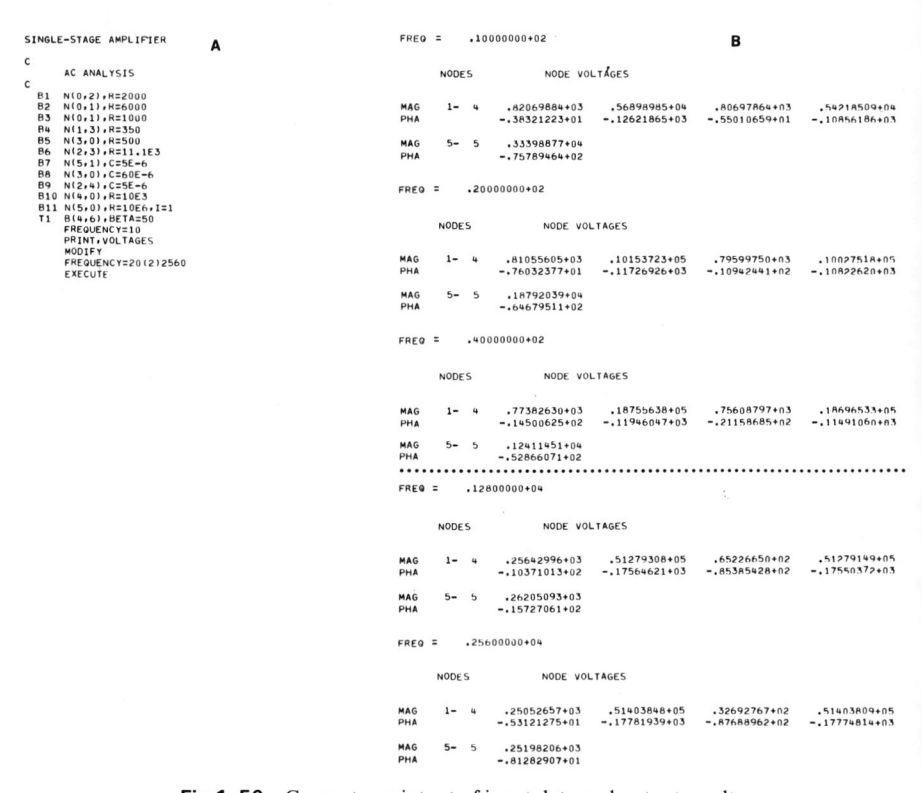

```
SINGLE-STAGE AMPLIFIER         A
C
      AC ANALYSIS
C
  B1  N(0,2),R=2000
  B2  N(0,1),R=6000
  B3  N(0,1),R=1000
  B4  N(1,3),R=350
  B5  N(3,0),R=500
  B6  N(2,3),R=11.1E3
  B7  N(5,1),C=5E-6
  B8  N(3,0),C=60E-6
  B9  N(2,4),C=5E-6
  B10 N(4,0),R=10E3
  B11 N(5,0),R=10E6,I=1
  T1  B(4,6),BETA=50
      FREQUENCY=10
      PRINT,VOLTAGES
      MODIFY
      FREQUENCY=20(2)2560
      EXECUTE
```

```
FREQ =   .10000000+02                         B

         NODES           NODE VOLTAGES

MAG   1-  4    .82069884+03   .56898985+04   .80697864+03   .54218509+04
PHA            -.38321223+01  -.12621865+03  -.55010659+01  -.10856186+03

MAG   5-  5    .33398877+04
PHA            -.75789464+02

FREQ =   .20000000+02

         NODES           NODE VOLTAGES

MAG   1-  4    .81055605+03   .10153723+05   .79599750+03   .10027518+05
PHA            -.76032377+01  -.11726926+03  -.10942441+02  -.10822620+03

MAG   5-  5    .18792039+04
PHA            -.64679511+02

FREQ =   .40000000+02

         NODES           NODE VOLTAGES

MAG   1-  4    .77382630+03   .18755638+05   .76608797+03   .18696533+05
PHA            -.14500625+02  -.11946047+03  -.21158685+02  -.11491060+03

MAG   5-  5    .12411451+04
PHA            -.52866071+02
************************************************************
FREQ =   .12800000+04

         NODES           NODE VOLTAGES

MAG   1-  4    .25642996+03   .51279308+05   .65226650+02   .51279149+05
PHA            -.10371013+02  -.17564621+03  -.85385428+02  -.17550372+03

MAG   5-  5    .26205093+03
PHA            -.15727061+02

FREQ =   .25600000+04

         NODES           NODE VOLTAGES

MAG   1-  4    .25052657+03   .51403848+05   .32692767+02   .51403809+05
PHA            -.53121275+01  -.17781939+03  -.87688962+02  -.17774814+03

MAG   5-  5    .25198206+03
PHA            -.81282907+01
```

Fig 1-50 Computer printout of input data and output results

20 Hz, 40 Hz, ..., 1.28 kHz, 2.56 kHz. (The output data for frequencies 80 Hz to 640 Hz are omitted to save space in fig 1-50b.) The ECAP is treated in greater detail in the IBM user's manual and Ref 7.

CORNAP

*CO*Rnell *N*etwork *A*nalysis *P*rogram uses the algorithm based on the state-space approach. The program obtains state equations and the desired transfer function together with frequency and time responses of an *n*-port general linear active network. The program is written entirely in FORTRAN language. The program as available is virtually machine-independent and can be used in many machines, such as IBM 7094, 360's, GE 635, CDC 6400, Univac 1108. The input to the program provides a topological description of the network together with element values. The output consists of the A, B, C, and D matrices of the state equations. Transfer functions between desired outputs and assigned inputs in the form of gain constant, poles, and zeros are given as outputs. This program will also compute time and frequency responses, magnitude and phase responses, impulse and step responses. As an example, consider the active linear network shown in fig 1-51. The

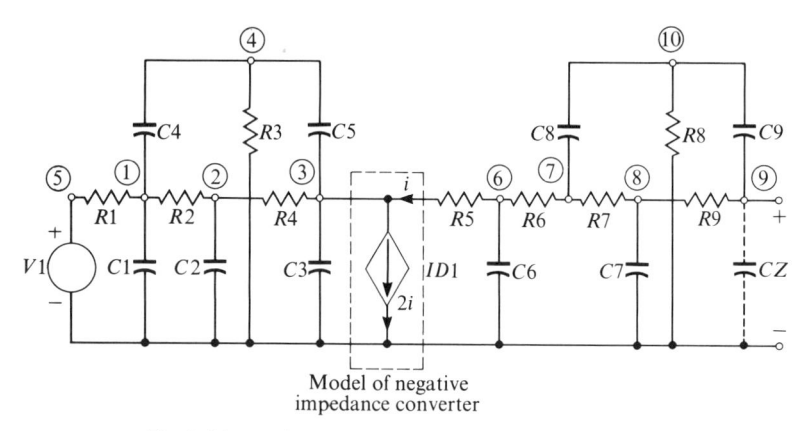

Fig 1-51 Active *RC* network used in CORNAP example

network has one dependent and one independent source. The zero-valued capacitor *CZ* is used to establish a branch at the output across which the output voltage (i.e., open-circuit output) is taken. The following pages (fig 1-52 and fig 1-53) show the required input cards and the computer-generated outputs for these data. Specifically, these are the A, B, C, D matrices for the state variable, the poles and zeros of the desired transfer function, and the frequency responses and time responses of the transfer function under consideration.[6] These

---

[6] If the ECAP and/or CORNAP packaged programs are available, it is instructive to devote more time to this topic and have students use the computer to solve some problems which are discussed and assigned later on in the text.

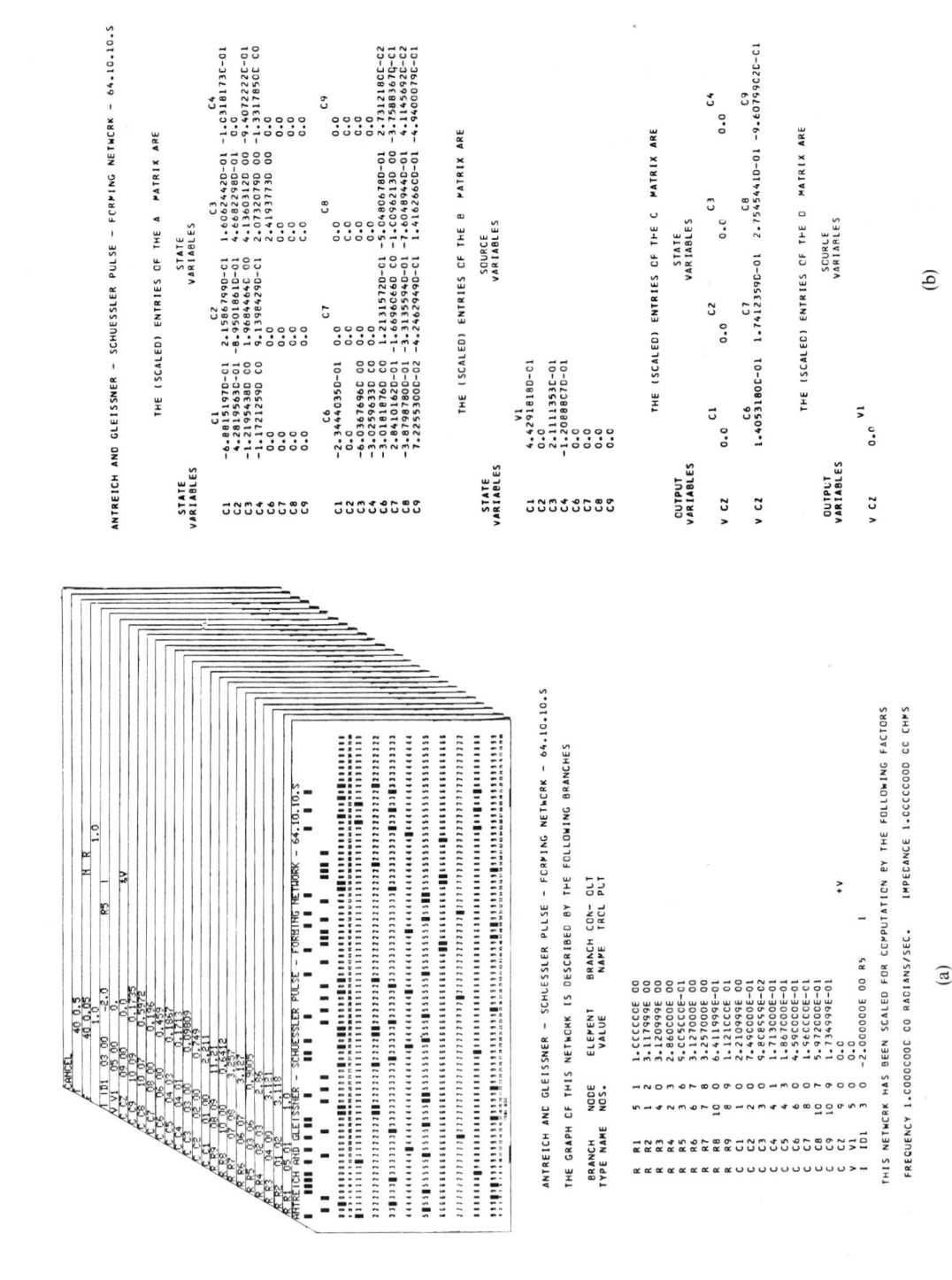

50   1/Review of Selected Topics from Circuit Theory

Fig 1-52 (a) Card input and computer listing for example; (b) State matrices for example

ANTREICH AND GLEISSNER - SCHUESSLER PULSE - FORMING NETWORK - 64.10.10.S

TRANSFER FUNCTION CRITICAL FREQUENCIES (SCALED)

OUTPUT VARIABLE - V CZ
SOURCE VARIABLE - V1

GAIN CONSTANT IS 7.1778486D-02

POLE POSITIONS REAL PART	POLE POSITIONS IMAGINARY PART	ORDER	ZERO POSITIONS REAL PART	ZERO POSITIONS IMAGINARY PART	ORDER
-4.4859237D-02	9.1549720D-01	1	-8.9500642D-01	0.0	1
-3.6668824D-01	6.2457207D-01	1	-2.0220165D 00	0.0	1
-8.9500642D-01	0.0	1	-6.0962475D-06	-1.2945755D 00	1
-2.0220163D 00	0.0	1	1.9923657D-04	1.0199096D 00	1
-4.9054554D-01	2.0810346D-01	1			

THIS NETWORK HAS BEEN SCALED FOR COMPUTATION BY THE FOLLOWING FACTORS
FREQUENCY 1.0000000D 00 RADIANS/SEC. IMPEDANCE 1.0000000D 00 OHMS

SCALED FREQ RESPONSE

FREQUENCY RPS	MAGNITUDE	MAGNITUDE DB	REAL PART	IMAGINARY PART	PHASE FRAC OF PI	DELAY SEC
0.0	1.0000E 00	-3.6240E-06	1.0000E 00	-6.1283E-17	-0.00000	4.9605E 00
5.0000E-02	9.9527E-01	-4.1147E-02	9.6481E-01	-2.4437E-01	-0.07896	4.9627E 00
1.0000E-01	9.8116E-01	-1.6523E-01	8.6277E-01	-4.6722E-01	-0.15799	4.9682E 00
1.5000E-01	9.5787E-01	-3.7387E-01	7.0418E-01	-6.4934E-01	-0.23711	4.9751E 00
2.0000E-01	9.2589E-01	-6.6877E-01	5.0506E-01	-7.7601E-01	-0.31635	4.9818E 00
2.5000E-01	8.8603E-01	-1.0510E 00	2.8518E-01	-8.3888E-01	-0.39569	4.9892E 00
3.0000E-01	8.3935E-01	-1.5211E 00	6.5375E-02	-8.3680E-01	-0.47518	5.0012E 00
3.5000E-01	7.8705E-01	-2.0799E 00	-1.3516E-01	-7.7536E-01	-0.55494	5.0231E 00
4.0000E-01	7.3024E-01	-2.7307E 00	-3.0081E-01	-6.6540E-01	-0.63515	5.0585E 00
4.5000E-01	6.6984E-01	-3.4806E 00	-4.2048E-01	-5.2142E-01	-0.71602	5.1054E 00
5.0000E-01	6.0669E-01	-4.3407E 00	-4.8819E-01	-3.6019E-01	-0.79767	5.1536E 00
5.5000E-01	5.4184E-01	-5.3226E 00	-5.0378E-01	-1.9949E-01	-0.87999	5.1873E 00
6.0000E-01	4.7690E-01	-6.4314E 00	-4.7362E-01	-5.5844E-02	-0.96264	5.1950E 00
6.5000E-01	4.1426E-01	-7.6546E 00	-4.1007E-01	5.8709E-02	0.95474	5.1885E 00
7.0000E-01	3.5688E-01	-8.9495E 00	-3.2839E-01	1.3972E-01	0.87195	5.2325E 00
7.5000E-01	3.0808E-01	-1.0227E 01	-2.4164E-01	1.9111E-01	0.78700	5.5088E 00
8.0000E-01	2.7158E-01	-1.1322E 01	-1.5458E-01	2.2329E-01	0.69274	6.5504E 00
8.5000E-01	2.5166E-01	-1.1984E 01	-5.0736E-02	2.4649E-01	0.56462	1.0389E 01
9.0000E-01	2.2461E-01	-1.2971E 01	1.2491E-01	1.8668E-01	0.31229	2.2851E 01
9.5000E-01	8.1652E-02	-2.1761E 01	8.0777E-02	-1.1919E-02	-0.04663	1.6656E 01
1.0000E 00	1.0027E-02	-3.9977E 01	7.6425E-03	-6.4911E-03	-0.22413	7.7333E 00
1.0500E 00	7.2598E-03	-4.2782E 01	-4.1409E-03	5.9630E-03	0.69321	4.5359E 00
1.1000E 00	9.8536E-03	-4.0128E 01	-4.0241E-03	8.9945E-03	0.63391	3.1461E 00
1.1500E 00	8.1541E-03	-4.1772E 01	-2.2647E-03	7.8333E-03	0.58958	2.4841E 00
1.2000E 00	5.3018E-03	-4.5511E 01	-8.8780E-04	5.2270E-03	0.55355	2.0710E 00
1.2500E 00	2.3710E-03	-5.2501E 01	-1.7136E-04	2.3648E-03	0.52303	1.7794E 00
1.3000E 00	2.6771E-04	-7.1447E 01	-3.3122E-06	-2.6769E-04	-0.50394	1.3576E 00
1.3500E 00	2.5091E-03	-5.2010E 01	-2.1339E-04	-2.5000E-03	-0.52710	1.3911E 00
1.4000E 00	4.3519E-03	-4.7226E 01	-6.5523E-04	-4.3023E-03	-0.54811	1.2525E 00
1.4500E 00	5.8339E-03	-4.4681E 01	-1.2205E-03	-5.7048E-03	-0.56709	1.1361E 00
1.5000E 00	7.0042E-03	-4.3093E 01	-1.8347E-03	-6.7597E-03	-0.58437	1.0371E 00
1.5500E 00	7.9122E-03	-4.2034E 01	-2.4492E-03	-7.5236E-03	-0.60018	9.5191E-01
1.6000E 00	8.6025E-03	-4.1308E 01	-3.0338E-03	-8.0498E-03	-0.61472	8.7778E-01
1.6500E 00	9.1138E-03	-4.0806E 01	-3.5712E-03	-8.3850E-03	-0.62817	8.1270E-01
1.7000E 00	9.4790E-03	-4.0465E 01	-4.0530E-03	-8.5689E-03	-0.64063	7.5514E-01
1.7500E 00	9.7255E-03	-4.0242E 01	-4.4760E-03	-8.6343E-03	-0.65224	7.0391E-01
1.8000E 00	9.8758E-03	-4.0109E 01	-4.8408E-03	-8.6080E-03	-0.66307	6.5803E-01
1.8500E 00	9.9486E-03	-4.0045E 01	-5.1502E-03	-8.5117E-03	-0.67321	6.1676E-01
1.9000E 00	9.9591E-03	-4.0036E 01	-5.4080E-03	-8.3628E-03	-0.68272	5.7945E-01
1.9500E 00	9.9199E-03	-4.0070E 01	-5.6187E-03	-8.1752E-03	-0.69167	5.4559E-01
2.0000E 00	9.8412E-03	-4.0139E 01	-5.7871E-03	-7.9598E-03	-0.70010	5.1475E-01

(a)

ANTREICH AND GLEISSNER - SCHUESSLER PULSE - FORMING NETWORK - 64.10.10.S

TRANSFER FUNCTION CRITICAL FREQUENCIES (SCALED)

OUTPUT VARIABLE - V CZ
SOURCE VARIABLE - V1

GAIN CONSTANT IS 7.1778486D-02

POLE POSITIONS REAL PART	POLE POSITIONS IMAGINARY PART	ORDER	ZERO POSITIONS REAL PART	ZERO POSITIONS IMAGINARY PART	ORDER
-4.4859237D-02	9.1549720D-01	1	-8.9500642D-01	0.0	1
-3.6668824D-01	6.2457207D-01	1	-2.0220165D 00	0.0	1
-8.9500642D-01	0.0	1	-6.0962475D-06	-1.2945755D 00	1
-2.0220163D 00	0.0	1	1.9923657D-04	1.0199096D 00	1
-4.9054554D-01	2.0810346D-01	1			

THIS NETWORK HAS BEEN SCALED FOR COMPUTATION BY THE FOLLOWING FACTORS
FREQUENCY 1.0000000D 00 RADIANS/SEC. IMPEDANCE 1.0000000D 00 OHMS

UNSCALED TIME RESPONSE

INTEGRATION STEP SIZE 4.5455E-02 SEC

TIME	STEP	IMPULSE
0.0	1.9058D-21	-3.2682D-21
5.0000D-01	6.8488D-03	2.4166D-02
1.0000D 00	2.2513D-02	3.8009D-02
1.5000D 00	4.5070D-02	5.2872D-02
2.0000D 00	7.6198D-02	7.2474D-02
2.5000D 00	1.1825D-01	9.6273D-02
3.0000D 00	1.7276D-01	1.2176D-01
3.5000D 00	2.3979D-01	1.4590D-01
4.0000D 00	3.1797D-01	1.6591D-01
4.5000D 00	4.0466D-01	1.7965D-01
5.0000D 00	4.9634D-01	1.8577D-01
5.5000D 00	5.8905D-01	1.8371D-01
6.0000D 00	6.7872D-01	1.7370D-01
6.5000D 00	7.6157D-01	1.5664D-01
7.0000D 00	8.3444D-01	1.3405D-01
7.5000D 00	8.9503D-01	1.0790D-01
8.0000D 00	9.4211D-01	8.0419D-02
8.5000D 00	9.7561D-01	5.3888D-02
9.0000D 00	9.9651D-01	3.0363D-02
9.5000D 00	1.0067D 00	1.1448D-02
1.0000D 01	1.0089D 00	-1.8933D-03
1.0500D 01	1.0058D 00	-9.4358D-03
1.1000D 01	1.0003D 00	-1.1685D-02
1.1500D 01	9.9483D-01	-9.7573D-03
1.2000D 01	9.9102D-01	-5.1745D-03
1.2500D 01	9.8982D-01	4.0417D-04
1.3000D 01	9.9134D-01	5.4539D-03
1.3500D 01	9.9499D-01	8.8187D-03
1.4000D 01	9.9977D-01	9.8670D-03
1.4500D 01	1.0045D 00	8.5437D-03
1.5000D 01	1.0080D 00	5.3146D-03
1.5500D 01	1.0096D 00	1.0196D-03
1.6000D 01	1.0090D 00	-3.3331D-03
1.6500D 01	1.0064D 00	-6.7886D-03
1.7000D 01	1.0025D 00	-8.6399D-03
1.7500D 01	9.9810D-01	-8.5590D-03
1.8000D 01	9.9423D-01	-6.6425D-03
1.8500D 01	9.9169D-01	-3.3687D-03
1.9000D 01	9.9097D-01	5.2198D-04
1.9500D 01	9.9217D-01	4.1924D-03
2.0000D 01	9.9500D-01	6.8863D-03

(b)

Fig 1-53 (a) Frequency response for example; (b) Time response for example

figures are reproduced directly from the CORNAP user's manual, to which the reader should refer for further details (see footnote 5, near beginning of this section).

Time-Sharing Systems

Time-sharing consists of a big central computer system with a number of input–output terminals, i.e., simultaneous access to the computer is provided from many remote individual smaller stations. The small remote terminals are made available, for example, to small organizations employing only a few engineers that may not use a computer all the time. Individuals using the input–output terminals (usually through teleprinters or special typewriters) "share" the use of the central computer in such a way that each appears to have sole use of the computer. This type of service gives immediate access to the computer and is very useful for rapid solutions to problems. As an example, the General Electric time-sharing service uses the BASIC (*B*eginner's *A*ll-Purpose *S*ymbolic *I*nstruction *C*ode) language, which is a user-oriented language and can be learned very quickly. In fact, with some of the library of programs available, such as dc linear network analysis, Laplace transform inversion, etc., the user need only furnish data in a simple form and the reader can learn to solve problems the same day even if he has never used a computer before. The size limitations for the GE and IBM time-sharing systems are different (see their manuals). It should be noted that there are also several other time-sharing systems for circuit analysis available. In the future, no doubt, the number will grow further.

With a time-shared computer, the turnaround time is very small (usually measured in seconds or minutes). Thus, changes in the response by varying circuit parameters can be very quickly observed by on-line use of computers. Time-sharing is efficient for problems which are not very large, i.e., those that do not require long on-line execution time. It is also appropriate to mention at this point that computer graphics also play an important role in circuit design. For example, on a graphic console, the circuit designer can enter the data on a cathode-ray tube by means of a light pencil, in the form of a circuit diagram. The graphic output is viewed as he makes changes in the circuit. The process is repeated several times until a satisfactory design is achieved. This process is analogous to making changes in the breadboard and observing the results on an oscilloscope.

Computer-aided analysis and design play a significant role in electronic circuit development. Design by computer is just an iterative analysis until certain prescribed specifications are met. Without the use of some of these packaged programs, accurate analysis of most of the linear and nonlinear integrated circuits is virtually impossible. Detailed analysis, using more accurate device models in a relatively complex subsystem, which could not have been performed without a computer, can now be handled. In fact, in integrated circuits where several transistors, diodes, resistors, and capacitors are used in the realizations of electronic functional blocks, the final design is usually analyzed in detail by a computer prior to fabrication for mass production.

REFERENCES AND SUGGESTED READING

1. Desoer, C. A., and Kuh, E. S., *Basic Circuit Theory*, New York: McGraw-Hill, 1969.
2. Cruz, J., and Van Valkenberg, M. E., *Introductory Signals and Circuits*, Waltham: Blaisdell, 1967.
3. Hayt, W. H., and Hughes, G., *Introduction to Electrical Engineering*, New York: McGraw-Hill, 1968.
4. Close, C. M., *The Analysis of Linear Circuits*, New York: Harcourt, Brace and World, 1966.
5. Van Valkenberg, M. E., *Network Analysis*, 2nd edition, Englewood Cliffs: Prentice-Hall, 1964.
6. DeRusso, P. M., Roy, R. J., and Close, C., *State Variables for Engineers*, New York: John Wiley, 1965. See Chapter 4 for a lucid discussion of matrices.
7. Jensen, R. W., and Leiberman, M. D., *IBM Electronic Circuit Analysis Program*, Englewood Cliffs: Prentice-Hall, 1968. Also *IBM 1620 Electronic Circuit Analysis Program* (Application Program 1620 EE-02X), White Plains: IBM, 1965.

PROBLEMS

1-1 Consider the following differential equation:

$$C_3 \frac{d^3 x}{dt^3} + (C_2 x + C_1 t)\frac{d^2 x}{dt^2} + C_0(1 + xt) = C_4 \sin t$$

where C_4, C_3, C_2, C_1, C_0 are real constants. What are the conditions on the value of the C's if the above is to represent a linear time-invariant system; a linear time-varying system; a nonlinear time-invariant system; a non-linear time-varying system?

1/Review of Selected Topics from Circuit Theory

1-2 Solve the following differential equation:
$$\frac{d^2x}{dt^2} + 2\frac{dx}{dt} + x = e^{-t} \qquad x(0) = 0 \qquad x'(0) = 1$$

1-3 Find the Laplace transform of
(a) $(2t + 1)u(t)$ \qquad (b) $\cosh(2t)u(t)$

1-4 Show that
(a) $\mathscr{L}[e^{-\lambda t}f(t)] = F(s + \lambda)$ \qquad (b) $\mathscr{L}[-tf(t)] = dF(s)/ds$

1-5 Use the convolution theorem (table 1-1) to obtain $f_1 * f_2$:
(a) $f_1(t) = u(t) \qquad f_2(t) = tu(t)$ \qquad (b) $\mathscr{L}^{-1}\{[1/(s + 1)] \cdot [1/(s + 2)]\}$

1-6 Find the inverse Laplace transform of
(a) $1/[(s + 1)(s^2 + s + 1)]$ \qquad (b) $(s^2 + 2s + 4)/(s^2 + 2s + 1)$
(c) $e^{-s}/(s + 1)$ \qquad (d) $(s^2 - 1)/s(s^2 + 1)$

1-7 A Laplace transformable periodic function $f(t)u(t)$ with period τ is given. Show that
$$\mathscr{L}[f(t)u(t)] = \frac{F_0(s)}{1 - e^{-s\tau}}$$
where
$$F_0(s) = \mathscr{L}[f_0(t)u(t)]$$
and
$$f_0(t) = \begin{cases} f(t) & 0 \le t \le \tau \\ 0 & \tau < t \end{cases}$$
(Hint: $1/(1 - e^{-s\tau}) = 1 + e^{-s\tau} + e^{-2s\tau} + \cdots$)

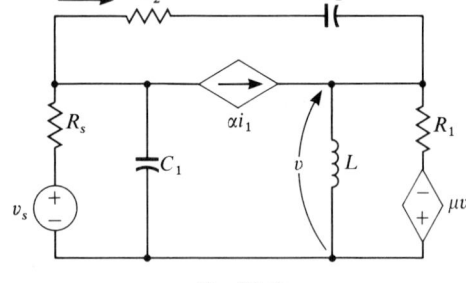

1-8 Write the equilibrium equations for the circuit in fig P1-8 which can be used to solve for the voltage across C_2. Choose whichever method is most convenient. *Do not solve.*

1-9 The linear time-invariant circuit shown in fig P1-9 is in the zero state.
(a) Set up the nodal equations in matrix form.
(b) If i_s is the excitation, set up the solution for v_0 in determinant form. *Do not solve.*

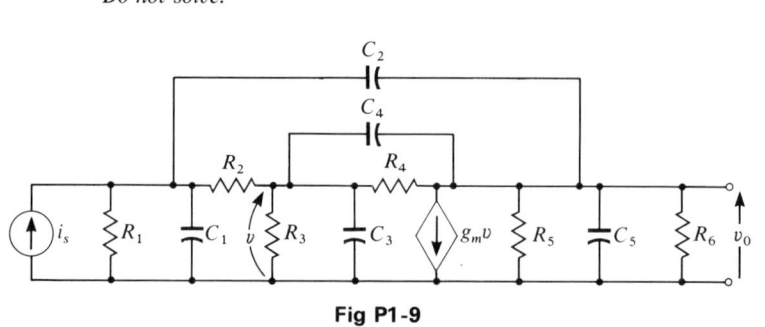

Fig P1-8

Fig P1-9

1-10 The linear time-invariant network shown in fig P1-10 is in the zero state.
  (a) Set up the loop equations in matrix form.
  (b) Set up the solution for v_o in the determinant form. *Do not solve.*

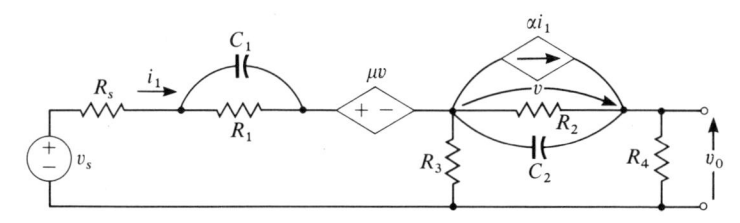

Fig P1-10

1-11 Consider the linear time-invariant circuit shown in fig P1-11.
  (a) Set up the state equations. Let v_1 and v_o be the state variables.
  (b) Derive the expression for the natural frequencies of the network, using the result of (a).

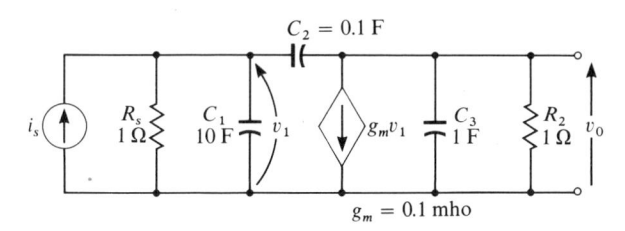

Fig P1-11

1-12 Set up the state equations for the circuit in fig 1-24.

1-13 The following state equations are given:

$$\frac{d}{dt}\begin{bmatrix} x_1 \\ x_2 \end{bmatrix} = \begin{bmatrix} -1 & 0 \\ 0 & -2 \end{bmatrix} \begin{bmatrix} x_1 \\ x_2 \end{bmatrix}$$

with

$$\begin{bmatrix} x_1(0) \\ x_2(0) \end{bmatrix} = \begin{bmatrix} 2 \\ 2 \end{bmatrix}$$

  (a) Find $[x(N\,\Delta t)]$ with $\Delta t = 0.2$ sec. Plot the trajectory, i.e., x_1 vs x_2 from above, stopping at any convenient value of N.
  (b) Plot the exact trajectory by solving the above equations.

1-14 The transfer function of a linear time-invariant circuit is given by

$$\frac{V_o}{V_i}(s) = \frac{30}{(s+1)(s+3)}$$

Fig P1-15

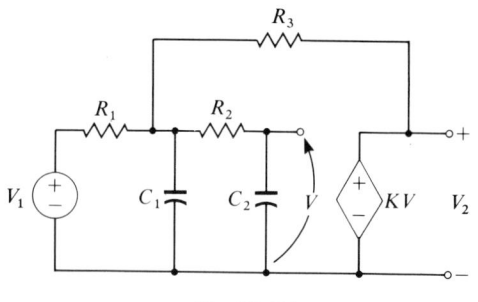

$f(v) = -2v + v^3$

Fig P1-17

Determine the state equations by the following two methods:
(a) $V_0(s) = 30V_i/[(s+1)(s+3)] = [15V_i/(s+3)] - [15V_i/(s+1)]$. Let $X_1(s) = 15V_i/(s+3)$ or $sX_1 = 15V_i(s) - 3X_1(s)$. Recall that $s = d/dt$, and so on.
(b) $(s^2 + 4s + 3)V_0(s) = 30V_i(s)$. Let $v_0(t) = x_1(t), dv_0/dt = dx_1/dt = x_2(t)$, and so on.

1-15 Consider the circuit shown in fig P1-15, where one resistor is a nonlinear device. If $i_L(0) = 2$ A and $v_c(0) = (4/3)$ V, find $i_L(N\,\Delta t)$ and $v_c(N\,\Delta t)$ for $\Delta t = 0.2$. Draw the trajectory in the state space, i.e., i_L vs v_c, with t as a parameter.

1-16 Evaluate e^{At} if A is given by

(a) $\begin{bmatrix} 0 & 1 \\ 2 & 3 \end{bmatrix}$ (b) $\begin{bmatrix} 2 & 1 \\ -1 & 4 \end{bmatrix}$

1-17 For the circuit shown in fig P1-17, derive the expression for $\dfrac{V_2}{V_1}(s)$.

1-18 For the circuit shown in fig P1-18, show that
$$\frac{I_0}{I_s} = \frac{K_2(Y_A - K_1 Y_B)}{Y_1 - (K_2 - 1)Y_2}$$

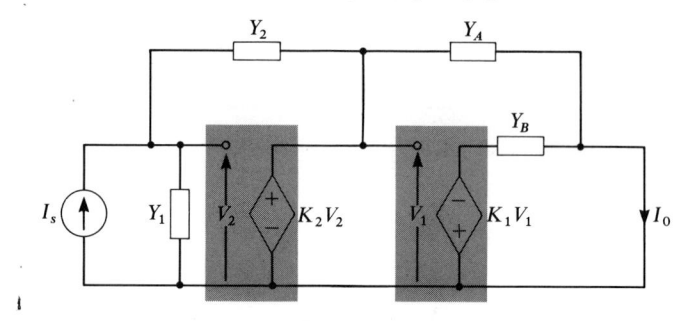

Fig P1-18

1-19 For the NIC circuits shown in fig P1-19, derive the expressions for the driving-point impedance looking into terminals 1–1'.

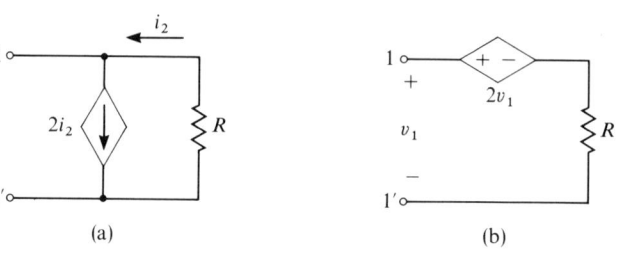

Fig P1-19

1-20 Determine the natural frequencies of the circuit shown in fig P1-20. (Hint: Use the result of Problem 1-19.)

1-21 Repeat Problem 1-20 for the circuit shown in fig P1-21.

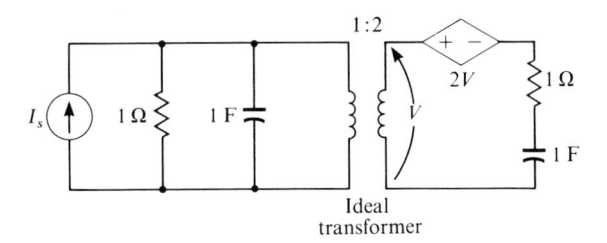

Fig P1-21

Fig P1-20

1-22 For the circuit shown in fig P1-22, determine the transfer impedance and the natural frequencies.

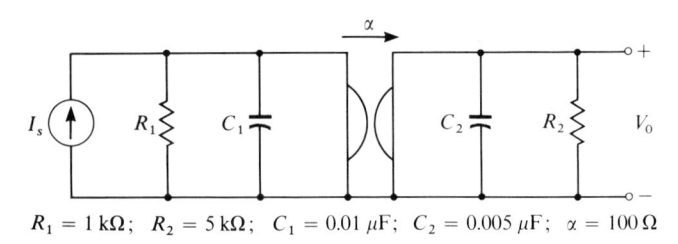

$R_1 = 1\ k\Omega;\ R_2 = 5\ k\Omega;\ C_1 = 0.01\ \mu F;\ C_2 = 0.005\ \mu F;\ \alpha = 100\ \Omega$

Fig P1-22

1-23 Determine the natural frequencies of fig 1-42 by the two methods (3) and (4) not used in the example in Section 1-11.

1-24 The driving-point function of a network is given by

$$Z(s) = \frac{(s+1)(s-1)}{(s+2)(s^2+4)}$$

What are the open circuit and short circuit natural frequencies of the network?

1-25 Prove the four properties of network functions as given in Section 1-11.

1-26 In the circuit shown in fig P1-26, the capacitor has an initial voltage $v_c(0^-) = V_1$ at the same time that $i_L(0^-) = 0$. Determine the expression for $v_2(t)$ for $t \geq 0$ if switch S is closed at $t = 0$. Let $R_1 = R_2 = 2\Omega$, $C = 1F$, and $L = 1H$. Use any method you prefer.

1-27 Rewrite the differential equation in Problem 1-1 in the normal form.

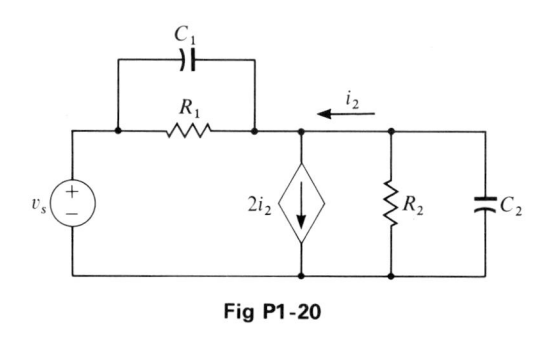

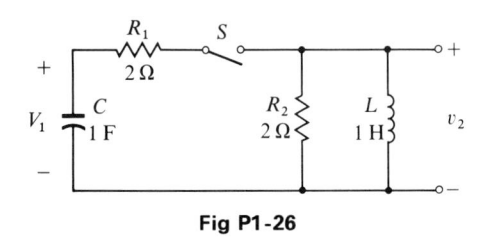

Fig P1-26

1-28 Obtain the Thévenin equivalent circuit to the left of terminal 1-1' in fig P1-28.

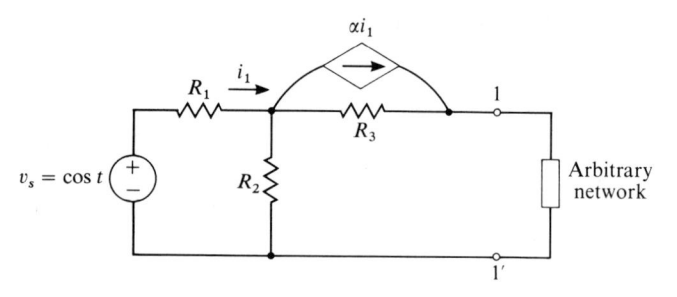

Fig P1-28

1-29 For the network shown in fig P1-29, show that the Norton equivalent admittance and current generator at terminals 1–1' are given by

$$Y_N = \frac{y_{22}(Y_s + y_{11}) - y_{12}y_{21}}{y_{11} + Y_s} \qquad I_N = \frac{-y_{21}Y_s E_{in}}{y_{11} + Y_s}$$

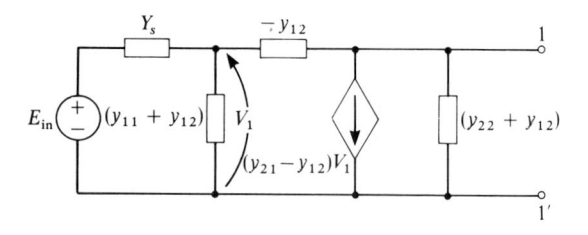

1-30 For the circuit shown in fig P1-30, determine the current through the nonlinear device $\mathcal{D}$. The v-i relationship of the device is given by $i = 0.5\, v^{3/2}$ (mA) for $v > 0$ and $i = 0$ for $v \leq 0$. Graphical solution is acceptable.

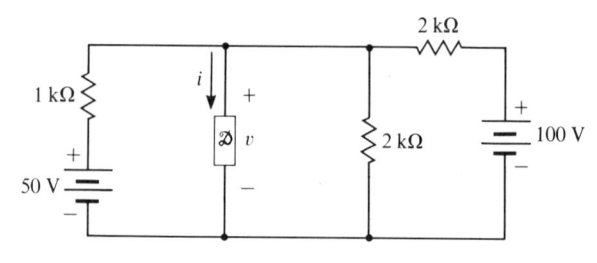

Fig P1-30

1-31 The following equivalence is referred to as the *Miller theorem.* Consider an arbitrary circuit configuration with N nodes, as shown in fig P1-31a. It is assumed that $V_2/V_1 = A$ is known. Show that the circuit is entirely equivalent to the one shown in fig P1-31b.

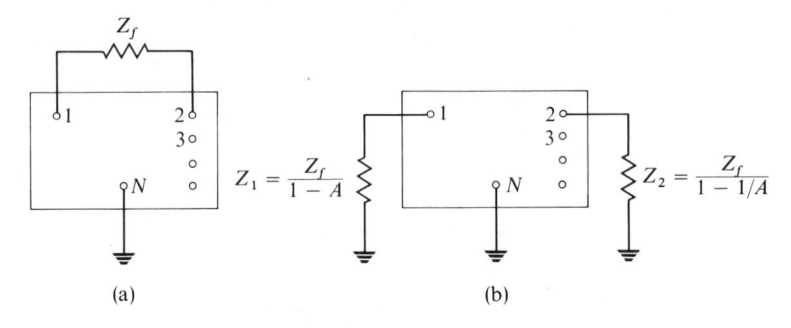

Fig P1-31

1-32 Consider the time-varying circuit shown in fig P1-32. Derive the differential equation that governs the output $v_o(t)$ and the input $v_i(t)$. The resistors are all linear time invariant.

Fig P1-32

1-33 For the nonlinear circuit shown in fig P1-33,
(a) derive the differential equation from which v can be obtained. Given that

$$C = f(v) = C_0\left(1 - \frac{v}{V_0}\right)^n$$

where C_0 and V_0 are constants. The element values of the battery V_1 and the resistors R_1 and R_2 are known.

(b) If the normalized values are $R_1 = R_2 = 2$, $C_0 = 1$, $n = -\frac{1}{2}$, $V_0 = V_1 = 2$, find the voltage across the nonlinear capacitor vs time by using the method of Section 1-9. Use $\Delta t = 0.1$ and assume zero initial state.

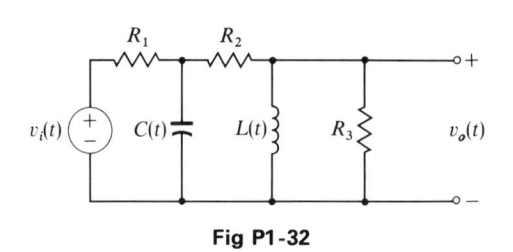

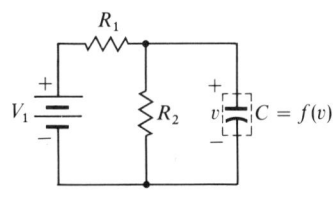

Fig P1-33

1-34 If the ECAP and/or CORNAP packaged programs are accessible write a program to analyze the circuit of fig 4-37. The circuit parameters are as follows:

$R_f = 1\,\text{k}\Omega$, $R_L = 100\,\Omega$, $R_s = 1\,\text{M}\Omega$, $r_b = 100\,\Omega$, $G_1 = 4 \times 10^{-3}$ mho,

$g_m = 0.2$ mho, $r_o = h_{oe}^{-1} = 50\,\text{k}\Omega$, $C_\pi = 74.6$ pF, $C_c = 5$ pF

Specifically the current gain I_o/I_s at $f = 10$ Hz is to be determined. Write the input data for ECAP such that the frequency response up to $f = 100$ MHz is also calculated.

2
Two-Port Networks and Related Topics

In this chapter we shall review the basic properties of two-port networks and discuss briefly some other related topics. A two-port network is a circuit which has only *two pairs* of accessible terminals. A pair of terminals in which the current leaving one terminal is equal to the current flowing into the other terminal is called a *port*. Two-port networks constitute a very important class of networks in signal processing and, therefore, merit special study. We often encounter two-port systems where one port represents the input and the other the output. The analysis and design of the transmission characteristics of signals from the input to the output of two-port circuits constitute the main subject of this book.

The various formulas and definitions discussed in this chapter are quite general, regardless of the nature of the two-port networks, provided the networks are linear and time-invariant.

Almost all active electronic devices, such as transistors, vacuum tubes, semiconductor diodes, etc., are nonlinear. Under small-signal conditions, however (described in the next chapter), nonlinear devices can be adequately characterized by linear time-invariant models. In a number of applications, such as the amplification of video and audio signals, linearity of signal processing must be maintained and, in such applications, the nonlinear device is deliberately operated and analyzed under small-signal conditions. The Laplace transform is useful and convenient for the analysis of linear time-invariant networks, and we shall, therefore, use the transform domain throughout this chapter. The transform quantities are denoted, as usual, by capital letters.

2-1 Two-Port Characterizations

A linear time-invariant two-port network is shown symbolically by the rectangle marked with the conventional reference directions for the voltages and currents, in fig 2-1. There are no independent voltage or current sources within the rectangle, and the network is in the zero state, i.e., the initial conditions are set to zero. There are four possible types of variables in such a two-port network: currents I_1 and I_2, and voltages V_1 and V_2. Among these four variables are two linear constraints; thus, there are only two independent variables. There are six ways of picking two quantities out of a set of four quantities; thus, six two-port descriptions are possible, depending on the choice of the dependent and independent quantities. The various possibilities are as follows.

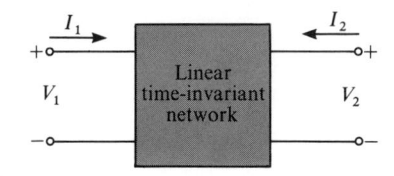

Fig 2-1 Linear time-invariant two-port network

Short-Circuit Admittance Parameters (y_{ij})

Consider V_1 and V_2 to be the independent variables, i.e., the network in fig 2-1 is excited by two independent voltage sources. We wish, therefore, to determine the currents I_1 and I_2. Such a situation is best illustrated by a simple example.

EXAMPLE

Let us calculate the responses I_1 and I_2 in fig 2-2, where V_1 and V_2 are independent voltage excitations.

The two nodal equations for fig 2-2 are

$$I_1 = V_1\left(sC_1 + \frac{1}{R+sL}\right) + V_2\left(-\frac{1}{R+sL}\right)$$

$$I_2 = V_1\left(-\frac{1}{R+sL}\right) + V_2\left(sC_2 + \frac{1}{R+sL}\right)$$

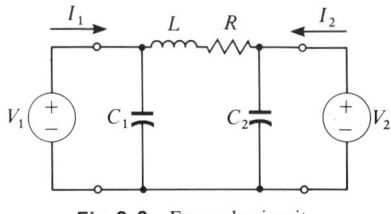

Fig 2-2 Example circuit

The currents I_1 and I_2 are thus determined for any given V_1 and V_2 and circuit element values.

For a general linear time-invariant two-port network, we can write, by superposition,

$$I_1 = y_{11}V_1 + y_{12}V_2 \qquad (2\text{-}1a)$$

$$I_2 = y_{21}V_1 + y_{22}V_2 \qquad (2\text{-}1b)$$

where

$$y_{11} = \left.\frac{I_1}{V_1}\right|_{V_2=0} \qquad y_{12} = \left.\frac{I_1}{V_2}\right|_{V_1=0}$$
$$y_{21} = \left.\frac{I_2}{V_1}\right|_{V_2=0} \qquad y_{22} = \left.\frac{I_2}{V_2}\right|_{V_1=0} \qquad (2\text{-}2)$$

The coefficients $y_{ij}(s)$ are called the *short-circuit admittance parameters*. Notice that the "short circuit" is applied to the port at which the excitation is *not* being applied. Note that for a reciprocal network $y_{12} = y_{21}$, which is always true for RLCM circuits. If there are gyrators and/or dependent sources in the circuit, the network in general is not reciprocal, i.e., $y_{12} \neq y_{21}$. Only in special cases can the circuit be reciprocal.

Equations (2-1) may be represented by the equivalent circuit with two dependent current sources, as shown in fig 2-3. The circuit in fig 2-3 is a Norton equivalent at both ports.

By adding and subtracting $y_{12}V_1$ in (2-1b), we obtain

$$I_2 = y_{12}V_1 + y_{22}V_2 + (y_{21} - y_{12})V_1 \qquad (2\text{-}3)$$

From (2-3) and (2-1a), an alternate network representation with one dependent current source, usually referred to as the π *equivalent circuit*, is obtained, as shown in fig 2-4.

In circuit analysis, the y-parameters of a network sometimes can be directly obtained by using fig 2-4.

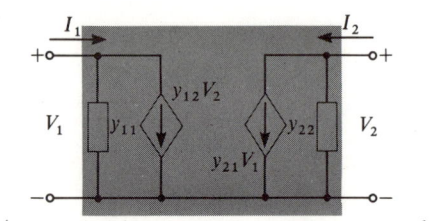

Fig 2-3 Norton equivalent of a two-port network

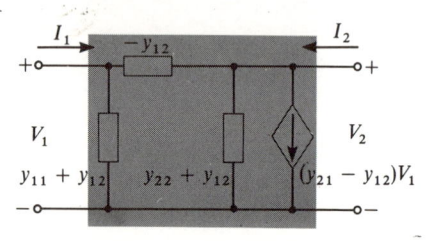

Fig 2-4 π-equivalent of an active two-port network

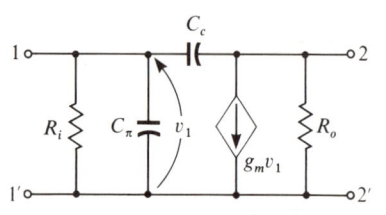

Fig 2-5 Example circuit

EXAMPLE

Let us calculate the short-circuit admittance parameters of the network shown in fig 2-5. From a comparison of fig 2-5 and fig 2-4, we have

$$y_{12} = -sC_c$$

$$y_{11} + y_{12} = G_i + sC_\pi \quad \text{hence} \quad y_{11} = G_i + s(C_\pi + C_c)$$

$$y_{22} + y_{12} = G_o \qquad \text{hence} \quad y_{22} = G_o + sC_c$$
$$y_{21} - y_{12} = g_m \qquad \text{hence} \quad y_{21} = g_m - sC_c$$

where $G_i = R_i^{-1}$ and $G_o = R_o^{-1}$. Note that $y_{12} \neq y_{21}$ for this circuit. The network is, therefore, not reciprocal.

Open-Circuit Impedance Parameters (z_{ij})

Let us consider I_1 and I_2 as the independent variables (i.e., the network of fig 2-1 excited by two independent current sources) and V_1 and V_2 as the responses to be determined. The circuit responses in terms of the excitations can, in this case, be written in the following matrix form:

$$\begin{bmatrix} V_1 \\ V_2 \end{bmatrix} = \begin{bmatrix} z_{11} & z_{12} \\ z_{21} & z_{22} \end{bmatrix} \begin{bmatrix} I_1 \\ I_2 \end{bmatrix} \qquad (2\text{-}4)$$

where z_{ij} are called the *open-circuit impedance parameters*, since

$$z_{11} = \left.\frac{V_1}{I_1}\right|_{I_2=0} \qquad z_{12} = \left.\frac{V_1}{I_2}\right|_{I_1=0}$$
$$z_{21} = \left.\frac{V_2}{I_1}\right|_{I_2=0} \qquad z_{22} = \left.\frac{V_2}{I_2}\right|_{I_1=0} \qquad (2\text{-}5)$$

Note that the "open circuit" is at the port at which excitation is *not* being applied.

EXAMPLE

Let us determine the open-circuit impedance matrix of the circuit depicted in fig 2-6.

The two loop equations for fig 2-6 are

$$V_1 = \left(sL_1 + R + \frac{1}{sC}\right)I_1 + \left(R + \frac{1}{sC}\right)I_2$$

$$V_2 = \left(R + \frac{1}{sC}\right)I_1 + \left(sL_2 + R + \frac{1}{sC}\right)I_2$$

In the circuit, I_1 and I_2 are independent current excitations. The impedance matrix $[z_{ij}]$ can thus be read directly.

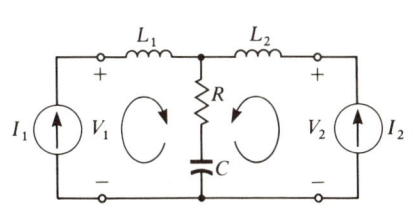

Fig 2-6 Example circuit

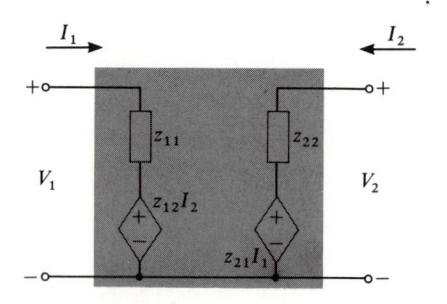

Fig 2-7 Thévenin equivalent of a two-port network

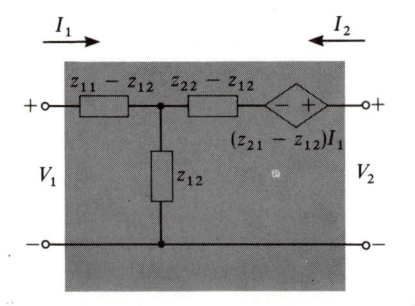

Fig 2-8 T-equivalent of an active two-port network

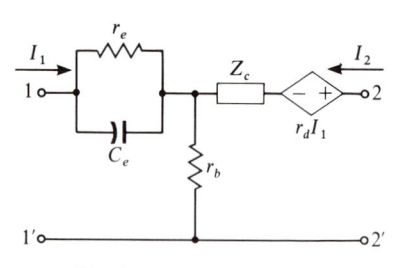

Fig 2-9 Example circuit

Equation (2-4) may be represented by the equivalent circuit shown in fig 2-7. The circuit has a Thévenin equivalent representation at both ports.

By adding and subtracting $z_{12}I_1$ in the expression for V_2 in (2-4), we obtain

$$V_2 = z_{12}I_1 + z_{22}I_2 + (z_{21} - z_{12})I_1 \quad (2\text{-}6)$$

From (2-6) and (2-4), an alternate network representation with one dependent voltage source, referred to as the *T-equivalent* circuit of a two-port, is obtained, as shown in fig 2-8. The use of fig 2-8 sometimes yields the z-parameters directly.

EXAMPLE

Let us calculate the open-circuit impedance parameters of the circuit shown in fig 2-9. From a comparison of fig 2-9 with fig 2-8, we can at once write:

$$z_{12} = r_b$$

$$z_{22} - z_{12} = Z_c \qquad \text{hence} \quad z_{22} = r_b + Z_c$$

$$z_{21} - z_{12} = r_d \qquad \text{hence} \quad z_{21} = r_d + r_b$$

$$z_{11} - z_{12} = \frac{r_e}{1 + r_e C_e s} \qquad \text{hence} \quad z_{11} = \frac{r_e}{1 + r_e C_e s} + r_b$$

In a reciprocal network such as that shown in fig 2-6, $z_{12} = z_{21}$. In general, when there are gyrators and/or dependent generators in a network, $z_{12} \neq z_{21}$. Only in special cases are these two quantities equal. Note that the z-parameters can be obtained directly from the y-parameters if $\det [y_{ij}] \neq 0$. Since

$$\begin{bmatrix} I_1 \\ I_2 \end{bmatrix} = \begin{bmatrix} y_{11} & y_{12} \\ y_{21} & y_{22} \end{bmatrix} \begin{bmatrix} V_1 \\ V_2 \end{bmatrix} \quad (2\text{-}7)$$

we can solve for V_1 and V_2 in terms of I_1 and I_2. The solutions are

$$V_1 = \frac{\begin{vmatrix} I_1 & y_{12} \\ I_2 & y_{22} \end{vmatrix}}{\begin{vmatrix} y_{11} & y_{12} \\ y_{21} & y_{22} \end{vmatrix}} = \frac{y_{22}}{\Delta y}I_1 - \frac{y_{12}}{\Delta y}I_2 \quad (2\text{-}8a)$$

$$V_2 = \frac{\begin{vmatrix} y_{11} & I_1 \\ y_{21} & I_2 \end{vmatrix}}{\Delta y} = \frac{y_{11}}{\Delta y}I_2 - \frac{y_{21}}{\Delta y}I_1 \quad (2\text{-}8b)$$

where

$$\Delta y = y_{11}y_{22} - y_{12}y_{21} \neq 0 \tag{2-9}$$

Hence, from (2-4) and (2-8a)

$$z_{11} = \frac{y_{22}}{\Delta y} \qquad z_{12} = -\frac{y_{12}}{\Delta y}$$
$$z_{22} = \frac{y_{11}}{\Delta y} \qquad z_{21} = -\frac{y_{21}}{\Delta y} \tag{2-10}$$

Similarly, y_{ij} may be obtained in terms of z_{ij} if det $[z_{ij}] \neq 0$. Note that $[z] = [y]^{-1}$ and the product of $\Delta y \Delta z = 1$.

The Hybrid Parameters (h_{ij}) and (g_{ij})

If I_1 and V_2 are chosen as the independent variables, then V_1 and I_2 are the dependent variables. Note that we have mixed (current and voltage) dependent and independent quantities. In other words, the excitations are an independent current source at port 1 and an independent voltage source across port 2, while the dependent responses are V_1 and I_2.

In this case, we have the h-parameter description

$$\begin{bmatrix} V_1 \\ I_2 \end{bmatrix} = \begin{bmatrix} h_{11} & h_{12} \\ h_{21} & h_{22} \end{bmatrix} \begin{bmatrix} I_1 \\ V_2 \end{bmatrix} \tag{2-11}$$

where

$$h_{11} = \frac{V_1}{I_1}\bigg|_{V_2=0} \qquad h_{21} = \frac{I_2}{I_1}\bigg|_{V_2=0}$$
$$h_{12} = \frac{V_1}{V_2}\bigg|_{I_1=0} \qquad h_{22} = \frac{I_2}{V_2}\bigg|_{I_1=0} \tag{2-12}$$

Note that h_{11} has the dimension of impedance, h_{22} has the dimension of admittance, and h_{21} and h_{12} are dimensionless, since they are ratios of currents and voltages, respectively. In particular, note that h_{21} is the *short-circuit current ratio*. As we shall see in Chapter 3, this characterization is often convenient in transistor circuit analysis. An equivalent circuit in terms of the hybrid h-parameters is shown in fig 2-10. In this

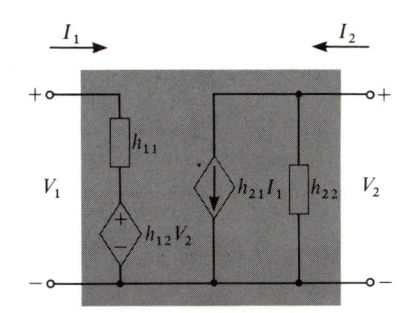

Fig 2-10 Hybrid representation of a two-port (input Thévenin, output Norton equivalent)

characterization, note that the input is a Thévenin equivalent, while the output is a Norton equivalent representation.

One could obtain the *h*-parameters in terms of *y*-parameters or *z*-parameters from (2-4) or (2-7). For example, solving for V_1 and I_2 in terms of I_1 and V_2 in (2-7), we have

$$\begin{bmatrix} V_1 \\ I_2 \end{bmatrix} = \begin{bmatrix} \dfrac{1}{y_{11}} & -\dfrac{y_{12}}{y_{11}} \\ \dfrac{y_{21}}{y_{11}} & \dfrac{\Delta_y}{y_{11}} \end{bmatrix} \begin{bmatrix} I_1 \\ V_2 \end{bmatrix} \tag{2-13}$$

hence

$$\begin{aligned} h_{11} &= \frac{1}{y_{11}} & h_{12} &= -\frac{y_{12}}{y_{11}} \\ h_{21} &= \frac{y_{21}}{y_{11}} & h_{22} &= \frac{\Delta_y}{y_{11}} \end{aligned} \tag{2-14}$$

Note that, for reciprocal networks (since $y_{12} = y_{21}$), we have $h_{12} = -h_{21}$.

Similarly, if V_1 and I_2 are chosen as the independent variables, then I_1 and V_2 are the dependent variables. In this case, we have the g-parameter description

$$\begin{bmatrix} I_1 \\ V_2 \end{bmatrix} = \begin{bmatrix} g_{11} & g_{12} \\ g_{21} & g_{22} \end{bmatrix} \begin{bmatrix} V_1 \\ I_2 \end{bmatrix} \tag{2-15}$$

where

$$\begin{aligned} g_{11} &= \left.\frac{I_1}{V_1}\right|_{I_2=0} & g_{21} &= \left.\frac{V_2}{V_1}\right|_{I_2=0} \\ g_{12} &= \left.\frac{I_1}{I_2}\right|_{V_1=0} & g_{22} &= \left.\frac{V_2}{I_2}\right|_{V_1=0} \end{aligned} \tag{2-16}$$

Note that g_{11} and g_{22} have the dimensions of admittance and impedance, respectively, while g_{12} and g_{21} are dimensionless quantities. In particular, note that g_{21} is an *open-circuit voltage ratio*. An equivalent circuit, in terms of the hybrid g-parameters, is shown in fig 2-11. In this case, the input has a Norton equivalent while the output has a Thévenin equivalent circuit.

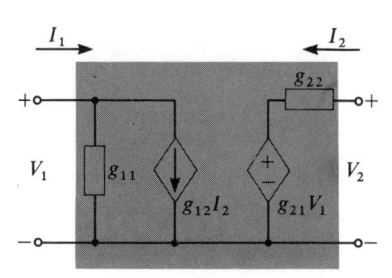

Fig 2-11 Hybrid representation of a two-port (input Norton, output Thévenin equivalent)

The Transmission (Chain) Parameters ($ABCD$ and $\mathscr{ABCD}$)

The transmission parameters, also referred to as the chain parameters, relate the input and output quantities. If the independent variables are chosen as V_2 and $-I_2$, then the dependent parameters are V_1 and I_1. The minus sign preceding I_2 is used for convenience, as we shall see. The relationship between the dependent and the independent variables in this description is given by the $ABCD$ parameters

$$\begin{bmatrix} V_1 \\ I_1 \end{bmatrix} = \begin{bmatrix} A & B \\ C & D \end{bmatrix} \begin{bmatrix} V_2 \\ -I_2 \end{bmatrix} \quad (2\text{-}17)$$

where

$$A = \left.\frac{V_1}{V_2}\right|_{I_2=0} \qquad C = \left.\frac{I_1}{V_2}\right|_{I_2=0}$$
$$B = -\left.\frac{V_1}{I_2}\right|_{V_2=0} \qquad D = -\left.\frac{I_1}{I_2}\right|_{V_2=0} \quad (2\text{-}18)$$

Note that all the preceding types of parameters (z, y, g, h) were defined as network functions, whereas the reciprocals of transmission parameters (e.g., A^{-1}) are network functions.

Similarly, if the independent variables are chosen as V_1 and $-I_1$, then the dependent parameters are V_2 and I_2; this relationship is given by $\mathscr{ABCD}$ parameters, namely

$$\begin{bmatrix} V_2 \\ I_2 \end{bmatrix} = \begin{bmatrix} \mathscr{A} & \mathscr{B} \\ \mathscr{C} & \mathscr{D} \end{bmatrix} \begin{bmatrix} V_1 \\ -I_1 \end{bmatrix} \quad (2\text{-}19)$$

The descriptions in (2-17) and (2-19) are very useful in the analysis of cascaded two-ports, and it is for this reason that the negative sign is associated with the independent current in each description.

For example, consider the cascaded two-ports shown in fig 2-12. For two-port N_a we have

$$\begin{bmatrix} V_1 \\ I_1 \end{bmatrix} = \begin{bmatrix} A^a & B^a \\ C^a & D^a \end{bmatrix} \begin{bmatrix} V_2 \\ -I_2 \end{bmatrix} \quad (2\text{-}20)$$

For two-port N_b we have

$$\begin{bmatrix} V_2 \\ -I_2 \end{bmatrix} = \begin{bmatrix} A^b & B^b \\ C^b & D^b \end{bmatrix} \begin{bmatrix} V_3 \\ -I_3 \end{bmatrix} \quad (2\text{-}21)$$

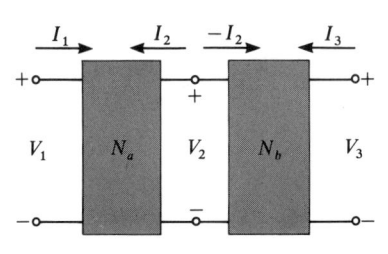

Fig 2-12 Cascaded two-port networks

From (2-20) and (2-21), we can immediately write

$$\begin{bmatrix} V_1 \\ I_1 \end{bmatrix} = \begin{bmatrix} A^a & B^a \\ C^a & D^a \end{bmatrix} \begin{bmatrix} A^b & B^b \\ C^b & D^b \end{bmatrix} \begin{bmatrix} V_3 \\ -I_3 \end{bmatrix} = \begin{bmatrix} A & B \\ C & D \end{bmatrix} \begin{bmatrix} V_3 \\ -I_3 \end{bmatrix} \quad (2\text{-}22)$$

where

$$\begin{aligned} A &= A^a A^b + B^a C^b & B &= A^a B^b + B^a D^b \\ C &= C^a A^b + D^a C^b & D &= C^a B^b + D^a D^b \end{aligned} \quad (2\text{-}23)$$

For convenience and subsequent reference, the matrix conversions from one parameter set to another parameter set are listed in table 2-1.

The choice of parameter characterization depends on the problem at hand. In certain cases, some characterizations may not exist. For example, the ideal transformer has no z- or y-parameter description but the other characterizations can be used. In some other cases, a particular choice may yield a very simple representation. As an example, consider the various idealized controlled-source two-port networks.

For the voltage-controlled voltage source (fig 1-9a) (VCVS), which is also referred to as the ideal voltage amplifier, the g-parameters give a simple description

$$\begin{bmatrix} I_1 \\ V_2 \end{bmatrix} = \begin{bmatrix} 0 & 0 \\ \mu & 0 \end{bmatrix} \begin{bmatrix} V_1 \\ I_2 \end{bmatrix} \quad (2\text{-}24)$$

i.e., $g_{21} = \mu$, and $g_{11} = g_{22} = g_{12} = 0$. The voltage-controlled current source (VCCS) shown in fig 1–9b is most simply described in terms of the y-parameters

$$\begin{bmatrix} I_1 \\ I_2 \end{bmatrix} = \begin{bmatrix} 0 & 0 \\ g_m & 0 \end{bmatrix} \begin{bmatrix} V_1 \\ V_2 \end{bmatrix} \quad (2\text{-}25)$$

i.e., $y_{21} = g_m$, and $y_{11} = y_{22} = y_{12} = 0$. For fig 1-9c, a current-controlled current source (CCCS), also referred to as the ideal current amplifier, the simplest description in terms of the h-parameters

$$\begin{bmatrix} V_1 \\ I_2 \end{bmatrix} = \begin{bmatrix} 0 & 0 \\ \alpha & 0 \end{bmatrix} \begin{bmatrix} I_1 \\ V_2 \end{bmatrix} \quad (2\text{-}26)$$

i.e., $h_{21} = \alpha$, and $h_{11} = h_{22} = h_{12} = 0$. Finally, for the current-controlled voltage source (CCVS) shown in fig 1-9d, the simplest description provided by the z-parameters is

2-1 Two-Port Characterizations

TABLE 2-1 *Two-Port Parameter Matrix Conversions*[a]

	$[z_{ij}]$	$[y_{ij}]$	$[g_{ij}]$	$[h_{ij}]$	$\begin{bmatrix} A & B \\ C & D \end{bmatrix}$	$\begin{bmatrix} \mathscr{A} & \mathscr{B} \\ \mathscr{C} & \mathscr{D} \end{bmatrix}$
$[z_{ij}]$	$z_{11} \quad z_{12}$ $z_{21} \quad z_{22}$	$\dfrac{y_{22}}{\Delta_y} \quad -\dfrac{y_{12}}{\Delta_y}$ $-\dfrac{y_{21}}{\Delta_y} \quad \dfrac{y_{11}}{\Delta_y}$	$\dfrac{1}{g_{11}} \quad -\dfrac{g_{12}}{g_{11}}$ $\dfrac{g_{21}}{g_{11}} \quad \dfrac{\Delta_g}{g_{11}}$	$\dfrac{\Delta_h}{h_{22}} \quad \dfrac{h_{12}}{h_{22}}$ $-\dfrac{h_{21}}{h_{22}} \quad \dfrac{1}{h_{22}}$	$\dfrac{A}{C} \quad \dfrac{\Delta_A}{C}$ $\dfrac{1}{C} \quad \dfrac{D}{C}$	$\dfrac{\mathscr{D}}{\mathscr{C}} \quad \dfrac{1}{\mathscr{C}}$ $\dfrac{\Delta_{\mathscr{A}}}{\mathscr{C}} \quad \dfrac{\mathscr{A}}{\mathscr{C}}$
$[y_{ij}]$	$\dfrac{z_{22}}{\Delta_z} \quad -\dfrac{z_{12}}{\Delta_z}$ $-\dfrac{z_{21}}{\Delta_z} \quad \dfrac{z_{11}}{\Delta_z}$	$y_{11} \quad y_{12}$ $y_{21} \quad y_{22}$	$\dfrac{\Delta_g}{g_{22}} \quad \dfrac{g_{12}}{g_{22}}$ $-\dfrac{g_{21}}{g_{22}} \quad \dfrac{1}{g_{22}}$	$\dfrac{1}{h_{11}} \quad -\dfrac{h_{12}}{h_{11}}$ $\dfrac{h_{21}}{h_{11}} \quad \dfrac{\Delta_h}{h_{11}}$	$\dfrac{D}{B} \quad -\dfrac{\Delta_A}{B}$ $-\dfrac{1}{B} \quad \dfrac{A}{B}$	$\dfrac{\mathscr{A}}{\mathscr{B}} \quad -\dfrac{1}{\mathscr{B}}$ $-\dfrac{\Delta_{\mathscr{A}}}{\mathscr{B}} \quad \dfrac{\mathscr{D}}{\mathscr{B}}$
$[g_{ij}]$	$\dfrac{1}{z_{11}} \quad -\dfrac{z_{12}}{z_{11}}$ $\dfrac{z_{21}}{z_{11}} \quad \dfrac{\Delta_z}{z_{11}}$	$\dfrac{\Delta_y}{y_{22}} \quad \dfrac{y_{12}}{y_{22}}$ $-\dfrac{y_{21}}{y_{22}} \quad \dfrac{1}{y_{22}}$	$g_{11} \quad g_{12}$ $g_{21} \quad g_{22}$	$\dfrac{h_{22}}{\Delta_h} \quad -\dfrac{h_{12}}{\Delta_h}$ $-\dfrac{h_{21}}{\Delta_h} \quad \dfrac{h_{11}}{\Delta_h}$	$\dfrac{C}{A} \quad -\dfrac{\Delta_A}{A}$ $\dfrac{1}{A} \quad \dfrac{B}{A}$	$\dfrac{\mathscr{C}}{\mathscr{D}} \quad -\dfrac{1}{\mathscr{D}}$ $\dfrac{\Delta_{\mathscr{A}}}{\mathscr{D}} \quad \dfrac{\mathscr{B}}{\mathscr{D}}$
$[h_{ij}]$	$\dfrac{\Delta_z}{z_{22}} \quad \dfrac{z_{12}}{z_{22}}$ $-\dfrac{z_{21}}{z_{22}} \quad \dfrac{1}{z_{22}}$	$\dfrac{1}{y_{11}} \quad -\dfrac{y_{12}}{y_{11}}$ $\dfrac{y_{21}}{y_{11}} \quad \dfrac{\Delta_y}{y_{11}}$	$\dfrac{g_{22}}{\Delta_g} \quad -\dfrac{g_{12}}{\Delta_g}$ $-\dfrac{g_{21}}{\Delta_g} \quad \dfrac{g_{11}}{\Delta_g}$	$h_{11} \quad h_{12}$ $h_{21} \quad h_{22}$	$\dfrac{B}{D} \quad \dfrac{\Delta_A}{D}$ $-\dfrac{1}{D} \quad \dfrac{C}{D}$	$\dfrac{\mathscr{B}}{\mathscr{A}} \quad \dfrac{1}{\mathscr{A}}$ $-\dfrac{\Delta_{\mathscr{A}}}{\mathscr{A}} \quad \dfrac{\mathscr{C}}{\mathscr{A}}$
$\begin{bmatrix} A & B \\ C & D \end{bmatrix}$	$\dfrac{z_{11}}{z_{21}} \quad \dfrac{\Delta_z}{z_{21}}$ $\dfrac{1}{z_{21}} \quad \dfrac{z_{22}}{z_{21}}$	$-\dfrac{y_{22}}{y_{21}} \quad -\dfrac{1}{y_{21}}$ $-\dfrac{\Delta_y}{y_{21}} \quad -\dfrac{y_{11}}{y_{21}}$	$\dfrac{1}{g_{21}} \quad \dfrac{g_{22}}{g_{21}}$ $\dfrac{g_{11}}{g_{21}} \quad \dfrac{\Delta_y}{g_{21}}$	$-\dfrac{\Delta_h}{h_{21}} \quad -\dfrac{h_{11}}{h_{21}}$ $-\dfrac{h_{22}}{h_{21}} \quad -\dfrac{1}{h_{21}}$	$A \quad B$ $C \quad D$	$\dfrac{\mathscr{D}}{\Delta_{\mathscr{A}}} \quad \dfrac{\mathscr{B}}{\Delta_{\mathscr{A}}}$ $\dfrac{\mathscr{C}}{\Delta_{\mathscr{A}}} \quad \dfrac{\mathscr{A}}{\Delta_{\mathscr{A}}}$
$\begin{bmatrix} \mathscr{A} & \mathscr{B} \\ \mathscr{C} & \mathscr{D} \end{bmatrix}$	$\dfrac{z_{22}}{z_{12}} \quad \dfrac{\Delta_z}{z_{12}}$ $\dfrac{1}{z_{12}} \quad \dfrac{z_{11}}{z_{12}}$	$-\dfrac{y_{11}}{y_{12}} \quad -\dfrac{1}{y_{12}}$ $-\dfrac{\Delta_y}{y_{12}} \quad -\dfrac{y_{22}}{y_{12}}$	$-\dfrac{\Delta_g}{g_{12}} \quad -\dfrac{g_{22}}{g_{12}}$ $-\dfrac{g_{11}}{g_{12}} \quad -\dfrac{1}{g_{12}}$	$\dfrac{1}{h_{12}} \quad \dfrac{h_{11}}{h_{12}}$ $\dfrac{h_{22}}{h_{12}} \quad \dfrac{\Delta_h}{h_{12}}$	$\dfrac{D}{\Delta_A} \quad \dfrac{B}{\Delta_A}$ $\dfrac{C}{\Delta_A} \quad \dfrac{A}{\Delta_A}$	$\mathscr{A} \quad \mathscr{B}$ $\mathscr{C} \quad \mathscr{D}$

[a] All matrices appearing in the same row in the table are equivalent, e.g.,

$$h_{11} = \frac{\Delta_z}{z_{22}} = \frac{1}{y_{11}} \text{ and } \Delta_z = z_{11}z_{22} - z_{12}z_{21} \text{ etc.}$$

$$\begin{bmatrix} V_1 \\ V_2 \end{bmatrix} = \begin{bmatrix} 0 & 0 \\ r_m & 0 \end{bmatrix} \begin{bmatrix} I_1 \\ I_2 \end{bmatrix} \qquad (2\text{-}27)$$

i.e., $z_{21} = r_m$, and $z_{11} = z_{22} = z_{12} = 0$. In all the above ideal controlled source cases, the only other mode of representation of the six possible uses is the *ABCD* parameters.

For the negative-impedance converter, we may use the *ABCD*, *g*, *h*, or $\mathscr{ABCD}$ parameters. The *ABCD*-parameter description is given by (see Section 8-4)

$$\begin{bmatrix} V_1 \\ I_1 \end{bmatrix} = \begin{bmatrix} \pm 1 & 0 \\ 0 & \mp 1 \end{bmatrix} \begin{bmatrix} V_2 \\ -I_2 \end{bmatrix} \qquad (2\text{-}28)$$

i.e., $A = \pm 1$, $D = \mp 1$, and $B = C = 0$. If $A = +1$, $D = -1$, the NIC is called unity gain INIC (fig 1-12a) and if $A = -1$, $D = +1$, the NIC is called unity gain VNIC (fig 1-12b) (since $B = C = 0$, $[z_{ij}]$, $[y_{ij}]$, conversion does not exist, see table 2-1).

The gyrator (fig 1-6) may be conveniently described using the *z*, *y*, *ABCD*, or $\mathscr{ABCD}$ parameters. The *z*- and *y*-parameter characterizations are

$$\begin{bmatrix} V_1 \\ V_2 \end{bmatrix} = \begin{bmatrix} 0 & \alpha \\ -\alpha & 0 \end{bmatrix} \begin{bmatrix} I_1 \\ I_2 \end{bmatrix} \quad \text{or} \quad \begin{bmatrix} I_1 \\ I_2 \end{bmatrix} = \begin{bmatrix} 0 & -1/\alpha \\ 1/\alpha & 0 \end{bmatrix} \begin{bmatrix} V_1 \\ V_2 \end{bmatrix} \qquad (2\text{-}29)$$

When two-ports are interconnected, the overall two-port parameters of the combined two-port can be obtained simply by direct addition of the respective parameters of the original two-ports (z_{ij}, y_{ij}, h_{ij}, g_{ij}) provided that the dependent variable is common to both ports and also the interconnection does not change the parameter sets. In other words, direct addition of the respective parameters is allowable if the current which enters one terminal of a port has the same value as the current which leaves the other terminal of the same port. It is usually clear in any given problem whether or not direct addition is permissible.

For example, consider the two-port networks shown in fig 2-13. If the two circuits are connected in shunt at both ports (i.e., a shunt-shunt interconnection) the total *y*-parameters are given by $y_{ij}^T = y_{ij}^a + y_{ij}^b$. If the two circuits are series-series interconnected, $z_{ij}^T \neq z_{ij}^a + z_{ij}^b$ since the element R_a will be short-circuited in a series-series interconnection. If an ideal transformer is used to provide isolation, then, in any of the

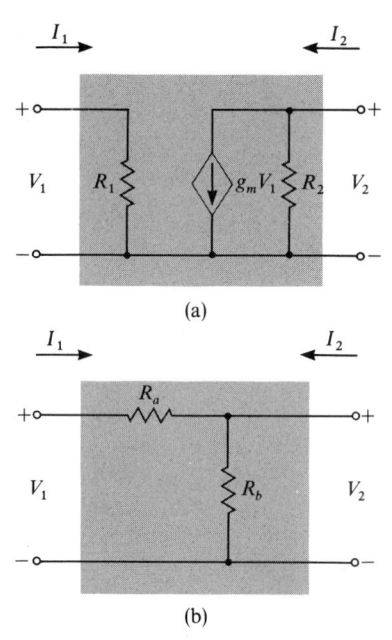

Fig 2-13 (a) and (b) Individual two-port networks

series-series, shunt-shunt, series-shunt, and shunt-series interconnections, the overall two-port parameters are always given by the respective addition of the parameters of the individual two-ports. For example, in a series-series interconnection, we may use the ideal transformer, as shown in fig 2-14. In this case, the equalities $z_{ij}^T = z_{ij}^a + z_{ij}^b$ always hold. Hence, if N^a and N^b are the networks in fig 2-13, we simply add each element of the $[z]$ matrix of N^a to those of N^b to get the overall z_{ij} parameters.

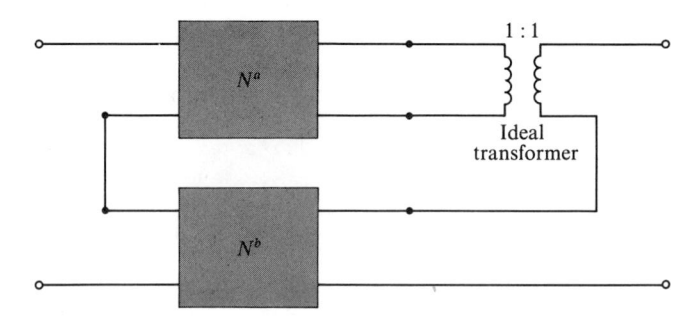

Fig 2-14 Series-series interconnected two-port where $z_{ij}^T = z_{ij}^a + z_{ij}^b$

2-2 Terminated Two-Port Networks

In many circuit applications, a two-port network is used to process and transmit a signal from a source to a load. Consider a linear time-invariant two-port network with initial conditions equal to zero, as represented by the box in fig 2-15. The terminating source and load resistances are R_s and R_L, respectively. For this circuit, we are interested in determining the input impedance Z_{in}, output impedance Z_o, the

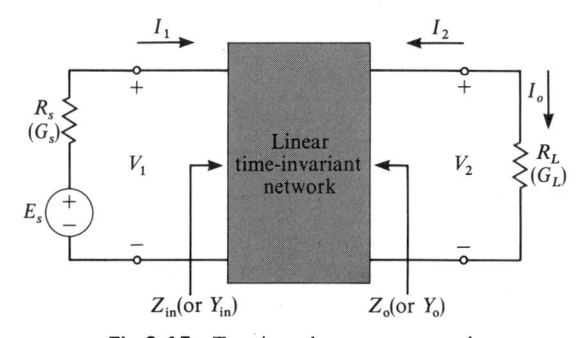

Fig 2-15 Terminated two-port network

voltage ratio V_2/V_1, and the current ratio I_2/I_1. (The input impedance, Z_{in}, is the impedance seen at the input port of the two-port network when the terminating source resistance is removed. The output impedance Z_o is the impedance seen at the output port of the two-port network when the terminating load resistance is removed.) The voltage gain V_2/E_s and the current gain $I_o/I_s = I_o/(E_s/R_s)$ can then be readily determined from the above results. The power gain calculations are discussed in the next section. Consequently, we shall consider only the admittance parameters (y_{ij}) for the sake of clarity. The reader can carry out the steps in terms of the generalized parameters—for any other parameter characterizations.[1]

The two-port description in terms of the y-parameters is

$$I_1 = y_{11}V_1 + y_{12}V_2 \tag{2-30a}$$

$$I_2 = y_{21}V_1 + y_{22}V_2 \tag{2-30b}$$

The source and load connections yield the following additional relations:

$$-I_2 = G_L V_2 \tag{2-31}$$

$$I_1 = G_s E_s - G_s V_1 \tag{2-32}$$

Note that G_s and G_L are the reciprocals of R_s and R_L, respectively. To obtain the input admittance, Y_{in}, we remove R_s, and use (2-31) to eliminate I_2 in (2-30b) and solve for I_1/V_1. This yields

$$Y_{in} = \frac{I_1}{V_1} = y_{11} - \frac{y_{12}y_{21}}{y_{22} + G_L} \tag{2-33}$$

Similarly to find the output admittance we remove R_L, set $E_s = 0$, and use (2-32) to eliminate I_1 in (2-30a) and solve for I_2/V_2:

$$Y_0 = \frac{I_2}{V_2} = y_{22} - \frac{y_{21}y_{12}}{y_{11} + G_s} \tag{2-34}$$

---

[1] To avoid repetition of identical steps in manipulating the equations, it is often helpful to use the generalized parameter k_{ij} which relates the dependent and the independent variables by the following matrix equation,

$$\begin{bmatrix} \phi_{d1} \\ \phi_{d2} \end{bmatrix} = \begin{bmatrix} k_{11} & k_{12} \\ k_{21} & k_{22} \end{bmatrix} \begin{bmatrix} \phi_{i1} \\ \phi_{i2} \end{bmatrix}$$

where k_{ij} represents y_{ij}, z_{ij}, h_{ij}, or g_{ij} depending upon the choice of the dependent and the independent variables. For example, if ϕ_{i1} and ϕ_{i2} denote I_1 and V_2, then ϕ_{d1} and ϕ_{d2} represent V_1 and I_2, and k_{ij} represents h_{ij}, respectively. The source will be represented by an impedance R_s and the load by an admittance G_L.

The voltage ratio V_2/V_1 is found from (2-31) and (2-30b) by eliminating I_2. Thus

$$\frac{V_2}{V_1} = \frac{-y_{21}}{y_{22} + G_L} \qquad (2\text{-}35)$$

Similarly, the current ratio is found by eliminating V_1 and V_2. The result is

$$\frac{-I_2}{I_1} = \frac{-y_{21}G_L}{\Delta_y + y_{11}G_L} \qquad (2\text{-}36)$$

If we are interested in finding the current gain I_o/I_s, where $I_s = E_s G_s$, and $I_o = -I_2$, we can incorporate G_s in the two-port to obtain a new two-port described by $[y]_s$ and then use (2-36). The modified two-port admittance matrix is

$$[y_{ij}]_s = \begin{bmatrix} (y_{11} + G_s) & y_{12} \\ y_{21} & y_{22} \end{bmatrix} \qquad (2\text{-}37)$$

The current gain I_o/I_s can then be written directly from (2-36) and (2-37):

$$\frac{-I_2}{I_s} = \frac{-y_{21s}G_L}{\Delta y_s + y_{11s}G_L} = \frac{-y_{21}G_L}{(y_{11} + G_s)y_{22} - y_{12}y_{21} + (y_{11} + G_s)G_L}$$
$$= \frac{-y_{21}G_L}{(y_{11} + G_s)(y_{22} + G_L) - y_{12}y_{21}} \qquad (2\text{-}38)$$

The voltage gain V_2/E_s is given by

$$\frac{V_2}{E_s} = \frac{-R_L I_2}{R_s I_s} = \frac{-y_{21}G_s}{(y_{11} + G_s)(y_{22} + G_L) - y_{12}y_{21}} \qquad (2\text{-}39)$$

Similarly, if we use the z-parameters, we obtain

$$Z_{in} = \frac{V_1}{I_1} = z_{11} - \frac{z_{12}z_{21}}{z_{22} + R_L} \qquad (2\text{-}40)$$

$$Z_o = \frac{V_2}{I_2} = z_{22} - \frac{z_{12}z_{21}}{z_{11} + R_s} \qquad (2\text{-}41)$$

$$\frac{V_2}{V_1} = \frac{z_{21}R_L}{\Delta_z + z_{11}R_L} \qquad (2\text{-}42)$$

$$-\frac{I_2}{I_1} = \frac{z_{21}}{z_{22} + R_L} \tag{2-43}$$

$$\frac{V_2}{E_s} = \frac{z_{21}R_L}{(z_{11} + R_s)(z_{22} + R_L) - z_{12}z_{21}} \tag{2-44}$$

and

$$\frac{I_o}{I_s} = \frac{z_{21}R_s}{(z_{11} + R_s)(z_{22} + R_L) - z_{12}z_{21}} \tag{2-45}$$

Note that $I_s = E_s/R_s$. The various gain and impedance relations in terms of the other parameters are given in table 2-2. It should be noted that the voltage gain can also be obtained from

$$\frac{V_2}{E_s} = \frac{V_2}{V_1} \cdot \frac{V_1}{E_s}$$

where $V_1/E_s = Z_{in}/(Z_{in} + R)$. The impedance relations, the voltage ratio, and the current ratio expressions in table 2-2 are convenient for gain calculations in cascaded two-port circuits (see Section 5-1 and Problem 2-8).

When considering active circuits, we are sometimes interested in the voltage insertion gain. The voltage insertion gain is defined as the ratio

TABLE 2-2 *Gain and Impedance Relations*

	$[z_{ij}]$	$[y_{ij}]$	$[g_{ij}]$	$[h_{ij}]$	$\begin{bmatrix} A & B \\ C & D \end{bmatrix}$	$\begin{bmatrix} \mathscr{A} & \mathscr{B} \\ \mathscr{C} & \mathscr{D} \end{bmatrix}$
Input impedance, Z_i	$\dfrac{\Delta_z + z_{11}Z_L}{z_{22} + Z_L}$	$\dfrac{y_{22} + Y_L}{\Delta_y + y_{11}Y_L}$	$\dfrac{g_{22} + Z_L}{\Delta_g + g_{11}Z_L}$	$\dfrac{\Delta_h + h_{11}Y_L}{h_{22} + Y_L}$	$\dfrac{AZ_L + B}{CZ_L + D}$	$\dfrac{\mathscr{D}Z_L + \mathscr{B}}{\mathscr{C}Z_L + \mathscr{A}}$
Output impedance, Z_o	$\dfrac{\Delta_z + z_{22}Z_s}{z_{11} + Z_s}$	$\dfrac{y_{11} + Y_s}{\Delta_y + y_{22}Y_s}$	$\dfrac{\Delta_g + g_{22}Y_s}{g_{11} + Y_s}$	$\dfrac{h_{11} + Z_s}{\Delta_h + h_{22}Z_s}$	$\dfrac{DZ_s + B}{CZ_s + A}$	$\dfrac{\mathscr{A}Z_s + \mathscr{B}}{\mathscr{C}Z_s + \mathscr{D}}$
Current ratio, $-\dfrac{I_2}{I_1}$	$\dfrac{z_{21}}{z_{22} + Z_L}$	$\dfrac{-y_{21}Y_L}{\Delta_y + y_{11}Y_L}$	$\dfrac{g_{21}}{\Delta_g + g_{11}Z_L}$	$\dfrac{-h_{21}Y_L}{h_{22} + Y_L}$	$\dfrac{1}{D + CZ_L}$	$\dfrac{\Delta_{\mathscr{A}}}{\mathscr{A} + \mathscr{C}Z_L}$
Voltage ratio, $\dfrac{V_2}{V_1}$	$\dfrac{z_{21}Z_L}{\Delta_z + z_{11}Z_L}$	$\dfrac{-y_{21}}{y_{22} + Y_L}$	$\dfrac{g_{21}Z_L}{g_{22} + Z_L}$	$\dfrac{-h_{21}}{\Delta_h + h_{11}Y_L}$	$\dfrac{Z_L}{B + AZ_L}$	$\dfrac{\Delta_{\mathscr{A}}}{\mathscr{B}Y_L + \mathscr{D}}$

$$\text{Voltage gain} = \frac{V_2}{E_s} = \left(\frac{Z_i}{Z_i + Z_s}\right)\frac{V_2}{V_1} \qquad \text{Current gain} = \frac{I_o}{I_s} = \frac{I_o}{E_i/R_s} = \left(\frac{Z_s}{Z_s + Z_i}\right)\left(-\frac{I_2}{I_1}\right)$$

of the voltage gain, when the terminated two-port device is connected to the source, to the voltage gain with the device removed and the source feeding the load directly (fig 2-16). In other words,

$$\text{Insertion Voltage Gain (IVG)} = \frac{V_2/E_s \text{ as obtained from fig 2-15}}{V_2/E_s \text{ as obtained from fig 2-16}} \quad (2\text{-}46)$$

Hence

$$\text{IVG} = \frac{-(G_s + G_L)y_{21}}{(y_{11} + G_s)(y_{22} + G_L) - y_{12}y_{21}} \quad (2\text{-}47)$$

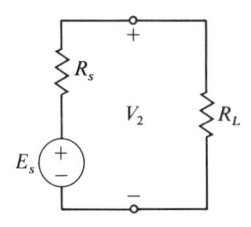

Fig 2-16 Two-port network removed; source feeding the load directly

In terms of the z-parameters,

$$\text{IVG} = \frac{(R_s + R_L)z_{21}}{(z_{11} + R_s)(z_{22} + R_L) - z_{12}z_{21}} \quad (2\text{-}48)$$

EXAMPLE

Consider the circuit shown in fig 2-17. From the circuit, the y-parameters are readily determined to be: $y_{12} = 0$, $y_{21} = g_m = 10^{-1}$, $y_{11} = \frac{1}{800}$, $y_{22} = 10^{-4}$ and $G_L = \frac{1}{50}$, $G_s = \frac{1}{100}$. From (2-47), we obtain IVG $= -13.32$. Note, however, that it is simple in this case to apply the definitions directly. The voltage gain and the insertion voltage gain are determined as follows:

$$V_1 = \frac{800}{900} E_s$$

$$V_2 = -(g_m V_1) \left[\frac{10^4(50)}{10^4 + 50} \right] = -V_1 \frac{50}{(10.05)}$$

Now

$$\frac{V_2}{E_s} = \frac{V_2}{V_1} \frac{V_1}{E_s}$$

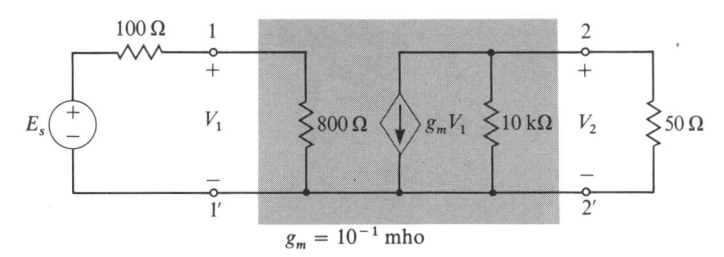

$g_m = 10^{-1}$ mho

Fig 2-17 Example circuit

Hence

$$\frac{V_2}{E_s} = -\frac{50}{10.05} \cdot \frac{800}{900} = -4.44 = \text{Gain with two-port connected.}$$

When the two-port is removed, $\dfrac{V_2}{E_s} = \dfrac{50}{150} = 0.333$. The insertion voltage gain $\text{IVG} = -\dfrac{4.44}{0.333} = -13.32$.

2-3 Various Power Gains in Active Two-Ports

In this section, we shall consider the definitions for various power gains and their corresponding expressions when the linear time-invariant active two-port is terminated at both ends by passive one-ports. It should be emphasized that the quantities defined in this section apply only to sinusoidal steady-state conditions. In Chapter 7, we shall see why and when these different power gains are applied.

Consider the network shown schematically in fig 2-18. The various power definitions and their expressions, on the basis of the y-parameters, are as follows:

(1) *Power gain* G_P is defined as the ratio of the power delivered to the load to the power into the active two-port network, i.e., P_o/P_i in fig 2-18. By active two-port we mean that the box in fig 2-18 contains some form of controlled source such that power gain larger than unity is obtained. Specifically, the average power into the input and the output ports is *not* always larger than zero for all frequencies (see

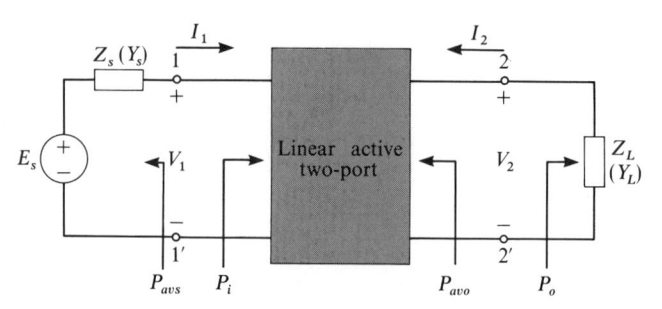

Fig 2-18 Terminated active two-port network

Problem 2-13) in active two-ports. By definition

$$G_P = \frac{P_o}{P_i} = \frac{|V_2|^2 \operatorname{Re} Y_L}{|V_1|^2 \operatorname{Re} Y_{in}} \quad (2\text{-}49a)$$

where Re designates the real part. G_L and G_s are replaced by Y_L and Y_s for generality. From (2-33), (2-35), and (2-49a), we obtain

$$G_P = \left|\frac{y_{21}}{y_{22} + Y_L}\right|^2 \frac{\operatorname{Re} Y_L}{\operatorname{Re}[y_{11} - (y_{12}y_{21})/(y_{22} + Y_L)]} \quad (2\text{-}49b)$$

(2) *Transducer power gain* G_T is defined as the ratio of the power delivered to the load to the power available from the source. The power available from the source P_{avs} is by definition the power delivered from the source to a conjugate matched load, as shown in fig 2-19. From fig 2-19 we have

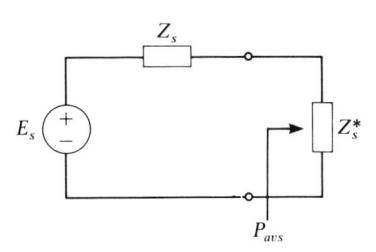

Fig 2-19 Circuit to determine P_{avs}

$$P_{avs} = \frac{|E_s|^2}{4\operatorname{Re} Z_s} = \frac{|E_s|^2 |Y_s|^2}{4\operatorname{Re} Y_s} \quad (2\text{-}50)$$

(Note that if $Z = R + jX = 1/Y$, $\operatorname{Re} Z = R = \operatorname{Re} Y|Z|^2 = \operatorname{Re} Y/|Y|^2$.) Since a conjugate matched load leads to a maximum power transfer, P_{avs} is the largest power deliverable from the source. From the definition, then, G_T is given by:

$$G_T = \frac{P_o}{P_{avs}} = \frac{|V_2|^2 \operatorname{Re} Y_L}{|E_s|^2/4\operatorname{Re} Z_s} = \left|\frac{V_2}{E_s}\right|^2 \frac{4\operatorname{Re} Y_s \operatorname{Re} Y_L}{|Y_s|^2} \quad (2\text{-}51)$$

But from (2-39), noting that the source and load are Y_s and Y_L, respectively, we have

$$\left|\frac{V_2}{E_s}\right| = \left|\frac{y_{21} Y_s}{[(y_{11} + Y_s)(y_{22} + Y_L) - y_{12}y_{21}]}\right| \quad (2\text{-}52)$$

Thus, from (2-51) and (2-52), we obtain

$$G_T = \frac{4|y_{21}|^2 \operatorname{Re} Y_L \operatorname{Re} Y_s}{|(y_{11} + Y_s)(y_{22} + Y_L) - y_{12}y_{21}|^2} \quad (2\text{-}53)$$

(3) *Available power gain* G_A is defined as the ratio of available power from the two-port to the available power from the source.

From fig 2-18 and the definition of G_A, we have

$$G_A = \frac{P_{avo}}{P_{avs}} \tag{2-54}$$

To determine P_{avo}, we can represent the linear two-port at terminals 2-2' with the source generator and source termination by the Norton equivalent circuit. Thus

$$P_{avo} = \frac{|I_N|^2}{4\text{Re } Y_N} \tag{2-55}$$

where I_N and Y_N are the Norton equivalent current and admittance looking into the output at terminals 2-2'. These quantities, which can be readily determined (Problem 1-29), are

$$I_N = \frac{-y_{21} Y_s E_s}{y_{11} + Y_s} \tag{2-56a}$$

$$Y_N = \frac{y_{22}(Y_s + y_{11}) - y_{12}y_{21}}{y_{11} + Y_s} = \frac{\Delta y + y_{22} Y_s}{y_{11} + Y_s} \tag{2-56b}$$

Hence, from (2-50), (2-55), (2-56a), and (2-56b) we obtain[2]

$$G_A = \frac{|y_{21}|^2 \text{Re } Y_s}{\text{Re}\,[(y_{11}y_{22} - y_{12}y_{21} + y_{22}Y_s)(y_{11}^* + Y_s^*)]} \tag{2-57}$$

where * denotes conjugate.

Other parameter characterizations, such as the z-, h-, and g-parameters may also be used to determine the expressions for the various power gains. These are similar to the above and the reader can write them by inspection using the generalized parameters. Due to the nature of the definitions for the various power gains, there is no simple interrelationship among them. However, a couple of observations can readily be made. Compare, for example, (2-49a) and (2-51) ($G_P = P_o/P_i$, $G_T = P_o/P_{avs}$, and $G_P/G_T = P_{avs}/P_i$); since $P_i \leq P_{avs}$, we know that

$$G_P \geq G_T \tag{2-58}$$

---

[2] Note that

$$Y_N = \frac{(\Delta_y + y_{22} Y_s)(y_{11}^* + Y_s^*)}{|y_{11} + Y_s|^2}$$

2-3 Various Power Gains in Active Two-Ports

Similarly, comparison of (2-51) with (2-54) ($G_T = P_o/P_{avs}$, $G_A = P_{avo}/P_{avs}$, and $G_A/G_T = P_{avo}/P_o$), noting that $P_{avo} \geq P_o$, yields

$$G_A \geq G_T \tag{2-59}$$

Only in the special case where the input and the output of an active two-port are conjugate matched do we have the simple relationships

$$(G_A)_{max} = (G_P)_{max} = (G_T)_{max} \tag{2-60}$$

The conjugate match condition is possible and meaningful only if the active two-port is absolutely stable. By absolutely stable we mean that no passive uncoupled terminations can make the system unstable. In other words, in an inherently stable two-port, at no frequency does $Y_{in}(j\omega_o)$ and/or $Y_{out}(j\omega_o)$ (for any $Z_s = R_s \pm jX_s$, $Z_L = R_L \pm jX_L$) ever exhibit negative real parts. Note that ω_o emphasizes a single frequency treatment and its value is arbitrary. If the two-port is not inherently stable, then the maximum power gain is infinite and the design with optimum terminations is meaningless.

Active two-ports may be either absolutely stable or potentially unstable, while passive networks are always absolutely stable. By *potential instability* we mean that the two-port in fig 2-18 can be made unstable for some choice of Z_s and/or Z_L.

The criteria for potential instability are stated, without proof, as follows (Ref 3). If any of the following conditions is *not* satisfied, the two-port is potentially unstable.

$$\text{Re } y_{11} > 0 \quad \text{Re } y_{22} > 0 \tag{2-61a}$$

$$\text{Re } y_{11} \text{Re } y_{22} > \tfrac{1}{2}|y_{21}y_{12}|(1 + \cos \theta) \tag{2-61b}$$

where

$$\theta = \tan^{-1} \frac{\text{Im}(y_{12}y_{21})}{\text{Re}(y_{12}y_{21})}$$

where *Im* and *Re* designate the imaginary part and the real part, respectively. Other parameters, such as the z-, g-, or h-parameters, may be used directly instead of y_{ij}. In other words, if the two-port description is available in terms of h_{ij}, use (2-61) but replace y_{ij} by h_{ij}. Do not convert to y-parameters in order to use (2-61). If any of the conditions in (2-61) is *not* satisfied, the two-port is said to be potentially unstable. In other words, passive terminations can be found which would make the terminated two-port unstable. If, however, (2-61) is satisfied, the system is inherently stable and no passive terminations whatsoever can cause instability. Only if the system is inherently stable, i.e., if (2-61) is satisfied, can we talk about optimum passive terminations for maximizing the power gain, and only for this case is the maximum power gain the same no matter what power definition is used.

2-4 Design of Optimum Passive Terminations for Inherently Stable, Active Two-Port Networks

In practice, such as in the design of narrow-band tuned amplifiers, we may wish to design the active network for maximum power gain at the center frequency of the tuned amplifier (see Chapter 7). In such cases we have the parameters of the active device measured (or given) at the center frequency ω_o, and we design the passive terminations for a given bandwidth and maximum power gain. Furthermore, since impedance matching transformers are used, maximization of voltage and current gain is not meaningful as it depends on the turn ratios of the transformers.

If the device is inherently stable, then such terminations exist and are determined as follows: From the definitions (2-53), (2-57), and (2-49b), it is seen that G_T is a function of the source and load impedance, G_A is a function of the source impedance only, and G_P is solely a function of the load impedance. We obtain maximum power gain when the input and the output ports are simultaneously conjugate-matched by the passive terminations. In this case (2-53), (2-57), and (2-49b) are equal:

$$(G_T)_{\max} = (G_A)_{\max} = (G_P)_{\max} \tag{2-62}$$

Thus, we may choose any of the following four methods to derive the expressions for the optimum passive terminations:

(1) Maximize G_A with respect to the real and the imaginary parts of the source impedance; then select the load impedance to conjugate-match the output port terminated with (Z_s) optimized.

(2) Maximize G_P with respect to the real and imaginary parts of the load impedance; then select the source impedance to conjugate-match the input port terminated with (Z_L) optimized.

(3) Maximize G_T with respect to the real and the imaginary parts of the source and load impedances.

(4) Simultaneously conjugate-match both ports.

2-4 Design of Optimum Passive Terminations

Methods (3) and (4) are quite involved if the two-port parameters are complex quantities, i.e., if there are frequency-dependent circuit elements in the box in fig 2-18. For such cases use methods (1) or (2) (see Ref 3). If, however, the parameters are real, i.e., $z_{ij} = r_{ij}$, method (4) leads to a simple solution. Using, for example, (2-40) and (2-41), we have

$$Z_{in} = r_{11} - \frac{r_{12}r_{21}}{r_{22} + R_L} = R_S \qquad (2\text{-}63a)$$

$$Z_0 = r_{22} - \frac{r_{12}r_{21}}{r_{11} + R_S} = R_L \qquad (2\text{-}63b)$$

Solving (2-63a) and (2-63b) for $r_{12}r_{21}$, we have

$$(r_{11} - R_S)(r_{22} + R_L) = r_{12}r_{21} \qquad (2\text{-}64a)$$

$$(r_{22} - R_L)(r_{11} + R_S) = r_{12}r_{21} \qquad (2\text{-}64b)$$

Subtracting (2-64a) from (2-64b), we obtain

$$2(r_{11}R_L - r_{22}R_S) = 0 \qquad (2\text{-}65)$$

From (2-65), we have

$$\frac{R_L}{r_{22}} = \frac{R_S}{r_{11}} \qquad (2\text{-}66)$$

Substituting (2-66) in (2-64a) or (2-64b), we have

$$R_S = \sqrt{\frac{r_{11}}{r_{22}}(r_{11}r_{22} - r_{12}r_{21})} \qquad (2\text{-}67)$$

$$R_L = \sqrt{\frac{r_{22}}{r_{11}}(r_{11}r_{22} - r_{12}r_{21})} \qquad (2\text{-}68)$$

Since, as we note from (2-61), for an absolutely stable active two-port we have $r_{11} > 0$, $r_{22} > 0$, and $r_{11}r_{22} - r_{12}r_{21} > 0$, the terminations are therefore positive and meaningful.

The maximum power gain under the conditions expressed in (2-67) and (2-68) is

$$G_{max} = \frac{(r_{21})^2}{(\sqrt{r_{11}r_{22} - r_{12}r_{21}} + \sqrt{r_{11}r_{22}})^2} \qquad (2\text{-}69)$$

The results, *for real parameters*, are listed below in terms of the generalized parameters.

$$(\Gamma_L)_{opt} = \sqrt{\frac{k_{22}}{k_{11}}(k_{11}k_{22} - k_{12}k_{21})} \qquad (2\text{-}70)$$

$$(\Gamma_s)_{opt} = \sqrt{\frac{k_{11}}{k_{22}}(k_{22}k_{11} - k_{21}k_{12})} \qquad (2\text{-}71)$$

$$(G_T)_{max} = \frac{(k_{21})^2}{(\sqrt{k_{11}k_{22} - k_{12}k_{21}} + \sqrt{k_{11}k_{22}})^2} \qquad (2\text{-}72)$$

where k_{ij} are the generalized parameters (i.e., any of the y_{ij}, z_{ij}, g_{ij}, or h_{ij} parameters), and all are assumed to be real quantities in this derivation.

$$\begin{aligned}\Gamma_L &= R_L \text{ for } z_{ij} \text{ and } g_{ij} & \Gamma_s &= R_s \text{ for } z_{ij}, \text{ and } h_{ij} \\ &= G_L \text{ for } y_{ij} \text{ and } h_{ij} & &= G_s \text{ for } y_{ij}, \text{ and } g_{ij}\end{aligned} \qquad (2\text{-}73)$$

EXAMPLE 1

Let

$$h_{11} = 500 \, \Omega \qquad h_{12} = 10^{-4}$$
$$h_{21} = 50 \qquad h_{22} = 10^{-4} \text{ mho}$$

Find the optimum passive terminations (if any) and the maximum power gain.

Solution: Since, from (2-61), $h_{11}h_{22} - h_{12}h_{21} = 450 \times 10^{-4} > 0$, there exist optimum terminations which are

$$(R_s)_{opt} = \sqrt{\frac{h_{11}}{h_{22}}(h_{22}h_{11} - h_{21}h_{12})} = 150 \, \Omega$$

$$(G_L)_{opt} = \sqrt{\frac{h_{22}}{h_{11}}(h_{22}h_{11} - h_{21}h_{12})} = 0.95 \times 10^{-2} \quad \text{mhos}$$

or

$$(R_L)_{opt} = 105 \, \Omega$$

$$(G_T)_{max} = \frac{(50)^2}{(\sqrt{450 \times 10^{-4}} + \sqrt{500 \times 10^{-4}})^2} = 1.32 \times 10^4$$

or

$$(G)_{max} = 10 \log (1.32 \times 10^4) = 41.2 \text{ dB}$$

EXAMPLE 2

The two-port parameters of an active device measured at 50 MHz are (all in mhos):

$$y_{11} = 10^{-6} \qquad y_{12} = -j6 \times 10^{-6}$$
$$y_{21} = 8 \times 10^{-3} \qquad y_{22} = 2 \times 10^{-6}$$

Find the maximum power gain if it is finite.

Solution: From (2-61) we have

$$\text{Re } y_{11} > 0 \qquad \text{Re } y_{22} > 0$$
$$\text{Re } y_{11} \text{Re } y_{22} - \tfrac{1}{2}|y_{12}y_{21}|(1 + \cos \theta) = (10^{-6})(2 \times 10^{-6})$$
$$- (\tfrac{1}{2})48 \times 10^{-9}(1 + 0) < 0$$

The inequality is not satisfied and, therefore, the device is potentially unstable at this frequency. The maximum power gain can be infinite and the device capable of oscillation at this particular frequency.

2-5 Frequency Response and the Bode Plot

The plot of magnitude and phase of a network function vs frequency is referred to as the frequency response. This information is very useful in the analysis and design of linear time-invariant networks. We shall briefly discuss the frequency response below.

Consider a simple one-pole network function:

$$H(s) = \frac{1/R}{s + 1/RC} = \frac{K}{s + p_1} \qquad (2\text{-}74)$$

To find the magnitude and the phase we let $s = j\omega$ in (2-74). Hence

$$H(j\omega) = \frac{K}{j\omega + p_1} \qquad (2\text{-}75)$$

2/Two-Port Networks and Related Topics

where $p_1 = 1/RC$ and

$$|H(j\omega)| = \frac{K}{|j\omega + p_1|} = \frac{K}{\sqrt{p_1^2 + \omega^2}} = \frac{K/p_1}{\sqrt{1 + (\omega/p_1)^2}} \quad (2\text{-}76)$$

A plot of $\log |H(j\omega)|$ vs frequency,[3] which is called magnitude response, is shown in fig 2-20a. The phase function from (2-75) is

$$\arg H(j\omega) = \arg K - \tan^{-1}\frac{\omega}{p_1} = -\tan^{-1}\frac{\omega}{p_1} \quad (2\text{-}77)$$

Note that at zero frequency $H(0) = K/p_1$ and at $\omega = p_1$ we have $|H(jp_1)| = (K/p_1)/\sqrt{2}$. In other words, $|H(jp_1)|$ is equal to 0.707 $H(0)$, or 3 dB lower than the dc value. The 3 dB frequency in this case is thus p_1. At the 3 dB frequency for this one-pole network function, the phase is $\arg[H(jp_1)] = -\tan^{-1}(p_1/p_1) = -45°$. The phase response is shown in fig 2-20b.

For an involved network function, the use of a computer is both desirable and recommended. For paper and pencil work, however, the graphical method is convenient. The graphical method is performed as follows: Consider the following normalized, for the sake of clarity, network function.

$$H(s) = 100\frac{(s + 1.0)}{s^2 + 2s + 2} = \frac{100(s + 1.0)}{(s + 1 + j1)(s + 1 - j1)}$$

The pole-zero plot is shown in fig 2-21a.

$$|H(j\omega)| = 100\frac{|j\omega + 1.0|}{|1 + j(1 + \omega)||1 + j(\omega - 1)|}$$

Hence, at any frequency ω_1 we have

$$|H(j\omega_1)| = 100\frac{l_1}{d_1 d_2}$$

where l_1 is the distance from the zero, and d_1 and d_2 are the distances from the poles, as shown in fig 2-21a. This symbology is used to keep

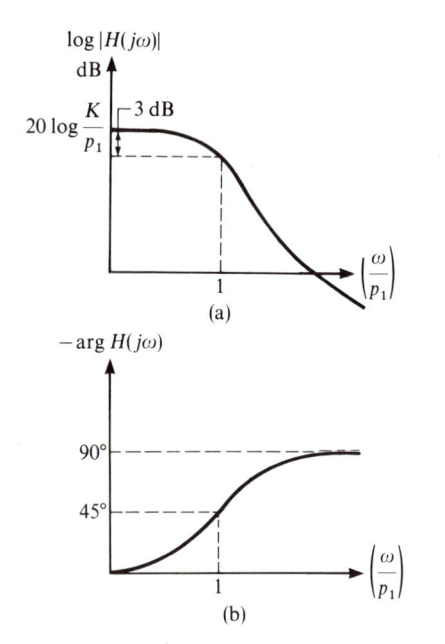

Fig 2-20 Frequency response: (a) Magnitude plot; (b) Phase plot

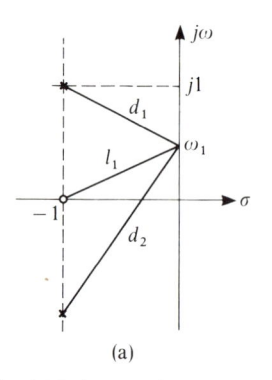

Fig 2-21 (a) Pole-zero plot of the example

---

[3] In this book, log refers to the base 10, while ln refers to the natural logarithm, i.e., to the base ε.

2-5 Frequency Response and the Bode Plot

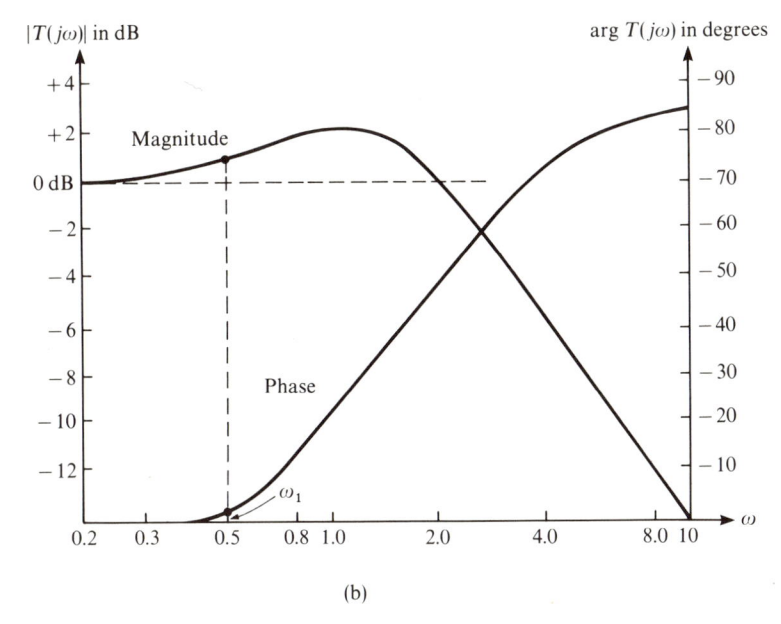

Fig 2-21 (b) Magnitude and phase plot vs ω

track of multiplication and division. Hence

$$\log |H(j\omega_1)| = \log 100 \left(\frac{l_1}{d_1 d_2}\right)$$

$$\arg H(j\omega) = \tan^{-1}\frac{\omega}{1.0} - \tan^{-1}\frac{1+\omega}{1.0} - \tan^{-1}\frac{\omega - 1}{1.0}$$

The magnitude and the phase plots for this case are shown in fig. 2-21b, where the transfer function is normalized, i.e., $T(j\omega) = H(j\omega)/H(0)$. The application of the above method when many poles and zeros are present in the network function should now be evident, and will not be discussed any further.

The Bode Plot

The Bode plot, or the asymptotic approximation, is essentially a break point approximation. This method of approximation is best suited for real poles and zeros. For complex critical frequencies, the error could be intolerable at the corner points. Since the Bode plot is a logarithmic plot of $H(j\omega)$ it has the advantage that product factors become additive. The plots are thus easy to construct.

In general, the network function for a lumped linear time-invariant system can be written in factored form as:

$$H(s) = K \frac{\prod_{i}^{l}(1 + a_i s)\prod_{k}^{m}(1 + b_k s + c_k s^2)}{s^p \prod_{r}^{q}(1 + d_r s)\prod_{u}^{n}(1 + e_u s + f_u s^2)} \tag{2-78}$$

where $a_i, b_k, c_k, d_r, e_u, f_u$ and K are real constants. The magnitude of $H(j\omega)$ in dB from (2-78) is

$$20 \log|H(j\omega)| = 20 \log K + 20\sum_{i}^{l} \log|1 + ja_i\omega| + 20\sum_{k}^{m} \log|1 - c_k\omega^2 \tag{2-79}$$

$$+ jb_k\omega| - 20\sum_{r}^{q} \log|1 + jd_r\omega| - 20p \log|\omega|$$

$$- 20\sum_{u}^{n} \log|1 - f_u\omega^2 + je_u\omega|$$

The phase of $H(j\omega)$ is

$$\arg T(j\omega) = \arg K + \sum_{i}^{l} \arg(1 + ja_i\omega) + \sum_{k}^{m} \arg(1 - c_k\omega^2 + jb_k\omega) \tag{2-80}$$

$$- p\left(\frac{\pi}{2}\right) - \sum_{r}^{q} \arg(1 + jd_r\omega) - \sum_{u}^{n} \arg(1 - f_u\omega^2 + je_u\omega)$$

From (2-79) and (2-78) it is seen that we have the sum of four simple types of factors:

(1) constant term K
(2) poles and zeros at the origin $\pm(j\omega)$
(3) real poles and zeros $\pm(1 + j\omega T)$
(4) complex poles and zeros $\pm(1 + j\omega B - C\omega^2)$

Each of the above factors is considered separately and all are then added or subtracted, depending upon whether the factors are due to the zeros or poles. The response is usually plotted on semilog paper with magnitude (dB) and phase on the linear scale and ω on the logarithmic scale. We shall now consider each factor separately.

(1) The constant term K yields a constant magnitude and zero phase contribution.

(2) The pole and zeros at the origin, for the magnitude response, represent terms which vary linearly with frequency when plotted on semilog paper, since

$$20 \log|(j\omega)^{\pm 1}| = \pm 20 \log \omega \quad \text{(in dB)} \tag{2-81}$$

The slope of the straight line is obtained by taking the derivative of (2-81):

$$\frac{d}{d \log \omega}[\pm 20 \log \omega] = \pm 20 \text{ dB/decade} \tag{2-82}$$

Note that a unit change in $\log \omega$ is equivalent to a decade change in frequency, i.e., 10 to 100, 1 to 10, etc., hence the name "per decade." Sometimes we may wish to know this slope per octave, i.e., a change in frequency by a factor of two, i.e., 1 to 2, 4 to 8, etc. This may be readily determined, since for $\omega_2 = 2\omega_1$ we have

$$\log \frac{\omega_2}{\omega_1} = \log 2 = 0.301 \tag{2-83}$$

Hence, the slope per octave is given by

$$0.301(\pm 20 \text{ dB}) \approx \pm 6 \text{ dB/octave} \tag{2-84}$$

It should be clear that if there are M more poles than zeros then the asymptotic high frequency slope is $-20M$ dB/decade or $-6M$ dB/octave. The phase shift for each extra pole or zero at the origin is minus or plus 90 degrees, since $\arg(\pm j\omega) = \pm \pi/2$.

(3) The magnitude and phase terms due to the real pole or zero are

$$\pm 20 \log|1 + j\omega T| = \pm 20 \log \sqrt{1 + \omega^2 T^2} \tag{2-85}$$

$$\pm \arg(1 + j\omega T) = \pm \tan^{-1} \omega T \tag{2-86}$$

To obtain the asymptotic behavior, we consider very low and very high frequencies; namely, for the magnitude

for

$$\omega \ll \frac{1}{T} \quad 20 \log|1 + j\omega T| \simeq 20 \log 1 \simeq 0 \text{ dB} \tag{2-87}$$

$$\omega \gg \frac{1}{T} \quad 20 \log|1 + j\omega T| \simeq 20 \log \omega T \tag{2-88}$$

Equations (2-87) and (2-88) represent two straight lines, one with a zero slope and the other with a slope of 20 dB/decade (or 6 dB/octave). The intersection of the low- and the high-frequency asymptotes is found by equating (2-87) and (2-88), yielding

$$20 \log \omega T = 20 \log 1 \Rightarrow \omega = \frac{1}{T} \qquad (2\text{-}89)$$

The frequency $\omega = 1/T$ is called the break, or corner frequency. The Bode plot for $\pm 20 \log |1 + j\omega T|$ and the actual magnitude response are shown in fig 2-22. The error between the asymptotic plot and the actual curve is maximum at the break frequency, with an error of ± 3 dB at $\omega = 1/T$.

The asymptotic phase behavior from $\pm \arg(1 + j\omega T)$ is readily seen to be as follows:

$$\omega T \gg 1 \qquad \pm \arg(1 + j\omega T) \simeq \pm \arg j\omega T = \pm 90° \qquad (2\text{-}90)$$

$$\omega T \ll 1 \qquad \pm \arg(1 + j\omega T) \approx \pm \arg 1 = 0° \qquad (2\text{-}91)$$

At the break frequency the phase is $\pm \arg(1 + j1) = \pm 45°$.

(4) The magnitude response due to the complex poles or zeros is

$$\pm 20 \log |1 + j\omega B - C\omega^2| \qquad (2\text{-}92)$$

The exact magnitude function is

$$\pm 20 \log \sqrt{(1 - C\omega^2)^2 + \omega^2 B^2} \qquad (2\text{-}93)$$

The low frequency asymptote is again $\approx \pm 20 \log 1 = 0$ dB. The high-frequency asymptote is determined by the ω^2 term; namely, from (2-92)

$$\pm 20 \log C\omega^2 = \pm 40 \log \sqrt{C}\omega \qquad (2\text{-}94)$$

Equation (2-94) represents a straight line with a slope of ± 40 dB/decade (or ± 12 dB/octave). The intersection of the low- and high-frequency asymptotes is given by

$$\pm 20 \log C\omega^2 = \pm 20 \log 1 \Rightarrow \omega = \frac{1}{\sqrt{C}} \qquad (2\text{-}95)$$

The error at the break frequency between the actual magnitude response and the asymptotic plot depends upon the value of $B/2C$ (i.e., the real

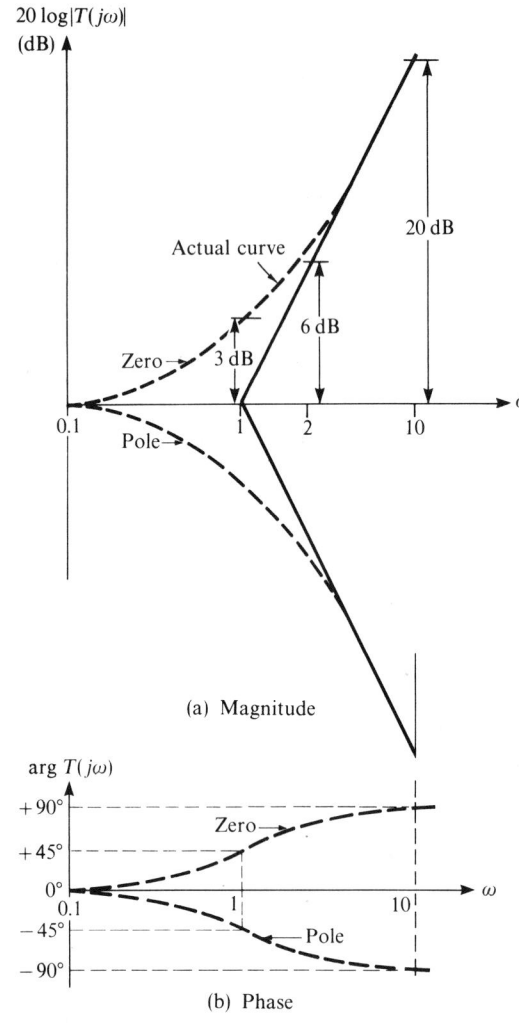

Fig 2-22 Actual and asymptotic plots of magnitude and phase vs ω for $T(s) = 1/(s + 1)$ and $(s + 1)$.

part of the poles or zeros.) This is readily seen, since the actual value of the magnitude from (2-93) at the break frequency is

$$\pm 20 \log \sqrt{(1 - C\omega^2)^2 + \omega^2 B^2} \bigg|_{\omega = 1/\sqrt{C}} = \pm 20 \log \frac{B}{\sqrt{C}} \quad (2\text{-}96)$$

while the Bode plot is independent of the value of B. Hence, from (2-96), it is seen that the closer the complex pole (zero) pairs are to the $j\omega$-axis, the larger the error. This is the reason why a Bode plot is not a good approximation for complex poles and zeros and is used mainly for real poles and zeros.

The phase behavior for the complex poles or zeros is determined from

$$\pm \arg(1 + j\omega B - C\omega^2) = \pm \tan^{-1} \frac{\omega B}{1 - C\omega^2} \quad (2\text{-}97)$$

The asymptotic behavior is readily seen to be as follows: The low-frequency asymptotic value of phase is zero, while the high-frequency asymptotic value is

$$\pm \tan^{-1} \left(\frac{B}{-C\omega} \right) \bigg|_{\omega \to \infty} = \pm 180° \quad (2\text{-}98)$$

Consider a quadratic term as poles of a transfer function:

$$H(s) = \frac{1}{1 + sB + Cs^2} = \frac{1}{1 + \frac{B}{\sqrt{C}}(\sqrt{C}s) + (\sqrt{C}s)^2} = \frac{1}{1 + 2\delta s_n + s_n^2} \quad (2\text{-}99)$$

where $s_n = \sqrt{C}s$ and $\delta = B/2\sqrt{C} = \cos \phi$ and ϕ is the angle of the pole from the negative real axis. The actual magnitude and phase response and the Bode plot for various values of δ are shown in fig 2-23.

EXAMPLE

We wish to sketch the Bode plot for the following transfer function:

$$H(s) = \frac{V_o}{V_s}(s) = -10^4 \frac{(s - 1)(s + 5)}{(s + 1)(s + 10)(s + 50)}$$

Note that at $s = 0$, the dc value is $H(0) = 10^2$. If we express the gain in decibels we have $20 \log H(0) = 40$ dB. (Note the factor of 20 for voltage or current ratio). For the magnitude response, the effects due

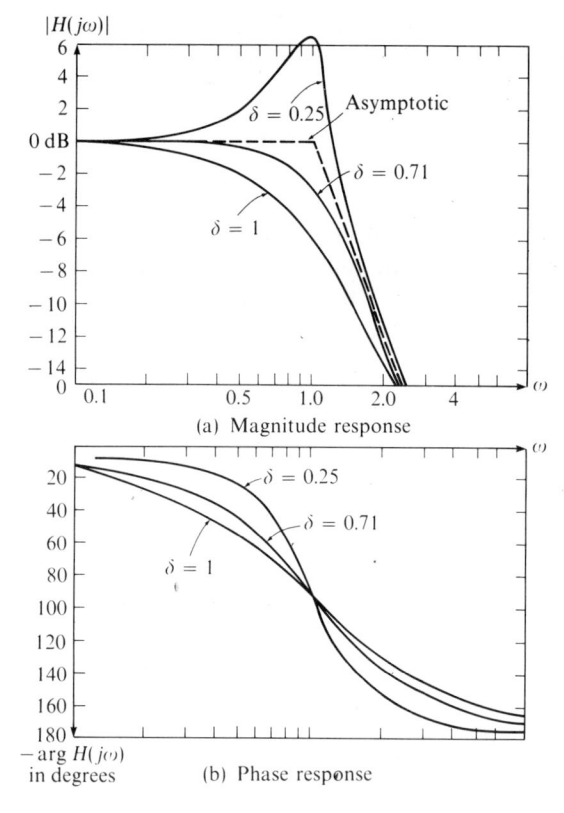

Fig 2-23 Magnitude and phase plot vs ω for $T(s) = 1/(1 + 2\delta s_n + s_n^2)$

to $(s - 1)$ and $(s + 1)$ cancel each other. The Bode plot is shown in fig 2-24.

Fig 2-24 Example Bode plot

2-6 Time-Domain Response

Consider the two-port linear time-invariant system shown in fig 2-25. The input is at port 1 and the output at port 2, and the system is in the zero state. The *network function* has already been defined in Section 1-11, and we repeat it here for convenience.

Fig 2-25 Schematic representation of a linear time-invariant two-port network

$$\begin{pmatrix}\text{Laplace transform of the}\\ \text{zero-state response}\end{pmatrix} = \begin{pmatrix}\text{Network}\\ \text{function}\end{pmatrix}\begin{pmatrix}\text{Laplace transform}\\ \text{of the input}\end{pmatrix}$$

2-6 Time-Domain Response

If the input and output quantities are voltages V_i and V_o, respectively, we then have

$$V_o(s) = H(s)V_i(s) \qquad (2\text{-}100)$$

If $v_i(t)$ is a unit impulse, then $V_i(s) = 1$ and the output $v_o(t)$, or *impulse response*, is given by

$$v_o(t) = \mathscr{L}^{-1}[V_o(s)] = \mathscr{L}^{-1}[H(s)] \equiv h(t) \qquad (2\text{-}101)$$

Thus

$$H(s) = \mathscr{L}[h(t)] = \text{Laplace transform of impulse response} \qquad (2\text{-}102)$$

If the impulse response $h(t)$ is known, then for the system in fig 2-25, the output for any other input is given by

$$v_o(t) = h(t) * v_i(t) = \int_0^\infty h(\lambda) v_i(t - \lambda) \, d\lambda \qquad (2\text{-}103)$$

where $*$ denotes convolution.

In electronic circuits, such as pulse amplifiers, one is often interested in the step response of the system. Thus, if $v_i(t) = u(t)$, then $V_i(s) = 1/s$, and from (2-100)

$$V_o(s) = H(s)\frac{1}{s} \qquad (2\text{-}104)$$

and

$$v_o(t) = \mathscr{L}^{-1}[V_o(s)] = \mathscr{L}^{-1}\left[\frac{H(s)}{s}\right] = \text{step response} \qquad (2\text{-}105)$$

A typical normalized step response for network functions with complex poles is shown in fig 2-26. Such a response is also called the transient response. Key features of the transient response are *overshoot* γ, *rise time* τ_R, and *delay time* τ_D. These quantities are illustrated in fig 2-26.

Overshoot is the percentage of the difference between the largest peak value and the final value to the final value. *Rise time* is the difference between the time when the response first reaches 90% of its final value and the time when the response is 10% of its final value.[4] *Delay time* is the time from $t = 0$ until the response reaches 50% of its final value.

[4] Other definitions of rise time are also used in the literature (see Refs 2 and 3).

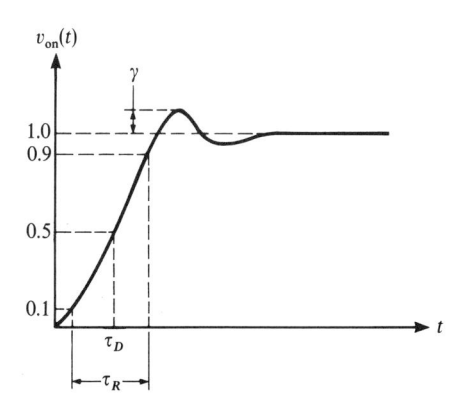

Fig 2-26 Time-domain response for a unit-step input

The evaluation of (2-105) to obtain the above quantities by partial fraction expansion is discussed in Section 1-5.

EXAMPLE

Let us calculate the step response of the following network function:

$$H(s) = \frac{A}{s^2 + \sqrt{2}s + 1}$$

From (2-105) we have

$$V_o(s) = \frac{1}{s} \frac{A}{s^2 + \sqrt{2}s + 1}$$

This may be expanded into a partial fractional expansion as follows:

$$V_o(s) = \frac{1}{s} \frac{A}{[s + 1/\sqrt{2} + j(1/\sqrt{2})][s + 1/\sqrt{2} - j(1/\sqrt{2})]}$$

$$= \frac{A}{s} + \frac{A(\sqrt{2}/2)\underline{/5\pi/4}}{s + 1/\sqrt{2} + j(1/\sqrt{2})} + \frac{A(\sqrt{2}/2)\underline{/-5\pi/4}}{[s + 1/\sqrt{2} - j(1/\sqrt{2})]}$$

From table 1-1, we obtain

$$v_o(t) = A\left[1 + \sqrt{2}e^{-(1/\sqrt{2})t}\cos\left(\frac{1}{\sqrt{2}}t + \frac{5\pi}{4}\right)\right]u(t)$$

The response is plotted in fig 2-27 ($\delta = \sqrt{2}/2$). The various parameters are

$$\gamma \approx 4.1\% \qquad \tau_R \approx 2.1 \qquad \tau_D \approx 1.4$$

Figure 2-27 also shows the step response for $H(s) = A/(s^2 + 2\delta s + 1)$ for $\delta = 1$ and $\delta = 0.25$. In pulse and video amplifiers, as discussed in later chapters, an overshoot of higher than 5 per cent is not desirable, since it causes distortion of the transient signal.

A distortionless ideal signal transmission is defined as follows: The output signal has the same waveform as the input signal (although it may be amplified and delayed by a constant time τ_0). In other words, $v_o(t) = A_o v_i(t - \tau_0)$. The network function for the ideal transmission

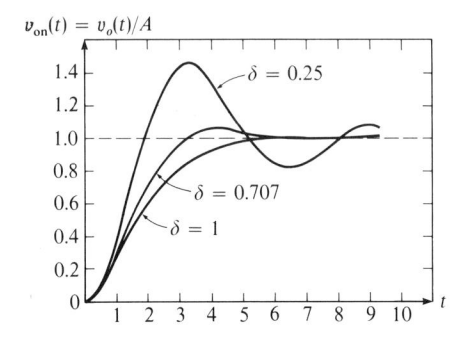

Fig 2-27 Step response for $H(s) = A/(s^2 + 2\delta s + 1)$

is

$$H(s) = \frac{\mathscr{L}[v_o(t)]}{\mathscr{L}[v_i(t)]} = \frac{\mathscr{L}[A_o v_i(t - \tau_o)]}{\mathscr{L}[v_i(t)]} \qquad (2\text{-}106)$$

From table 1-1, $\mathscr{L}[A_o v_i(t - \tau_o)] = A_o e^{-\tau_o s} V_i(s)$. Hence

$$H(s) = \frac{A_o e^{-\tau_o s} V_i(s)}{V_i(s)} = A_o e^{-\tau_o s} \qquad (2\text{-}107)$$

Thus, from (2-107), for ideal signal transmission:

$$|H(j\omega)| = A_0 \qquad (2\text{-}108a)$$

$$\arg H(j\omega) = -\tau_o \omega \qquad (2\text{-}108b)$$

The ideal characteristics expressed in (2-108a) and (2-108b) are not realizable in practice, and we can thus only try to approximate them. Since lumped networks will be used in the actual realizations, the approximation of $H(s)$ must be in the form of rational functions of the complex frequency.

Magnitude and phase functions can be approximated simultaneously within a band of frequencies. This type of approximation, however, is quite involved and normally is treated in advanced texts. We shall consider simple approximations by concentrating either on the phase or the magnitude characteristics. Obviously, such approximations are less accurate, but they are still useful in many practical applications.

2-7 Linear Phase Transfer Functions

We shall now consider the approximation of a linear phase characteristic by a rational transfer function of the form

$$H(s) = \frac{b_m s^m + b_{m-1} s^{m-1} + \cdots + b_1 s + b_0}{s^n + a_{n-1} s^n + \cdots + a_1 s + a_0}. \quad (2\text{-}109)$$

We assume that $n > m$ since for any physical system $|H(j\omega)|$ goes to zero as $\omega \to \infty$.

The transfer function (2-109) can also be written as

$$H(j\omega) = \frac{A(\omega^2) + j\omega B(\omega^2)}{C(\omega^2) + j\omega D(\omega^2)} \quad (2\text{-}110)$$

where A, B, C, and D are polynomials in ω^2. In other words, from (2-109) and (2-110)

$$\begin{aligned}
A(\omega^2) &= b_0 - b_2 \omega^2 + b_4 \omega^4 - \cdots \\
B(\omega^2) &= b_1 - b_3 \omega^2 + b_5 \omega^4 - \cdots \\
C(\omega^2) &= a_0 - a_2 \omega^2 + a_4 \omega^4 - \cdots \\
D(\omega^2) &= a_1 - a_3 \omega^2 + a_5 \omega^4 - \cdots
\end{aligned} \quad (2\text{-}111)$$

Before we consider the general case, it may be instructive to consider a simple example in order to understand the method.

EXAMPLE

Suppose we wish to approximate a linear phase characteristic with the following simple transfer function:

$$H(s) = \frac{K}{s^2 + a_1 s + a_0}$$

We wish to find the pole locations or, alternatively stated, the relation between a_1 and a_0 such that the above equation is an approximation for a linear phase transfer function. For this case

$$\phi(\omega) \equiv \arg H(j\omega) = -\tan^{-1} \frac{a_1 \omega}{a_0 - \omega^2} \quad (2\text{-}112)$$

2-7 Linear Phase Transfer Functions

The method is to expand $\phi(\omega)$ in a Taylor series about the origin; i.e., to express $\phi(\omega)$ for small ω as follows:

$$\phi(\omega) = \alpha_1 \omega + \alpha_3 \omega^3 + \alpha_5 \omega^5 + \cdots \quad (2\text{-}113)$$

Note that the even powers of ω are absent because the phase function is always an odd function of frequency. We set as many of $\alpha_3, \alpha_5, \ldots,$ as possible equal to zero. In order to write (2-110) in the form of (2-113), we use two series expansions

$$\tan^{-1} x = x - \frac{x^3}{3} + \frac{x^5}{5} - \cdots \quad |x| < 1 \quad (2\text{-}114)$$

and

$$(1 \pm \delta)^{-n} = 1 \mp n\delta + \frac{n(n+1)}{2!}\delta^2 \mp \frac{n(n+1)(n+2)}{3!}\delta^3 + \cdots \quad \delta < 1 \quad (2\text{-}115)$$

Hence for ω near zero, (2-112) can be written as

$$\phi(\omega) = -\left[\frac{a_1 \omega}{a_0 - \omega^2} - \frac{1}{3}\left(\frac{a_1 \omega}{a_0 - \omega^2}\right)^3 + \cdots\right]$$

or

$$\phi(\omega) = -\frac{a_1 \omega}{a_0}\frac{1}{(1 - \omega^2/a_0)} + \frac{a_1^3 \omega^3}{3a_0^3}\frac{1}{(1 - \omega^2/a_0)^3} - \cdots \quad (2\text{-}116)$$

Each term in (2-116) can be expressed in a Taylor series from (2-115), thus

$$\phi(\omega) = -\frac{a_1 \omega}{a_0}\left(1 + \frac{\omega^2}{a_0} - \frac{\omega^4}{a_0^2} + \cdots\right) + \frac{a_1^3 \omega^3}{3a_0^3}\left(1 + 3\frac{\omega^2}{a_0} + 6\frac{\omega^4}{a_0^2} + \cdots\right)$$

Rearranging the terms, we get

$$\phi(\omega) = -\frac{a_1}{a_0}\omega - \left(\frac{a_1^3}{3a_0^3} - \frac{a_1}{a_0^2}\right)\omega^3 - \cdots \quad (2\text{-}117)$$

For (2-117) to approximate a linear phase transfer function,

$$\phi(\omega) \simeq -\frac{a_1}{a_0}\omega$$

Hence, all the terms except those that are linear must be set equal to zero. In this special approximating function, we have only one degree of freedom, and we therefore set the first term of the series beyond the linear terms equal to zero. This yields

$$\frac{a_1^3}{3a_0^3} - \frac{a_1}{a_0^2} = 0$$

or

$$a_1^2 = 3a_0 \tag{2-118}$$

Thus, if we use a frequency normalization such that $a_0 = 1$, a two-pole network function for a linear phase function is

$$H(s_n) = \frac{A}{s_n^2 + \sqrt{3}s_n + 1}$$

The general case can now be treated similarly. From (2-110), we have

$$\phi(\omega) = \arg H(j\omega) = \tan^{-1}\frac{\omega B(\omega^2)}{A(\omega^2)} - \tan^{-1}\frac{\omega D(\omega^2)}{C(\omega^2)} \tag{2-119}$$

From (2-119) and (2-114), we have

$$\phi(\omega) = \left[\frac{\omega B(\omega^2)}{A(\omega^2)} - \frac{\omega^3 B^2(\omega^2)}{A^2(\omega^2)} + \cdots\right] - \left[\frac{\omega D(\omega^2)}{C(\omega^2)} - \frac{\omega^3 D^2(\omega^2)}{C^2(\omega^2)} + \cdots\right]$$

$$\tag{2-120}$$

Next we expand

$$\frac{1}{A(\omega^2)}, \quad \frac{1}{A^2(\omega^2)}, \quad \frac{1}{C^2(\omega^2)}$$

etc., using (2-115), and then rearrange and collect terms to rewrite (2-120) as

$$\phi(\omega) = \alpha_1\omega + \alpha_3\omega^3 + \alpha_5\omega^5 + \cdots \tag{2-121}$$

For (2-121) to approximate a linear phase characteristic, we must have

$$\alpha_1 \neq 0 \quad \text{and} \quad \alpha_i = 0, \tag{2-122}$$

where $i = 3, 5, 7, \ldots$.

For a given-order transfer function, we can set only a few of the α_i equal to zero, because of the available degrees of freedom. In such cases, low-order power coefficients, i.e., α_3, α_5, etc., take priority over the high-order ones, such as α_9, α_{11}, etc., since in (2-121) at ω near zero, i.e., ω very small, ω^3 contributes more than ω^5, and so on.

Thomson Filters

A class of transfer functions which has no finite transmission zero and which has a linear phase characteristic is often used in practical applications. This type of transmission network is referred to as the Thomson filter, linear-phase filter, or maximally flat delay filter. It is used when a simple network function with no distortion in the signal transmission is desired. Consider the following transfer function

$$H(s) = \frac{k}{s^n + a_{n-1}s^{n-1} + \cdots + a_1 s + a_0} \quad (2\text{-}123)$$

The procedure discussed above can be used to find the relation between the coefficients for a linear phase characteristic. For this special class, however, other simple methods are also available (Ref 3).

For convenience and later reference, we list the first four low-order polynomials of the Thomson filter in table 2-3. The 3 dB bandwidth for each case is also listed therein. Note that the second-degree polynomial is related to the denominator polynomial in the preceding example for a frequency scaling of $\sqrt{3}$. If the magnitude function varies slowly with frequency, and if the phase function over the frequency band of interest is approximately linear, we can arrive at a physically meaningful definition of the *delay function* (Ref 2).

$$\tau(\omega) = -\frac{d\phi(\omega)}{d\omega} \quad (2\text{-}124)$$

TABLE 2-3 *Polynomials for Linear Phase (Thomson) Response*

Various order polynomials	ω_{3dB}
$p_1(s) = s + 1$	1.00
$p_2(s) = s^2 + 3s + 3$	1.36
$p_3(s) = s^3 + 6s^2 + 15s + 15 = (s + 2.322)(s^2 + 3.678s + 6.46)$	1.75
$p_4(s) = s^4 + 10s^3 + 45s^2 + 105s + 105 = (s^2 + 5.79s + 9.13)(s^2 + 4.21s + 11.46)$	2.13

Since the phase characteristics of Thomson filters are approximately linear, the delay function from (2-121) and (2-124) is approximately constant, hence *maximally flat* near the origin. This is the reason for the alternative name "maximally flat delay filter." If the delay is to be normalized to unity from (2-124) and (2-121), or simply from (2-117) near $\omega = 0$, we have

$$\tau(0) = -\frac{d\phi(\omega)}{d\omega}\bigg|_{\omega=0} = \frac{a_1}{a_0} = 1 \qquad (2\text{-}125)$$

Note that the polynomials listed in table 2-3 are normalized such that the delay is unity at $\omega = 0$. The delay functions vs frequency for Thomson filters are shown in fig 2-28. Figure 2-29 shows the step response for Thomson filters up to the fourth order. The overshoot for Thomson filters is negligible for any order network function.

2-8 Maximally Flat Magnitude Functions

Consider the approximation of a constant magnitude within a band of frequencies by a rational transfer function of the form

$$H(s) = \frac{b_m s^m + b_{m-1} s^{m-1} + \cdots + b_1 s + b_0}{s^n + a_{n-1} s^{n-1} + \cdots + a_1 s + a_0} \qquad (2\text{-}126)$$

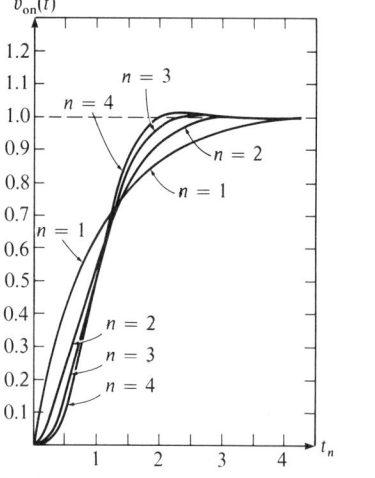

Fig 2-29 Step response of low-order Thomson filters

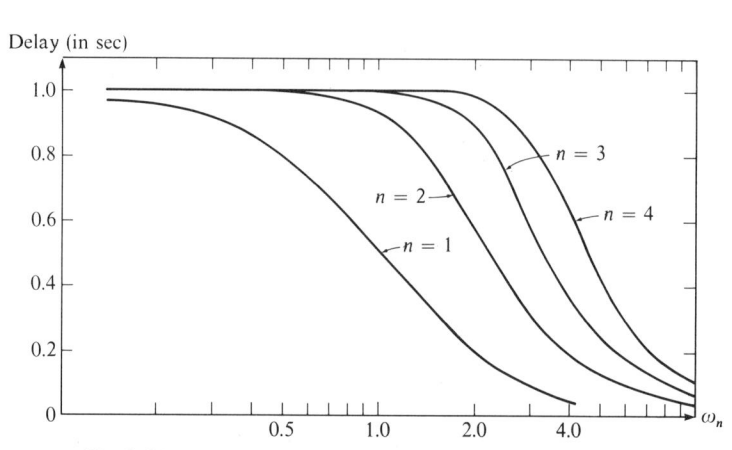

Fig 2-28 Delay characteristics of low-order Thomson filters

2-8 Maximally Flat Magnitude Functions

with $n > m$. Equation (2-126) may be rewritten as

$$|H(j\omega)|^2 = A^2 \frac{1 + c_2\omega^2 + c_4\omega^4 + \cdots c_{2m}\omega^{2m}}{1 + d_2\omega^2 + d_4\omega^4 + \cdots + d_{2n}\omega^{2n}} \quad (2\text{-}127)$$

One method of approximation is to have (2-127) represent a monotonically decreasing function of frequency and be as flat as possible near the dc (zero) frequency. Such an approximation is referred to as a maximally flat magnitude approximation. For (2-127) to be a maximally flat function near the origin, as many of its derivatives as possible must be equal to zero. Hence, one approach to find the relation between the coefficients is to do just that: set $(n-1)$ derivatives of (2-127) equal to zero, since there are $(n-1)$ relations for an nth-order function. Another, simpler, approach to finding the derivatives of (2-127) is to use the Maclaurin series expansion:

$$|H(j\omega)|^2 \equiv f(\omega^2) = A^2 + f'(0)\omega + \frac{f''(0)}{2!}\omega^2 + \frac{f'''(0)}{3!}\omega^3 + \cdots \quad (2\text{-}128)$$

where

$$f'(0) = \frac{d}{d\omega}|H(j\omega)|^2 \quad \text{at } \omega = 0$$

$$f''(0) = \frac{d^2}{d\omega^2}|H(j\omega)|^2 \quad \text{at } \omega = 0, \text{ etc.} \quad (2\text{-}129)$$

By long division, i.e., dividing the numerator by the denominator in (2-127), we get

$$|H(j\omega)|^2 = A^2\{1 + (c_2 - d_2)\omega^2 + [(c_4 - d_4) - d_2(c_2 - d_2)]\omega^4 + \cdots\} \quad (2\text{-}130)$$

Note that all the odd powers of ω are missing since $|H(j\omega)|$ is an even function of ω. Since a power series is unique within the region of convergence, (2-130) and (2-128) are equal. Therefore, one can identify the coefficients of (2-130) with the derivatives in (2-128). The derivatives are set to zero to obtain flatness; hence

$$\begin{aligned} c_i &= d_i & i &= 2, 4, \ldots, 2m \\ d_i &= 0 & i &= 2(m+1), \ldots, 2(n-1) \end{aligned} \quad (2\text{-}131)$$

Hence, to get a maximally flat magnitude function, we obtain $|H(j\omega)|^2$ from $H(s)$ and then use the conditions in (2-131). Note that a maximally flat magnitude function from (2-131) and (2-127) is of the form

$$|H(j\omega)|^2 = \frac{F_1(\omega^2)}{F_1(\omega^2) + d_{2n}\omega^{2n}} \qquad (2\text{-}132)$$

Butterworth Functions

Butterworth functions, which are also called maximally flat magnitude functions, are special cases of the general function in (2-126) in which there are no finite transmission zeros, i.e., $b_i = 0$ ($i \neq 0$). Hence, the Butterworth functions are of the form

$$|H(j\omega)|^2 = A^2 \frac{1}{1 + d_{2n}\omega^{2n}} = \frac{A^2}{1 + (\omega/\rho)^{2n}} \qquad (2\text{-}133)$$

where $\rho^{-2n} = d_{2n}$ is introduced for convenience, as shall be seen later.

EXAMPLE

Let

$$H(s) = \frac{K}{s^2 + a_1 s + a_0}$$

We want to find the relationship between a_1 and a_0 such that the transfer function has a maximally flat magnitude function.

The magnitude-squared function is

$$|H(j\omega)|^2 = \frac{K^2}{(a_0 - \omega^2)^2 + a_1^2 \omega^2}$$

$$= \frac{K^2/a_0^2}{1 + \omega^2[(a_1^2 - 2a_0)/a_0^2] + \omega^4/a_0^2}$$

Using the conditions expressed in (2-131) in the above, we obtain

$$a_1^2 - 2a_0 = 0, \text{ hence } a_1 = \sqrt{2a_0} \qquad (2\text{-}134)$$

A second-order Butterworth function may thus be written as

$$H(s) = \frac{K}{s^2 + \sqrt{2a_0}\,s + a_0} \qquad (2\text{-}135)$$

The pole locations are shown in fig 2-30a.

2-8 Maximally Flat Magnitude Functions

Note also that the 3-dB frequency as obtained from (2-135) is

$$\left.\frac{K^2/a_0^2}{1 + \omega^4/a_0^2}\right|_{\omega=\omega_{3\,\mathrm{dB}}} = \frac{1}{2}\left(\frac{k^2}{a_0^2}\right) \rightarrow \omega_{3\,\mathrm{dB}} = \sqrt{a_0}$$

The pole distributions for the general case in (2-133) are on a semicircle and may be written as follows:

$$n \text{ even:} \quad \prod_{k=1}^{n/2} (s^2 + 2\rho \cos \theta_k s + \rho^2) \quad \theta_k = \frac{k\pi}{2n} \quad (2\text{-}136a)$$

$$n \text{ odd:} \quad (s + \rho) \prod_{l=1}^{(n-1)/2} (s^2 + 2\rho \cos \theta_l s + \rho^2) \quad \theta_l = \frac{l\pi}{n} \quad (2\text{-}136b)$$

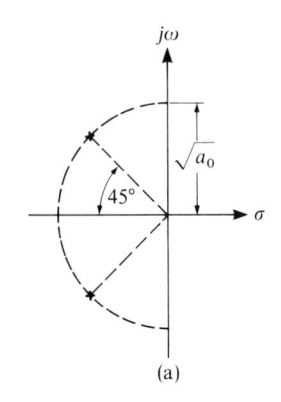

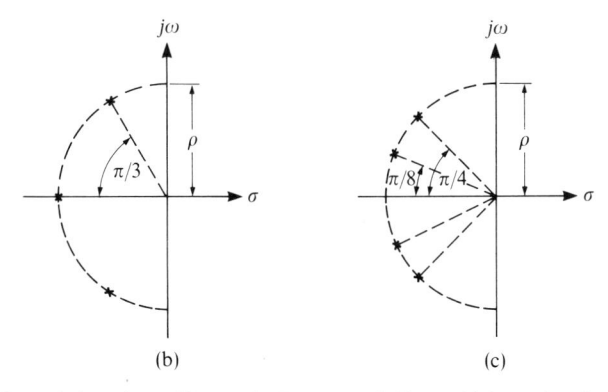

Fig 2-30 Pole locations of low-order Butterworth filters: (a) Second-order; (b) Third-order; (c) Fourth-order

For example, the pole locations for the third- and fourth-order Butterworth functions are shown in fig 2-30b and fig 2-30c, respectively. The 3-dB bandwidth for any order Butterworth function, as readily obtained from (2-133), is $\omega_{3\,\text{dB}} = \rho$.

Usually one normalizes frequency, i.e., sets $s_n = s/\rho$. For the normalized case, the Butterworth functions are of the form

$$H(s_n) = \frac{A}{s_n^n + a_{n-1}s_n^{n-1} + a_{n-2}s_n^{n-2} + \cdots + a_1 s_n + 1} \quad (2\text{-}137)$$

The values of a_i for the first four Butterworth polynomials are listed in table 2-4. Note that the second-degree polynomial corresponds to $a_0 = 1$ in (2-135). The magnitude and the step response of the Butterworth filters corresponding to (2-137) for n up to 4 are shown in fig 2-31 and fig 2-32, respectively. Note that the overshoot becomes larger as the order of the filter increases. By comparing fig 2-32 and fig 2-29

TABLE 2-4 *Butterworth Polynomials*

$s + 1$
$s^2 + \sqrt{2}s + 1$
$s^3 + 2s^2 + 2s + 1 = (s + 1)(s^2 + s + 1)$
$s^4 + 2.613s^3 + 3.414s^2 + 2.613s + 1 = (s^2 + 0.765s + 1)(s^2 + 1.848s + 1)$

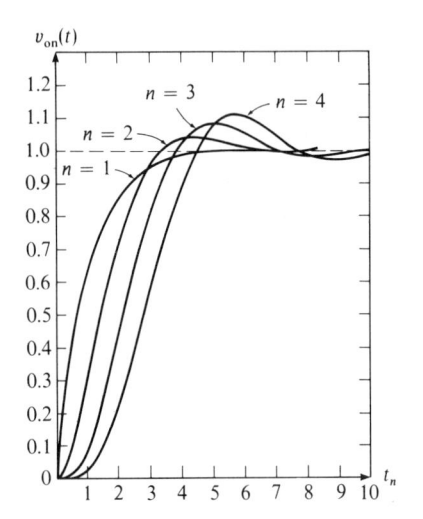

Fig 2-32 Step response of low-order Butterworth filters

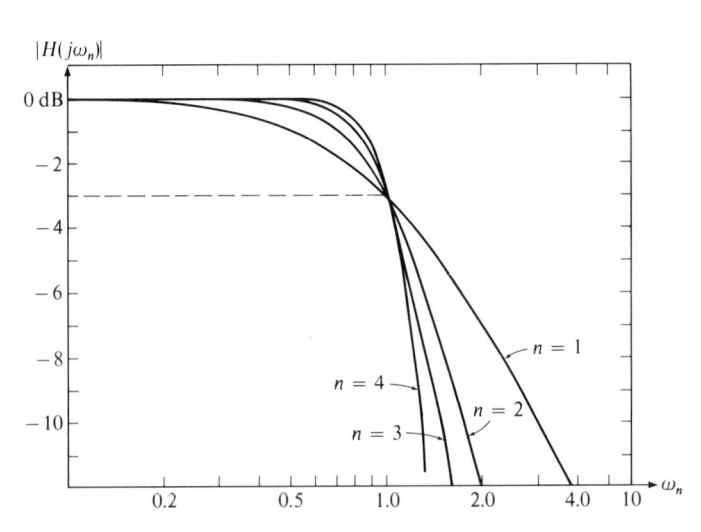

Fig 2-31 Magnitude response of low-order Butterworth filters

for, say, $n = 4$ or higher orders, we conclude that the approximation confined to the phase function provides a time response that is a better replica of the input than does the approximation restricted to the magnitude function.

In video signal processing, phase characteristics are important in distortionless signal transmission. In telephone communication, the magnitude characteristic is important, since the human ear is insensitive to phase.

2-9 Other Approximation Criteria

We mention at this point that the maximally flat magnitude criterion is not the only way to approximate a constant magnitude within a band. There are other types of approximations, one of which is the equiripple, or Chebyshev, filter. In the equiripple approximation, the constant magnitude is approximated within a band of frequencies such that the maximum error is minimized. This approximation distributes the error evenly throughout the pass-band in an oscillating manner, such that the magnitude function varies between equal maximum and minimum specified error around the desired constant-magnitude function. For example, an all-pole network function of order four, with an allowable error δ of 0.5 dB in magnitude within a band of frequencies $0 \leq \omega \leq \omega_0$, has with the Chebyshev approximation the magnitude shown in fig 2-33. The step response of a Chebyshev filter has a large overshoot. For a more detailed discussion of this topic, as well as other types of approximations, the reader is directed to Ref 2, 3, and 6.

An interesting class of all-pole filters, which gives a good compromise between the frequency response and the transient response, is described in Problem 2-35. In this class of filters, the overshoot for a step input is less than 6 per cent for any order filter, while the frequency response is better than the Thomson filters.

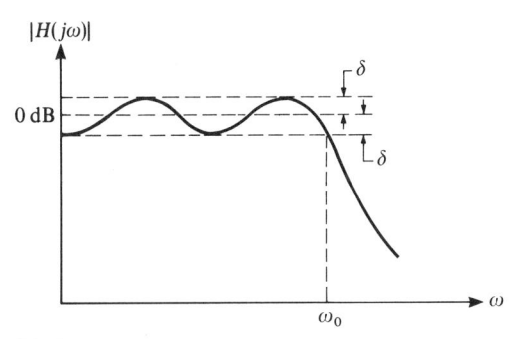

Fig 2-33 Fourth-order Chebyshev magnitude response

REFERENCES AND SUGGESTED READING

1. Desoer, C. and Kuh, E., *Basic Circuit Theory*, New York: McGraw-Hill, 1969, Chapter 17.
2. Kuh, E. and Pederson, D., *Principles of Circuit Synthesis*, New York: McGraw-Hill, 1959, Chapters 3, 4, and 5.

3. Ghausi, M. S., *Principles and Design of Linear Active Circuits*, New York: McGraw-Hill, 1965, Chapters 3 and 4.
4. Ruston, H. and Bordogna, J., *Electric Networks, Functions, Filters, Analysis*, New York: McGraw-Hill, 1966, Chapters 4, 5, and 8.
5. Skwirzynski, J. K., *Design Theory and Data for Electrical Filters*, New York: Van Nostrand Reinhold, 1965. A useful handbook for filter design.
6. Weinberg, L., *Network Analysis and Synthesis*, New York: McGraw-Hill, 1962, Chapters 11 and 13.

PROBLEMS

2-1 Determine the *h*-parameters of the two-port network shown in fig P2-1.

2-2 Determine the *z*-parameters of fig 1-11b.

2-3 Calculate *ABCD* parameters in terms of *y*-parameters.

2-4 Determine the *z*-parameters of the circuit shown in fig P2-4.

2-5 Find a convenient two-port description for the circuit shown in fig P2-5. If $Z_a = -R$ and $Z_b = R$, is the network reciprocal?

2-6 Determine the *z*- and *y*-parameters of the lattice network shown in fig P2-6.

2-7 Determine the voltage ratio expression in terms of the *ABCD* parameters for the circuit of fig 2-18.

2-8 For the cascaded linear two-port circuit shown in fig P2-8, determine the following:
(a) The expression for the overall voltage gain V_o/E_s in terms of the given $[z_{ij}]$ parameters.
(b) If $[z_{ij}^b]$ are the parameters of an ideal gyrator, (2-29), and $[z_{ij}^a]$ are those of an ideal current-controlled voltage source, (2-27), what is the overall voltage gain?

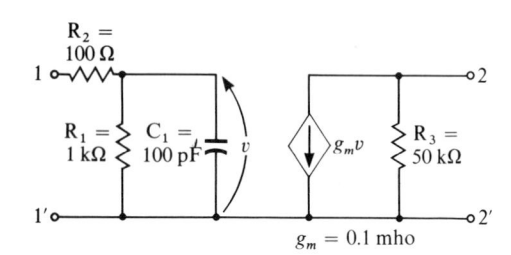

Fig P2-1

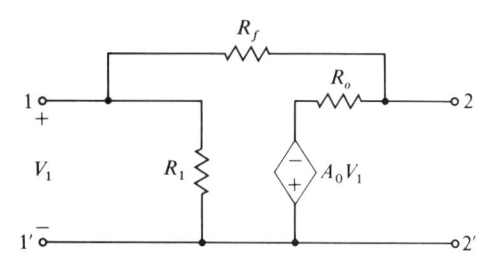

Fig P2-4

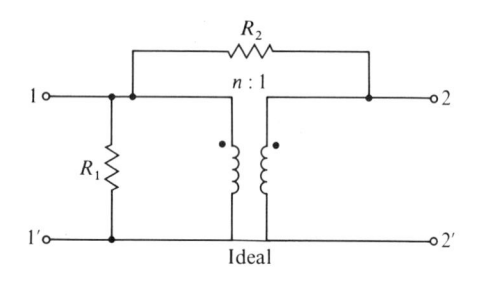

Fig P2-5

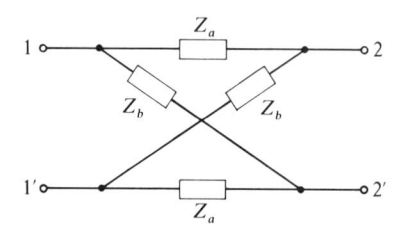

Fig P2-6

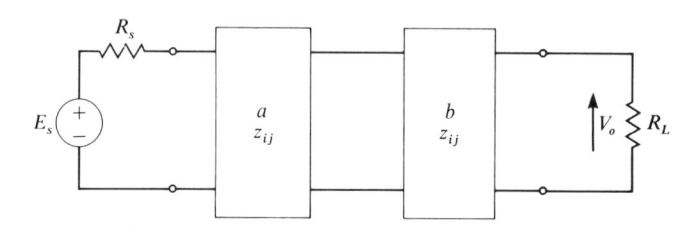

Fig P2-8

2-9 Derive the expressions in (2-56) and (2-57).

2-10 The two-port description of a transistor at 25 MHz is given by measurements to be

$$h_{11} = (60 - j60)\Omega \qquad h_{22} = (50 + j40) \times 10^{-5} \text{ mho}$$
$$h_{12} = 0.01 \, \Omega \qquad h_{21} = -j2.0 \text{ mho}$$

Is the transistor potentially unstable at this frequency? What is the tranducer power gain at 25 MHz if the transistor is used with $R_S = R_L = 100 \, \Omega$.

2-11 The two-port description of a device at 1 MHz is found to be

$$y_{11} = (1 + j2) \times 10^{-3} \text{ mho} \qquad y_{21} = 4 \times 10^{-2} \text{ mho}$$
$$y_{12} = -j10^{-5} \text{ mho} \qquad y_{22} = (2 + j10) \times 10^{-6} \text{ mho}$$

Is the device inherently stable? What is the actual power gain G_P of this device at 1 MHz with source and load terminations of $R_S = R_L = 1 \, k\Omega$?

2-12 If $y_{12} = 0$ in a linear two-port,
(a) What are the optimum terminations in terms of the y-parameters?
(b) What is the maximum power gain? (This power gain is called the unilateral power gain.)

2-13 Consider a linear two-port network described by the y-parameters. Show that the average ac power entering the two-port is

$$P_{av} = [V]^{t*}[\tfrac{1}{2}\{[y] + [y]^{t*}\}][V]$$

where t and $*$ denote transpose and conjugate, respectively. Hint:

$$P_{av} = \sum_{j=1}^{2} \text{Re } V_j^* I_j = \tfrac{1}{2}\{[V]^{t*}[I] + [V]^t[I^*]\}$$

2-14 Determine the y-parameters of the circuit shown in fig P2-14. Is the circuit reciprocal; unilateral? If it is not unilateral, what value of R can make the two-port unilateral?

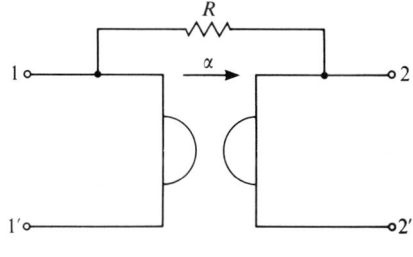

Fig P2-14

2-15 An active two-port has the following parameters at low frequencies (all in mhos):

$$y_{11} = 10^{-3} \qquad y_{21} = 5 \times 10^{-2}$$
$$y_{12} = 10^{-5} \qquad y_{22} = 10^{-6}$$

Is the two-port inherently stable? If $R_S = 1 \, k\Omega$ and $Y_L = G_L + sC_L$, for what values of G_L and C_L will the circuit exhibit instability? Hint: Examine the real part of the admittance of the two-port with R_S termination.

2-16 What is the simplest two-port network equivalent to a cascade of two gyrators with $\alpha_1 = \alpha_2 = 100\,\Omega$? What is the input impedance of this network if the output is terminated by an inductance L?

2-17 For the network shown in fig P2-17, the $ABCD$ parameters of $\mathcal{N}$ are as indicated. Show that

$$\left.\frac{V_2}{I_1}\right|_{I_2=0} = \frac{z_{21}^a z_{21}^b}{z_{22}^a - z_{11}^b}$$

Fig P2-17

2-18 If in the network of fig P2-17, $\mathcal{N}$ is a gyrator with the gyration impedance $\alpha_0\,\Omega$, determine V_2/I_1 in terms of the z-parameters of the individual networks.

2-19 For a linear time-invariant system S, shown schematically in fig P2-19 the impulse response is $h(t) = (e^{-t} + e^{-2t})\,u(t)$.
(a) What is the step response of the system?
(b) What is the output in time domain for an input $e^{-t}u(t)$?

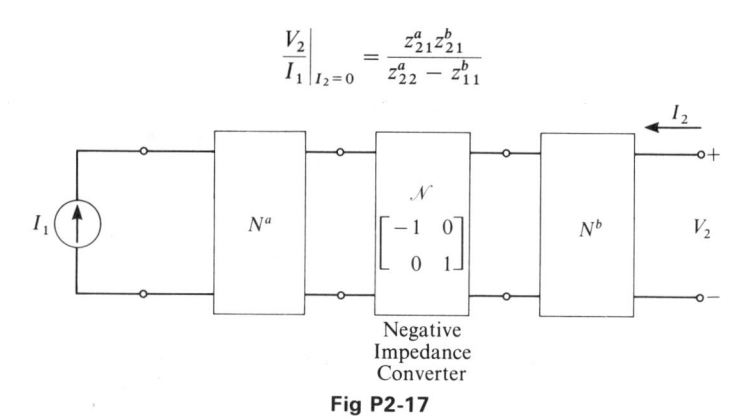

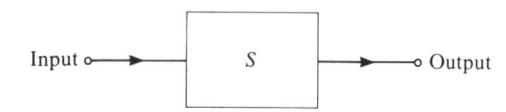

Fig P2-19

2-20 For a linear time-invariant one-port network, the following information is given: When the excitation is a current $i = u(t)$, the voltage across the terminals is $v(t) = (1 - e^{-2t})\,u(t)$. When the excitation is a voltage $v = u(t)$, the current flowing into the terminal is $i(t) = (1 - e^{-t})\,u(t)$. What is the driving-point impedance function? Show one possible circuit which has the above $Z(s)$.

2-21 For the system shown in fig P2-19, the transfer function is given by

$$\frac{V_o}{V_i}(s) = \frac{1}{s^2 + 3s + 3}$$

Determine the step response, i.e., $v_o(t)$ for $v_i(t) = u(t)$. What is the 10-per cent to 90-per cent rise time, overshoot, and delay of the system.

2-22 Determine the impulse response for Problem 2-21 and sketch the waveform.

2-23 For the linear time-varying circuit shown in fig P2-23, determine the impulse response and step response. Is the impulse response the derivative of the step response?

Fig P2-23

$G = 1 + 2t$
$L = 1$

2-24 Sketch the Bode plot for the magnitude of the following network functions (show the breakpoints and the slopes of each segment):

(a) $(s + 1)(s + 10)/(s + 2)(s^2 + s + 1)(s + 5)$

(b) $(s - 2)/(s + 2)(s^2 + 6s + 25)$

2-25 The transfer function of a network is given by

$$T(s) = \frac{(s + z)}{s^2 + as + b}$$

Show that for the transfer function to have a maximally flat magnitude, the relation between z, a, and b is given by:

$$a^2 - 2b = \left(\frac{b}{z}\right)^2$$

2-26 For the transfer function given in Problem 2-25, derive the relation between z, a, and b for a linear phase response.

2-27 The network shown in fig P2-27 is to have a maximally flat magnitude response, a 3-dB radian frequency $\omega_{3\,dB} = 10^7$ rad/sec, and a dc value of $V_o/I_S(0) = 1$ kΩ. Design the circuit.

Fig P2-27

2-28 Repeat Problem 2-27 for a maximally flat delay response.

2-29 Given the following network function:

$$N(s) = \frac{K}{s^3 + 5s^2 + a_1 s + a_0}$$

determine the constants a_1 and a_0 for the following approximation criteria:
(a) Linear phase.
(b) Maximally flat magnitude.

2-30 A maximally flat magnitude filter is desired for a cutoff frequency of 0.5 kHz and a minimum attenuation of 20 dB at 1 kHz. What minimum order Butterworth filter is necessary? Write the actual voltage ratio function for this filter if the dc level is 100.

2-31 If two ideal voltage-controlled current source two-port networks are available, show that, by their proper interconnection, a gyrator can be obtained.

2-32 For a reciprocal resistive linear two-port network, shown schematically in fig P2-32, the following data are given:
(a) When the excitation is a current I_s and $R_L = 0$, the current gain is

$$\frac{I_L}{I_s} = -\frac{1}{3}$$

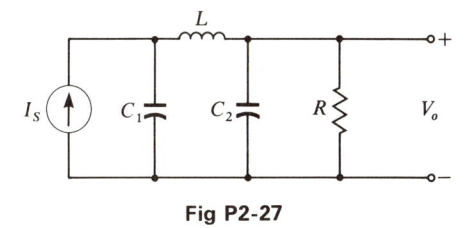

(b) When the excitation is a voltage V_s and $R_L = \infty$, the voltage gain is

$$\frac{V_L}{V_s} = \frac{1}{5}$$

(c) When $R_L = 1\ \text{k}\Omega$, $Z_{in} = 1.75\ \text{k}\Omega$.

Determine the z-parameters of the two-port, and show the circuit elements of the T-model.

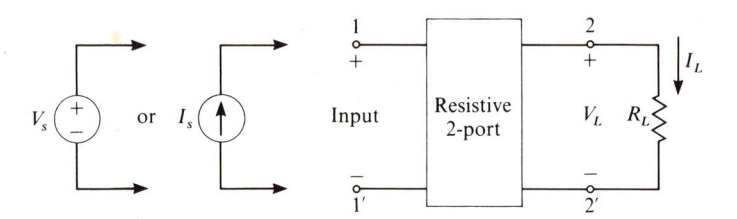

Fig P2-32

2-33 The circuit shown in fig P2-33 is to be designed for a maximally flat magnitude response with the 3 dB cutoff at $\omega_{3\ \text{dB}} = 10^7$ rad/sec. Determine the values of R and L and calculate the voltage gain at dc frequency.

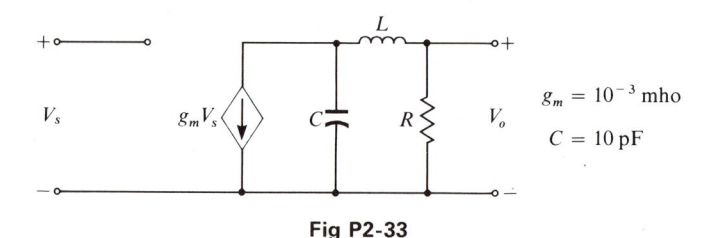

$g_m = 10^{-3}$ mho
$C = 10$ pF

Fig P2-33

2-34 Repeat Problem 2-33 for a maximally flat delay design.

2-35 A class of filters, with no finite transmission zeros, which has good frequency and transient response is described below. This class is referred to as catenary filters, since all the poles lie on a hyperbolic curve of the form

$$\sigma_k = \frac{1}{b-1}(\cosh \omega_k - b)$$

where b is real. The actual pole locations for various order filters are determined by the intersection of the above curve with radial straight lines from the origin.

$$\omega_k = \sigma_k \tan \theta_k$$

where θ_k are the angles corresponding to those used to locate the poles of the Butterworth filter. For example, for $b = 2.0$ and a third-order filter, the poles are as shown in fig P2-35. For this case, the polynomial is

$$P_3(s) = s^3 + 2.0728s^2 + 2.2237s + 1.1509$$
$$= (s + 1)(s + 0.5364 + j0.9291)(s + 0.5364 - j0.9291)$$

(a)[5] Show that for the above case $\omega_{3\,dB} = 1.04$, overshoot $\gamma = 7.24\%$, and rise time $\tau_R = 2.20$.

(b)[5] Repeat the above for a fourth-order filter and show that

$$\omega_{3\,dB} = 1.05, \gamma = 8.33\%, \text{ and } \tau_R = 2.34$$

It is interesting to note that for this particular value of b, i.e., $b = 2.0$, the overshoot is less than 9 per cent for any order filter. For example, for a tenth-order filter, $n = 10$ and the overshoot is about 8.6 per cent. Also, for $b = 2.50$, the overshoot is less than 6 per cent for any order filter. Maximum overshoot for $b = 2.50$ occurs for $n = 3$, in which case $\gamma = 5.61$ per cent and the overshoot then reduces monotonically as the order of the filter is increased. For $n = 10$, the overshoot is $\gamma \simeq 2.5$ per cent.

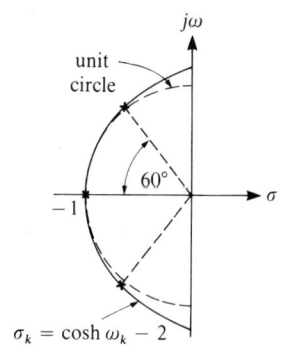

Fig P2-35

[5] Computer-aided analysis is recommended for this problem.

3
Devices, V-I Characteristics, Models, and Biasing

In this chapter, we shall consider the V-I characteristics and models of the various important electronic devices. These include junction diodes, tunnel diodes, breakdown (Zener) diodes, bipolar transistors, field-effect transistors, unijunction transistors, semiconductor controlled rectifiers, vacuum triodes and pentodes. In particular, we shall discuss briefly the measured voltage-current relationships and circuit models for large-signal and small-signal applications. For the latter, high-frequency circuit models also will be discussed. The concept of the small-signal model is very important, since a linear model is obtained from a nonlinear device under small-signal conditions. Under such conditions, the extensive theory of linear systems can be applied to the analysis and design of amplifiers, active filters, and other linear circuits.

Since this book is designed for a course in electronic circuits, the internal electronics of the devices will not be discussed. This aspect of electronics is important and should not be deemphasized, but it is anticipated that the student has been or will be exposed to a physical electronics course where these topics are covered in some detail. As a bare minimum, the reader should become familiar with Appendix A before reading this and the following chapters.

3-1 p-n Junction Diodes

A *p-n* junction diode consists of *p-* and *n*-type semiconductor materials joined together to form a junction. The *p-n* junction is essential to all semiconductor devices. The characteristics and properties of *p-n* junctions are reviewed in Appendix A.

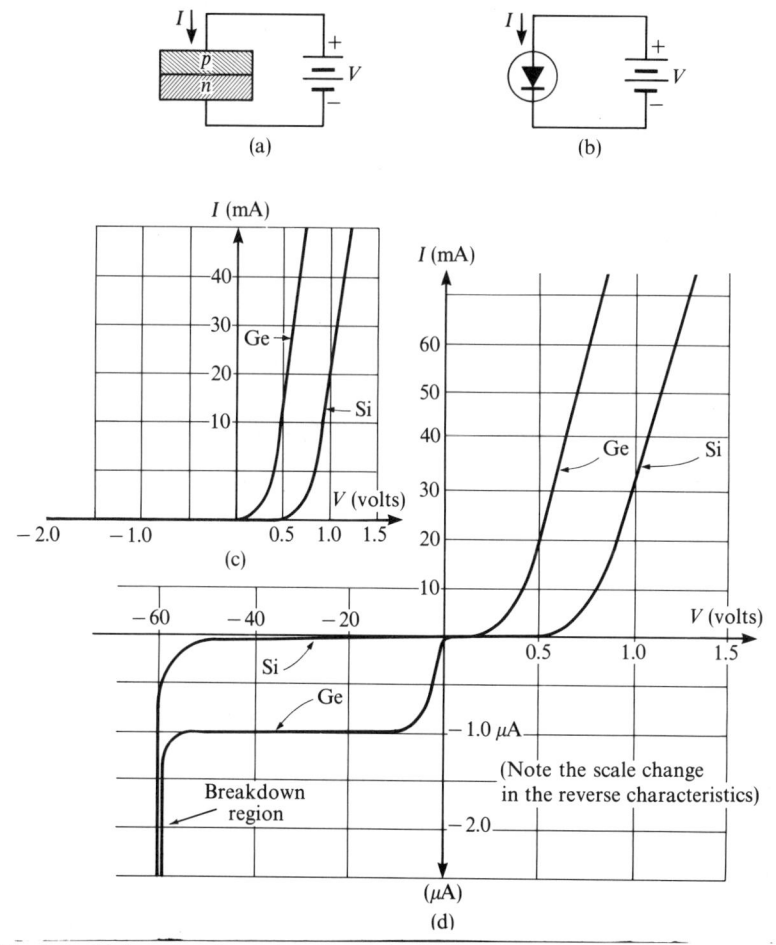

Fig 3-1 (a) A semiconductor junction diode; (b) Graphic symbol; (c) *v-i* characteristics of typical junction diodes; (d) *v-i* characteristics on expanded scale

Figure 3-1a is a schematic representation of a *p-n* junction. When the junction is forward biased (i.e., the positive terminal of the battery is connected to the *p* type material), the potential barrier is reduced and a large number of holes from the *p* region flow to the *n* region. Similarly, electrons from the *n* region flow to the *p* region and a current flow results in the direction of the arrow shown on the symbol for a junction diode, fig 3-1b. When the diode is reverse biased (i.e., the positive terminal of the battery is connected to the *n* type material), the potential barrier is increased and a very small number of holes in the *n* region flow into the *p* region. The current, therefore, is very small and in the direction opposite to the arrow.

The measured *V-I* characteristics of typical silicon (Si) and germanium (Ge) junction diodes at room temperature are shown in fig 3-1c. Figure 3-1d shows an expanded scale for the reverse region. The *V-I* relationship of a *p-n* junction diode, exclusive of the breakdown region, is given by

$$I = I_s[\exp(\lambda qV/kT) - 1] \qquad (3\text{-}1)$$

where

q = electronic charge (1.60×10^{-19} coulomb)
k = Boltzmann's constant (1.38×10^{-23} joules/°K)
T = absolute temperature (°K)
λ = empirical constant which has a nominal value of unity but which, in a practical device, lies between 0.5 and 1.

At room temperature (290°K),

$$\frac{q}{kT} \equiv \Lambda = 40 \text{ V}^{-1} \qquad \text{or} \qquad \frac{kT}{q} = 0.025 \text{ V} \qquad (3\text{-}2)$$

For a forward voltage slightly larger than 25 mV, the current increases exponentially; for a reverse voltage, the current is equal to the saturation current I_s. In Appendix A it is shown that I_s is strongly temperature-dependent. The reverse saturation current approximately doubles for every 10°C in germanium and every 6°C in silicon, near room temperature (Problem 3-3). At this temperature, the values of I_s of actual Si and Ge diodes are typically 10^{-9} and 10^{-6} amperes, respectively.

Because of the nonlinear nature of the *V-I* characteristics of diodes, circuit analysis for many diodes and RLC circuits is complicated. In simple circuits, the characteristics can be used in a graphical analysis.

Graphical Analysis

Consider a circuit consisting of independent sources, resistors, and a diode. This type of circuit can be readily analyzed graphically. Pull out the diode terminals and represent the rest of the circuit across the diode by the Thévenin equivalent, as shown in fig 3-2a.

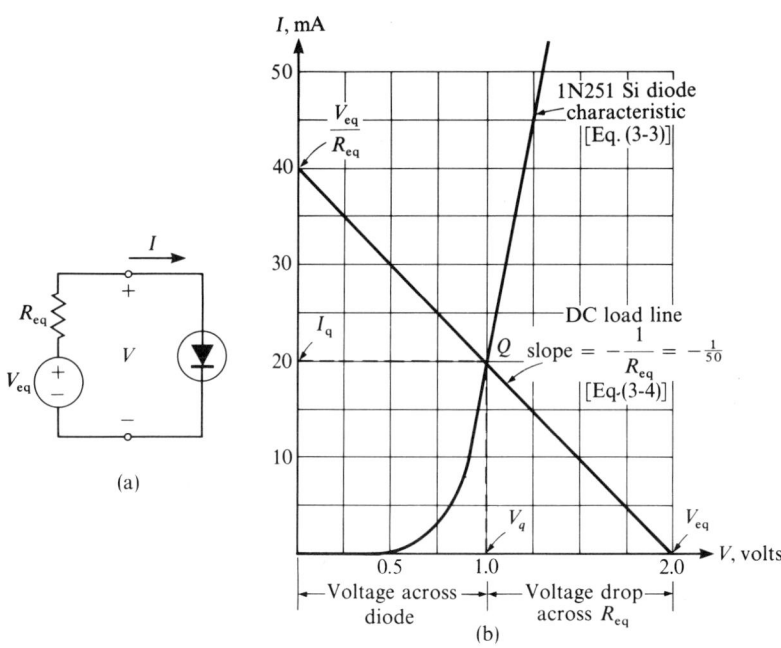

Fig 3-2 (a) Thévenin equivalent of a resistive circuit with batteries and a single diode; (b) Graphical solution

The equilibrium equations for the circuit are

for the nonlinear diode: $\quad I = f(V) = I_s[\exp(\lambda qV/kT) - 1]\quad$ (3-3)

for the Thévenin one-port: $\quad V = V_{eq} - R_{eq}I \quad$ (3-4)

The solution of (3-3) and (3-4) determines the values of I and V. In a simple case, such as the one shown in fig 3-2a, the analysis is best performed graphically, as shown in fig 3-2b. For example, let $V_{eq} = 2$ V and $R_{eq} = 50\,\Omega$. The straight line corresponding to (3-4) is drawn on the device V-I characteristic. The straight line is usually referred to as the *dc load line*. The intersection of the two curves yields the operating point Q. The operating point is often called the *quiescent* point. In the above example the quiescent point is $I_q = 19.7$ mA, $V_q = 1.02$ V.

For a general, large-signal analysis when the independent source is time varying (e.g., sinusoidal) and/or the device is a diode or some other nonlinear element, the reader is referred to Chapter 9.

Large-Signal Model

A general nonlinear circuit may alternatively be analyzed by piecewise-linear analysis. In piecewise-linear analysis, the nonlinear device is approximated by a piecewise-linear model and then linear analysis is used.

For the piecewise-linear analysis, it is convenient to introduce the *ideal diode*. The ideal diode is defined by the following equations:

$$I = 0 \quad \text{for } V \leq 0 \quad (3\text{-}5a)$$

$$V = 0 \quad \text{for } I \geq 0 \quad (3\text{-}5b)$$

The representation of the ideal diode and its characteristic is shown in fig. 3-3. (Note that we use a triangle with no circle envelope as a symbol for the *ideal* diode.)

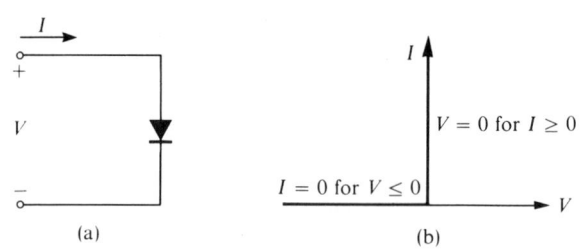

Fig 3-3 (a) Ideal diode; (b) Characteristic of the ideal diode

The characteristics of *p-n* junction diodes may be approximated by the piecewise-linear characteristic, as shown in fig 3-4a. The threshold voltage V_0 usually lies in the range from 0.2 to 0.5 V for Ge diodes, and from 0.5 to 0.9 V for Si diodes. The threshold voltage is also referred to as *offset voltage* or *cut-in voltage*. The temperature variation of this voltage ranges from -1 to -3 mV/°C for both Ge and Si.

The large-signal model corresponding to the piecewise-linear characteristics and the ideal diode is shown in fig 3-4b. (For further details on piecewise-linear analysis, see Chapter 9.) For the example considered above, piecewise-linear analysis yields the following results:

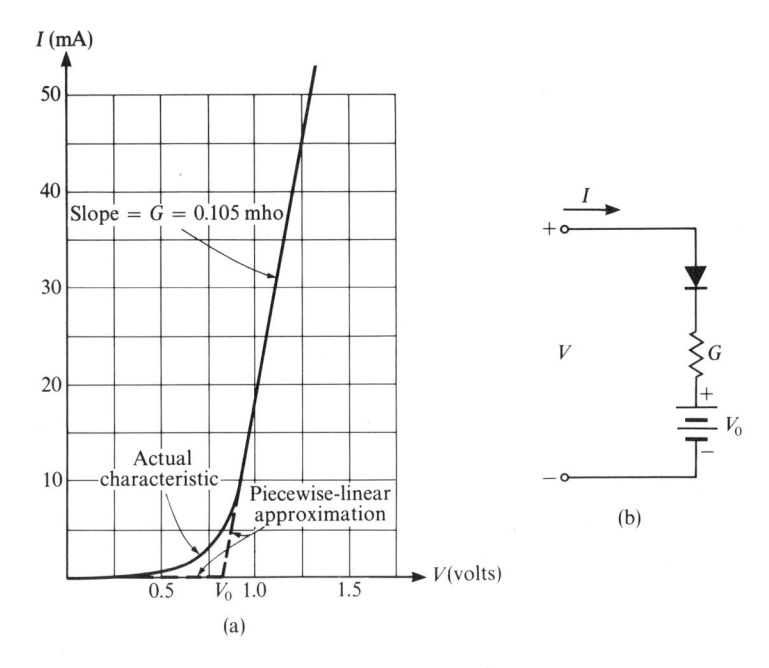

Fig 3-4 (a) Piecewise-linear approximation of the diode characteristic; (b) Large-signal model for the diode

The threshold voltage for the diode in this case is $V_0 = 0.82$ V; for $V < V_0$, $I = 0$, and (3-4) is not satisfied. Hence, the solution must have $V > V_0$. For $V > V_0$ we have to solve the two linear equations

$$I = G(V - V_0) = 0.105\,(V - 0.820) \tag{3-6a}$$

$$V = V_{eq} - R_{eq}I = 2 - 50I \tag{3-6b}$$

which gives the solution $I = I_q = 19.8$ mA, $V = V_q = 1.01$ V. The result is in good agreement with the graphical analysis.

Note that in the above models we have assumed that the reverse voltage is below the breakdown voltage; otherwise, another diode model is required for the breakdown region (Problem 3-12). For a very rough analysis, the ideal diode model may be used to examine the gross behavior of diode circuits.

Applications

Diodes are used in a variety of applications, including rectifiers, detectors, function generators, waveform shapers, and logic gates. The

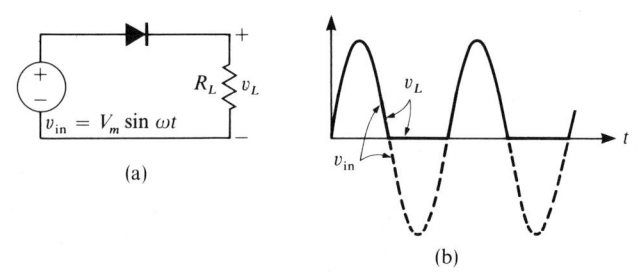

Fig 3-5 (a) Halfwave rectifier circuit; (b) Waveform of the halfwave rectifier

use of diodes for waveform shaping and logic gates is discussed in detail in Chapters 10 and 11, and nonlinear function generation is discussed in Chapter 9.

Rectification, i.e., changing an ac signal to a dc signal, is achieved by diode circuits. In the circuit of fig 3-5a, which is called a half-wave rectifier, the output voltage follows the input voltage when the diode conducts and is zero when the diode does not conduct, i.e., when $v_{in} < 0$. Full-wave rectification can be achieved by using a diode bridge circuit, as shown in fig P3-10. To reduce the ac ripple from the ac to dc conversion, filters are usually used across the load. Detailed discussion of this topic is found in Ref 7.

Small-Signal Model

Junction diodes are seldom operated solely in the linear region of the characteristic curve. This is because, as we shall presently show, under small-signal operation at low frequencies, the diode behaves as a resistor. Small-signal models play a very important role in simplifying the analysis and design of linear active circuits, however, since the extensive theory of linear systems can be utilized. Small-signal models are, therefore, widely used in the analysis of linear circuits employing transistors, vacuum tubes, and so on. We introduce the concept here because it is easier to understand it for a two-terminal device.

The circuit in fig 3-6a shows a diode biased by the dc voltage V_{DD} with the dc operating point at Q (i.e., I_q, V_q). A small-amplitude sinusoidal voltage is superimposed on V_{DD}, thereby varying V_{DD} by an incremental amount. As the supply voltage across the diode changes to maximum limits, $V_{DD} \pm V_m$ (with $V_m \ll V_{DD}$), the operating point I_q and V_q changes to maximum limits of $I_q \pm \Delta I$ and $V_q \pm \Delta V$. These are illustrated in fig 3-6b, where the ac diode current is also shown. If we are interested in ac

3-1 p-n Junction Diodes

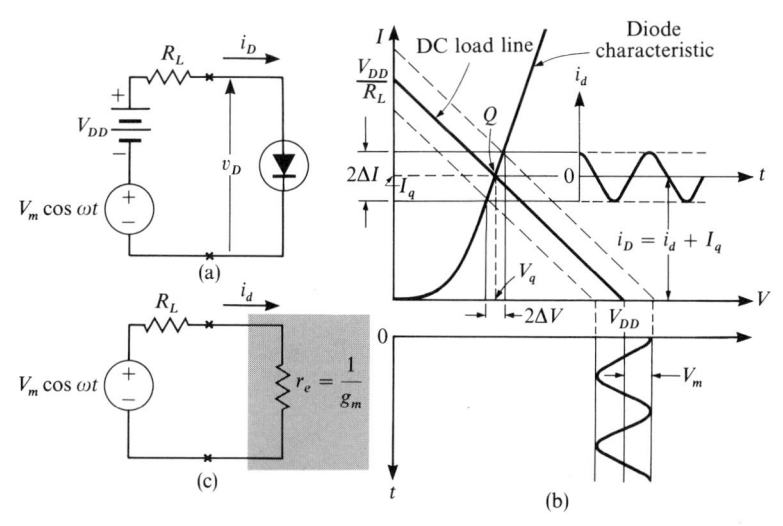

Fig 3-6 (a) Diode circuit for small-signal operation; (b) Load line and quiescent point; (c) Small-signal model at the quiescent point

signals only, then the diode can be modeled by a conductance g_m:

$$g_m = \frac{\Delta I}{\Delta V} = \frac{1}{r_e} \qquad (3\text{-}7)$$

Hence, the low-frequency model for the forward-biased junction diode is a resistance r_e, as shown in fig 3-6c. For small signals, the value of the resistance depends on the slope of the characteristic at the quiescent point.

Actually, the value of g_m may also be derived analytically by making a Taylor series expansion of $i_D = f(v_D)$ at the operating point (I_q, V_q). We shall do this here for illustrative purposes; these concepts will be used later for transistors.

Consider the total diode voltage, which is the sum of the small-signal variation ΔV and the quiescent point V_q. That is,

$$v_D = V_q + \Delta V \qquad (3\text{-}8)$$

In (3-8), v_D varies as the sinusoidal voltage ΔV varies (see fig 3-6b). Similarly,

$$i_D = I_q + \Delta I \qquad (3\text{-}9)$$

The nonlinear V-I relationship of the diode from (3-1) is

$$i_D = I_s\left[\exp\left(\frac{\lambda q v_D}{kT}\right) - 1\right] = I_s[\exp(\Lambda' v_D) - 1] \qquad (3\text{-}10)$$

where $\Lambda' = \lambda q/kT \equiv \lambda \Lambda$, i.e., the empirical constant is included. We expand (3-10) in a Taylor series about V_q to obtain

$$i_D = I_s[\exp(\Lambda' V_q) - 1] + I_s[\exp(\Lambda' V_q)]\left[\Lambda' \Delta V + \frac{1}{2!}(\Lambda')^2(\Delta V)^2 + \cdots\right] \quad (3\text{-}11)$$

If we assume *small-signal* operation, i.e., $\Delta V \ll V_q$ or, in particular, $\Lambda'\Delta V \leq 1/5$, the nonlinear terms within the last bracket can be ignored without any significant error. Noting that $I_q = I_s[\exp(\Lambda' V_q) - 1]$, if we ignore the nonlinear terms we get, from (3-9) and (3-11),

$$\Delta I = I_s \exp(\Lambda' V_q)\Lambda' \Delta V \quad (3\text{-}12)$$

The small-signal conductance of the forward-biased diode is

$$\frac{1}{r_e} \equiv g_m \equiv \frac{\Delta I}{\Delta V} = I_s \Lambda' \exp(\Lambda' V_q) \quad (3\text{-}13)$$

But $\exp(\Lambda' V_q) \gg 1$ so $I_s \exp(\Lambda' V_q) \approx I_q$, hence

$$\frac{1}{r_e} = g_m = \Lambda' I_q = \frac{\lambda q}{kT} I_q \quad (3\text{-}14)$$

Recall that λ is a numerical constant, $0.5 \leq \lambda \leq 1$, and $q/kT = 40$ mhos at room temperature. For the nominal value of $\lambda = 1$, we have

$$r_e = \frac{1}{g_m} = \frac{kT}{qI_q} = \frac{0.025}{I_q} \quad \text{at } T = 290°\text{K} \quad (3\text{-}15)$$

Note that r_e can be varied by varying the dc current I_q.

At high frequencies, where the carrier transit time must be taken into consideration, the diode model becomes a shunt RC circuit where $R = r_e$ and C is the total capacitance of a forward-biased p-n junction.

In the reverse region, the model consists of a leakage resistance r_c, which is very high ($10^6 \Omega$ or larger), and a junction capacitance of the reverse-biased p-n junction.

3-2 Breakdown (Zener) Diodes

The breakdown voltage of a p-n junction in the reverse-bias condition can be controlled by controlling the concentration of impurities on each side of the junction. Breakdown, or Zener, diodes are such devices and they find application in constant-voltage reference and regulator circuits.

Typical Zener *V-I* characteristics of a low-voltage reference are shown in fig 3-7. Note that the forward characteristic is similar to that

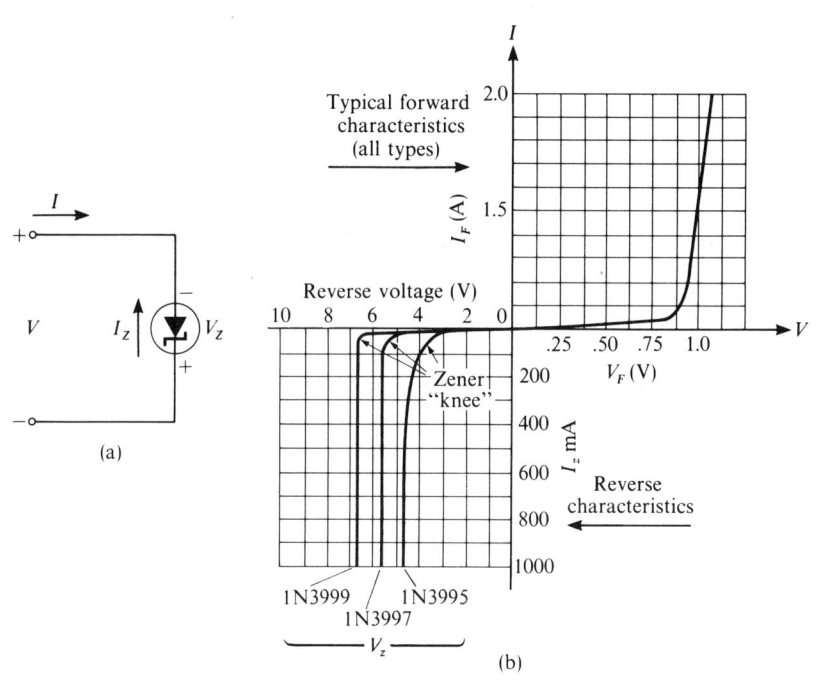

Fig 3-7 (a) Breakdown (Zener) diode; (b) Typical breakdown (Zener) diode *v-i* characteristic (courtesy Motorola Semiconductor Products)

of the conventional *p-n* junction diode. The reverse characteristic shows a breakdown voltage V_z, at which it is independent of the diode current. A wide range of Zener diodes are commercially available; values from 2 to 200 volts are typical, with power ratings from a fraction of a watt to 100 watts.

A typical application of a Zener diode for voltage regulation is shown in fig 3-8. The dc voltage V is unregulated and can vary, as can the load resistance. In spite of these variations, by proper design the voltage across the load can be held constant to the nominal Zener value V_z. The operation of the circuit is as follows (assume the value of R is such that the operation is beyond the "knee" of the breakdown curve): If V changes, the voltage drop across R changes, since V_z is a constant. The change in the drop across R results in a change in current

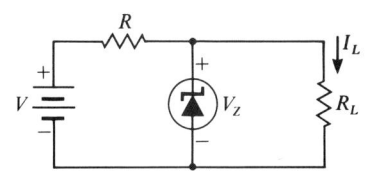

Fig 3-8 Simple voltage-regulator circuit

through R. This current flows through the Zener diode, and the load current remains the same, because the voltage across it has not changed. If the input voltage is constant and the load resistance is varied, the variation in the load resistance causes a change in the load current. Since V and V_z are unchanged, the current variation of the load results only in changes in the Zener diode current. Thus, the voltage across the load remains constant and equal to V_z, in spite of these variations. It should be pointed out, however, that changes in temperature generally cause a change in the Zener reference voltage. The typical temperature coefficient of the Zener diode is specified by the manufacturer.

3-3 Tunnel Diodes

Tunnel diodes are *p-n* junction diodes in which both sides of the *p-n* junction are very heavily doped. As a result of this high doping, the space-charge layer is very thin and additional current flow occurs because of a phenomenon known as *tunneling*. When the forward voltage is increased enough, the tunneling effect ceases to dominate and the normal diode conduction takes over. The tunnel diode has an interesting and unusual *V-I* characteristic. The measured *V-I* characteristic of a typical tunnel diode is shown in fig 3-9. Note that the *V-I*

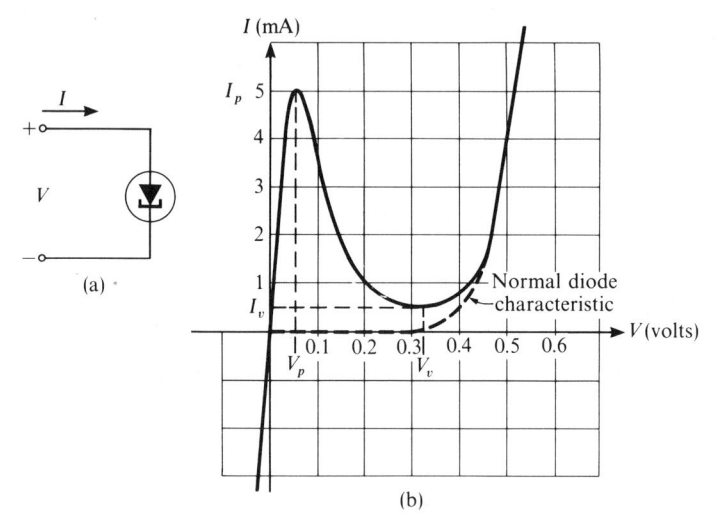

Fig 3-9 (a) Tunnel diode; (b) Typical tunnel diode v-i characteristic

characteristic exhibits a negative slope between the peak and valley points; hence, in small-signal applications, when operated at this region, it can be modeled as a *negative* resistance. This negative resistance characteristic makes it useful as a high-frequency amplifier and/or oscillator. A small-signal equivalent circuit for the tunnel diode at moderate frequencies is shown in fig 3-10a, where C is the junction

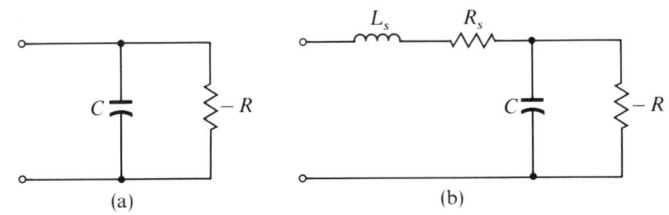

Fig 3-10 (a) Small-signal model of tunnel diode biased in negative-resistance region; (b) A more accurate small-signal model

capacitance. Typical values of $|R|$ and C are of the order of 50 Ω and 5 pF, respectively. Since the RC time constant of the tunnel diode is very small (of the order 10^{-10} second), these diodes are useful in very-high-frequency applications. At such high frequencies (of the order of 10^9 Hz or 1 GHz), the lead inductance must also be included in the circuit model. A more accurate equivalent circuit for the tunnel diode, which includes the lead inductance and the bulk series resistance, is shown in fig 3-10b. Typical values of L_s and R_s are 5×10^{-9} H and 1 Ω, respectively.

In large-signal applications, the tunnel diode can be used in computer circuitry because of its bistable state. The bistable state is more easily understood if a load line is drawn in the *V-I* characteristic. The load line intersects the curve at three points. The two intersecting points in the positive slope region are the stable points. The circuit is designed so that the operation can be shifted to either of these points. Bistable operation is discussed in more detail in Chapter 10.

3-4 Bipolar Junction Transistors

Bipolar *junction transistors* (BJT's) are very useful active devices which play an important role in today's electronic age, especially in wideband electronic amplifiers and fast digital circuitry, where they dominate the field. A BJT is comprised of two *p-n* junctions back-to-back, and its operation depends upon both the majority and minority carriers; hence, they are called bipolar junction transistors, or simply *transistors*.

The essential parts of the physical structure of a *pnp* planar transistor are shown in fig 3-11. It is made by taking a *p*- (or *n*-) type material and diffusing onto it an *n*- (or *p*-) type material, respectively, and then another *p*- (or *n*-) type material. The construction of integrated-circuit transistors is discussed in Appendix B.

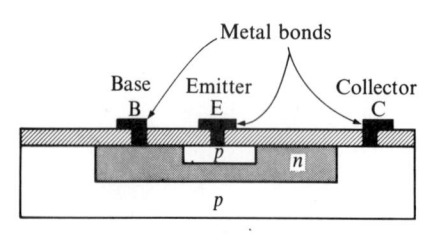

The current-flow mechanism of the transistor can be described in terms of the *p-n* junction theory, with some minor modification. Figure 3-12 shows a schematic representation of a *pnp* transistor. The three parts of the transistor are called *emitter* (*E*), *base* (*B*), and *collector* (*C*), with the base region being very thin (of the order of 10^{-4} cm). The discussion applies equally to *npn* transistors but, in order not to duplicate similar arguments, we shall consider only *pnp* transistors in the following discussion. Consider the biasing arrangement shown in fig 3-12, where the emitter-base junction is forward biased and the collector-base junction is reverse biased. This biasing arrangement is used in amplifiers. As a result of the forward bias, the emitter junction injects many majority carriers across the emitter junction, i.e., holes into the base region and electrons into the emitter region, resulting in the emitter current. Most of the holes injected into the base region are swept across the very thin base region and are collected by the collector. As a result of the finite transit time across the base, some recombination of electrons and holes takes place and a small portion (1 per cent or less) is lost in the base region. The collector current, therefore, consists of two terms, the dominant term being a fraction of I_e, which may be written as αI_e, and the second term being the reverse saturation current I_{CO}. The proportionality constant α depends on the doping and construction of the transistor. An approximate expression used to calculate α at low frequencies is $\alpha \approx 1 - \frac{1}{2}(W/L_b)^2$, where W is the base width and L_b is the diffusion length of the injected carriers in the base (i.e., $L_b = L_p$ for *pnp* and $L_b = L_n$ for *npn*) and $W/L_b \ll 1$. Typical values of α range from 0.900 to 0.995. Because of the reverse bias across the collector junction, as shown in fig 3-12 the reverse saturation current I_{CO} is very small. It is of the order of μA in Ge and nA in Si transistors. Note that if the emitter junction is open-circuited, collector current I_{CO} flows. The collector current, according to the sign convention of fig 3-12, can be written as

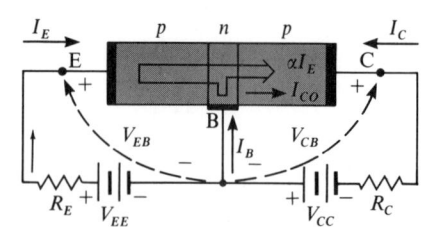

$$I_C = -\alpha I_E - I_{CO} \qquad (3\text{-}16)$$

From Kirchhoff's current law (KCL), at the external node we have

$$I_E + I_B + I_C = 0 \qquad (3\text{-}17)$$

The collector current, in terms of I_B and I_{CO}, is obtained by eliminating I_E in (3-16) and (3-17):

$$I_C = \frac{\alpha}{1-\alpha}I_B - \frac{I_{CO}}{1-\alpha} \qquad (3\text{-}18)$$

As we shall see, the quantities α and $\alpha/(1-\alpha)$ often appear in transistor circuit equations. The quantity $\alpha/(1-\alpha)$ is commonly called β. A subscript 0, such as in α_0 and β_0, is used to denote dc quantities. From the above discussion, it is seen that current in the low-resistance region (forward-biased emitter junction) is transferred at essentially the same value to the high-resistance region (reverse-biased collector junction), thus providing power gain. Note further that bipolar transistors are current-sensitive devices. The topic of current-sensitive and voltage-sensitive devices will be considered quantitatively and in more detail in Chapter 4.

Mathematically, a transistor may be described in terms of two diodes coupled to each other. This is not unreasonable since a transistor consists of two *p-n* junctions back-to-back. The base region, which is common to both, provides the coupling.

The dc behavior of the *pnp* transistor can be expressed mathematically as[1]

$$I_E = I_{ES}[\exp(\Lambda V_{EB}) - 1] - \alpha_R I_{CS}[\exp(\Lambda V_{CB}) - 1] \qquad (3\text{-}19\text{a})$$

$$I_C = -\alpha_F I_{ES}[\exp(\Lambda V_{EB}) - 1] + I_{CS}[\exp(\Lambda V_{CB}) - 1] \qquad (3\text{-}19\text{b})$$

where $\Lambda = q/kT$. The subscripts *F*, *R*, and *S* designate *forward*, *reverse*, and *short-circuit* conditions, and are used for convenience. The term I_{CS} defines the *reverse saturation* current of the collector diode, i.e., the collector current with $V_{EB} = 0$ and V_{CB} reverse biased such that $\exp(\Lambda V_{CB}) \ll 1$, as is seen in (3-19b). Similarly, I_{ES} is the *reverse saturation* current of the emitter diode. Equations (3-19a) and (3-19b) are called the *Ebers–Moll* equations.[2] These equations have been shown to

---

[1] For *npn*, a minus sign must be introduced for each component, viz:

$$I_E = -I_{ES}[\exp(-\Lambda V_{EB}) - 1] + \alpha_R I_{CS}[\exp(-\Lambda V_{CB}) - 1]$$

$$I_C = \alpha_F I_{ES}[\exp(-\Lambda V_{EB}) - 1] - I_{CS}[\exp(-\Lambda V_{CB}) - 1].$$

[2] The Ebers–Moll equations are also sometimes given (for *pnp*) as:

$$I_E = a_{11}[\exp(\Lambda V_{EB}) - 1] + a_{12}[\exp(\Lambda V_{CB}) - 1]$$

$$I_C = a_{21}[\exp(\Lambda V_{EB}) - 1] + a_{22}[\exp(\Lambda V_{CB}) - 1],$$

where $a_{12}/a_{11} = \alpha$, $(a_{11}a_{22} - a_{12}a_{21})/a_{11} = I_{CO}$, and $a_{12} = a_{21}$.

be true for any transistor regardless of geometry. They also show that in a transistor the following reciprocity relation holds:

$$\alpha_R I_{CS} = \alpha_F I_{ES} \tag{3-20}$$

Of the four parameters α_R, α_F, I_{CS}, and I_{ES} (which, by definition, are all positive) only three are independent, and the fourth can be obtained from (3-20).

Two additional parameters can also be defined which are related to the above parameters. These two are I_{CO} and I_{EO}, the former having already been mentioned. The parameter I_{CO} is defined in terms of open-circuited conditions, i.e., the reverse collector current when $I_E = 0$ and V_{CB} is reverse biased such that $\exp(\Lambda V_{CB}) \ll 1$. Similarly, I_{EO} is the reverse emitter current that flows when $I_C = 0$ and $\exp(\Lambda V_{EB}) \ll 1$. Note that both I_{CO} and I_{EO} are positive quantities. I_{CO} and I_{EO} may be obtained in terms of the other four parameters simply by applying the definition conditions. For example, to find I_{CO} put $I_C = I_{CO}$, set $I_E = 0$ in (3-19a), use the approximation $\exp(\Lambda V_{CB}) \ll 1$, and eliminate $[\exp(\Lambda V_{EB}) - 1]$ in (3-19a) and (3-19b) to get:

$$I_{CO} = (1 - \alpha_F \alpha_R) I_{CS} \tag{3-21a}$$

Similarly,

$$I_{EO} = (1 - \alpha_F \alpha_R) I_{ES} \tag{3-21b}$$

The Ebers–Moll equations are important and useful in large-signal applications where the device nonlinearities must be considered. The equations apply for any bias or mode of operation.

3-5 *V-I* Characteristics of Bipolar Junction Transistors

The circuit symbols for *pnp* and *npn* transistors are shown in fig 3-13. Note that the emitter arrow in the symbol points in the actual direction of current flow from emitter to base or base to emitter in each case. In fig 3-13, which has the base as the common terminal, the circuit is described as being in the *common-base (CB)* configuration. Two other configurations, *common-emitter (CE)* and *common-collector (CC)*, are also possible and are shown, for a *pnp* in transistor, in fig 3-14.

3-5 V-I Characteristics of Bipolar Junction Transistors

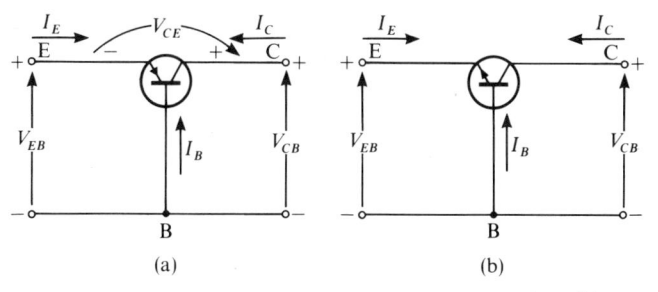

Fig 3-13 (a) *pnp* graphic symbol in the common-base connection; (b) *npn* graphic symbol in the common-base connection

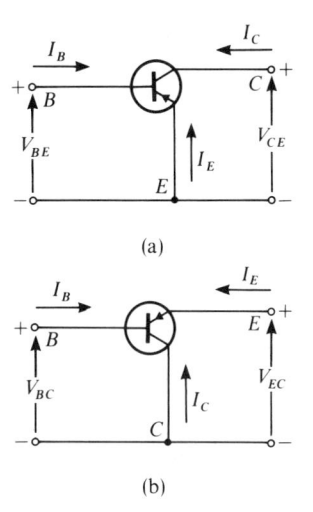

Fig 3-14 (a) *pnp* Transistor in the common-emitter connection; (b) *pnp* Transistor in the common-collector connection

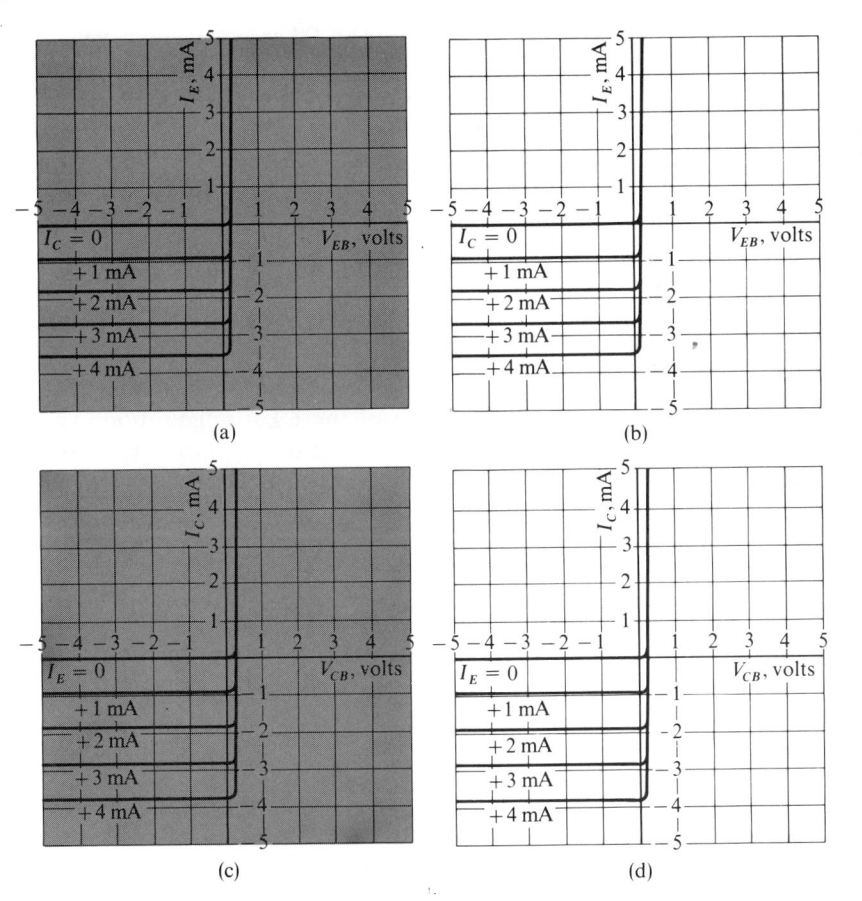

Fig 3-15 Common-base *pnp* characteristics: (a) Ebers-Moll input *v-i* characteristic; (b) Corresponding measured curves; (c) Ebers-Moll output *v-i* characteristic; (d) Corresponding measured curves (after Searle; v 3, p 43, Ref 5)

The measured *CB input* characteristics of an actual transistor, i.e., a plot of the input current I_E vs the input voltage V_{EB}, with the output current I_C as a parameter, are shown in fig 3-15b. We could also use the output voltage V_{CB} as a parameter, but the choice of I_C turns out to be more convenient. Note that the input characteristics in fig 3-15b may be described mathematically by the relation $I_E = f(V_{EB}, I_C)$. For the theoretical model, this relation is found by eliminating $[\exp(\Lambda V_{CB}) - 1]$ in (3-19) and using (3-21). The result is

$$I_E = -\alpha_R I_C + I_{EO}[\exp(\Lambda V_{EB}) - 1] \qquad (3\text{-}22)$$

The plot of (3-22) is shown in fig 3-15a. For an actual transistor, the empirical constant (λ) of the diode should be included in a detailed analysis and comparison of experimental and theoretical equations.

The *output* characteristics of a transistor in the *CB* connection are a plot of I_C vs V_{CB}, with I_E as a parameter. A typical measured plot is shown in fig. 3-15d. The mathematical relation is $I_C = f(V_{CB}, I_E)$. The relation is found by eliminating $[\exp(\Lambda V_{EB}) - 1]$ in (3-19) and using (3-21) to obtain

$$I_C = -\alpha_F I_E + I_{CO}[\exp(\Lambda V_{CB}) - 1] \qquad (3\text{-}23)$$

The plot of (3-23) for a typical transistor is shown in fig 3-15c.

In the majority of applications, the *CE* configuration is used. The reasons for this are given in Chapter 4. Manufacturers usually give the input and output *V-I* characteristics of the *CE* configuration. The measured input and output characteristics of a typical transistor in the *CE* configuration are shown in fig 3-16b and 3-16d, respectively.

The relation for the theoretical model in the *CE* configuration may be obtained from (3-19) and (3-17) by expressing $I_B = f(V_{BE}, V_{CE})$ and $I_C = f(I_B, V_{CE})$. To get the equations in this form, two other relations (see fig 3-14a) may be used to eliminate I_E and V_{CB}:

$$I_B = -(I_C + I_E) \qquad (3\text{-}24)$$

$$V_{CB} = V_{CE} + V_{EB} = V_{CE} - V_{BE} \qquad (3\text{-}25)$$

Substituting (3-24) in (3-19) we obtain:

$$\begin{aligned} I_B = &-I_{ES}(1 - \alpha_F)[\exp(-\Lambda V_{BE}) - 1] \\ &- I_{CS}(1 - \alpha_R)\{\exp[\Lambda(V_{CE} - V_{BE})] - 1\} \end{aligned} \qquad (3\text{-}26)$$

From (3-26) we may solve for $\exp(\Lambda V_{BE})$ in terms of I_B and V_{CE} and then substitute in (3-19b) to get $I_C = f(I_B, V_{CE})$. The expressions,

3-5 V-I Characteristics of Bipolar Junction Transistors

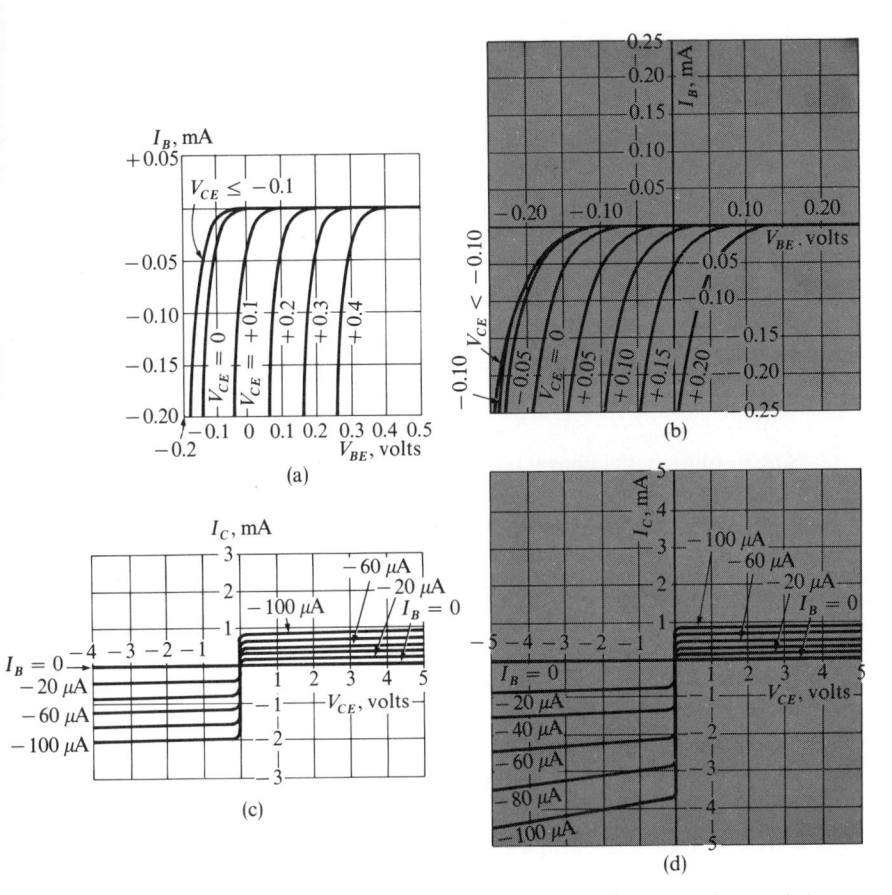

Fig 3-16 Common-emitter *pnp* characteristics: (a) Ebers-Moll input *v-i* characteristic; (b) Corresponding measured curves; (c) Ebers-Moll output *v-i* characteristic; (d) Corresponding measured curves (after Searle; v 3, p 46, Ref 5)

however, become involved and therefore, less useful. The curves derived from these equations for a typical transistor are shown in figs 3-16a and 3-16c, in order to show the reasonableness and validity of the theoretical model. A useful form is obtained directly from (3-23) by expressing I_C in terms of I_B and V_{CE}. Since $I_E = -I_C - I_B$, we have

$$I_C = \frac{\alpha_F}{1-\alpha_F} I_B + \frac{I_{CO}}{1-\alpha_F} \{\exp[\Lambda(V_{CE} - V_{BE})] - 1\} \quad (3\text{-}27)$$

Transistor manufacturers usually give the *CE* output *V-I* characteristics for the normal active region, as shown in fig 3-17. For a *pnp* transistor, if $V_{CE} - V_{BE} < -0.1$ V, the exponential part of the second term

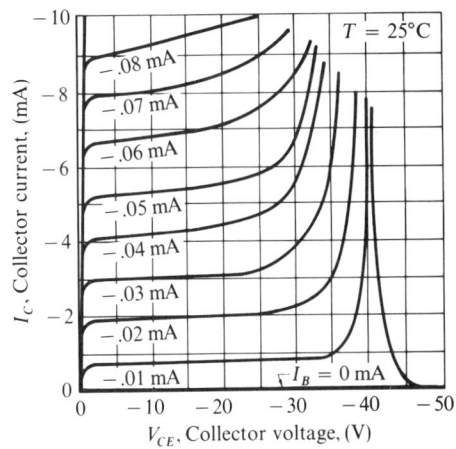

Fig 3-17 Output *v-i* characteristic of *pnp* transistor in the *CE* configuration

$$I_C = \frac{\alpha_F}{1-\alpha_F}I_B - \frac{I_{CO}}{1-\alpha_F} \quad (3\text{-}28)$$

Note that (3-28) confirms the shape of the curves in fig 3-17. The result in (3-28) was obtained qualitatively in (3-18).

3-6 Regions of Operation in Bipolar Transistors

The nonlinear V-I characteristics of the transistor describe the device under all bias voltages. In most circuit applications, however, the transistor is not subjected to all of the bias conditions. There are four possible bias conditions, depending on the state of the bias voltages of the two junctions. These bias conditions define the region of operation of the transistors. The four possible combinations for a *pnp* transistor are shown in fig 3-18. These are

(1) *The forward-active region* corresponds to forward bias in the emitter junction and reverse bias in the collector junction ($V_{EB} > 0$ and $V_{CB} < 0$ for a *pnp* transistor).

(2) *The reverse-active region* corresponds to reverse bias in the emitter junction and forward bias in the collector junction ($V_{EB} < 0$ and $V_{CB} > 0$ for a *pnp* transistor).

(3) *The cutoff region* corresponds to both junctions being reverse biased ($V_{EB} < 0$ and $V_{CB} < 0$ for a *pnp* transistor).

(4) *The saturation region* corresponds to both junctions being forward biased ($V_{EB} > 0$ and $V_{CB} > 0$ for a *pnp* transistor).

In linear circuit applications, where the transistor is used as an amplifier, the operating conditions are confined to the forward-active region. In pulse and digital circuits, where the transistor is used as a switch, the transistor is ON when it is in the saturated region and OFF when it is in the cutoff region. During circuit operation, when the state of the transistor changes, the transistor will be driven between these two regions.

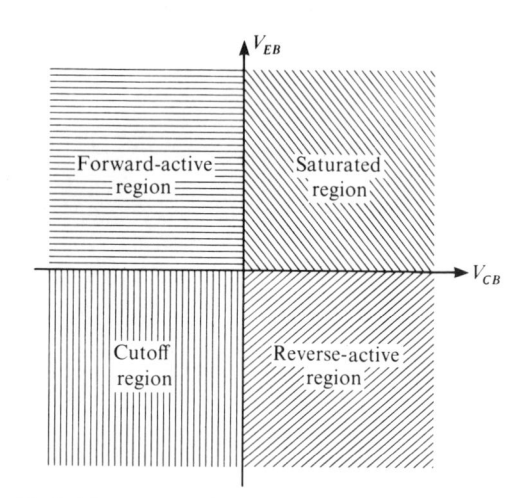

Fig 3-18 Regions of operation for a *pnp* transistor (for *npn*, use negative signs for V_{EB} and V_{CB})

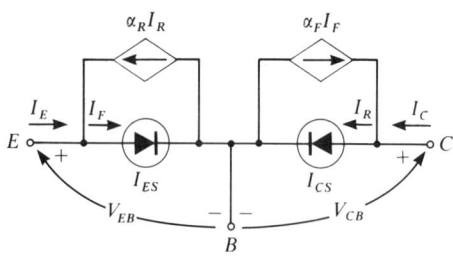

Fig 3-19 Large-signal circuit model for *pnp* transistor, in terms of short-circuited terminals

$$\begin{pmatrix} I_F \equiv I_{ES}[\exp(\Lambda V_{EB}) - 1)] \\ I_R \equiv I_{CS}[\exp(\Lambda V_{CB}) - 1)] \end{pmatrix}$$

3-7 Large-Signal Models for Bipolar Transistors

A transistor model for circuit applications may be obtained directly from the Ebers–Moll equations (3-19). The resulting model for a *pnp*

3-7 Large-Signal Models for Bipolar Transistors

transistor is shown in fig 3-19. The model is commonly referred to as the Ebers–Moll model. This model describes the dc behavior of the transistor. For an actual transistor, the empirical constant λ may be used in (3-19). In circuit analysis it is usually desirable to have the model in a form where the dependent sources are controlled by terminal currents rather than by the diode currents, as is the case in fig 3-19. The alternative form may be obtained as follows: From (3-22) and (3-23) we have

$$I_E = -\alpha_R I_C + I_{EO}[\exp(\Lambda V_{EB}) - 1] \quad (3\text{-}29a)$$

$$I_C = -\alpha_F I_E + I_{CO}[\exp(\Lambda V_{CB}) - 1] \quad (3\text{-}29b)$$

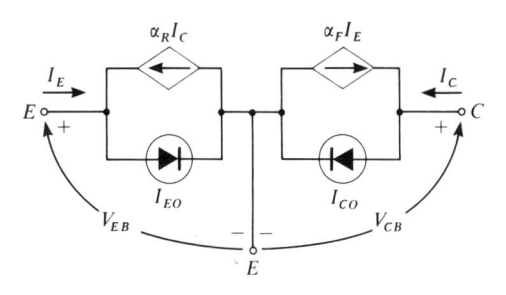

Fig 3-20 Alternative large-signal circuit model for *pnp* transistor, in terms of open-circuited terminals

The model corresponding to (3-29) is shown in fig 3-20. Note that in this case the dependent current sources are controlled by the terminal currents. The model in fig 3-19 is in terms of short-circuited terminals, that is, when $V_{CB} = 0, I_E = I_{ES}[\exp(\Lambda V_{EB}) - 1]$, while the model in fig 3-20 is in terms of the open-circuited terminals, that is, when $I_C = 0, I_E = I_{EO}[\exp(\Lambda V_{EB}) - 1]$, and so on. The currents I_{ES} and I_{CS} are usually called short-circuited saturation currents, while I_{EO} and I_{CO} are called open-circuited saturation currents. These two models, which are completely interchangeable, can be used for transistor operation in all four regions. As we might expect, if we restrict the operation of the transistor to one region only, then simplication in the model is achieved. Let us examine such a simplified model for a *pnp* transistor (the change in signs for an *npn* transistor should pose no problem).

For the *forward-active* region, which is also referred to as the *normal* region, $V_{CB} < 0$ and $V_{EB} > 0$. For $V_{CB} \ll -kT/q$ (e.g., more negative than -0.1 V), the exponential part of the second term in (3-19) is approximately zero and the equations reduce to

$$I_E = I_{ES}[\exp(\Lambda V_{EB}) - 1] + \alpha_R I_{CS} \quad (3\text{-}30a)$$

$$I_C = -\alpha_F I_{ES}[\exp(\Lambda V_{EB}) - 1] - I_{CS} \quad (3\text{-}30b)$$

Equation (3-30b) may also be written from (3-29b) as

$$I_C = -\alpha_F I_E - I_{CO} \quad (3\text{-}31)$$

The model corresponding to the above equations for the active region is shown in fig 3-21a. I_{CO} is a small current of the order 10^{-6} A for Ge, and 10^{-9} A for Si at room temperature. The quantity α_F of a transistor in the active mode of operation is usually written as α_0 where the subscript *F* is omitted and 0 is added to denote the dc value. The range of

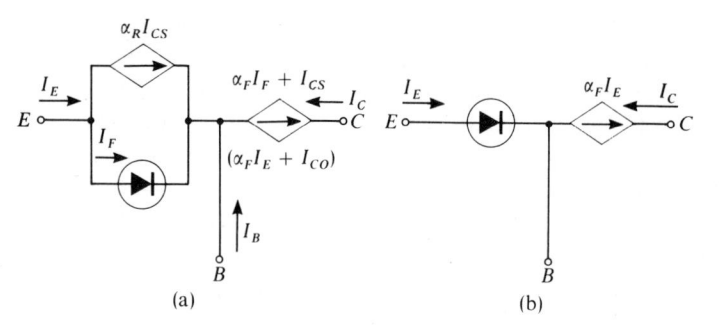

Fig 3-21 (a) Normal active-region model; (b) Approximate model of (a)

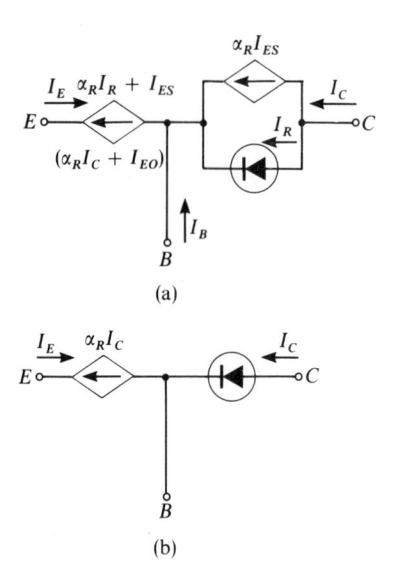

Fig 3-22 (a) Inverse active-region model; (b) Approximate model of (a)

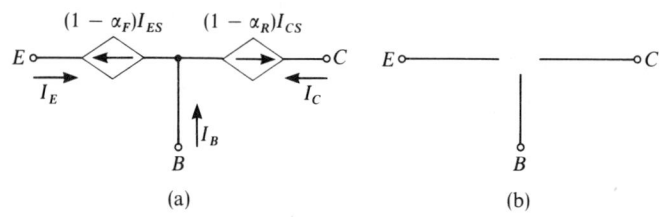

Fig 3-23 (a) Cutoff-region model; (b) Approximate model of (a)

values of α_0 is from 0.900 to 0.995. As we have already mentioned, the narrower the base width, the smaller is the recombination of the minority carriers in the base region. In fact, $\alpha \approx 1 - \frac{1}{2}(W/L_B)^2$, where L_B is the diffusion length of the minority carriers in the base region and W is the base width. The model of fig 3-21a, if I_{CO} and I_{CS} are assumed to be much smaller than I_E, can then be further approximated, as shown in fig 3-21b.

The model for the *inverted-active* region is similar to the normal-active region, except that the collector and emitter terminals are interchanged. In this region, $V_{EB} < 0$ and $V_{CB} > 0$ for *pnp* transistors. Thus, if $V_{EB} \ll -kT/q$ the exponential terms $\exp(\Lambda V_{EB}) \ll 1$ and from (3-29) the model is as shown in fig 3-22a. Further approximation of the model is shown in fig 3-22b.

The *cutoff region* has both junctions reverse biased. Thus, for a *pnp* transistor, if $V_{EB} \ll -kT/q$ and $V_{CB} \ll -kT/q$ the exponential terms in (3-19) can be ignored and the equations reduce to

$$I_E = -I_{ES} + \alpha_R I_{CS} = -I_{ES}(1 - \alpha_F) \tag{3-32a}$$

$$I_C = +\alpha_F I_{ES} - I_{CS} = -I_{CS}(1 - \alpha_R) \tag{3-32b}$$

Note that (3-20) has been utilized in writing the alternate forms for (3-32). The model corresponding to (3-32) is shown in fig 3-23. Since

I_{ES} and I_{CS} are very small quantities, the model can be further approximated, as shown in fig 3-23b. A reverse voltage of the order of 0.1 V is usually sufficient to cut off the transistor. In a silicon transistor, it can be even closer to zero. Note that in this mode, the transistor can be viewed as an OFF switch.

In the *saturation region* both junctions are forward biased. In this case, the nonlinear model of figs 3-19 and 3-20 cannot be simplified if the exponential nature of the junction cannot be ignored. In many switching applications, however, the simplified model of fig 3-24a can be used for the saturation region for $V_{BE}, V_{CE} \gg kT/q$. In fact, in some cases the saturation voltages can also be ignored. In this mode, the transistor can be considered as an ON switch. The saturation voltages, in the CE configuration, $(V_{CE})_{sat}$ and $(V_{BE})_{sat}$ are usually supplied by the manufacturer. These voltages are quite temperature-dependent; the temperature dependence of $(V_{BE})_{sat}$ is approximately equal to -2.5 V/°C and that of $(V_{CE})_{sat}$ is about $+0.2$ V/°C for both Ge and Si transistors. At saturation $I_B \geq I_C/\beta_0$, where $\beta_0 = \alpha_0/(1 - \alpha_0)$ as defined earlier.

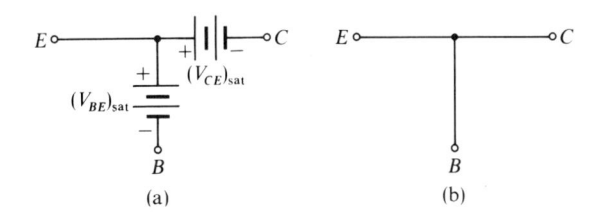

Fig 3-24 (a) Saturation-region model; (b) Approximate model of (a)

3-8 Low-Frequency Small-Signal Models

Amplification of small signals is one of the most important functions in electronic circuits. Bipolar transistors play a dominant role at high frequencies. When used as an amplifier, the transistor is usually biased (i.e., operated) in the linear portion of the normal-active region. The small-signal circuit operation is on the load line about the quiescent point.

The concept of a small-signal model developed for a *p-n* junction diode can be readily applied to the transistor. For example, in fig 3-21b, which is the model for the normal-active region, if the small-signal condition is imposed on the forward-biased diode (with the dc emitter current I_E), the diode can be replaced by its equivalent circuit, which is a resistance. The value of the resistance, from (3-14), is

$$r_e = \frac{kT}{\lambda q I_E} = \frac{kT}{q I_E} = \frac{0.025}{I_E} \Omega \quad \text{(at room temperature)} \quad (3\text{-}33)$$

The collector-base diode, which is reverse biased, is equivalent, for a small signal, to a leakage resistance r_c which is large (of the order of one megohm). The small-signal model for the intrinsic transistor (a transistor characterized by the diffusion mechanism) is then as shown in fig 3-25a. To this low-frequency model, an extrinsic parameter, such as a base-spreading resistance r_b, may be added in the base terminal (as shown in fig 3-25b) to obtain an improved model for the transistor. Note that in fig 3-25b, a typical element value for r_b is 100 Ω, and for α_0 is 0.980.

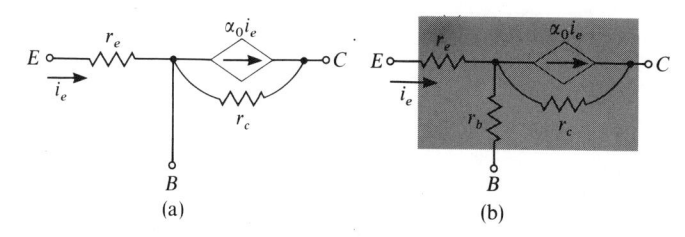

Fig 3-25 (a) Intrinsic low-frequency small-signal model for the bipolar transistor; (b) Low-frequency small-signal T-model for the bipolar transistor

A bipolar transistor is a nonlinear two-port device which can be represented schematically with its associated biasing circuitry as shown in fig 3-26. For small-signal operation the nonlinear two-port can be approximated by a linear two-port circuit.

A transistor is most commonly used in the CE configuration (the reasons will be discussed in Chapter 4), so we shall consider a CE configuration in the following discussion.

The input and output characteristics of the CE transistor in the normal region are usually given by the manufacturer. A simple CE stage with its characteristics is shown in fig 3-27. Because of the availability of these data the hybrid parameter representation for the linearized two-port is most convenient and useful. In fact, typical h-parameters for the transistors are often specified by the manufacturer. If these are unavailable, they can be readily obtained from the input and output characteristics. Even if the characteristics are unavailable, the h-parameters can be readily measured.

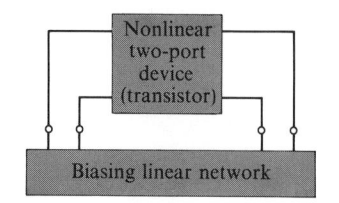

Fig 3-26 General form of a biasing network

3-8 Low-Frequency Small-Signal Models

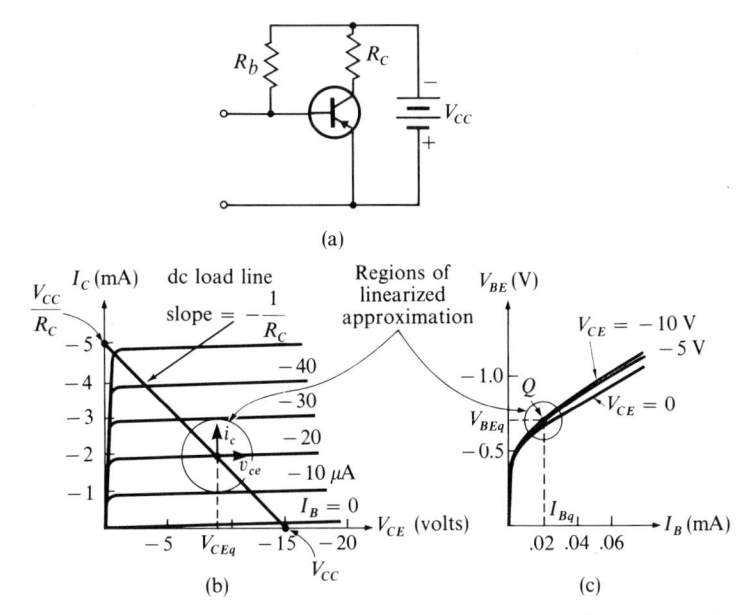

Fig 3-27 (a) A very simple biasing circuit; (b) *pnp* common-emitter output characteristics; (c) *pnp* common-emitter input characteristics

The nonlinear relations for the *CE* transistor may be written in the form:

$$v_{BE} = f(i_B, v_{CE}) \quad (3\text{-}34a)$$

$$i_C = f(v_{CE}, i_B) \quad (3\text{-}34b)$$

where $i_C = i_c + I_{Cq}$, $v_{CE} = v_{ce} + V_{CEq}$, and so on; i_C is the total current and i_c is the ac current superimposed on the dc quiescent value of I_{Cq}, similar to fig 3-6.

The variations of the parameters about the quiescent point Q may be expressed as follows:

$$dv_{BE} = \left.\frac{\partial v_{BE}}{\partial i_B}\right|_{v_{CE}=\text{const}} di_B + \left.\frac{\partial v_{BE}}{\partial v_{CE}}\right|_{i_B=\text{const}} dv_{CE} + \cdots \quad (3\text{-}35a)$$

$$di_C = \left.\frac{\partial i_C}{\partial i_B}\right|_{v_{CE}=\text{const}} di_B + \left.\frac{\partial i_C}{\partial v_{CE}}\right|_{i_B=\text{const}} dv_{CE} + \cdots \quad (3\text{-}35b)$$

For small incremental variations we may write (3-35) as

$$\Delta v_{BE} \approx \left.\frac{\partial v_{BE}}{\partial i_B}\right|_{v_{CE}=V_{CEq}} \Delta i_B + \left.\frac{\partial v_{BE}}{\partial v_{CE}}\right|_{i_B=I_{Bq}} \Delta v_{CE} \qquad (3\text{-}36a)$$

$$\Delta i_C \approx \left.\frac{\partial i_C}{\partial i_B}\right|_{v_{CE}=V_{CEq}} \Delta i_B + \left.\frac{\partial i_C}{\partial v_{CE}}\right|_{i_B=I_{Bq}} \Delta v_{CE} \qquad (3\text{-}36b)$$

To simplify the notation (as in the diode case) we shall use lower-case letter subscripts:

$$\Delta i_C = i_c, \quad \Delta i_B = i_b, \quad \Delta v_{CE} = v_{ce}, \quad \Delta v_{BE} = v_{be} \qquad (3\text{-}37)$$

Thus (3-36) can be written as

$$v_{be} = \left.\frac{\partial v_{BE}}{\partial i_B}\right|_{v_{ce}=0} i_b + \left.\frac{\partial v_{BE}}{\partial v_{CE}}\right|_{i_b=0} v_{ce} \qquad (3\text{-}38a)$$

$$i_c = \left.\frac{\partial i_C}{\partial i_B}\right|_{v_{ce}=0} i_b + \left.\frac{\partial i_C}{\partial v_{CE}}\right|_{i_b=0} v_{ce} \qquad (3\text{-}38b)$$

The partial derivatives in the linearized two-port description in (3-38) can be identified as the h-parameters. Thus[3]

$$v_{be} = h_{11} i_b + h_{12} v_{ce} \qquad (3\text{-}39a)$$

$$i_c = h_{21} i_b + h_{22} v_{ce} \qquad (3\text{-}39b)$$

where

$$h_{11} = h_{ie} = \left.\frac{v_{be}}{i_b}\right|_{v_{ce}=0} = \left.\frac{\partial v_{BE}}{\partial i_B}\right|_{v_{ce}=0} = \left.\frac{\Delta v_{BE}}{\Delta i_B}\right|_{V_{CEq}} \qquad (3\text{-}40a)$$

$$h_{12} = h_{re} = \left.\frac{v_{be}}{v_{ce}}\right|_{i_b=0} = \left.\frac{\partial v_{BE}}{\partial v_{CE}}\right|_{i_b=0} = \left.\frac{\Delta v_{BE}}{\Delta v_{CE}}\right|_{I_{Bq}} \qquad (3\text{-}40b)$$

$$h_{21} = h_{fe} = \left.\frac{i_c}{i_b}\right|_{v_{ce}=0} = \left.\frac{\partial i_C}{\partial i_B}\right|_{v_{ce}=0} = \left.\frac{\Delta i_C}{\Delta i_B}\right|_{V_{CEq}} \qquad (3\text{-}40c)$$

[3] In the literature one often encounters the notation
$$v_{be} = h_{ie} i_b + h_{re} v_{ce}$$
and
$$i_c = h_{fe} i_b + h_{oe} v_{ce},$$
where the first subscripts i, r, f, o denote the input, reverse, forward, and output, respectively, and the second subscript e denotes the common emitter configuration.

$$h_{22} = h_{oe} = \left.\frac{i_c}{v_{ce}}\right|_{i_b=0} = \left.\frac{\partial i_C}{\partial v_{CE}}\right|_{i_b=0} = \left.\frac{\Delta i_C}{\Delta v_{CE}}\right|_{I_{Bq}} \quad (3\text{-}40\text{d})$$

Note the I_{Bq}, V_{CEq} are the quiescent point values.

The h-parameters are determined from the characteristic curves as follows: To determine h_{fe}, draw a vertical line through Q in the output characteristic curve. The ratio of the current increments in the neighborhood of Q from (3-40c) is then the required value of h_{fe}. This is a very important small-signal parameter of the transistor and is usually denoted by β. It is to be noted that h_{fe} and β are not exactly equal since, from (3-28) and neglecting I_{CO}, we have

$$i_C \approx \frac{\alpha}{1-\alpha} i_B = \beta i_B \quad (3\text{-}41)$$

Thus

$$\frac{\Delta i_C}{\Delta i_B} = \beta + \frac{\Delta \beta}{\Delta i_B} i_B \quad (3\text{-}42)$$

If the second term in (3-42) is much smaller relative to β, then the values $h_{fe} \simeq \beta \equiv h_{FE}$ from (3-40c) and (3-42). Usually this is the case under normal operation and we will assume it to hold throughout this text.

The parameter h_{ie} is determined from (3-40a). Draw a tangent line at Q in the input characteristic; the slope of the line gives h_{ie}.

The parameter h_{re} is determined from the ratio of the voltage increments, from (3-40b), as read from a vertical line drawn through Q at the input characteristics.

The parameter h_{oe} is determined from (3-40d) by drawing a tangent line through Q at the output characteristics and finding its slope.

For the characteristic curves shown in figs 3-27b and 3-27c, and the quiescent point $I_{CQ} = -2$ mA, $V_{CEq} = 7.5$ V, $I_{Bq} = 0.02$ mA, we determine the following h-parameters:

$$\begin{array}{ll} h_{ie} = 1.4 \text{ k}\Omega & h_{re} = 1.0 \times 10^{-4} \\ h_{fe} = 100 & h_{oe} = 1.5 \times 10^{-5} \text{ mho} \end{array} \quad (3\text{-}43)$$

The parameters listed in (3-43) are typical for a variety of types of transistors.

The characteristic curves vary appreciably with changes in the environmental temperature; such data are often provided by the manufacturer. The transistor parameters vary with changes in the temperature as well as with bias-point variations. It is important, therefore, to

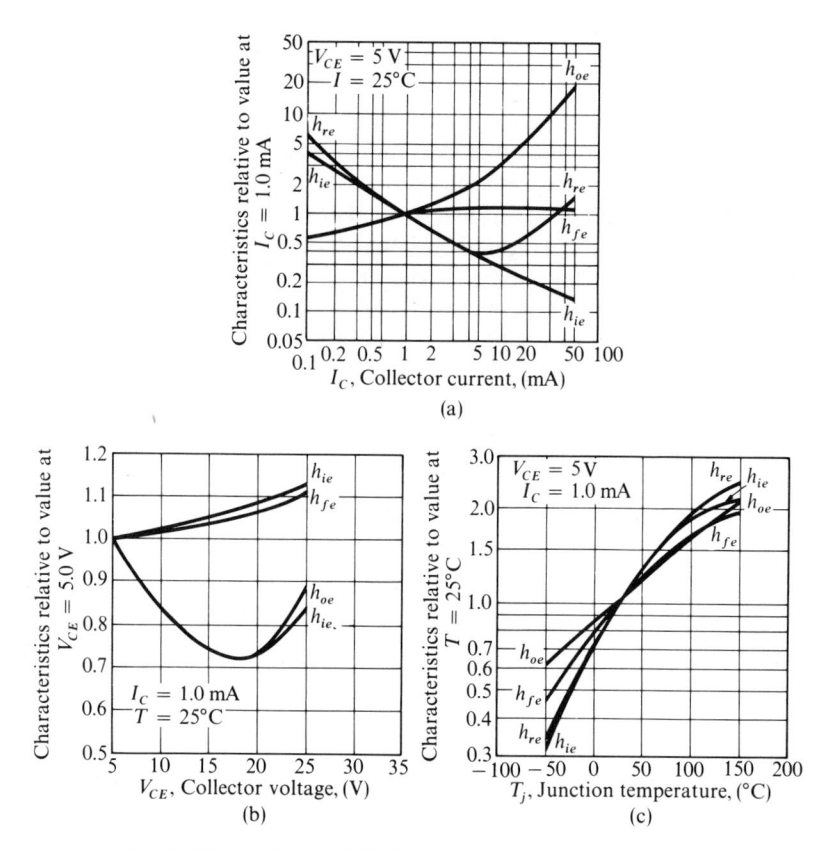

Fig 3-28 (a), (b) and (c) Typical common-emitter characteristics

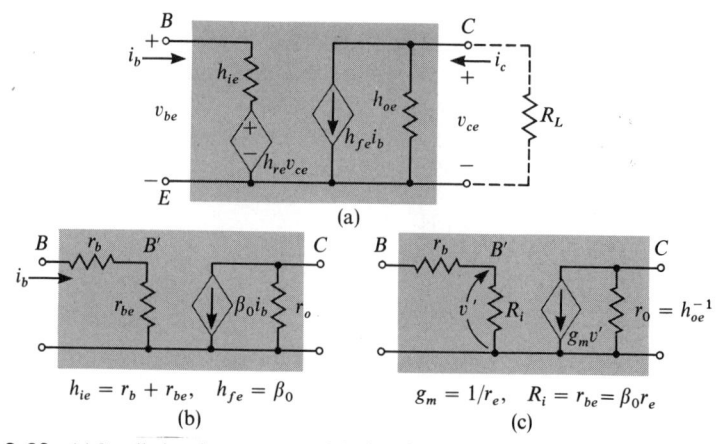

Fig 3-29 (a) Small-signal two-port model of *CE* transistor configuration in terms of the *h*-parameters; (b) Approximate model of (a) for $R_L \ll h_{ie}/h_{fe}h_{re}$; (c) Model of (b), but in terms of voltage-controlled current source

stabilize the bias point in bipolar transistors. This topic is discussed in Section 3-17.

Typical variations of the h-parameters for an *npn* silicon planar transistor as a function of the operating point and the temperature are shown in fig 3-28.

The small-signal *CE* model for the linearized two-port, in terms of the h-parameters, is shown in fig 3-29a (see also fig 2-10). From the typical values given in (3-43), it is seen that h_{oe}^{-1} is a very large impedance, and if the load resistance R_L (in shunt with h_{oe}^{-1}) is small, that is, $R_L h_{oe} \leq 0.1$, then we may ignore h_{oe} in the model. Also, under these conditions we have

$$|h_{re}v_{ce}| \approx h_{re}R_L i_c = h_{re}R_L h_{fe} i_b \qquad (3\text{-}44)$$

Generally, $h_{re}h_{fe} \leq 10^{-2}$, the voltage $h_{re}v_{ce}$ may be neglected in comparison with $h_{ie}i_b$ if $R_L \ll h_{ie}/h_{fe}h_{re}$. For the typical values listed in (3-43), $R_L \ll 140 \text{ k}\Omega$, which is a more relaxed requirement than $R_L \ll 0.1\ h_{oe}^{-1} \approx 6 \text{ k}\Omega$. Hence, in some cases we may leave h_{oe} in the model but ignore the controlled voltage source at the input of the model. The approximation made when we ignore the voltage source provides considerable simplification in analysis and design. The fact that V_{CE} has a negligible effect on the input characteristic (fig 3-27c) also validates this approximation. In transistor literature, the input resistance h_{ie} is usually split into two series resistances, one indicating the base-spreading resistance r_b, which is typically 100 ohms, and the other, the input of the intrinsic transistor (which is usually denoted as r_{be}), as shown in fig 3-29b. Note that in the model $h_{re}v_{ce}$ is ignored and $h_{fe} = \beta$ is used. In the circuit, the symbol B' is used to denote the base junction which is not an accessible terminal. In a detailed analysis, it can be shown that $r_{be} = \beta_0 r_e \equiv R_i$, where the subscript 0 is added to denote the low-frequency value of β. The model of fig 3-29b can thus be represented as in fig 3-29c. The value of r_e depends on the operating point (3-33), and β_0 is typically 50.

CB Configuration

The model shown in fig 3-29b is a reasonably accurate model of the transistor and may be used for small-signal transistor circuit analysis in any of the three configurations at low frequencies. For the *CB* and *CC* configurations, the model in fig 3-25b is more convenient than that of fig 3-29b, because when the circuit of fig 3-29b is rearranged to show the *CB* and *CC* connections, the dependent sources will not be explicit functions of terminal voltages or currents.

The h-parameters of the CB configuration may also be obtained directly from the corresponding input and output characteristics. The approach is quite similar to that used for the CE configuration and will not be repeated here. Typical h-parameter values for the CB configuration are

$$h_{ib} \simeq 20\ \Omega \qquad h_{fb} = -\alpha_0 \approx -0.99$$
$$h_{rb} \simeq 10^{-4} \qquad h_{ob} \simeq 0.2 \times 10^{-6}\ \text{mho} \tag{3-45}$$

The h-parameter model of the CB configuration, shown in fig 3-30a, may be further approximated to a simpler model, shown in fig 3-30b, since $h_{rb}h_{fb} \lesssim 10^{-3}$ and $R_L \ll h_{ob}^{-1}$ in almost all practical cases.

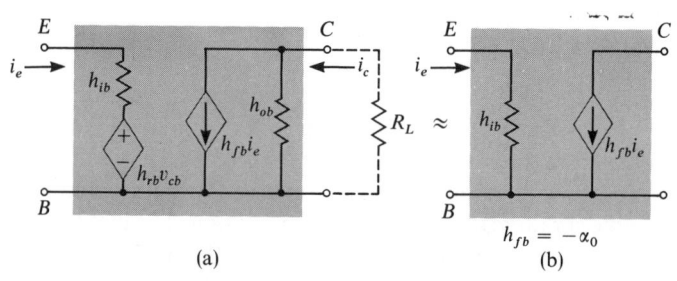

Fig 3-30 (a) Small-signal two-port model of CB transistor configuration in terms of the h-parameters; (b) Approximate model of (a)

3-9 High-Frequency Circuit Models for Bipolar Transistors

$$\alpha = \frac{\alpha_0}{1 + j\omega/\omega_\alpha} \qquad C_e = \frac{1}{r_e \omega_\alpha}$$

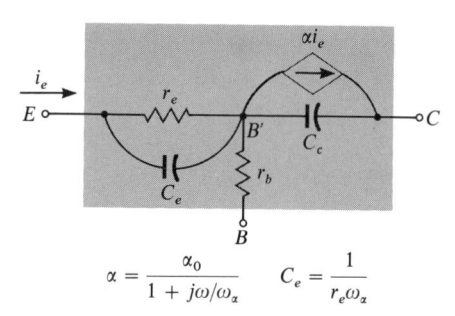

Fig 3-31 Small-signal high-frequency $(f \leq f_\alpha)$ Tee-equivalent circuit of bipolar transistor

When transistors are used to amplify high-frequency signals, the model for analysis and design must include the appropriate energy-storage elements in order to describe the dynamic behavior of the transistor. In any model, one must make some compromise between accuracy and complexity. The simpler the model, the easier is the analysis and design but, in order to obtain a realistic simple model, some approximations must be made.

A high-frequency equivalent circuit which is applicable for transistors in all configurations is shown in fig 3-31. This model is commonly called the high-frequency T equivalent circuit. The derivation of the model from the physics of the device is given in Ref 6. In the model

3-9 High-Frequency Circuit Models for Bipolar Transistors

shown in fig 3-31, C_c is the junction capacitance of the reverse-biased collector junction. At high frequencies, the reactance of C_c is much smaller than r_c, so r_c is ignored in the high-frequency model. C_e consists of two parts: one is the junction capacitance of the forward-biased emitter junction, and the other is the diffusion capacitance (which is the dominant component). The alpha cutoff frequency ω_α is the radian frequency at which the value of α_0 is down to $0.707\,\alpha_0$. Actually, the frequency dependence of $\alpha(j\omega)$ is a transcendental function (Ref 6), but its magnitude and frequency response can be approximated with reasonable accuracy (see Section 8-10) by

$$\alpha(j\omega) \approx \frac{\alpha_0}{1 + j(\omega/\omega_\alpha)} \exp\left[-j\left(\frac{m\omega}{\omega_\alpha}\right)\right] \approx \frac{\alpha_0}{1 + j(\omega/\omega_\alpha)} \quad (3\text{-}46)$$

where m is the excess phase factor and its value ranges from 0.2 to 1.2, depending on the type of transistor. In many applications, where the phase response of $\alpha(j\omega)$ does not play a crucial role, the exponential term may be ignored and the one-pole approximation for $\alpha(j\omega)$ may be used. The alpha cutoff frequency, in terms of the physics of the device, may be written as (Ref. 6):

$$\omega_\alpha = \frac{\kappa D_b}{W^2} \quad (3\text{-}47)$$

where D_b is the diffusion constant of the minority carrier in the base region (i.e., for a pnp transistor, $D_b = D_p$ and for an npn transistor $D_b = D_n$) and W is the base width. The proportionality constant κ ranges roughly from 2.5 to 20, depending on the type of transistor. The base-spreading resistance is roughly proportional to ρ_b/W, where ρ_b is the base resistivity. A typical range of values for the elements of the T circuit is given below:

$$\begin{aligned}r_e &= \frac{kT}{\lambda q I_e} \approx 5\,\Omega \text{ (at } 25°\text{C and 5 mA)}, \quad \alpha_0 \simeq 0.95\text{-}0.99 \\ r_b &\simeq 20\text{-}100\,\Omega, \quad C_c = 1\text{-}20\text{ pF}, \quad C_e \simeq 20\text{-}5000\text{ pF}\end{aligned} \quad (3\text{-}48)$$

Small values for capacitors correspond to high-frequency transistors; e.g., for $\omega_\alpha = 2\pi f_\alpha = 4 \times 10^9$ rad/s, $C_e \approx (1/r_e\omega_\alpha) = 50$ pF, $C_c \approx 1$ pF.

Note that at low frequencies, i.e., when $\omega \ll \omega_\alpha$, the model of fig 3-31 reduces to that of fig 3-25. The high-frequency model is valid only for frequencies up to f_α; beyond this frequency, the transistor is not useful as an amplifier.

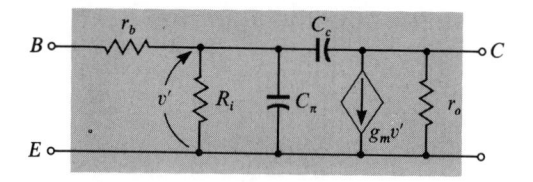

Fig 3-32 Small-signal high-frequency ($f \leq f_T/3$) hybrid π model of bipolar transistor

In the majority of applications, transistors are used in the *CE* configuration. The most useful high-frequency model of the transistor is shown in fig 3-32. This model is very widely used in the literature and is called the *hybrid π equivalent circuit*. Different symbols for the elements are used by various authors.[4] Note that at low frequencies, where the capacitance impedances are so high they can be ignored, the model reduces to the low-frequency equivalent circuit shown in fig 3-29c.

For the *CE* configuration, a commonly used and convenient cutoff frequency f_β is defined as follows: Consider the short-circuited current gain of the transistor in the *CE* configuration shown in fig 3-33. The short-circuited current gain is readily obtained. From the circuit

$$I_s = v'\left(\frac{1}{\beta_0 r_e} + sC_\pi + sC_c\right) \qquad (3\text{-}49)$$

and

$$I_0 = \frac{v'}{r_e} - v' s C_c \qquad (3\text{-}50)$$

Hence

$$\frac{I_0}{I_s} = h_{21} = \frac{\beta_0}{1 + s(C_\pi + C_c)\beta_0 r_e} \quad \text{for } \frac{1}{r_e} \gg |sC_c| \qquad (3\text{-}51)$$

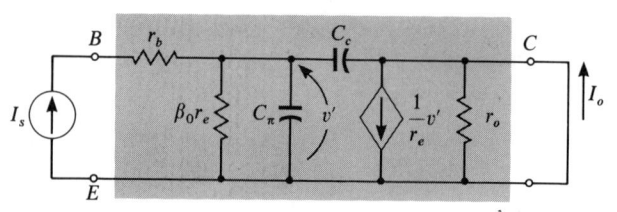

Fig 3-33 Short-circuited *CE* stage with current-source drive

[4] The following symbols are commonly used:

$r_b \to r'_{bb}, r'_b, r_x \qquad C_c \to C_{ob}, C_\mu$
$R_i \to r'_{be}, r_\pi \qquad r_0 \to r_{ce}, h_{oe}^{-1}$
$C_\pi \to C_e, C'_{be} \qquad g_m \to \alpha_0/r'_e, 1/r_e$

For example, the SEEC series (Refs 4 and 5) use $r_x, r_\pi, C_\pi, C_\mu, r_0, g_m$, for the hybrid π parameters.

3-9 High-Frequency Circuit Models for Bipolar Transistors

The 3-dB cutoff frequency of the current gain for *short-circuited CE* stage, driven by a current source, is defined as f_β. Thus, from (3-51)

$$\omega_\beta \equiv \frac{1}{\beta_0 r_e(C_\pi + C_c)} \quad \text{or} \quad f_\beta = \frac{1}{2\pi R_i(C_\pi + C_c)} \quad (3\text{-}52)$$

Hence (3-51) can be written as

$$h_{21} = \frac{I_0}{I_s} = \frac{\beta_0}{1 + j(\omega/\omega_\beta)} \quad (3\text{-}53)$$

The useful upper frequency limit of a transistor in the *CE* configuration is commonly denoted by f_T, or $\omega_T/2\pi$. Note that ω_T is defined as the radian frequency at which the short-circuit *CE* current gain is unity. From (3-53) and using the definition of ω_T, we obtain

$$1 = \frac{\beta_0}{|1 + j(\omega_T/\omega_\beta)|} \quad (3\text{-}54a)$$

or

$$\left(\frac{\omega_T}{\omega_\beta}\right)^2 = (\beta_0^2 - 1) \quad (3\text{-}54b)$$

Since, in almost all transistors $\beta_0 \equiv \alpha_0/(1 - \alpha_0) > 10$, (3–54b) reduces to

$$\omega_T = \beta_0 \omega_\beta \quad \text{or} \quad f_T = \beta_0 f_\beta \quad (3\text{-}55)$$

The Bode plot based on (3-53) and the actual short-circuit gain of a typical transistor is shown in fig 3-34. Note that the *actual* transistor gain is not unity at ω_T. This is because the model of fig 3-32 is not valid for $\omega \geq \omega_T/2$. Thus, ω_T for a transistor is based on the model of fig 3-32, and h_{fe} is measured at any high frequency $f_1 = \omega_1/2\pi$ where $3\omega_\beta \leq \omega_1 \leq \omega_T/3$, and $h_{fe}\omega_1 = \omega_T$ is as shown in fig 3-34. The frequency f_T is a very important parameter of the transistor and is almost always given by the manufacturer. It is also sometimes called the gain-bandwidth product of the transistor. The frequencies f_T and f_α are related as one would expect. The relationship is approximately as follows (Ref 6):

$$f_T \simeq \frac{f_\alpha}{1 + m} \quad (3\text{-}56)$$

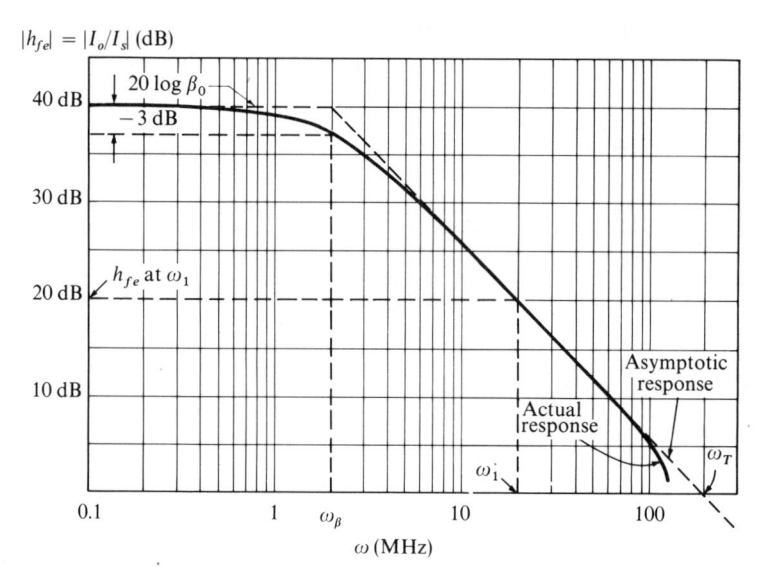

Fig 3-34 Frequency response of current gain from fig 3-33

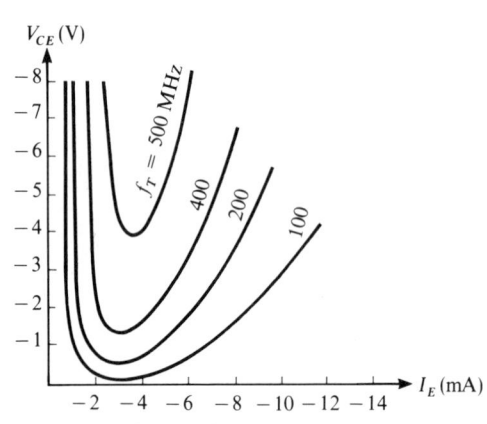

Fig 3-35 Variation of f_T vs I_E and V_{CE}

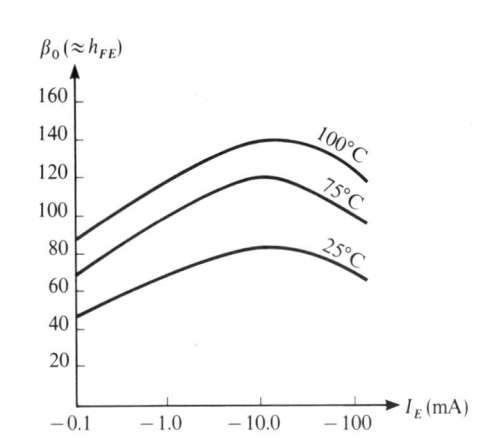

Fig 3-36 Variation of β_0 vs I_E and T for a silicon transistor

where m is the excess phase factor in (3-46) and $0.2 \leq m \leq 1.2$. From (3-56) it is seen that the excess phase factor has a larger degradation effect on the frequency performance of a CE transistor. This is because f_α determines the 3-dB cutoff frequency of a CB stage, while f_T determines the cutoff frequency of a CE stage.

From (3-52) and (3-55) we have

$$C_\pi = \frac{1}{r_e \omega_T} - C_c \qquad (3-57)$$

Usually $(1/r_e\omega_T) \gg C_c$ so that $C_\pi \approx (1/r_e\omega_T)$. The value of r_b could be determined from the fact that $r_b \approx h_{ie} - R_i$ at low frequencies. Since h_{ie} and R_i are both much larger than r_b, however, the difference between these two nearly equal quantities could lead to a meaningless result. A better way is to find $Re(h_{ie})$ at $\omega \gg \omega_\beta$. A typical range of values for the elements of the hybrid π circuit is given in table 3-1. Typical variation of f_T vs I_e and V_{ce} is shown in fig 3-35; the variation of β_0 vs I_e and temperature is shown in fig 3-36.

The high-frequency models given above for transistors are approximate equivalent circuits which represent the best compromise between accuracy and circuit complexity. They are valid for all transistor types such as alloyed junction, grown junction, planar, mesa, epitaxial

TABLE 3-1 *Element Values of the Hybrid π Equivalent Circuit*

$g_m = 1/r_e = \lambda q I_e/kT \approx 40 I_e = 0.4$ mho (for $I_e = 1$ mA)
$R_i = \beta_0 r_e \qquad \beta_0 = 30\text{–}100$
$C_c = 1\text{–}20$ pF $\qquad r_b = 20\text{–}100\ \Omega$
$C_\pi = 1/r_e \omega_T - C_c = 10\text{–}1000$ pF
$r_0 = K/I_e \qquad r_0 = 10\text{–}100$ kΩ $\qquad$ (K is a proportionality constant.)

planar, epitaxial mesa, lateral, and integrated circuits.[5] The various names designate the method of fabrication of the transistor. The fabrication process for *npn* epitaxial planar transistors and integrated circuits (*IC*'s) is discussed briefly in Appendix B.

More elaborate, accurate, and complicated circuit models for the transistor appear in the literature (Ref 6) and may be used in computer-aided analysis and design. In special cases, even more complicated models than those cited may be used. We should be careful, however, not to overtax the computer unnecessarily, and it should be noted that the parameter values of transistors vary appreciably from unit to unit, even within the same type of transistors. This fact also suggests that *only* in special cases, and for a detailed analysis, is the use of a very complicated model justified.

3-10 Field-Effect Transistors

The unipolar or *f*ield-*e*ffect *t*ransistor (FET) is a semiconductor device whose operation depends upon one type of carrier, the majority carrier. The current in the FET is controlled by an electric field. FET's are voltage-sensitive devices and have extremely high input impedances (of the order of $10^{12}\ \Omega$ or higher).

There are two basic types of FET's. One type is called the *j*unction *f*ield-*e*ffect *t*ransistor (JFET) and the other type is called the *i*nsulated-gate *f*ield-*e*ffect *t*ransistor (IGFET, or MOSFET). The term MOS is an abbreviation for *m*etal, *o*xide, and *s*emiconductor, which are the materials used in the fabrication of a MOSFET. The importance of the MOSFET is rapidly increasing in integrated circuits because of low

---

[5] If the integrated circuit has no buried layer, a series resistance r_{cs} is added to the collector terminal (r_{cs} is usually about 10–40 Ω in such cases). The collector-to-ground parasitic junction capacitance may also be added in a detailed analysis of *IC*.

fabrication cost and certain improved performance characteristics at low frequencies. The principles of operation of both JFET's and MOSFET's are basically the same.

JFET

The schematic and the graphic symbol for a JFET are shown in figs 3-37a and 3-37b, respectively. The device consists of an *n*-type channel with two ohmic contacts, called the *source* and the *drain*, and two *p*-type rectifying junctions with ohmic contacts, called the *gate*.

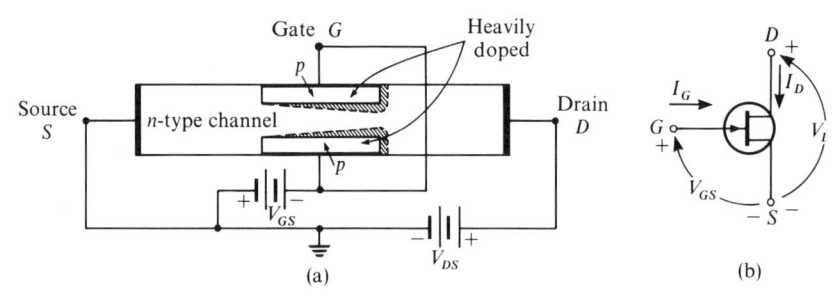

Fig 3-37 (a) *n*-channel JFET schematic circuit; (b) Graphic symbol with voltage polarity and current reference directions

The reason for the names source, drain, and gate, will become apparent when we consider the operation of the device. FET's are also made with *p*-type channels and *n*-type gates. They operate in the same manner as *n*-type channels, except that the polarities of all the voltages and currents are reversed for *p*-type channels. To avoid unnecessary repetition, we will consider the *n*-type channel unless otherwise stated.

The operation of the device is as follows: Let us assume that V_{GS} is zero so that the source and gate terminals are connected to ground potential, and V_{DS} is some dc voltage with the positive polarity on the drain, as indicated in fig 3-37a. Under these conditions, the *p-n* junctions are reverse biased and essentially no current flows through the gate lead. The *n*-type material acts like a resistor for small values of V_{DS} and electrons flow from the source to the drain. The initial amount of current depends on V_{DS} and the resistance of the *n*-type material.

As we increase the voltage V_{DS}, the junctions are further reverse biased and the depletion region becomes wider. Note that the drain side of the junction has a wider depletion width because the reverse voltage is higher at the drain end than at the source end. The depletion region is shown by the dashed area in fig 3-37a. As the depletion region becomes wider the channel becomes narrower, and thus the resistance increases and the slope of the V-I characteristic becomes smaller. This is illustrated in fig 3-38. The voltage V_{PO} is called the *pinchoff* voltage at zero gate voltage; at this voltage the depletion region on each side of the channel joins together and the channel connection between the source and the drain is pinched off, as shown in fig 3-38a. In typical devices,

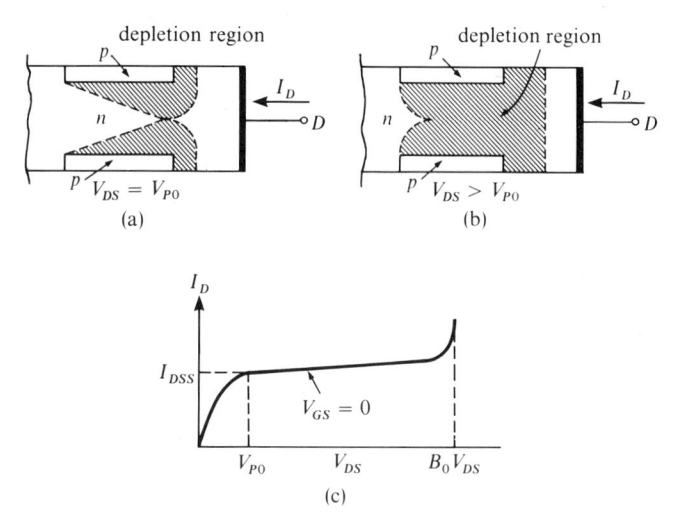

Fig 3-38 (a) Channel of JFET for $V_{DS} = V_{PO}$; (b) Channel of JFET for $V_{DS} > V_{PO}$; (c) $I_D - V_{DS}$ curve for $V_{GS} = 0$

V_{PO} ranges from 1 to 6 volts. The onset of pinchoff changes the physical process occurring in the device and the current I_D remains almost constant and independent of V_{DS} for $V_{DS} > V_{PO}$. Increasing V_{DS} beyond $B_0 V_{DS}$ causes avalanche breakdown. The breakdown voltage is typically in the range of 20 to 50 volts.

The measured output or drain characteristics and the transfer characteristic of a typical *n*-channel JFET are shown in fig 3-39. The reference directions are indicated in fig 3-37b. (For a *p* channel JFET, the signs of all the voltages and currents are reversed.)

If V_{DS} is held constant and a gate voltage V_{GS} (which reverse biases the *p-n* junctions in fig 3-37) is applied, the depletion region is increased and thus the channel resistance is increased, thereby decreasing I_D. The effect of increasing the magnitude of V_{GS} (i.e., making V_{GS} more negative) also reduces the value of V_{DS} at the threshold of pinchoff, as shown in fig 3-39. The value of V_{DS} at the threshold of pinchoff is given

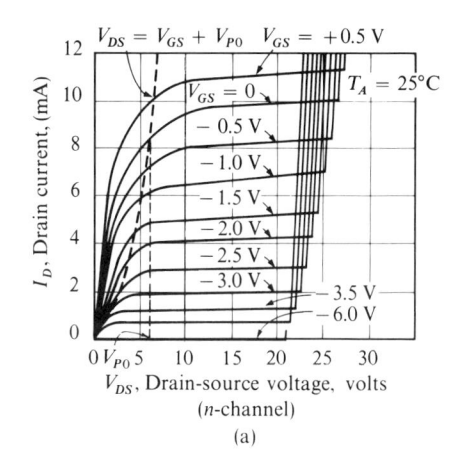

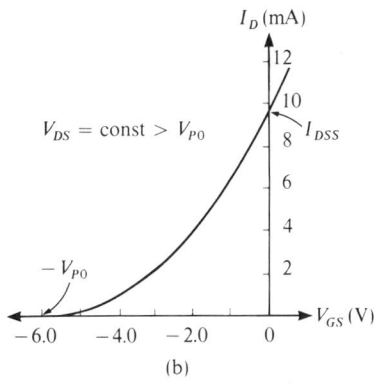

Fig 3-39 (a) Common-source FET drain or output characteristic; (b) Common-source FET transfer characteristic

approximately by

$$V_{DS} \text{ (at pinchoff)} = V_{P0} + V_{GS} = V_P \tag{3-58}$$

Note that $V_{P0} > 0$ for *n*-channel FET. In practice V_{GS} is not made more positive than 0.5 volt in order to avoid the gate to source current. The breakdown voltage for an arbitrary V_{GS} is given approximately by

$$BV_{DS} \text{ (for arbitrary } V_{GS}) = B_0 V_{DS} \text{ (at } V_{GS} = 0) + V_{GS} \tag{3-59}$$

and reduces as V_{GS} becomes more negative (fig 3-39).

Note that in fig 3-39 a small gate-voltage variation causes appreciable changes in the drain current; thus, the FET is a voltage-sensitive device. In integrated circuits, FET's can be used as voltage-controlled resistors for $V_{DS} < V_{P0} + V_{GS}$. When an FET is used as an amplifier, it is usually operated in the constant-current region, i.e., where $V_{DS} > V_{P0} + V_{GS}$.

For analytic purposes, the *V-I* characteristics of FET's may be conveniently divided into two regions of operation. The two regions and their *V-I* relationships are:

(a) *Region below pinchoff*: When the drain voltage is low and the channel is not pinched off, the *V-I* relationship is given approximately by

$$I_D \approx K[2(V_{GS} + V_{P0})V_{DS} - V_{DS}^2] \tag{3-60}$$

for $0 < V_{DS} \leq V_{GS} + V_{P0}$, where K is a constant depending on the electrical properties and geometry of the channel.

(b) *Region between pinchoff and breakdown* (also called the saturation-current region): When the drain voltage is greater than the pinchoff value, the *V-I* relationship is given approximately by

$$I_D \approx K[V_{GS} + V_{P0}]^2 \tag{3-61}$$

for $V_{GS} + V_{P0} \leq V_{DS} < BV_{DS}$. If we substitute (3-58) into (3-60) or (3-61), we obtain the locus of the boundary between the two regions, which is a parabola,

$$I_D = KV_{DS}^2 \tag{3-62}$$

The locus separating the two boundaries is shown by the dashed line in fig 3-39a.

Equation (3-61), which describes the region of operation of FET as an amplifier, may be put in a convenient form as[6]

$$I_D = KV_{PO}^2\left(1 + \frac{V_{GS}}{V_{PO}}\right)^2 = I_{DSS}\left(1 + \frac{V_{GS}}{V_{PO}}\right)^2 \qquad (3\text{-}63)$$

The current I_{DSS} is temperature-dependent and decreases for increasing temperature. The parameters I_{DSS} and V_{PO} are important in FET's and are usually specified by the manufacturer. The value of V_{GS} for a given I_D may then be determined from (3-63).

MOSFET (IGFET)

The structures shown in fig 3-40 and fig 3-41 are alternative ways of realizing field-effect transistors. These structures consist of a lightly doped p-type substrate into which two heavily doped n-type regions are diffused. In fig 3-40, a shallow channel of lightly doped n-type material is diffused between the source and the drain. The region between the source and the drain is covered by an oxide layer over which is deposited a metal plate which serves as a gate. The *m*etal-*o*xide-*s*emiconductor construction gives the FET the name MOSFET. Since the gate is insulated from the semiconductor by the oxide layer, the name IGFET is also used.

The operation of the MOSFET is basically similar to that of the JFET. In the structure of fig 3-40, if a voltage is applied between the gate and the source, making the gate negative relative to the source, positive charges are induced in the channel through the SiO_2 of the gate capacitor. The induced positive charge causes a depletion of the majority carriers (electrons in the n-channel) and thus the channel becomes less conductive.

If the gate is made sufficiently negative, the depletion region extends completely across the channel and the channel cannot conduct current. This condition is the pinchoff and is generally within the same range of values as in the JFET. The V-I characteristics of the depletion-mode MOSFET are quite similar to those of JFET.

If the gate voltage is made positive relative to the source, negative charges are induced in the channel through the gate capacitor, thus enhancing the concentration of the majority carriers in the channel. The conductivity of the channel is therefore increased and the current can exceed I_{DSS}.

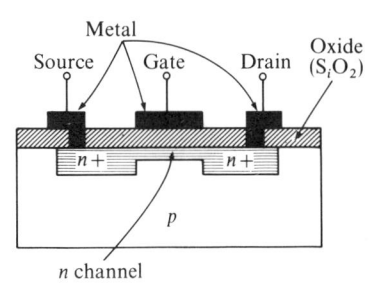

Fig 3-40 Depletion-type MOSFET schematic with n-channel (can also be operated in the enhancement mode)

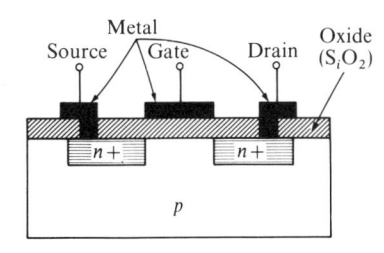

Fig 3-41 Enhancement-type MOSFET with induced n-channel (cannot be operated in the depletion mode)

[6] The classical result (Ref. 1) is $I_D = I_{DSS}[1 + 3(V_{GS}/V_{P0}) + 2(-V_{GS}/V_{P0})^{3/2}]$. The results of the two equations agree very closely; (3-63) is preferable because of its simpler form.

Thus, a depletion-type MOSFET can be operated in the enhancement mode or the depletion mode. The *V-I* characteristics of a typical depletion-type MOSFET are shown in fig 3-42. Frequently the graphic symbol for the JFET is used for the MOSFET *when it is operated in the depletion mode.*

Another simple MOSFET structure is shown in fig 3-41. In this case, no channel exists between the drain and the source if there is no applied voltage on the gate-to-source terminals. This structure, usually the *p*-channel, is becoming very common for integrated circuits (see Appendix B) because of its ease of fabrication and some other performance characteristics. This type of MOSFET is useful *only* in the *enhancement mode* of operation, i.e., when the applied voltage makes the gate positive in relation to the source (for the *n*-channel MOSFET). Only when this situation exists, as in fig 3-41, is *n*-channel induced and the device operates properly. The gate voltage at which the channel forms is called the threshold voltage V_T. The value of V_T depends upon the electrical properties of the substrate and oxide and the thickness of the oxide. Typical values of V_T range from 1 to 5 volts. Thus, when the gate potential is less than V_T, no channel exists and the drain current is approximately zero. For higher values of V_{DS}, i.e., $V_{DS} > V_{GS} - V_T > 0$, the channel is pinched off and the device operates with essentially constant drain current. Note that the quantity I_{DSS} has no meaning in the enhancement-type MOSFET because $I_D = 0$ when $V_{GS} = 0$. For this reason, a different graphic symbol, shown in fig 3-43, is used for the enhancement-type MOSFET. Usually another terminal is available which designates the substrate (or bulk) and is connected to the source or to a negative potential. This is done to reverse bias the *p-n* junction and ensure isolation for the device. Typical *V-I* characteristics of an enhancement-type MOSFET are shown in fig 3-44. Note

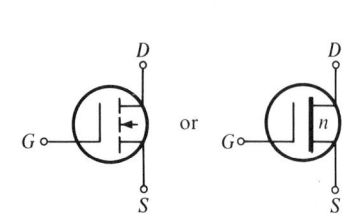

Fig 3-43 Graphic symbols for *n*-channel enhancement-type MOSFET (for a *p*-channel, change the direction of arrow and use *p* instead of *n*)

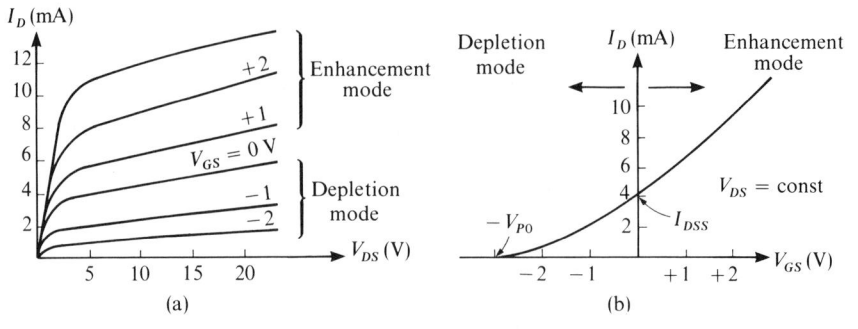

Fig 3-42 (a) Drain characteristics of *n*-channel depletion-type MOSFET; (b) Transfer characteristic

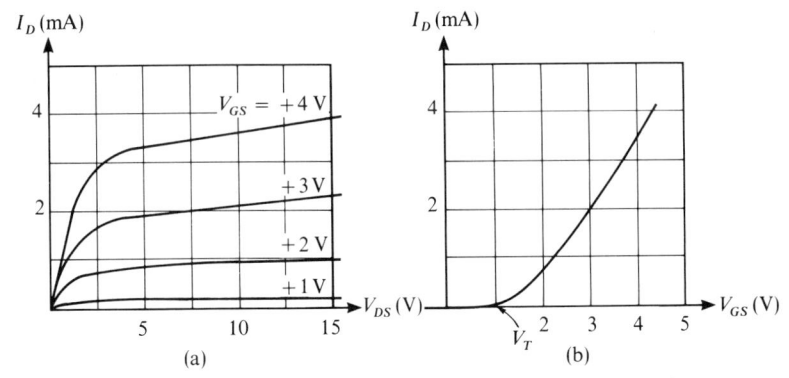

Fig 3-44 (a) Drain characteristics of *n*-channel enhancement-type MOSFET; (b) Transfer characteristic

that for the enhancement-type MOSFET, increasing the gate voltage increases the conductivity, while in the JFET and depletion-type MOSFET, increasing the gate voltage decreases the conductivity of the channel.

The analytic V-I relationship in (3-63) also gives an excellent fit with the experimentally determined characteristics of MOSFET devices. The V-I characteristics of the three types of FET's are alike, so the model obtained in the next section applies equally to all types of FET's and other devices having similar characteristics, such as vacuum-tube pentodes, which will be discussed in Section 3-16.

3-11 FET Small-Signal Model

When an FET is used as a small-signal amplifier, the device is operated in the current-saturation region, i.e., beyond pinchoff. A basic common-source FET amplifier circuit is shown in fig 3-45. The quiescent point and the load line for $V_{DD} = 20$ V and $V_{GG} = -0.7$ V are shown on the output characteristic curve. We wish to determine the small-signal model around the operating point for the FET, as we did in Section 3-8 for bipolar transistors.

From the V-I characteristics of the FET, it is seen that in the current-saturation region the slope is not zero, so increasing V_{DS} also increases I_D. We may write the dependence of I_D on V_{DS} and V_{GS} in the following functional form:

$$i_D = f(v_{GS}, v_{DS}) \tag{3-64}$$

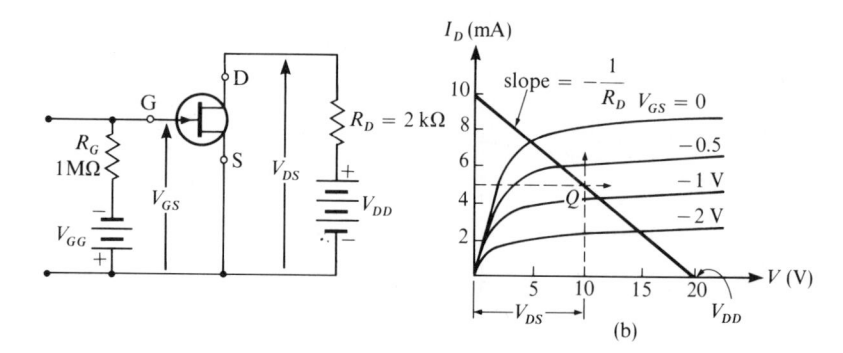

Fig 3-45 (a) Common-source JFET biasing circuit; (b) Output characteristics with load line and quiescent point

where the symbols denote instantaneous quantities, e.g., $i_D = I_D + i_d$.

We proceed as in (3-34), by obtaining a Taylor series expansion of (3-64) near the operating point Q. Thus we have

$$\Delta i_D \approx \frac{\partial i_D}{\partial v_{GS}}\bigg|_{V_{DS}} \Delta v_{GS} + \frac{\partial i_D}{\partial v_{DS}}\bigg|_{V_{GS}} \Delta v_{DS} \qquad (3\text{-}65)$$

In small-signal notation we write

$$i_d = g_m v_{gs} + g_d v_{ds} \qquad (3\text{-}66)$$

where

$$g_m = \frac{\partial i_D}{\partial v_{GS}}\bigg|_{v_{ds}=0} = \frac{\Delta i_D}{\Delta v_{GS}}\bigg|_{V_{DS}} \qquad (3\text{-}67)$$

$$g_d = \frac{\partial i_D}{\partial v_{DS}}\bigg|_{v_{gs}=0} = \frac{\Delta i_D}{\Delta v_{DS}}\bigg|_{V_{GS}} \qquad (3\text{-}68)$$

The input current $i_G = 0$ for all values of v_{GS} and v_{DS} under small-signal conditions. Using this fact and (3-66), we have the two-port model for an FET in the common-source (CS) configuration, as shown in fig 3-46. The model is, of course, valid for all types of FET's, although the element values may be different in different types of FET's.

The current-controlled source may be changed to a voltage-controlled source by using the Thévenin equivalent form. For such cases, it is convenient to define

$$\mu \equiv \frac{g_m}{g_d} = \frac{\Delta v_{DS}}{\Delta v_{GS}}\bigg|_Q \qquad (3\text{-}69)$$

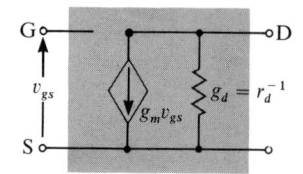

Fig 3-46 Low-frequency small-signal model of an FET

where μ is the amplification factor of the device.

3-11 FET Small-Signal Model

The high-frequency model, which must be used when the incremental variables change rapidly, is shown in fig 3-47. This π model has been experimentally verified by many workers in the field and has been shown to give excellent agreement with measured results. The capacitances for high-frequency FET's are typically in the range from 0.5 to 2 pF and their dependence of the operating point is very weak. Typical values of the parameters for JFET's and MOSFET's are shown in table 3-2.

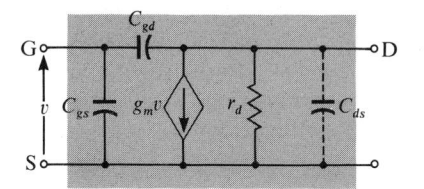

Fig 3-47 High-frequency small-signal model of an FET (C_{ds} can usually be ignored)

TABLE 3-2 *Typical Parameter Values for JFET's and MOSFET's*

Parameter	JFET	MOSFET
g_m(mhos)	0.1×10^{-3}–10×10^{-3}	0.5×10^{-3}–10×10^{-3}
$r_d^{-1} = g_d$(mhos)	10^{-6}–10^{-5}	10^{-5}–10^{-4}
C_{gd}, C_{ds} (pF)	0.1–2	0.1–2
C_{gs} (pF)	2–10	2–10

The circuit element values can be readily obtained from manufacturers' data sheets. The data are usually given in terms of the *y*-parameters, which are most convenient for FET's (see Appendix D). From the high-frequency model of fig 3-47 and the manufacturer's *y*-parameters we obtain

$$C_{gd} \simeq C_{rss}, \quad C_{gs} \simeq C_{iss} - C_{rss}, \quad C_{ds} \simeq C_{oss} - C_{rss}$$

$g_m \simeq |y_{fs}|$ at low frequencies and $g_d \simeq \text{Re}(y_{os})$ at low frequencies. Note that at very high frequencies $\text{Re}(y_{is})$ is not negligible. Capacitance C_{ds} is very small and may usually be neglected in circuit calculations. Capacitance C_{gd}, though small, cannot be ignored because of the feedback through it at high frequencies (see Section 4-8).

The variation of g_m, r_d, and μ as a function of the operating point, for a typical FET, is shown in fig 3-48. The transconductance g_m is directly proportional to the quiescent V_{GS}, as can readily be seen from (3-63) and (3-67).

In spite of the fact that the capacitances of FET's are small, as we shall see in Chapters 4 and 5, the bipolar transistor has a higher gain-bandwidth capability than the FET devices and is, therefore, used in high-speed applications. In high-frequency applications, the load resistance is necessarily small and capacitance C_{ds}, which is in shunt with R_L, can be ignored, just as in the bipolar transistor (BJT) model.

Note that the FET model is similar to the BJT model *except* that r_b and R_i are not present in the FET model. The input properties of the two devices are quite different; in the FET, the low-frequency input impedance is infinite and it must, therefore, be driven by a voltage source. The input impedance of the BJT is moderate and it can be driven by a voltage or a current source. The latter is most often used because the BJT is a current-sensitive device.

Finally, it should be clear that the model we have shown in fig 3-47 can be used for FET's in any of the three configurations, common-source (*CS*), common-drain (*CD*), and common-gate (*CG*), by merely reorienting the model. The properties of FET's in the various configurations are discussed in Chapter 4. Circuit applications of transistors are discussed in later chapters throughout the text.

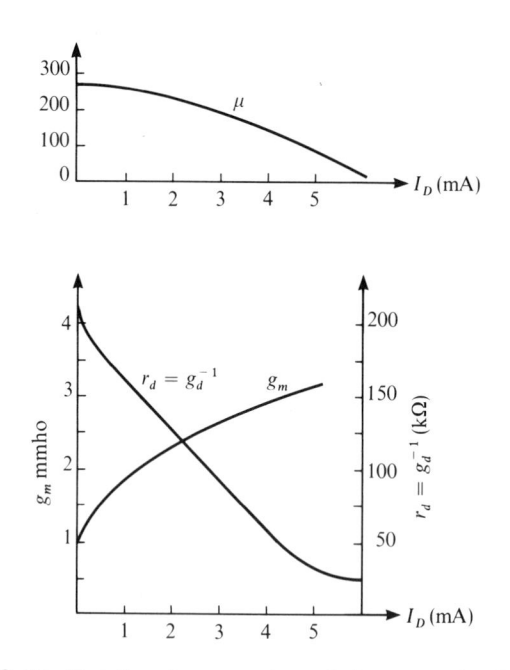

Fig 3-48 Variation of g_m, r_d, and μ vs drain current of an FET

3.12 Unijunction Transistors

Another useful device which is constructed somewhat similarly to the FET is the so-called *unijunction transistor* (UJT). The emitter junction of the UJT is very much smaller than the junction area of the JFET. It is a three-terminal device with one junction, hence the name unijunction, as shown schematically in fig 3-49a. Because of the two bases, it is sometimes called a *double-base diode*. The graphic symbol for the device and its *V-I* characteristics are shown in figs 3-49b and 3-49c, respectively. Note that the emitter arrow in fig 3-49b is angled and points toward *B*1. As we shall see, the junction terminals *E* and *B*1 play an important role in determining the *V-I* characteristic of the UJT.

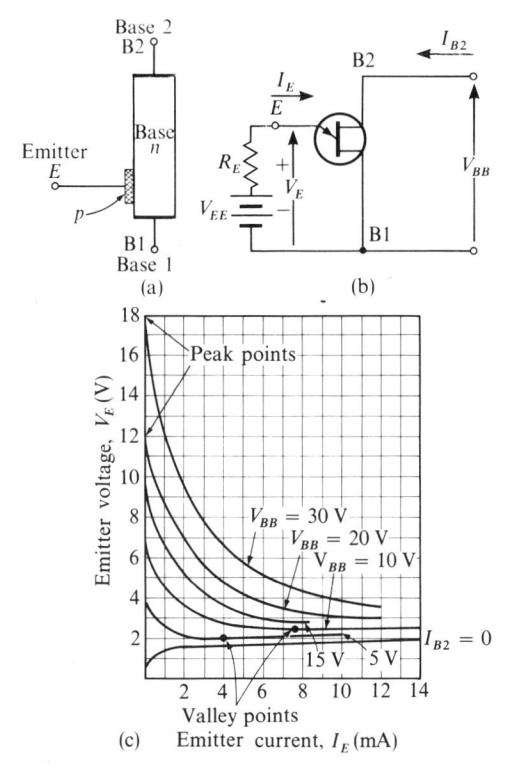

Fig 3-49 (a) Schematic of a unijunction transistor (UJT); (b) Graphic symbol for a UJT with voltage polarity and current reference directions; (c) Typical UJT emitter input characteristics (Courtesy General Electric Co)

For proper operation of the device, the emitter is forward biased and a supply voltage V_{BB} is used across the two base terminals such that *B*2 is made positive relative to *B*1. If terminal *B*2 is open-circuited, so that $I_{B2} = 0$, we have the exponential curve of the *p-n* junction diode, as in fig 3-49b. Note, however, that when I_{B2} flows, the impedance between *E* and *B*1 has a negative-resistance characteristic, which makes it very useful in a number of applications, such as the relaxation oscillator discussed in Chapters 9 and 10.

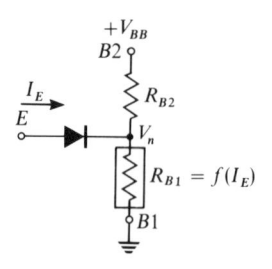

Fig 3-50 Circuit model for a UJT

The circuit model for the UJT is shown in fig 3-50. The *V-I* characteristics may be explained qualitatively in terms of the circuit model of fig 3-50 as follows: For small values of the emitter voltage

$$V_E < \left(\frac{R_{B1}}{R_{B1} + R_{B2}}\right) V_{BB},$$

the *p-n* junction is reverse biased and thus negligible reverse leakage current flows. When

$$V_E \geq V_0 + \left(\frac{R_{B1}}{R_{B1} + R_{B2}}\right) V_{BB},$$

the diode conducts and holes are injected into the lightly doped *n*-type base. V_0 is the threshold voltage of the *E-B*1 diode, $V_0 \approx 0.6$ V. Because of the electric field due to V_{BB}, the injected holes are swept into *B*1, so the resistance of R_{B1} is greatly reduced while that of R_{B2} is negligibly affected. The *decrease* in R_{B1} causes a regenerative action, that is, the more R_{B1} decreases, the more the voltage V_n (see fig 3-50) decreases and the more the emitter current increases. The increase in current while the voltage is decreasing is the cause of negative resistance for the device. The value of the emitter current is limited by the value of V_{EE}/R_E. Note that in the characteristic, the orientation of the plot is such that I_E is the independent variable.

Since the UJT is used only in large-signal applications, we shall not develop a small-signal model for the device.

3-13 Semiconductor Controlled Rectifiers

The *semiconductor controlled rectifier* (SCR) is a three-junction semiconductor device, shown schematically in fig 3-51a. The graphic symbol for the SCR is shown in fig 3-51b. The device is sometimes called a *pnpn triode*. The *V-I* characteristics of a typical SCR are shown in fig 3-51c. This device also has a negative-resistance characteristic. The negative-resistance characteristic and the high anode-current capability in the ON state (*pnpn* SCR's with 200-ampere current ratings are commerically available) make it very useful in industrial applications.

The *V-I* characteristic of the SCR may be described with the aid of a two-transistor analog, as shown in fig 3-52. Operation is as follows:

3-13 Semiconductor Controlled Rectifiers

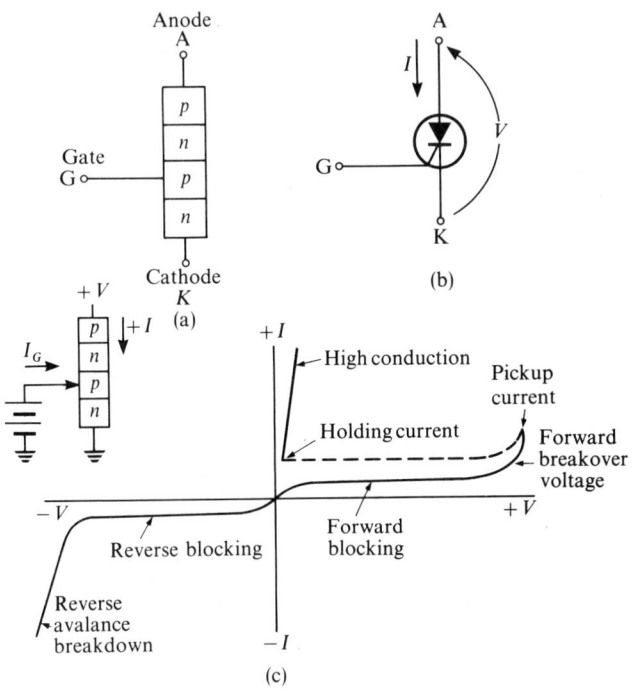

Fig 3-51 (a) Schematic of a semiconductor controlled rectifier (SCR); (b) Graphic symbol for an SCR with voltage polarity and current reference directions; (c) Typical SCR voltage-current characteristics (Courtesy General Electric Co)

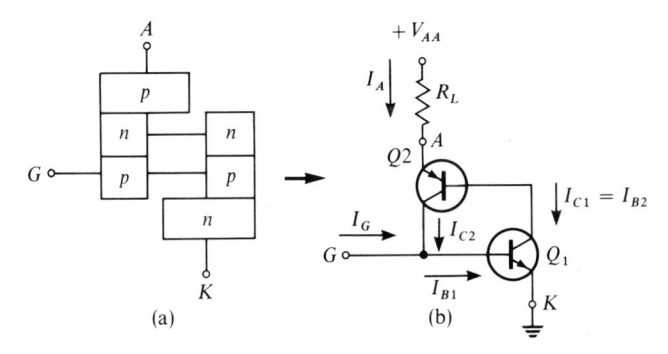

Fig 3-52 (a) Schematic of two-transistor analog of an SCR; (b) Bipolar transistor equivalent of an SCR

The two transistors are connected in a positive feedback arrangement, that is, a current I_{B1} is amplified by transistor Q_1 and again by Q_2, and returned in phase to the base of Q_1. This regenerative feedback action (feedback is discussed in detail in Chapter 6) increases I_{B1} more and more until both transistors are saturated. This corresponds to the ON state of the device with V_{AK} having the lowest value and I_A the maximum value. The maximum value of current is limited by the load resistance R_L. If transistor Q_1 is kept off by controlling the gate voltage (i.e., reverse biasing the emitter junction of Q_1), then Q_2 will also be off and the device will be in the OFF state. The fact that both transistors cannot be kept in the active mode of operation is exhibited by the negative resistance region of the *V-I* characteristic. Specifically, positive feedback causes the negative resistance (see Problem 3-30). Note that, once the device is in the ON state, the gate voltage has no control over the SCR anode current. In order to restore the SCR to the nonconducting state, the anode current I_A must be made smaller than the holding current. Of course, the SCR will shut off when the anode potential becomes negative. The manufacturer usually specifies the holding current, the maximum gate trigger voltage, and the maximum gate current.

Like the UJT, the SCR is used exclusively in large-signal applications, so there is no need to develop a small-signal equivalent circuit.

Finally, it should be mentioned that other *pnpn* devices are also commercially available. For example, in the structure shown in fig 3-51a, if only the cathode and anode leads are accessible, the device is called a *Shockley diode* or a *four-layer pnpn diode*. The schematic and the graphic symbol for the four-layer *pnpn* diode are shown in fig 3-53. The *V-I* characteristic of this device has similar negative-resistance characteristics and is discussed, with circuit applications, in Ref 2.

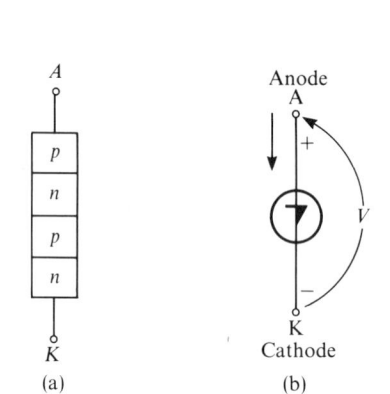

Fig 3-53 Four-layer *pnpn* (Shockley) diode: (a) Schematic; (b) Graphic symbol

3-14 Vacuum-Tube Devices

Vacuum-tube devices once played an important role in the electronics industry, but, for low-power applications, they have been virtually replaced by transistors. Only at very high frequencies where a very high power level is required, do vacuum tube devices still play a dominant role. These microwave devices, such as klystrons, traveling-wave tubes, and backward-wave oscillators, are used in applications where the frequency and power requirements are of the order of GHz and kilowatts, respectively. Since, in this text, we are interested in signal processing at frequencies below the GHz range, and further because these microwave devices are somewhat specialized, we will not discuss them here.

3-15 The Triode

In television receivers, radio transmitters, radar systems, etc., some vacuum triodes and pentodes still are used. Although these are being replaced by transistors, vacuum tubes are nonetheless commercially available and some familiarity with their behavior is helpful in circuit design. As we shall see, these vacuum-tube devices behave very much like FET's in circuit applications. Some of the various vacuum-tube devices are the diode, triode, tetrode, and pentode. Among these, the triode and pentode are still in use and will be considered briefly in the following sections.

3-15 The Triode

A vacuum triode is a thermionic device with three electrodes. These are the plate (or anode), the cathode, and the grid. The pictorial representation and the graphic symbol for the triode are shown in figs 3-54a and 3-54b, respectively. The cathode is usually made of nickel alloy coated with a barium compound, or other similar alloy. When the cathode is heated, large quantities of electrons are emitted from its surface. The cathode is heated indirectly by the heater, which is usually connected to a 6-volt supply. The heater is usually not shown in the graphic symbol, but it is understood that it is connected to a heater (or filament) supply. The plate is connected to a positive supply voltage which will attract the emitted electrons. The grid is a wire mesh which is placed close to the cathode. The grid plays the role of a control electrode in the following manner: When the grid potential is negative, some of the electrons from the cathode are repelled by the grid and return to the cathode; only those electrons having sufficient energy pass through the grid and reach the plate. Thus, if the grid potential is

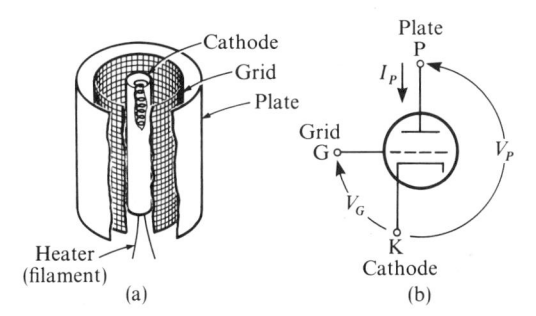

Fig 3-54 Vacuum-tube triode: (a) Pictorial representation; (b) Graphic symbol

made more negative, a smaller number of electrons reach the plate and the plate current is reduced. The grid potential variation thus exerts controlling action on the plate current.

The V-I characteristics of a typical triode are shown in fig 3-55. The grid current is zero as long as the grid potential is negative.[7]

The V-I relationship of an idealized triode is approximately given by (Ref 6 and note that I_p, V_G, V_p instead of I_b, E_c, E_b are used):

$$I_p = K(V_G + \mu^{-1}V_p)^n \qquad V_G \leq 0; \quad V_G + \mu^{-1}V_p \geq 0 \qquad (3\text{-}70)$$

where μ is called the *amplification factor* and K is called the *perveance*, and both depend on the geometry of the tube. The exponent n is a constant and is approximately equal to $\frac{3}{2}$. This is the reason for the nomenclature "3/2 power-law device." Note that the triode is a voltage-sensitive device, similar to the FET.

Small-Signal Model of a Triode

The small-signal model of a triode may be developed identically to that in Section 3-11 for the FET. We may write the dependence of I_p on V_G and V_p in the functional form

$$i_P = f(v_{GK}, v_{PK}) \qquad (3\text{-}71)$$

where the symbol i_P denotes an instantaneous quantity, such as $i_P = I_p + i_p$.

Similar to (3-66) we obtain

$$i_p = g_m v_{gk} + g_p v_{pk} \qquad (3\text{-}72)$$

where g_m and $r_p \equiv g_p^{-1}$ are called the transconductance and the incremental plate resistance, respectively. These parameters are given by the following relation:

$$g_m = \left.\frac{\partial i_P}{\partial v_{GK}}\right|_{v_{pk}=0} = \left.\frac{\Delta i_P}{\Delta v_{GK}}\right|_{V_{PK}} \qquad (3\text{-}73)$$

$$g_p = \left.\frac{\partial i_P}{\partial v_{PK}}\right|_{v_{gk}=0} = \left.\frac{\Delta i_P}{\Delta v_{PK}}\right|_{V_{GK}} \qquad (3\text{-}74)$$

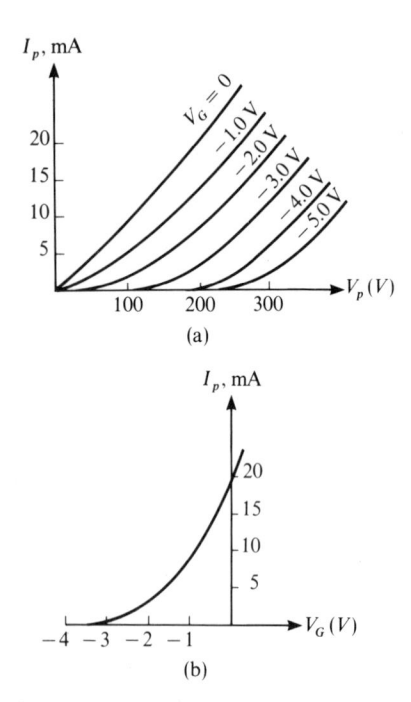

Fig 3-55 (a) Output characteristics of a typical vacuum triode; (b) Transfer characteristic

---

[7] A vacuum tube without the grid electrode, i.e., one having only an anode and a cathode, is called a vacuum diode. The V-I relationship of such a device is given approximately by $I_p \approx K V_p^{3/2}$ for $V_p > 0$, and $I_p = 0$ for $V_p < 0$.

Note that the amplification factor is

$$\mu = \frac{g_m}{g_p} = g_m r_p = \frac{\Delta v_{PK}}{\Delta v_{GK}}\bigg|_Q \qquad (3\text{-}75)$$

The low- and the high-frequency incremental models of a triode are shown in fig 3-56. Note that the model of a triode is identical to that of an FET, shown in fig 3-47. Typical range of values for the parameters of actual triodes are given in table 3-3. The variation of μ, g_m, and r_p as a function of the operating point for a typical triode is shown in fig 3-57. Note that μ is approximately constant.

TABLE 3-3 *Typical Parameter Values for Triodes*

μ	2–100	g_m(mhos)	0.5×10^{-3}–10×10^{-3}
$r_p(\Omega)$	10^3–5×10^4	C_{gk}, C_{pk}, C_{gp}(pF)	1–10

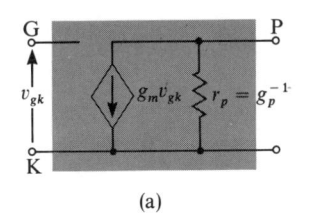

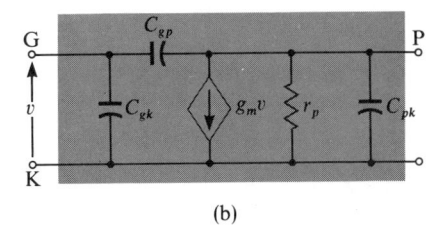

Fig 3-56 (a) Low-frequency small-signal model of a vacuum triode; (b) High-frequency small-signal model of a vacuum triode

3-16 Pentode Characteristics and Model

Multigrid vacuum tubes are designed to obtain better performance than triodes. As we shall show in the next chapter, the grid-to-plate capacitance of a triode limits its high-frequency amplification capability. In order to improve the frequency response, other grid structures

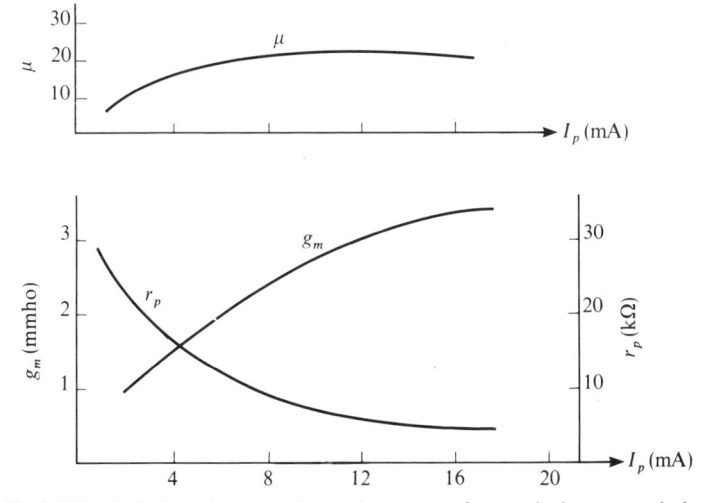

Fig 3-57 Variation of μ, g_m, and r_p vs plate current for a typical vacuum triode

are inserted between the grid and the plate. In this case, the shielding between the grid and the plate reduces the grid-to-plate capacitance, thereby reducing the high-frequency coupling of the grid and plate circuits. Among the multigrid vacuum tubes, tetrodes and pentodes are the most useful devices. In a tetrode, one additional grid is added, called the *screen grid*. In a pentode, two additional grids are added. These are called the screen and the *suppressor grid*, respectively. We shall consider only pentodes because of their superior performance for small-signal amplification as compared with tetrodes.

The pentode graphic symbol is shown in fig 3-58a. The pentode is normally operated with fixed suppressor-grid and screen-grid potentials. (The suppressor grid is usually connected to the cathode, as shown in fig 3-58a.) The V-I characteristics of a typical pentode are shown in fig 3-58b. A pentode is characterized by the same small-signal model as a triode since the signal can appear on two electrodes and $i_P = f(v_{PK}, v_{GK})$. Of course, the element values are different (some by orders of magnitude); specifically, the value of C_{gp}, which is extremely small (on the order of 0.01 pF) because of shielding, and the value of r_p, which is very large (on the order of 1 MΩ) as seen by the slope in the V-I characteristics for pentodes. Typical range of values for pentode

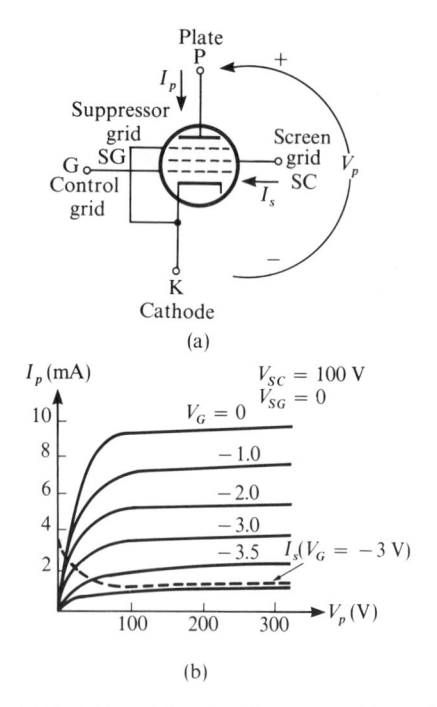

Fig 3-58 (a) Graphic symbol for vacuum-tube pentode with voltage polarity and current reference directions; (b) Output characteristics of a typical vacuum pentode

TABLE 3-4 *Typical Pentode Parameters*

μ	100–1000	C_{gk}, C_{pk}	1–10 pF
r_p(MΩ)	0.1–5	C_{gp}	0.005–0.05 pF
g_m(mmhos)	0.2–10		

parameters are given in table 3-4. Note that the amplification factor μ of a pentode has a much higher value than that of a triode. Usually C_{gk} and C_{pk} of a pentode are specified by the manufacturer and denoted by C_i and C_o, respectively. The high-frequency small-signal model of a pentode is shown in fig 3-59. Note that C_{gp} and r_p are ignored in the model because they have a negligible effect on circuit operation due to their values, as given in table 3-4. For low frequencies, the pentode approximates very closely an ideal voltage-controlled current source active element (fig 1-9b).

3-17 Bipolar Transistor Biasing

For transistor amplifiers to function properly, we must bias the transistor in the normal-active region. The common-emitter output characteristic of a typical low-power *pnp* silicon transistor is shown in

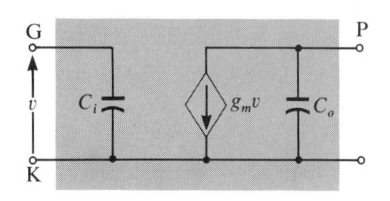

Fig 3-59 High-frequency small-signal model of a vacuum pentode

3-17 Bipolar Transistor Biasing

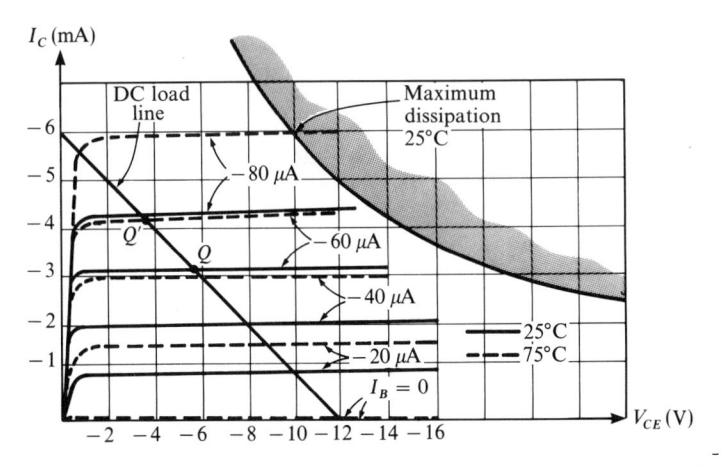

Fig 3-60 Typical low-power *pnp* silicon transistor output characteristics at two different temperatures. The shift in the operating point is also shown

fig 3-60. The region beyond the area of safe operation is marked by shading. The limit of the safe operating region is the maximum-dissipation hyperbola, defined by

$$P_C = V_{CE} I_C \qquad (3\text{-}76)$$

The limit is lowered in an increased-temperature environment. In low-power, high-frequency amplification we operate the transistor much below the dissipation hyperbola. Only in power amplifiers is the transistor operated close to the dissipation curve. In such cases power transistors, which have power ratings from a few watts to over one-hundred watts, are used. Power amplifiers are discussed in Section 5-14. In the following discussion, we consider low-power amplifiers.

A very simple biasing circuit is shown in fig 3-61 to illustrate some problems in biasing and the shift of operating point. The dc load line and the quiescent operating point are shown in fig 3-60. At room temperature, the quiescent point is at Q ($I_{Cq} = -3.2$ mA, $V_{Cq} = -5.6$ V). If the temperature is increased to 75°C, however, the operating point will shift to Q' ($I_{Cq} = -4.2$ mA, $V_{Cq} = -3.6$ V); thus, the transistor parameters will be different. Had the original operating point at 25°C been at $I_B = 80$ μA, the new operating point at 75°C would be in the saturation region—a very undesirable situation.

The aim of the biasing circuit is to establish the quiescent point independent of variations in transistor characteristics (which may result from temperature changes, transistor replacement, or other sources).

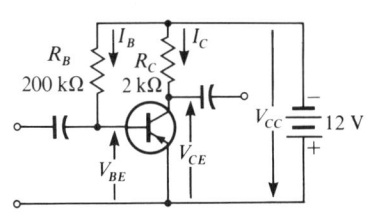

Fig 3-61 Very simple biasing circuit for *pnp* transistors (for *npn*, the voltage polarity is reversed)

If the Q-point is fixed, differences in transistor dc characteristics will not have a significant effect on circuit behavior.

We have seen that transistor characteristics are strongly temperature-dependent. This can also be readily seen by examining the characteristic curves of transistors at different temperatures, as provided by the manufacturers. Further, the transistor parameters vary as the operating point changes. The question is how to choose and stabilize the operating point. The answer to this question depends on the circuit application. In the common simple case considered in the following, the junction and ambient temperatures are essentially the same. This is the case when the supply voltage and the resistive load are such that the load line always falls well below the maximum power dissipation curve.

The transistor operating point is usually selected on the basis of optimal device performance in terms of gain (β_0), frequency response f_T, large dynamic swing without distortion, and minimal noise considerations. Usually the manufacturer provides such operating point data and the choice of the Q-point is not too difficult. Once the operating point is chosen, it must be stabilized by external circuits against temperature changes and transistor replacement. We shall now consider stabilization of the operating point.

It should be emphasized at the outset that proper stabilization is that which maintains I_C and V_{CE} constant against environmental changes. Note that by maintaining I_B constant we cannot achieve operating-point stability. For example, consider fig 3-60, which shows the common-emitter output characteristics of the same transistor for two different temperatures. (It could just as well have been the output characteristics of two different transistors of the same type.) If I_B is maintained constant by external biasing, the operating point will move to a new point, as indicated in fig 3-60. In some cases, it is possible for such a shift to bring the transistor operating point near the non-linear region or to bring it fully into the saturation region. We must design the circuit so that I_C, V_{CE} is constant for various temperature conditions as well as to provide for interchangeability of transistors in assembly-line production.

The main sources of shift in the operating point of a transistor due to temperature changes are the variation of I_{CO}, V_{BE}, and β_0. Another important source is the variation of β_0 from unit to unit, which can vary by a factor of three or more among different specimens of the same type of transistor. The I_{CO} approximately doubles for a 6°C increase in temperature above 25°C in silicon transistors, and it roughly doubles for about every 10°C increase above 25°C in germanium transistors. At room temperature, V_{BE} is approximately 0.6 V for Si and 0.2 V for

Ge. The magnitude of the emitter-junction bias for both Si and Ge decreases approximately as 2.5 mV/°C. The increase in β_0 due to temperature increase is approximately linear. β_0 roughly doubles for an increase of 60°C in Ge and 80°C in Si. The variation of β_0 is difficult to show analytically, but can be read directly from the CE output characteristics for various temperatures. Interchanging transistors, however, further complicates the problem, since variations can only be treated statistically.

In the active region, the relationship between I_C and I_B is given by (3-28):

$$I_C = -(\beta_0 + 1)I_{CO} + \beta_0 I_B \tag{3-77}$$

From (3-77), and using an ideal diode model for the forward-biased emitter-base junction, a temperature-dependent model for the *pnp* transistor is shown in fig 3-62. For *npn* transistors, the direction of the current and voltage sources should be reversed. Typical values for Ge and Si transistor parameters are given in table 3.5a. The approximate analytical expressions of Table 3-5b may be used for a rough estimate of I_{CO}, β_0, and V_{BE} at different temperatures if more accurate data are not available.

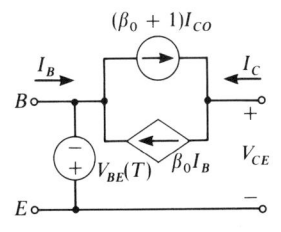

Fig 3-62 Temperature-dependent models for *pnp* transistors

TABLE 3-5a *Typical Values of β_0, I_{CO}, and V_{BE} for Silicon and Germanium Transistors at Various Temperatures*

	Silicon			Germanium		
$T(°C)$	−50	25	150	−30	25	75
$I_{CO}(\mu A)$	10^{-5}	1×10^{-3}	8	10^{-3}	1	32
β_0	20	50	120	10	50	100
$V_{BE}(V)$	0.78	0.60	0.25	0.33	0.20	0.08

TABLE 3-5b *Approximate Temperature Variation Expressions for β_0, I_{CO} and V_{BE} for Silicon and Germanium Transistors*

	Silicon	Germanium
$I_{CO}(25° + \Delta T)$	$[I_{CO}(25°)](2)^{\Delta T/6 °C}$	$[I_{CO}(25°)](2)^{\Delta T/10 °C}$
$\beta_0(25° + \Delta T)$	$\beta_0(25°)(1 + \Delta T/80°C)$	$\beta_0(25°)(1 + \Delta T/60°C)$
$V_{BE}(25° + \Delta T)$	$(0.60 - 0.002 \Delta T)$V	$(0.20 - 0.002 \Delta T)$V

From table 3-5a it is seen that Si transistors are to be used when the device is to be subjected to high temperatures. Note also that for

Si transistors the effect of I_{CO} is negligible while for Ge transistors the effect of V_{BE} is usually negligible. Silicon transistors are more often used industrially.

Let us examine now the simple biasing circuit of fig 3-61. The design is derived from the following equations. Note that for dc, the capacitors are open-circuited.

The collector-emitter and the base-emitter loop equations are

$$V_{CC} + V_{CE} + I_C R_C = 0 \qquad (3\text{-}78)$$

$$V_{CC} + I_B R_B + V_{BE} = 0 \qquad (3\text{-}79)$$

Normally, we design the biasing circuit so the operating point is nearly in the center of the active region, that is, $V_{CE} \simeq -V_{CC}/2$. Hence, for a desired I_{Cq}, V_{Cq} we obtain the circuit parameters from (3-78), namely,

$$I_C = -\frac{V_{CC} + V_{CE}}{R_C} = -\frac{V_{CC}}{2R_C} \qquad (3\text{-}80)$$

from (3-77),

$$I_B = \frac{I_C + (\beta_0 + 1)I_{CO}}{\beta_0} \approx \frac{I_C}{\beta_0} = -\frac{V_{CC}}{2R_C \beta_0} \qquad I_C \gg \beta_0 I_{CO} \qquad (3\text{-}81)$$

and finally, from (3-79),

$$R_B = -\frac{(V_{CC} + V_{BE})}{I_B} \approx -\frac{V_{CC}}{I_B} \qquad V_{CC} \gg V_{BE} \qquad (3\text{-}82)$$

The biasing circuit of fig 3-61 is not satisfactory for maintaining I_C and V_{CE} constant. For example, if the transistor is replaced by another of the same type but with a β_0 twice as large as the first one (not an uncommon situation), the operating point will shift considerably. Consider the circuit of fig 3-61 for a Ge transistor and the following numerical parameters:

$$V_{CC} = 12 \text{ V}, \quad R_C = 1.2 \text{ k}\Omega, \quad R_B = 120 \text{ k}\Omega$$

If the Ge transistor parameters at room temperature (25°C) are $\beta_0 = 50$, $I_{CO} = 1 \text{ }\mu\text{A}$, $V_{BE} = 0.2 \text{ V}$, the operating point will be approximately at

$$I_C = -5 \text{ mA} \qquad V_{CE} = -6 \text{ V}$$

3-17 Bipolar Transistor Biasing

If the second Ge transistor has the same parameters except that $\beta_0 = 100$, the operating point will be approximately at

$$I_C = -10 \text{ mA} \qquad V_{CE} \approx 0$$

Obviously the new transistor will not function properly in the circuit.

Let us pursue the analysis of the circuit somewhat further and consider the original transistor in the circuit, that is, the one with $\beta_0 = 50$, and increase the temperature from 25°C to 55°C. Because of the temperature increase, the transistor parameters will change to $V_{BE} \approx 0.12 \text{ V}$, $\beta_0 \approx 80$, $I_{CO} \approx 8 \text{ } \mu\text{A}$. The operating point in this case, as calculated above, is

$$I_C = -8.6 \text{ mA} \qquad V_{CE} \approx -1.7 \text{ V}$$

Again notice the considerable shift in operating point.

In order to compare the various biasing circuits, we define a stability factor[8] S_I, which is the rate of change of the collector current with respect to the reverse saturation current, holding V_{BE} and β_0 constant,

$$S_I \equiv \frac{\partial I_C}{\partial I_{CO}} \qquad (3\text{-}83)$$

The larger the S_I, the poorer the stabilization of I_C due to changes in I_{CO}.

From the general expression of (3-77), which is valid for all biasing circuits, we have

$$\frac{dI_C}{dI_{CO}} = -(\beta_0 + 1) + \beta_0 \frac{dI_B}{dI_{CO}} \qquad (3\text{-}84a)$$

or

$$S_I = -\frac{(\beta_0 + 1)}{1 - \beta_0(dI_B/dI_{CO})} \qquad (3\text{-}48b)$$

For the circuit of fig 3-61, I_B is independent of I_{CO} because I_B is fixed and determined from (3-81) so that $dI_B/dI_{CO} = 0$. Hence, for this circuit $S_I = -(\beta_0 + 1)$, which is rather poor, since β_0 is usually a large number.

---

[8] Other stability factors, such as $S_V \equiv \partial I_C/\partial V_{BE}$ and $S_{\beta_0} \equiv \partial I_C/\partial \beta_0$ [or $S_\alpha \equiv \partial I_C/\partial \alpha$], have also been used in the literature on transistor biasing. Since $I_C \simeq f(I_{CO}, V_{BE}, \beta_0)$ the variations of I_C can be written approximately as $\Delta I_C = S_I \Delta I_{CO} + S_V \Delta V_{BE} + S_{\beta_0} \Delta \beta_0$. We can thus find ΔI_C if the changes in the independent variables are small.

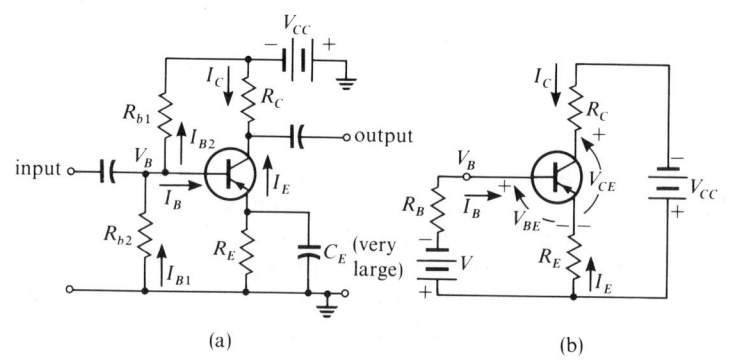

Fig 3-63 (a) Stabilized biasing circuit for discrete *pnp* transistors (for *npn*, the voltage polarity is reversed); (b) Simplified equivalent of (a) at dc

The circuit shown in fig 3-63a, which is referred to as the self-bias or emitter-bias circuit, is one of the most frequently used biasing circuits for single-stage and *RC*-coupled stages in discrete electronic applications. In this case, the addition of R_E in the emitter lead provides stabilization of the operating point. Physically, this improvement is provided as follows: If β_0 and/or I_{CO} is increased due to either interchangeability or temperature increase, then I_C increases. Since I_E is approximately the same as I_C, the increase in I_E causes an increase in the voltage drop across R_E, hence a decrease in I_B. The decrease in I_B tends to decrease I_C, so a stabilization effect is achieved. This stabilization due to feedback is one of the important properties of negative feedback and is discussed further in Chapter 6.

If we replace the circuit to the left of the base-to-ground terminal by its Thévenin equivalent circuit, as shown in fig 3-63b, we obtain

$$V = \frac{R_{b2} V_{CC}}{R_{b2} + R_{b1}} \quad (3\text{-}85)$$

and

$$R_B = R_{b1} \| R_{b2} = \frac{R_{b1} R_{b2}}{R_{b1} + R_{b2}} \quad (3\text{-}86)$$

Kirchhoff's voltage law applied around the base-emitter loop gives

$$V_{BE} = -V + I_E R_E - I_B R_B \quad (3\text{-}87)$$

Substituting $I_E = -(1/\alpha_0)(I_{CO} + I_C)$ and $I_B = -(I_E + I_C)$ in (3-87) and rearranging yields

$$\frac{R_{b2}}{R_{b1} + R_{b2}} V_{CC} + V_{BE} = -I_C \left[\frac{R_E}{\alpha_0} + \frac{R_B(1 - \alpha_0)}{\alpha_0} \right] \quad (3\text{-}88)$$
$$- \frac{I_{CO}}{\alpha_0}(R_E + R_B)$$

From (3-88) and (3-83) we obtain

$$S_I = \frac{\partial I_C}{\partial I_{CO}} = -\frac{1 + R_E/R_B}{(1 - \alpha_0) + R_E/R_B} \quad (3\text{-}89a)$$

Or, in terms of β_0,

$$S_I = -(1 + \beta_0) \frac{(1 + R_B/R_E)}{1 + \beta_0 + R_B/R_E} \quad (3\text{-}89b)$$

In (3-89a), if $R_E/R_B \to 0$, we have $S_I = -1/(1 - \alpha_0) = -(\beta_0 + 1)$, the situation in fig 3-61. If $R_E/R_B \gg 1$, then $S_I \to -1$. Also, for $R_B/R_E \ll \beta_0$, S_I is essentially independent of β_0. It should be noted, however, that choosing R_B as small as possible conflicts with the ac gain of the stage because the resistance R_B shunts the input of the amplifier. Usually R_B is such that it is much larger than the input impedance, that is, $R_B \gg (r_b + \beta_0 r_e)$. The requirement that R_E be as large as possible is limited by the available or specified voltage supply. Thus the design of a biasing circuit is a compromise between conflicting requirements. In certain cases, specified tight variations in I_C and V_{CE} may have to be relaxed in order to achieve the design with the biasing circuit of fig 3-63. Note also that R_E is bypassed by a large capacitance ($> 10\ \mu F$) in order to avoid the loss of ac signal gain due to R_E. The choice of the value of C_E depends upon the desired low-frequency cutoff of the stage; this is discussed in Chapter 4. The static and dynamic load lines are shown in fig 3-64. The very large value of capacitor C_E short-circuits R_E under the dynamic conditions, hence the slope of the load line for the ac case is $-1/R_C$, which is larger than for the dc case.

Design of the Biasing Circuit

If all the pertinent factors are included in the design, the design will not be simple. For a simple design, in order to obtain adequate stability, we usually choose $R_B/R_E \leq (\beta_0)_{min}/5$. As a rule of thumb, the ratio R_B/R_E is generally chosen to be 5 or less. R_E and R_C are chosen to be of the same order of magnitude with R_E somewhat smaller than R_C so that the dc voltage across R_C is at least as much as V_{CE}.

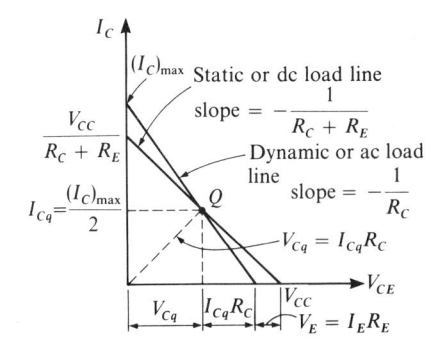

Fig 3-64 Static and dynamic load lines

EXAMPLE

Consider the design of the biasing circuit of fig 3-63 for a silicon transistor with the manufacturer's specifications at 25°C as follows: $(\beta_0)_{min} = 50$; $(\beta_0)_{max} = 150$; typical value for transistor type, $\beta_0 = 100$. The supply voltage $V_{CC} = 20$ V. The desired operating point is $I_C = -5$ mA, $V_{CE} = -5$ V.

For the collector-emitter loop, using Kirchhoff's voltage law, we have

$$V_{CC} = -V_{CE} - I_C R_C + I_E R_E \simeq -V_{CE} - I_C(R_C + R_E) \quad (3\text{-}90)$$

Substitution of numerical values yields

$$20 = 5 + 5 \times 10^{-3}(R_C + R_E) \quad \text{or} \quad R_C + R_E = 3 \text{ k}\Omega.$$

We choose $R_C = 2$ kΩ and then $R_E = 1$ kΩ. For proper stability we choose $R_B/R_E = 5$, hence $R_B = 5$ kΩ. Now, neglecting I_{CO}, we have

$$I_C \approx \beta_0 I_B \quad \text{or} \quad I_B = \frac{5 \times 10^{-3}}{100} = 50 \ \mu\text{A}$$

From (3-85), (3-86) we can now solve for R_{b1} and R_{b2}. From (3-85),

$$R_{b1} = \frac{R_B V_{CC}}{V} \quad \text{and} \quad R_{b2} = \frac{R_{b1} V}{V_{CC} - V} = \frac{R_B V_{CC}}{V_{CC} - V}$$

where V is determined from (3-87):

$$V = -V_{BE} - \frac{R_B}{\beta_0} I_C - \frac{R_E}{\alpha_0} I_C$$

$$V = 0.6 + (0.05)5 + (5 + 0.05)1 = 5.9 \text{ V}$$

and

$$R_{b1} = (5 \times 10^3)\frac{20}{5.9} = 17 \text{ k}\Omega$$

$$R_{b2} = \frac{17(5.9)}{20 - 5.9} = 7.1 \text{ k}\Omega$$

We now determine the operating point for interchangeability, that is, $(\beta_0)_{min}$ and $(\beta_0)_{max}$ rather than the typical value used in the design.

The expressions for the operating points are

$$I_C = -\left[\frac{V - V_{BE}}{R_E + (R_B + R_E)/\beta_0}\right],$$

$$V_{CE} = -\left[V_{CC} + \left(R_C + R_E + \frac{R_E}{\beta_0}\right)I_C\right]$$

For the $(\beta_0)_{min}$ case, the operating point is calculated to be

$$I_C = -4.74 \text{ mA} \qquad V_{CE} = -5.8 \text{ V}$$

For the $(\beta_0)_{max}$ case, the operating point will be

$$I_C = -5.11 \text{ mA} \qquad V_{CE} \approx -4.7 \text{ V}$$

Hence we have adequate stability for interchangeability of transistors.

Next, let us determine the new operating point of the designed example for a temperature of 100°C, instead of 25°C. For the increased temperature the transistor parameters are approximately estimated by the rule in table 3-5b.

$$(\beta_0)_{nominal} \approx 160, \quad I_{CO} \approx 0.2 \text{ }\mu\text{A}, \quad V_{BE} \approx -0.4 \text{ V}$$

The analysis of the circuit with the new transistor parameters yields

$$I_C = -5.3 \text{ mA} \qquad V_{CE} = -4.1 \text{ V}$$

It is seen that β_0 variations, because of interchangeability, produce the same general effect as temperature variations. Thus, stabilizing for temperature variations also stabilizes unit-to-unit variations in β_0. The biasing circuit of fig 3-63 is, therefore, very useful and the simple design method outlined above provides adequate stability for the operating point.

In order not to reduce the amplification at signal frequencies, the emitter resistance R_E is usually bypassed by a large capacitor. The effect of this capacitor on the frequency response of the amplifier is discussed in Chapter 4. Obviously, C_E has no effect on the dc biasing.

It should be noted that it is sometimes desirable to locate the Q point so as to obtain a maximum symmetrical collector current swing. To obtain the maximum symmetrical collector current swing or, alternatively stated, a large dynamic range, the Q point must bisect the ac load line (see fig 3-64) such that $I_{Cq} = \frac{1}{2}(I_c)_{max}$. The pertinent segments for this condition are labeled in fig 3-64.

Worst-Case Analysis

The extreme values of the operating point under worst-case conditions may be determined as follows: Substitute the temperature-dependent model of fig 3-62 into the circuit of fig 3-63. Then, by straightforward analysis of the circuit,

$$I_E = \frac{I_{CO} + (V_{CC}/R_{b1}) - (V_{BE}/R_B)}{[1/(1 + \beta_0)] + (R_E/R_B)} \tag{3-91}$$

where $R_B = R_{b1} \| R_{b2}$ as above.

The maximum value of I_E occurs when I_{CO} and β_0 are maximum and V_{BE} is minimum. We must include the temperature range as well as the manufacturer's tolerances. The smallest value of I_E will occur when I_{CO} and β_0 have their minimum values and V_{BE} its maximum. Although such conditions rarely occur together, in some applications (such as in space and military electronics) such a pessimistic design approach is used.

3-18 Integrated-Circuit (IC) Biasing

In addition to the circuit discussed above, there are other stabilized biasing circuits for discrete electronic circuits (see, for example, Problem 3-37). In some cases, compensation techniques, using temperature-sensitive devices, such as diodes, transistors, and thermistors (temperature-sensitive resistors), are also used. We shall not discuss these points in detail here, except to point out that in integrated circuits, the circuit of fig 3-63 is not suitable because of the necessity for large capacitors, particularly a very large capacitor across $R_E (>10 \, \mu F)$. Further, in integrated circuits (see Appendix B) (a) close matching of active and passive devices over a large temperature range is achievable, (b) the active elements are cheaper than the passive elements, and (c) all the elements are fabricated on a very small chip, hence the same temperature environment exists for the entire circuit. The design philosophy must take an altogether different viewpoint.

We no longer avoid active elements but indeed use more transistors than capacitors so that the occupied area on the chip, the so-called "real estate cost," is less than that required by the large bypass and coupling capacitors.

3-18 Integrated-Circuit (IC) Biasing

There are several ingenious biasing circuits given in the literature for the biasing of integrated circuits. We shall consider only one commonly used biasing arrangement for integrated circuits (shown in fig 3-65), in order to examine some of the basic concepts. Since the active and passive elements are fabricated in an identical manner in a monolithic structure, we can assume identical transistors and resistors over a wide temperature range. In the circuit of fig 3-65, the two bases are driven from a common voltage node through essentially equal resistances, so the base currents are the same:

$$I_{B1} = I_{B2} = I_B$$

Hence, the collector currents are matched for matched or identically constructed transistors:

$$I_{C1} = I_{C2} = I_C$$

Also, both devices are driven by the same base-emitter voltage.

The collector current is determined from Kirchhoff's voltage law,

$$V_{CC} = I_1 R_1 + (V_{CE})_1 \qquad (3\text{-}92)$$

But

$$(V_{CE})_1 = I_B R_B + V_{BE} \quad \text{and} \quad I_1 = I_{C1} + 2I_B \qquad (3\text{-}93)$$

Solving (3-92) and (3-93) for I_C, we get

$$I_{C1} = I_{C2} = I_C = \frac{V_{CC} - V_{BE}}{R_1} - \left(\frac{2R_1 + R_B}{R_1}\right) I_B \qquad (3\text{-}94)$$

If $V_{CC} \gg V_{BE}$ and $[(2R_1 + R_B)/R_1] I_B \ll V_{CC}/R_1$, then

$$I_C \approx \frac{V_{CC}}{R_1} \qquad (3\text{-}95)$$

Similarly,

$$(V_{CE})_2 = V_{CC} - I_C R_C \simeq V_{CC}\left(1 - \frac{R_C}{R_1}\right) \qquad (3\text{-}96)$$

If R_C and R_1 are fabricated in the same diffusion step, the ratio R_C/R_1 is temperature insensitive. Usually $R_C = R_1/2$ so that $(V_{CE})_2 = V_{CC}/2$.

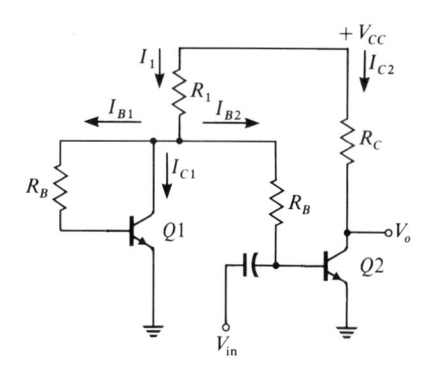

Fig 3-65 Biasing circuit for integrated circuits

Note that the operating point is independent of temperature variation and depends only on the matching of the transistors. Since matching in integrated circuits can be achieved with very close tolerances, the circuit has a very well stabilized quiescent point. For improved performance, small equal resistances (of the order of 100 Ω) are often inserted in series with the emitter leads of each transistor to obtain feedback. Note that no large bypass capacitors are required in this biasing.

3-19 Biasing Techniques for FET and Vacuum-Tube Devices

The biasing of vacuum tubes and FET's is usually simpler than that for bipolar transistors. In the vacuum-tube case, the filament temperature is usually over 1000°C so the ambient temperature changes are negligible. In FET's the gate current is usually very small (of the order of 1 nA) and does not vary appreciably with temperature so it can be neglected in the design.

The biasing circuit for JFET and depletion-mode MOSFET devices is shown in fig 3-66. Note that for an *n*-channel device, the arrowhead of the gate symbol points toward the source and the drain terminals of the FET.

The selection of an operating point (I_D, V_{DS}) for an FET amplifier stage is usually suggested by the manufacturer. The design is as follows: For an FET amplifier stage having the quiescent point

$$I_D = 2.0 \text{ mA} \qquad V_{DS} = 10 \text{ V}$$

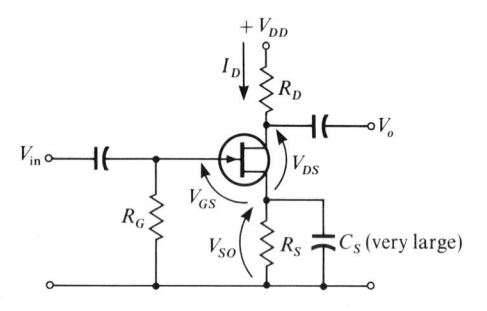

Fig 3-66 Biasing circuit for a JFET and depletion-type MOSFET

(at which point $V_{GS} = -3.0$ V) we first choose a large value for R_G, say 1 megohm. The voltage drop across R_G is $I_G R_G$, which is of the order of a millivolt and is negligible as compared to V_{GS}. From KVL, at the input we have

$$V_{GS} = I_G R_G - I_D R_S \approx -I_D R_S \qquad (3\text{-}97)$$

$$R_S = -\frac{V_{GS}}{I_D} = \frac{3.0}{2.0} \times 10^3 = 1.5 \text{ k}\Omega$$

3-19 Biasing Techniques for FET and Vacuum-Tube Devices

The value of R_D is determined from KVL applied to the source-drain circuit,

$$V_{DD} - V_{DS} - V_{SO} = I_D R_D \qquad (3\text{-}98)$$

For a supply voltage $V_{DD} = 20$ volts, the value of R_D is

$$R_D = \frac{V_{DD} - V_{DS} - V_{SO}}{I_D} = \frac{20 - 10 - 1.5(2.0)}{2.0 \times 10^{-3}} = 3.5 \text{ k}\Omega$$

If the FET characteristic curve is available, we can determine $R_S + R_D$ by drawing the load line corresponding to the desired operating point and the available supply voltage. The value of V_{GS} at the operating point determines R_S. The choice of the large bypass capacitance C_S depends upon the desired low-frequency cutoff of the stage and is discussed in the next chapter.

The biasing circuit of a triode (or pentode) amplifying stage is shown in fig 3-67. For a selected operating point (I_P, E_C) and supply voltage (E_{BB}), suggested by the manufacturer, the design is as before.

For example, consider a 6CG7 (miniature) or 6SN7 (octal) triode, which is to be biased at $E_C = -6$ V and $I_p = 8$ mA (at this point $E_p = 200$ V). The supply voltage E_{BB} is 300 V.

The value of R_K is determined from

$$R_K = -\frac{E_C}{I_p} = \frac{6}{8} \times 10^3 = 750 \ \Omega$$

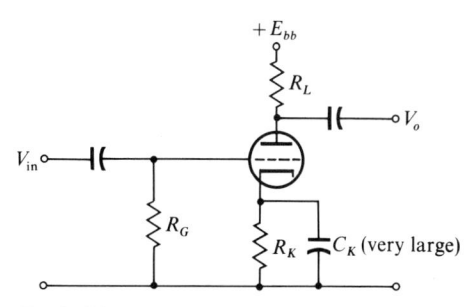

Fig 3-67 Biasing circuit for vacuum-tube triode

Graphically, the slope of the load line drawn on the characteristic curve determines the value of $R_L + R_k$. Analytically, this is equivalent to

$$R_L + R_k = \frac{E_{BB} - E_p}{I_p} \qquad (3\text{-}99)$$

or

$$R_L = \frac{300 - 200}{8 \times 10^{-3}} - 750 \approx 12 \text{ k}\Omega$$

The design of pentode biasing is essentially similar to that for triodes except that the screen current must be added to the plate current in computing the value of the cathode-bias resistor. The design of pentode biasing is left as an exercise (Problem 3-43).

The biasing circuit for an *n*-channel enhancement-type MOSFET is given in fig P3-41 (Problem 3-41). Note that the gate is to have a positive potential with respect to the source, and the graphic symbol,

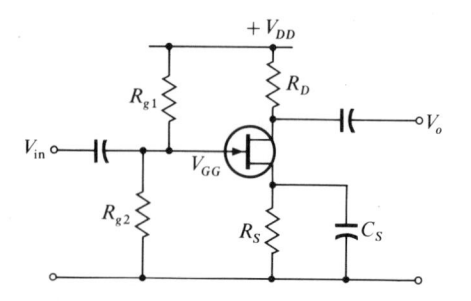

Fig 3-68 Modified biasing circuit to stabilize the operating point for an FET

therefore, is different. The enhancement-type MOSFET biasing requirements are the following: (a) The gate voltage is of the same polarity as the drain voltage. (b) $|V_{DS}| \geq |V_{GS} - V_T|$ to ensure that the operating point is in the current-saturation region. Note that absolute values are used to make the result applicable to both p- and n-channel enhancement-type MOSFET's.

Finally, it should be noted that FET and vacuum-tube devices also differ from unit to unit. Even though the manufacturers do not supply such information, variability from unit to unit exists and the operating point shifts as a result of the interchangeability of devices. A common biasing circuit which circumvents manufacturing variations is shown in fig 3-68. The circuit applies for FET and vacuum-tube devices (the supply voltage and resistor sizes would be different). To see how the operating point can be stabilized despite interchangeability, consider the modified FET biasing circuit shown in fig 3-68. Due to the nature of the manufacturing process, good control over I_{DSS} and V_{PO} cannot be maintained. Within the same type of FET, these parameters may exhibit a spread of 4 to 1 in the above characteristics. For example, the transfer characteristics of two FET's of the same type might be as shown in fig 3-69. If the FET is biased as in fig 3-66, the bias line is as shown in fig 3-69a. In order for I_D to be approximately the same for both FET's, the slope of the bias line must be considerably decreased (i.e., R_S is considerably increased). But if R_S is made very high the drain current will be very low, so that the transistor might be biased in the nonlinear region, which is undesirable.

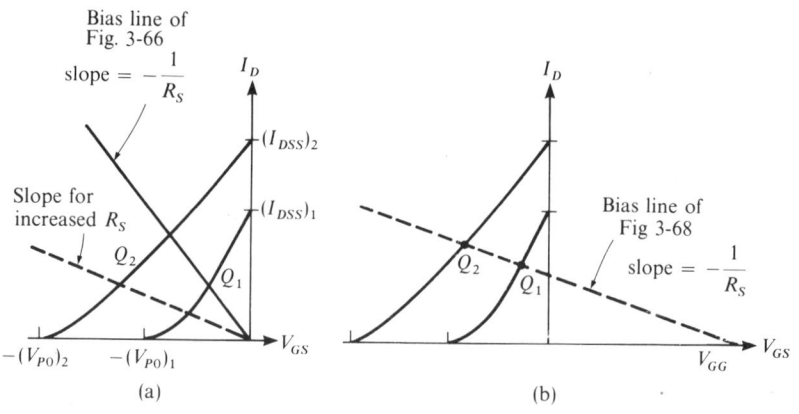

Fig 3-69 (a) Bias line and quiescent drain-current shift for fig 3-66 shown on the transfer characteristic; (b) Bias line and quiescent drain current shift for fig 3-68 shown on the transfer characteristic

Now consider the modified biasing circuit of fig 3-68. In this case, under quiescent conditions

$$V_{GS} = V_{GG} - R_S I_D \quad \text{or} \quad I_D = \frac{V_{GG}}{R_S} - \frac{1}{R_S} V_{GS} \quad (3\text{-}100)$$

The bias line corresponding to (3-100) is shown in fig 3-69b. A reasonably large drain current is maintained when R_s is increased, and since the slope of the bias line is small, the variation of the quiescent point is also relatively small from unit to unit.

3-20 Summary

This chapter presents several important devices along with their voltage-current relationships. From their nonlinear V-I characteristics, incremental linear models are obtained which represent the device for a restricted set of operating conditions: specifically, the small-signal conditions. Both low- and high-frequency models are given for BJT's, FET's, and vacuum tubes. In general, the models can be of varying degrees of complexity; only simple practical models are given which approximate the device with a reasonable degree of accuracy. The hybrid π model is the best compromise between accuracy and simplicity for a BJT in the CE configuration. The π model is the best compromise for FET's, vacuum triodes, and pentodes. The models of FET's and vacuum tubes differ only in the values of their parameters. FET's and vacuum tubes are voltage-sensitive devices; the BJT is a current-sensitive device. The parameters of the models vary with the operating point and temperature. Stabilization of the quiescent operating point against these changes as well as against interchangeability is also discussed.

Chapters 4 through 8 are concerned with linear circuits, including the analysis and design of different types of amplifiers, active filters, etc. In these chapters, small-signal models are utilized in the analysis and design. The powerful theory of linear circuits covered in Chapters 1 and 2 is directly applied.

Chapters 9 through 11 are concerned with nonlinear circuits. In such circuits, the devices are usually subjected to large-signal operation; thus, the small-signal model cannot be used. Some basic nonlinear circuits are discussed in which the analysis is performed either graphically or by piecewise-linear approximation.

REFERENCES AND SUGGESTED READING

1. Gray, P., *Introduction to Electronics*, New York: Wiley, 1967.
2. Gibbons, J., *Semiconductor Electronics*, New York: McGraw-Hill, 1969.
3. Angelo, J., *Electronics: BJT's, FET's and Microcircuits*, New York: McGraw-Hill, 1969.
4. Gray, P., *Physical Electronics and Circuit Models of Transistors*, New York: Wiley, 1965, SEEC series, Vol. 2.
5. Searle, C., *Elementary Circuit Properties of Transistors*, New York: Wiley, 1964, SEEC series, Vol. 3.
6. Ghausi, M. S., *Principles and Design of Linear Active Circuits*, New York: McGraw-Hill, 1965, Chapters 5–9.
7. Alley, C., and Atwood, K., *Electronic Engineering*, New York: Wiley, 1966.

PROBLEMS

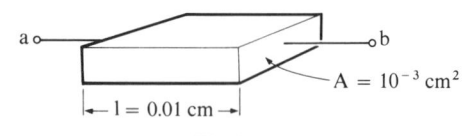

*3-1 A crystal is shown in fig P3-1. What is the resistance, R_{ab}, of the bar at room temperature (a) if it is an intrinsic Ge crystal; (b) if it is an intrinsic Si crystal? (c) Can the resistance be increased by doping the crystal but keeping the temperature constant? If so, what is the maximum resistance in each case?

*3-2 An abrupt *p-n* junction Ge diode has the same physical parameters as in the example Section A-3. Determine the corresponding parameters for a reverse bias of -5 volts and a forward bias of $+0.1$ volt.

*3-3 Show that the saturation current I_s of a diode approximately doubles near room temperature for every 10°C increase in Ge and every 6°C in Si. (Hint: $I_s = K_1 n_i^2 = KT^3 e^{-C/T}$, where $C \approx 7000°K$ for Si and $C \approx 4500°K$ for Ge.)

*3-4 Derive an expression for the ratio of the hole current to the electron current crossing a forward-biased $(V_a \gg \Lambda^{-1})$ junction. If $N_d = 10^{16}$ cm^{-3}, what should N_a be so that the electron current is 1% of the total current in a Si diode? Assume that $\tau_n = \tau_p = 10 \,\mu\text{sec}$.

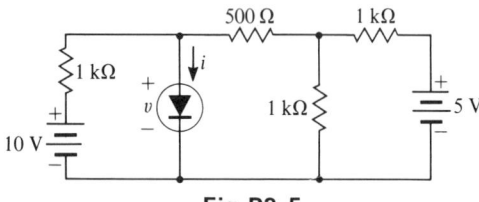

3-5 The *v-i* relationship of a Si diode is given in fig 3-2. If the diode is used as in fig P3-5, determine the voltage across and the current through the diode. (Hint: Use the Thévenin theorem and load line.)

3-6 Repeat Problem 3-5, but solve analytically, using a piecewise-linear approximation.

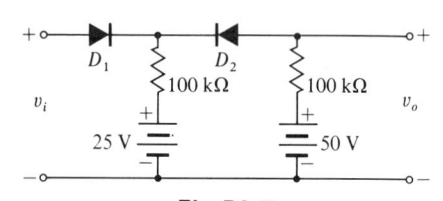

3-7 The input voltage to the circuit shown in fig P3-7 varies linearly from 0 to 100 volts. Sketch the output voltage to the same time scale as the input voltage. Assume ideal diodes.

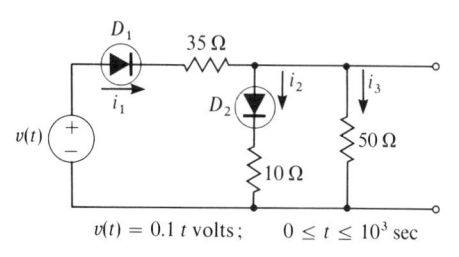

3-8 Assume that diodes D_1 and D_2 in fig P3-8 are defined by their piecewise-linear characteristics. D_1 and D_2 are Si diodes and their threshold voltage

* Consult Appendix A to solve any problem marked with an asterisk.

is $V_{01} = V_{02} = 0.6$ V. Their forward resistance is $(r)_1 = (r)_2 = 15 \Omega$. Find the times at which D_1 and D_2 are triggered (i.e., start to conduct) as well as the currents I_1, I_2, and I_3 from $t = 0$ until the moment at which both diodes have been triggered.

3-9 The circuit shown in fig P3-9 is a full-wave rectifier.
(a) Assume ideal diodes and sketch the output voltage $v_0(t)$.
(b) If a capacitor is connected in shunt across R, how is the voltage $v_0(t)$ affected? Sketch $v_0(t)$.

3-10 Repeat Problem 3-9 for the full-wave rectifier circuit shown in fig P3-10, but assume ideal diodes.

3-11 For the circuit shown in fig P3-11, plot the output if $v_{in}(t)$ is as shown and the Si diode has the characteristics given in fig 3-1.

3-12 For the v-i characteristics of the Si diode shown in fig 3-1, show a piecewise-linear model which includes the reverse breakdown region, using ideal diodes, resistors, and batteries.

3-13 The v-i characteristics of a Si diode are shown in fig 3-1. Sketch the v-i characteristics for $v \geq 0$, if two such diodes are connected (a) in series and (b) in parallel.

3-14 The ac circuit of a tunnel diode is shown in fig P3-14.
(a) Using the simple model of fig 3-10a, what is the constraint on R_g so that no natural frequency of the circuit is in the right-half s plane?
(b) What are the constraints for stability if the model of fig 3-10b is used?

3-15 For the circuit shown in fig P3-15a, draw the load line and comment on the operating point(s). The v-i characteristic of the device $\mathcal{N}$ is shown in fig P3-15b.

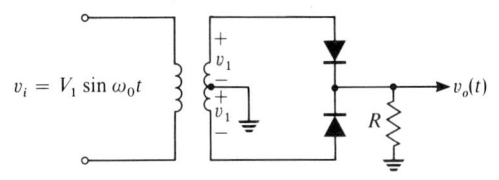

Fig P3-9

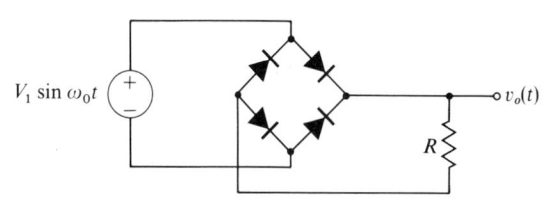

Fig P3-10

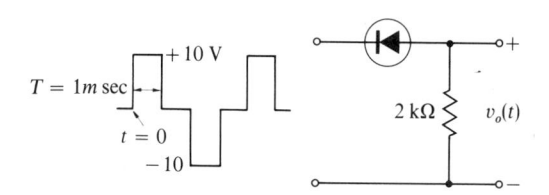

Fig P3-11

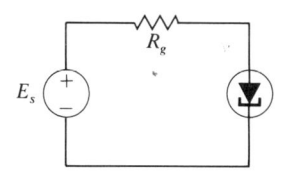

Fig P3-14

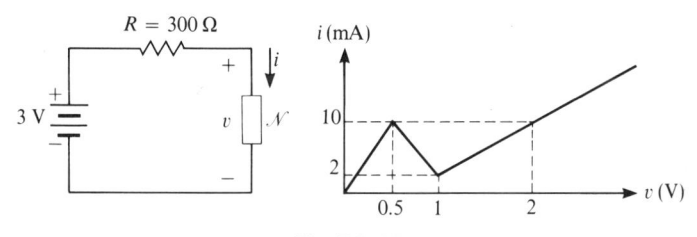

Fig P3-15

3-16 For the circuit shown in fig P3-16, determine the base voltage V_{BB} to saturate the transistor. Given that $(V_{CE})_{sat} = 0.1$ V, $(V_{BE})_{sat} = 0.6$ V, and $h_{FE} = 50$.

3-17 Show that the magnitude of $V_{BE}(T)$ near room temperature decreases approximately by 2.5 mV/°C for both Ge and Si transistors. (Hint: From the idealized diode equation (3-1) and $I \gg I_s$, solve for V and evaluate $(\partial V/\partial T)|_{I=\text{const}}$. Note that $I_s = KT^3 e^{-C/T}$.)

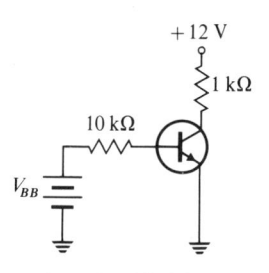

Fig P3-16

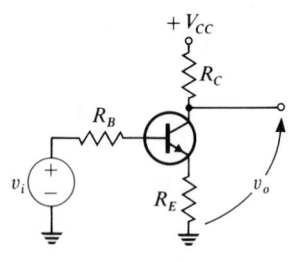

Fig P3-18

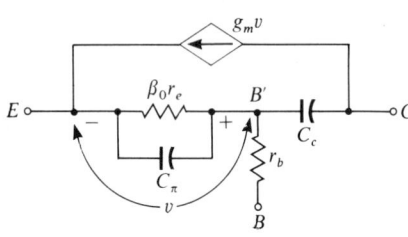

Fig P3-25

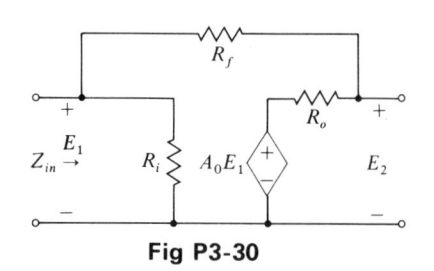

Fig P3-30

3-18 For the circuit shown in fig P3-18, sketch the static transfer characteristic, i.e., v_o vs v_i for all v_i. Assume the transistor to be open-circuited when in the cutoff mode and perfectly short-circuited when in the saturation mode. What value of v_i will saturate the transistor?

3-19 Suppose that a nonlinear resistive two-port is characterized by

$$i_1 = 2v_1^2 + v_1 v_2^2$$
$$i_2 = v_2(v_1 + v_1 e^{-v_2})$$

Calculate the incremental admittance matrix about the operating point $v_1 = v_2 = 1$ V. Show the small-signal model.

3-20 The low-frequency v-i description of a certain device is given by

$$v_1 = 0.1 e^{0.1 i_1}$$
$$v_2 = 0.1 i_1^2 (1 + e^{0.1 i_2})$$

Calculate the incremental impedance matrix about the operating point $i_1 = 10$, $i_2 = 5$ A. Show the small-signal model.

*3-21 The hole lifetime in the base region of a bipolar transistor is $\tau_p = 1$ μsec. The basewidth $W = 10^{-3}$ cm. The impurity concentration of carriers in the base region, in the absence of bias, is exponential such that $\kappa = 10$. Determine the approximate low-frequency common-base current gain α_0, and the alpha cutoff frequency ω_α, for the following cases:
(a) a *pnp* Si transistor;
(b) a *pnp* Ge transistor.

*3-22 Repeat Problem 3-21, if $\tau_n = 1$ μsec, $W = 10^{-3}$ cm, and $\kappa = 10$ for
(a) an *npn* Si transistor;
(b) an *npn* Ge transistor.

3-23 The manufacturer's data for Fairchild transistor type 2N3011 is given in Appendix D. Compute all the circuit-element values of the hybrid-π model at $I_e = 10$ mA.

3-24 Repeat Problem 3-23 for Texas Instruments transistor type 2N3251 as described in Appendix D.

3-25 The hybrid-π model is shown in the *CB* configuration in fig P3-25. Use the source transformation theorem by shifting the controlled source (from *C* to node *B'* and from *B'* to *E*) and then make circuit manipulations to obtain the *T*-equivalent circuit. Note that the results are not exactly the same as in fig 3-31. Comment briefly on why.

3-26 If the *CE* stage *h*-parameters are as given in (3-43), find the values of r_b, r_e, r_c, and α_0 of the *T*-model.

3-27 If the *T*-model parameters are $\alpha_0 = 0.98$, $r_e = 25\,\Omega$, $r_b = 200\,\Omega$, and $r_c = 1$ MΩ, find the *h*-parameters of a *CB* stage.

Problems   179

3-28 Find the y-parameters of an FET high-frequency model and determine the circuit values of the n-channel Si FET type 2N3823 (data sheets are given in Appendix D).

3-29 (a) Determine the high-frequency circuit-element values for RCA MOSFET type 40461 (data sheets are given in Appendix D).
(b) Compare the transfer characteristic in (3-63) with the classical result given in footnote 6.

3-30 Consider the circuit shown in fig P3-30. Determine the range of values for A_0 in terms R_0, R_i, and R_f such that the input impedance Z_{in} exhibits a negative resistance value.

3-31 For the circuit given in fig P3-31,
(a) calculate I_C, V_{CE} for a Ge transistor with $V_{BE} = +0.20$ V, $\beta_0 = 50$, and $I_{CO} = 1$ μA at 25°C.
(b) If the temperature is increased to 75°C, use table 3-5a to determine the new operating point.

3-32 For the circuit shown in fig P3-32, a silicon transistor is used. The transistor β variation from unit to unit ranges from 50 to 100, $V_{BE} \approx 0.6$ V and is approximately the same for all units.
(a) Determine the values R_E, R_{b1}, and R_{b2} such that the quiescent current does not change by more than 20 per cent.
(b) Determine the low-frequency ac current signal loss due to the biasing resistors R_{b1} and R_{b2}. Assume $C_E = \infty$.

3-33 Consider the circuit shown in fig P3-33.
(a) If a Si transistor is used with $V_{BE} = 0.7$ V, $h_{FE} = 60$, and $I_{CO} \approx 0$, determine the quiescent point V_{CE}, I_C.
(b) If a Ge transistor is used with $V_{BE} = 0.2$ V, $h_{FE} = 60$, and $I_{CO} \approx 5$ μA, what is V_{CE}, I_C?

3-34 For the circuit shown in fig P3-34, determine the values of R_{b1} and R_{b2} such that a maximum symmetrical collector current swing is obtained. What is the quiescent point under this condition? The transistor has $\beta_0 = 50$, $I_{CO} = 1$ μA, and $V_{BE} \approx 0.3$ V.

3-35 For the MOSFET (enhancement-mode) amplifier stage shown in fig P3-35 the operating point is to be at $I_D = 10$ mA and $V_{DS} = 10$ V ($V_{GS} = +5$ V). At the operating point, the following data are given: $|y_{fs}|$ at low frequencies = 4000 μmhos, $C_{iss} = 12$ pF, $C_{rss} = 2$ pF, and $r_d = 10$ kΩ.
(a) Determine the resistor values R_G and R_S. $R_G = R_{g1} || R_{g2}$.
(b) Show the incremental high-frequency model for the MOSFET and indicate the parameters.
(c) Find the midband voltage gain.

3-36 Design the biasing circuit of fig 3-63 using an npn Ge transistor such that the operating point is in the center of the active region. The supply voltage $V_{CC} = 20$ volts. Given that $40 < \beta_0 < 100$, $I_{CO} \approx 5$ μA, $V_{BE} \simeq$

Fig P3-31

Desired quiescent point
$I_C = 2$ mA; $V_{CE} = 5$ V

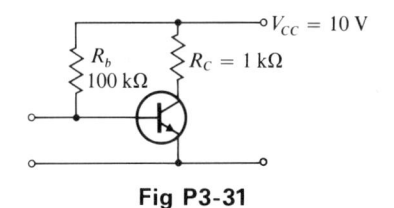

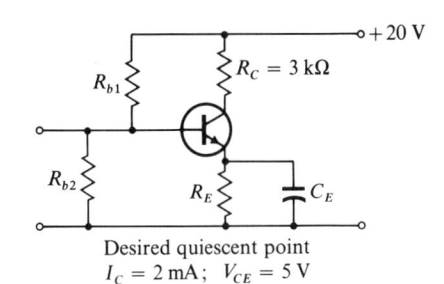

Fig P3-32

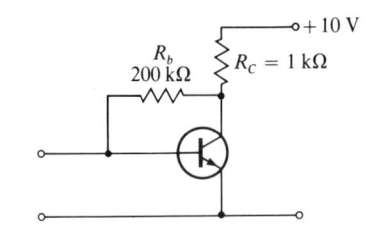

Fig P3-33

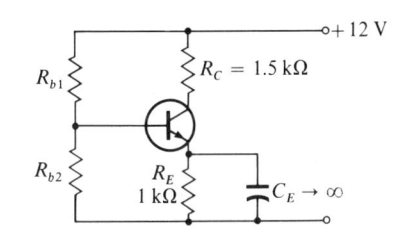

Fig P3-34

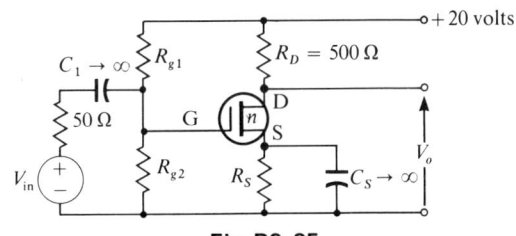

Fig P3-35

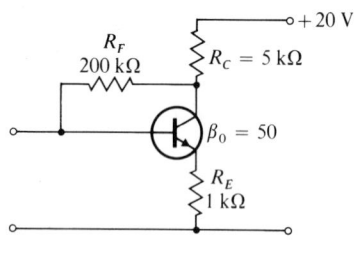

Fig P3-37

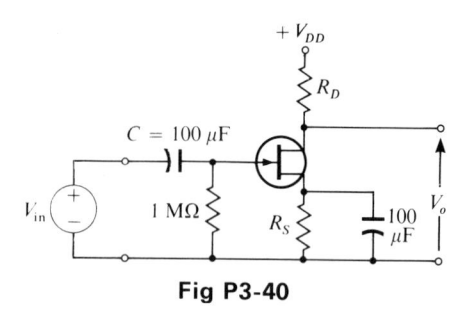

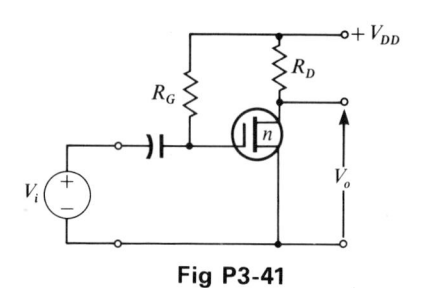

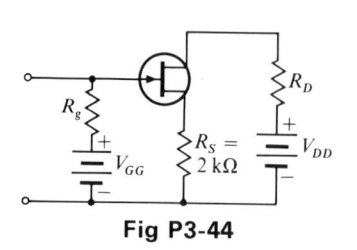

(Circuit: $R_F = 200\text{ k}\Omega$, $R_C = 5\text{ k}\Omega$, $\beta_0 = 50$, $R_E = 1\text{ k}\Omega$, $+20$ V supply)

Fig P3-40

(Circuit: $C = 100\ \mu F$, $1\text{ M}\Omega$, R_D, R_S, $100\ \mu F$, $+V_{DD}$, V_{in}, V_o)

Fig P3-41

(Circuit: R_G, R_D, n-channel MOSFET, $+V_{DD}$, V_i, V_o)

Fig P3-44

(Circuit: R_g, V_{GG}, $R_S = 2\text{ k}\Omega$, R_D, V_{DD})

0.2 V. Use $R_B \simeq 0.1(\beta_0)_{min}R_E$, $R_C = 2R_E$. Note that the center of the active region corresponds to

$$V_{CE} = \frac{(V_C)_{max} + (V_C)_{min}}{2} = \frac{1}{2}\left[V_{CC} + \left(\frac{R_E}{R_E + R_C}\right)V_{CC}\right]$$

3-37 A Si transistor is used in the circuit shown in fig P3-37. Calculate I_{CE} and V_{CE}. Determine the new operating point if another transistor of the same type, but with $\beta_0 = 100$, is used.

3-38 A Ge transistor is used in the circuit shown in fig P3-37.
(a) Derive the expression for S_I.
(b) What is the approximate shift in the operating point if I_{CO} changes from 2 μA to 10 μA, but the other parameters remain constant?

3-39 The I_D–V_{DS} characteristic curve of a Si FET is given in Appendix D. Design the biasing circuit of fig 3-66 such that the quiescent operating point is at $I_D = 4$ mA and $V_{DS} = 12$ V. The available supply voltage $V_{DD} = 20$ V and $R_G = 1$ MΩ.

3-40 The parameters of an FET are $I_{DSS} = 8$ mA, $V_{PO} = 5.0$ V. The FET stage is to be biased as in fig P3-40 at $I_D = 2$ mA, $V_{DS} = 8$ V.
(a) If the supply voltage V_{DD} is 20 V, determine the values of R_D and R_S.
(b) Find the voltage gain V_o/V_{in} at $\omega = 10^4$ rad/sec if $g_d = 3 \times 10^{-4}$ mhos, $g_m = 4 \times 10^{-3}$ mhos at the operating point.

3-41 The MOSFET of Problem 3-39 is to be biased in the enhancement mode. The desired operating point is $I_D = 12$ mA, $V_{DS} = 8$ V. The circuit is shown in fig P3-41. Determine the value of R_D and R_G if $V_{DD} = 20$ V. What is the voltage gain of the circuit at low frequencies?

3-42 The V–I characteristic of a triode is shown in fig 3-55. Determine the biasing circuit-element values corresponding to fig 3-67 such that the quiescent point is at $I_p = 5$ mA, $V_G = -2.0$ V. The available supply voltage is $E_{BB} = 300$ V. Let $R_G = 100$ kΩ and ignore the capacitors.

3-43 The V–I characteristics of a pentode are given in fig 3-58. Show the complete biasing circuit and the element values such that the operating point is at $I_p = 4$ mA, $V_P = 150$ V. Bias supply $V_{BB} = 300$ V.

3-44 The biasing circuit shown in fig 3-68 is to be used for an FET amplifier. The circuit shown in fig P3-44 is a Thévenin equivalent circuit for R_{g1}, R_{g2}, and V_{DD}. The desired operating point is $I_D = 2$ mA, $V_{DS} = 6$ V. The FET transistor parameters are $I_{DSS} = 8$ mA, $V_{PO} = 5.0$ V. The supply voltage is $V_{DD} = 20$ V. For operating-point stability the value of R_S is chosen to be 2 kΩ.
(a) Determine the value of R_D.
(b) If $R_g = 1$ MΩ, determine the values of R_{g1}, R_{g2}, and V_{GG}.
(c) Find the new operating point if the FET is replaced by another FET with $I_{DSS} = 16$ mA, $V_{PO} = 7$ V.

3-45 For the circuit shown in fig P3-45, the operating point is desired to be $V_{CE} = 5\,V$, $I_C = 10\,mA$. The following data are given about the transitor at this operating point:

$$h_{FE} = 100, \quad C_c = 5\,pF, \quad \omega_T = 4 \times 10^9\,rad/sec$$

at low frequencies $h_{ie} = 300$, $h_{oe} = 50\,\mu mhos$

$$V_{BE} = 0.2\,V, \quad I_{CO} = 1\,\mu A$$

(a) Determine the values of R_B and R_C.
(b) Determine the high-frequency element values of the hybrid π model.
(c) Determine the voltage gain V_o/V_{in} at $\omega = 10^4$ rad/sec.

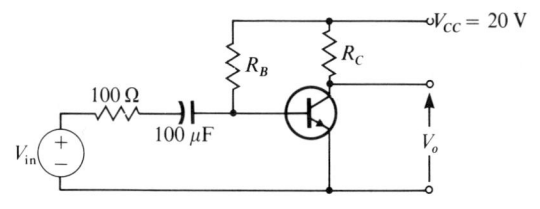

Fig P3-45

3-46 A transitor, operated in the ω_β region, in the circuit shown in fig P3-46 is sometimes used to simulate an inductor in an integrated circuit. Show that the input impedance is inductive and find the equivalent inductance and resistance. Given: $\omega_\alpha = 10^9$ rad/sec, $\alpha_0 = 0.98$, $I_E = 10\,mA$, $C_c = 2\,pF$, $r_b + R_b = 1\,k\Omega$. [Hint: Use the T-model with α given by (3-46), and note that $\omega_\beta \ll \omega_\alpha$.]

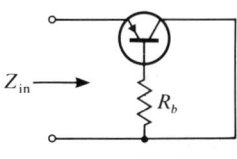

Fig P3-46

4
Analysis and Design of Single-Stage Amplifiers

Amplification is one of the most important functions in electronic circuits. In this chapter we will discuss low-pass small-signal amplification in a single-stage amplifier. Though amplifiers invariably comprise more than one stage, we will specifically consider a single-stage amplifier in this chapter to focus our attention on the salient characteristics of its performance; particularly the gain, impedance, frequency response, and transient response, without concerning ourselves with interstage interactions and other problems which arise in a multistage amplifier. The analysis and design of cascaded and multistage amplifiers is discussed in Chapter 5.

We will discuss the bipolar junction transistor (BJT) in some detail because of the prominent role it plays in wide-band amplifiers. Field-effect transistors (FET's) are also becoming more and more important in discrete and integrated circuits. In fact, the combination of FET's and BJT's provides considerable versatility to the designer. FET's and vacuum tubes are voltage-amplifying devices. They have the same circuit models and their mathematical treatment is identical; hence

4-1 Low-Pass Amplifier Frequency Characteristics

we need discuss only one device in detail. Certain approximations used for BJT's can also be used for FET's and vacuum tubes. In the following discussion, the frequency characteristics of the various devices and their midband properties in various configurations, including the gain and impedance relations, are briefly analyzed. The concept of dominant poles and zeros and the usefulness of the Miller approximation in determining the dominant natural frequency of the circuit is described, together with its limitations. Frequency response and transient response calculations and their interrelationships are then examined.

4-1 Low-Pass Amplifier Frequency Characteristics

The normalized frequency response of a typical low-pass ac amplifier is shown in fig 4-1b. This response represents the frequency characteristics of many of the devices considered in Chapter 3. In other words, although shown for a BJT, the frequency response could represent the response of a field-effect transistor, or a triode or pentode vacuum tube. The normalization scale, of course, depends upon the device parameters and the external circuit elements. The complete amplifier frequency response in fig 4-1b can be obtained from the circuit of fig 4-1a by replacing the device with its small-signal equivalent circuit and then calculating the magnitude response of the desired gain function. This calculation, if done by paper and pencil methods (i.e., noncomputer methods), is very involved and is not recommended. A simple approximation method for obtaining the frequency response of an amplifier is to consider the frequency response at various frequency ranges, namely the low-frequency, midband, and high-frequency ranges as discussed in the following. In fig 4-1, A corresponds to either voltage gain or current gain, ω_h is the high-frequency cutoff or upper 3-dB radian frequency, and ω_l is the low-frequency cutoff or lower 3-dB radian frequency. These frequencies are also called *half-power* frequencies. In other words, at $\omega = \omega_h$ and $\omega = \omega_l$, the power gain is lower than at its midband value by a factor of one half or, alternatively stated, the voltage gain or current gain is lower than its midband value by a factor of $1/\sqrt{2}$. Note that, since, for a resistive load power $P \alpha V^2$ or I^2 we have $10 \log 1/2 = 20 \log 1/\sqrt{2} = 3$ dB. Usually ω_h is much larger than ω_l by several orders of magnitude. For example, in a video amplifier ω_h may be equal to $2\pi (10^7 \text{ Hz})$ while ω_l may be $2\pi (10^2 \text{ Hz})$. The amplifier bandwidth $f_{3\text{ dB}}$ or $\omega_{3\text{ dB}} = 2\pi(f_{3\text{ dB}})$ is defined as

$$f_{3\text{ dB}} = f_h - f_l \approx f_h \quad \text{or} \quad \omega_{3\text{ dB}} = \omega_h - \omega_l \simeq \omega_h \quad (4\text{-}1)$$

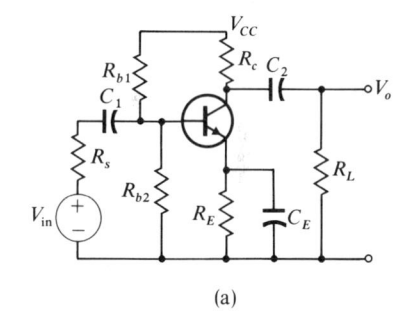

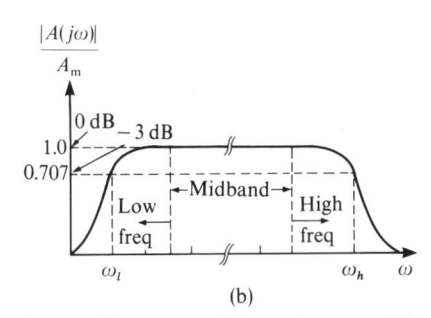

Fig 4-1 (a) Single-stage *CE* transistor amplifier; (b) Typical frequency response of the magnitude of the gain function

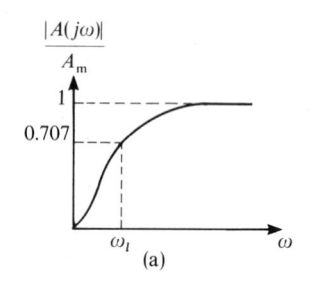

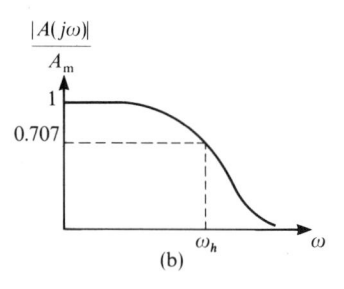

Fig 4-2 (a) Highpass frequency response; (b) Lowpass frequency response

In view of the fact that the upper-edge and lower-edge frequency responses are separated by a wide range of frequencies (at which frequencies the gain is constant) considerable simplification results. We can break up the frequency response into three regions; the low-frequency, the midband, and the high-frequency regions, as indicated in fig 4-1b. The loss of gain in the low-frequency response is due to the coupling and bypass capacitors of the biasing circuit. It has the high-pass (i.e., high frequencies are not attenuated) response characteristics shown in fig 4-2a. Note that because of capacitors C_1 and C_2 no signal is transmitted at dc, while at high frequencies the relatively large capacitances are essentially short-circuited. In a design, the appropriate value of ω_l is obtained by the proper selection of values for these capacitors, as discussed in Section 4-4. The high-frequency response and the attenuation are a result of the interelectrode capacitances and other capacitances of the devices, as depicted in their high-frequency models. This response has a low-pass (i.e., low frequencies are not attenuated) characteristic, as shown in fig 4-2b. Since the capacitance values of the high-frequency models are not entirely under our control for a given device, we cannot do much to improve the high-frequency cutoff without paying for it in the midband gain. We will discuss this problem in some detail in Sections 4-7 through 4-14. In the midband region, the frequencies are high enough so that the coupling and bypass capacitors are essentially short-circuited; and low enough so that the device capacitances can be neglected. The devices can thus be represented by their low-frequency models. In other words, at midband frequencies we have no reactive elements in the circuit, so that the gain is independent of the frequency. We shall now consider the midband gain and impedance relations in transistors in order to acquire insight into the operation of single-stage devices.

4-2 Midband Gain and Impedance Relations in Bipolar Junction Transistors

The bipolar junction transistor is a three-terminal device that can be used for signal amplification in three possible configurations, the common-emitter (CE), common base (CB), and common collector (CC) connections. Which configuration to use depends upon the desired property in a specific circuit application. An examination of the midband properties of the BJT in the three configurations will help us in making a choice.

4-2 Midband Gain and Impedance in BJT's

To begin the discussion in general terms, we consider the transistor as a two-port device, wherein a signal is fed in at an input port having a source resistance R_s and is subsequently transmitted from the output port to a load resistance R_L. This arrangement, regardless of which terminal of the transistor is common to the input and output ports, is shown in fig 4-3. The two-port formalism discussed in Chapter 2, which is applicable to any linear network, may now be used to obtain the various expressions for the network properties of the device. Any set of parameters may be used, though in certain cases a particular choice may be more convenient because the parameters may be easily measured or may be obtained directly from the manufacturer's data sheets. For example, let the transistor in fig 4-3 be characterized by the h-parameters. We then have the following matrix relation:

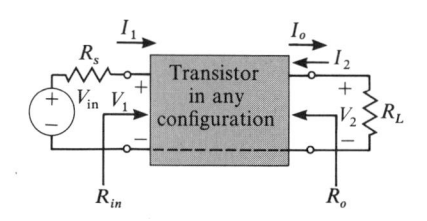

Fig 4-3 Transistor amplifier stage used as an active two-port circuit

$$\begin{bmatrix} V_1 \\ I_2 \end{bmatrix} = \begin{bmatrix} h_i & h_r \\ h_f & h_o \end{bmatrix} \begin{bmatrix} I_1 \\ V_2 \end{bmatrix} \qquad (4\text{-}2)$$

where the additional subscript e, b, or c may be used if the common terminal is known; e.g., if the device is in the CE configuration, we would use h_{ie}, h_{re}, etc.

From (4-2) and the relations for the source and load circuits

$$-I_2 = G_L V_2 \qquad V_1 = V_{in} - I_1 R_s \qquad (4\text{-}3)$$

we can obtain the network properties of the two-port as discussed in Section 2-2. Hence, the signal transmission characteristics from a source to a load for the device in a given connection may be determined.

EXAMPLE

Let us assume that the transistor in fig 4-3 is in the CE connection. Suppose that at $I_C = 1$ mA, $V_{CE} = 5$ V, the h-parameters are

$$\begin{aligned} h_{ie} &= 1.3 \times 10^3 \, \Omega & h_{fe} &= 50 \\ h_{re} &= 10^{-4} & h_{oe} &= 2 \times 10^{-5} \text{ mhos} \end{aligned} \qquad (4\text{-}4)$$

Let the source and load resistances be $R_L = R_S = 10^3 \, \Omega$. We shall determine the input impedance R_{in}, output impedance R_o, current gain A_i, voltage gain A_v, and the power gain G_P of the circuit.

From the results derived in Chapter 2 and summarized in table 2-2, we have

$$R_{in} = h_{ie} - \frac{h_{re}h_{fe}}{h_{oe} + G_L} = 1.3 \times 10^3 - \frac{5 \times 10^{-3}}{2 \times 10^{-5} + 10^{-3}} \simeq 1.3 \, k\Omega$$

$$R_o^{-1} = h_{oe} - \frac{h_{re}h_{fe}}{h_{ie} + R_s} = 2 \times 10^{-5} - \frac{5 \times 10^{-3}}{1.3 \times 10^3 + 10^3}$$

$$\simeq 1.8 \times 10^{-5} \, \text{mho}$$

$$\frac{I_0}{I_1} = -\frac{I_2}{I_1} = -\frac{h_{fe}}{1 + h_{oe}R_L} = -\frac{50}{1 + 2 \times 10^{-2}} \approx -49$$

Hence

$$A_i \equiv \frac{I_o}{I_s} = \left(\frac{I_o}{I_1}\right)\left(\frac{R_s}{R_s + R_{in}}\right) = (-49)\left(\frac{1}{1 + 1.3}\right) = -21.3$$

Similarly,

$$\frac{V_2}{V_1} = \frac{h_{fe}}{h_{re}h_{fe} - h_{ie}(G_L + h_{oe})}$$

$$= \frac{50}{5 \times 10^{-3} - 1.3 \times 10^3(10^{-3} + 2 \times 10^{-5})} \simeq -37.8$$

and the voltage gain is

$$A_v \equiv \frac{V_2}{V_{in}} = \frac{V_2}{V_1}\left(\frac{R_{in}}{R_{in} + R_s}\right) = (-37.8)\left(\frac{1.3}{1.3 + 1}\right) = -21.4$$

Finally, the power gain, since all parameters are real, is given by

$$G_P = -\frac{V_2 I_2}{V_1 I_1} = \left(\frac{V_2}{V_1}\right)^2 \frac{G_L}{G_{in}} = (37.8)^2(10^{-3})(1.3 \times 10^3) = 1.85 \times 10^3$$

$$= 32.6 \, \text{dB}$$

Note that, in the above, once $R_{in}(=G_{in}^{-1})$ and I_0/I_1 are determined, we can also obtain

$$G_P = \left(\frac{I_o}{I_1}\right)\left(\frac{V_2}{V_1}\right) = \left(\frac{I_o}{I_1}\right)^2 \frac{R_L}{R_{in}} = (-49)^2\left(\frac{1}{1.3}\right) = 1.85 \times 10^3$$

We now examine the midband properties of the various connections in terms of the transistor parameters in the various configurations. To do so, we must use an appropriate model for the device. For convenience of comparison we shall use the Tee model for the transistor shown in fig 4-4, and assume that the biasing resistors have negligible loading effects on the stage.

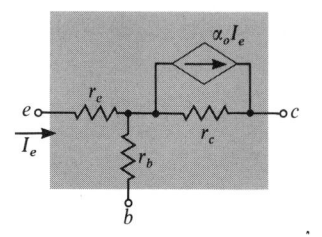

Fig 4-4 Low-frequency small-signal model for the bipolar transistor

The Common Base Configuration

The transistor stage in the *CB* configuration is shown with the biasing circuit in fig 4-5a. Capacitor *C* is a very large bypass capacitor and its impedance at midband is essentially zero. Since the large value of the capacitor is of no concern in midband calculations, we label it $C \to \infty$ to indicate a short circuit at midband frequency. The small-signal midband equivalent circuit is shown in fig 4-5b.

In order to find the network properties of the circuit, several approaches may be taken. We may (1) find the *h*-parameters of the *CB* stage directly or in terms of the *CE* parameters[1] and then use the previous formulas of this section to determine the midband properties; (2) read directly the *z*-parameters (see fig 2-7 and change the dependent current source of fig 4-5b to a dependent voltage source) and then use the formulas of table 2-2 in terms of the *z*-parameters; or (3) use any other parameter characterization we wish to choose. By method (2):

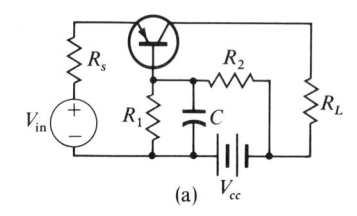

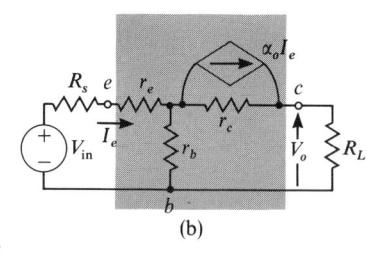

$$z_{12} = r_b \qquad z_{11} = r_e + r_b$$
$$z_{22} = r_c + r_b \qquad z_{21} = r_b + \alpha_0 r_c \qquad (4\text{-}5)$$

Fig 4-5 (a) Single-stage *CB* transistor amplifier (capacitor *C* is very large and is essentially short-circuited at midband and higher frequencies); (b) Equivalent circuit of (a)

A knowledge of the typical parameters is of great help in making suitable approximations. Typical parameters for the junction transistor at room temperature are as follows:

$$r_e \approx \frac{25}{I_e(\text{mA})} \text{(e.g., at } I_e = 1 \text{ mA}, r_e = 25 \, \Omega\text{)}, g_m = \frac{1}{r_e}$$

$$\alpha_0 \geq 0.97, \frac{1}{1-\alpha_0} \approx \beta_0 = 30 \text{ to } 150$$

$$r_c = 1 \text{ to } 10 \text{ M}\Omega, r_b = 50 \text{ to } 200 \, \Omega$$

---

[1] The results (see Problem 4-5): $h_{fb} = -h_{fe}/(1 + h_{fe}) \simeq -1$ (for $h_{fe} \gg 1$), $h_{rb} \simeq 0$, $h_{ib} = h_{ie}/(1 + h_{fe})$, and $h_{ob} = h_{oe}/(1 + h_{fe})$.

4/Analysis and Design of Single-Stage Amplifiers

Thus, with no significant error we can use the approximation

$$(1 - \alpha_0)r_c \gg r_e, r_b; \quad \beta_0 \gg 1.$$

The various network properties of the CB stage are then determined directly from (4-5) and the results of table 2-2:

$$Z_{in} = R_{in} = z_{11} - \frac{z_{12}z_{21}}{z_{22} + Z_L} = r_e + r_b \frac{r_c(1 - \alpha_0) + R_L}{r_b + r_c + R_L} \quad (4\text{-}6a)$$

$$\approx r_e + r_b(1 - \alpha_0) \qquad (1 - \alpha_0)r_c \gg R_L \quad (4\text{-}6b)$$

$$Z_o = R_o = z_{22} - \frac{z_{12}z_{21}}{z_{11} + R_S} = r_c - r_b \frac{\alpha_0 r_c - r_e - R_S}{r_e + r_b + R_S} \quad (4\text{-}7a)$$

$$\approx r_c \qquad r_c \gg R_S \gg r_b \quad (4\text{-}7b)$$

$$\frac{I_0}{I_1} = \frac{z_{21}}{z_{22} + Z_L} = \frac{\alpha_0 r_c + r_b}{r_c + r_b + R_L} \quad (4\text{-}8a)$$

$$\approx \alpha_0 \qquad R_L \ll r_c \quad (4\text{-}8b)$$

$$\frac{V_0}{V_1} = \frac{z_{21} Z_L}{-z_{12}z_{21} + z_{11}(Z_L + z_{21})}$$

$$= \frac{R_L(r_b + \alpha_0 r_c)}{r_b[(1 - \alpha_0)r_c + r_e + R_L] + r_e(r_c + R_L)} \quad (4\text{-}9a)$$

$$\approx \frac{\alpha_0 R_L}{r_e + r_b(1 - \alpha_0)} \qquad R_L \ll r_c \quad (4\text{-}9b)$$

It can be seen from (4-8) and (4-9) that the current gain (i.e., the ratio of the output current to the input current for a current-source input) of a common-base stage is almost unity, and that the voltage gain (i.e., the ratio of the output voltage to the input voltage for a voltage-source input) is approximately equal to $g_m R_L$. Note that for finite values of R_s, the voltage gain is

$$A_v \equiv \frac{V_{out}}{V_{in}} = \left(\frac{V_0}{V_1}\right)\left(\frac{V_1}{V_{in}}\right) = \left(\frac{V_0}{V_1}\right)\left(\frac{R_{in}}{R_{in} + R_s}\right)$$

and the current gain is

$$A_i \equiv \frac{I_o}{I_s} = \left(\frac{I_o}{I_1}\right)\left(\frac{I_1}{I_s}\right) = \left(\frac{I_o}{I_1}\right)\left(\frac{R_s}{R_s + R_{in}}\right)$$

The results in the form of (4-8) and (4-9) are, however, useful for multistage calculations, as shown in Section 5-1. The input impedance is very low and the output impedance is very high. As a result of this tremendous mismatch between the input and the output circuits, the voltage gain and the current gain of cascaded CB stages are less than unity and no power gain can be obtained unless a transformer is used between the stages. This fact is readily seen by setting $R_L = R_{in}$ and then finding A_i and A_v, which will both be approximately equal to α_0 and hence $G_P \simeq \alpha_0^2 < 1$.

The CB stage is used to change the impedance levels in a circuit and to provide isolation for CE stages in tuned amplifiers. It is also used as a noninverting amplifier with a voltage gain. Sometimes it is used as a current-controlled current source element. The comparison of the midband properties of a typical CB stage with those of the other configurations is shown in table 4-1.

TABLE 4-1 *Comparison of Network Properties of Transistor Configurations for a Typical Transistor*[a]

	CB	CE	CC
R_{in} (ohms)	29	1450	50,200
R_o (ohms)	1 MΩ	4500	49
$\dfrac{I_o}{I_1}$	0.98	−49	−50
$\dfrac{V_o}{V_1}$	33.8	−33.8	1
$G_P = \dfrac{V_o I_o}{V_1 I_1}$	33.1	1650	50
$A_v = \dfrac{V_o}{V_s} = A_i$ (since $R_S = R_L$)	0.95	20	0.98

[a] $\alpha_0 = 0.98$, $r_e = 25\,\Omega$, $r_b = 200\,\Omega$, $r_c = 1\,\text{M}\Omega$, $R_L = R_S = 1\,\text{k}\Omega$.

The Common Emitter Configuration

The transistor stage in the CE configuration and its midband small-signal Tee model are shown in fig 4-6. Usually the h-parameters of the CE stage are given and the midband properties are determined as described earlier in this section. However, the h-parameters in terms of the Tee model parameters can be obtained by straightforward

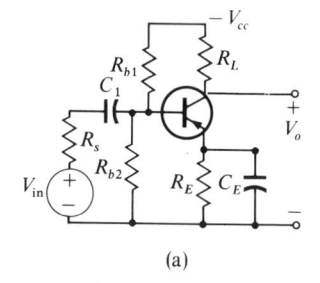

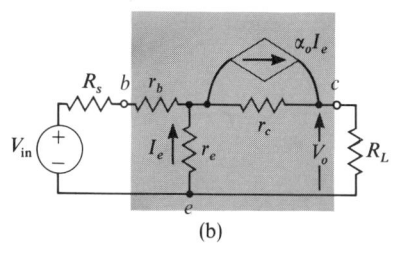

Fig 4-6 (a) Single-stage CE transistor amplifier (capacitors C_1 and C_E are very large and are essentially short-circuited at midband and higher frequencies); (b) Equivalent circuit of (a)

analysis (see Problem 4-4). The *h*-parameters are

$$h_{11} \approx r_b + \frac{r_e}{1 - \alpha_0} \qquad h_{12} \simeq \frac{r_e}{r_c(1 - \alpha_0)}$$

$$h_{21} \simeq \frac{\alpha_0}{1 - \alpha_0} \qquad h_{22} \approx \frac{1}{r_c(1 - \alpha_0)}$$

(4-10)

The various network properties of the *CE* stage are then determined directly from (4-10) and table 2-2:

$$Z_{in} = R_{in} = r_b + r_e \frac{r_c + R_L}{r_c(1 - \alpha_0) + R_L + r_e} \qquad (4\text{-}11\text{a})$$

$$\simeq r_b + \frac{r_e}{1 - \alpha_0} \qquad R_L \ll r_c(1 - \alpha_0) \qquad (4\text{-}11\text{b})$$

$$Z_o = R_o = r_c(1 - \alpha_0) + r_e \frac{R_S + r_b + \alpha_0 r_c}{R_S + r_b + r_e} \qquad (4\text{-}12\text{a})$$

$$\approx r_c \left[(1 - \alpha_0) + \frac{r_e}{R_S} \right] \qquad r_b \ll R_S \ll r_c \qquad (4\text{-}12\text{b})$$

$$\frac{I_0}{I_1} = -\frac{\alpha_0 r_c - r_e}{r_c(1 - \alpha_0) + r_e + R_L} \qquad (4\text{-}13\text{a})$$

$$\approx -\frac{\alpha_0}{1 - \alpha_0} = -\beta_0 \qquad (4\text{-}13\text{b})$$

$$\frac{V_0}{V_1} = -\frac{R_L(\alpha_0 r_c - r_e)}{r_b[r_c(1 - \alpha_0) + r_e + R_L] + r_e(r_c + R_L)} \qquad (4\text{-}14\text{a})$$

$$\simeq -\frac{\alpha_0 R_L}{r_e + r_b(1 - \alpha_0)} \qquad r_c(1 - \alpha_0) \gg R_L \qquad (4\text{-}14\text{b})$$

From the above expressions it is seen that for the *CE* configuration, the current gain is $-\beta_0$ for a current-source input, and the voltage gain is roughly $-g_m R_L$ for a voltage-source input. Note that, in this configuration, both current and voltage gains of greater than unity can be achieved. The input impedance is medium and is essentially independent of the load impedance, if R_L is not too high. The output impedance is high ($\approx r_c/\beta_0$) and fairly independent of the source impedance. The mismatch of the input and output impedances of cascaded *CE* stages is the least of the three configurations (see table 4-1), and power amplification is maintained in this configuration even if

The Common Collector Configuration

The transistor stage in the CC configuration and its midband small-signal Tee model are shown in fig 4-7. This connection is commonly called the emitter follower.

The various network properties of this circuit are obtained in exactly the same manner as before, that is, we find the h-parameters in terms of the Tee model (see Problem 4-2) and then use the previously obtained results. The expressions are

$$R_{in} = r_b + r_c \frac{R_L + r_e}{(1 - \alpha_0)r_c + R_L + r_e} \qquad (4\text{-}15a)$$

$$\approx r_b + \frac{R_L + r_e}{1 - \alpha_0} \approx \beta_0(R_L + r_e) \qquad (1 - \alpha_0)r_c \gg R_L \qquad (4\text{-}15b)$$

$$R_O = r_e + \frac{(1 - \alpha_0)r_c(R_S + r_b)}{R_S + r_b + r_c} \qquad (4\text{-}16a)$$

$$\approx r_e + (R_S + r_b)(1 - \alpha_0) \qquad r_c \gg R_S \qquad (4\text{-}16b)$$

$$\frac{I_0}{I_1} = \frac{r_c}{r_c(1 - \alpha_0) + R_L + r_e} \qquad (4\text{-}17a)$$

$$\simeq \frac{1}{1 - \alpha_0} \approx \beta_0 \qquad r_c(1 - \alpha_0) \gg R_L \qquad (4\text{-}17b)$$

$$\frac{V_0}{V_1} = \frac{r_c R_L}{r_b[(1 - \alpha_0)r_c + R_L + r_e] + r_c(R_L + r_e)} \qquad (4\text{-}18a)$$

$$\approx \frac{R_L}{R_L + r_e} \approx 1 \qquad R_L \gg r_e, r_b(1 - \alpha_0) \qquad (4\text{-}18b)$$

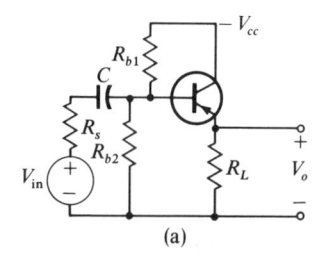

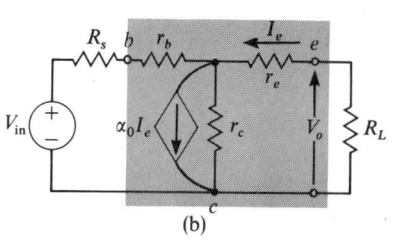

Fig 4-7 (a) Single-stage CC transistor amplifier (capacitor C is very large and is essentially short-circuited at midband and higher frequencies); (b) Equivalent circuit of (a)

From the above expressions, it is seen that for the CC configuration the current gain, for a current-source input, is almost equal to β_0 and the voltage gain, for a voltage-source input, is almost unity. For the CC configuration, the input resistance is the highest ($\approx \beta_0 R_L$) and the output resistance the lowest ($\approx R_S/\beta_0$) among the three configurations. Because of these properties, the CC configuration is used as a buffer stage; since the output of a *CC* stage is very low, it can also be used to drive a variable-load impedance. The midband properties of a *CC* configuration for a typical transistor as compared with the other configurations are summarized in table 4-1.

4-3 Midband Gain and Impedance Relations in Field-Effect Transistors

Since the models of FET's and vacuum tubes are the same, as shown in fig 4-8, the results would naturally apply to vacuum tubes. We shall discuss only the FET stage to avoid duplication. Only the symbols need be changed for vacuum tubes.

The method of finding the midband properties are the same as in the previous section. Since the model of the FET is simpler, however, considerable simplification in the analysis results. Consider, for example, the FET stage in the common-source (*CS*) connection (this corresponds to a common-cathode connection in vacuum tubes, as shown in fig 4-9).

The input impedance is, by inspection, found to be infinite. Because of this, no current can flow at the input and thus input signal must be a voltage source at the input, i.e., $R_S \approx 0$. This is the reason why FET's and vacuum tubes are referred to as voltage-amplifying devices and we do not speak of a current gain.

The output impedance, by inspection, is

$$R_o = g_d^{-1} = r_d \qquad (4\text{-}19)$$

The voltage gain is given by[2]

$$A_v = \frac{V_o}{V_i} = -g_m(r_d \| R_L) = -\frac{g_m R_L r_d}{R_L + r_d} \qquad (4\text{-}20a)$$

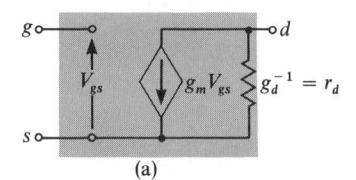

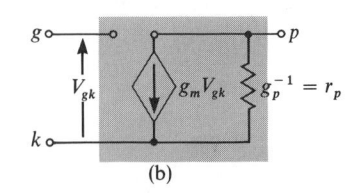

Fig 4-8 (a) Low-frequency small-signal model for an *FET*; (b) Low-frequency small-signal model for vacuum tubes

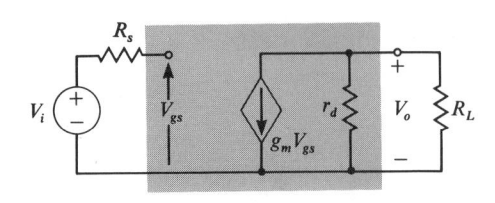

Fig 4-9 Common-source FET single-stage amplifier at midband frequencies

[2] The ∥ sign designates "in parallel with."

$$\approx -g_m R_L \qquad R_L \ll r_d \qquad (4\text{-}20b)$$

In a similar manner, we can obtain the network properties of the common-gate (*CG*) and the common-drain (*CD*) configurations. These properties are summarized in table 4-2.

TABLE 4-2 *Gain and Impedance Relations for Single-Stage FET's (or Vacuum Tube[a])*

FET (vacuum tube)	Common Source (grounded cathode)	Common Gate (grounded grid)	Common Drain (grounded plate)
R_{in}	∞	$\dfrac{r_d + R_L}{\mu + 1} \approx \dfrac{1}{g_m}$	∞
R_o	r_d	$r_d + (\mu + 1)R_S \approx r_d + \mu R_S$	$\dfrac{r_d}{\mu + 1} \approx \dfrac{1}{g_m}$
A_v	$-\dfrac{\mu R_L}{r_d + R_L} \approx -g_m R_L$	$\dfrac{(\mu+1) R_L}{r_d + R_L} \approx g_m R_L$	$\dfrac{\mu R_L}{(\mu+1)R_L + r_d}$ $\approx \dfrac{g_m R_L}{1 + g_m R_L}$

[a] $\mu = g_m/g_d$. For tubes, use g_p instead of g_d, where $g_p = 1/r_p$.

Typical values for different types of FET's and vacuum tubes are given in Chapter 3. Typical FET parameters are $g_m = 10^{-3}$ mho and $r_d = 50 \text{ k}\Omega$. Notice that an FET in the *CD* connection (also referred to as the source follower) has a voltage gain of less than unity, and the lowest output impedance as compared with the other configurations. This configuration is analogous to the emitter follower in BJT circuits. Similarly, the *CS* (or grounded cathode) is roughly analogous to the *CE* configuration, while the *CG* (or grounded grid) is roughly analogous to the *CB* configuration. The analogies are drawn from the voltage gain, polarity inversion property, and impedance characteristic of the devices. As in all analogies, however, we should be careful of their usage. For example, the input impedance of a *CE* stage is relatively medium, while that of a grounded-cathode stage is infinite. Further, because of the availability of *pnp* and *npn* transistors, certain circuit configurations and designs are possible which have no counterparts in vacuum-tube circuitry (see fig 5-53).

High-input-impedance devices should preferably be driven by a voltage source and low-impedance devices by a current source. This is because a practical voltage source has a low series source resistance.

If the amplifier input impedance is low, a considerable portion of the source signal is lost across the source resistance. On the other hand, a practical current source has a high shunt source resistance. If the input impedance of the amplifier is high, a fair amount of the source signal is lost across the source resistance. Thus, the natural input parameter for very low-input-impedance devices is a current source, while for very large-input-impedance devices it is a voltage source. This is the reason why a CB is almost always driven by a current source, and FET's and vacuum tubes are always driven by voltage sources. It should be remembered, however, that if maximum power transfer is to be attained, the input impedance of the device should match the source impedance.

4-4 Low-Frequency Response Characteristics Due to Bias and Coupling Circuitry in BJT's

Consider the single-stage CE amplifier circuit shown in fig 4-10a. We are now interested in determining the low-frequency response of the single-stage amplifier.[3] For low and midband frequencies, we use the low-frequency model of the transistor, since at these frequencies the effects of the device capacitances are insignificant. The complete low-midband frequency equivalent circuit is shown in fig 4-10b.

We can now determine the current-gain function $A_i(s)$ for fig 4-10b, by either nodal or loop analysis. The equilibrium equations of the circuit by nodal analysis in the transformed frequency domain are

$$I_{in} = V_1 G_1 + (V_1 - V_2)sC_1 \tag{4-21a}$$

$$(V_1 - V_2)sC_1 = V_2 G_b + (V_2 - V_3)\left(\frac{1}{r_b + R_i}\right) \tag{4-21b}$$

$$(V_2 - V_3)\left(\frac{1}{r_b + R_i}\right) + g_m V' = V_3(G_E + sC_E) \tag{4-21c}$$

$$g_m V' + V_4 G_C + V_4\left(\frac{sC_2}{R_2 C_2 s + 1}\right) = 0 \tag{4-21d}$$

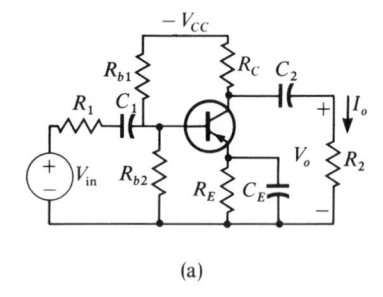

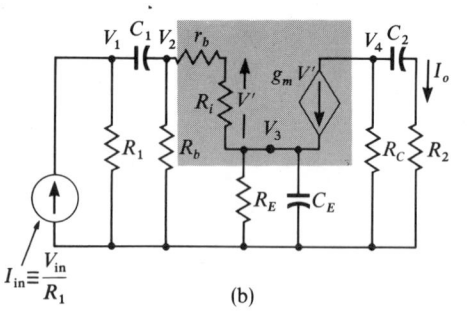

Fig 4-10 (a) Single-stage CE amplifier; (b) Low-frequency model of (a) ($R_b = R_{b1} \| R_{b2}$)

---
[3] In a computer-aided analysis, the entire frequency response is calculated without subdividing the frequency response into low-, mid-, and high-frequency responses.

4-4 Low-Frequency Response Characteristics in BJT's

Note that

$$V' = R_i \left[(V_2 - V_3) \left(\frac{1}{r_b + R_i} \right) \right] \quad \text{and} \quad I_o = V_4 \left(\frac{sC_2}{R_2 C_2 s + 1} \right)$$

The solution of (4-21) after some manipulation is

$$A_i(s) = \frac{I_o}{I_{in}} = \frac{-a_m s^2 (s + 1/R_E C_E)}{[s + 1/(R_2 + R_C)C_2](s^2 + a_2 s + a_1)} \quad (4\text{-}22)$$

where

$$a_2 = \frac{\dfrac{1}{C_1}(R_b + r_b + R_i) + \dfrac{R_b + R_1}{R_E C_E}(R_i + \beta_0 R_E + r_b) + \dfrac{R_1 R_b}{R_E C_E}}{R_b R_1 + (R_b + R_1)(r_b + R_i)} \quad (4\text{-}23\text{a})$$

$$\approx \frac{1}{C_1}\left(\frac{1}{R_1 + r_b + R_i}\right) + \frac{1}{C_E}\left(\frac{\beta_0}{R_1 + r_b + R_i}\right) \qquad R_b \gg R_i, R_1$$

$$a_1 = \frac{R_b + R_i + \beta_0 R_E + r_b}{R_E C_E C_1 [R_b R_1 + (R_b + R_1)(r_b + R_i)]} \quad (4\text{-}23\text{b})$$

$$= \frac{R_b + \beta_0 R_E}{R_b R_E (R_1 + r_b + R_i) C_E C_1} \qquad R_b \gg R_i, R_1$$

$$a_m = \beta_0 \left(\frac{R_1 \| R_b}{(R_1 \| R_b) + r_b + R_i}\right)\left(\frac{R_C}{R_C + R_2}\right) \quad (4\text{-}23\text{c})$$

$$\approx \beta_0 \frac{R_1}{R_1 + R_i + r_b} \qquad R_b \gg R_1, \quad R_C \gg R_2$$

Note that in (4-23) we have used the approximation $\beta_0 \gg 1$, $\beta_0 R_E \gg r_b$.

Note that the gain function in (4-22) has a high-pass characteristic, and that as $s \to \infty$, the magnitude of the current gain is equal to the midband value $-a_m$. From (4-22) it is also seen that the output coupling circuit (C_2) has no interaction with the emitter bypass capacitance (C_E) and the input coupling circuit (C_1), because R_2, C_2 and R_C do not appear anywhere in the other terms. The emitter bypass circuit and the input coupling circuit exhibit strong interactions as seen by their appearance in both terms a_1 and a_2. A numerical example will illustrate this point.

EXAMPLE

In the circuit of fig 4-10, let the parameters be as follows:

$$R_C = 2 \text{ k}\Omega, R_b = 6 \text{ k}\Omega, R_E = 2.7 \text{ k}\Omega, R_1 = R_2 = 1 \text{ k}\Omega$$
$$C_1 = 2 \text{ }\mu\text{F}, C_2 = 10 \text{ }\mu\text{F}, \text{ and } C_E = 50 \text{ }\mu\text{F}$$
$$\beta_0 = 50, g_m = 0.2 \text{ mho}, r_b = 100 \text{ }\Omega$$

Substitution of the numerical values in (4-22) yields

$$A_i(s) = -23.7 \frac{s^2(s + 7.4)}{(s + 33)(s + 1.14 \times 10^3)(s + 55)} \quad (4\text{-}24)$$

The pole-zero plot and the asymptotic plots are shown in fig 4-11.

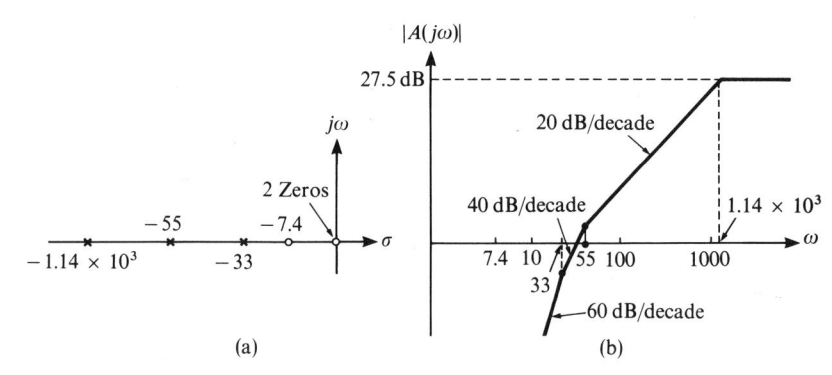

Fig 4-11 (a) Pole-zero plot of the gain function (not to scale); (b) Frequency response of the gain function

The low-frequency cutoff point of this example is approximately equal to the farthest left pole, namely $[(1.14 \times 10^3)/2\pi]$Hz.

Note that if we assume no interaction between the emitter bypass and input coupling circuits, the approximate location of the poles will be in considerable error. For example, if we assume C_E and C_2 to be perfect short circuits, the pole due to the input coupling circuit alone is determined from fig 4-10b. By inspection, using the concept of natural frequencies, we have

$$P_a|_{\text{due to } C_1 \text{ alone}} = -\frac{1}{C_1[R_1 + R_b \| (r_b + R_i)]} \quad (4\text{-}25)$$
$$= -3.76 \times 10^2 \text{ rad/sec (for this example)}$$

4-4 Low-Frequency Response Characteristics in BJT's

The pole due to C_E, assuming C_1 and C_2 are short circuits, is determined from fig 4-10 by short-circuiting C_1 and C_2. The natural frequencies as determined by any of the methods of Section 1-11 are given by

$$P_b|_{\text{due to }C_E \text{ alone}} = -\frac{1 + \beta_0 R_E/[(R_1 \| R_b) + R_i + r_b]}{R_E C_E} \quad (4\text{-}26)$$

$$= -8.31 \times 10^2 \text{ rad/sec}$$

From (4-26) it is seen that the input resistance $(R_1 \| R_b)$ has a strong influence on the poles due to C_E even though C_1 is short-circuited. Note further that the calculation of the pole locations by this method, that is, considering the pole due to each capacitance alone, is appreciably in error. The actual pole locations are at $p_1 = -55$ and $p_2 = -1.14 \times 10^3$. The effect of the interaction is to split the poles apart, as shown in fig 4-12. In a high-pass circuit, when the poles are on the negative real axis, the pole farthest to the left plays a dominant role in determining the low-frequency cutoff point of the amplifier. This is because the 3-dB cutoff frequency is the frequency at which the gain is 0.707 of its midband value. The midband value is reached asymptotically as the frequency increases. Thus, if $|p_2| \gg |p_1|, |p_c|$, where p_c is the pole due to C_c, and p_2 and p_1 are the actual poles due to the combined effect of C_1 and C_E, then $\omega_l \approx |p_2|$.

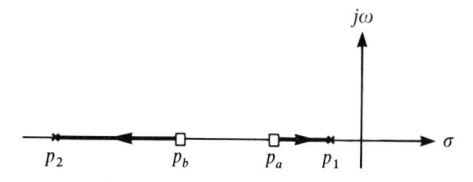

Fig 4-12 Pole splitting due to interaction

In a design for a specified ω_l, we have three variables to choose, namely C_1, C_2, and C_E. The external resistor values are already determined and known from the biasing design. In some cases C_2 may not be present, and even if present causes no problem, since the output coupling circuit does not interact with the other circuits. Thus we consider the poles due to the combined effect of C_E and C_1. As we have pointed out (see the example above), these poles are usually far enough apart so that one pole is much larger than the other. (In the following discussion, we use the negative of the values of the poles so that magnitude signs are not carried along.) In other words, we can approximate the actual roots by the following simple expressions:

$$(s^2 + a_2 s + a_1) = (s + p_2)(s + p_1) \quad \text{where } p_2 \gg p_1$$

$$\approx (s + a_2)\left(s + \frac{a_1}{a_2}\right) \quad \text{where } a_2 \gg \frac{a_1}{a_2} \quad (4\text{-}27)$$

From (4-27) and (4-23) we thus have

$$p_2 \approx a_2 \approx \frac{1}{(R_1 + r_b + R_i)}\left(\frac{1}{C_1} + \frac{\beta_0}{C_E}\right) \quad (4\text{-}28a)$$

$$p_1 \approx \frac{a_1}{a_2} \approx \frac{R_b + \beta_0 R_E}{R_b R_E (C_E + \beta_0 C_1)} \qquad (4\text{-}28\text{b})$$

The condition $p_2 \geq 10\, p_1$ (i.e., $p_2 \gg p_1$) enables us to use (4-27) and (4-28). In an analysis problem, this condition can be readily checked.

In a design problem, however, the values of C_1 and C_E are unknown. If we wish to design on the basis of widely separated poles, we must meet the condition described above. Let us examine the condition $p_2/p_1 \geq 10$. From (4-28) we have

$$\begin{aligned}\frac{p_2}{p_1} &= \frac{(1/R_x)[(C_E + \beta_0 C_1)/C_1 C_E]}{(R_b + \beta_0 R_E)/[R_b R_E (C_E + \beta_0 C_1)]} \\ &= \frac{(1/R_x)[(C_E + \beta_0 C_1)/C_1 C_E]}{(1/R_p)[\beta_0/(C_E + \beta_0 C_1)]}\end{aligned} \qquad (4\text{-}29)$$

where $R_x \equiv R_1 + r_b + R_i$ and $R_p = R_b \| \beta_0 R_E$. By setting $p_2/p_1 \geq 10$ we obtain from (4-29)

$$\frac{R_p}{R_x} \frac{(C_E + \beta_0 C_1)^2}{\beta_0 C_1 C_E} \geq 10 \qquad (4\text{-}30\text{a})$$

or

$$\frac{R_p}{R_x}\left(\frac{C_E}{\beta_0 C_1} + 2 + \frac{\beta_0 C_1}{C_E}\right) \geq 10 \qquad (4\text{-}30\text{b})$$

The term in parentheses of (4-30b) is of the form $(x + 2 + 1/x)$ which has a minimum value of 4. Hence the most pessimistic condition is that $R_p/R_x \geq 2.5$. This guarantees the separation of poles by at least a factor of 10. The pole-separation condition, written explicitly in terms of the circuit parameters, is

$$(R_b \| \beta_0 R_E) \geq 2.5\,(R_1 + r_b + R_i) \qquad (4\text{-}31)$$

Hence, if (4-31) is met, no matter what the values of C_1 and C_E are, the poles are separated by at least a factor of 10. If R_1 is not too large, this condition is often satisfied. In the example of this section, where R_1 is not small, we have $R_b \| \beta_0 R_E = 6\text{ k}\Omega$ and $R_1 + r_b + R_i = 1.35\text{ k}\Omega$, which clearly satisfies the condition in (4-31). It is seen that the poles are actually separated by a factor of 27.3. The poles as obtained by the approximate expressions (4-28a) and (4-28b) are

$$p_2 \simeq a_2 = 1.2 \times 10^3 \text{ rad/sec} \qquad \text{and} \qquad p_1 \simeq \frac{a_1}{a_2} = 51 \text{ rad/sec}$$

4-4 Low-Frequency Response Characteristics in BJT's

which clearly shows a very simple and fairly accurate method of predicting the pole locations due to the interaction of C_E and C_1. In fact, the error in calculation of this approximate method is always less than 10 per cent.

In a design problem there is a great deal of freedom in the choice of C_1, C_2, and C_E for a given ω_l. A simple design procedure is as follows: First insure the satisfaction of the pole separation condition (4-31). Under this condition recall that no matter what the values of C_1 and C_E, the pole $p_2 \geq 10p_1$. Also if $p_c \leq p_2/10$, then $p_2 \approx \omega_l$. In an actual design, make $\omega_l \simeq p_2 \approx a_2$ and arbitrarily select C_1 and C_E in (4-23a) such that:

$$C_1 = \frac{2}{\omega_l}\left[\frac{R_b + r_b + R_i}{R_b R_1 + (R_b + R_1)(r_b + R_i)}\right] \quad (4\text{-}32a)$$

$$C_E = \frac{2}{\omega_l}\left[\frac{(R_b + R_1)(\beta_0 + R_i/R_E)}{R_b R_1 + (R_b + R_1)(r_b + R_i)}\right] \quad (4\text{-}32b)$$

The output coupling circuit in this case must be chosen such that the output circuit pole $p_c \ll \omega_l$. Specifically, we may use the following relation

$$p_c = \frac{1}{(R_2 + R_C)C_2} \leq \frac{\omega_l}{10}$$

or

$$C_2 \geq \frac{10}{\omega_l}\left(\frac{1}{R_2 + R_c}\right) \quad (4\text{-}33)$$

The pole-zero plot for this design is shown in fig 4-13. The zero at $1/R_E C_E$ may or may not be larger than p_1, depending on the element values.

A design example is worked out in Section 4-15. It should be noted that the low-frequency response can be improved by splitting R_C such

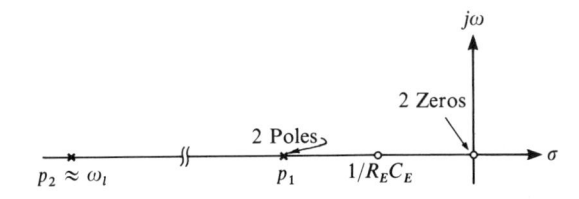

Fig 4-13 Pole-zero plot for simple design

that the main part of it is bypassed by a low-frequency compensating capacitor C_d such as is shown in fig 4-18. The effect of this capacitor is to increase the gain at low frequencies, thus compensating for the loss of gain due to the other bias and coupling capacitors (see Problem 4-18).

In integrated circuits, the large values of the capacitors in the bypass and coupling circuitry are undesirable, since they require a large area on the chip. Most integrated-circuit amplifiers employ different biasing schemes (fig 3-65) and are often dc coupled in a differential configuration. These are discussed in Chapter 5.

4-5 Low-Frequency Response of FET's and Vacuum Tubes Due to Bias and Coupling Circuitry

Calculations of the low-frequency response due to biasing and coupling circuitry for FET's and vacuum tubes are similar to those for the bipolar transistor. For example, from the low-frequency equivalent circuit of the devices, we have the circuit shown in fig 4-14b. In fig 4-14a, appropriate values for an FET or vacuum triode are obtained, as discussed in Chapter 3. The circuit of fig 4-14b is similar to that shown in fig 4-10b and is, in fact, simpler. The analysis and design considerations are, therefore, essentially as in the previous section.

Note that in the FET and vacuum tube cases $R_i \simeq \infty$ and we always apply a voltage-source input. Because of the open circuit due to R_i, the circuit divides into two noninteracting subcircuits. Thus, if we solve for the voltage gain function $V_o/V_i(s)$ in fig 4-14b, we will find that C_1 and C_3 do not interact and the pole due to each capacitor is determined separately and independent of the other. To provide adequate by-passing for the FET over the same frequency range as for the BJT, the value of the source bypass capacitor is much smaller than the emitter bypass capacitor, i.e., $C_E \gg C_S$ (C_3 in fig 4-14), because of the drastically different values of g_m in each case. Usually the size of the capacitors is not so important in discrete circuits, in that we can use bigger capacitances at essentially no extra cost, and we can use a simple design because of the absence of interaction.

Thus the pole-zero locations due to various circuits, considering them one at a time (i.e., considering one capacitance at a time and assuming perfect short-circuits for the other capacitors), are determined by straightforward analysis as follows:

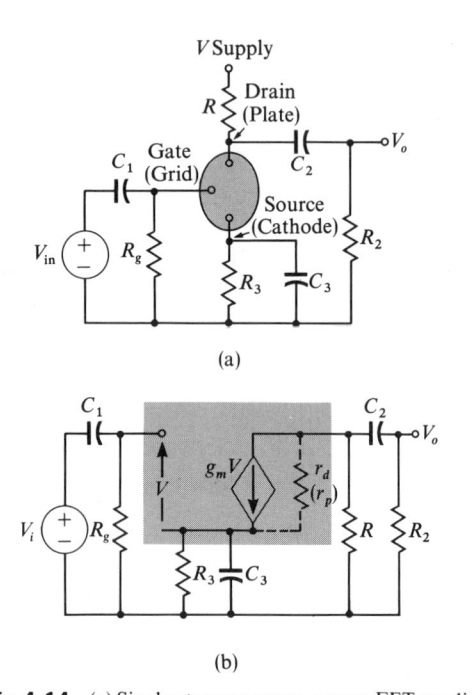

Fig 4-14 (a) Single-stage common-source FET amplifier (or a common-cathode vacuum tube); (b) Low-frequency small-signal equivalent circuit of (a)

4-5 Low-Frequency Response of FET's and Vacuum Tubes

We write the general gain expression in the form of

$$A_v(s) = A_m \frac{(s + z_1)}{s + p_1} \qquad (4\text{-}34)$$

where $p_1 > z_1$, since the low-frequency response has a high-pass characteristic. Of course, z_1 can be equal to zero, as it is for the coupling capacitor circuits. The value of A_m is simply the midband gain and is readily determined by short-circuiting all the bypass and coupling capacitors. Thus, from fig 4-14b, the midband voltage gain A_m is

$$A_m = -g_m(r_d \| R \| R_2) = -g_m(R_L \| r_d) \approx -g_m R_L \qquad (4\text{-}35)$$

where $R_L = R \| R_2$. The pole p_1 is the natural frequency of the network and is determined by any of the methods discussed in Section 1-12. For C_1 and C_2 it can be written by inspection. The value of the zero is determined by making sure that (4-34) is correct at dc. For C_1 and C_2 it is obviously zero, since the signal path is open at dc and $A_v(0) = 0$.

Thus, for the drain-coupling (or plate-coupling) circuit, the low-frequency voltage-gain function is given by

$$\frac{V_o}{V_{in}} \approx A_m \frac{s}{s + 1/[(R + R_2)C_2]} \qquad (4\text{-}36)$$

The voltage-gain function of the source-bypass (or cathode-bypass) circuit is given by[4]

$$\frac{V_o}{V_{in}} \approx A_m \frac{(s + 1/R_3 C_3)}{s + (1 + g_m R_3)/R_3 C_3} \qquad (4\text{-}37)$$

Finally, for the gate-coupling (or grid-coupling) circuit we have:

$$\frac{V_o}{V_{in}} = A_m \frac{s}{s + 1/R_g C_1} \qquad (4\text{-}38)$$

---

[4] A more accurate expression including r_d is:

$$A_v = A_m \frac{(s + 1/R_3 C_3)}{s + 1/(R_3 C_3 \{[(R \| R_2) + r_d]/[R_3(\mu + 1) + r_d + R \| R_2]\})}$$

which reduces to (4-37) for $r_d \gg R \| R_2$, $g_m \gg \dfrac{1}{r_d}$

The low-frequency cutoff is approximately equal to the largest of the poles in (4-36), (4-37), and (4-38) if they are not too close to each other. If the poles are close to each other but are not of identical value, an approximate expression for the low-frequency cutoff is given by

$$\omega_l \simeq \sqrt{p_1^2 + p_2^2 + p_3^2} \qquad (4\text{-}39)$$

(see Section 5-3), where p_1, p_2, and p_3 are the poles in (4-36), (4-37), and (4-38). In the extreme case when all three poles in (4-39) are equal, the low-frequency cutoff radian frequency will be about twice the value of the pole.

In an FET stage, the source bypass capacitor ($C_s \equiv C_3$, fig 4-14) is usually the determining factor in establishing the low-frequency cutoff point. For the same low-frequency cutoff point, the value of C_s is much lower than that of the emitter bypass capacitor C_E of a bipolar transistor circuit because of the difference in the magnitudes of g_m. The following example illustrates the above.

EXAMPLE

Consider an FET stage, such as shown in fig 4-14, with the following parameters:

$$R_D \equiv R = 3.5\,\text{k}\Omega \qquad R_g \equiv R_2 = 1\,\text{M}\Omega \qquad R_S \equiv R_3 = 1.5\,\text{k}\Omega$$

$$C_1 = C_2 = C_s = 10\,\mu\text{F} \qquad g_m = 2 \times 10^{-3}\,\text{mhos} \qquad r_d = 20\,\text{k}\Omega$$

We wish to determine the approximate low-frequency cutoff point of the amplifier.

Solution: We determine the value of the poles due to each coupling and bypass circuit. From (4-36), the pole of the drain-coupling circuit alone p_d is given by

$$p_d = \frac{1}{(R_D + R_g)C_2} \simeq \frac{1}{(3.5 \times 10^3 + 10^6)(10 \times 10^{-6})} = 0.1\,\text{rad/sec}$$

Since $r_d \gg (R\|R_2)$, $1/g_m$, the pole due to the source bypass capacitance p_s from (4-37) is given by

$$p_s = \frac{1 + g_m R_S}{R_S C_s} = \frac{4}{(1.5 \times 10^3)(10 \times 10^{-6})} = 2.66 \times 10^2\,\text{rad/sec}$$

From (4-38), the pole due to the gate-coupling circuit p_g is given by

$$p_g = \frac{1}{R_g C_1} = \frac{1}{10^6(10 \times 10^{-6})} = 0.1\,\text{rad/sec}$$

4-6 Step Response of a Typical High-Pass Circuit and Sag Calculation

Since $p_s \gg p_d$, and $p_s \gg p_g$, the low-frequency cutoff point is given by the pole of the source bypass circuit;

$$\omega_l = p_s = 2.66 \times 10^2 \text{ rad/sec} \quad \text{or} \quad f_l = \frac{2.66 \times 10^2}{2\pi} = 42 \text{ Hz}$$

Note that for the same values of capacitors, the source bypass capacitance pole is larger than the other poles by three orders of magnitude. Note also that, for the same cutoff frequency, the value of C_E for a bipolar transistor is much larger than the value of C_s for an FET.

4-6 Step Response of a Typical High-Pass Circuit and Sag Calculation

The transfer function of an individual high-pass circuit, as shown in Section 4-5, is of the form of (4-34) rewritten as

$$A = \frac{I_o}{I_i} = A_m \frac{(s + z_1)}{s + p_1} \quad (4\text{-}40)$$

where $p_1 > z_1$ and both are real quantities, and I_o and I_i are the transforms of the output and input signals. In (4-40), we could just as well use transforms of the output and input voltages, which would be the case for FET's and vacuum tubes. The magnitude and phase response of (4-40) are shown in figs 4-15a and 4-15b, respectively. We shall now

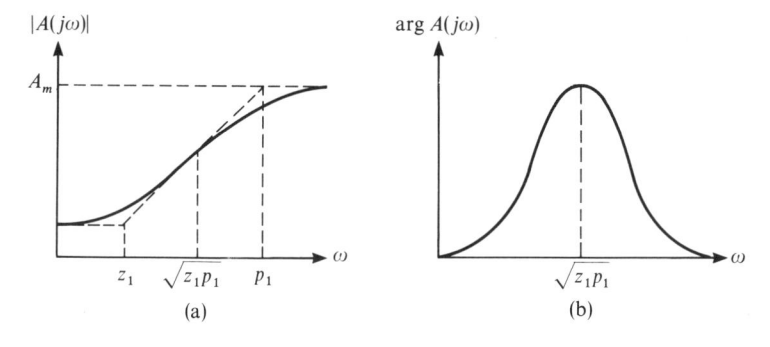

Fig 4-15 (a) Magnitude response of $A_m(s + z_1)/(s + p_1)$; (b) Phase response of $A_m(s + z_1)/(s + p_1)$

determine the response of a typical high-pass circuit in the time domain. This is important in pulse amplifiers where we wish to amplify linearly a pulse without distortion.

The transform of the signal for a unit step input [i.e., $v_i(t) = u(t)$] is $1/s$. Thus, the transform of the output signal is

$$V_o(s) = \frac{A_m}{s} \frac{(s + z_1)}{(s + p_1)} \qquad (4\text{-}41)$$

The inverse Laplace transform of (4-41), obtained by simple partial fraction expansion, is

$$v_o(t) = A_m \frac{z_1}{p_1}\left[1 + \left(\frac{p_1}{z_1} - 1\right)e^{-p_1 t}\right] \qquad (4\text{-}42)$$

The magnitude of the output step at $t = 0$ is equal to A_m, in accordance with the initial value theorem of Laplace transforms (see table 1-1). The output response decays with a time constant $(1/p_1)$, and the final value is $(A_m z_1/p_1)$, as shown in fig 4-16. For linear pulse amplification, the input pulse has a pulse duration time of, say τ_p. We define a pulse sag δ as

$$\delta \equiv \frac{A_m - v_o(\tau_p)}{A_m} = 1 - \frac{v_o(\tau_p)}{A_m} \qquad (4\text{-}43)$$

The above quantities are illustrated in fig 4-17. To calculate the sag, we then use (4-43) and (4-42). Usually $p_1 \tau_p \ll 1$ so that we can use an approximation by retaining the linear term of the exponential expansion, i.e.,

$$e^{-p_1 \tau_p} \approx 1 - p_1 \tau_p \qquad (4\text{-}44)$$

Then from (4-44), (4-43), and (4-42), we have

$$\delta \approx (p_1 - z_1)\tau_p \qquad (4\text{-}45)$$

Note that if we normalize $v_o(t)$ such that $v_o(0^+) = 1$ and then differentiate (4-42) we obtain:

$$\frac{dv_{on}}{dt} = -(p_1 - z_1)e^{-p_1 t} \qquad (4\text{-}46a)$$

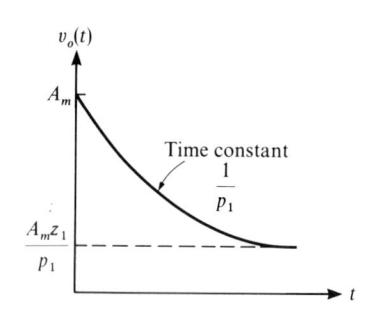

Fig 4-16 Unit step response of $A_m(s + z_1)/(s + p_1)$

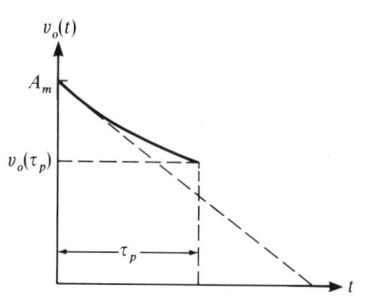

Fig 4-17 Pulse response of $A_m(s + z_1)/(s + p_1)$, illustrating the sag (exaggerated)

4-6 Step Response of a Typical High-Pass Circuit and Sag Calculation

and

$$\left.\frac{dv_{on}}{dt}\right|_{t=0^+} = z_1 - p_1 \qquad (4\text{-}46b)$$

Hence the slope of the normalized output at $t = 0^+$ yields the sag/sec, that is, from (4-45) and (4-46b)

$$\text{sag/sec} = \frac{\delta}{\tau_p} \simeq z_1 - p_1 = \left.\frac{dv_{on}}{dt}\right|_{t=0^+} \qquad (4\text{-}47)$$

For many sources of sag, if the sags of the individual coupling bypass circuits are small (so that the initial slope $v_o(0^+)$ can be used to represent the sag), it can be shown that (see Problem 4-19)

$$\frac{d}{dt}v_o(0^+) = \sum_{i=1}^{N}(z_i - p_i) \qquad (4\text{-}48)$$

or alternatively stated,

$$\text{total sag} = \tau_p \sum \text{individual sags (slopes)} \qquad (4\text{-}49)$$

EXAMPLE

Let a voltage pulse of low amplitude A and a pulse duration of 1 millisecond be applied to the FET example of the previous section. (Note that we specifically mention low amplitude, so that the amplifier is not saturated and thus operates linearly. When the pulse amplitude is very large, saturation and nonlinear effects occur. This topic is discussed in detail in Chapter 10.) We wish to calculate the total sag.

Since $\tau_p p_1$ for each circuit (see Section 4-5) is much less than unity, we can use (4-45) and (4-49). The gate and drain coupling circuits each contribute a per cent sag of

$$\delta_g = \delta_d = (p_1 - z_1)\tau_p = (0.1)(10^{-3}) = 10^{-4}$$

The source bypass circuit contributes a sag of

$$\delta_s = (2.66 \times 10^2 - 0.66 \times 10^2)(10^{-3}) = 0.2$$

Hence the total sag is $\delta = \delta_s + \delta_g + \delta_d \simeq 0.2$. In other words (see fig 4-17), the output pulse at τ_p will be lower in amplitude than at $t = 0^+$ by 20 per cent.

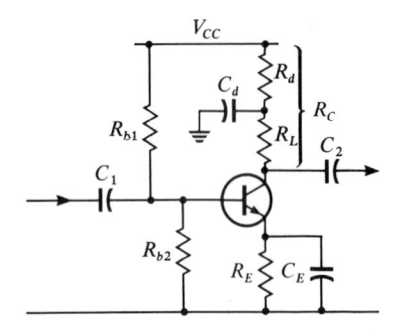

Fig 4-18 Circuit for sag compensation

The sag can be improved by a low-frequency compensation network $R_d C_d$, as shown in fig 4-18. This circuit provides a positive initial slope (see Problem 4-18), thereby compensating the negative initial sag of the circuit (if appropriately adjusted). Improving low-frequency response thus improves the sag. The $R_d C_d$ circuit also has an additional useful property, in that it allows the total collector resistance to be divided into two parts. The high-frequency load resistance is R_L, the low-frequency load resistance is $R_C = R_d + R_L$. For a wide bandwidth stage we generally want R_L small (of the order of 200 Ω). For bias stability, we want $R_d + R_L$ large (of the order of 2 kΩ), as was discussed in the section on biasing design.

4-7 High-Frequency Response Characteristics of BJT's

Consider the single-stage CE amplifier in fig 4-10a. The midband and high-frequency circuit model, using the hybrid π equivalent circuit for the transistor, is shown in fig 4-19. Note that in this frequency range, all the coupling and bypass capacitors are short-circuited. The frequency response of the circuit at high frequencies can now be

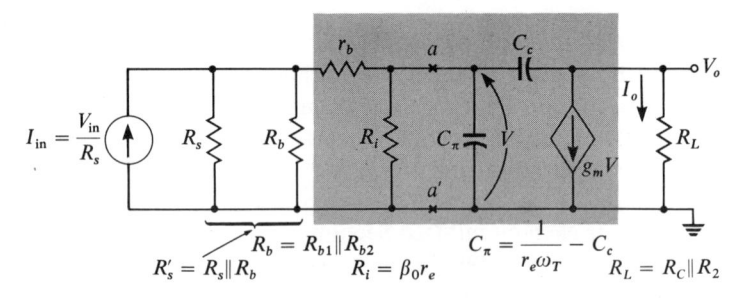

Fig 4-19 High-frequency ($f \leq f_T/3$) small-signal equivalent circuit of a single-stage CE transistor amplifier

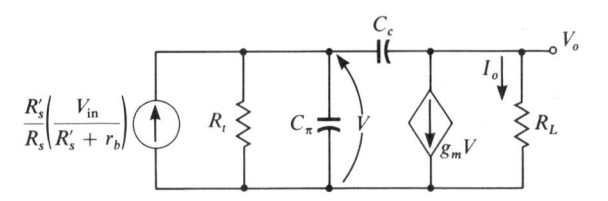

Fig 4-20 Simplified equivalent of fig 4-19

4-7 High-Frequency Response Characteristics of BJT's

determined by an analysis of the circuit in fig 4-19. The algebra can be simplified if we replace the circuit to the left of terminals a–a' by its Norton equivalent circuit. This equivalence is shown in fig 4-20. Nodal analysis of fig 4-20 yields the following equilibrium equations in the transformed domain:

$$\frac{R'_s V_{in}}{R_s(R'_s + r_b)} + V(G_t + sC_\pi) = (V - V_o)sC_c \qquad (4\text{-}50)$$

$$(V - V_o)sC_c = g_m V + V_o G_L$$

If we eliminate V in (4-50) and solve for the voltage-gain function, we obtain:

$$\frac{V_o}{V_{in}} = \frac{R'_s(s - g_m/C_c)}{(R'_s + r_b)R_s C_\pi \{s^2 + s[(g_m + G_t)/C_\pi + G_L(1/C_c + 1/C_\pi)] + (G_t G_L)/(C_c C_\pi)\}} \qquad (4\text{-}51)$$

where $R_t = R_i \| (r_b + R'_s)$ has been substituted in (4-51). The current-gain function is obtained using the relations $I_{in} \equiv V_{in} G_s$ and $I_o = V_o G_L$, and (4-51). Thus

$$\frac{I_o}{I_{in}} = \left(\frac{R_s}{R_L}\right)\frac{V_o}{V_{in}} = \frac{R'_s(s - g_m/C_c)}{R_L(R'_s + r_b)C_\pi \{s^2 + s[1/C_\pi(g_m + G_t) + G_L(1/C_c + 1/C_\pi)] + (G_t G_L)/C_c C_\pi\}} \qquad (4\text{-}52)$$

In (4-51) we find again a wide separation between the two poles for typical transistor parameters. Recall that since the model is not accurate near and beyond the frequency f_T, the expression in (4-51) is valid only for $f \leq f_T/3$. We shall illustrate the analysis and determine the order of magnitudes of the poles and zeros for a typical transistor in the example below, and then make use of the dominant-pole approximation concept in the preliminary design of transistor amplifiers.

EXAMPLE

Let the transistor parameters at the quiescent operating point be $I_C = 5$ mA, $V_{CE} = 5$ V, and the element values of the circuit in fig 4-10a be

$$f_T = 400 \text{ MHz}, \quad r_b = 100 \text{ }\Omega, \quad C_c = 5 \text{ pF}, \quad \beta_0 = 50$$

$$R_b = 6 \text{ k}\Omega, \quad R_1 = 1 \text{ k}\Omega, \quad R_L = R_C \| R_2 = 500 \text{ }\Omega \qquad (4\text{-}53)$$

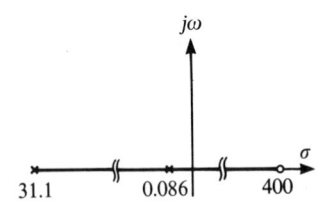

Fig 4-21 Pole-zero plot of the example gain function

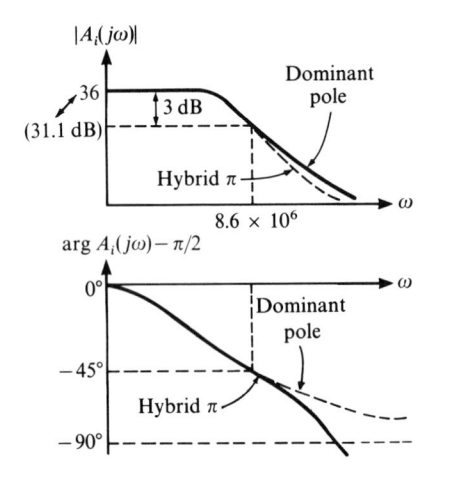

Fig 4-22 (a) Magnitude response of the example gain function; (b) Phase response of the example gain function

From the above data we obtain

$$R_i = \beta_0 r_e = 250\,\Omega \qquad C_\pi = \frac{1}{r_e \omega_T} - C_c = 74.6 \text{ pF}$$

Substitution of the numerical values in (4-52) yields

$$\frac{I_o}{I_{in}} = (2.4 \times 10^7)\frac{s - 4 \times 10^{10}}{s^2 + s(31.1 \times 10^8) + 2.67 \times 10^{16}} \qquad (4\text{-}54)$$

For convenience we normalize the frequency by a scale of 10^8, and determine the roots of (4-54) to be

$$\left(\frac{I_o}{I_{in}}\right) = (0.24)\frac{s_n - 400}{(s_n + 31.1)(s_n + 0.086)} \qquad (4\text{-}55)$$

The pole-zero plot and the magnitude and phase responses are shown in fig 4-21 and fig 4-22, respectively. Remember that the expression in (4-55) is valid for frequencies up to $f_T/3$, i.e., about 1.3×10^8 Hz. Note further that the magnitude of the zero is much larger than the poles and that the 3-dB bandwidth from (4-55) is $\omega_{3\text{ dB}} \simeq 8.6 \times 10^6$ rad/sec, which is the dominant pole of the circuit (i.e., in the low-pass frequency range of interest the smallest pole dominates the response as discussed in Sec 4-10).

The 3-dB cutoff frequency is $f_{3\text{ dB}} = \omega_{3\text{ dB}}/2\pi = 1.37$ MHz. The low-frequency current gain is -36. Note that the low frequency here refers to the actual midband gain, since the bypass and coupling circuitry are assumed to be short circuits and the frequency range starts from midband and extends to higher frequencies. This is understood throughout this chapter.

The current-gain function in (4-52) can also be written as

$$A_i(s) = \frac{I_o}{I_{in}} = a_o \left(\frac{b_1}{z_1}\right)\frac{(s - z_1)}{s^2 + b_2 s + b_1} \qquad (4\text{-}56)$$

where

$$a_o = \beta_0 \frac{R_s \| R_b}{R_i + r_b + R_s \| R_b}$$

$$z_1 = \frac{g_m}{C_c} = \frac{1}{r_e C_c}$$

$$b_2 = \frac{1}{C_\pi}(g_m + G_t) + \left(\frac{1}{C_c} + \frac{1}{C_\pi}\right)G_L \qquad (4\text{-}57)$$

$$b_1 = \frac{G_t G_L}{C_c C_\pi}$$

Note that at $s = 0$, $A_i(0) = -a_o$, the midband current gain. If a dominant pole situation exists, then the two roots of (4-56) are far apart. For two widely separated poles, as in (4-27), we have the condition $b_1/b_2 \ll b_2$, and (4-56) can be approximated as follows:

$$A_i(s) \simeq a_o \left(\frac{b_1}{z_1}\right) \frac{(s - z_1)}{(s + b_2)(s + b_1/b_2)} \quad (4\text{-}58)$$

Thus the dominant pole expression is b_1/b_2. It is further understood that $|z_1| \gg b_1/b_2$, which is almost always the case. Under these conditions (4-58) may be written in the simple approximate form

$$A_i(s) = \frac{-a_o b_1}{b_2(s + b_1/b_2)} \quad (4\text{-}59)$$

The upper 3-dB radian frequency is given by

$$\omega_h \equiv \omega_{3\,\text{dB}} \approx \frac{b_1}{b_2} = \frac{G_t}{C_\pi + C_c[1 + g_m R_L + R_L G_t]} \quad (4\text{-}60)$$

The reader is reminded that the expression in (4-59) is valid for frequencies up to roughly three or four times the cutoff frequency. The magnitude and phase responses of (4-59), as compared with the more accurate expression in (4-56) for the example, are shown in fig 4-22.

We may wonder at this time if one could make some simplifications in the circuit at the outset, such that the resultant gain function of the modified circuit is one-pole, with the pole location given approximately by (4-59). If this approximation can be achieved, what are the conditions and the limitations of the new model? This approximation can be achieved, and is often applicable for gain and bandwidth calculations up to a frequency range of approximately $f_T/10$ if we use the Miller approximation. We must be aware, of course, of the conditions under which it cannot be used. But first let us find out about the Miller effect and the unilateral approximation.

4-8 The Miller Effect and the Unilateral Approximation

Consider the circuit shown in fig 4-23. We use general symbols so that we can apply the results obtained in this section to bipolar transistors, FETs, and vacuum tubes. Note that in the bipolar transistor

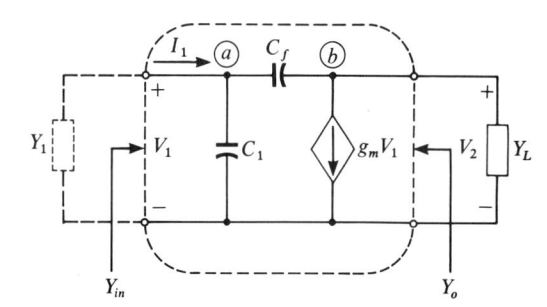

Fig 4-23 Circuit used to illustrate the Miller effect and to obtain the unilateral approximation. Circuit is applicable for BJT, FET, and vacuum triodes

case, R_i is included in Y_1, this admittance is not of direct concern here. Similarly, the output capacitance may be included in Y_L. What we wish to do is to "unilateralize" the circuit such that the effect of the feedback capacitance C_f and the load admittance Y_L is included at the input, while still retaining some features of the input–output decoupling.

This process, of course, requires approximation. As we shall soon see, however, tremendous simplification can result, and without significant error, if the approximation is applied properly. It should be pointed out, however, that in a unilateral circuit, the input admittance is independent of the load admittance. Since

$$Y_{in} = y_{11} - \frac{y_{12}y_{21}}{y_{22} + Y_L} = y_i - \frac{y_r y_f}{y_o + Y_L} \tag{4-61}$$

for a unilateral two-port, $y_{12} = y_r = 0$ and thus Y_{in} is independent of Y_L. In a BJT, FET, or triode at high frequencies, the input is not independent of the load, as is readily seen in fig 4-23. Therefore any good approximation must retain this feature. Because of this dependence of input on output even in a decoupled circuit, we do not actually have a unilateral model. For convenience, however, we shall use the term loosely in the following.

We determine the expression for the input admittance by nodal analysis. The two nodal equations in the transformed domain are

$$I_1 = s(C_1 + C_f)V_1 - sC_f V_2 \tag{4-62a}$$

$$0 = (g_m - sC_f)V_1 + (Y_L + sC_f)V_2 \tag{4-62b}$$

From (4-62a) and (4-62b) we determine the input admittance Y_{in}, which is given by

$$Y_{in} \equiv \frac{I_1}{V_1} = s(C_1 + C_f) + \frac{g_m - sC_f}{Y_L + sC_f} sC_f \tag{4-63}$$

If we assume that in the range of frequencies of interest

$$\omega C_f \ll g_m, |Y_L| \tag{4-64}$$

then (4-63) can be approximated by

$$\frac{I_1}{V_1} \approx s\{C_1 + C_f[1 + g_m Z_L(s)]\} = sC_1 + Y_{eq} \tag{4-65}$$

where

$$Y_{eq} = sC_f[1 + g_m Z_L(s)]$$

4-8 The Miller Effect and the Unilateral Approximation

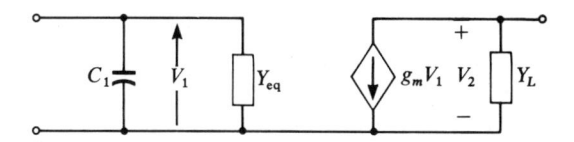

Fig 4-24 Unilateral approximate equivalent of fig 4-23, valid only for input admittance and forward-gain function calculations (provided $\omega C_f \ll g_m, |Y_L|$)

Using the assumption of (4-64) and the approximation of (4-65), the circuit model of fig 4-23 reduces to the one shown in fig 4-24 *as far as the input admittance is concerned.*

The forward gain function of the circuit in fig 4-23, as determined from (4-62b), is given by

$$A_v(s) \equiv \frac{V_2}{V_1} = -\left(\frac{g_m - sC_f}{Y_L + sC_f}\right) \quad (4\text{-}66a)$$

Using the assumption of (4-64), the above expression reduces to

$$A_v(s) \approx -\frac{g_m}{Y_L} = -g_m Z_L \quad (4\text{-}66b)$$

From the "unilateralized" circuit of fig 4-24 we also obtain the same gain function as in (4-66b). Hence if the approximation (4-64) is valid, the "unilateralized" model gives fairly correct results as far as the input admittance and the forward gain are concerned. Note, however, that the frequency range is further reduced due to the restriction in (4-64). Note also that the circuit *does not* give the correct *output admittance* and *reverse transmission.* The conclusion that these two quantities are incorrect is readily seen in fig 4-23 and fig 4-24. In fig 4-24 Z_o is infinite and the reverse transmission is zero while in the more accurate model of fig 4-23, Z_o is finite and the reverse transmission is not zero. Thus in situations where these two parameters play an important role, the model of fig 4-24 should not be used. As an example, when the load admittance is inductive, as in tuned amplifiers (see Chapter 7), or when the load resistance is very high, the model of fig 4-24 should not be used.

The decoupling of the input and output circuits is the most attractive simplifying feature of the circuit in the analysis and design. With this model, which we shall henceforth refer to as the *unilateralized model*, the analysis and design of cascaded multistage low-pass amplifiers are simplified in that the overall circuit can be broken into single "noninteracting" stages, as will be seen later in Chapters 5 and 6.

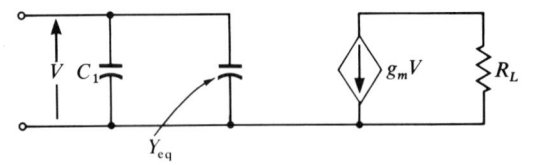

Fig 4-25 For a purely resistive load $Y_{eq} = C_f(1 + g_m R_L)$

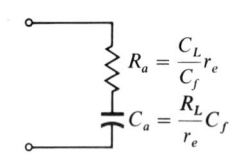

Fig 4-27 Realization of $Y = g_m R_L(C_f/1 + sR_L C_L)$

Now let us consider two important special cases of Y_L in fig 4-24. If Y_L is a purely resistive load, then $Y_L = G_L = 1/R_L$. In this case the circuit model of fig 4-24 reduces to the one shown in fig 4-25. See also the Miller theorem in Problem 1-31. The apparent increase in the input capacitance of the circuit caused by the voltage gain of the amplifying device is known as the *Miller effect*. The Miller effect may increase the effect of the feedback capacitor C_f by a very large factor, since $g_m R_L$ is approximately the midband voltage gain and can be very large. For example, if $C_1 = 90 \text{ pF}$, $C_f = 5 \text{ pF}$, and $g_m R_L = 50$, the total input capacitance will be 345 pF. The unilateral approximation thus incorporates the Miller effect. Note also that the shunt conductance G_c across C_f, which we have ignored here, increases G_i in the same manner. In other words, in fig 4-19, $G'_i = G_i + G_c(1 + g_m R_L)$. This effect is still negligible provided G_i is not small, since G_c is usually of the order of 10^{-6} mhos.

Next consider the special case where $Y_L = G_L + sC_L$, i.e., a shunt RC load. For this case, the circuit model of fig 4-24 reduces to that in fig 4-26. This fact for the special case can readily be seen as follows:

$$C_f g_m Z_L(s) = C_f g_m \left(\frac{1}{G_L + sC_L} \right) = g_m R_L \left(\frac{C_f}{1 + sR_L C_L} \right) \quad (4\text{-}67)$$

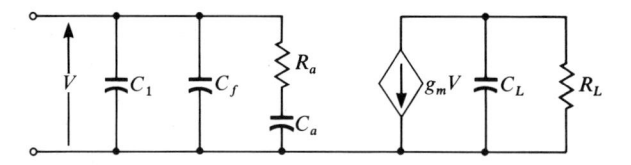

Fig 4-26 For a simple shunt RC load Y_{eq} is an RC admittance function and not only a capacitor

Equation (4-67) is recognized as the admittance of a series RC network as shown in fig 4-27, where

$$R_a = \frac{C_L}{C_f g_m} = r_e \left(\frac{C_L}{C_f} \right) \quad (4\text{-}68a)$$

$$C_a = g_m R_L C_f = \left(\frac{R_L}{r_e} \right) C_f \quad (4\text{-}68b)$$

We repeat that this approximation of a "unilateralized" model is applicable not only to bipolar *CE* transistors but also to common-cathode vacuum tubes and *CS* field-effect transistors. This is due to the very nature of the circuit used in fig 4-23, where C_f is C_c for a CE bipolar transistor, C_{gd} for a CS field-effect transistor, or C_{gp} for a vacuum tube. Of course Y_1 and Y_L of fig 4-23 will be different in BJT's than in FET's and tubes. Even for a given device, Y_1 can be different, depending upon whether the device is used at the input stage or in an interior stage.

The central theme in this section has been to make an approximation by removing C_f, i.e., decoupling the input and output circuits and yet including its effect at the input and on forward transmission. This approximation presents no serious problem in calculating the forward gain and the 3-dB bandwidth of low-pass amplifiers, provided the load resistance is not very large. In bandpass amplifiers, however, or when examining the stability of two-ports or the stability of some feedback amplifiers, the effect of the output impedance and the reverse transmission cannot be ignored, and the unilateral model must not be used. Note further that the unilateral model is valid for a smaller range of frequencies ($\leq f_T/10$) than the hybrid π model ($\leq f_T/3$).

4-9 Gain and Bandwidth Calculation for BJT's Using the Unilateral Model

To bring clearly into focus the simplicity, limitations, and usefulness of the unilateral model, we consider the analysis of the single-stage resistively loaded amplifier of Section 4-7. For convenience we again show the unilateralized model of fig 4-25 in fig 4-28.

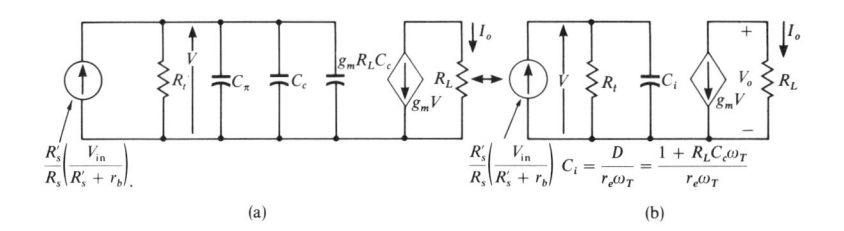

Fig 4-28 (a) Unilateral approximate model of fig 4-25; (b) Simplified equivalent circuit of (a)

For the circuit of fig 4-28 and from (3-52), we have

$$C_\pi = \frac{1}{r_e \omega_T} - C_c \qquad (4\text{-}69)$$

From fig 4-28a we define $C_i \equiv C_\pi + C_c + g_m R_L C_c$, hence

$$C_i = \frac{1}{r_e \omega_T} + g_m R_L C_c = \frac{D}{r_e \omega_T} \qquad (4\text{-}70)$$

where

$$D = 1 + R_L C_c \omega_T \qquad (4\text{-}71)$$

D is the degradation factor in the bandwidth due to the feedback effect of C_c. From the circuit of fig 4-28, it is a simple matter to write the gain function:

$$\frac{V_o}{V_{in}} = \frac{-g_m R_L R'_s}{R_s(R'_s + r_b)} \frac{1}{G_t + s[C_\pi + C_c(1 + g_m R_L)]} = \frac{-g_m R_L R'_s}{R_s(R'_s + r_b)} \frac{1}{G_t + sC_i} \qquad (4\text{-}72)$$

The current-gain function is

$$A_i = \frac{I_o}{I_{in}} = -\frac{g_m R'_s}{R'_s + r_b} \frac{1}{G_t + sC_i} = -\frac{g_m R'_s}{(R'_s + r_b)C_i} \frac{1}{[s + (1/R_t C_i)]} \qquad (4\text{-}73)$$

Since the gain function above has only one pole, the 3-dB radian frequency is given by the pole, namely

$$\omega_{3\,\text{dB}} = \frac{1}{R_t C_i} = \frac{G_t}{C_\pi + C_c(1 + g_m R_L)} \qquad (4\text{-}74)$$

Comparison of (4-74) with (4-60) shows that the expressions are nearly the same if R_L is not too large. In fact, if $R_L G_t \ll (1 + g_m R_L)$, the expression in (4-60) reduces to the result of the unilateral model given in (4-74). The unilateral model is thus very useful and sufficiently accurate for gain and bandwidth calculations in the analysis and design of low-pass amplifiers.

Similarly, for the shunt RC load in the unilateral model, analysis shows (Problem 4-21) that a pole-zero cancellation always occurs at $1/R_L C_L$ and that the other two-poles of the circuit in fig 4-26 for $C_L \gg C_c$ are the same as the natural frequencies of the hybrid π model with a shunt RC load.

Let us now consider the numerical example of Section 4-7 for the resistive load and the unilateral CE model. From (4-73) the low-frequency current gain is given by

$$A_i(0) = \frac{I_o}{I_{in}} = \frac{-g_m R_s}{G_t(R'_s + r_b)} = -36 \qquad (4\text{-}75)$$

The bandwidth is then determined from (4-74) to be

$$\omega_{3\,dB} = \frac{G_t}{C_\pi + C_c(1 + g_m R_L)} = 0.086 \times 10^8 \text{ rad/sec} \quad (4\text{-}76)$$

Hence $f_{3\,dB} = (8.6 \times 10^6)/2\pi = 1.37$ MHz, which is the same result as was obtained using the hybrid π model.

For future reference, let us rewrite the expressions for gain and bandwidth in *CE* stages with a resistive load. The biasing resistance R_b is usually large such that it can be ignored. Thus the gain and bandwidth are given by

$$A_v = \frac{V_o}{V_{in}} = -\beta_0 \frac{R_L}{R_s + r_b + R_i} \quad (4\text{-}77)$$

$$A_i = \frac{I_o}{I_{in}} = -\beta_0 \frac{R_s}{R_s + r_b + R_i} = -\beta_0 \frac{1}{1 + [(r_b + R_i)/R_s]} \quad (4\text{-}78)$$

$$\omega_{3\,dB} = \frac{\omega_\beta}{D} \frac{R_s + r_b + R_i}{R_s + r_b} = \frac{\omega_\beta}{(1 + R_L C_c \omega_T)} \frac{R_s + r_b + R_i}{R_s + r_b} \quad (4\text{-}79)$$

where $R_i = \beta_0 r_e$ and $\omega_\beta = \omega_T/\beta_0$.

From (4-78) and (4-79), it is seen that when R_s is very large the current gain approaches β_0 and the bandwidth approaches ω_β/D. If, in addition, $R_L \to 0$, $\omega_{3\,dB} \to \omega_\beta$.

The effect of R_1 and R_L on the gain and bandwidth of a *CE* stage can be seen in (4-78) and (4-79). From (4-79), as R_L increases, so does the degradation factor D, and thus the bandwidth decreases, while a decrease in R_L increases the bandwidth. Note also that for transistors with a high f_T and/or where the load resistance is not too small, the *D*-factor is significant and cannot be ignored in bandwidth calculations. Lowering R_s increases the bandwidth and decreases the current gain. For a very small R_s, the effect of r_b on the bandwidth will be pronounced, and due to inaccuracies in the modeling of r_b, bandwidth calculation under this condition may not give reliable results. Note the condition of validity of the unilateral model as given in (4-64), i.e., $R_L \omega_{3\,dB} C_c \ll 1$. Thus if R_L is large, the unilateral model should not be used in circuit applications where the reverse transmission and/or the output impedance plays a dominant role, since the model is inaccurate for these purposes, as was discussed in the previous section.

4-10 The Dominant Pole-Zero Concept and Approximation

We digress briefly to discuss the dominant pole-zero concept. This concept is very important when the network functions have widely separated pole-zero patterns and the analysis is to be performed manually. In the analysis of electronic circuits, we may often not need to go into great detail, as approximate results may be satisfactory, at least in the initial design phase. If a detailed analysis is needed, computer methods of circuit analysis are used. A design based on the simple approximate methods can be very useful, however, as an initial iteration in a computer-aided design.

An approximation based on the dominant pole-zero concept simply ignores second-order effects due to far-removed poles and zeros of the low pass network functions. The inclusion of only first-order effects yields considerable simplification in the analysis and design of electronic circuits. We have already used such an approximation in Sections 4-4 and 4-7, where we ignored second-order effects due to the nondominant poles. In most cases, the inclusion of only dominant poles yields satisfactory results in determining the magnitude response; hence 3 dB calculations. When the role of the phase shift is important, however, such as in a stability study of feedback amplifiers (Chapter 6), one should be extremely careful. For example, the phase-shift contribution due to the nondominant poles and zeros may cause instability, and ignoring them completely would give meaningless results.

Let us first consider the magnitude approximation by ignoring the far removed, or the so-called nondominant, poles and zeros. For the sake of clarity, we consider the following simple example

$$H(s) = \frac{1 + s/z_1}{(1 + s/p_1)(1 + s/p_1^*)(1 + s/p_2)} \quad (4\text{-}80)$$

Let p_1 and p_1^* be the dominant poles (* denotes the conjugate pole), as shown in fig 4-29, where $|p_2|, |z_1| \gg |p_1|$. The magnitude function from (4-80) is

$$|H(j\omega)| = \frac{\sqrt{1 + (\omega/z_1)^2}}{|(1 + j\omega/p_1)(1 + j\omega/p_1^*)|\sqrt{1 + (\omega/p_2)^2}} \quad (4\text{-}81)$$

For the range of frequencies where $|\omega/z_1| \leq \frac{1}{4}$ and $|\omega/p_2| < \frac{1}{4}$, (4-81) can be approximated as

$$|H(j\omega)| \approx \frac{1}{|1 + j\omega/p_1||1 + j\omega/p_1^*|} \quad (4\text{-}82)$$

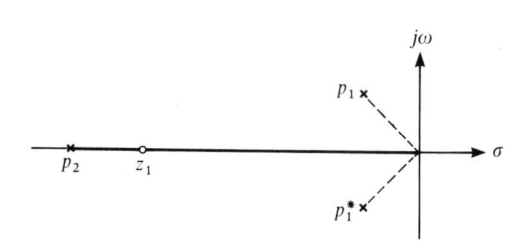

Fig 4-29 Pole-zero plot for a transfer function with a dominant complex pole pair and nondominant pole and zero

4-10 The Dominant Pole-Zero Concept and Approximation

Hence the magnitude function is essentially that of the dominant pair of poles in (4-82), neglecting the nondominant poles and zeros. In some cases, the nondominant poles may be complex. Note, however, that a pole (or a zero) as far apart as by a factor of five has a negligible effect on the magnitude response. For example, if

$$H(s) = \frac{p_1 p_2}{(s + p_1)(s + p_2)} \quad \text{with } p_2 = 5p_1 \qquad (4\text{-}83)$$

then

$$|H(j\omega)| = \frac{1}{|1 + j\omega/p_1||1 + j\omega/5p_1|}$$

$$= \left\{ \frac{1}{[1 + (\omega/p_1)^2][1 + \omega^2/25p_1^2]} \right\}^{1/2} \qquad (4\text{-}83a)$$

The 3-dB bandwidth of (4-83) is obviously less than p_1; thus, at p_1 (i.e., a frequency slightly higher than the 3-dB frequency), the contribution of the farther pole p_2 to the magnitude response is less than 2 per cent. This error decreases monotonically for decreasing frequencies. Note that the phase contribution due to the far pole, p_2, is 25 per cent, which is appreciable. Magnitude responses for various locations of nondominant poles and zeros are shown in fig 4-30. Also included is an approximate pole-zero cancellation case, which may be of interest. Note that the magnitude responses are essentially the same, while the phase response has considerable error.

Thus for 3-dB frequency calculations there is negligible error if the nondominant poles are completely ignored. In stability studies, where one must examine the behavior of the network function (magnitude and phase) for all frequencies (see Section 6-6), ignoring the nondominant poles may yield disastrous results. This is because the phase shift due to these so called nondominant poles at higher frequencies may be large.

In passive distributed *RC* networks, such as are encountered in integrated circuits (see Section 8-8), the network functions have an infinite number of poles, all lying on the negative real axis. The poles of the network function are often widely separated. Thus, the magnitude response can be approximated simply by ignoring all the poles except the one nearest the origin (or, if more accuracy is desired, we may consider only the first two poles near the origin). The phase responses of the actual distributed *RC* network function, and that of the one-pole approximation, are obviously quite different. Phase correction is usually

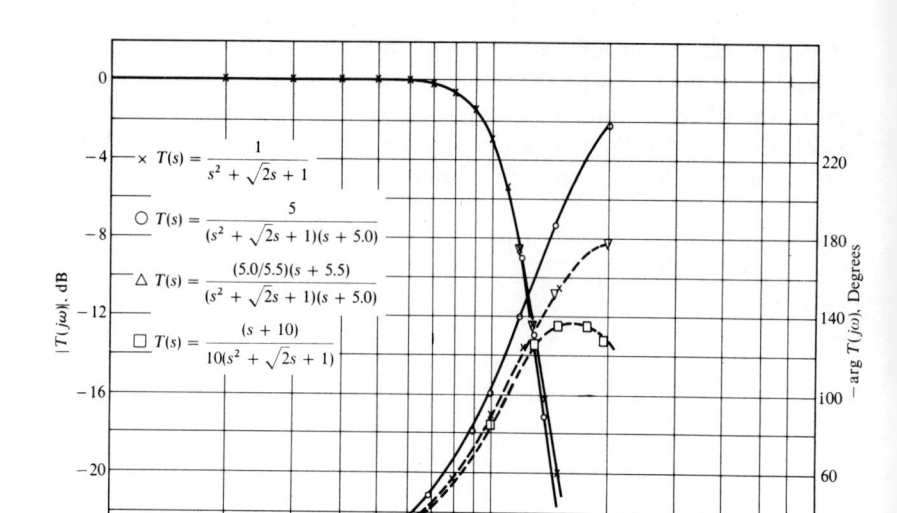

Fig 4-30 Magnitude and phase response of a dominant pole pair with nondominant pole and/or zero

Curves shown:
- × $T(s) = \dfrac{1}{s^2 + \sqrt{2}s + 1}$
- ○ $T(s) = \dfrac{5}{(s^2 + \sqrt{2}s + 1)(s + 5.0)}$
- △ $T(s) = \dfrac{(5.0/5.5)(s + 5.5)}{(s^2 + \sqrt{2}s + 1)(s + 5.0)}$
- □ $T(s) = \dfrac{(s + 10)}{10(s^2 + \sqrt{2}s + 1)}$

made by introducing a delay time in the approximation in the form of

$$H(s) \approx \frac{H_o}{1 + s/p_1} e^{-s\tau} \qquad (4\text{-}84)$$

This approximation is delineated in the following example:

EXAMPLE

Consider the short-circuit CB current gain of a bipolar diffusion transistor as given by (Ref 6)

$$\alpha = \frac{1}{\cosh\left[(W/L)^2 + j\omega(W^2/D)\right]^{1/2}}$$

$$= \frac{\alpha_0}{(1 + j\omega/\omega_\alpha)(1 + j\omega/9\omega_\alpha)(1 + j\omega/25\omega_\alpha)\cdots} \qquad (4\text{-}85)$$

4-10 The Dominant Pole-Zero Concept and Approximation

where $\omega_\alpha = \pi^2 D/4W^2$, $\alpha_0 = 1/\cosh(W/L)$, D is the diffusion constant of the minority carriers in the base region, and W is the effective bandwidth. Equation (4-85) can be approximated by (see also Section 8-11)

$$\alpha(j\omega) \approx \frac{\alpha_0}{1 + j\omega/\omega_\alpha} \exp\left(-\frac{0.2j\omega}{\omega_\alpha}\right) \tag{4-86}$$

where the delay factor $\exp[-0.2j(\omega/\omega_\alpha)]$ is introduced for phase correction and has no effect on the magnitude response. The actual magnitude and phase response of (4-85) and (4-86) are compared in fig 4-31.

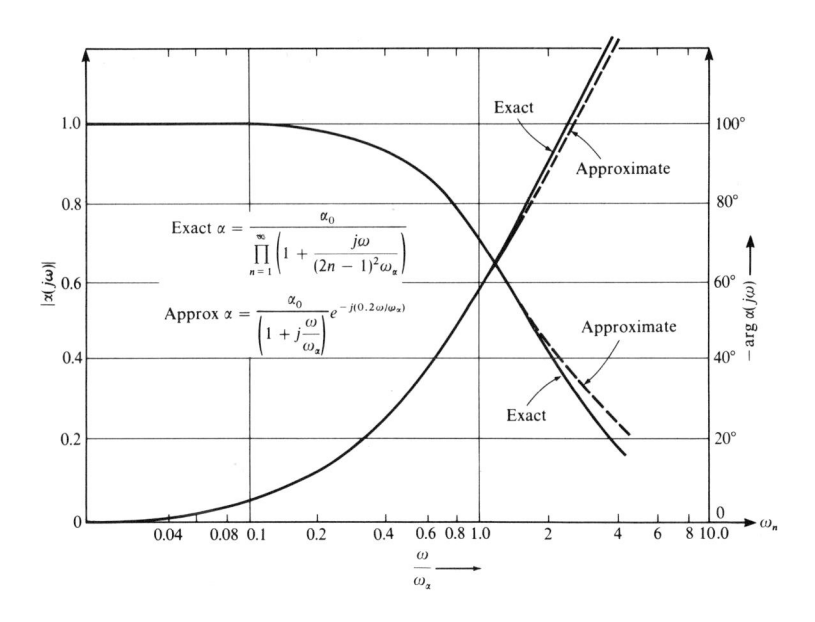

Fig 4-31 Magnitude and phase approximation of $\alpha(j\omega)$

Note that, for $\omega/\omega_\alpha \leq 4$, the two results are very close in phase and magnitude. Beyond this frequency, the error, especially in the magnitude, is not negligible.

The concept of dominant natural frequencies is of considerable help in obtaining simplified approximate expressions and can be used for both frequency and time-domain analysis. For frequency-response calculations, this amounts to ignoring the far-removed poles. For calculations where the phase contribution is important this effect must be included. In distributed parameter systems, the phase correction can generally be made by introducing a delay factor.

4-11 High-Frequency Response of FET's and Vacuum Tubes

For the single-stage common-source FET amplifier and the common-cathode vacuum-tube amplifier, the midband and high-frequency circuit model is shown in fig 4-32. A comparison of the circuit of fig 4-32

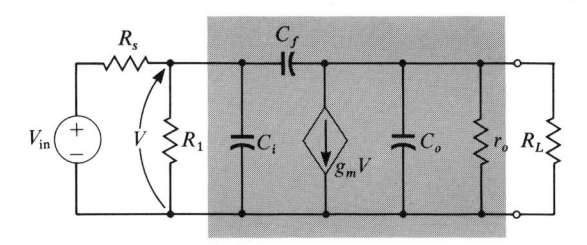

Fig 4-32 High-frequency equivalent circuit of a single-stage common-source FET amplifier (or common-cathode vacuum triode amplifier). For FET: $C_i = C_{gs}$, $C_f = C_{gd}$, $C_o = C_{ds}$, and $r_o = r_d$ (for vacuum triode: $C_i = C_{gk}$, $C_f = C_{gp}$, $C_o = C_{pk}$, and $r_o = r_p$)

with that of fig 4-20 for bipolar transistors shows that the same methods of analysis, unilateralization, and simplification can be employed if R_s is not very small. In FET's and vacuum tubes, however, the input signal is always a voltage source so that $R_s \approx 0$. Thus, for a voltage-source input, the input capacitance is in shunt with a zero impedance and has no effect whatever on bandwidth calculation in a single stage. The bandwidth in such cases is determined by the output circuit capacitance $C_o + C_f$ in shunt with the load including r_d. In multistage amplifiers, however, the input capacitance of every interior and final stage plays a significant part in bandwidth calculations. This is due to the fact that each input capacitance amplified by the Miller effect is in shunt with the load of the preceding stage. The method of analysis for BJT's can be carried over to FET's or vacuum tubes by simply making $r_b \to 0$, $R_i \to \infty$, $C_c \to C_{gd}$, $r_o \to r_d$, $g_m \to g_m$, and $C_\pi \to C_{gs}$. This method is discussed in the next chapter, where multistage amplifiers are considered (Section 5-2).

4-12 Gain-Bandwidth Relations in BJT's

Let us briefly consider the single-stage pentode amplifier. Recall that in this case C_{gp} is very small (of the order of 0.01 pF) and r_o is very large (of the order of 1 MΩ). For a voltage-source input, the high-frequency pentode model is shown in fig 4-33. The voltage-gain function of the circuit is given by

$$A_v = \frac{V_o}{V_{in}} = -\frac{g_m}{C_o} \frac{1}{s + 1/R_L C_o} \qquad (4\text{-}87)$$

where the midband voltage gain and bandwidth are

$$A_v = -g_m R_L \quad \text{and} \quad \omega_{3db} = \frac{1}{R_L C_o} \qquad (4\text{-}88)$$

The gain-bandwidth product of a pentode is, therefore,

$$A_v \omega_{3db} = \frac{g_m}{C_o} = \text{constant for a given pentode} \qquad (4\text{-}89a)$$

For an interior pentode stage

$$A_v \omega_{3db} = \frac{g_m}{C_o + C_i} = \text{constant depending on the pentode} \qquad (4\text{-}89b)$$

Thus, gain and bandwidth can be traded, leaving their product unchanged. In this case, C_{gp} is so small that degradation due to the Miller effect is negligible. The maximum voltage gain that can be achieved in a single-stage amplifier occurs when $R_L = \infty$, and since R_L is in parallel with r_p, the maximum gain is $g_m r_p = \mu$. In the design of a single stage, either gain or bandwidth can be specified, the other is determined by (4-88). Usually the gain, the bandwidth, and the load impedance are specified. In such cases, multistage cascaded amplifiers are used. To improve the gain-bandwidth product capability of a stage, a compensating network is usually used.

4-12 Gain-Bandwidth Relations in BJT's

A single-stage CE transistor amplifier, exclusive of biasing circuitry, is shown in fig 4-34. We shall now show that the gain-bandwidth product in a transistor stage is not constant. This can readily be seen

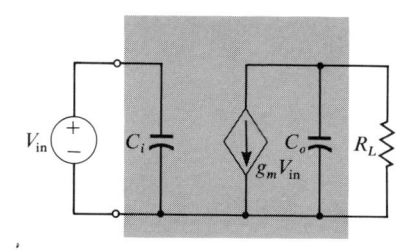

Fig 4-33 High-frequency small-signal equivalent circuit of a single-stage vacuum pentode amplifier

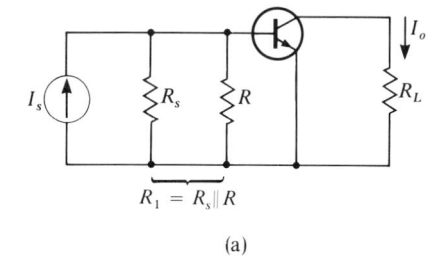

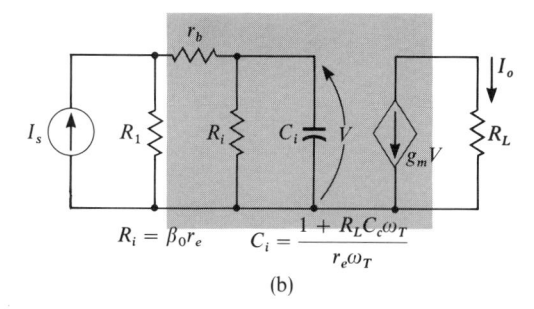

Fig 4-34 (a) Single-stage CE amplifier exclusive of dc biasing circuitry; (b) High-frequency ($f \leq f_T/3$) small-signal equivalent circuit of (a)

$R_1 = R_s \| R$

$R_i = \beta_0 r_e \qquad C_i = \dfrac{1 + R_L C_c \omega_T}{r_e \omega_T}$

4/Analysis and Design of Single-Stage Amplifiers

from the gain and bandwidth expressions in (4-78) and (4-79). For $R_1 = 0$ the current gain is zero while the bandwidth is finite. Thus the gain-bandwidth product (GBP) is zero at $\omega = [(R_i + r_b)/r_b]\omega_\beta/D$. If the value of r_b could be made zero, the GBP would remain constant. Since r_b is never zero, however, it has a serious degradation effect on the GBP at high frequencies.

The current gain-bandwidth product, in general, is given by

$$\text{GBP} \equiv |A_i \omega_{3\,\text{dB}}| = \frac{\beta_0 \omega_\beta}{(1 + R_L C_c \omega_T)} \left[\frac{R_1}{(R_1 + r_b)}\right] = \frac{\omega_T}{D} \frac{R_1}{R_1 + r_b} \quad (4\text{-}90)$$

Note that the voltage gain-bandwidth product is equal to

$$\frac{\omega_T}{D}\left(\frac{R_L}{R_1 + r_b}\right)$$

From (4-90), it is seen that the GBP decreases with increasing R_L and decreasing R_1.

For example, consider the transistor parameters of Section 4-7:

$$\begin{aligned}\omega_T &= 2\pi(4 \times 10^8)\,\text{rad/sec} & r_b &= 100\,\Omega, \\ C_c &= 5\,\text{pF}, & \beta_0 &= 50, & g_m &= 0.2\,\text{mhos}\end{aligned} \quad (4\text{-}91)$$

Let the input signal be from a current source. The gain vs bandwidth ratio for various values of R_1 and two different values of R_L, namely $R_L = 500\,\Omega$ and $R_L = 100\,\Omega$, are shown in fig 4-35a. Note that the actual source resistance of a practical current source and the effect of biasing resistance R_b may be included in R_1. From fig 4-35a, it is seen that for small values of R_1, the gain is drastically reduced while the increase in bandwidth is insignificant. Therefore, this type of amplifier circuit is very inefficient for wide-band amplifiers, and thus broad-banding methods must be used in order to utilize the circuit efficiently. By efficient, we mean that the gain-bandwidth product of the broad-banded stage is larger than ω_T/D, preferably close to ω_T. Figure 4-35b shows the normalized GBP vs normalized $\omega_{3\,\text{dB}}$ for resistive broad-banded stage ($R_L = 100\,\Omega$, i.e., $D = 2.25$). It is seen that GBP is not constant and that it decreases for increasing bandwidth. It should be pointed out that both β_0 and ω_T are functions of the operating point I_C. Usually the operating point is chosen such that these parameters are optimized. Even so, however, for a given transistor, the gain and bandwidth requirements of the amplifier often necessitate broadband multistage amplification.

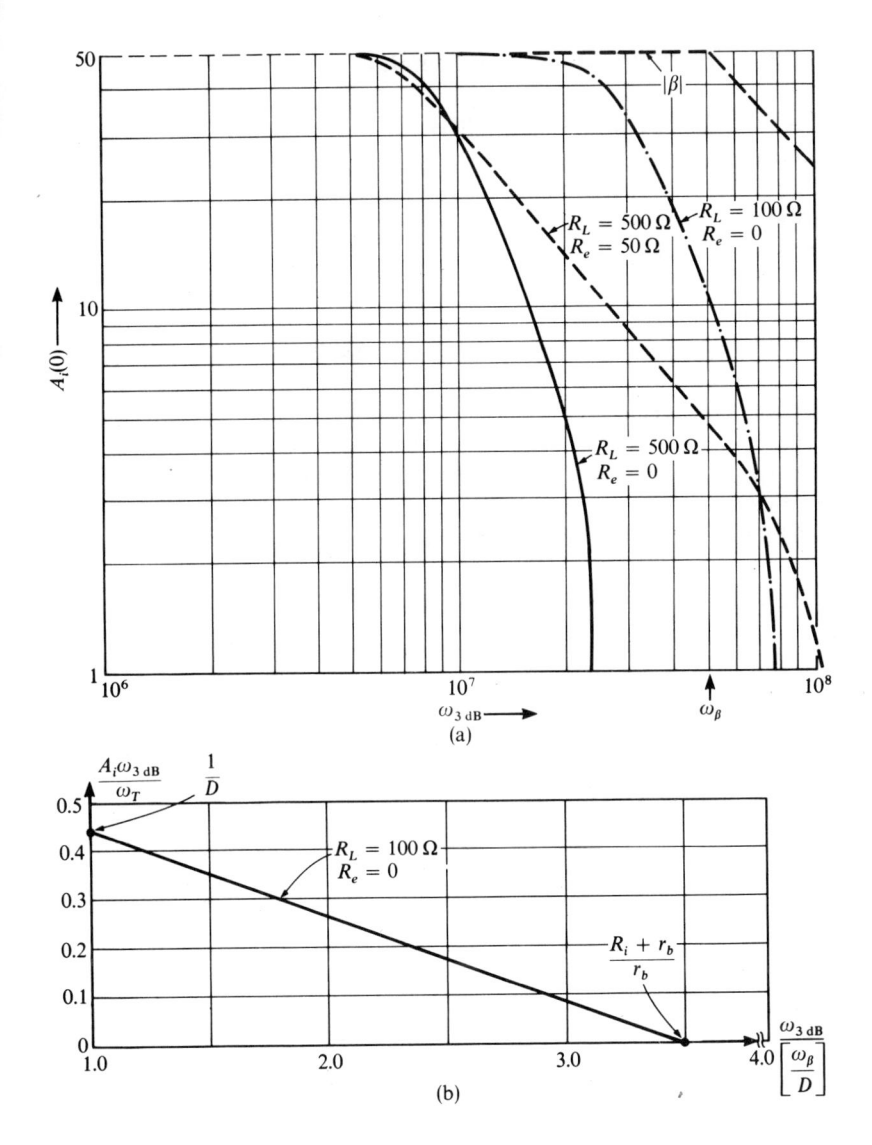

Fig 4-35 (a) Plot of $A_i(0)$ vs $\omega_{3\,dB}$ for the example with and without the series emitter resistance for two different values of R_L; (b) Plot of normalized GBP vs normalized bandwidth for the example with no broadbanding

There are many broadbanding schemes that can improve the gain-bandwidth product of a stage. We shall consider a few simple practical cases here to illustrate this point and more will be said in the next chapter, where we consider multistage amplifiers.

4-13 Simple Broadbanding Techniques

From the previous section, it is quite clear that if the bandwidth requirements of an amplifier stage are such that $\omega_{3\,\text{dB}} \geq (\omega_\beta/D)[(r_b + R_i)/(r_b + r_e)]$, we cannot use the simple resistive broadbanded circuit, since, at the equality sign, the gain of the stage will be approximately unity. For example, for the transistor parameters listed in (4-31) and $R_2 = 500\,\Omega$, suppose we wish to design the stage such that its bandwidth is $\omega_{3\,\text{dB}} = 2.5 \times 10^7$ rad/sec. The quantity $(\omega_\beta/D)[(R_i + r_b)/(r_b + r_e)] = 2.31 \times 10^7$, which is less than the required bandwidth. The circuit of fig 4-34 cannot meet this specification. Hence, we must use some other sort of broadbanding to be able to meet this specification with the given transistor. Let us now consider the use of an additional resistor in a feedback arrangement. In addition to increasing the bandwidth, feedback has other desirable features, which are discussed in detail in Chapter 6.

Addition of a Series Resistance in the Emitter Lead

For large bandwidth requirements, i.e., $\omega_{3\,\text{dB}} \gg \omega_\beta/D$, the addition of a small series feedback resistance (of the order of 50 Ω) at the emitter terminal, as shown in fig 4-36a, improves the gain-bandwidth product.

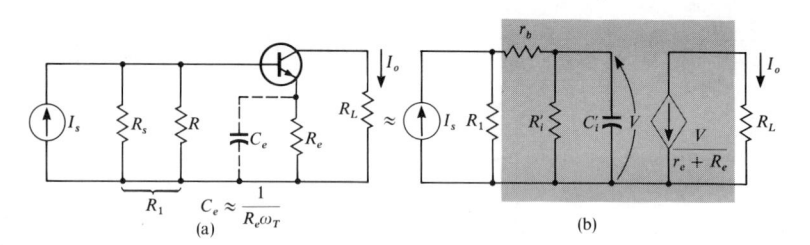

Fig 4-36 (a) Simple series-feedback broadbanded amplifier stage; (b) Small-signal high-frequency ($f \leq f_T/3$) equivalent circuit of (a)

We shall show that in this arrangement the effect of r_b on the bandwidth is small, and that this type of amplifier has a much better GBP capability than the simple resistive broadbanded stage. Actually, the same effect (i.e., increasing r_e) can be achieved by decreasing the emitter current (since $r_e = kT/qI_e$). This approach leads to some complications, however, since β_0 and f_T are functions of the current and will not remain constant, as was seen in fig 3-35.

The circuit of fig 4-36a can be analyzed by replacing the transistor by its hybrid π model and then using nodal or loop analysis to solve for

4-13 Simple Broadbanding Techniques

the current-gain function. For specified R_e and the specific transistor parameters, the GBP vs bandwidth can be examined as in fig 4-35. Thus we need not make any further approximation. For typical transistor parameters, however, the zero and the poles are widely separated and the dominant-pole situation exists (see Problem 4-26). Thus, suitable approximation can also be made in this circuit to simplify the analysis and design, when considering only the forward gain and the bandwidth of the amplifier.

The circuit model of fig 4-36b is an approximation which includes only the dominant natural frequency of the circuit. This circuit is readily obtained by using the source transformation (see Section 1-11). In fact, if a shunt capacitance $C_e \approx 1/R_e\omega_T$ is added across R_e, the zero at $1/R_e C_e \approx \omega_T$ will be close to one of the nondominant poles, suggesting pole-zero cancellation for nondominant poles (see Problem 4-27). The stray capacitance across moderate values of emitter resistance provides approximately such a capacitance.

Thus the model of fig 4-36b would also have a good phase approximation of the circuit of fig 4-36a with C_e across R_e, provided C_e is small (roughly in the range of 1 to 10 pF). If a capacitor C_e is used, its value usually is chosen to be much larger than $1/R_e\omega_T$ so as to achieve high-frequency compensation. In such cases, the zero and the poles are not far apart and the model of fig 4-36a cannot be used. For this situation we use the hybrid π model in our analysis and design. An interesting design situation is to select C_e such that a maximally flat magnitude response is obtained. The GBP would then be better than the case where R_e is not shunted by C_e (Ref 6).

The circuit in fig 4-36b is the same as that in fig 4-34b, so that the expressions for the gain and bandwidth are also similar. These are given by[5]

[5] Note that
$$\text{GBP}|_{R_e = 0} = \frac{\omega_T}{D}\left(\frac{R_1}{R_1 + r_b}\right) \quad \text{and} \quad \text{GBP}|_{R_e \neq 0} = \frac{\omega_T}{D'}\left(\frac{R_1}{R_1 + r_b}\right)$$

One might wrongly conclude that always $\text{GBP}|_{R_e = 0} > \text{GBP}|_{R_e \neq 0}$, since $D' > D$. The fallacy in this argument is that, even though R_1 is the same for both cases, the gain and 3-dB frequencies in each case are different and, since GBP vs 3-dB frequency is not constant, this comparison is not meaningful. In fact, for the same gain, the bandwidth in each case is given by

$$\omega_{3\,\text{dB}}|_{R_e = 0} = \frac{\omega_\beta}{D}\frac{r_b + R_1 + R_i}{r_b + R_1}$$

$$\omega_{3\,\text{dB}}|_{R_e \neq 0} = \frac{\omega_\beta}{D'}\frac{(r_b + R_i')(r_b + \beta_0 r_e + R_1)}{r_b(r_b + \beta_0 r_e) + R_1(r_b + R_i')}$$

Further note that, for R_1 large, $\omega_{3\,\text{dB}}|_{R_e \neq 0} < \omega_{3\,\text{dB}}|_{R_e = 0}$, and for R_1 small, $\omega_{3\,\text{dB}}|_{R_e \neq 0} > \omega_{3\,\text{dB}}|_{R_e = 0}$. In fact, the value of R_1 for which the two quantities, for a given value of R_e, are equal, can be readily determined from above.

$$A_i = \frac{I_o}{I_s} = -\beta_0 \frac{1}{1 + (r_b + R_i')/R_1} \simeq -\beta_0 \frac{1}{1 + \beta_0(R_e + r_e)/R_1} \quad (4\text{-}92)$$

$$\omega_{3\,\text{dB}} \approx \frac{\omega_\beta}{D'} \frac{R_1 + r_b + R_i'}{R_1 + r_b} \simeq \frac{\omega_\beta}{1 + (R_L + R_e)C_c\omega_T} \left(\frac{R_1 + R_i'}{R_1 + r_b} \right) \quad (4\text{-}93)$$

where

$$R_i' = \beta_0(r_e + R_e) \quad \text{and} \quad D' = 1 + (R_L + R_e)C_c\omega_T$$

The effectiveness of R_e in this series feedback arrangement in improving the GBP can best be illustrated by a numerical example. Consider the example of this section where $R_L = 500\,\Omega$, $R_e = 50\,\Omega$, and a current source excitation at the input. The plots of gains vs bandwidth for various values of R_1 based on (4-92) and (4-93) for the transistor are shown in fig 4-35a. Note that beyond a certain bandwidth ($\omega_{3\,\text{dB}} \geq 9 \times 10^6$ rad/sec in this example) the gain is higher for the same bandwidth using the additional series resistance R_e. For a bandwidth $\omega_{3\,\text{dB}} = 2.5 \times 10^7$ rad/sec, which could not be realized by the simple resistive broadbanded stage, the gain using this broadbanding scheme is a little over 10. Calculation of gain and bandwidth using the hybrid π model shows that the results are not far from those given in fig 4-35a. The addition of R_e also provides a degree of freedom in the design of an amplifier stage for a specified bandwidth. In fact, there exists an optimum value for R_e when the load resistance and the 3 dB bandwidth are specified. This optimum value is determined as follows: For a specified $\omega_{3\,\text{dB}}$ and R_L, we solve R_1 in terms of R_e in (4-93). We substitute the resulting expression in (4-92) and then perform the differentiation with respect to R_e. The result is a quadratic equation for optimum R_e,

$$(R_e)_{\text{opt}}^2 + K_1(R_e)_{\text{opt}} + K_2 = 0 \quad (4\text{-}94)$$

where

$$K_1 = -\frac{2[r_b\omega_{3\,\text{dB}}(1 + R_L C_c\omega_T) - (\beta_0 r_e + r_b)\omega_\beta]}{\omega_T(\omega_{3\,\text{dB}} r_b C_c - 1)} \quad (4\text{-}95a)$$

and

$$K_2 = \frac{-r_b\beta_0 r_e C_c\omega_T - r_b(1 + R_L C_c\omega_T) + r_b(\omega_{3\,\text{dB}}/\omega_\beta)(1 + R_L C_c\omega_T)^2}{C_c\omega_T(\omega_{3\,\text{dB}} r_b C_c - \beta_0)} \quad (4\text{-}95b)$$

As a numerical example, for $R_L = 500\,\Omega$, $\omega_{3\,\text{dB}} = 4 \times 10^7$ rad/sec, and the same transistor parameters used in the examples of this section, we obtain $K_1 = 9.6$, $K_2 = -5.0 \times 10^3$, and $(R_e)_{opt} = 68 \, \Omega$. These values of R_e and ω_{3dB} yield $R_1 = 565 \, \Omega$ and the gain $A_i(0) = 6.6$, which is slightly higher than that for $R_e = 50 \, \Omega$ (see fig 4-35a). It has been found, in general, that $(R_e)_{opt}$ has a rather broad range, and that its value is not critical, i.e., the loss of gain near the optimum value is usually not significant. In practice, a somewhat lower value of R_e than that obtained from (4-94) is desirable in cascaded stages, because a lower value of R_e lowers the D-factor of the preceding stage, thus improving the overall gain-bandwidth product. Thus, for wideband stages, we shall use a convenient value of R_e, usually in the range $25 \, \Omega < R_e < 100 \, \Omega$, without using (4-94).

Resistive Shunt Feedback

Broadbanding is also achieved by placing a resistor in shunt, as shown in fig 4-37a. Biasing and coupling circuitry have been omitted for

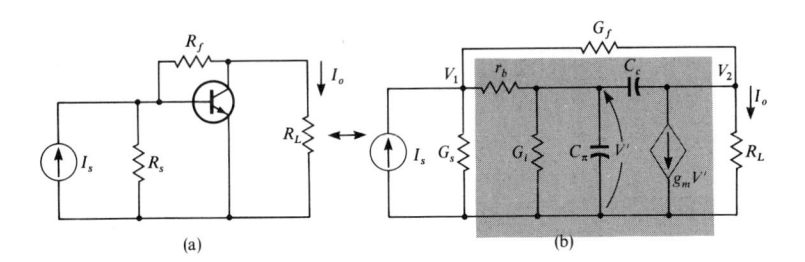

Fig 4-37 (a) Simple shunt-feedback broadbanded amplifier stage; (b) Small-signal high-frequency ($f \leq f_T/2$) equivalent circuit of (a)

simplicity. This configuration is referred to as a shunt-feedback stage, as contrasted to the previous case, which is known as a series feedback configuration. The effect of R_f in stabilizing the operating point has already been noted in Problem 3-33. Here we examine the gain-bandwidth capability of the stage.

The analysis and design of the circuit by the feedback method are quite simple. Since we have not yet covered this topic, however (as we shall in Chapter 6), we shall use more familiar methods of analysis. The equivalent circuit of fig 4-37a, using the hybrid π model for the transistor, is shown in fig 4-37b.[6] The equilibrium nodal equations in the transformed s-domain are

$$I_s = V_1 G_s + (V_1 - V_2)G_f + (V_1 - V')g_b$$
$$(V - V')g_b = V'(G_i + sC_\pi) + (V' - V_2)sC_c \quad (4\text{-}96)$$
$$(V_1 - V_2)G_f + (V' - V_2)sC_c = g_m V' + V_2 G_L$$

From (4-96), we solve for V_2 in terms of I_s and then substitute $V_2 = I_o R_L$. The current-gain function after some manipulation is found to be

$$A_i(0) = \frac{I_o}{I_s} = -\frac{A_o b_o}{z_1} \frac{(s + z_1)}{s^2 + b_1 s + b_o} \quad (4\text{-}97)$$

$$A_o = \frac{-G_L[(g_b + G_i) + g_m g_b]}{g_m G_f g_b + g_b(G_f G_s + G_L G_s + G_f G_L) + G_L(g_b G_f + g_b G_L + G_f G_s + G_L G_s + G_f G_L)}$$

$$\approx \frac{g_m g_b G_L}{G_f g_b(g_m + G_L) + G_L G_i(g_b + G_f)}$$

$$b_1 = \left(\frac{C_c + C_\pi}{C_c C_\pi}\right)\left(\frac{g_b G_f + g_b G_L + G_f G_s + G_L G_s + G_f G_L}{g_b + G_s + G_f}\right)$$
$$+ \frac{1}{C_\pi}\left(g_m - \frac{2g_b G_f + g_b^2}{g_b + G_s + G_f}\right)$$

$$\approx \frac{[g_b(G_f + G_L) + G_f G_L]}{g_b C_c} + \frac{(g_m - 2G_f - g_b)}{C_\pi} \quad (4\text{-}98)$$

$$b_o = \frac{1}{C_c C_\pi}\left\{\frac{g_m g_b G_f + g_b(G_f G_s + G_L G_s + G_f G_L) + G_i[g_b G_f + g_b G_L + G_f G_s - G_L G_s + G_f G_L]}{g_b + G_s + G_f}\right\}$$

$$\approx \frac{1}{C_c C_\pi}[g_m G_f + G_L(G_f + G_s) + G_i(G_f + G_L)]$$

$$z_1 = \frac{g_b(g_m + G_f) + G_f G_i}{C_c(G_f - g_b) + C_\pi G_f}$$

$$\approx \frac{g_b g_m}{G_f C_\pi - g_m C_c}$$

[6] Computer-aided analysis of fig 4-37b (with $r_o = h_{oe}^{-1} = 50\,\text{k}\Omega$ added) via ECAP yields $A_i(0) = 7.69$ and $f_{3\,\text{dB}} = 19.9$ MHz (see Problem 1-34).

4-13 Simple Broadbanding Techniques

In the above we have used the following often-valid approximations to arrive at the simplified expressions

$$C_\pi \gg C_c, \quad g_b \gg G_f, G_s, \quad g_m \gg G_f, G_s, G_i$$

For the transistor parameters listed in (4-91), assuming a current-source input ($R_s \to \infty$), and $R_L = 100\,\Omega$, $R_f = 1000\,\Omega$, the approximate numerical results are

$$A_i(s) \simeq -2.65 \times 10^9 \frac{s + 2.1 \times 10^9}{s^2 + 4.60 \times 10^9 s + 6.96 \times 10^{17}} \qquad (4\text{-}99a)$$

$$\approx -2.65 \times 10^9 \frac{s + 2.1 \times 10^9}{(s + 4.58 \times 10^9)(s + 1.50 \times 10^8)} \qquad (4\text{-}99b)$$

Note that the dominant-pole situation exists. The gain and the bandwidth from (4-99) are[7]

$$A_i(0) = 8.0, \quad f_{3\,\text{dB}} = \frac{1.5 \times 10^8}{2\pi} \approx 24\,\text{MHz} \qquad (4\text{-}99c)$$

The much higher bandwidth in this example than that in the series feedback case is due mainly to the smaller load resistance, hence lower D-factor for the shunt feedback. The GBP of the shunt feedback can be further improved if an appropriate value of inductance L_f is used in series with R_f (Ref 6). The midband input impedance of the shunt-feedback circuit is low, as can be readily seen using the Miller approximation (see Problem 4-34). This is in contrast to the series feedback circuit of the previous section, which had a very large input impedance.

The Shunt-Peaked Stage

Another simple and useful broadbanding method for discrete-circuit design is the use of an inductor in shunt with the input, hence the

---

[7] We shall show in Chapter 6, by using feedback technique, that the expressions for the midband gain and bandwidth of the shunt feedback circuit are approximately given by:

$$\left. \begin{array}{l} A_i(0) \approx \dfrac{\beta_0}{1 + \beta_0 R_L G_f} \\[2mm] f_{3\,\text{dB}} \approx \dfrac{f_\beta}{D}(1 + \beta_0 R_L G_f) \end{array} \right\} \quad \text{for } R_f \gg R_i, R_L$$

For this example these simple expressions yield $A_i(0) \approx 8.3$, and $f_{3\,\text{dB}} \simeq 21\,\text{MHz}$.

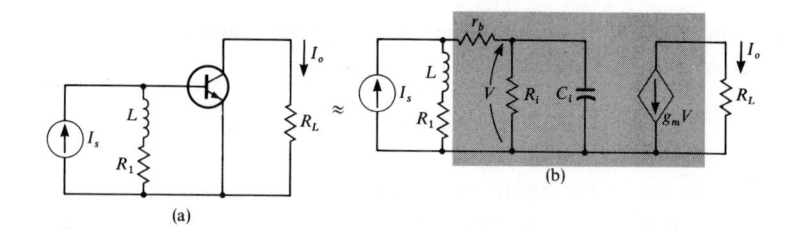

230 4/Analysis and Design of Single-Stage Amplifiers

Fig 4-38 (a) Shunt-peaked amplifier stage; (b) Small-signal high-frequency ($f \leq f_T/3$) equivalent circuit of (a)

name shunt-peaked, as shown in fig 4-38a. Biasing and coupling circuitry are again omitted for the sake of simplicity and clarity.

For a more accurate frequency response analysis of the circuit, we may use the hybrid π model for the transistor. For gain and bandwidth calculations, however, the "unilateral" model utilizing the Miller approximation is sufficiently accurate and quite simple. The dominant poles are obtained at the outset, as discussed earlier in this chapter. From the equivalent circuit of fig 4-38b, the input circuit equilibrium equations in the s-plane are

$$I_s = \frac{V_1}{sL + R_1} + \frac{V_1 - V}{r_b}$$
$$(V_1 - V)g_b = V(G_i + sC_i) \tag{4-100}$$

Solving (4-100) for V/I_s we get

$$\frac{V}{I_s} = \frac{R_1 + sL}{s^2 L C_i + s(C_i R_1 + G_i L_1 + r_b C_i) + r_b G_i + R_1 G_i + 1} \tag{4-101}$$

The output current I_o is equal to $-g_m V$. Hence the current-gain function is given by

$$\frac{I_o}{I_s} = -g_m \frac{(R_1 + sL)}{s^2 L C_i + s(C_i R_1 + G_i L + r_b C_i) + r_b G_i + R_1 G_i + 1} \tag{4-102}$$

We may rewrite (4-102) as follows

$$\frac{I_o}{I_s} = -\frac{\beta_0}{R_i C_i} \frac{(s + R_1/L)}{s^2 + s[(R_1 + r_b)/L + 1/R_i C_i] + (R_1 + r_b + R_i)/(R_i C_i L)} \tag{4-103}$$

4-13 Simple Broadbanding Techniques

Equation (4-103) can also be written for convenience as

$$\frac{I_o}{I_s} = a_o\left(\frac{b}{z}\right)\frac{s+z}{s^2 + as + b} \qquad (4\text{-}104)$$

where

$$a_o = -\beta_0 \frac{R_1}{R_1 + r_b + R_i}, \qquad z = \frac{R_1}{L}$$

$$a = \frac{R_1 + r_b}{L} + \frac{1}{R_i C_i} \qquad b = \frac{R_1 + r_b + R_i}{R_i C_i L} \qquad (4\text{-}105)$$

The effect of the interaction of L with C_i is to move the poles to the left and to create complex poles, as we shall soon see. For a given low-frequency gain specification and load resistance, the gain function can be designed for a maximally flat magnitude function (see Section 2-8) or a linear-phase function (see Section 2-7) by selection of the proper value of inductance L.

EXAMPLE

Let the circuit of fig 4-38 be designed for a current gain of 10 and a maximally flat magnitude function. The transistor is assumed to have the same parameters as (4-91) and $R_L = 500\ \Omega$. The design proceeds as follows.

From the low-frequency gain specification, we determine R_1, i.e., from the expression for a_o in (4-105)

$$R_1 = (r_b + R_i)\left(\frac{a_o}{\beta_0 - a_o}\right) = 350\left(\frac{10}{40}\right) = 87.5\ \Omega$$

From (4-105) and the transistor parameters, we obtain

$$z = \frac{86}{L}, \quad b = \frac{386}{L}(6.89 \times 10^6), \quad a = \frac{136}{L} + 6.89 \times 10^6$$

The maximally flat magnitude condition in (2-25) requires that

$$a^2 - 2b = \left(\frac{b}{z}\right)^2 \qquad (4\text{-}106)$$

From (4-106) we determine the value of L, which is $L = 3.5\ \mu\text{H}$. The 3 dB bandwidth from (4-106) and (4-104) is given by the expression

(see Problem 4-30)

$$\omega_{3\,dB} = \sqrt{\frac{1}{2}\left(\frac{b}{z}\right)^2 + \left[\frac{1}{4}\left(\frac{b}{z}\right)^4 + b^2\right]^{1/2}} \qquad (4\text{-}107)$$

Hence from the values of L, b, z and (4-107) we obtain $\omega_{3\,dB} = 4 \times 10^7$ rad/sec. Note the improvement of the gain-bandwidth product as compared to the resistive broadbanding in fig 4-35a. Compare this value also with the series feedback as shown in fig 4-35a.

If the 3 dB bandwidth is specified for a maximally flat magnitude response rather than the gain specification, the use of (4-106) and (4-107) is very cumbersome. A simple method of design for this case is the iterative approach. Assume a GBP $\approx \omega_T/D$, and determine the gain A_1 from the specified bandwidth (i.e., $A_1 = \omega_T/\omega_{3\,dB}D$) and then follow the design equations for a specified gain A_1. Calculate the bandwidth from (4-107), say it is ω_1. Compare this value with the specified bandwidth, and if the result is not close enough reiterate by designing for $A_2 = A_1\omega_1/\omega_{3\,dB}$ and so on. Usually no more than two iterations are necessary if the initial guess is not too far off. Figure 4-39, which shows A_i vs $\omega_{3\,dB}$, may be used for such purposes, i.e., to determine the initial value for iteration.

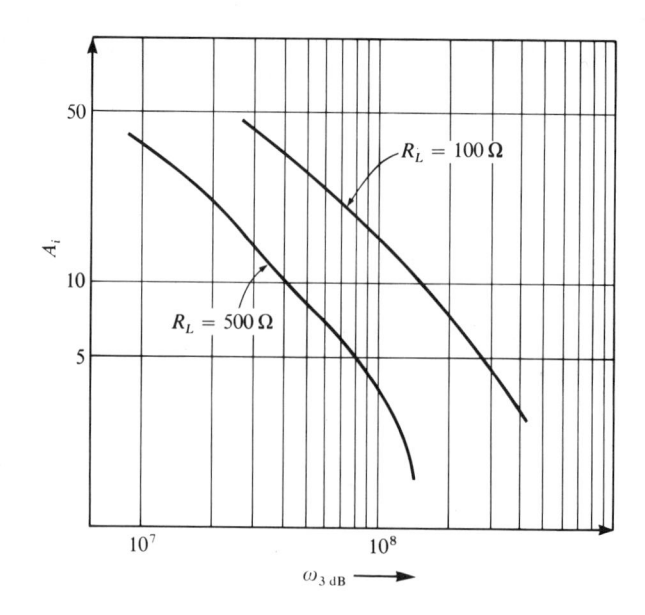

Fig 4-39 Current gain vs bandwidth for a shunt-peaked maximally flat magnitude design for two values of load resistors

4-14 Step Response of an Amplifier Stage

present in the case of the FET amplifier. The Miller effect, which appreciably reduces the GBP of a stage, will be present in the FET amplifier and is taken into consideration in the same manner as for the BJT amplifier. These topics are discussed further in the next chapter, where multistage amplifiers are considered.

4-14 Step Response of an Amplifier Stage

Consider the gain function of a simple resistively loaded amplifier stage in the midband and high-frequency regions. We have already seen this to be a low-pass function; in many cases, the approximate expression for the gain function is of the form

$$A = \frac{\Phi_o}{\Phi_i}(s) = \frac{A_o p_o}{s + p_o} \qquad (4\text{-}114)$$

where p_o is the dominant pole of the high-frequency circuit, A_o is the midband gain, and Φ_o and Φ_i are the transforms of the output and input signals. These may be either currents or voltages. The magnitude and phase responses of (4-114) are shown in fig 4-41.

If the input is a unit step voltage, $v_i(t) = u(t)$, its transform is $1/s$ and the transform of the output voltage is

$$V_o(s) = \frac{A_o p_o}{s}\left(\frac{1}{s + p_o}\right) \qquad (4\text{-}115)$$

The inverse Laplace transform of (4-115) yields the time-domain response of the output signal, $v_o(t)$. Hence

$$v_o(t) = \mathcal{L}^{-1}[V_o(s)] = A_o(1 - e^{-tp_o}) \qquad (4\text{-}116)$$

The unit step response is shown in fig 4-42.

In the time-domain response, the key features of a step response are the 10 to 90 per cent rise time τ_R and the 50 per cent delay time τ_D. (Other quantities, such as overshoot, which results from the gain function having complex poles are also of importance (see Section 2-6), but will not be present in this case). The rise time and the delay time can be readily determined from (5-81) and their definitions (see fig 4-42).

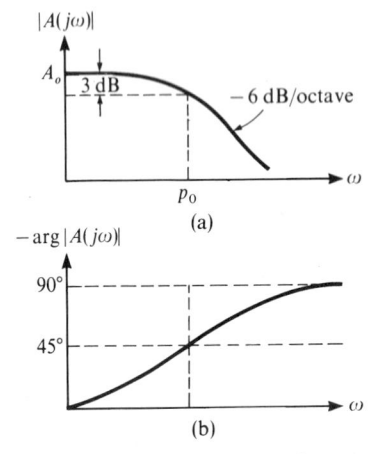

Fig 4-41 (a) Magnitude response of a gain function having one pole; (b) Phase response of a gain function having one pole

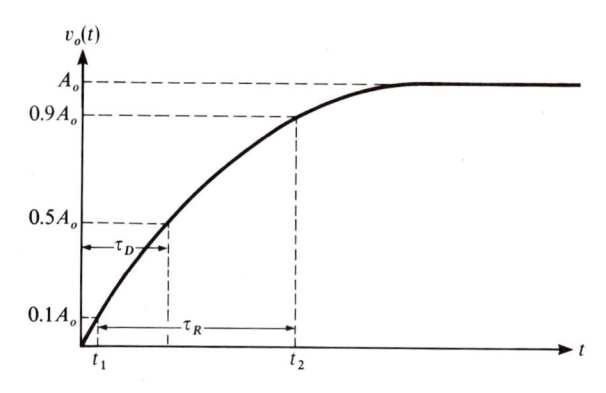

Fig 4-42 Unit step response for a function with single pole

To find τ_D, we set $v_o(t) = 0.5\,A_o$ and solve for t. Thus

$$0.5\,A_o = A_o(1 - e^{-\tau_D p_o}) \quad \text{or} \quad \tau_D = \frac{0.69}{p_o} \qquad (4\text{-}117)$$

To determine τ_R, we set $v_o(t) = 0.1\,A_o$ to find t_1, and $v_o(t) = 0.9\,A_o$ to find t_2, and then $\tau_R = t_2 - t_1$. Hence

$$0.1\,A_o = A_o(1 - e^{-t_1 p_o}) \quad \text{or} \quad t_1 = \frac{0.1}{p_o}$$

$$0.9\,A_o = A_o(1 - e^{-t_2 p_o}) \quad \text{or} \quad t_2 = \frac{2.3}{p_o}$$

The 10 to 90 per cent rise time is

$$\tau_R = t_2 - t_1 = \frac{2.2}{p_o} \qquad (4\text{-}118)$$

Note that the rise time and the 3 dB bandwidth for the gain function given in (4-114) are completely related. Since $\omega_{3\,\text{dB}} = p_o$, we have

$$\tau_R = \frac{2.2}{\omega_{3\,\text{dB}}} \qquad (4\text{-}119\text{a})$$

or

$$\tau_R f_{3\,\text{dB}} = \frac{2.2}{2\pi} = 0.35 \qquad (4\text{-}119\text{b})$$

The relationship in (4-119) has also been found to be empirically true for a much wider class of gain functions. Specifically, if the gain functions are monotonically increasing vs time, the relationship in (4-119) can be used. We shall discuss this matter in the next chapter in connection with multistage amplifiers. In passing, we note that when the network function of a distributed network is approximated by a dominant pole and a delay factor, as in (4-86), the response to a step input is similar to (4-116), except that it is delayed in time by τ_o. For example, for (4-86), the time response is

$$v_o(t) = \mathscr{L}^{-1}\left[\frac{A_o p_o}{s}\left(\frac{1}{s + p_o}\right)e^{-s\tau_o}\right]$$
$$= A_o[1 - \exp{-(t - \tau_o)p_o}]u(t - \tau_o) \qquad (4\text{-}120)$$

where $\tau_o = 0.2/\omega_\alpha$ and we have used the shifting property of the Laplace transform (table 1-1). Note that $u(t - \tau_o)$ is the Heaviside unit function, which has a value of unity for $t > \tau_o$ and is identically zero for $t < \tau_o$. For (4-120), the rise time is the same as in (4-116), but the delay time τ_D is increased by the dead time τ_o, so that the delay time is equal to $(0.69/p_o) + \tau_o$.

4-15 Summary

We have considered the analysis and design of single-stage amplifiers in order to focus our attention on the salient performance parameters of the amplifier in the frequency and time domains. Amplifiers almost invariably use more than one stage. Some of the basic features of the design of single-stage amplifiers carry over to the design of practical multistage amplifiers used in electronic circuits.

Knowledge of the midband network properties of the device is of considerable help in deciding what configurations to use and whether the desired gain can be achieved. The formulas of the gains and impedances developed in Section 4-2 and Section 4-3 can be of considerable help in calculating, in a simple manner, the midband properties of a multistage amplifier.

For frequency-response calculations, such as those of ac coupled stages, it can be shown that analysis and design may be subdivided into two simple cases: a low-frequency region and a high-frequency region. The low-frequency behavior is determined by the coupling and bypass capacitors. If the amplifier is to amplify the dc signal as well, then dc coupling is necessary, as discussed in the next chapter. The high-frequency response is determined by the device parameters and the external circuit resistors. The gain-bandwidth product of the device imposes a limitation on the achievable upper 3-dB frequency cutoff point of the amplifier. The accuracy of results in analysis and design depends on the models chosen for the device. A compromise between the accuracy and the complexity of the model is often made, especially if a preliminary paper-and-pencil analysis or design is performed. A simple unilateralized model based on the dominant-pole concept often gives reasonably accurate results for gain and bandwidth calculations if properly applied. In a real design situation, these could be interpreted as a preliminary design. Computer-aided iteration may then be performed using a more adequate model, such as the hybrid π model for the transistor. Even if computer-aided design is to be used, we should not overtax the computer with unnecessarily complicated models. In a discrete circuit design, the final design may be adjusted experimentally in order to achieve the actual specifications.

In order to summarize and clarify some of the concepts discussed in this chapter, we will now illustrate the simple preliminary design of a complete single-stage amplifier.

Design Example

Let us consider the complete design of a single-stage CE amplifier. We have available in stock transistors which we must use. The quiescent operating point suggested by the manufacturer for an optimum gain-frequency response is approximately $I_C = 5\,\text{mA}$, $V_{CE} = 5\,\text{V}$. At this operating point, the typical transistor parameters at room temperature are as follows:

$$\beta_0 = 100, \qquad f_T = 500\,\text{MHz}, \qquad r_b = 100\,\Omega, \qquad C_c = 4\,\text{pF}$$

For this particular design, we also have the following specifications: the supply voltage is 20 V, the source and load terminations are $R_s = 10\,\text{k}\Omega$, and $R_L = 500\,\Omega$, respectively.

We wish to achieve the following approximate performance parameters in the single-stage amplifier. The low- and high-frequency cutoff points, respectively, are to be $f_l \leq 1\,\text{kHz}$, $f_h \simeq 5\,\text{MHz}$. The stage must have a current gain of approximately 20 dB (i.e., $A_i \geq 10$) or more with a one-pole rolloff near the upper band-edge frequency. The operating point is to be reasonably stable for normal environmental temperature changes and interchangeability of transistors.

Solution: We make a quick rough check to see if the specifications can be met in a single stage. The maximum current gain for any single stage is β_0, which, in this case, is 100. The GBP from the specification is $A_i f_{3dB} \geq 50$ MHz, which is less than f_T. At this point there is no obvious indication of not being able to meet the specifications. We now proceed with the design, starting with dc biasing, high-frequency, and low-frequency design, respectively.

Dc Biasing

From the biasing design procedure of Section 3-17 (see example) we have already determined the following circuit resistor values for reasonably stabilized biasing (note that we would usually select standard size resistors closest to the calculated values):

$$R_c = 2 \text{ k}\Omega, \qquad R_E = 1 \text{ k}\Omega$$
$$R_{b1} = 17 \text{ k}\Omega, \qquad R_{b2} = 7 \text{ k}\Omega, \qquad R_b = R_{b1} \| R_{b2} = 5 \text{ k}\Omega$$

High-Frequency Design

First, we design the high-frequency circuit, because the broadbanding resistor R_1 needed at the input affects the low-frequency design. Hence, the value of R_1 must be determined. This will become clear when we come to the low-frequency design. We calculate $(f_\beta/D)[(R_i + r_b)/(r_b + r_e)] = 3.9 \times 10^6$, a quantity which is less than the desired high-frequency cutoff. Hence, simple resistive broadbanding at the input cannot do the job. We must, therefore, choose some other broadbanding scheme. Let us try the series feedback configuration. We could use the expression in (4-94) to obtain $(R_e)_{\text{opt}}$. For simplicity, however, let us arbitrarily choose $R_e = 35 \ \Omega$.

From (4-93), we determine the value of R_1 as

$$2\pi(5 \times 10^6) = \frac{2\pi(5 \times 10^6)}{7.28}\left(\frac{R_1 + 4000}{R_1 + 100}\right)$$

or

$$R_1 = 520 \ \Omega$$

The current gain is determined from (4-92):

$$A_i \approx -100\left(\frac{1}{1 + 4000/520}\right) = -11.5$$

which meets the desired specification. Hence the preliminary high-frequency design is complete.

Low-Frequency Design

We proceed according to the design method given in Section 4-4. First we check the pole separation condition as in (4-31). We find that $R_b \| \beta_0 R_E = 5 \text{ k}\Omega$ is larger than $2.5(R_i + r_b + R_1) = 3.1 \text{ k}\Omega$ and thus the design method will be valid. Note that we have used $\beta_0 r_e$ and not $\beta_0(r_e + R_e)$ for R_i since, for frequencies near or below the low-frequency cutoff point, the emitter Z_E due to R_E (and C_E) is much larger than the value of R_e. Since these impedances are in series, the value of R_e can be ignored in comparison with Z_E.

Thus from (4-32) and (4-33) we obtain:

$$C_1 = \frac{2}{2\pi(10^3)} \left[\frac{5.6 \times 10^3}{(2.6 \times 10^6) + (3.3 \times 10^6)} \right] = 0.303 \text{ }\mu\text{F} \rightarrow 0.33 \text{ }\mu\text{F}$$

$$C_E = \frac{2}{2\pi(10^3)} \left[\frac{5.52 \times 10^3}{(2.6 \times 10^6) + (3.3 \times 10^6)} \right] = 25 \text{ }\mu\text{F} \rightarrow 27 \text{ }\mu\text{F}$$

$$C_2 \geq \frac{10}{2\pi(10^3)(2.5 \times 10^3)} = 0.64 \text{ }\mu\text{F} \rightarrow 0.68 \text{ }\mu\text{F}$$

In the above we have selected standard-size capacitors. Note the particularly large value of the emitter bypass capacitor C_E. This capacitor is almost always very large. Because of their relatively large values ($> 25 \text{ }\mu\text{F}$) vs physical size, electrolytic capacitors are sometimes used. Electrolytic capacitors have an associated series resistance in the range of 1 to 5 Ω. This small resistance can affect the gain and bandwidth performance of the stage, especially if R_e is zero. Also recall that in the relation expressed in (3-14), i.e., $r_e = kT/\lambda q I_e$, we have used the nominal value $\lambda = 1$. But in a real transistor, λ varies from 0.5 to 1. This variation can certainly affect the gain and bandwidth of the amplifier (see Problem 4-37).

The preliminary design example may be analyzed more accurately by any of the available packaged computer programs. For example, via ECAP (see Section 1-13), using the hybrid π model (with $r_0 = h_{oe}^{-1} = 50 \text{ k}\Omega$) for the transistor and the circuit parameters of fig 4-43, we would obtain (Problem 4-44) the following results:

$$f_l = 0.14 \text{ kHz}, \quad f_h = 4.8 \text{ MHz}, \quad A_i = -14.9$$

The above results meet the design specifications of the amplifier example.

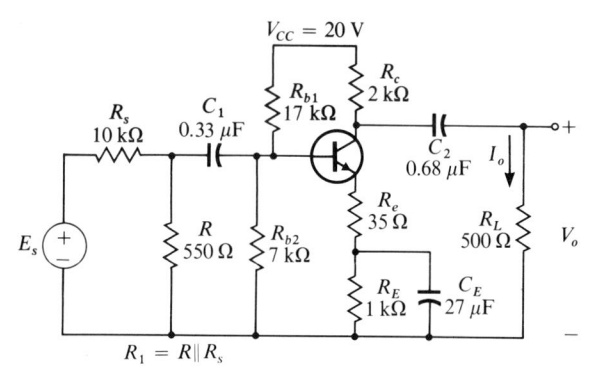

Fig 4-43 Complete single-stage amplifier of the design example

REFERENCES AND SUGGESTED READING

1. Searle, C. L., et al, *Elementary Circuit Properties of Transistors*, New York: Wiley, 1964, Chapters 6 and 7.
2. Pederson, D. O., *Electronic Circuits* (Prelim. ed.), New York: McGraw-Hill, 1965, Chapters 5 and 7.
3. Schilling, D. L. and Belove, C., *Electronic Circuits: Discrete and Integrated*, New York: McGraw-Hill, 1968, Chapters 12 and 13.
4. Comer, D., *Introduction to Semiconductor Design*, Reading, Mass: Addison-Wesley, 1968, Chapters 5 and 7.
5. Millman, J., and Halkias, C., *Electronic Devices and Circuits*, New York: McGraw-Hill, 1967, Chapters 12, 13, and 14.
6. Ghausi, M. S., *Principles and Design of Linear Active Circuits*, New York: McGraw-Hill, 1965, Chapter 12.

PROBLEMS

4-1 Show that the midband h-parameters of the CB configuration in terms of the Tee-model parameters are given approximately by

$$h_{ib} \approx r_e + \frac{r_b}{\beta_0} \qquad h_{ob} \approx \frac{1}{r_c}$$

$$h_{fb} \simeq -\alpha_o \qquad h_{rb} \approx \frac{r_b}{r_c}$$

4-2 Show that the midband h-parameters of the CC configuration in terms of the Tee-model parameters are given approximately by

$$h_{ic} \approx r_b + \beta_0 r_e \qquad h_{oc} \simeq \frac{\beta_0}{r_c}$$

$$h_{fc} \simeq -\beta_0 \qquad h_{rc} \simeq 1$$

4-3 Show that the Tee-model parameters in terms of the h-parameters for the CE configuration are given by

$$\alpha_0 = \frac{h_{fe}}{1 + h_{fe}} \qquad r_c = \frac{1 + h_{fe}}{h_{oe}}$$

$$r_e = \frac{h_{re}}{h_{oe}} \qquad r_b = h_{ie} - \frac{h_{re}}{h_{oe}}(1 + h_{fe})$$

Determine the numerical values for the transistor described in (4-4).

4-4 Derive (4-10).

4-5 Derive the expressions in footnote 1.

4-6 If the h-parameters of the CB configuration are known, determine the h-parameters of the CE configuration.

4-7 For the paraphase circuit shown in fig P4-7, obtain the midband Thévenin equivalents (a) for V_1 to ground; (b) for V_2 to ground. Assume that $R_b \gg h_{ie}$.

4-8 In integrated circuits a constant current source is usually obtained using the transistor circuit shown in fig P4-8. Determine the midband Norton equivalent of the circuit looking into terminals ab. What are the numerical values of the current and the parallel resistance if the transistor has the Tee-model parameters of table 4-1, and $R_1 = R_2 = 1\ \text{k}\Omega$ and $R_3 = 2\ \text{k}\Omega$.

4-9 In an amplifier, it is sometimes useful to know a quantity referred to as the voltage isolation, V.I., which is defined as follows:

$$\text{V.I.} = \frac{\text{voltage gain in the forward direction}}{\text{voltage gain in the reverse direction}}$$

For the transistor parameters and terminations given in table 4-1, find the midband V.I. for the three configurations.

4-10 For the transistor parameters given in table 4-1, (a) determine the midband available power gain, G_A, for the three configurations. [Hint: Use (2-57) in terms of the generalized parameters k_{ij} and the results of Section 4-2.] (b) Is the device potentially unstable? If not, what is the maximum power gain in each case?

4-11 Determine the approximate midband y-parameters of a CE stage from the low-frequency hybrid π model in an effortless manner.

4-12 The transistor described in (4-4) has the following h-parameters at $I_C = 5\ \text{mA}$, $V_{CE} = 5\ \text{V}$:

$$h_{ie} = 300\ \Omega \qquad h_{fe} = 60$$
$$h_{re} = 5 \times 10^{-5} \qquad h_{oe} = 5 \times 10^{-5}\ \text{mhos}$$

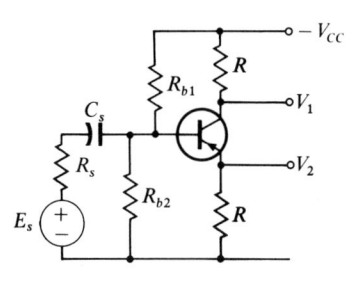

Fig P4-7

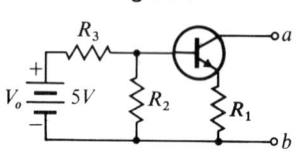

Fig P4-8

The source and load terminations are unchanged from (4-4). Determine the current, voltage, and power gain of the stage and compare it with those of the text referring to (4-4) at $I_C = 1$ mA.

4-13 (a) Determine the midband R_{in}, R_o, and A_v of a common-gate FET stage.
(b) Repeat the above calculations for an FET in the common drain connection.

4-14 If transformers are used at the input and output to obtain the maximum power gain, determine the power gain, the transformer turns ratios, and the configuration, if the transistor parameters are given in Problem 4-12. Assume ideal transformers. The source and load terminations are $R_s = 50\,\Omega$ and $R_L = 1\,\text{k}\Omega$. What configuration yields the maximum power?

4-15 For the circuit shown in fig P4-15, determine C_1 so that the lower 3-dB frequency of the current-gain function is $f_l = 50$ Hz. Consider the following two cases:
(a) $R_e = 0$
(b) $R_e = 50\,\Omega$.
(Given $\beta_0 = 100$, $r_b = 100\,\Omega$, and $h_{ie} = 1\,\text{k}\Omega$ in both cases.)

4-16 For the circuit shown in fig P4-16, determine the values of C_1 and C_E such that the low-frequency cutoff is at $f_l = 100$ Hz. Determine the midband voltage gain and the upper cutoff frequency f_h. Plot V_o/V_{in} (asymptotic plot). Note that $I_C \approx 1$ mA. Assume the following parameters:

$$\beta_0 = 90, \quad r_b \simeq 0, \quad f_T = 200 \text{ MHz}, \quad C_c = 6 \text{ pF}$$

(Hint: Since R_b is not much larger than R_i (4-31) is not satisfied. Therefore, use either (4-32a) and (4-32b), or (4-25) and (4-26) as an iteration for (4-22).)

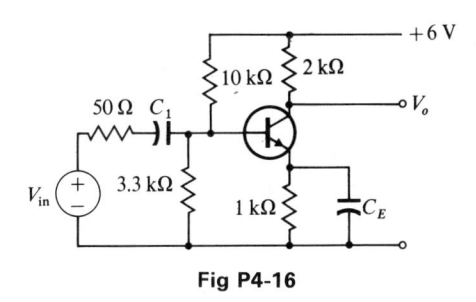

Fig P4-15

Fig P4-16

4-17 For the FET amplifier shown in fig P4-17, $g_m = 2 \times 10^{-3}$ mho, and $r_d = 20\,\text{k}\Omega$
(a) Determine the complete low- and mid-frequency voltage-gain function.
(b) If $C_1 = C_s = 10\,\mu\text{F}$, plot V_o/V_{in} (asymptotic plot) and find ω_l.

Fig P4-17

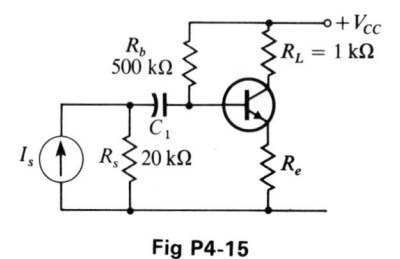

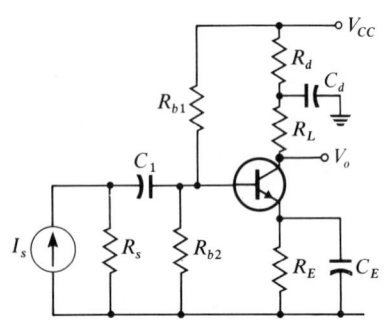

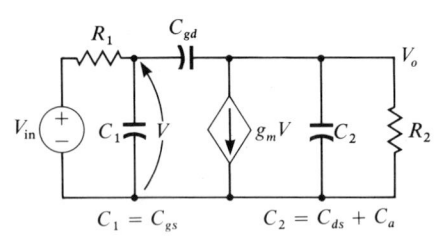

4/Analysis and Design of Single-Stage Amplifiers

Fig P4-18

Fig P4-21
$C_1 = C_{gs}$ $C_2 = C_{ds} + C_a$

4-18 A *CE* amplifier circuit with sag compensation is shown in fig P4-18. Let $R_b \gg R_i$ and assume that C_1 and C_E are short-circuited in the frequency range of interest. Use the low-frequency hybrid π model to show that the initial slope of the output voltage is positive, i.e., $dv_2/dt|_{t=0^+} > 0$. Determine the low-frequency pole-zero location of the stage under the above conditions. Note that $R_b = R_{b1} \| R_{b2}$.

4-19 Derive (4-48).

4-20 For the circuit shown in fig 4-23, determine the input impedance if the load impedance is a shunt *LC* circuit, i.e., $Y_L = sC + 1/sL$. Under what condition is the real part of Y_{in} negative? Repeat the preceding for $Z_L = sL + 1/sC$. Use the Miller approximation and comment on your results.

4-21 For the FET circuit shown in fig P4-21, use the Miller approximation and then find the gain function V_o/V_{in}. Note that C_L is to be included, since additional capacitance C_a is added at the output. Observe the pole-zero cancellation in your results.

4-22 The high-frequency gain function of a certain amplifier is given by

$$A_i = K \frac{(s_n + 50)}{(s_n + 25)(s_n + 55)(s_n + 5)}$$

where the frequency is normalized by a factor of 10^6, i.e., $s_n = s/10^6$.
(a) Determine the approximate 3 dB bandwidth of the amplifier.
(b) What is the phase angle at $\omega_n = 5$?
(c) Sketch the asymptotic plot.

4-23 The high-frequency gain function of a cascaded FET amplifier is given by

$$A_v = \frac{K(s_n + 12)(s_n + 15)}{(s_n + 1)(s_n + 13)(s_n + 1.5)(s_n + 16)}$$

where $s_n = s/10^6$. Determine the approximate 3 dB bandwidth of the amplifier.

4-24 The low-frequency gain function of a transistor amplifier is given by

$$A_i = \frac{10^3 s^2 (s + 1)}{(s + 30)(s + 30)(s + 4)}$$

Determine the low-frequency cutoff f_l and sketch the asymptotic plot.

4-25 Determine the 3-dB bandwidth of an amplifier which has the following gain function:

$$A_i(s) = \frac{A_o}{(1 + s/p_o)^3}$$

4-26 For the series-feedback circuit shown in fig P4-26, find the current-gain function; use the hybrid π model. If the transistor parameters are as given in (4-91) and $R_e = 50\,\Omega$, $C_e = 10\,\text{pF}$, and $R_1 = R_L = 500\,\Omega$, find the pole-zero locations of the current-gain function, the 3-dB bandwidth, and the midband gain of the amplifier.

4-27 (a) Show that if $1/R_e C_e \approx \omega_T$, $C_\pi \gg C_c$, and $\beta_0 \gg 1$, the circuit in fig P4-26 is equivalent to the hybrid π model shown in fig P4-27.
(b) Obtain the approximate unilateral model (compare with fig 4-36b) and repeat the calculation of Problem 4-26.

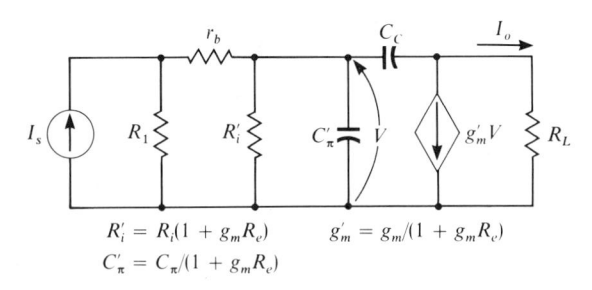

Fig P4-26

$R'_i = R_i(1 + g_m R_e)$
$C'_\pi = C_\pi/(1 + g_m R_e)$
$g'_m = g_m/(1 + g_m R_e)$

Fig P4-27

4-28 The manufacturer of a certain type of transistor gives the following information at room temperature: At the operating point (which is at $I_E = 2.5\,\text{mA}$, $V_{CE} = 5\,\text{V}$), h_{fe} measured at 10 kHz is 100, h_{fe} measured at 10 MHz is 10, h_{ie} measured at 10 kHz is approximately 1 kΩ, and $R_e(h_{ie})$ measured at 10 MHz is 100 Ω. For a current-source input and a load resistance of 1 kΩ, $f_{3\,\text{dB}} = 200\,\text{kHz}$. Determine the values of g_m, r_e, r_b, β_0, C_π, C_c, R_i, f_T, and f_β for the transistor.

4-29 For the simple resistive stage shown in fig P4-29, the parameters are as given in (4-28).
(a) Find R_I for a current gain of 10 and determine the bandwidth.
(b) Find R_I for a 3-dB bandwidth of 2 MHz and determine the current gain.

4-30 Derive (4-107).

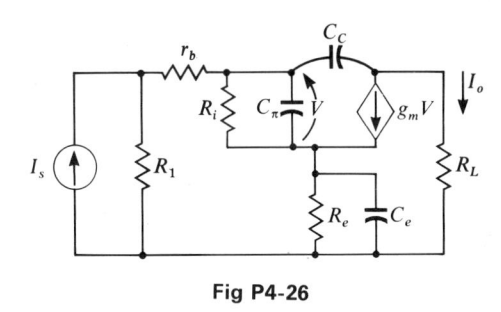

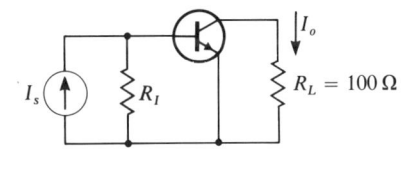

Fig P4-29

4-31 For the shunt-peaked transistor stage shown in fig 4-38, the transistor parameters are given in (4-91). The load resistance $R_L = 500\,\Omega$. Design the circuit for a bandwidth of $\omega_{3\,\text{dB}} = 10^7\,\text{rad/sec}$, and an approximate maximally flat magnitude function. Determine the midband gain of the designed circuit. (Hint: Use iterative design, taking the text example as the first iteration.)

4-32 For the shunt-peaked stage in fig 4-38, $R_L = 500\,\Omega$. Design the circuit for a constant-resistance input. Find L, R_1, $A_i(0)$, and $f_{3\,\text{dB}}$. Use the transistor parameters in (4-91).

4/Analysis and Design of Single-Stage Amplifiers

Fig P4-33

4-33 A shunt-peaked pentode stage is shown in fig P4-33. Given that $C_i = C_o = 5$ pF, and $g_m = 5 \times 10^{-3}$ mhos.
Design the circuit for a gain $A_v(0) = 10$ and a maximally flat magnitude response. What is the 3-dB bandwidth of the designed circuit? Compare the bandwidth with the case when $L = 0$.

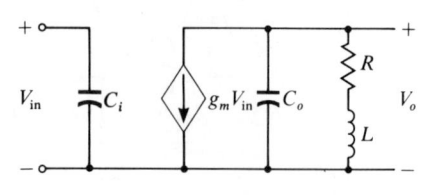

4-34 For the shunt feedback circuit shown in fig 4-37, determine the midband input impedance using the Miller approximation.

4-35 A transistor series-peaked approximate equivalent circuit is shown in fig P4-35.

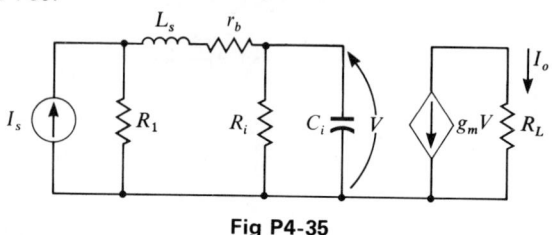

Fig P4-35

(a) Determine the current-gain function.
(b) If the transistor has the parameters given in (4-91) and $R_L = 100\,\Omega$, design the circuit (i.e., find R_1 and L_s) to obtain a maximally flat magnitude function with $\omega_{3\,\text{dB}} = 2 \times 10^7$ rad/sec. What is the midband gain of the circuit?

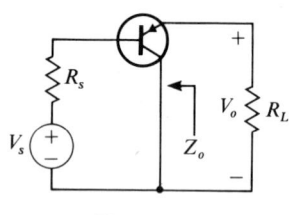

4-36 For a CC stage with $R_L = 500\,\Omega$ and $R_s = 50\,\Omega$, as shown in fig P4-36, perform the following:
(a) Determine the upper 3-dB bandwidth of the amplifier if the transistor parameters are as in (4-91). What is the midband gain of the circuit? (Use the hybrid π model and suitable approximations.)
(b) Determine the output impedance $Z_o(s)$.

Fig P4-36

4-37 For the design example given in Section 4-15, consider the following situations.
(a) If the real transistor used in the circuit has a constant of proportionality other than 25 mV (i.e., $r_e = kT/\lambda q I_e = 30/I_e = 6\,\Omega$), what is the midband gains, f_l, and f_h, for this case?
(b) If R_e of the circuit is zero, comment on the effects of (a) if the emitter bypass capacitor C_E has a series resistance of 1 Ω.

4-38 A single-stage CE amplifier is to be designed such that the 10 to 90 per cent rise time is 0.1 μsec, the sag per second is to be less than $0.1/\tau_p$, where $\tau_p = 10^{-4}$ sec. The transistor parameters are as given in (4-91). The signal input is a current source and the load is $R_L = 200\,\Omega$. Show a completely designed circuit (i.e., give the values of R's and C's). Use the dc biasing of fig 4-43. Let $C_E = 10C_2 = 10C_1$.

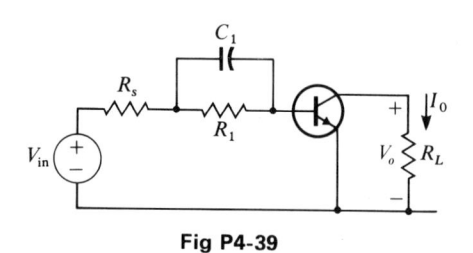

Fig P4-39

4-39 For the base-compensated broadbanded stage shown in fig P4-39, the transistor parameters are as given in (4-91). For proper compensation,

C_1 is chosen such that $R_1 C_1 = R_i C_i$. Given that $R_s = R_L = 100\,\Omega$, perform the following.

(a) Find the expressions for $A_i(s)$, and $\omega_{3\,\mathrm{dB}}$, and show that

$$\mathrm{GBP} = \frac{\omega_T}{D}\left(\frac{R_s}{R_s + r_b}\right)$$

(Note that $I_{in} \equiv V_{in}/R_s$.)

(b) Design the circuit for a bandwidth of $\omega_{3\,\mathrm{dB}} = 8 \times 10^7$ rad/sec. What are the gains $A_i(0)$ and $A_v(0)$?

(c) If $R_s = 0 = r_b$, can the unilateral model be used for the bandwidth calculation? Why?

4-40 For the MOSFET source-follower circuit shown in fig P4-40, determine the following:

(a) The expression for the low-frequency cutoff, f_l.
(b) The expression for the midband voltage gain.
(c) The expression for the high-frequency cutoff, f_h.

4-41 A single-stage MOSFET amplifier is shown in fig P4-41. The MOSFET parameters and the circuit element values are:

$$|y_{fs}| = 5\,\mathrm{mmho}, \quad C_{iss} = 10\,\mathrm{pF}, \quad C_{rss} = 2\,\mathrm{pF}, \quad r_d = 20\,\mathrm{k\Omega},$$

$$R_i = 50\,\Omega, \quad R_D = R_S = 2\,\mathrm{k\Omega}, \quad R_g = R_{g1} \| R_{g2} = 50\,\mathrm{k\Omega},$$

$$C_g = C_s = 1\,\mu\mathrm{F}$$

(a) Find the midband gain V_o/V_i, the low-frequency cutoff ω_l, and the high-frequency cutoff ω_h, of the amplifier.
(b) Sketch $v_o(t)$, if v_i is a unit pulse input with a pulse width equal to $0.2/\omega_l$.

4-42 For the resistive shunt feedback circuit of fig 4-37, $R_L = 100\,\Omega$ and $R_f = 1\,\mathrm{k\Omega}$. The transistor parameters are as given in (4-91), except assume $r_b = 0$. Determine the gain and the bandwidth of the circuit, using the Miller approximation for R_f and C_c.

4-43 For the circuit shown in fig P4-43, determine the midband voltage gain (V_o/V_s); the input impedance Z_{in}, and the output impedance Z_o, both at

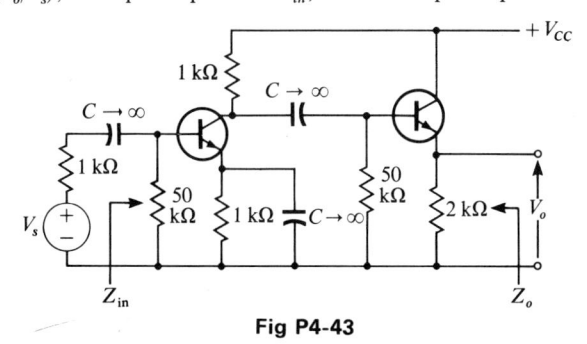

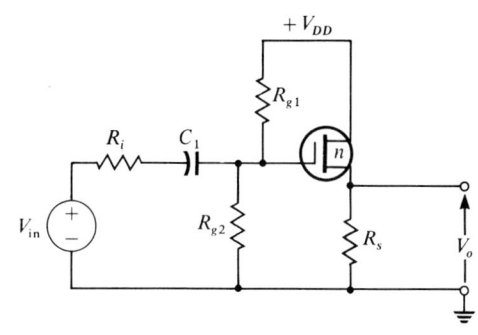

Fig P4-40

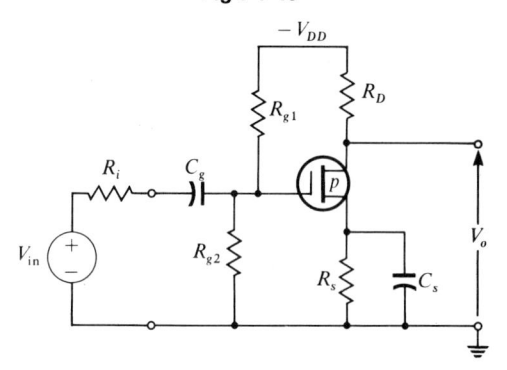

Fig P4-41

Fig P4-43

midband. (You may use the results of Section 4-2, i.e., you need not derive any equation you may use from the text.) Assume identical transistors and the following parameters: $h_{fe} = \beta_0 = 200$, $h_{ie} = 1 \text{ k}\Omega$, $h_{re} = 0 = h_{oe}$, $I_E = 5 \text{ mA}$, $V_{CE} = 5 \text{ V}$.

4-44 If ECAP package and computer facilities are available, analyze the circuit of fig 4-43, using the hybrid π model, and confirm the results obtained in the text.

5
Analysis and Design of Multistage Amplifiers

5-1 Introduction

Amplifiers used in electronic circuits are invariably multistage because of the demands on the gain and bandwidth. Hence more than one stage of amplification is almost always included in the amplifying functional block. Different types of stages are often used at the output it may be desired for an amplifier to have a large current or voltage gain, a wide bandwidth, and the capability to drive a variable high impedance load. In this case, the output impedance of the amplifier must be very low, so as to simulate a voltage source to the load. The input impedance of the amplifier must be much smaller than the signal source impedance, if we are interested in obtaining maximum current from the source. On the other hand, if we are interested in obtaining a maximum voltage from the source, the input impedance of the amplifier should be much larger than the source impedance. For maximum power transfer of course, we match the input impedance to the source impedance. The bandwidth of the input and output stages must also be sufficiently large so as not to impair the overall bandwidth of the amplifier.

The requirements made on the amplifier may be quite different, however. For example, the noise consideration may be important so that the input stage will have an altogether different requirement than when gain and bandwidth are the main criteria. It may be desired for the output stage to deliver an appreciable amount of power to drive a very small load resistance, thus requiring power amplifiers and quite different design considerations. Since the latter two categories require special and new design considerations, they are treated separately in some detail later in this chapter. Small signal models cannot be used for large signal amplifiers, since the signal swings are very large and we must, therefore, use nonlinear, or graphical analysis.

We first focus our attention on the analysis and design of small signal, high-frequency, multistage amplifiers, where the desired gain and bandwidth mandate multistage amplification. As pointed out earlier, we may think of these amplifiers as consisting of three parts: the main amplifier which consists of cascaded iterative *CE* stages (or *CS* stages for FET), the input stage, which couples the signal source to the main amplifier, and the output stage, which couples the main amplifier to the load. In some special cases, the input and output stages may not be much different than the individual stages of the main amplifier.

In small signal linear amplifier design, usually the overall gain and bandwidth, and sometimes the transmission shape, are also specified. Depending upon the particular application, the transistor type, the signal source resistance, and the load impedances, etc., may also be specified. Since the latter specifications, specifically the source and load terminations, can be so different for different applications, the design for a particular situation often becomes an individual problem. The former specifications, that is gain and bandwidth, however, are at the root of every amplifier design and can be treated in general terms. Here again, one cannot prescribe a general design procedure. There are, however, certain common rules that are encountered in the design of the majority of cascaded amplifiers. Often, for reasons of economy and the availability of particular types of transistors, we also specify the transistor types to be used in the design. Since transistor parameters vary from unit to unit in any selected type of transistors, we may either measure the parameters or use typical parameters in the design, and leave provisions in the design for alignment. In a discrete circuit design, the alignment provisions to realize a transmission shape can readily be provided by using variable inductors and/or capacitors. The use of variable resistors is usually costly, and should be avoided. Furthermore, by varying resistors, we not only affect the transmission shape, but also the gain of the amplifier.

5-1 Introduction

In a monolithic integrated circuit design, the use of variable components is not allowed. In this case, however, transistors can be produced to be nearly identical. In the design of integrated circuits (IC), one usually resorts to a computer-aided design examining a variety of situations and alterations. This is important, since the circuit may be complex and no changes can be made once the integrated circuit mask and fabrication are completed. Sometimes a discrete counterpart of an IC is designed and then thoroughly tested on a breadboard. After achieving complete satisfaction in the design, the designer uses some trade-off of the stray wiring capacitance of the discrete circuit to the junction isolation capacitances of the integrated circuit, and goes ahead with the integrated circuit design. This trade-off is very difficult to establish analytically and is usually done from experience on the part of the circuit designer.

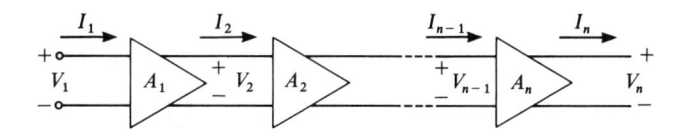

Fig 5-1 Cascaded stages of the main amplifier

The cascaded stages of the *main amplifier* are shown schematically in fig 5-1. The voltage gain and the current gain of the structure are given by

$$A_v = \frac{V_n}{V_1} = \frac{V_2}{V_1}\frac{V_3}{V_2}\cdots\frac{V_n}{V_{n-1}} = A_{v1}A_{v2}\cdots A_{vn} = \prod_{j=1}^{n} A_{vj} \qquad (5\text{-}1)$$

$$A_i = \frac{I_n}{I_1} = \frac{I_2}{I_1}\frac{I_3}{I_2}\cdots\frac{I_n}{I_{n-1}} = A_{i1}A_{i2}\cdots A_{in} = \prod_{k=1}^{n} A_{ik} \qquad (5\text{-}2)$$

From (5-1) and (5-2) it is seen that the overall gain is the product of the individual stage gains, actually voltage ratios and current ratios. It should be pointed out that in calculating overall voltage and current gains as products of individual stage voltage and current gains, the individual stage gains must be determined under the same conditions of termination as are to be used in the final overall cascade, unless noninteracting models are used. Of course, the product amplification can be effective only if the gain of the individual stage is larger than unity.[1]

---

[1] If the bandwidth requirements are such that the individual gains are less than or very close to one, then additive amplification, or the so-called distributed amplifier can be used (see Ref 3, Chapter 13).

252 5/Analysis and Design of Multistage Amplifiers

This is the reason why stages are cascaded so that efficient amplification is achieved. The price paid for obtaining a large gain is bandwidth reduction and vice versa. These topics are discussed in subsequent sections.

5-2 Gain and Bandwidth Calculations for an RC-Coupled FET Multistage Amplifier

A three-stage, RC-coupled, FET amplifier is shown in fig 5-2a. The source and load terminations are also given as shown. Note that the n-channel FET devices are biased as described in Section 3-19, namely

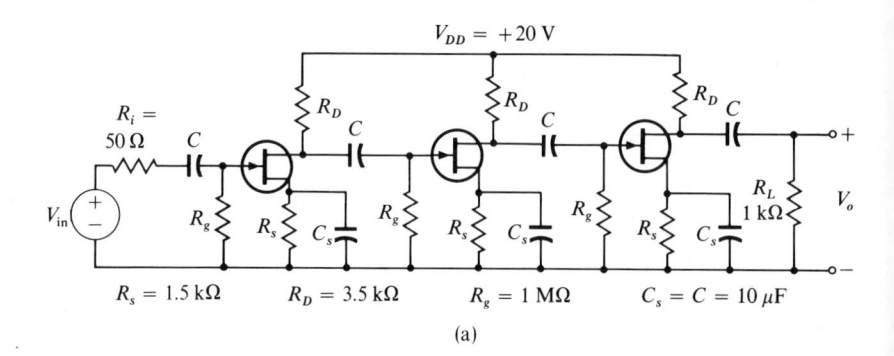

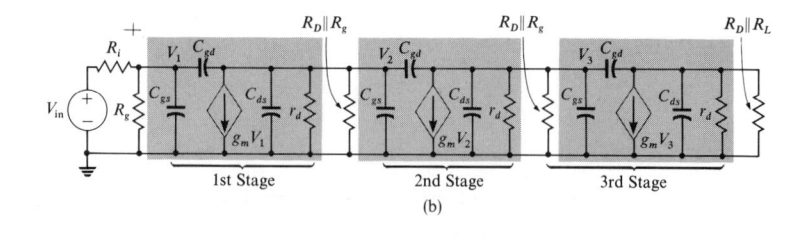

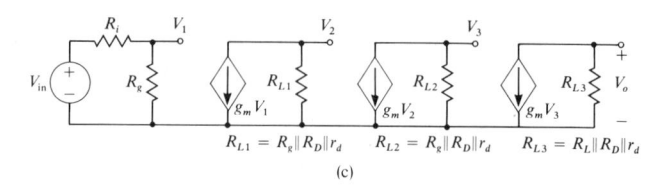

Fig 5-2 (a) Three-stage RC-coupled FET amplifier; (b) Small-signal high-frequency model of (a); (c) Small-signal dc model of (b) for midband calculation

5-2 Gain and Bandwidth Calculations

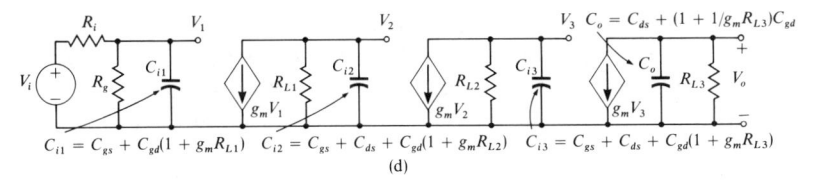

Fig 5-2 (d) Approximate unilateral model of (b)

the biasing point is $I_D = 2.0$ mA, and $V_{DS} = 10$ V. We wish to calculate the midband voltage gain, and the upper and lower 3-dB cutoff frequencies of the amplifier. The JFET parameters at the operating-point are given (or measured) as

$$g_m = 2 \times 10^{-3} \text{ mho} \qquad C_{gd} = 2 \text{ pF}$$
$$C_{gs} = 3 \text{ pF} \qquad (5\text{-}3)$$
$$r_d \equiv g_d^{-1} = 20 \text{ k}\Omega \qquad C_{ds} = 1 \text{ pF}$$

For midband and high frequency calculation the effect of biasing and coupling circuitry capacitors can be ignored, since the capacitors are essentially short-circuited at mid-high frequencies (e.g., for this circuit at $\omega \geq 100$ kHz, $1/\omega C \leq 1\,\Omega$). The midband and high-frequency equivalent circuit of the amplifier is shown in fig 5-2b.

For midband voltage gain calculation we can consider the device capacitances to be open-circuited, since their reactances are very high at midband frequencies (e.g., at $\omega \leq 100$ kHz, $1/\omega C_{gs} \geq 3$ MΩ) and have negligible effect on the midband gain. The model for midband gain calculation is shown in fig 5-2c. The voltage gain is easily determined. Since

$$V_1 = \frac{R_g V_i}{R_i + R_g} \approx V_i, \qquad V_2 = g_m R_{L1} V_1 \qquad \text{etc.}$$

we have

$$A_V = \frac{V_o}{V_i} = \frac{V_o}{V_3} \cdot \frac{V_3}{V_2} \cdot \frac{V_2}{V_1} \cdot \frac{V_1}{V_i}$$
$$A_V = (g_m R_{L3})(g_m R_{L2})(g_m R_{L1})(1) \qquad (5\text{-}4)$$

For this example

$$A_V = (1.50)(5.97)(5.97) \approx 53$$

For the upper cutoff frequency calculation, we can use the high-frequency model shown in fig 5-2b (combining the resistors as in fig 5-2c) and write the four nodal equations in the s-domain and then solve for $V_o/V_i(s)$. This is a very simple task if a computer-aided circuit analysis is used. However, for paper and pencil calculation this is rather tedious, especially the determination of the 3-dB frequency. Furthermore, such a detailed analysis is often not necessary in a preliminary design. We, therefore, use approximations to get the approximate answers. We use the Miller approximation developed in Section 4-8, for an approximate dominant-pole analysis of the circuit. We also note that, since the load of each stage is not a simple resistance, the effect of the load on each input stage is frequency dependent. Beginning with the last stage we see that the load for the third stage is a shunt RC. The Miller approximation for the input of stage 3 is then as in fig 4-26. For the second stage the load is not a shunt RC, but a shunt RC in parallel with a series RC admittance. The Miller approximation for the input of stage 2 will thus be complicated, and similarly for stage 1 it will be even more complicated. In order to simplify the approximate analysis and the preliminary design, we make one more simplifying approximation. We assume a resistive load for each stage with the load resistance given by $Z_L(0)$. Since for a resistive load the Miller approximation yields a pure capacitance at the input of value $C_{gd}(1 + g_m R_L)$, the approximate unilateral model, shown in fig 5-2d, is obtained. Since for each stage $Z_L(0) > Z_L(j\omega)$, this approximation will be slightly pessimistic in that the approximate Miller contribution in fig 5-2d will be larger than the actual values. Note also that the approximation in (4-64), namely, $\omega C_{gd} \ll g_m$, $Y_1(0)$, must be valid. In this example the approximation is valid for frequencies up to about 5 MHz. Hence, if our 3-dB frequency calculation yields a value higher than 5 MHz we have to reexamine our analysis.

For the circuit of fig 5-2d, the overall voltage-gain function can be readily determined as follows:

$$\frac{V_o}{V_3} = \frac{g_m}{G_{L3} + sC_o} \qquad \frac{V_2}{V_1} = \frac{g_m}{G_{L1} + sC_{i2}}$$

$$\frac{V_3}{V_2} = \frac{g_m}{G_{L2} + sC_{i3}} \qquad \frac{V_1}{V_i} = \frac{G_i}{G_i + G_g + sC_{i1}} \qquad (5\text{-}5a)$$

Note that at the output of the last stage the Miller effect (Problem 1-31) is included so that $C_o = C_{ds} + C_{gd}[1 - (1/A)]$. For this example we obtain

$$\frac{V_o}{V_3} = \frac{5 \times 10^9}{s + (3.33 \times 10^9)} = \frac{1.50}{1 + s/(3.33 \times 10^9)}$$

$$\frac{V_3}{V_2} = \frac{22.2 \times 10^7}{s + (3.75 \times 10^7)} = \frac{5.97}{1 + s/(3.75 \times 10^7)}$$

$$\frac{V_2}{V_1} = \frac{11.1 \times 10^7}{s + (1.87 \times 10^7)} = \frac{5.97}{1 + s/(1.87 \times 10^7)} \quad (5\text{-}5b)$$

$$\frac{V_1}{V_i} = \frac{1.18 \times 10^9}{s + (1.18 \times 10^9)} = \frac{1}{1 + s/(1.18 \times 10^9)}$$

From the numerical results in (5-5b) it is readily seen that there are only two dominant poles in the gain function. Hence, the approximate gain function can be written as

$$\frac{V_o}{V_s} \approx \frac{53}{[1 + s/(3.75 \times 10^7)][1 + s/(1.87 \times 10^7)]} \quad (5\text{-}6)$$

The reader is reminded that the numerator constant in (5-6) is the product of all four numerator constants of (5-5b), even those for which the poles are nondominant. From (5-6) the 3-dB bandwidth can be readily calculated.

For convenience, and later reference, we have plotted in fig 5-3 the 3-dB bandwidth for a gain function for two poles, real or complex. For two real poles if the smaller of the two poles is labeled p_1, the bandwidth is determined in terms of p_1 for any ratio $p_2/p_1 \leq 5$. For larger ratios than 5 the second pole is nondominant and $\omega_{3\text{dB}} \approx p_1$. For a complex pole pair, if the magnitude is p_1 and the angle with the negative real axis is ψ, the 3-dB bandwidth can be determined for any $\psi \leq 45$ degrees. Note that $\psi = 45$ degrees corresponds to a maximally flat magnitude function, and for $\psi > 45$ degrees the magnitude response will exhibit peaking. Also $\psi = 30$ degrees corresponds to the linear phase case.

For the gain function in (5-6), $p_2/p_1 \approx 2.0$, and thus from fig 5-3, $\omega_{3\text{dB}} = 0.835 \, p_1 = 1.56 \times 10^7$ rad/sec. Hence

$$f_{3\text{dB}} = \frac{\omega_{3\text{dB}}}{2\pi} = \frac{1.56 \times 10^7}{2\pi} \approx 2.5 \text{ MHz}$$

(For more than two dominant poles in the gain function, an approximate method of calculating $\omega_{3\text{dB}}$ is given in Section 5-3.) Note that $f_{3\text{dB}}$ is less than 5 MHz and the approximation in (4-64) is satisfied.

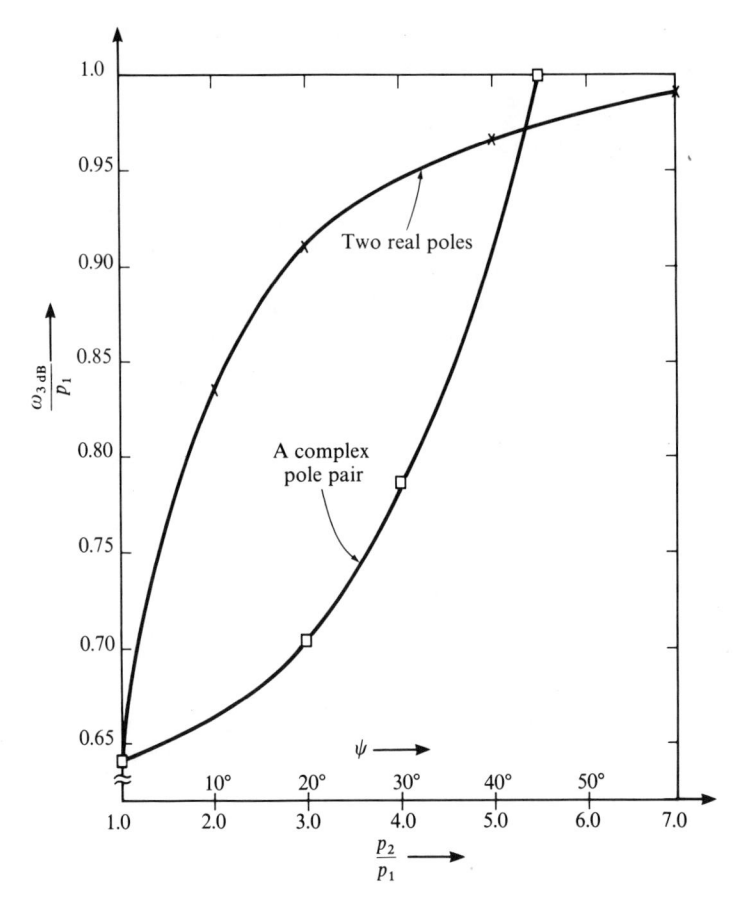

Fig 5-3 Normalized 3 dB bandwidth of $A_1(s) = K_1/(s + p_1)(s + p_2)$ and $A_2(s) = K_2/[s^2 + (2p_1 \cos \psi)s + p_1^2]$

Note that even though the individual stage voltage gain is low, the Miller effect is still appreciable and must be taken into consideration. Because of the Miller effect degradation, the gain-bandwidth product (GBP) capability of the individual stage is greatly reduced. The Miller degradation effect on the GBP is less pronounced for MOSFET's than for JFET's, because C_{gd} is usually much smaller for the former type (see Problems 5-3 and 5-4).

Needless to say, the method of analysis discussed above can be identically carried over to the analysis of triode vacuum tube circuits. For pentodes, since the value of C_{gp} is extremely small (of the order of 0.01 pF), the Miller effect becomes insignificant and can be altogether ignored.

The approximate low-frequency cutoff point of the FET amplifier is readily determined by using the method of Section 4-5. Since the parameters of the first and second stages are the same as those of the example in Section 4-5, the low-frequency cutoff points of the first and second stages are f_{l1} and $f_{l2} = 42$ Hz (which is determined mainly by the source bypass circuit). For the third stage, since R_L is different from R_g, the pole due to the drain coupling circuit from (4-36) is

$$p\Big|_{\substack{\text{drain}\\\text{circuit}}} = \frac{1}{(3.5 \times 10^3 + 10^3)(10 \times 10^{-6})} \approx 22 \text{ rad/sec}$$

This is still much smaller than the pole due to the source bypass circuit, namely, $p_s = 2\pi(42) \approx 264$ rad/sec. Hence, the third stage will also have the same dominant pole as the other stages and the low-frequency cutoff $f_{l3} = 42$ Hz. For three stages, each having identical one dominant pole low-frequency response (remember that for a high-pass circuit the largest poles are dominant) it will be shown in Section 5-3 that the overall low-frequency cutoff of the amplifier for 3 identical dominant poles is

$$f_l = \frac{42}{0.51} = 82 \text{ Hz}$$

The frequency response calculations for vacuum tubes are similar to those above.

It is interesting to note that if ECAP is used for the complete circuit of fig 5-2a, which includes the high-frequency as well as the low-frequency capacitances, the results are (see Problem 5-41a)

$$A_m = 52.8, f_l = 75 \text{ Hz}, f_h = 2.0 \text{ MHz}$$

The simplified analysis is seen to provide close agreement with those obtained from a more complicated and accurate circuit model via computer-aided analysis.

Before we consider the analysis and design of multistage bipolar transistor amplifiers, it is expedient at this point to develop an approximate method of calculating the 3-dB cutoff frequency of an amplifier where there is a cluster of dominant poles in the network function. The cluster of dominant poles in the high-frequency gain function leads to a bandwidth shrinkage in the overall frequency response. This factor must be taken into consideration in the design of an amplifier.

5-3 The Effect of a Number of Dominant Poles on the Bandwidth, Frequency Response, and Step Response of an Amplifier

In order to appreciate the effect of dominant pole-zero locations on the frequency response of an amplifier, we consider in this section simple cases where we have only a set of dominant real poles. The effect of complex poles and/or zeros is discussed later in Section 5-7.

Identical Real Poles, Low-Pass Case

Let the amplifier gain function be given by

$$A(s) = \frac{A_0 p_0^n}{(s + p_o)^n} \tag{5-7}$$

where A_0 is the overall midband gain, $A(s)$ is either a current gain or a voltage gain function, p_o is the dominant pole location of each stage, and n is the number of stages having the dominant pole at p_o. Note that (5-7) does not imply identical stages. The stages may have different gains and values of R and C, but the dominant pole of each stage is assumed to be at p_o.

The overall upper 3-dB cutoff frequency or the 3-dB bandwidth of the amplifier can now be determined as follows. From the definition of the 3-dB radian frequency, we have

$$|A(\omega_{3\,dB})| = \frac{A_0}{|1 + j\omega_{3\,dB}/p_o|^n} = \frac{A_0}{\sqrt{2}} \tag{5-8a}$$

or

$$\left[1 + \left(\frac{\omega_{3\,dB}}{p_o}\right)^2\right]^{n/2} = \sqrt{2} \tag{5-8b}$$

Solving (5-8b) we get

$$\omega_{3\,dB} = p_o\sqrt{2^{1/n} - 1} = p_o S_n \tag{5-9}$$

where

$$S_n \equiv \sqrt{2^{1/n} - 1} \tag{5-10}$$

5-3 The Effect of a Number of Dominant Poles

The factor S_n, which is always less than unity, is called the *identical pole shrinkage factor*. This factor tells us by how much the overall bandwidth will be reduced (for a given number of stages) from that of the individual stage bandwidth, which is equal to p_o.

An approximate, but more convenient expression for S_n is obtained as follows. Since

$$2^{1/n} = \exp\left(\frac{\ln 2}{n}\right) = \exp\left(\frac{0.693}{n}\right) = 1 + \frac{0.693}{n} + \left(\frac{0.693}{n}\right)^2 + \cdots$$

$$\approx 1 + 0.693/n \qquad n \geq 4$$

$$S_n = \sqrt{2^{1/n} - 1} \simeq \sqrt{\frac{0.693}{n}} = \frac{0.83}{\sqrt{n}} \qquad n \geq 4 \qquad (5\text{-}11)$$

Table 5-1 gives the value of S_n for n up to 5. Note that for $n \geq 4$ the approximate expression is accurate to within 5 per cent and is much simpler to use. For $n < 4$ we must use the exact values given in table 5-1.

TABLE 5-1 *Real Pole Shrinkage Factors*

n	1	2	3	4	5
$S_n = \sqrt{2^{1/n} - 1}$	1	0.64	0.51	0.44	0.39
$S_n \approx 0.83/\sqrt{n}$	—	—	—	0.42	0.37

The shrinkage factor is used in a design as follows. Suppose that the gain and bandwidth requirements of an amplifier are such that a three-stage RC-coupled amplifier must be used for the main amplifier. If the desired overall bandwidth is designated by $(\omega_{3\,\text{dB}})_o$, then the individual stages (assuming identical and one dominant pole per stage) must have a bandwidth $(\omega_{3\,\text{dB}})_i$

$$(\omega_{3\,\text{dB}})_i = \frac{(\omega_{3\,\text{dB}})_o}{0.51} = 1.96(\omega_{3\,\text{dB}})_o$$

where 0.51 is the shrinkage factor due to three identical stages. Note that if a cascade of stages is to be down 3 dB, then the individual stages must be down less than 3 dB. Therefore, they must have a greater 3-dB bandwidth.

Transient Response Consideration

At this point let us consider the time-domain response, and examine the output for a unit step input for an amplifier having the gain function given by (5-7), namely,

$$A(s) = \frac{\Phi_o(s)}{\Phi_i(s)} = \frac{A_0 p_o^n}{(s + p_o)^n} \tag{5-12}$$

where Φ_o and Φ_i are the transforms of the output and input voltages or currents. If the input is a unit step voltage, i.e., $\phi_i(t) \equiv u(t)$, then $\Phi_i(s) = 1/s$ and from (5-12)

$$\Phi_o(s) \equiv V_o(s) = \frac{A_0 p_o^n}{s(s + p_o)^n} \tag{5-13}$$

The normalized output voltage $v_{on}(t)$ (i.e., $V_{on}(s) = V_o(s)/A_0$) is given by

$$v_{on}(t) = \mathscr{L}^{-1}[V_{on}(s)] = \mathscr{L}^{-1} \frac{p_o^n}{s(s + p_o)^n} \tag{5-14}$$

From a table of Laplace transforms (see table 1-1) we obtain

$$v_{on}(t) = 1 - e^{-tp_o} \sum_{k=0}^{n-1} \frac{(p_o t)^k}{k!} \tag{5-15}$$

The plot of (5-15) for n up to 5 is shown in fig 5-4. The 10 per cent to 90 per cent rise time, τ_R, as read from fig 5-4 is shown in table 5-2. In table 5-2 we have also included the product $f_{3\,dB}\,\tau_R$, where $f_{3\,dB}$ is

TABLE 5-2 $p_o \tau_R$ and $f_{3\,dB} \tau_R$ *for Identical Real Poles*

n	1	2	3	4	5
$p_o \tau_R$	2.2	3.3	4.2	4.9	5.5
$f_{3\,dB}\,\tau_R$	0.350	0.338	0.342	0.343	0.342

given by (5-9). Note that this product is approximately the same for each case, i.e.,

$$f_{3\,dB}\,\tau_R \approx 0.35 \tag{5-16}$$

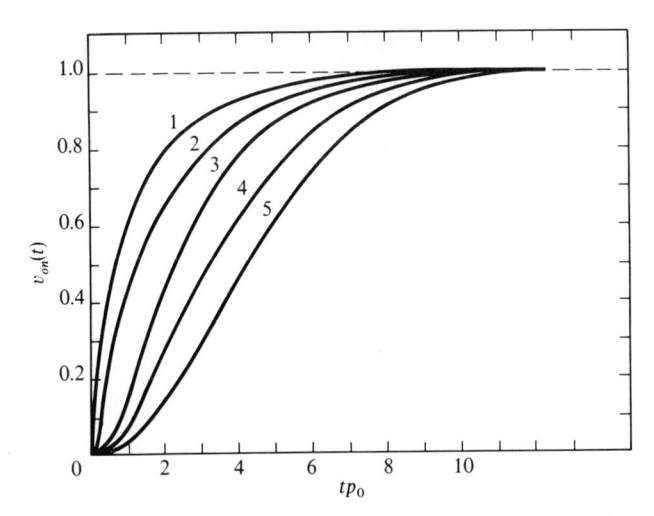

Fig 5-4 Normalized step response of transfer functions with n identical real poles for $n = 1$ to $n = 5$

The result in (5-16) is also found to be true for higher n. Note that this result is in agreement with the empirical relation given in (4-119).

Nonidentical Real Poles, Low-Pass Case

The poles of a cascaded amplifier, in general, may not be identical. Let the gain-function of the amplifier be given by

$$A(s) = \frac{A_0}{(1 + s/p_1)(1 + s/p_2)(1 + s/p_3)\cdots} \qquad (5\text{-}17)$$

This situation arises often in BJ transistors, FET, and triodes, where due to the interaction of stages, the dominant poles are not identical even though identical stages are used. The approximate gain-function of an RC coupled transistor CE amplifier, using the unilateral model, will be of the form of (5-17).

The magnitude function for (5-17) is

$$A(j\omega) = \frac{A_0}{\{[1 + (\omega/p_1)^2][1 + (\omega/p_2)^2][1 + (\omega/p_3)^2]\cdots\}^{1/2}} \qquad (5\text{-}18)$$

We equate (5-18) with $A_0/\sqrt{2}$ to obtain the 3-dB radian frequency, namely

$$2 = \left[1 + \left(\frac{\omega_{3\text{dB}}}{p_1}\right)^2\right]\left[1 + \left(\frac{\omega_{3\text{dB}}}{p_2}\right)^2\right]\left[1 + \left(\frac{\omega_{3\text{dB}}}{p_3}\right)^2\right]\cdots \qquad (5\text{-}19)$$

Since $\omega_{3\,dB}$ is less than any one of the poles, we can expand (5-19) and retain only the terms of the order ω^2, i.e.,

$$2 \approx 1 + \omega_{3\,dB}^2 \left(\frac{1}{p_1^2} + \frac{1}{p_2^2} + \frac{1}{p_3^2} + \cdots \right) \quad (5\text{-}20)$$

Hence

$$\omega_{3\,dB} \approx \frac{1}{\sqrt{(1/p_1)^2 + (1/p_2)^2 + (1/p_3)^2 + \cdots}} \quad (5\text{-}21)$$

From (5-21) it is noted that if any of the poles $p_i \geq 4p_1$, where the poles are numbered such that $p_1 < p_2 < \cdots < p_n$, its contribution in calculating the bandwidth can be neglected, since $(1/p_1)^2(1 + 1/4^2) \approx (1/p_1)^2$. For the worst case, where the poles are all identical, the error between (5-21) and (5-9) is about 20 per cent. Hence the error will always be less than 20 per cent by using (5-21).

In the time domain the rise time τ_R of the output for a step input is determined by using the approximate relation

$$\tau_R^2 \approx \tau_{R1}^2 + \tau_{R2}^2 + \tau_{R3}^2 + \cdots = \sum_{i=1}^{N} \tau_{Ri}^2 \quad (5\text{-}22)$$

where τ_{R1}, τ_{R2}, etc., correspond to the rise time due to poles p_1, p_2, p_3, etc. In other words, $\tau_{Ri} = 2.2/p_i$. The approximate relation, (5-22), follows directly from (5-16) and (5-21).

High-Pass Case

In multistage amplifiers, the gain function also has a set of dominant poles due to the coupling and bypass circuits of the various stages. In this case, if we assume identical dominant poles, the gain function can be written as

$$A(s) = \frac{A_0 s^n}{(s + p_l)^n} \quad (5\text{-}23)$$

where p_l is the dominant pole (s) of the low frequency circuits and A_0 is the overall midband gain. In this case the low frequency 3-dB cutoff radian frequency ω_l is determined from

$$|A(j\omega_l)| = \frac{A_0 \omega_l^n}{(p_l^2 + \omega_l^2)^{n/2}} = \frac{A_0}{\sqrt{2}} \quad (5\text{-}24)$$

Solution of (5-24) yields

$$\omega_l = \frac{p_l}{\sqrt{2^{1/n} - 1}} = \frac{p_l}{S_n} \qquad (5\text{-}25a)$$

or using the more convenient expression for S_n we have

$$\omega_l \approx \frac{p_l\sqrt{n}}{0.83} = 1.2\sqrt{n p_l} \qquad n \geq 4 \qquad (5\text{-}25b)$$

Note that since $S_n \leq 1$, $\omega_l \geq p_l$, the equality sign obviously holds for $n = 1$ in (5-25a). For example, if in a three-stage amplifier each stage contributes one dominant pole (remember that for low-frequency response functions, the farthest pole from the origin is dominant), we then have

$$A_l(s) = \frac{A_0 s^3}{(s + p_l)^3} \qquad (5\text{-}26)$$

Note that (5-26) could just as well describe the low-frequency gain function of a single stage where the three poles of the various coupling and bypass circuits are equal, and the zero is smaller than p_l so that it is ignored. From (5-26) and (5-25), for $n = 3$, we obtain

$$\omega_l = \frac{p_l}{\sqrt{2^{1/3} - 1}} = \frac{p_l}{0.51} = 1.96\, p_l$$

(For example, in Section 4-5 if $C_1 = C_2 = 0.003\ \mu F$, $C_3 = 10\ \mu F$, $f_l = 42/0.51 = 82$ Hz.)

If the low frequency circuit dominant poles are not identical we can use an expression similar to (5-21), namely

$$\omega_l \approx \sqrt{(p_1)^2 + (p_2)^3 + (p_3)^2 + \cdots} \qquad (5\text{-}27)$$

where p_1, p_2, p_3, etc., are the pole locations of the low-frequency equivalent circuit of the amplifier. Again note that the approximation in (5-27) for the worst case, where $p_1 = p_2 = p_3 = p_i$, has an error of 20 per cent. Hence the error by using (5-27) is always less than 20 per cent. When the poles are identical we, of course, use (5-25) instead of (5-27).

5-4 Gain and Frequency Response Calculations for RC-Coupled Cascaded CE Transistor Stages

We turn our attention now to the analysis of cascaded, *RC*-coupled, *CE* multistage amplifiers. In particular, we shall obtain expressions for the midband gain and the upper 3-dB cutoff frequency of such an

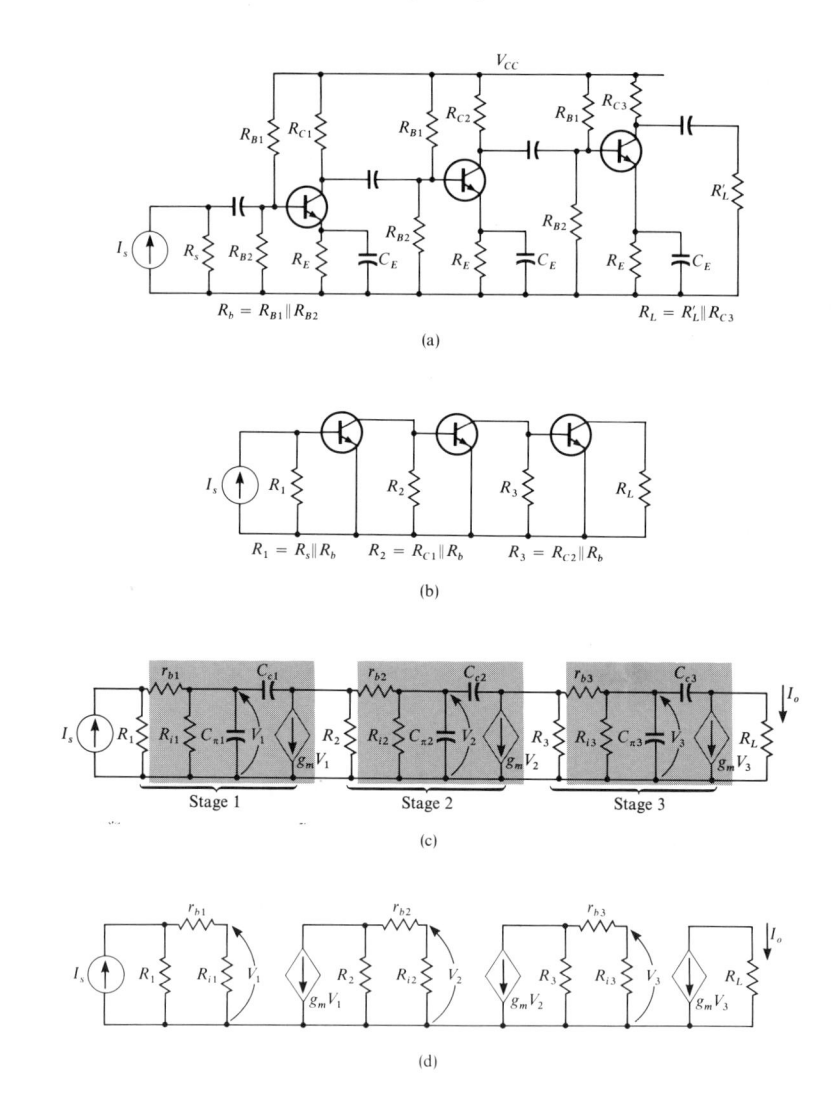

Fig 5-5 (a) Three-stage RC-coupled transistor amplifier; (b) Signal circuit of (a) i.e., the amplifier circuit exclusive of biasing and dc coupling circuitry; (c) Small-signal hybrid π equivalent circuit of (b); (d) Low-frequency or midband model of (b)

amplifier. Familiarity with these expressions will be of aid when we consider design procedures for such amplifiers.

A simple three-stage cascaded RC-coupled CE transistor amplifier is shown in fig 5-5a. The circuit is redrawn in fig 5-5b, where all the low-frequency coupling and bypass capacitors are short-circuited, since they do not affect the high-frequency response of the amplifier. The

5-4 Gain and Frequency Response Calculations

collector resistances and the biasing resistances are adsorbed in the interstage resistances R_1, R_2, R_3, i.e. $(R_2 = R_b \| R_{C1})$. The equivalent circuit of the amplifier using the hybrid π model is shown in fig 5-5c. The midband gain calculation is very simple. For midband gain calculation, the high-frequency capacitances are open-circuited and the model of fig 5-5c reduces to that of fig 5-5d. Thus the midband current gain from fig 5-5d is given by

$$A_I(0) = \frac{I_o}{I_s} = \left(\frac{I_o}{V_3}\right)\left(\frac{V_3}{V_2}\right)\left(\frac{V_2}{V_1}\right)\left(\frac{V_1}{I_s}\right) \tag{5-28a}$$

Substitution of the expressions from fig 5-5d yields

$$A_I(0) = \frac{I_o}{I_s}(0) = -(\beta_{01})\left(\frac{R_1}{R_1 + r_{b1} + R_{i1}}\right)(\beta_{02})\left(\frac{R_2}{R_2 + r_{b2} + R_{i2}}\right)$$

$$\times (\beta_{03})\left(\frac{R_3}{R_3 + r_{b3} + R_{i3}}\right) \tag{5-28b}$$

The midband voltage gain is given by

$$A_V(0) \equiv \frac{V_o}{V_s} = \frac{R_L I_o}{R_s I_s} = \frac{R_L}{R_s} A_I(0) \tag{5-29}$$

The calculation of the 3-dB frequency response from fig 5-5c is quite tedious and involved. The determinant of the nodal analysis, which is the most convenient method for this analysis, will be a 6 by 6 array. In other words, the circuit has six natural frequencies. Of these three may be dominant. For a computer-aided analysis the solution to this problem poses no difficulty. However, for preliminary design purposes, or quick paper-and-pencil calculation, the approximation based on the Miller theorem has been found to provide a reasonably accurate means of calculating the 3-dB bandwidth of an amplifier.

In order to use the Miller theorem, consider a typical stage, say stage 2 in fig 5-5c, as redrawn in fig 5-6a. The unilateral model for fig 5-6a is shown in fig 5-6b. In this case also, since $Z_L(s)$ is not a pure resistance, Y_M is not simply a capacitance. However, if we make a further approximation by using $Z_L(0)$ and assuming that $\omega C_c \leq g_m$, $Y_L(0)$, then Y_M becomes a pure capacitance. This approximation has already been used in Section 5-2, and provides considerable simplification in the analysis and design of amplifiers. From fig 5-5c, for example,

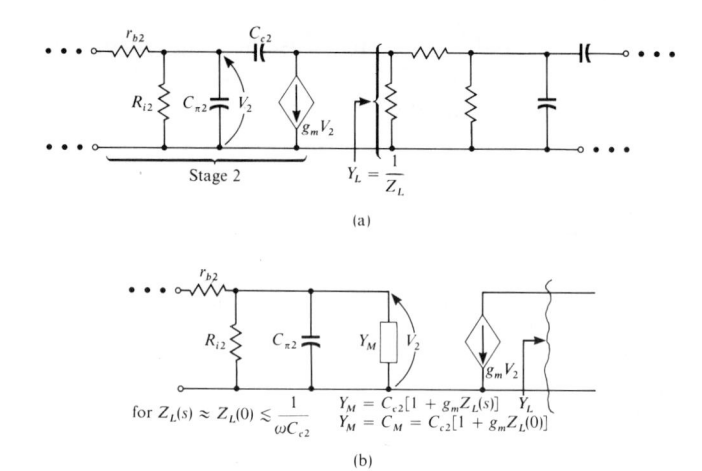

Fig 5-6 (a) Load of an iterative amplifier stage; (b) Unilateral approximate model of (a)

the Miller capacitance for stage 2 is given by

$$C_{M2} = C_{c2}[1 + g_{m2}Z_L(0)] = C_{c2}\{1 + g_{m2}[R_3\|(r_{b3} + R_{i3})]\} \quad (5\text{-}30a)$$

$$= C_{c2}[1 + g_{m2}R_{L2}] \quad (5\text{-}30b)$$

where

$$R_{L2} = R_3\|(r_{b3} + R_{i3})$$

Under this approximation the model of fig 5-5c reduces to that of fig 5-7. Note that

$$C_{i2} = C_{M2} + C_{\pi2} = C_{c2}(1 + g_{m2}R_{L2}) + \frac{1}{r_{e2}\omega_{T2}} - C_{c2}$$

$$= \frac{1 + R_{L2}C_{c2}\omega_{T2}}{r_{e2}\omega_{T2}} = \frac{D_2}{r_{e2}\omega_{T2}} \quad (5\text{-}31)$$

and similarly for C_{i1} and C_{i3}.

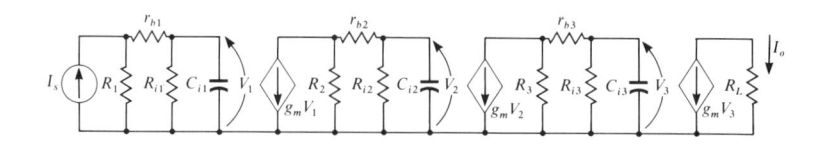

Fig 5-7 Unilateral approximate model of fig 5-5

5-4 Gain and Frequency Response Calculations

Bandwidth calculation based on (5-30) will be slightly pessimistic, since at the 3-dB frequency Z_L will be smaller than $Z_L(0)$.

The current-gain function of the circuit in fig 5-7 can now be easily determined to be

$$A_I = \frac{I_o}{I_s}(s) = \left(\frac{I_o}{V_3}\right)\left(\frac{V_3}{V_2}\right)\left(\frac{V_2}{V_1}\right)\left(\frac{V_1}{I_s}\right) \quad (5\text{-}32a)$$

In (5-32a) each ratio contributes one pole except I_o/V_3, which is equal to $-g_m$. Hence, from (5-32a) and fig 5-7 we can write the gain function in the following convenient form:

$$A_I = \frac{I_o}{I_s} = \frac{A_0 p_1 p_2 p_3}{(s+p_1)(s+p_2)(s+p_3)} \quad (5\text{-}32)$$

where

$$p_1 = \frac{1}{R_{i1} C_{i1}} \frac{R_1 + R_{i1} + r_{b1}}{R_1 + r_{b1}} \quad (5\text{-}33a)$$

$$p_2 = \frac{1}{R_{i2} C_{i2}} \frac{R_2 + R_{i2} + r_{b2}}{R_2 + r_{b2}} \quad (5\text{-}33b)$$

$$p_3 = \frac{1}{R_{i3} C_{i3}} \frac{R_3 + R_{i3} + r_{b3}}{R_3 + r_{b3}} \quad (5\text{-}33c)$$

and A_0 is equal to $A_I(0)$ and is given by (5-28). The approximate 3-dB bandwidth is then determined from (5-33) and (5-21).

EXAMPLE

To illustrate the midband gain and the 3-dB frequency calculations of RC-coupled, cascaded CE stages of the type shown in fig 5-5, we consider the following numerical example.

For simplicity, we assume identical transistors with the identical operating point: $V_{CE} = 5$ V, and $I_C = 5$ mA. The biasing resistor $R_{Ci} = 1$ kΩ, $R_b = 3.5$ kΩ. The transistor parameters at the operating point are as follows

$$f_T = 300 \text{ MHz} \qquad r_b = 50 \, \Omega$$
$$C_c = 4 \text{ pF} \qquad \beta_0 = 50 \qquad (5\text{-}34)$$

Let $R_s = R_L = 50 \, \Omega$.

From the above information we find

$$r_e = \frac{1}{g_m} = 5\,\Omega, \qquad R_i = \beta_0 r_e = 250\,\Omega, \qquad \frac{1}{r_e \omega_T} = 1.06 \times 10^{-10}$$

Note that the interstage resistance values are

$$R_1 \simeq 50\,\Omega, \; R_2 = R_3 = 1\,\text{k}\Omega \| 3.5\,\text{k}\Omega = 780\,\Omega$$

From (5-28) the magnitude of the midband current and voltage gains are

$$A_I(0) = (50)\frac{50}{50 + 50 + 250}(50)\frac{780}{780 + 50 + 250}(50)\frac{780}{780 + 50 + 250}$$
$$= 9320 = 79.4\,\text{dB}$$

$$A_V(0) = \frac{R_L}{R_s} A_I(0) = 9.32 \times 10^3 = 79.4\,\text{dB}$$

The input capacitances of the three stages are

$$C_{i3} = (1.06 \times 10^{-10})D_3 = (1.06 \times 10^{-10})(1 + R_{L3}C_c\omega_T)$$
$$= 1.44 \times 10^{-10}\,\text{F}$$

$$C_{i2} = (1.06 \times 10^{-10})D_2 = (1.06 \times 10^{-10})(1 + R_{L2}C_c\omega_T)$$
$$= 2.76 \times 10^{-10}\,\text{F}$$

$$C_{i1} = (1.06 \times 10^{-10})D_1 = (1.06 \times 10^{-10})(1 + R_{L1}C_c\omega_T)$$
$$= 2.76 \times 10^{-10}\,\text{F}$$

From (5-33) we obtain

$$p_3 = 3.61 \times 10^7\,\text{rad/sec}, \; p_2 = 1.88 \times 10^7\,\text{rad/sec},$$
$$p_1 = 5.08 \times 10^7\,\text{rad/sec}$$

We normalize the frequency by a factor of 10^7 (i.e., $\Omega_0 = 10^7$) for convenience. The 3-dB bandwidth is then determined from (5-21) to be

$$(\omega_{3\,\text{dB}})_n \approx \frac{1}{\sqrt{(1/p_{1n})^2 + (1/p_{2n})^2 + (1/p_{3n})^2}}$$

$$= \frac{1}{\sqrt{(1/5.08)^2 + (1/1.88)^2 + (1/3.61)^2}} \approx 1.59$$

5-5 Midband Gain and Impedance Calculation in Multistage Circuits

Hence, the denormalized 3-dB radian frequency is $\omega_{3\,dB} = 1.59 \times 10^7$ rad/sec or

$$f_{3\,dB} = \frac{1.59 \times 10^7}{2\pi} = 2.54 \text{ MHz}$$

The student is encouraged to perform this analysis using the hybrid π model for the transistor as shown in fig 5-5c, using a computer-aided circuit analysis package and the circuit values of this example (see Problem 5-41b).

5-5 Midband Gain and Impedance Calculation in Multistage Circuits

In the previous sections we considered midband gain and frequency response calculations for bipolar and FET devices. The expressions for FET devices are identical to those for the corresponding vacuum tube circuits, and, therefore, the latter were not separately discussed.

The method discussed in the previous section, namely that of replacing the device by its low-frequency model and ignoring the reactive elements due to the high- and low-frequency circuitry for midband calculations, is quite general and can always be used. However, for midband calculations for multistage circuits having devices in the same or other configurations, it may often be expedient and relatively easy if use is made of the results in Section 4-2, i.e., using the gain and impedance relations for the device in the three configurations. In some cases the h-parameters may be given, and it may be more convenient to work directly with these parameters. In others we may use the gain and impedance relations in terms of the circuit parameters such as those of the Tee model. The following examples illustrate these points.

EXAMPLE

The Darlington pair consists of two devices in a compound configuration. The various configurations are shown in fig 5-8. These configurations yield overall α_0 close to unity, and also have other interesting properties. In fact, some manufacturers package the Darlington pair as a single composite device to obtain a very high β_0 (on the order of 1000), the package having only three external leads.

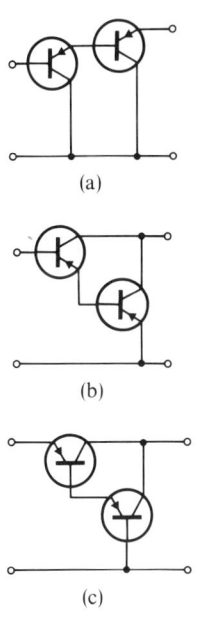

Fig 5-8 Mixed pairs or Darlington compound configurations: (a) *CC-CC* pair; (b) *CE-CE* pair; (c) *CB-CB* pair

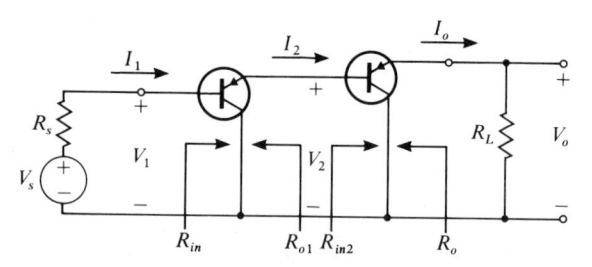

5/Analysis and Design of Multistage Amplifiers

Fig 5-9 *CC-CC* cascade amplifier

We shall develop expressions for the gain and impedance of the circuit shown in fig 5-8a (redrawn in fig 5-9 for convenience) using the results of Section 4-2.

For the *CC-CC* cascade circuit of fig 5-9, the overall current gain is

$$A_I = \frac{I_o}{I_1} = \left(\frac{I_o}{I_2}\right)\left(\frac{I_2}{I_1}\right)$$

But from applying (4-17) in fig 5-9

$$\frac{I_o}{I_2} = \frac{r_{c2}}{r_{c2}(1-\alpha_{02}) + r_{e2} + R_L} \simeq \frac{1}{1-\alpha_{02}} \simeq \beta_{02}$$

Similarly from (4-17), and using (4-15b), we have

$$\frac{I_2}{I_1} = \frac{r_{c1}}{r_{c1}(1-\alpha_{01}) + r_{e1} + R_{in2}} \approx \frac{r_{c1}}{\frac{r_{c1}}{\beta_{01}} + \beta_{02}(R_L + r_{e2})}$$

Hence

$$A_I \simeq \frac{\beta_{01}\beta_{02}}{1 + \beta_{01}\beta_{02}[(r_{e2} + R_L)/r_{c1}]}$$

The overall voltage gain is

$$A_V = \frac{V_o}{V_s} = \left(\frac{V_o}{V_2}\right)\left(\frac{V_2}{V_1}\right)\left(\frac{V_1}{V_s}\right)$$

But from applying (4-18), in fig 5-9

$$\frac{V_o}{V_2} = \frac{R_L}{R_L + r_{e2}} \simeq 1 - \frac{r_{e2}}{R_L} \quad \text{for } R_L \gg r_{e2}$$

5-5 Midband Gain and Impedance Calculation in Multistage Circuits

Similarly

$$\frac{V_2}{V_1} = \frac{R_{in2}}{R_{in2} + r_{e1}} \simeq \frac{\beta_{02}(R_L + r_{e2})}{\beta_{02}(R_L + r_{e2}) + r_{e1}} \simeq 1 - \frac{r_{e1}}{\beta_{02}(R_L + r_{e2})}$$

and

$$\frac{V_1}{V_s} = \frac{R_{in}}{R_{in} + R_s} \simeq \frac{\beta_{01}[\beta_{02}(r_{e2} + R_L)]}{\beta_{01}[\beta_{02}(r_{e2} + R_L)] + R_s} \simeq 1, \text{ where } R_s \ll \beta_{01}\beta_{02}R_L$$

Hence

$$A_V \approx \left(\frac{1}{1 + r_{e2}/R_L}\right)\left\{\frac{1}{1 + r_{e1}/[\beta_{02}(R_L + r_{e2})]}\right\} \simeq 1 - \frac{r_{e2}}{R_L}$$

$$- \frac{r_{e1}}{\beta_{02}(R_L + r_{e2})} \tag{5-36}$$

The input impedance from (4-15) is

$$R_{in} = r_{b1} + \frac{R_{in2} + r_{e1}}{1 - \alpha_{01}} \simeq (\beta_{01})(r_{e1} + R_{in2})$$

$$= \beta_{01}[r_{e1} + \beta_{02}(r_{e2} + R_L)] \simeq \beta_{01}\beta_{02}(r_{e2} + R_L) \tag{5-37}$$

The output impedance from (4-16) is

$$R_o \simeq r_{e2} + (R_{01} + r_{b2})(1 - \alpha_{02})$$

$$= r_{e2} + [r_{e1} + (R_s + r_{b1})(1 - \alpha_{01}) + r_{b2}](1 - \alpha_{02}) \tag{5-38}$$

From the above properties we see that the Darlington amplifier has a much higher current gain than any of the single-stage configurations. The circuit of fig 5-9 is in some respects similar to a single CC stage but with a much greater current gain.

The Cascode Configuration

The cascode configuration is simply a CE-CB cascade as shown, exclusive of biasing circuitry, in fig 5-10. We shall see later that it is very useful as an input stage when the noise consideration is important, and it is therefore given a special name. We shall determine both the

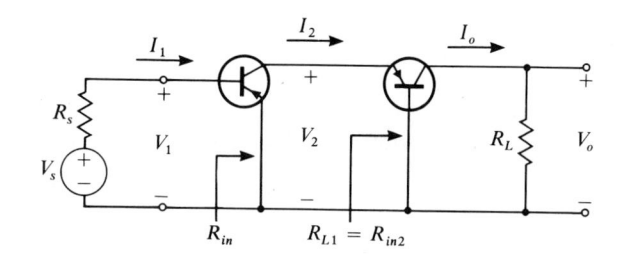

Fig 5-10 Cascode amplifier (*CE-CB* cascade)

voltage and current gains in terms of the circuit parameters, as well as find the overall *h*-parameters.

The overall voltage gain is given by

$$A_V = \frac{V_o}{V_s} = \frac{V_o}{V_2}\frac{V_2}{V_1}\frac{V_1}{V_s}$$

But from (4-9b) and (4-14b) we have

$$\frac{V_o}{V_2} = \frac{\alpha_{02} R_L}{r_{e2} + r_{b2}(1 - \alpha_{02})} \tag{5-39a}$$

$$\frac{V_2}{V_1} = -\frac{\alpha_{01} R_{L1}}{r_{e1} + r_{b1}(1 - \alpha_{01})} = -\frac{\alpha_{01}[r_{e2} + r_{b2}(1 - \alpha_{02})]}{r_{e1} + r_{b1}(1 - \alpha_{01})} \tag{5-39b}$$

$$\frac{V_1}{V_s} = \frac{R_{in}}{R_{in} + R_s}\frac{R_{i1} + r_{b1}}{R_{i1} + r_{b1} + R_s} \tag{5-39c}$$

Assuming identical parameters (for simplicity), (5-39b) reduces to $-\alpha_0$ and we have

$$A_V = -\frac{\alpha_o^2 R_L}{r_e + r_b(1 - \alpha_0)}\left(\frac{R_i + r_b}{R_s + R_i + r_b}\right) \tag{5-40}$$

Similarly the current gain is given by

$$A_I = \frac{I_o}{I_1} = \left(\frac{I_o}{I_2}\right)\left(\frac{I_2}{I_1}\right)$$

where from (4-8b) and (4-13b) we have

$$\frac{I_o}{I_2} \approx \alpha_{02}, \quad \frac{I_2}{I_1} \simeq -\beta_{01}$$

$$A_i \simeq -\alpha_{02}\beta_{01} \simeq -\beta_{01} \tag{5-41}$$

Note that this circuit as a unit acts very much like a single CE stage. The current gain and the voltage gain are approximately the same as those of the single CE stage. For the same large voltage gain, however, it has a greater bandwidth than can be obtained from a single CE stage. This fact can be readily demonstrated as follows.

A CB stage has a very large bandwidth (of the order of ω_α) and the CE stage has a small bandwidth (of the order of ω_β). Since R_L of the CE stage is the input impedance of the CB stage, which is very low, the Miller effect is appreciately reduced and the D-factor is close to unity. Thus the overal bandwidth of the cascode circuit, for a current source input, is approximately equal to ω_β even though R_L is very large. On the other hand, a very large value of R_L increases the D factor of a single CE stage, thus reducing the bandwidth. Note that in the cascode circuit accurate bandwidth calculations can be made without resorting to a computer-aided analysis. This circuit has an additional desirable feature in that its reverse transmission is extremely low. In fact the overall two-port h-parameters of the entire circuit are given approximately by (see Problem 5-18)

$$h_{11} \simeq h_{ie} \qquad h_{12} \simeq h_{re}h_{rb} \Rightarrow 0$$
$$h_{21} \simeq h_{fe} \qquad h_{22} \simeq h_{ob} \Rightarrow 0 \qquad (5\text{-}42)$$

Note that since in (5-42) $h_{12} \simeq (\beta_0 r_e/r_c)(r_b/r_c) \Rightarrow 0$, the reverse transmission, or internal feedback, is almost completely negligible, hence the interaction of the succeeding stage can be ignored without causing any significant error. The CE-CB cascade can be modeled more accurately as a unilateral two-port than a CE stage. The CE-CB composite stage is, therefore, less prone to instability problems and is used for tuned amplifiers where instability poses a severe problem (see Section 7-8).

The emitter-coupled pair (CC-CB cascade) also has similar useful properties, and is often used in integrated circuits in various forms. The analysis of this configuration is left as an exercise in Problem 5-11.

5-6 Design of *RC*-Coupled *CE* Transistor Amplifiers for Specified Gain and Bandwidth

In a wide-band multistage amplifier design, the circuit shown in fig 5-5a is not often used because of the inefficient gain-bandwidth product of the individual stages, as was demonstrated in Section 4-12.

Usually a broadbanding scheme is utilized in the design, as discussed in the next section. Nonetheless, we will provide a design example for the circuit of fig 5-5, in order to illustrate the use of shrinkage factor and other ideas important in a design, and furthermore to bring out the need for broadbanding techniques.

Let us assume that the proper operating point for obtaining high f_T, β_0, etc., for the available transistors is $V_{CE} = 5$ V, and $I_C = 5$ mA. The parameters of the available transistors are assumed to be the same as those given in (5-34). The design requires an approximate bandwidth $f_{3\,dB} = 5$ MHz, and the overall voltage gain is to be larger than, or approximately equal to, 1000 (60 dB). The source and load resistances are $R_s = R_L = 100\,\Omega$.

First we design the biasing circuit in accordance with the design procedure given in Section 3-17. For the operating point $V_{CE} = 5$ V, $I_C = 5$ mA, and a supply voltage of 12 V, the following design values were obtained:

$$R_C = 1\,\Omega \quad R_E = 400\,\Omega \quad R_{b1} = 15\,k\Omega \quad R_{b2} = 4.7\,k\Omega$$

The low-frequency cutoff is not specified, hence we may use any reasonably large values of C.

We then proceed with the design of the high-frequency circuitry. For the given transistor parameters, gain, and bandwidth specifications, if we use two stages, the individual stage gain must be greater than 30 and the individual stage bandwidth about 7.7 MHz ($= f_{3\,dB}/0.64$), which is twice as large as $f_\beta/D = 3.5$ MHz. Since the GBP for simple resistive broadbanding decreases drastically for frequencies larger than f_β/D, we cannot realize the required specification with two stages when using the given transistors.[2] We can thus estimate that three stages will probably be sufficient.

From the specified bandwidth we now determine the gain and bandwidth of each of the individual stages when using three stages of amplification. For an amplifier with three identical-pole stages, the pole for each stage is given by (5-9), that is

$$p_o = \frac{\omega_{3\,dB}}{S_n} = \frac{\omega_{3\,dB}}{\sqrt{2^{1/3} - 1}} = \frac{\omega_{3\,dB}}{0.51} \tag{5-43}$$

---

[2] The proof of this lies in the fact that if $p_1 = p_2 = 2\pi(7.7 \times 10^6)$ rad/sec for two identical amplifier stages, then from the individual stage bandwidth expression, $p_2 = (\omega_\beta/D) \times (R_I + R_i + r_b)/(R_I + r_b) \Rightarrow R_I = 158\,\Omega$. The current gain of each of the individual stages, a_i, is found from $a_i = \beta_0 R_I/(R_I + R_i + r_b) \Rightarrow a_i \approx 17$. Therefore, since the product of the individual stage gains $(17)^2 < 1000$, two stages of amplification are insufficient to meet the design requirement.

5-6 Design of RC-Coupled CE Transistor Amplifiers

Since $\omega_{3\,dB} = 2\pi(5\text{ MHz}) = 3.14 \times 10^7$, the individual poles have the value of $p_o = 6.15 \times 10^7$ rad/sec, as obtained from (5-43). Now from (5-33), starting with last stage so that the D-factor be known, we obtain

$$D_3 = 1 + R_L C_c \omega_T = 1.7 \qquad C_{i3} = \frac{D_3}{r_e \omega_T} = 1.82 \times 10^{-10}$$

$$R_3 = \frac{R_{i3}}{p_3 R_{i3} C_{i3} - 1} - r_b = \frac{250}{1.8} - 50 = 90\,\Omega$$

Also

$$R_{L2} = R_3 \| (r_b + R_i) = 90\|300 = 70\,\Omega, \qquad D_2 = 1 + R_{L2} C_c \omega_T = 1.49$$

$$R_2 = \frac{R_{i2}}{p_2 R_{i2} C_{i2} - 1} - r_b = 123\,\Omega$$

and

$$R_{L1} = R_2 \| (r_b + R_i) = 123\|300 = 88\,\Omega, \qquad D_1 = 1.62$$

$$R_1 = \frac{R_{i1}}{p_1 R_{i1} C_{i1} - 1} - r_b = 100\,\Omega$$

In this case, coincidentally, $R_1 = R_s$. If the specified value of R_s is larger than that of R_1, then a resistor R_a of value $R_b \| R_a \| R_s = R_1$ may be added in shunt with R_b (see fig 5-5a). As mentioned previously, however, the input stage may be altogether different, depending on what we wish to achieve at the input, in addition to realizing the gain and bandwidth requirements of the design specification.

The overall voltage gain is given by

$$A_V = \frac{V_o}{V_s} = \frac{I_o R_L}{R_s I_s} = A_I \frac{R_L}{R_s}$$

where A_I is given by (5-28). For the resistor values of this design, $R_L = R_s = 100\,\Omega$, the magnitudes of the gains are

$$A_V = A_I = (\beta_0)^3 \left(\frac{R_s}{R_s + r_b + R_i} \right) \left(\frac{R_2}{R_2 + r_b + R_i} \right) \left(\frac{R_3}{R_3 + r_b + R_i} \right) \quad (5\text{-}44)$$

$$A_V = A_I = 2100$$

The actual gain will be slightly lower than above, since the biasing resistance $R_b = 3.5\,k\Omega$ is in shunt with the input of each stage, thus

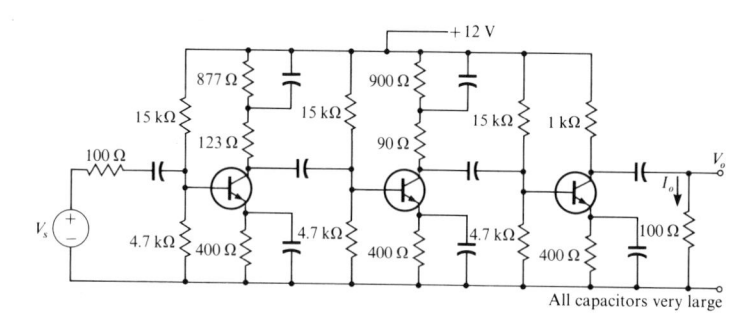

Fig 5-11 Amplifier of the design example

lowering slightly the values of R_1, R_2, and R_3 to $R_1 \| R_b$, etc. Now that we have met the gain specifications, the preliminary design is complete. The designed circuit is shown in fig 5-11, where R_C is split into low- and high-frequency collector load resistors by the use of a shunt capacitor. When discrete circuits are used, the preliminary design is experimentally tested and, if necessary, adjusted to exactly meet the design specifications.

Note that if the design specifications had called for the same gain, the same load termination, and $f_{3\,\text{dB}} = 10$ MHz, similar calculation yields the value of $R_3 = 4.5\,\Omega$. The current gain of the third stage would be 0.7, and the circuit would thus be useless for this and greater bandwidths. In fact, in simple resistive broadbanded stages, we would not be able to meet the specifications with the given transistors, unless another type of broadbanding were used. It is quite clear from this example that the transitor cascade amplifier circuit shown in fig 5-11 does not make efficient use of the gain-bandwidth product. We thus need broadbanding techniques, some of which were described in Section 4-13. We further explore, in the next section, some of these broadbanded stages in conjunction with the multistage amplifier configuration.

5-7 Multistage Broadband Transistor Amplifiers

In multistage amplifier design there are many factors to be taken into consideration. These factors vary, depending upon the particular application of the amplifier. In this section we will briefly consider a few typical cascade amplifier circuits and examine their gain-bandwidth capabilities.

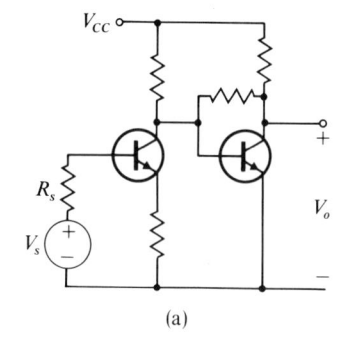

Fig 5-12 Feedback broadbanded amplifiers: (a) Series-feedback stage in cascade with a shunt-feedback stage

5-7 Multistage Broadband Transistor Amplifiers

Feedback Configurations

Transistor amplifiers must often use feedback in order to achieve desensitivity, as well as broadbanded frequency response. Since feedback amplifiers are extremely useful, the topic is discussed in detail in Chapter 6. A few such circuit arrangements are shown in fig 5-12. Biasing, low-frequency coupling, and bypass circuitry have been omitted for simplicity. Figure 5-12a shows a series-feedback amplifier with a voltage source input in cascade with a shunt-feedback amplifier. For a current source input, the shunt-feedback stage would precede the series-feedback stage so as not to lose signal current. More stages may be cascaded, of course, as alternate shunt-series-shunt, etc., if the gain specifications cannot be met with two stages.

The circuits of figs 5-12b and 5-12c are the so-called series-shunt feedback pair and shunt-series feedback pair, respectively. The circuits of figs 5-12d and 5-12e are the shunt-shunt and series-series feedback triples, respectively. These circuits are named in accordance with their feedback connections. The analysis and design of these circuits, using feedback techniques, are discussed in detail in Chapter 6, and will not be considered here. We will, however, briefly consider the circuit of fig 5-12a, using the familiar *approximate* methods discussed in the previous chapter.

The circuit to be analyzed is redrawn for convenience in fig 5-13, wherein the circuit element values are indicated. The transistor parameters are the same as those in (5-34). We shall calculate the approximate gain and bandwidth of the circuit.

The midband input impedance of a shunt-feedback stage, using the Miller approximation, is given by

$$(R_{in})_2 \approx [G_f(1 + g_m R_L)]^{-1} \| (r_b + R_{i2})$$

$$= \frac{R_f}{1 + g_m R_L} \| (r_b + \beta_0 r_{e2}) \simeq 75 \, \Omega$$

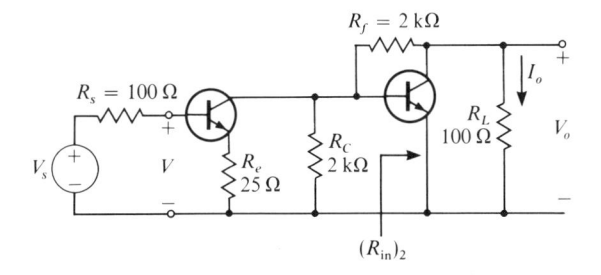

Fig 5-13 Amplifier having a series-feedback stage cascaded to a shunt-feedback stage

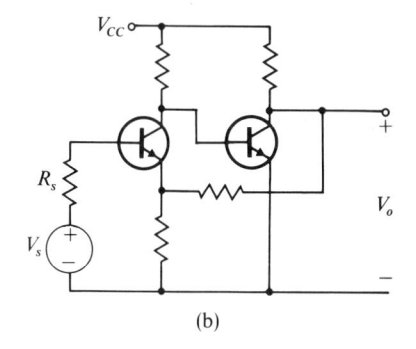

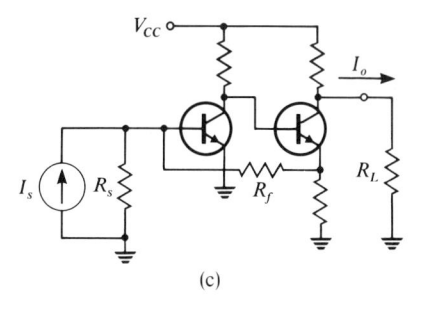

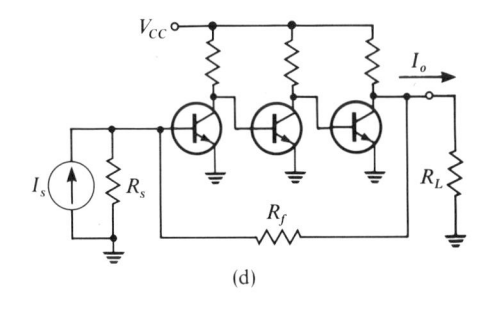

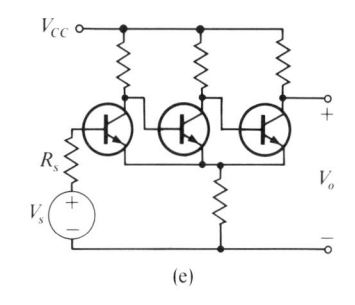

Fig 5-12 (b) Series-shunt feedback pair; (c) Shunt-series feedback pair; (d) Shunt-shunt feedback triple; (e) Series-series feedback triple

Since $(R_{in})_2$ is much smaller than $R_C \| R_0$ (where R_0 is the output impedance of stage 1, which is very high), we can consider the second stage to be driven by a current source, and analyze the circuit for its gain and bandwidth separate of the first stage. The gain and bandwidth of the second stage may now be determined as in Section 4-7, or very closely using the Miller approximation (or using the expression in footnote 6 of Chapter 4). Thus, without going into the details of the calculation, we find that the current gain and the bandwidth of the second stage are

$$A_{i2}(0) = 14 \qquad (\omega_{3\,dB})_2 \simeq 2\pi(12 \text{ MHz})$$

The first stage may now be separately analyzed as a series-feedback configuration having a load resistance $(R_L)_1 = 75\,\Omega$. For this load value, the given transistor parameters, and $R_s = 100\,\Omega$, we can use the expressions in (4-92) and (4-93) to find the current gain and the bandwidth. Substitution of the numerical values into these expressions yields

$$A_{i1}(0) = \frac{I_1}{I_s} \simeq 3 \qquad (\omega_{3\,dB})_1 = 2\pi(36 \text{ MHz})$$

Thus the overall current and voltage gains are given by

$$A_i = A_{i1} A_{i2} = 42 = A_V, \quad \text{since } R_L = R_s$$

The overall 3-dB bandwidth is determined from fig 5-3 or (5-21). For $p_2/p_1 = 3$ we read:

$$\omega_{3\,dB} = 0.91\,p_1 \quad \text{or} \quad f_{3\,dB} \simeq 11 \text{ MHz}$$

The above results were obtained in a very approximate manner just to illustrate the gain-bandwidth capability of configurations such as shown in fig 5-13. Note that a bandwidth of 10 MHz, with some gain, could not be realized using simple cascaded resistive broadbanded stages. The reader is encouraged to perform a more precise analysis of the circuit of fig 5-13, using the hybrid π model and, preferably, a computer-aided analysis. The series-shunt cascade can also be designed to realize approximately the voltage-controlled voltage source element in fig 1-9a. The shunt-series cascade, on the other hand, can be designed to realize approximately the current-controlled current source element in fig 1-9c.

Mixed Pairs

Transistors in an amplifier need not all be connected in the same configuration. Mixed pairs can also be used as a building block unit of amplification. There are many possible combinations of such pairs. Two such arrangements are shown in fig 5-9 and fig 5-10. The circuit of fig 5-9 is a cascade of CC-CC stages while the circuit of fig 5-10 is a cascade of CE-CB stages. Both of these circuits were discussed in Section 5-5. The emitter-coupled pair (see fig P5-11) is also a useful configuration, which as a two transistor block has certain desirable features. One of its attractive properties is the isolation due to the CB stage in the same manner as in the cascode configuration. Another configuration useful for wide-band amplification is a cascade of CC-CE stages, as shown in fig 5-14. Biasing and coupling circuitry in this figure have been omitted for simplicity. In the following, we shall consider the analysis of the CC-CE pair in some detail to see why it has a good GBP capability. Since the output impedance of a CC stage is inductive (see Problem 4-36) in nature, it interacts with the capacitive input impedance of the CE stage to provide complex poles and thus leads to a broadbanding situation.

In order to simplify the analysis and design, we neglect C_c of the CC stage. This omission is often permissible for wide-band applications, since C_c is in shunt with $(R_I + r_b)$, which is a low resistance, and the resulting error is thus negligible in most cases.[3] The high-frequency equivalent circuit for fig 5-14, with the above approximation, is shown in fig 5-15. The resistors R_{i1} and R_{i2} need not be the same, and thus provide some flexibility in the design. Even if the operating points are identical, the addition of a small external emitter resistance provides a degree of freedom, as was seen in the series-feedback stage. We therefore distinguish the parameters of each stage by appropriate subscripts.

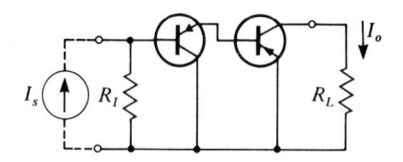

Fig 5-14 Amplifier with the CC-CE pair configuration

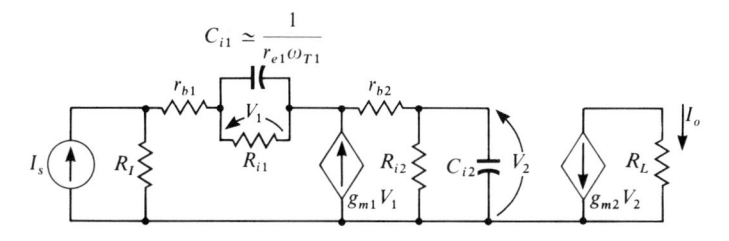

Fig 5-15 Equivalent circuit of the CC-CE pair

---

[3] If $R_L \gg (R_s + r_b)$ in a single CC stage, then include C_c of the hybrid π model, or use the high-frequency Tee model. Otherwise, erroneous results may be obtained (see Problem 4-36 where $R_L = 1$ kΩ).

We write the nodal equation for the circuit in the *s*-domain, and then determine the current-gain function. After some manipulation (see Problem 5-19), and using the approximations $\beta_0 \gg 1$, and $\beta_0 r_e \gg r_b$, we obtain (note that we have used $r_{b1} = r_{b2} = r_b$ and $\omega_{T1} = \omega_{T2} = \omega_T$)

$$A_I = \frac{I_o}{I_s} = \frac{R_I}{(R_I + 2r_b)D}$$

$$\times \frac{(s/\omega_T + 1)}{\left(\dfrac{s}{\omega_T}\right)^2 + \left(\dfrac{s}{\omega_T}\right)\left[\dfrac{R_I + \beta_0(r_b + r_{e1})}{\beta_0(R_I + 2r_b)} + \dfrac{R_I + \beta_0 r_{e2}}{\beta_0(R_I + 2r_b)D}\right] + \dfrac{r_{e2}}{D(R_I + 2r_b)}} \quad (5\text{-}45)$$

where $D = 1 + R_L C_c \omega_T$. Equation (5-45) can be rewritten for convenience in the normalized form

$$A_I(s_n) = \frac{b A_I(0)(s_n + 1)}{s_n^2 + a s_n + b} \quad (5\text{-}46)$$

where

$$A_I(0) = \frac{R_I}{r_{e2}} \qquad s_n = \frac{s}{\omega_T} \quad (5\text{-}47a)$$

$$a = \frac{R_I + \beta_0(r_b + r_{e1})}{\beta_0(R_I + 2r_b)} + \frac{R_I + \beta_0 r_{e2}}{\beta_0(R_I + 2r_b)D} \quad (5\text{-}47b)$$

$$b = \frac{r_{e2}}{(R_I + 2r_b)D} \quad (5\text{-}47c)$$

Note that the zero is located at ω_T, and is therefore nondominant in most cases. The denominator poles will be complex if $a^2 < 4b$. An example will illustrate the gain-bandwidth capability of the *CC-CE* pair.

EXAMPLE

For the operating point and transitor parameters given in (5-34), we calculate the gain and the 3-dB bandwidth. Let $R_I = 500\,\Omega$, $R_L = 100\,\Omega$. The current gain is given by (5-47a), that is

$$A_I(0) = \frac{R_I}{r_{e2}} = \frac{500}{5} = 100$$

From (5-47b) and (5-47c) we determine the values of a and b

$$a = \frac{500 + 50(55)}{50(500 + 100)} + \frac{500 + 50(5)}{50(500 + 100)(1.7)} = 0.122$$

$$b = \frac{5}{(500 + 100)(1.7)} = 0.0049$$

Hence (5-46) can be written as

$$A(s_n) = \frac{0.49(s_n + 1)}{s_n^2 + 0.122s + 0.0049}$$

$$= \frac{0.49(s_n + 1)}{(s_n + 0.061 + j0.034)(s_n + 0.061 - j0.034)}$$

Note that the zero is nondominant, since the magnitude of the complex pole pair is $|p_1| = 0.07 \ll 1$, and can be ignored for purposes of bandwidth calculation. The normalized 3-dB frequency for the complex pole-pair corresponding to $\psi = 31$ degrees from fig 5-3 is $0.79\, p_1$, hence $(\omega_{3\,dB})_n = 0.055$. Hence the actual 3-dB bandwidth is $\omega_{3\,dB} = 0.055 \times \omega_T = 2\pi(16.5 \text{ MHz})$. Note that this bandwidth could not be realized by a cascade of resistive broadbanded stages. It is interesting to note also that the response shape is approximately the same as that of a two-pole, linear phase characteristic. A smaller value of R_I can achieve a two pole, maximally flat, magnitude response characteristic (see Problem 5-19) with a wider bandwidth.

As a final comment on the circuit of fig 5-14, note that the current gain (5-47) is dependent only on R_I and r_{e2}. If an external resistance R_{e2} is added such that $R_{e2} > r_{e2}$, then the current gain of the amplifier will be essentially the ratio of the two resistors and will be independent of the transistor parameters. This is a very desirable feature in assembly line production, and the ratio feature is also attractive in integrated circuits.

Multistage Gain-Bandwidth Product

For cascaded multistage amplifiers the overall gain-bandwidth product is

GBP = (overall midband gain) × (overall 3-dB bandwidth)   (5-48)

It should be noted, however, that (5-48) is *not* a valid basis for comparing multistage amplifiers. A valid and meaningful basis of comparison in a multistage amplifier is the so-called mean gain-bandwidth product, MGBP, namely if the number of stages is n

$$\text{MGBP} = (\text{overall midband gain})^{1/n}(\text{overall 3-dB bandwidth}) \qquad (5\text{-}49)$$

To see that (5-48) is not a valid basis of comparison, consider the simple case of a two-stage, RC-coupled, pentode amplifier. Increasing the load resistance of each stage by a factor K increases the overall gain by K^2, while the bandwidth is reduced only by a factor of $1/K$. Thus the GBP in (5-48) can be made arbitrarily large without using any broadbanding techniques. It is thus seen that (5-48) does not give a meaningful comparison of multistage amplifiers. If (5-49) is used, the increase in load resistance by a factor of K increases the mean gain [i.e., (overall gain)$^{1/n}$] by K, while the bandwidth is reduced by a factor $1/K$. Hence the MGBP is unchanged, which is a meaningful conclusion.

5-8 Stagger-Tuning and Synchronous-Tuning in Amplifier Design

In wide-band amplifier design, the bandwidth shrinkage effect can be reduced or eliminated by stagger-tuning, and thus for a given gain level, a high GBP can be realized. A stagger tuned circuit is designed such that the desired overall network function pole-zero pattern is broken into simple pole-zero locations, which are assigned to individual noninteracting stages. For example, an overall four-pole Butterworth function is achieved by a cascade of two noninteracting stages having complex pole pairs as indicated in fig 5-16. This simple procedure can not be used in transistor stages where interaction occurs. Analytically, for such cases stagger tuned design is achieved by obtaining the overall gain-function of the amplifier, and then equating the coefficients of the overall gain-function with those of the polynomial resulting from the desired overall pole-zero pattern. Because approximate models are used for the transistors, however, calculated values frequently fail to achieve the design specifications, necessitating the varying of some of the values of the circuit elements on an experimental basis until the desired results are achieved. Such an alignment for interacting stages can be difficult, because varying one component may change many coefficients of the network function, and thus many circuit elements may have to be varied by trial and error to achieve the overall desired response. The interaction in some applications, such as tuned amplifiers, is intentionally reduced by a mismatch between the input and the output of the

5-8 Stagger-Tuning and Synchronous-Tuning in Amplifier Design

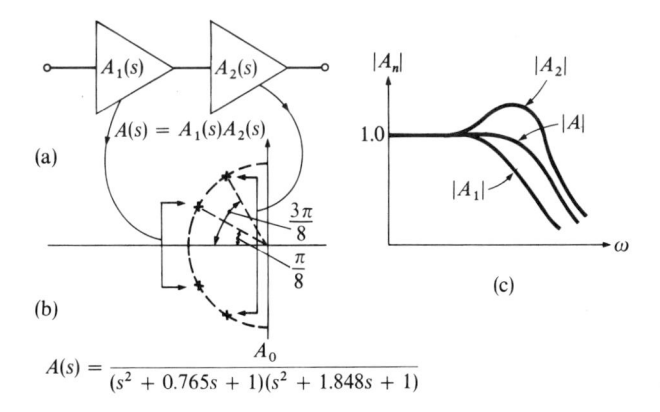

$$A(s) = \frac{A_0}{(s^2 + 0.765s + 1)(s^2 + 1.848s + 1)}$$

Fig 5-16 Stagger-tuned system with two noninteracting blocks (a) Individual transfer functions in cascade; (b) Pole locations of the individual transfer functions; (c) Magnitude responses of the individual transfer functions and the overall transfer function

individual stages, thus sacrificing power gain at the expense of alignability (see Chapter 7). The cascading of shunt-feedback amplifiers and series-feedback amplifiers, or vice versa, is another example of mismatch that reduces interaction between stages. The cascode circuit, as a unit, can be considered a noninteracting block, and thus stagger-tuned with another stage at its output (see Problem 5-21).

Experimentally, the alignment of transistor amplifiers in a stage-by-stage manner to achieve the widest overall bandwidth with no frequency peaking in the overall gain function is a form of stagger-tuning to obtain a maximally flat magnitude response.[4] In such cases, the alignment is performed by starting with the last stage and progressing to the preceding stage. The gain function of each unit (which may be a single-stage or multistage) is adjusted by varying around the preliminary design values so as to get the desired response from the unit. Usually the individual block, or unit, is arranged such that the interaction is small. As an example consider the shunt-series feedback pair of fig 5-12c as a unit. It is shown in Chapter 6 that this circuit can be designed to obtain a complex pair of dominant poles. Its input impedance is shown to be low while its output impedance is very high. If we cascade two such units, the interaction can be very small. Suppose we wish to design the circuits to realize an approximate four-pole Butterworth response. We may initially design the circuits by

---

[4] Stagger-tuning can be used to obtain an approximately linear phase characteristic, by aligning an amplifier in the time domain to provide a minimum rise time, with no overshoot, in the step response.

ignoring the interaction and realize one complex pole pair with $\psi = \pi/8$, say $A_1(s)$ in fig 5-16a, using one shunt-series pair and $A_2(s)$ to have $\psi = 3\pi/8$ by another shunt-series pair. The initial design may then be adjusted, starting with A_2, such that $A_2(s)$ and $A_1(s)$ in cascade yield the widest overall bandwidth with no frequency peaking in the overall response. For interacting stages, the sequence of realization can make some difference, i.e., in the above example if the interaction is not very small the results of assigning the complex pole pair with $\psi = \pi/8$ to A_1 or A_2 can give different results, because of the effect of the termination impedance in each stage. Only in the case of noninteracting stages is the assignment arbitrary.

A *synchronously tuned* design is one in which the individual noninteracting stages have identical pole-zero patterns. In noninteracting stages the overall pole-zero pattern has a certain multiplicity, depending on the form of the gain function of the individual stage. If the interaction is very small, an identical pole-zero pattern may be assigned to the individual stage or unit. In all synchronously tuned circuits the effect of the shrinkage factor must be included in the design. The CE, cascaded, identical one-pole per stage design described in Section 5-6 is one example of synchronous tuning. In the case of shunt-series pairs, which were considered for illustrative purposes earlier in this section, if the individual units are assigned such that each unit has a two-pole maximally flat magnitude response, the resulting design is a synchronous tuned design. In such a case, both units will have complex poles with $\psi = \pi/4$ as shown in fig 5-17. The shrinkage factor for this

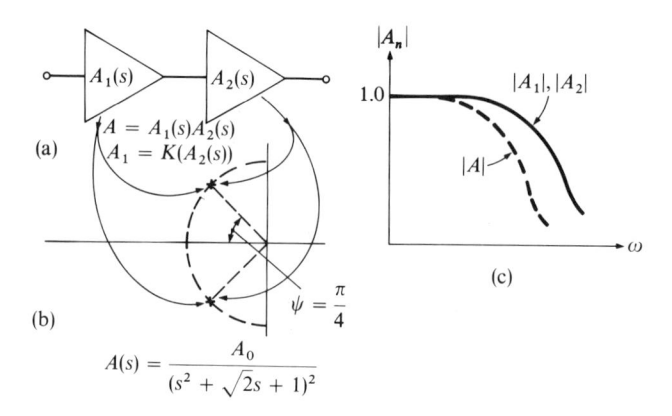

$$A(s) = \frac{A_0}{(s^2 + \sqrt{2}s + 1)^2}$$

Fig 5-17 Synchronously tuned system with two noninteracting blocks: (a) Individual transfer functions: $A_1(s) = KA_2(s)$; (b) Pole locations of the individual transfer functions; (c) Magnitude response of the individual transfer functions and the overall transfer function

case is (see Problem 5-22)

$$S = (\sqrt{2} - 1)^{1/4} = 0.804$$

Hence the overall bandwidth will be smaller than the individual stage bandwidth by the above shrinkage factor.

In a synchronously tuned design, the overall GBP is generally smaller than that of the staggered tuned design. However, the advantage gained through using synchronous tuning is ease of alignment, i.e., each stage or unit is aligned such as to have the same response characteristics as the other stages or units, and the alignment procedure is simpler than in the stagger tuned case. In the above example, if the interaction is small, each unit is aligned to have the widest bandwidth with no frequency peaking (beginning with the last stage) in order to achieve a synchronously-tuned alignment.

5-9 Dc and Differential Amplifiers

There are a number of applications where we must provide amplification for dc signals and for signals which vary very slowly with time. In the first category, we cannot use coupling and bypass capacitors, while in the second category, the values of these capacitors will be so prohibitively large as to make ac coupled amplifiers impractical.

A simple direct-coupled amplifier is shown in fig 5-18, where the collector of the first stage is tied directly to the base of the second stage. This circuit obviously has no low-frequency cutoff point, and can amplify very low-frequency signals, including dc. It has the obvious advantage of saving components by eliminating the coupling and bypass capacitors. The disadvantage of this circuit lies in instability of the operating point and the bias conditions. The operating point of the second stage depends upon the operating point of the first stage, thus making temperature stabilization a difficult problem. We cannot use bypassed emitter resistors in this case to achieve stabilization. Note also, that in this circuit the quiescent collector-to-emitter voltage of the first stage is equal to the emitter-to-base voltage of the second stage, and is very small (of the order of a fraction of a volt). Also, the collector current of the first stage is smaller than that of the second stage. In order to reduce the leakage current I_{CO} at low values of I_C, silicon transistors are often used in practical designs. For silicon transistors, $V_{BE} \gtrsim 0.6$ V, hence the transistor can be operated in the linear active region. However,

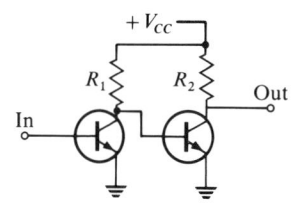

Fig 5-18 Simple dc coupled amplifier

the very low values of the operating point yield low gain and low f_T. In addition, drift problems usually require other compensation techniques. There are many clever compensation methods and circuits available for dc-coupled ac amplifiers in the literature. Most of these focus attention to one or another particular feature of the amplifier for the specific problems they are intended to solve. These are too specialized to consider here.

One design method, which uses transistors in a differential scheme, is particularly important and has extremely wide application in integrated, as well as discrete circuits. We shall, therefore, subsequently discuss this topic in some detail.

A *differential amplifier* is also called a *difference amplifier* and, as the name implies, its function is to amplify the difference between two signals. The differential amplifier finds wide usage in modern electronics. It is widely used in integrated circuits because of its balanced nature. A differential amplifier usually requires a minimum number of external capacitors, and can be used without the bypass and coupling capacitors. Hence it is used for dc and ac amplification.

In order to illustrate the basic principles of operation of a differential amplifier, we will consider the schematic representation shown in fig 5-19. In an ideal differential amplifier, the output signal would be given by

$$v_o = A(v_1 - v_2) \quad (5\text{-}50)$$

where A is the voltage gain of the amplifier. In an actual differential amplifier the output signal is given by

$$v_o = A_1 v_1 + A_2 v_2 \quad (5\text{-}51)$$

where A_1 and A_2 depend upon the difference and the sum of the signals. In symmetrical circuits it is convenient to talk about the in-phase signals, which are termed *common mode* (*CM*) signals, and the difference or antiphase signals, which are called *differential mode* (*DM*) signals. Thus according to this terminology, the *CM* signal v_c is defined by

$$v_c = \tfrac{1}{2}(v_1 + v_2) \quad (5\text{-}52)$$

and the *DM* signal, v_d, is defined by

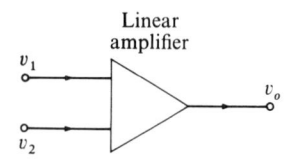

$$v_d = \tfrac{1}{2}(v_1 - v_2) \quad (5\text{-}53)$$

Fig 5-19 Schematic of a differential amplifier

Then from (5-51), (5-52), and (5-53), we obtain

$$v_o = A_1(v_d + v_c) + A_2(v_c - v_d) \tag{5-54a}$$

$$v_o = A_d v_d \left[1 + \left(\frac{v_c}{v_d}\right)\left(\frac{A_c}{A_d}\right) \right] \tag{5-54b}$$

where

$$A_d = (A_1 - A_2) \quad \text{and} \quad A_c = A_1 + A_2 \tag{5-55}$$

Note that A_d and A_c can be readily measured in a differential amplifier. For example, from (5-54b) we have

$$A_d = \left.\frac{v_o}{v_d}\right|_{v_c = 0} \tag{5-56}$$

But $v_c = 0$ implies $v_1 = -v_2$. Hence we apply equal magnitude signals of opposite polarity and measure the output voltage. The resulting voltage gain for equal magnitude antiphase input signals yields A_d. Similarly, from (5-54a) we have

$$v_o = A_c v_c \left[1 + \left(\frac{A_d}{A_c}\right)\left(\frac{v_d}{v_c}\right) \right] \tag{5-57}$$

and

$$A_c = \left.\frac{v_o}{v_c}\right|_{v_d = 0} \tag{5-58}$$

But $v_d = 0$ implies $v_1 = v_2$. Hence we apply equal magnitude signals of the same polarity and measure the output voltage. The voltage gain of the equal magnitude in-phase signal yields A_c. Generally, the desired signals in a differential amplifier are *DM*, and the undesired signals, such as those caused by temperature drift, etc., are *CM*.

We define a quantity called the *common mode rejection ratio* ρ_{CM} as follows:

$$\rho_{CM} = \left|\frac{A_d}{A_c}\right| \tag{5-59}$$

The quantity ρ_{CM} is usually used as a figure of merit for a differential amplifier. It is also sometimes referred to as the *discrimination factor* of a differential amplifier.

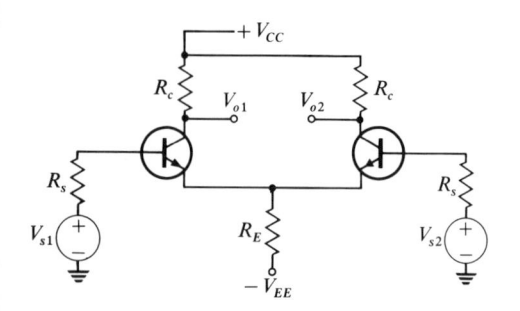

Fig 5-20 Simple differential amplifier

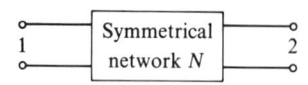

Fig 5-21 General symmetrical linear network shown schematically

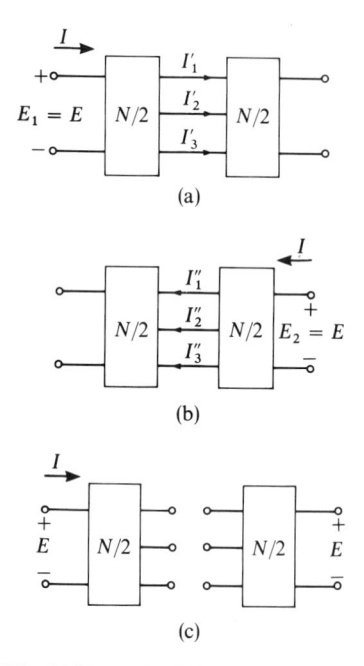

Fig 5-22 (a) Network of fig 5-21 redrawn to exhibit vertical symmetry and a voltage E applied at port 1; (b) Same network but the voltage E applied in phase at port 2; (c) Simplified network under the in-phase excitation conditions

Ideally $A_c = 0$. However, in most practical amplifiers A_c is nonzero, but very small, while A_d is very large. Thus, ρ_{CM} is a very large number in a well designed differential amplifier (of the order of $10^4 = 80$ dB, for a multistage differential amplifier). Substitution of (5-59) into (5-54b) yields

$$v_o = A_d v_d \left[1 + \left(\frac{1}{\rho_{CM}} \right) \left(\frac{v_c}{v_d} \right) \right] \quad (5\text{-}60)$$

Since ρ_{CM} is a very large number, and for v_c and v_d of the same order of magnitude $(1/\rho_{CM})(v_c/v_d) \ll 1$, (5-60) is approximately equal to (5-50) with $A_d/2 = A$. Note that in a differential amplifier with $\rho_{CM} = 10^4$, a 10 μV differential input would give the same output as a 100 mV common mode signal.

A basic differential amplifier circuit is shown in fig 5-20. A complete analysis of this circuit can be performed using the equivalent circuit for the transistors. From the resulting equivalent circuit, the output signal v_o for the difference mode and common mode signals can be solved by straight-forward analysis. This type of analysis, however, is very tedious and is not recommended. Due to the symmetry of the circuit, we can obtain considerable simplification if the transistors and resistors are assumed to be identical. We repeat that this is not a difficult condition to attain with integrated circuit technology. Precise matching between the active and passive components is obtained because they are simultaneously fabricated in adjacent areas on a small chip. Because of this, the transistor pair also has a very good temperature tracking characteristic. In fact even in the case of discrete circuits, a number of manufacturers sell matched transistors made specifically for differential amplifier applications. In some cases critical passive components are matched within 1 per cent. Thus the analysis of the circuit in fig 5-20 can be considerably simplified if we make use of its *symmetry* and the bisection theorem.

The Bisection Theorem

We digress here to briefly discuss a useful theorem known as Bartlett's bisection theorem. The use of this theorem provides considerable simplification in the analysis of *symmetrical circuits*. To see how we can use the bisection theorem, consider the linear symmetrical network shown schematically in fig 5-21. We assume, for the time being, vertical symmetry only, i.e., the two halves of the network on each side of a vertical line of symmetry are identical. We split the network into two sections as shown in fig 5-22a. We apply a voltage source $E_1 = E$ as shown, and the resulting currents in the wires between the two sections are I'_1, I'_2, I'_3, etc. We then apply the same excitation with the same polarity as shown in fig 5-22b at the other side, i.e., $E_2 = E$, and the currents in the wires between the two sections are I''_1, I''_2, I''_3, etc. Since the network is linear, we can apply the superposition theorem, that is $I_j = I'_j + I''_j$. Because of symmetry, $I'_j = -I''_j$ and hence $I_j = 0$. Therefore, we can open-circuit the wires without changing any of the parameters of the network, as shown in fig. 5-22c.

Now consider some wires cross-connected as shown in fig 5-23a. Because of symmetry of the network and the excitations, the potentials $V'_1 = V''_1$ and the currents $I'_j = I''_j$. Hence the crossed wires can be short-circuited and the network can be split into two identical sections, as shown in fig 5-23b.

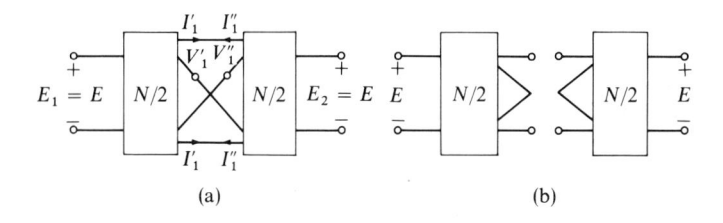

Fig 5-23 (a) Symmetrical network with crossed wires and in-phase excitations applied; (b) Simplified network under the in-phase excitation conditions

We next consider the symmetrical network of fig 5-21 excited by a voltage source $E_1 = E$, as in fig 5-24a. The potentials on the wires between the two sections are V'_1, V'_2, V'_3, etc. We then apply the same excitation with opposite polarity at the other port, as shown in fig 5-24b, which results in potentials V''_1, V''_2, V''_3, etc. By superposition, the total potential on the wires is $V_j = V'_j + V''_j$. But because of symmetry $V'_j = -V''_j$, hence $V_j = 0$. Thus the wires can be short-circuited and the network separated into two sections as shown in fig 5-24c. If we have some crossed wires as shown in fig 5-25a, the currents in each cross-

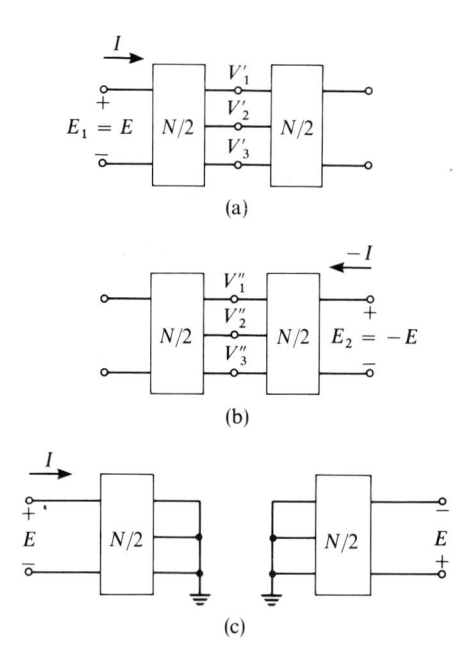

Fig 5-24 (a) Symmetrical network with a voltage E applied at port 1; (b) Same network but the voltage E applied in anti-phase at port 2; (c) Simplified network under the anti-phase excitation condition

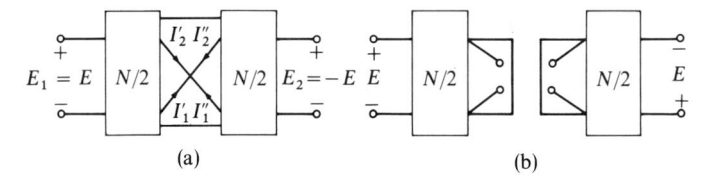

Fig 5-25 (a) Completely symmetrical network (i.e., a network with both vertical and horizontal symmetry) with crossed wires and anti-phase excitations applied; (b) Simplified network under the anti-phase excitation conditions

connected wire are zero *only* if we have in addition to vertical symmetry a horizontal symmetry. Thus, under the horizontal symmetry condition $I_2' = I_1''$, $I_1' = I_2''$, and the potentials on the uncrossed wires will be the same. The network can then be split into two identical sections as shown in fig 5-25b.

The use of the above theorem in the analysis of the differential amplifier is illustrated in the following: Note that in differential amplifiers we have vertical symmetry only, but with no crosswires, hence the bisection theorem applies.

Analysis of Differential Amplifier Circuit

We wish to determine the gain and bandwidth performance of the circuit of fig 5-20 using the bisection theorem. Specifically, we shall determine the voltage-gain functions for equal in-phase signals and equal antiphase signals.

To find A_d, we set $V_{s1} = -V_{s2} = V_s$. For this case the circuit of fig 5-20 reduces to the two sections shown in fig 5-26. The single section circuit of fig 5-27 is a simple CE stage, which has the approximate equivalent circuit shown in fig 5-27b. This circuit has been analyzed before, and the voltage-gain function of the circuit found to be

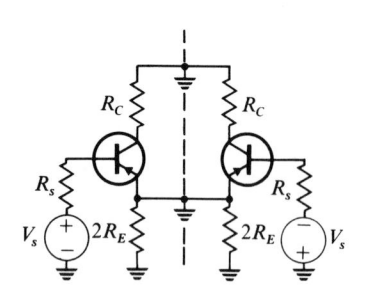

Fig 5-26 Differential amplifier with the anti-phase excitations (differential-mode signals) applied

$$A_d = \frac{V_o}{V_s} = -\frac{A_d(0)p_o}{s + p_o} \tag{5-61}$$

where

$$A_d(0) = \frac{\beta_0 R_C}{R_s + r_b + R_i} \tag{5-62a}$$

$$p_o = \left(\frac{\omega_\beta}{1 + R_C C_c \omega_T}\right)\left(1 + \frac{\beta_0 r_e}{r_b + R_s}\right) \tag{5-62b}$$

Note that if R_C is very large, the hybrid π model should be used.

$$R_i = \beta_0 r_e \qquad C_i = \frac{1 + R_C C_c \omega_T}{r_e \omega_T}$$

(a)   (b)

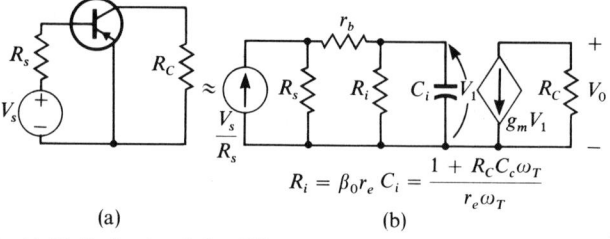

Fig 5-27 (a) Half circuit of the differential amplifier for the differential mode of operation; (b) Approximate model of the differential amplifier for the differential-mode calculation

5-9 DC and Differential Amplifiers

To find A_C, we set $V_{s1} = V_{s2} = V_s$. For this case, the circuit of fig 5-20 reduces to the two sections shown in fig 5-28. The half-circuit is just a CE stage with a large series emitter resistor. The approximate equivalent circuit of the half-circuit in fig 5-29a is shown in fig 5-29b. The voltage-gain function of this circuit has also been already obtained in (4-92), which is of the form

$$A_c(s) = -\frac{A_c(0)p_1}{s + p_1} \tag{5-63}$$

where

$$A_c(0) = \frac{\beta_0 R_C}{R_s + r_b + R_i + 2\beta_0 R_E} = \frac{R_C}{r_e + 2R_E + (R_s + r_b)/\beta_0} \tag{5-64a}$$

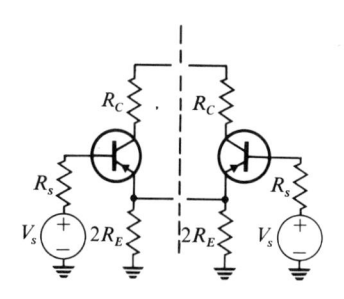

Fig 5-28 Differential amplifier with the in-phase excitations (common-mode signals) applied

$R_i \approx \beta_0(2R_E + r_e) \quad g'_m = \dfrac{1}{r_e + 2R_E}$

$C'_i \approx \dfrac{1 + (R_c + 2R_E)C_c\omega_T}{(r_e + 2R_E)\omega_T}$

(a)   (b)

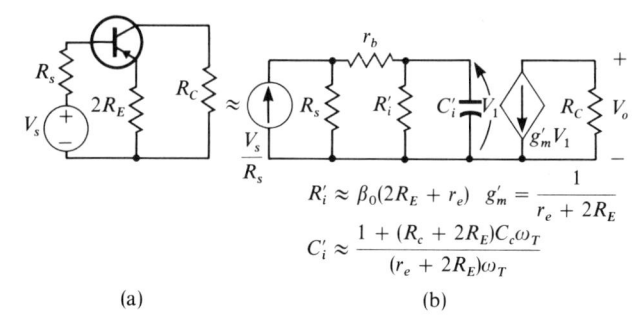

Fig 5-29 (a) Half circuit of the differential amplifier for the common mode of operation; (b) Approximate model of the differential amplifier for the common mode of operation

$$p_1 \approx \left(\frac{\omega_\beta}{1 + (R_C + 2R_E)C_c\omega_T}\right)\left(1 + \frac{2\beta_0 R_E + \beta_0 r_e}{R_s + r_b}\right) \tag{5-64b}$$

The low frequency common mode rejection ratio, ρ_{CM}, from (5-54), (5-62a), and (5-64a) is given by

$$\rho_{CM} = \frac{A_d(0)}{A_c(0)} = \frac{2\beta_0 R_E + \beta_0 r_e + R_s + r_b}{(R_s + r_b + \beta_0 r_e)} \tag{5-65}$$

From (5-65) it is seen that the common mode rejection ratio increases with decreasing R_s, increasing R_E, and increasing β_0. Since a very large value of R_E necessitates a large V_{EE} supply voltage, in order to achieve the chosen operating point, in practice other schemes are usually used to

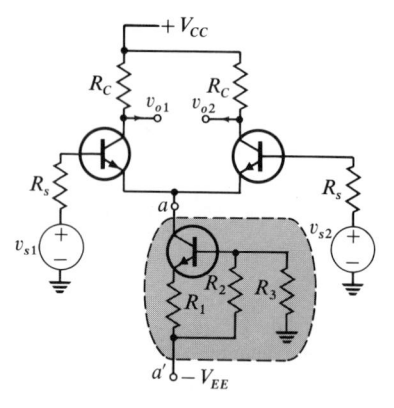

Fig 5-30 Differential amplifier with simulated large emitter resistance

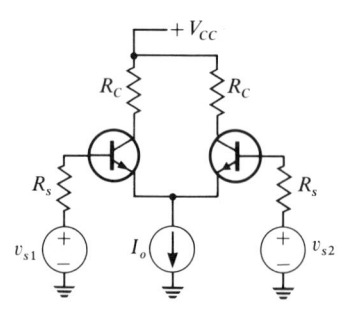

Fig 5-31 Differential amplifier with emitters coupled to a current source

obtain a very high value of R_E. A high value of β_0 can be obtained by using a Darlington compound transistor. A high value of R_E is obtained as shown in fig 5-30, where the third transistor and its associated circuitry (shown enclosed in dashed lines) provide a constant current source I_E. In this case, the effective value of R_E is very high, and of the order of several MΩ for typical cases (see Problem 4-8). The two transistors may be depicted as being fed by a current source as shown in fig 5-31. Note that if a larger amplification is desired, differential amplifiers may be cascaded by feeding the output signals v_{o1} and v_{o2} of fig 5-30 directly to the two base circuits of another differential amplifier. The circuits of practical integrated circuit differential amplifiers are shown in Appendix D. (See the operational amplifier circuits μ702C and MC1522.)

Differential amplifiers are used as a gain block in general purpose ac–dc amplification in integrated circuits. In fact, they are also used extensively as operational amplifiers, which are discussed in the next chapter.

5-10 Input Stage Design Considerations: Noise

In addition to the gain and bandwidth considerations in an amplifier, as pointed out at the very beginning of this chapter, there are other factors that may have to be considered in the design. If many stages of amplification are used to obtain a very high gain, the progressive increase of signal level at the last stage may cause distortion. This is because the signal level at the input of the final stage may make large excursions over the output load line, and may be so large as to violate the small signal assumption and the linear operation of the transistor used. The distortion problem arising as such is considered later in this chapter. If, on the other hand, we use an extremely small amplitude signal at the input so that the signal level at the final stage is small and in the linear region of the characteristic curve, we have solved the distortion problem, but we may have created another problem. The signal level at the input stage may be so small (of the order of the noise voltage), that we may not be able to distinguish the signal from the noise at the output. We shall now consider how this noise voltage (or current) comes about, and what should be done to minimize this problem.

The output signal of an amplifier, in the absence of an input signal, is called *noise*. Noise is exhibited in the amplifier output by hiss or crackle (in the case of an audio amplifier connected to a loudspeaker), when there is no input and the gain control is set to maximum. The magnitude of the noise voltage generated within the circuit, in general, limits the minimum signal that can be amplified by an amplifier. We will briefly consider these aspects and show that noise consideration is of utmost importance in the input stage of an amplifier. The various sources of noise are given in the following without any derivation.

Thermal or Johnson Noise

Noise in a resistor is due to the random motion produced by thermal agitation of the electrons. This noise is temperature-dependent, and is referred to as *thermal noise* or *Johnson noise*. The noise voltage in a resistor has a mean-square value given by

$$\overline{v_n^2} = 4kTR\Delta f \qquad (5\text{-}66)$$

where
T = absolute temperature, K($= °C + 273°$)
k = Boltzmann's constant (1.37×10^{-23} joules/K)
Δf = noise bandwidth in Hz
R = resistance in ohms

The noise equivalent circuit for a resistor, as a noiseless resistor, connected to a noise voltage source is shown in fig 5-32b, or alternatively the equivalent circuit for a resistor, as a noiseless resistor, connected to a current noise source is shown in fig 5-32c. Note that polarity is not shown for the sources, and that it makes no difference which way it is assigned for independent noise sources. To get an idea of the order of magnitude of the thermal noise voltage, consider at room temperature ($T = 300°$K) $R = 1$ kΩ, and $\Delta f = 10$ kHz. Substitution of these values in (5-66) and taking its square root yields an rms noise voltage $v_n = 0.41$ μV. Note that a wider bandwidth and/or a higher temperature will produce a larger noise voltage.

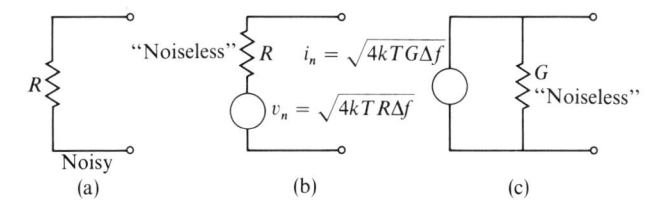

Fig 5-32 (a) An actual resistor (noisy resistor); (b) Noiseless resistor with a noise-voltage source; (c) Noiseless conductance with a noise-current source

Shot Noise

The random fluctuation in the number of charged carriers when emitted from a surface, or diffused across a junction, is referred to as *shot noise*. This noise is usually of the form

$$\overline{i_n^2} = 2qI_o\Delta f \qquad (5\text{-}67)$$

where I_o is the dc current flow, Δf is the noise bandwidth, and q is the magnitude of the electronic charge (1.6×10^{-19} coulomb).

Excess Noise

Another source of noise in a semiconductor device is the excess noise, or $1/f$ noise, which is due to surface imperfections resulting from the emission and diffusion processes. This noise is important at low frequencies, and can be reduced considerably by proper fabrication processes.

Partition Noise

Another source of noise is the partition noise. In BJT's it is due to the random recombination process in the base region. Because of the randomness of emission and recombination, two sources of noise must be introduced to account for this; hence the name partition noise.

5-11 Noise Models

From a knowledge of the presence of these noise sources, the simple noise models for the bipolar junction transistor using the Tee and the π models are derived and shown in figs 5-33a and 5-33b, respectively. In order to concentrate on the basic ideas and techniques, we use bipolar transistors exclusively in the following. The methods, however, are also applicable for other devices, albeit there are some differences in the models.

In the models of figs 5-33a and 5-33b, the noise generators have the following expressions:

$$\overline{i_e^2} = 2qI_e\Delta f \qquad (5\text{-}68\text{a})$$

$$\overline{v_b^2} = 4kTr_b\Delta f \qquad (5\text{-}68\text{b})$$

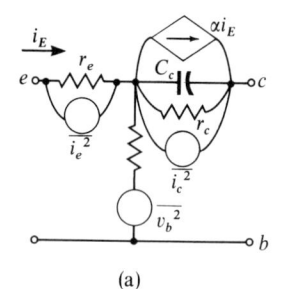

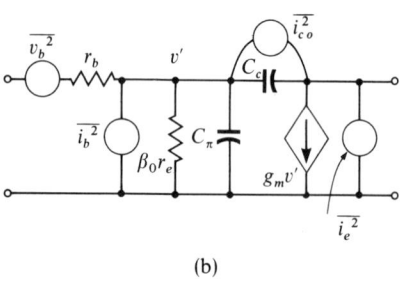

Fig 5-33 Noise models of the bipolar transistor: (a) T-model of the transistor with the various sources of noise; (b) Hybrid π model of the transistor with various sources of noise

5-11 Noise Models

$$\overline{i_c^2} = 2q\Gamma_c I_c \Delta f \qquad (5\text{-}68c)$$

$$\overline{i_b^2} = 2qI_b\Delta f \simeq 2q\frac{I_e}{\beta_0}\Delta f = \frac{\overline{i_e^2}}{\beta_0} \qquad (5\text{-}68d)$$

$$\overline{i_{CO}^2} = 2qI_{CO}\Delta f \qquad (5\text{-}68e)$$

Actually Γ_c is frequency-dependent, $1 - \alpha_0 \leq \Gamma_c \leq 1$. It has its minimum value at low frequencies, and increases as the frequency is increased. The contribution due to $\overline{i_{CO}^2}$ can usually be neglected. We shall assume in the following that the noise sources are uncorrelated. This final assumption is found to be quite reasonable, and enables one to use superposition of power in noise calculations—a factor that greatly simplifies such calculations.

The model shown in fig 5-33b is very useful for transistors in the CE configuration. For $\beta_0 \geq 10$ and $I_C \gg I_{CO}$, the noise source due to I_{CO} may be entirely neglected without any significant error. This is particularly true in the case of silicon transistors. We shall next consider an example to illustrate noise calculation in an amplifier circuit.

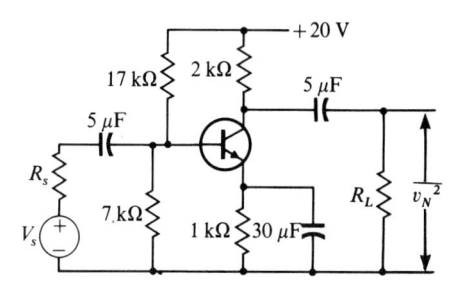

Fig 5-34 Example circuit for noise calculations

EXAMPLE

The amplifier circuit shown in fig 5-34 uses a silicon transistor. We wish to calculate the midband rms noise voltage at the output of the circuit at room temperature. The bias conditions are $V_{CE} = 5$ V, $I_E = 5$ mA, and $I_{CO} = 10^{-9}$ A. For simplicity we assume that[5] $\Delta f \approx f_{3\,dB}$. This is a common practice. The transistor parameters are assumed to be $\beta_0 = 50$, $r_b = 50\,\Omega$, $C_c = 4$ pF, and $f_T = 300$ MHz. The source and load terminations are $R_s = R_L = 100\,\Omega$.

For the circuit using the unilateral model, we can readily determine the approximate 3-dB bandwidth, $f_{3\,dB}$. For $R_L = 100\,\Omega$, $D = 1 + R_L C_c \omega_T = 1.7$, $C_i = D/r_e\omega_T = 1.8 \times 10^{-10}$ and the upper cutoff frequency $f_{3\,dB} = 1/\{2\pi C_i[R_i\|(r_b + R_s)]\} \simeq 9.3$ MHz $\simeq \Delta f$, according to our assumption.

The midband noise model for the circuit, ignoring the noise due to I_{CO}, is shown in fig 5-35. Since the circuit is linear and all the sources are assumed to be independent, we can separately determine the noise output due to each source and then add the results. In other words,

---

[5] Actually Δf is determined from

$$\Delta f = \int_0^\infty \frac{G_A(f)}{G_A(f_o)} df$$

where G_A is the available power gain, and f_o is the center frequency (for low-pass amplifiers f_o is zero).

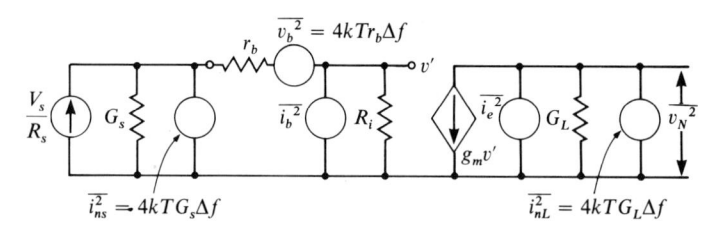

Fig 5-35 Noise model of fig 5-34 at midband

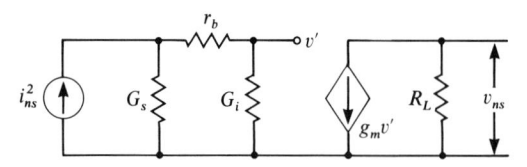

Fig 5-36 Circuit for noise calculation due to the source resistor only

to calculate the output noise due to the source resistance noise, i_{ns}, we ignore all the other sources (short-circuit the voltage source and open-circuit the current source) and find its contribution to $\overline{v_N^2}$ alone. We repeat the calculation for each of the other sources and then add the mean-square voltages.

Thus the output noise contribution due to the source resistance noise, $\overline{v_{ns}^2}$, is found from fig 5-36 to be

$$v' = \left(\frac{1}{G_s r_b G_i + G_i + G_s}\right) i_{ns}$$

$$v_{ns} = g_m v' R_L = \left(\frac{g_m R_L}{G_s r_b G_i + G_i + G_s}\right) i_{ns} \qquad (5\text{-}69a)$$

From (5-69a), and substitution of the expression for $\overline{i_{ns}^2}$, we obtain

$$\overline{v_{ns}^2} = \left(\frac{g_m R_L}{G_s r_b G_i + G_i + G_s}\right)^2 (4kTG_s \Delta f) \qquad (5\text{-}69b)$$

$$= 24.4 \times 10^{-10}$$

Similarly, the output noise contribution due to $\overline{v_b^2}$, i.e., $\overline{v_{nb}^2}$, is found to be

$$v' = R_i\left(\frac{v_b}{r_b + R_s + R_i}\right)$$

$$v_{nb} = g_m v' R_L = \left(\frac{g_m R_L R_i}{r_b + R_s + R_i}\right) v_b \qquad (5\text{-}70a)$$

$$\overline{v_{nb}^2} = \left(\frac{g_m R_L R_i}{r_b + R_s + R_i}\right)^2 (4kTr_b \Delta f) \qquad (5\text{-}70b)$$

$$= 12.2 \times 10^{-10}$$

The output noise contribution due to $\overline{i_b^2}$, i.e., $\overline{v_{ni}^2}$, is as follows:

$$v' = \left(\frac{r_b + R_s}{r_b + R_s + R_i}\right) i_b \beta_0 r_e$$

$$v_{ni} = g_m v' R_L = \frac{R_L(r_b + R_s)}{r_b + R_s + R_i} \beta_0 i_b \qquad (5\text{-}71a)$$

$$\overline{v_{ni}^2} = \left[\frac{R_L(r_b + R_s)}{r_b + R_s + R_i}\right]^2 \beta_0^2 \left(\frac{2qI_e\Delta f}{\beta_0}\right) \qquad (5\text{-}71b)$$

$$= 10.5 \times 10^{-10}$$

Finally the output noise due to $\overline{i_e^2}$, i.e., $\overline{v_{ne}^2}$, and the noise due to R_L, i.e., $\overline{v_{nL}^2}$, are

$$\overline{v_{ne}^2} = R_L^2 \overline{i_e^2} = R_L^2 (2qI_e\Delta f) \qquad (5\text{-}72a)$$

$$= 1.49 \times 10^{-10}$$

and

$$\overline{v_{nL}^2} = R_L^2 \overline{i_{nL}^2} = 4kTR_L\Delta f \qquad (5\text{-}72b)$$

$$= 0.16 \times 10^{-10}$$

The total ouput noise voltage is the sum of the above quantities, that is

$$\overline{v_N^2} = \overline{v_{ns}^2} + \overline{v_{nb}^2} + \overline{v_{ni}^2} + \overline{v_{ne}^2} + \overline{v_{nL}^2} \qquad (5\text{-}73)$$

$$\overline{v_N^2} = (24.4 + 12.2 + 10.5 + 1.49 + 0.16) \times 10^{-10} = 48.8 \times 10^{-10}$$

$$v_N \simeq 7 \times 10^{-5} = 70\,\mu V$$

In other words, in the absence of a signal, i.e., $v_s = 0$, we will have an undesired signal of 70 μV amplitude at the output of the amplifier. In a cascaded amplifier this voltage will be further increased and amplified by the succeeding stages.

5-12 Noise Figure

The noise figure provides a quantitative measure of how noisy a circuit is. The noise figure, F_n, is defined as follows:

$$F_n = 10 \log \frac{\text{total noise power output}}{\text{noise power output due to } R_s \text{ alone}} \qquad (5\text{-}74a)$$

A more convenient form of (5-74a) is one of the following:

$$F_n = 10 \log \left[\frac{\text{total mean-square noise current at output}}{\text{mean-square noise current at output due to } R_s \text{ alone}} \right]_{R_L = 0} \quad (5\text{-}74\text{b})$$

or

$$F_n = 10 \log \left[\frac{\text{total mean-square noise voltage at output}}{\text{mean-square noise voltage at output due to } R_s \text{ alone}} \right]_{R_L = \infty} \quad (5\text{-}74\text{c})$$

In some cases the value of R_L is chosen to match the output impedance of the amplifier in order to maximize the power gain. Furthermore, the contribution of R_L to the noise figure is usually insignificant. Note the factor of 10 for calculating the noise figure in dB, instead of 20, because power ratios are involved, rather than voltage or current ratios. Since the numerator is equal to the denominator plus additional terms we may also write (5-74a) as

$$F_n = 1 + \frac{R_{neq}}{R_s} \quad (5\text{-}75\text{a})$$

or

$$R_{neq} = (F_n - 1)R_s \quad (5\text{-}75\text{b})$$

where R_{neq} is a function of kT and the circuit parameters, and its determination is explained in the following: For the example of this section from (5-74) and (5-73) we have

$$F_n = \frac{\overline{v_N^2}}{\overline{v_{ns}^2}} = 1 + \frac{\overline{v_{nb}^2} + \overline{v_{ni}^2} + \overline{v_{ne}^2} + \overline{v_{nL}^2}}{\overline{v_{ns}^2}} \quad (5\text{-}76)$$

$$= \frac{48.8}{24.4} = 2 = 3 \text{ dB}$$

Thus the noise figure of the one-stage amplifier of the example is 3 dB. From (5-75) and (5-76) the value of $R_{neq} = R_s = 100 \, \Omega$.

The expression for the noise figure in terms of the circuit parameters may easily be found by using (5-74b) or (5-74c). For the CE configuration, from (5-74c), (5-69b), (5-70b), (5-71b), and (5-72) we obtain

$$F_n = 10 \log \left[1 + \frac{r_b}{R_s} + \frac{(r_b + R_s)^2}{4kTR_s} \left(\frac{2qI_e}{\beta_0} \right) \right.$$
$$\left. + \frac{2qI_e}{4kTR_s\beta_0} \left(\frac{r_b + R_s + \beta_0 r_e}{\beta_0} \right)^2 \right] \quad \text{(dB)} \quad (5\text{-}77)$$

Substituting $kT/qI_e = r_e$ in (5-77), and using the approximation $\beta_0 \gg 1$, we can simplify (5-77) to

$$F_n = 10 \log \left[1 + \frac{r_b}{R_s} + \frac{r_e}{2R_s} + \frac{(R_s + r_b + r_e)^2}{2\beta_0 r_e R_s} \right] \quad \text{(dB)} \quad (5\text{-}78)$$

From the noise expression (5-78) it is seen that F_n depends both on R_s and r_e. One can optimize F_n with respect to these two quantities, because F_n increases for high and very low values of r_e and R_s. If r_e is fixed, i.e., fixed I_e, then F_n can be optimized with respect to R_s. The result is

$$(R_s)_{opt} = r_b \sqrt{1 + \frac{\beta_0 r_e}{r_b} \left(2 + \frac{r_e}{r_b} \right)} \quad (5\text{-}79)$$

For this example $(R_s)_{opt} = 170 \, \Omega$ and $F_{min} = 10 \log (1.90) = 2.8$ dB. Note that the noise figure for $R_s = 100 \, \Omega$ is not far off from the minimum value. Also note from (5-78) that for low noise figure, transistors with low r_b and high β_0 are preferable. A typical plot of F_n with respect to both I_c and R_s is shown in fig 5-37. Note that the optimum region is reasonably broad, and that F_n is poor for very high and very low I_c. F_n is also high for very low or very high source resistance. Since the CE stage has a moderate input impedance, it can best match the optimum source resistance for low noise performance. Thus, these considerations are to be taken into account when a minimal noise amplifier is to be designed. From the typical plot, such as in fig 5-37, which is usually provided by the manufacturer, the selection of R_s and I_c can be made.

The noise figure of a transistor, in addition to being dependent on R_s and I_c, also depends on frequency. Typical variation of F_n vs frequency is shown in fig 5-38. Note that at midband frequencies F_n is fairly constant, and that it increases at both the low and the high frequencies. At low frequencies the $1/f$ noise dominates, and it is for this reason that one must be careful in dc-coupled ac amplifiers, when noise consideration is an important factor. The high-frequency increase of F_n is essentially due to the increase in Γ_c in (5-68c), and also the decrease of power gain vs frequency.

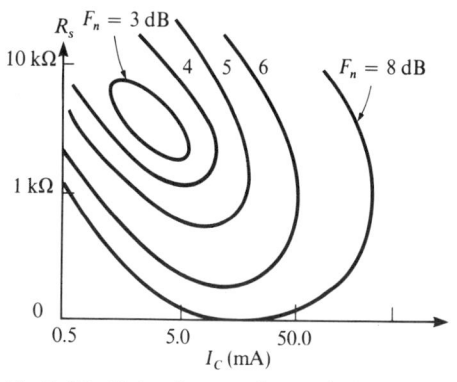

Fig 5-37 Noise figures of a typical transistor for various source resistors and collector currents

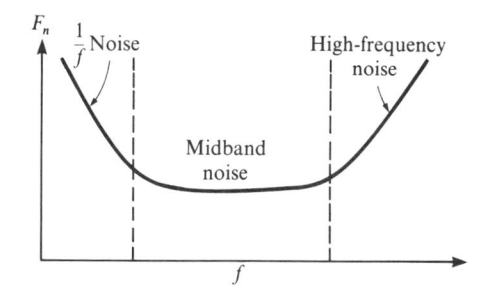

Fig 5-38 Noise figure of a typical transistor vs frequency

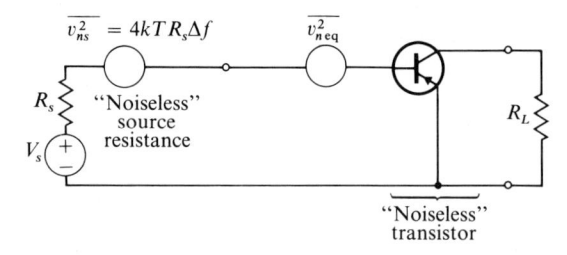

5/Analysis and Design of Multistage Amplifiers

Fig 5-39 Noise representation of a device

The noise generated in a device can also be represented by placing an equivalent input noise voltage at the input of the device as shown for a transistor in fig 5-39. In this case, the relation between $\overline{v_{neq}^2}$ and F_n is defined as follows:

$$F_n = 1 + \frac{\overline{v_{neq}^2}}{\overline{v_{ns}^2}} \qquad (5\text{-}80)$$

From (5-80), or (5-75), the value of R_{neq}, or $\overline{v_{neq}^2}$, determines the noise figure, F_n of the stage, and vice versa. If there are external resistors, the noise contribution due to them should be added to that due to $\overline{v_{neq}^2}$.

The FET amplifier also has $1/f$ noise at low frequencies. The thermal and shot noise sources are also present, but generally speaking, the noise performance of a FET is better than that of vacuum tubes of

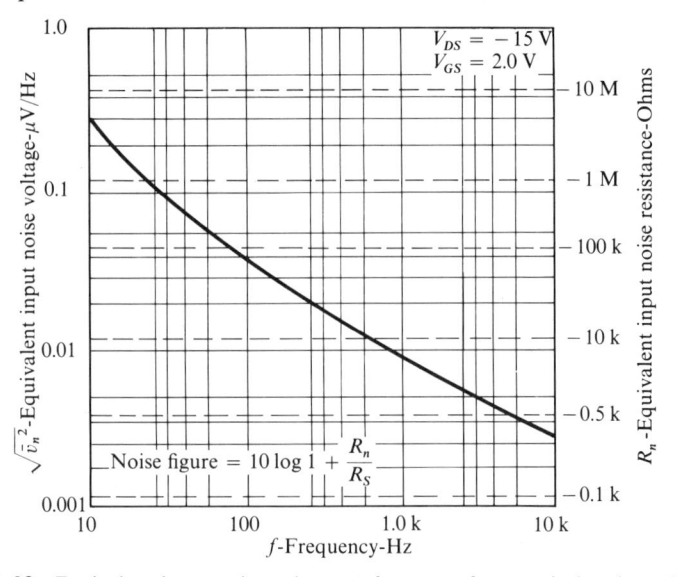

Fig 5-40 Equivalent input noise voltage vs frequency for a typical p-channel FET (Courtesy Fairchild Semiconductor)

comparable transconductance. For very large values of R_s, the noise figure of a FET is also better than that of a bipolar junction transistor, since the noise figure of a FET is small when it is connected to a very large source resistance. Furthermore, the noise figure of a FET is essentially independent of bias conditions. A typical noise figure for a FET is about 2 dB. The noise representation shown in fig 5-39 may also be used for FET's and vacuum tubes. The manufacturers of FET's usually give the values of v_{neq} or R_{neq} vs frequency. A typical plot is shown in fig 5-40 where v_n and R_n are used for short.

Noise Figure of Cascaded Networks

We next examine the noise figure of cascaded networks. In order to do so, it is convenient to think in terms of the available power gain. For example, consider the two-stage cascaded network shown schematically in fig 5-41, where for simplicity we assume that Δf is the

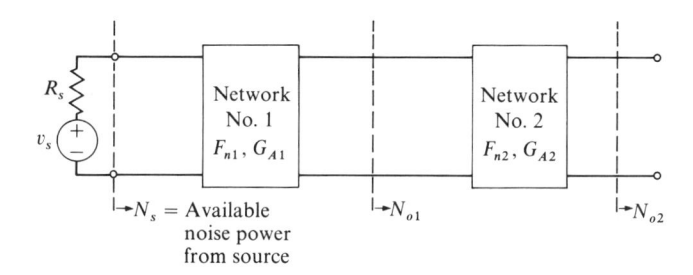

Fig 5-41 Noise-figure representation of cascaded amplifiers

same for both networks. The available noise power from the source is

$$N_s \equiv \frac{\overline{v_s^2}}{4R_s} = \frac{4R_s kT\Delta f}{4R_s} = kT\Delta f \tag{5-81}$$

The available noise power at the outputs of networks 1 and 2 is

$$N_{o1} = F_{n1} G_{A1} kT\Delta f \tag{5-82a}$$

$$N_{o12} = F_n G_{A1} G_{A2} kT\Delta f \tag{5-82b}$$

The noise generated by network 2 alone is

$$N_{n2} = (F_{n2} - 1) G_{A2} kT\Delta f \tag{5-83}$$

If we assume the noise to uncorrelated, we can add noise power; then

$$N_{o12} = N_{o1} G_{A2} + N_{n2} \tag{5-84}$$

From (5-81), (5-82), and (5-84) we have

$$F_n G_{A1} G_{A2} kT\Delta f = F_{n1} G_{A1} G_{A2} kT\Delta f + (F_{n2} - 1) G_{A2} kT\Delta f \quad (5\text{-}85)$$

From (5-85) we get

$$F_n = F_{n1} + \frac{F_{n2} - 1}{G_{A1}} \quad (5\text{-}86)$$

From (5-86) it is seen that in the overall noise figure of the cascaded networks, the noise contribution of the first stage is the most significant if $G_{A1} \gg 1$. In fact if the available power gain of the first stage is very high, the noise contribution due to F_{n2} will be negligible. It is for this reason that, in bipolar junction transistor circuitry, minimal noise cascaded networks use a *CE* stage as the input stage. The second stage, therefore, does not play an important role under the above condition, and can be any configuration. However, if the input *CE* stage is followed by a *CB* stage (see the cascode circuit in fig 5-10), the circuit will be nearly unilateral and less prone to instability, and thus the unit can be designed for an overall maximum power gain, with less instability problems than any of the other combinations.

If the input source resistance is very high, a hybrid cascode circuit using a FET common source in cascade with a *CB* bipolar junction transistor (see fig P5-16) may be used to obtain a minimal noise performance.

5-13 Output Stage Design Considerations: Power Amplifiers

Thus far we have considered small signal amplifiers, where the input and intermediate stages were used to obtain a large gain-bandwidth product, or to minimize the effect of noise in the signal. The stages were not required, however, to provide appreciable amounts of power. The operation of the amplifying stages was in the linear active region, and hence distortion was negligible.

In this section we will consider power amplifiers, where the requirements are quite different. The output stage may drive the loudspeaker of an audio amplifier, a servomotor, etc. In such cases appreciable amounts of power are required at the output stage. Other requirements are to keep the signal distortion low and the dc power requirements as small as possible. In other words, we wish to furnish the required power to the load as economically as possible, while meeting the other design specifications.

For power amplifiers, the signal swings are often very large. Since the nonlinear performance of a power amplifier cannot be represented by a linear model, analysis must be performed graphically using the characteristics of the amplifying device. Simplifying assumptions are usually made to make the analysis and design tractable. We will consider such an analysis for audio frequency power amplifiers, where the frequency range is below 20 kHz. In particular, we shall discuss distortion calculations and the efficiency of conversion of power amplifiers in the various modes of operations.

The various modes of operation are classified in accordance with the location of the operating point, i.e., the bias condition, as shown in fig 5-42. Class A refers to power amplifiers for which the bias condition and the input signals are such that the output signal is never zero. In other words the operating point always remains in the active region and the output signal current flows during the complete cycle. Class B refers to operation where the bias is at the edge of cutoff, so that the output current is zero during only half of the cycle. Class AB refers to the case where the operating point is between that of class A and class B, in other words, output current flows and is non-zero for more than half the cycle but not the complete cycle. Finally, class C operation refers to the case where the bias is beyond the cutoff point, so that the output current is zero for more than one half-cycle.

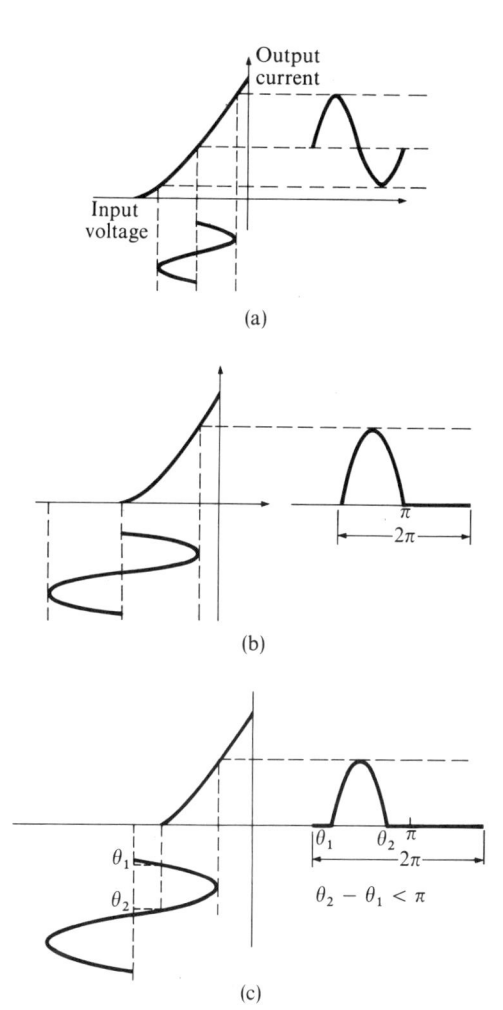

Fig 5-42 Power amplifier input and output signals: (a) Class A operation; (b) Class B operation; (c) Class C operation

5-14 Class A Power Amplifiers

When a single device is used in the output stage for linear amplification, the operation must be in Class A mode. A simple such amplifier, using a transistor, is shown in fig 5-43. The operating point bias conditions are such that the operation is not outside the hyperbola of the

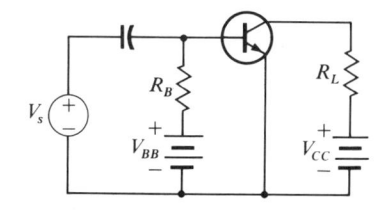

Fig 5-43 Simple class A power amplifier

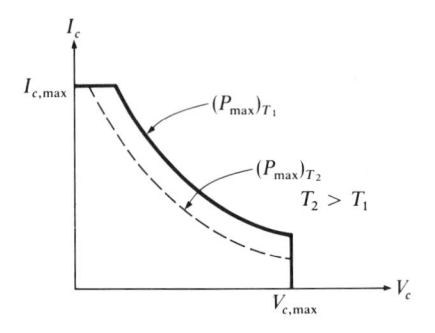

Fig 5-44 Allowable region of operation and maximum power dissipation hyberbola

maximum power dissipation. The limitations on the operating region of the transistor are shown in fig 5-44. Note that the maximum power dissipation of a transistor depends upon the ambient temperature. As temperature increases, this value decreases. The maximum average power that a transistor can dissipate depends upon the type and construction of the transistor. The maximum power is limited by the temperature of the collector-to-base junction. The junction temperature can rise because of environmental temperature change, or self-heating due to collector junction power dissipation. Collector junction power dissipation raises the junction temperature, which in turn increases the collector current, hence a further increase in power dissipation. This phenomenon which is referred to as *thermal runaway* should be avoided. Otherwise the transistor may be permanently damaged. In the design of power amplifiers, the transistor maximum collector dissipation is usually lowered for a case temperature greater than room temperature. Such derating information is often provided by the manufacturer (see Appendix D). In other words, for temperatures beyond some specified value, the dissipation is reduced linearly as specified by the data sheets.

Assume the output characteristics and the load line for a transistor amplifier as shown in fig 5-45. Initially we assume that the output characteristics are equidistant for equal increments of input signal, so that the output signals will be sinusoidal for sinusoidal excitation. In other words, we assume a linear dynamic transfer characteristic.

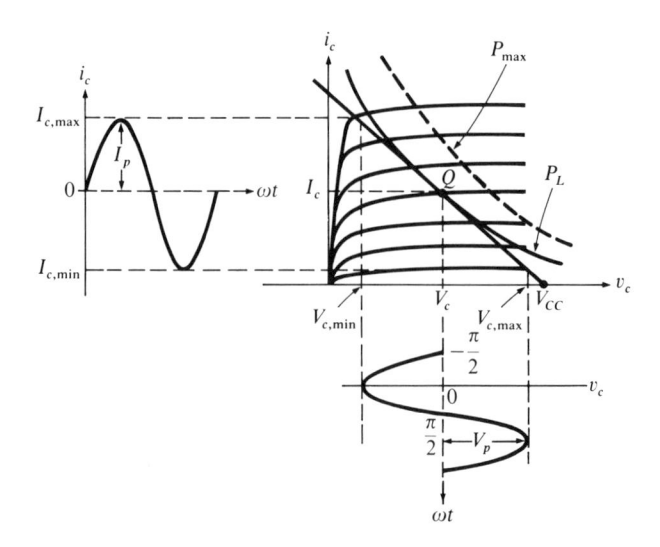

Fig 5-45 Output characteristics, input and output signals of a class A power amplifier

5-14 Class A Power Amplifiers

In this case nonlinear distortion is negligible, and the various power calculations are straightforward. The dc power required from the power supply is

$$P_{dc} = V_{CC}I_c \qquad (5\text{-}87)$$

The ac output power is

$$P_{ac} = \left(\frac{I_p}{\sqrt{2}}\right)\left(\frac{V_p}{\sqrt{2}}\right) = \frac{I_p^2 R_L}{2} \qquad (5\text{-}88)$$

where V_p and I_p are the peak values of the output circuit voltage and current, respectively. Under the linear assumption, the rms values of the load current and voltage from the characteristic curve, in terms of the peak values, are

$$\frac{I_p}{\sqrt{2}} = \frac{1}{\sqrt{2}}\left(\frac{I_{c,max} - I_{c,min}}{2}\right) \qquad (5\text{-}89)$$

$$\frac{V_p}{\sqrt{2}} = \frac{1}{\sqrt{2}}\left(\frac{V_{c,max} - V_{c,min}}{2}\right) \qquad (5\text{-}90)$$

Hence

$$P_{ac} = \frac{I_p V_p}{2} = \frac{(V_{c,max} - V_{c,min})(I_{c,max} - I_{c,min})}{8} \qquad (5\text{-}91)$$

We define the conversion efficiency of the dc power to the total ac power by

$$\eta \equiv \frac{\text{ac power delivered to load}}{\text{dc power of the output circuit}} \times 100\% = \frac{P_{ac}}{P_{dc}} \times 100\% \qquad (5\text{-}92)$$

From (5-87), (5-91), and (5-92)

$$\eta = \frac{(V_{c,max} - V_{c,min})(I_{c,max} - I_{c,min})}{8(V_{CC}I_c)} \qquad (5\text{-}93)$$

Under ideal conditions for maximum efficiency of class A power amplifiers

$$V_{c,min} = 0, \quad I_{c,min} = 0; \quad V_{c,max} = V_{CC}, \quad I_{c,max} = 2I_c$$

and (5-93) reduces to

$$\eta_m = \frac{P_{ac}}{P_{dc}} = \frac{I_p V_p}{2(V_{CC}I_c)} \approx \frac{(V_{c,\max})(I_{c,\max})}{4(V_{c,\max})(I_{c,\max})} = 25\% \quad (5\text{-}94)$$

Hence the maximum theoretical efficiency for class A operation is 25%. The actual conversion efficiency is, of course, less than 25%, since all the ac power is not delivered to the load if the device low-frequency parameters are included. The maximum collector power dissipation in a transistor, which sets the lowest bounds on the transistor power rating, is also a very important consideration in the design of power amplifiers.

The transistor collector dissipation P_c is given by

$$P_c = P_{dc} - P_L \quad (5\text{-}95)$$

where P_{dc} is the supplied power, and is equal to $V_{CC}I_c$, and P_L is the output power to the load.

A figure of merit, F_p, for power amplifiers is defined as

$$F_p = \frac{\text{maximum collector dissipation}}{\text{maximum output power}} = \frac{P_{CM}}{P_{LM}} \quad (5\text{-}96)$$

For the resistive-coupled circuit of fig 5-43, we have

$$P_{LM} = \frac{I_p^2 R_L}{2} = \frac{V_{CC}^2}{8R_L} \quad (5\text{-}97)$$

The maximum collector dissipation occurs when P_L is minimum. Hence

$$P_{CM} = V_{CC}I_c/2 - 0 = \frac{V_{CC}^2}{4R_L} \quad (5\text{-}98)$$

Thus from (5-95), (5-97), and (5-98) we obtain

$$F_p = \frac{V_{CC}^2/4R_L}{V_{CC}^2/8R_L} = 2 \quad (5\text{-}99)$$

Thus if the maximum output load power is specified as 10 watts, the collector must be able to dissipate at least 20 watts. This is an undesirable feature of class A amplifiers.

Transformer-Coupled Amplifier

The efficiency of conversion can be improved considerably by preventing the quiescent current from flowing into the load. In the case of the series-fed amplifier shown in fig 5-43, the flow of the quiescent current is the main cause of the poor efficiency. Often it is also undesirable to pass the dc component of the current through the output device. Therefore, a more desirable circuit for the output stage is that shown in fig 5-46. The transformer at the output circuit also provides an impedance match in order to transfer maximum power to the load. The load, such as the impedance of the voice coil of a loudspeaker, is usually very small, e.g., 4 to 15 ohms. Sometimes the input circuit also utilizes a transformer for matching the output impedance of the driver to the input impedance of the output stage.

Recall that for an ideal transformer, the voltage-current relations are

$$V_1 = \frac{n_1}{n_2} V_2 \qquad I_1 = \frac{n_2}{n_1} I_2 \qquad (5\text{-}100)$$

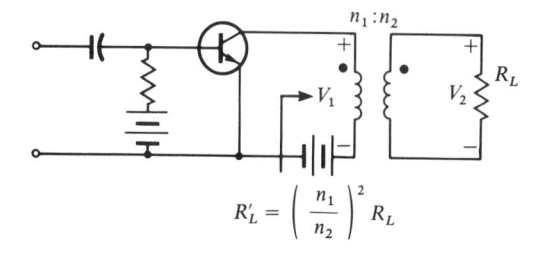

Fig 5-46 Transformer-coupled class A power amplifier

$$R'_L = \left(\frac{n_1}{n_2}\right)^2 R_L$$

Where the subscript 1 indicates the primary winding and the subscript 2 the secondary winding (current, voltage, and turns). From (5-100)

$$R'_L = \frac{V_1}{I_1} = \left(\frac{n_1}{n_2}\right)^2 \frac{V_2}{I_2} = \left(\frac{n_1}{n_2}\right)^2 R_L \qquad (5\text{-}101)$$

Of course, in a more complete analysis of efficiency and power loss, one must include the other parameters of the transformer, such as the core loss, primary and secondary resistances, etc.

For a transformer-coupled load, the static and dynamic load lines are shown in fig 5-47. The static load line is almost vertical, because of the very small primary resistance of the transformer. The maximum efficiency of a transformer-coupled stage is readily determined as follows. The maximum peak-to-peak voltage swing $2V_p = V_{max} = 2V_{CC}$. Also, the peak-to-peak current swing $2I_p = I_{c,max} = 2I_c$. The maximum ac power is then given by

$$(P_{ac})_{max} = \frac{V_p I_p}{2} = \frac{V_{CC} I_c}{2} \qquad (5\text{-}102)$$

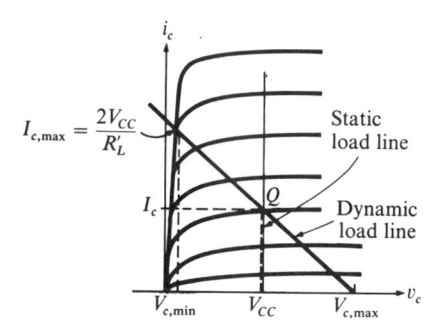

Fig 5-47 Static and dynamic load lines of a transformer-coupled class A power amplifier

The maximum efficiency is

$$\eta_m = \frac{(P_{ac})_{max}}{P_{dc}} = \frac{V_{CC} I_c / 2}{V_{CC} I_c} = 50\% \qquad (5\text{-}103)$$

Hence the maximum efficiency is doubled by using transformer coupling to the load.

The figure of merit of transformer-coupled class A amplifiers is the same as that of the series-fed case. This is readily seen to be the case, since

$$P_{LM} = \frac{I_p^2 R_L'}{2} = \frac{(V_{CC})^2}{2R_L'} \tag{5-104}$$

$$P_{CM} = V_{CC}I_c - 0 = \frac{(V_{CC})^2}{R_L'} \tag{5-105}$$

and

$$F_p = \frac{P_{CM}}{P_{LM}} = 2 \tag{5-106}$$

DESIGN EXAMPLE

Consider the following specifications for the design of a class A power output stage. The audio output power to the load is to be 2 watts. The output load resistance is 4 ohms, and the available supply voltage is 12 volts.

Solution: Since the load resistor is very small, and the efficiency of a transformer-coupled stage is better than that of the series-fed class A amplifier, we shall use a transformer-coupled class A amplifier. The transistor to be chosen must have the following ratings:

$$(V_{CE})_{max} > 2V_{CC} = 2(12) = 24 \text{ V}$$

From (5-106), the collector power dissipation must be at least twice the power output delivered to the transformer primary. Thus[6]

$$P_c \geq 2(2) = 4 \text{ watts}$$

---

[6] In a real transformer there are some losses, since the efficiency of the transformer is not 100 per cent. Hence the output power to the load must be greater by this factor. Finally, if the operating temperature is high, the transistor rating should be lowered to consider this factor. In other words, the derated value should be higher than $2(P_{ac}/\kappa)$ where κ is the transformer efficiency.

There are many commercially available transistors that meet these specifications, and have frequency capabilities suitable for audio frequencies. For the selected transistor, we plot the 4 watt dissipation hyperbola on the typical characteristic curve of the transistor. A load line is then drawn tangent to the hyperbola passing through the quiescent point $V_{CE} = 12$ V, as shown in fig 5-48. Hence the value of the quiescent current $I_c = \frac{4}{12} = 0.3$ A is known.[7]

From the slope of the load line we determine the ac load resistance R'_L, for the transistor. For this example

$$R'_L = \frac{12}{0.33} = 36\,\Omega$$

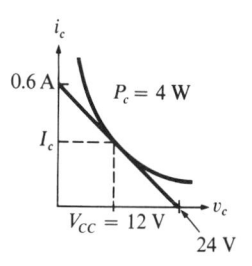

Fig 5-48 Load line of the design example

The required turns ratio of the transformer is then found from

$$R'_L = \left(\frac{n_1}{n_2}\right)^2 R_L \Rightarrow \frac{n_1}{n_2} = \sqrt{\frac{R'_L}{R_L}} = \sqrt{\frac{36}{4}} = 3.0$$

This preliminary design is then experimentally tested and modified, as necessary, to achieve minimum distortion, maximum efficiency, etc.

5-15 Distortion Characterization and Calculation

In the previous section we assumed a perfectly linear device, i.e., we assumed a linear dynamic transfer characteristic. However, we have seen in Chapter 3 that the device transfer characteristics are nonlinear. For large signal swings around Q, the nonlinear portion of the transfer characteristics is utilized, and a linear assumption can cause significant error in the analysis. If the device dynamic curve is nonlinear over the operating range, the waveform of the output differs from that of the input and is hence distorted. This type of distortion is known as *nonlinear* or *amplitude distortion*. A linear assumption of the dynamic curve $i = gv$, of course, cannot reveal this distortion. Thus, instead of assuming a linear dynamic characteristic, we express the characteristic curve with respect to the quiescent point Q by a

---

[7] The biasing technique of Chapter 3 can also be used here. However, an important difference is in the size of the resistances. The value of R_E is usually selected to be very small, of the order of a few ohms, or zero so that its power dissipation is negligible, i.e., $I_c^2 R_E \ll V_{CE} I_c$.

power series of the form

$$y - y_0 = g_1 x + g_2 x^2 + g_3 x^3 + \cdots \qquad (5\text{-}107)$$

where y is the output current and x the input voltage. We use x and y so as to keep the generality of the treatment, and thus avoid considering a specific device. In some cases the simple expression $y - y_0 = g_1 x + g_2 x^2$ is acceptable, e.g., for FET. This fact is borne out by a comparison of (3-61) and (3-107). In order to characterize the distortion, we have the graphical waveforms, and thus graphical Fourier analysis must be performed. Since the output curve is an even function of time (see fig 5-49), the Fourier series expansion of (5-107) can be expressed as a series of cosine terms only:

$$(y - y_0) = a_0 + a_1 \cos \omega t + a_2 \cos 2\omega t + a_3 \cos 3\omega t + \cdots \qquad (5\text{-}108)$$

The Fourier components must now be determined graphically. The graphical Fourier analysis, in conjunction with fig 5-49, is illustrated in the following. A very accurate Fourier series expansion can readily

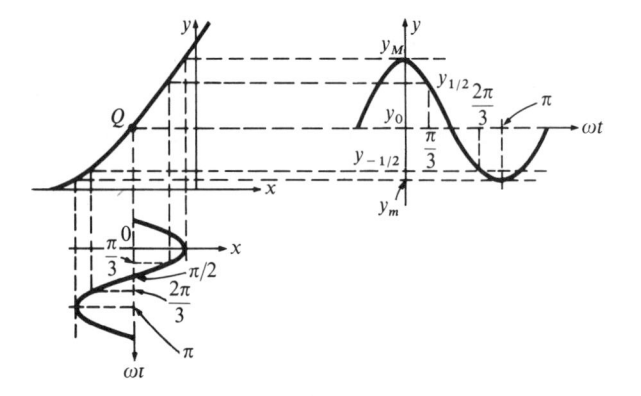

Fig 5-49 Transfer characteristic, input and output signals for distortion calculations

be made by using a computer. However, to illustrate the method by a simple approximate technique, we assume a truncated Fourier series with 5 terms, up to $a_4 \cos 4\omega t$. The input is assumed to be a sinusoidal voltage signal:

$$x = x_m \cos \omega t \qquad (5\text{-}109)$$

5-15 Distortion Characterization and Calculation

To evaluate the five components, the values of y at 5 different values of x are needed. We choose these to be as follows:

$$\begin{aligned} \text{at } \omega t &= 0 & y &= y_M \\ \text{at } \omega t &= \frac{\pi}{3} & y &= y_{1/2} \\ \text{at } \omega t &= \frac{\pi}{2} & y &= y_0 \\ \text{at } \omega t &= \frac{2\pi}{3} & y &= y_{-1/2} \\ \text{at } \omega t &= \pi & y &= y_m \end{aligned} \qquad (5\text{-}110)$$

From (5-110) and the following equation for the output current

$$y = \sum_{n=0}^{4} a_n \cos n\omega t \qquad (5\text{-}111)$$

We then solve for the values of a_n, which results in (see Problem 5-37)

$$\begin{aligned} a_0 &= \tfrac{1}{6}(y_M + 2y_{1/2} + 2y_{-1/2} + y_m) - y_0 \\ a_1 &= \tfrac{1}{3}(y_M + y_{1/2} - y_{-1/2} - y_m) \\ a_2 &= \tfrac{1}{4}(y_M - 2y_0 + y_m) \\ a_3 &= \tfrac{1}{6}(y_M - 2y_{1/2} + 2y_{-1/2} - y_m) \\ a_4 &= \tfrac{1}{12}(y_M - 4y_{1/2} + 6y_0 - 4y_{-1/2} + y_m) \end{aligned} \qquad (5\text{-}112)$$

If there is no distortion $y_0 = (y_M + y_m)/2 = (y_{1/2} + y_{-1/2})/2$, and $y_M - y_m = 2(y_{1/2} - y_{-1/2})$, and thus $a_0 = a_2 = a_3 = a_4 = 0$, and we have a perfectly sinusoidal output signal. We define the nth harmonic distortion as the ratio of the nth harmonic amplitude to the fundamental component:

$$D_n \equiv \left| \frac{a_n}{a_1} \right| \qquad (5\text{-}113)$$

If the distortion is not negligible, the total output power can be expressed as

$$P_{ac} = (a_1^2 + a_2^2 + a_3^2 + \cdots)\frac{R_L}{2} = P_1(1 + D_2^2 + D_3^2 + \cdots) \qquad (5\text{-}114)$$

where $P_1 = \frac{1}{2}a_1^2 R_L$ is the output power due to the fundamental component. We can rewrite (5-114) as

$$P_{ac} = (1 + D^2)P_1 \qquad (5\text{-}115)$$

where D is called the total distortion:

$$D \equiv \sqrt{D_2^2 + D_3^2 + D_4^2 + \cdots} \qquad (5\text{-}116)$$

For a total distortion of 10 per cent, which is usually more than can be tolerated, the total output power is only 1 per cent higher than that contributed by the fundamental component. For a good high fidelity system, the distortion is usually limited to 1 per cent or less.

5-16 Class B Push-Pull Power Amplifiers

When two devices are used in the class B push-pull configuration shown in fig 5-50, the collector dissipation can be reduced from that of a single-ended stage, and the efficiency considerably improved. The operation of the circuit in class B mode, assuming identical ideal transistors, is as follows. The input transformer supplies base currents

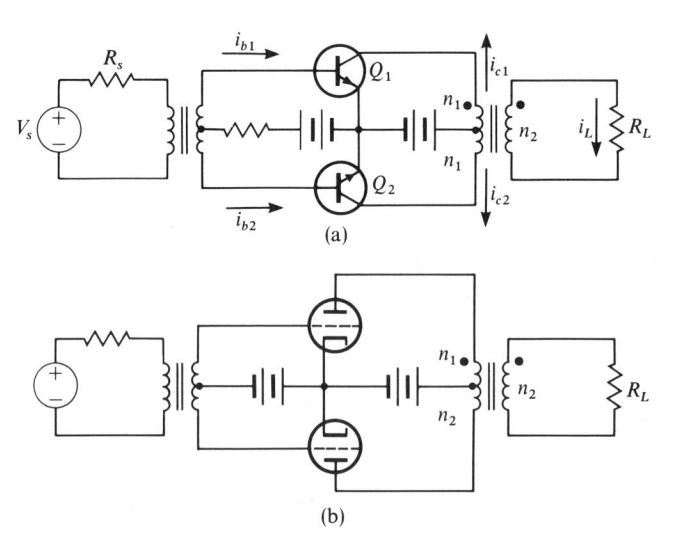

Fig 5-50 Class B push-pull amplifiers: (a) Transistor push-pull amplifier; (b) Vacuum triode push-pull amplifier

5-16 Class B Push-Pull Power Amplifiers

of equal amplitude and opposite phase to the two transistors. In the first half-cycle i_{b2} and i_{c2} are zero, and the operation of Q_1 can be expressed by a Fourier series as shown in (5-117) (if nonlinearities of the transfer characteristic are included).

$$i_{c1} = a_0 + a_1 \cos \omega t + a_2 \cos 2\omega t + a_3 \cos 3\omega t + \cdots \quad (5\text{-}117)$$

For linear variation, of course, the harmonic components are zero. On the second half-cycle Q_1 is cutoff and Q_2 is conducting as shown in fig 5-50. Thus, in general

$$\begin{aligned} i_{c2} &= a_0 + a_1 \cos(\omega t + \pi) + a_2 \cos 2(\omega t + \pi) + \cdots \\ &= a_0 - a_1 \cos \omega t + a_2 \cos 2\omega t - a_3 \cos \omega t + \cdots \end{aligned} \quad (5\text{-}118)$$

From fig 5-50, the load current i_L is related to the individual currents by

$$i_L = K(i_{c1} - i_{c2}) = K(a_1 \cos \omega t + a_3 \cos 3\omega t + \cdots) \quad (5\text{-}119)$$

where K is a real constant.

Since the dc and even harmonics cancel out, the distortion can be reduced in this arrangement from that of a single-ended power amplifier. The nonlinearity of the device, however, produces odd harmonics in the output signal. In FET devices since the transfer characteristics are square law, there are no odd harmonics (in fact in (5-108) $a_i = 0$ for $i \geq 3$) and the output current in a push-pull arrangement is thus

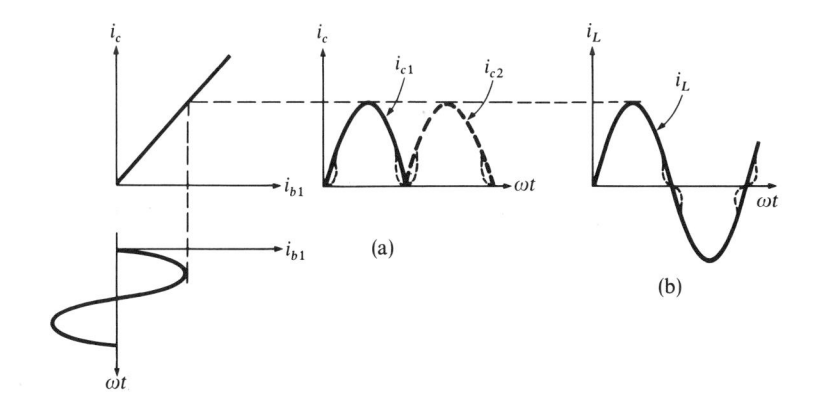

Fig 5-51 (a) Input–output signals of individual transistors in class B operation; (b) Load current

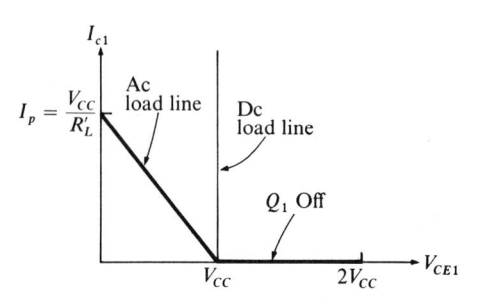

Fig 5-52 Load line of a single transistor in push-pull circuit

virtually distortion free. In bipolar transistors because of the non-linearity of the input $I_b - V_{BE}$ characteristic, however, the load current will be distorted near the zero crossings, as shown in fig 5-51c by the dashed curve. This distortion is usually referred to as *crossover distortion*. It arises because with no input signal (i.e., $i_b = 0$), V_{BE} is zero, and hence i_c is zero. This distortion is almost entirely eliminated by operating the push-pull circuit such that V_{BE} is at the break voltage and i_b is never zero. The operation is therefore in the class AB mode. However, the break voltage of the $I_b - V_{BE}$ curve is very small (0.6 V for silicon devices, and 0.3 V for germanium devices), so that the operation is essentially in class B mode.

The load line for a single transistor of the class B push-pull circuit is shown in fig 5-52, where $R'_L = (n_1/n_2)^2 R_L$. The dc current for each transistor is found from the Fourier expansion. For example, from fig 5-51b, the dc current for Q_1 is

$$(I_{dc})_1 = \frac{1}{2T} \int_{-T}^{T} i_{c1}\, dt = \frac{\omega}{2\pi} \int_0^{\pi} I_p \sin \omega t\, dt = \frac{I_p}{\pi} \qquad (5\text{-}120)$$

Thus the dc input power is

$$P_{dc} = I_{dc} V_{CC} = \frac{2}{\pi} I_p V_{CC} \qquad (5\text{-}121)$$

The factor of 2 is used to account for the dc current of both transistors. For maximum output power swing we have

$$(P_{ac})_{max} = \frac{I_p^2 R'_L}{2} = \frac{I_p V_{CC}}{2} \qquad (5\text{-}122)$$

Thus

$$\eta_m = \frac{V_{CC} I_p/2}{(2/\pi) I_p V_{CC}} = \frac{\pi}{4} = 78 \text{ per cent} \qquad (5\text{-}123)$$

Note that the maximum efficiency is much higher than that of class A operation.

The collector power dissipation for both transistors is given by

$$2P_c = P_{dc} - P_L = \frac{2}{\pi} I_p V_{CC} - \frac{I_p^2 R'_L}{2} \qquad (5\text{-}124)$$

To find the maximum power dissipation, we differentiate P_c with respect to I_p, and set the resulting expression equal to zero, which yields

$$I_p = \frac{2}{\pi} \frac{V_{CC}}{R'_L} \qquad (5\text{-}125)$$

From (5-125) and (5-124) we obtain

$$2P_{CM} = \frac{2V_{CC}^2}{\pi^2 R'_L} \qquad (5\text{-}126)$$

The maximum output power, P_{LM}, is

$$P_{LM} = (P_{ac})_{max} = \frac{I_p^2 R'_L}{2} = \frac{V_{CC}^2}{2R'_L} \qquad (5\text{-}127)$$

The figure of merit for each transistor is

$$\frac{P_{CM}}{P_{LM}} = \frac{V_{CC}^2/\pi^2 R'_L}{V_{CC}^2/2R'_L} = \frac{2}{\pi^2} \approx \frac{1}{5} \qquad (5\text{-}128)$$

The improvement in comparison with class A amplifiers is a factor of 10. In other words, the class B push-pull amplifier operating at the ideal maximum efficiency may be designed for a power output approximately five times the power dissipation rating of the individual transistors. This is indeed quite an advantage.

EXAMPLE

Consider the same specifications as in the design example of the class A amplifier. We shall now consider the use of a class B push-pull circuit.

For the specified output load power of 2 watts and the supply voltage of 12 volts, from (5-127) we have

$$R'_L = \frac{V_{CC}^2}{2P_{LM}} = \frac{(12)^2}{2(2)} = 36 \, \Omega$$

The peak current

$$I_p = \frac{V_{CC}}{R'_L} = \frac{12}{36} = 0.33 \text{ A}$$

The collector dissipation of each transistor in this case is

$$P_c \geq \tfrac{1}{5}(2) = 0.4 \text{ W}$$

Thus transistors with 2 or 1 watt ratings, which could not be used in the previous example, can be used in this case and have an extra margin of safety. The turns ratio is found from

$$\frac{n_1}{n_2} = \sqrt{\frac{R'_L}{R_L}} = \sqrt{\frac{36}{4}} = 3.0$$

Similarly, if the output impedance of the driver stage is known, the input coupling transformer turns ratio is selected such that the input impedance of the power amplifier is matched to that of the driver.

Complementary Pairs

In transistor circuitry, because of the availability of *pnp* and *npn* devices, one can altogether avoid using transformers for class B push-pull operation. The input transformer may be replaced by a modified paraphase amplifier (see fig P4-7), or simply by using *RC* circuits with supply voltages of both polarities, as shown in fig 5-53. The principle of operation of this circuit is quite similar to the one discussed earlier. When the ac signal is positive, *npn* transistor Q_1 conducts, while *pnp* transistor Q_2 is cutoff. During the negative portion

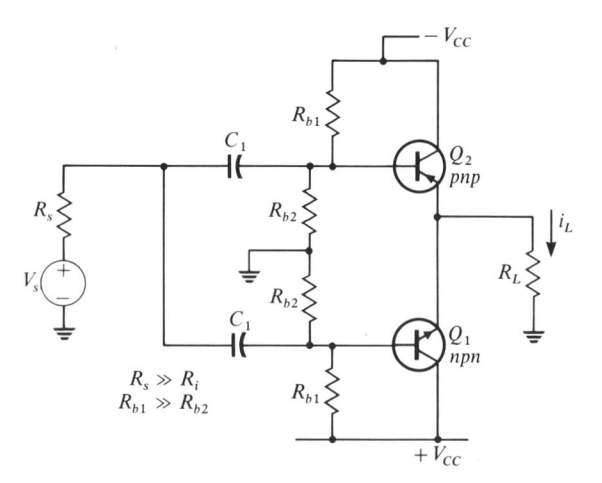

Fig 5-53 Complementary pair push-pull circuit

of the input signal, the *pnp* transistor conducts and the *npn* transistor is cutoff. The load current, which is equal to the difference of the collector currents, is thus sinusoidal and is an amplified version of the input signal. The analysis of the circuit in fig 5-53 may be performed by considering each stage a class B emitter follower.

In a complementary pair circuit, if the transistors are not matched, the output signal will be distorted. Thus, matching a pair of *pnp-npn* transistors should be done in both discrete and integrated circuit designs. In the latter case, a lateral *pnp* transistor should be used in order to obtain matching in the low-frequency performance. If other types of *pnp* are to be used in IC designs, because of their low gains, a *pnp* Darlington arrangement is usually used with an *npn* to get a complementary push-pull circuit.

5-17 Summary

The design of an amplifier most often requires more than one stage of amplification. First the main amplifier stage is designed for the given gain and bandwidth specifications, and then the input and output stages are designed to meet other requirements if any. The effect of the input and the output on the overall design must also be considered. If the gain-bandwidth product is the prime objective, the biasing points are selected to obtain high β_0 and f_T, and efficient broadband stages are often used. The design of such an amplifier is based on the small signal models for the device. Often simple models are used in the preliminary design, since active device parameters vary from unit to unit of the same type. For example, the Miller approximation used to simplify the model for forward gain and input impedance calculations is a good simplification for initial analysis and design purposes. If a more accurate and detailed analysis is desired, a computer-aided analysis using a more accurate model, such as the hybrid π model for the bipolar junction transistor, or the measured two-port parameters of the device may be used.

When the amplifier is to amplify very weak signals, the input stage must be designed for minimal noise and the operating point of the first stage must be chosen to obtain the smallest possible noise figure. The input stage is also usually designed for a large power gain, so as to reduce the effect of the noise contribution due to the other cascaded stages.

If the amplifier is to provide a large power output to the load, of the order of a few hundred milliwatts or more, then the output stage is designed first to meet the power requirements. Since the signal swing is very large so as to violate small signal assumptions, graphical analysis is used. Other factors in the design of power amplifiers such as efficiency, distortion, frequency response, etc., may also have to be considered. Because the main interest in power amplifiers is maximization of output power with good efficiency and small, or negligible distortion, we are limited in the frequency response by the output stage. Fortunately, these are most often used in audio amplifiers, and hence the frequency limitation usually does not pose a problem.

REFERENCES AND SUGGESTED READING

1. Thornton, R. D., et al., *Multistage Transistor Circuits*, 2nd edition, New York: Wiley, 1965, Volume 5, Chapters 1, 2, 5, and 6.
2. Pederson, D. O., *Electronic Circuits*, New York: McGraw-Hill, 1965, (Prelim. ed.) Chapters 6, 7, 8, and 9.
3. Ghausi, M. S., *Principles and Design of Linear Active Circuits*, New York: McGraw-Hill, 1965, Chapters 12 and 18.
4. Schilling, D., and Belove, C., *Electronic Circuits: Discrete and Integrated*, New York: McGraw-Hill, 1968, Chapters 5, 7, and 13.
5. Alley, C., and Atwood, K., *Electronic Engineering*, 2nd edition, New York: Wiley, 1966, Chapters 9 and 10.
6. Gray, P., and Searle, C., *Electronic Principles: Physics, Models, and Circuits*, New York: Wiley, 1969, Chapters 15 and 16.

PROBLEMS

5-1 For the circuit shown in fig P5-1, silicon transistors are used. Determine the following at room temperature:
(a) Quiescent points for the transistors.

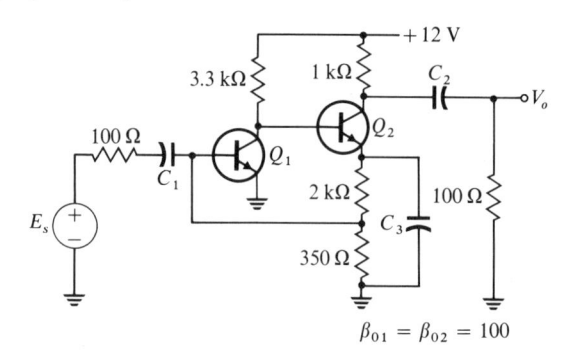

Fig P5-1

(b) The midband voltage gain of the amplifier. Assume that $r_b \ll R_i$. The capacitances are essentially short-circuited at midband frequencies. (Hint: Assume $V_{BE} = 0.7$ V and $I_C \gg I_B$ for each transistor.)

5-2 If the transistors in fig P5-1 have the following parameters:

$$f_T = 400 \text{ MHz}, \quad C_c = 2 \text{ pF}, \quad r_b = 50 \text{ }\Omega, \quad r_e = 13 \text{ }\Omega$$

determine the approximate 3-dB bandwidth of the amplifier.

5-3 For the two-stage FET amplifier shown in fig P5-3, determine the following:
(a) Midband voltage gain.
(b) The upper and lower 3-dB cutoff frequencies.

The JFET parameters are given as follows:

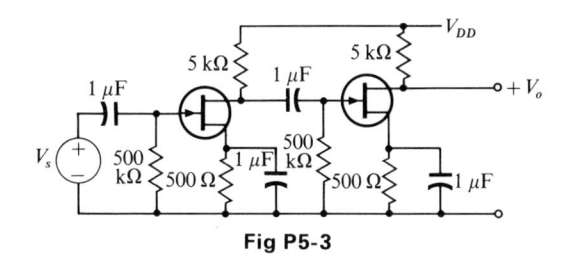

$$r_d = 10 \text{ k}\Omega, \qquad C_{gs} = 6 \text{ pF}, \qquad C_{ds} = 2 \text{ pF}$$
$$g_m = 4 \times 10^{-3} \text{ mho}, \qquad C_{gd} = 2 \text{ pF}$$

Fig P5-3

5-4 Repeat Problem 5-3 assuming MOSFET with the following parameters:

$$r_d = 20 \text{ k}\Omega, \qquad C_{gs} = 6 \text{ pF}, \qquad C_{ds} = 1 \text{ pF},$$
$$g_m = 4 \times 10^{-3} \text{ mho}, \qquad C_{gd} = 0.5 \text{ pF}$$

5-5 An amplifier has a voltage-gain function given by

$$A_v(s) = \frac{8 \times 10^{12} s^2 (s+3)}{(s+5)(s+10)(s+200)(s+10^5)(s+2 \times 10^5)}$$

Determine the approximate upper and lower 3-dB cutoff frequencies and the midband gain of the amplifier. Sketch the asymptotic plot.

5-6 (a) An amplifier has a gain function given by

$$A_i(s) = \frac{(5 \times 10^{14}) s^2 (s+5)}{(s+8)(s+100)^2 (s^2 + \sqrt{3} \times 10^6 s + 10^{12})}$$

Determine the midband gain, and the upper and lower 3-dB cutoff frequencies.

(b) What is the approximate 10 per cent to 90 per cent rise time of the above amplifier?

5-7 Estimate the f_T required to build an amplifier (with a minimum number of transistors) with a midband current gain of 5000 and a bandwidth $\omega_{3\,dB} = 2\pi \times 10^6$ rad/sec. The transistors have $\beta_0 \simeq 60$, and $C_c \approx 5$ pF. For simplicity, assume the CE stages to be identical interior stages and with a simple resistive broadbanding scheme. Let $r_b = 0$ and $r_e = 5\,\Omega$. What is the gain-bandwidth product if $r_b = 100\,\Omega$ at $I_E = 5$ mA?

5-8 An amplifier with a gain of 1000 uses three identical RC-coupled CS field effect transistors. Estimate the increase in bandwidth if a MOSFET instead of JFET is used. In other words, assume identical parameters except (C_{gd}) MOSFET $\approx \frac{1}{2}(C_{gd})$ JFET. For simplicity let $C_{gs} = 5C_{gd}$, and ignore C_{ds}.

5-9 (a) For a voltage gain of 10^4 or larger and $f_{3\,dB} = 1$ MHz, determine the minimum number of RC-coupled pentode stages that would be required. The pentodes to be used have the following typical parameters: $C_i = C_o = 4$ pF (including stray capacitances), and $g_m = 5 \times 10^{-3}$ mho. Assume identical stages for simplicity (i.e., the output capacitance of the last stage is $C_i + C_o$).
(b) For the same gain and bandwidth as in (a), estimate the number of RC-coupled triodes that would be required if $C_{gk} = C_{gp} = C_{pk} = 4$ pF (including stray capacitances), $g_p = 10^{-4}$ mho and $g_m = 5 \times 10^{-3}$ mho. Again assume identical stages.

5-10 Assume an iterative three-stage RC-coupled amplifier having identical stages as shown in fig P5-10. For a midband voltage gain of 10^4, compare the bandwidths if BJT's, JFET's, MOSFET's, pentodes, or triodes are used. Assume the parameters of each device are as in Problems 5-2, 5-3, 5-4, and 5-9, respectively.

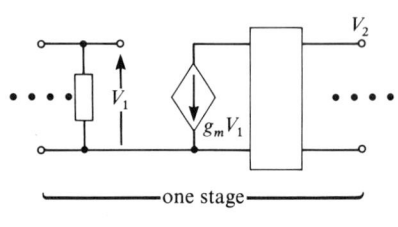

Fig P5-10

5-11 An emitter-coupled pair (CC-CB cascade), commonly found in integrated circuits, is shown in fig P5-11. Determine approximately the midband

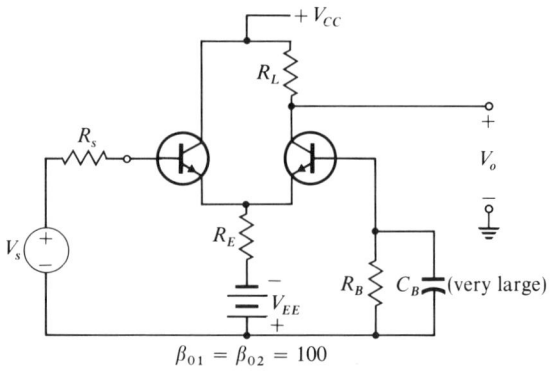

$\beta_{01} = \beta_{02} = 100$

Fig P5-11

gain and the upper 3-dB frequency of the amplifier. For simplicity, assume that the transistor parameters are the same as in Problem 5-2. Let $R_L = R_E = 1\,\text{k}\Omega$ and $R_s = 50\,\Omega$.

(Hint: The CB stage bandwidth is very much larger than that of the CC stage. Thus, use the low-frequency h model for the CB stage. Note also that for the CC stage the capacitor C_c of the hybrid π model can be ignored! Why?)

5-12 For the cascode amplifier shown in fig 5-10, $R_s = R_L = 10\,\text{k}\Omega$. The transistor parameters are as given in (5-34). Determine the voltage gain and the 3-dB bandwidth of the amplifier.

5-13 Design a transistor RC-coupled amplifier using the minimum number of stages to meet the following requirements:

$$A_V(0) = 2500,\ \omega_{3\,\text{dB}} = 2\pi(10\,\text{MHz}),\ R_L = 1\,\text{k}\Omega,\ R_s = 100\,\Omega$$

The operating point is $V_{CE} = 5\,\text{V}$, and $I_C = 5\,\text{mA}$. The transistor parameters at this quiescent point are as given in (5-34). For simplicity, assume that $r_b = 0$ and ignore the biasing resistors. Show the designed signal circuit exclusive of biasing circuitry.

5-14 Show that the midband properties of the circuit in fig P5-14 approximately realize an ideal current-controlled voltage source. Comment on the values of R and r_e for good approximation.

5-15 Show that the midband properties of the circuit in fig P5-15 approximately realize an ideal voltage-controlled current source. Comment on the values of R and r_e for a good approximation.

5-16 A hybrid cascode is shown in fig P5-16. Determine the midband voltage gain and the approximate 3-dB bandwidth of the amplifier if the transistor parameters are:

BJT	JFET
$f_T = 100\,\text{MHz}$	$g_m = 2 \times 10^{-3}\,\text{mho}$
$\beta_0 = 50$	$C_{gs} = 10\,\text{pF}$
$r_b = 100$	$C_{ds} = C_{gd} = 4\,\text{pF}$

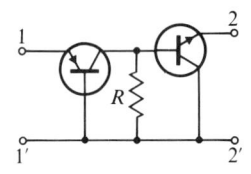

Fig P5-14

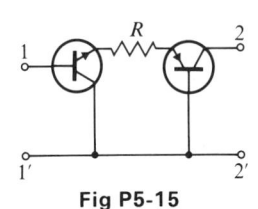

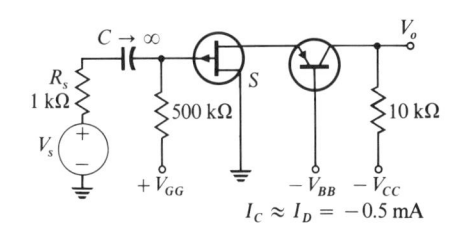

Fig P5-15

Fig P5-16

Hint: The 3-dB cutoff frequency is much lower than f_T, so that the CB stage internal capacitance can be ignored and the low frequency h-model can be used for the BJT.

5-17 For the pentode circuit shown in fig P5-17, design the circuit (i.e., determine R_f) such that a maximially flat magnitude gain function is obtained. Determine the 3-dB bandwidth and the midband gain of the designed circuit. If you have a choice in the values of R_f, select the value which will lead to a larger voltage gain.

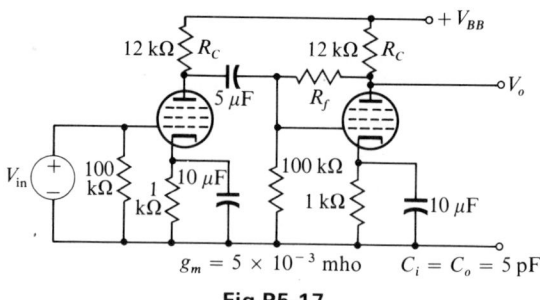

Fig P5-17

$g_m = 5 \times 10^{-3}$ mho, $C_i = C_o = 5$ pF

5-18 (a) Derive (5-42).
(b) Express the overall h-parameters in terms of the low-frequency Tee model parameters (you may use the results of Chapter 4).

5-19 (a) Derive (5-45).
(b) Determine R_I for the CC-CE cascade of the text for a maximally flat magnitude response characteristic. Determine the midband gain and the 3-dB bandwidth for this case.

5-20 The two-stage pentode amplifier shown in fig P5-20 is to be used in a stagger-tuned design to get a maximally flat magnitude response. Determine the values of R_1, R_L, and L to obtain a bandwidth $\omega_{3\,dB} = 2 \times 10^7$ rad/sec. What is the mid-band gain of this amplifier? The pentode parameters are $g_m = 5 \times 10^{-3}$ mho and $C_i + C_s = C_o + C_s = 5$ pF, where C_s is the stray wiring capacitance.

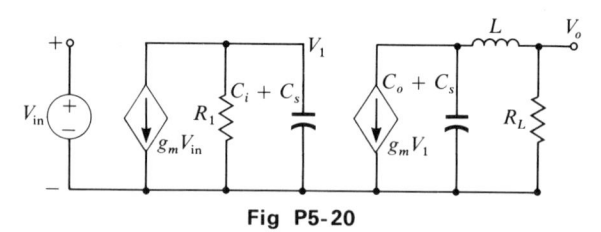

Fig P5-20

5-21 A cascode circuit is used in cascade with a shunt-peaked CE transistor stage as shown in fig P5-21 (biasing and coupling circuitry have been omitted for simplicity). Design the circuit (i.e., determine R_I and L) such that the voltage-gain function has an approximate dominant identical 2 pole response characteristic, and a midband voltage gain of 500. Calculate the 3-dB bandwidth of the designed circuit. (Hint: Use the result of (5-42). Cancel the zero of the shunt-peaked stage by the pole of the cascode.) Assume the following transistor parameters:

$\beta_0 = 100$, $\quad r_b \simeq 0$, $\quad r_e = 20\,\Omega$,

$f_T = 200$ MHz, $\quad C_c = 10$ pF

Can this circuit be designed for a two-pole maximally flat magnitude response, if the gain specification is removed?

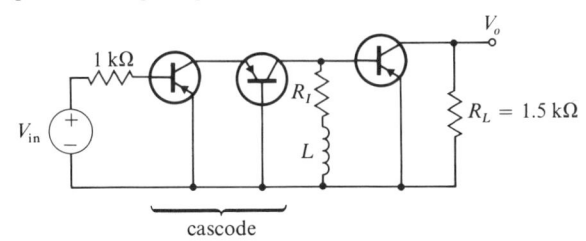

Fig P5-21

5-22 (a) Determine the shrinkage factor for a cascade of two stages, each stage having a two-pole maximally flat magnitude response.
(b) If the overall bandwidth is $\omega_{3\,dB} = 2\pi(1\text{ MHz})$, what is the individual stage bandwidth?

5-23 In dc amplifiers and integrated circuits, it is often of importance to provide a *level shifting* network without incurring a loss in the ac signal. Such a network is shown in fig P5-23. Where the dc value of V_{C1} can be shifted to $V_{C3} = 0$ without a loss in ac gain.
(a) Determine the values of R_b and R_c such that V_{C3} is at zero potential with respect to ground. List the quiescent points of each transistor.
(b) Show that the ac voltage gain from the collector of Q_1 to the collector of Q_3 is approximately unity.

5-24 For the difference amplifier shown in fig P5-24, find the quiescent operating point and also the expression for I_L in terms of the *CM* and *DM* signals. (Note that $V_{BE} \simeq 0.6$ V.) If $I_{s2} = 0$, determine I_L/I_{s1} at dc.

5-25 For a completely symmetrical two-port network, we define the *open-circuit half* section driving point impedance as

$$Z_{och} = \left.\frac{V_1}{I_1}\right|_{V_1=V_2}$$

and the *short-circuit half* section driving point impedance as

$$Z_{sch} = \left.\frac{V_1}{I_1}\right|_{V_1=-V_2}$$

where V_1, I_1, etc., are the conventional two-port voltage and current quantities.
(a) Show that $Z_{och} = z_{11} + z_{12}$, and $Z_{sch} = z_{11} - z_{12}$.
(b) Use the results in the above to find the z-parameters of the symmetrical two-port lattice network shown in fig P5-25.

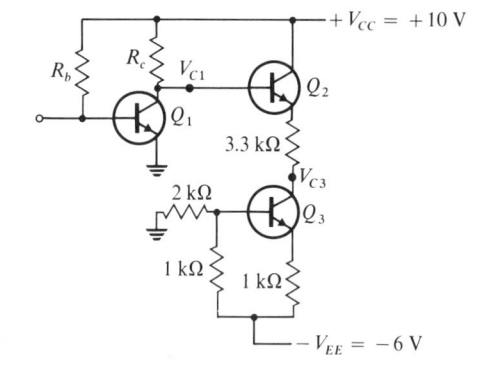

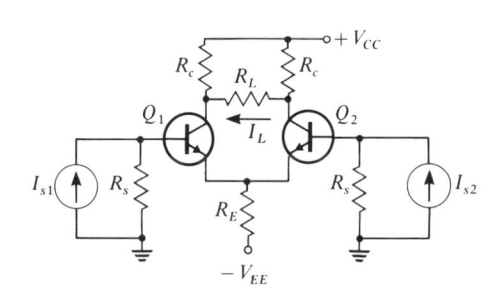

Fig P5-24

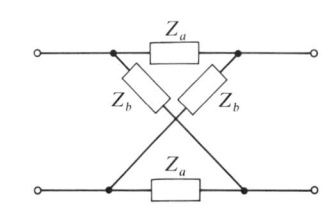

Fig P5-25

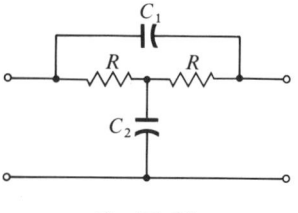

Fig P5-26

5-26 Find the two-port y-parameters of the bridged-Tee network shown in fig P5-26. Use the results of Problem 5-25a.

5-27 For the integrated circuit difference amplifier shown in fig P5-27, find the quiescent points and the midband common mode rejection ratio. Note that $Z_{aa'} \approx 1/h_{ob} \approx r_c$ (see Problem 4-8).

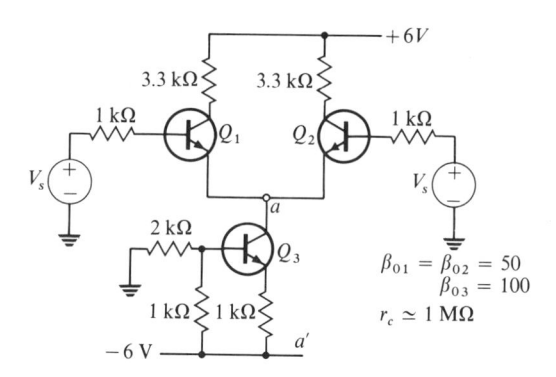

Fig P5-27

$\beta_{01} = \beta_{02} = 50$
$\beta_{03} = 100$
$r_c \simeq 1\ M\Omega$

5-28 An amplifier is shown within the dashed box in fig P5-28, as a noiseless amplifier and an equivalent input noise voltage that represents all the noises within the amplifier. The noise voltage, $\overline{v_1^2}$, is the thermal noise voltage due to the external resistance R_1. Determine the noise figure of the amplifier. If $R_L = R_0$ and $R_s = R_1 \| R_i$, obtain the expression for the available power gain of the amplifier.

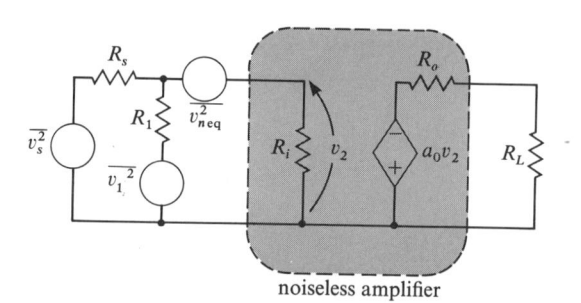

Fig P5-28

noiseless amplifier

5-29 Consider the single-stage triode amplifier shown in fig P5-29a. The noise model of a triode is shown in fig P5-29b. Determine the total noise voltage at the output and the noise figure at room temperature. Use the triode parameters of Problem 5-9b.

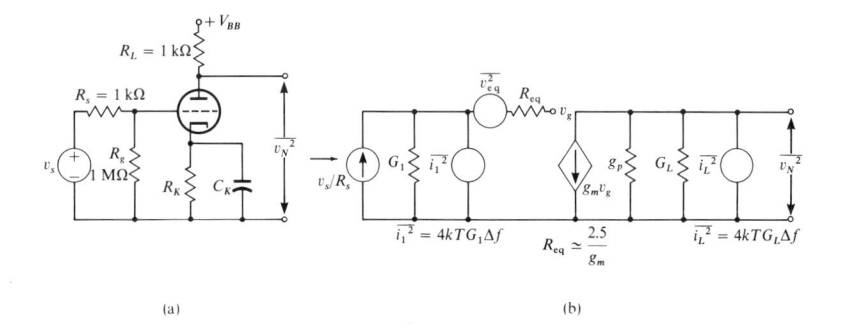

Fig P5-29

5-30 Derive the expressions for the midband noise figure of a transistor by using (5-74b) for the following connections (use the Tee model, and make reasonable approximations):
(a) A CE configuration.
(b) A CB configuration.
(c) A CC configuration.

5-31 The equivalent input noise voltage of a FET amplifier is as shown in fig 5-40. If the generator input impedance is 10 kΩ, what is the noise figure of the stage at a frequency of 1 kHz. What is the total noise output power if the stage has a power gain of 10 dB? What is the overall noise figure if the input stage is cascaded with another stage having the same noise figure? What is the total noise power at the output in this case?

5-32 Determine the noise figure of the silicon transistor cascode amplifier shown in fig 5-10. The transistor parameters are given in (5-34). Let $R_s = 250\,\Omega$, $R_L = 1\,k\Omega$. Comment why the cascode amplifier is useful as a low-noise amplifier.

5-33 Design a class A transformer-coupled power amplifier to supply a power of 10 watts to an 8 ohm load resistance. The supply voltage is 12 volts. Use a stabilized biased scheme, and show all the circuit element values, if the transistor has $\beta_0 = 50$. Also specify the transistor power dissipation rating.

5-34 A transistor is to be used in a class A power amplifier. The load resistance is 5 ohms. The transistor ratings are

$$P_{c,max} = 10 \text{ watts} \quad (V_{CE})_{sat} = 1 \text{ V}$$
$$(V_{CE})_{max} = 60 \text{ V}$$

(a) If no transformer is used, determine the maximum attainable voltage swing, the maximum power dissipated by the load, and the efficiency of the amplifier.

(b) Repeat (a) if transformer-coupling is used, with $n = 2$.

5-35 For a transformer-coupled class A power amplifier, the following specifications are given: the audio power to the load is to be 3 watts. The load is a loudspeaker with an 8 ohm impedance. The available supply voltage is 12 volts. The efficiency of the available transformers is 80 per cent. The driver stage has an output resistance of 1 kΩ. Specify the transistor ratings and the turns ratio of the input and output transformers if at the operating point $h_{ie} = 100\,\Omega$.

5-36 Show that for a class B push-pull amplifier, the power due to the fundamental component is given by

$$P_1 = \left[\frac{\sqrt{2}}{3}(I_{max} + I_{1/2})\right]^2 \left(\frac{n_1}{n_2}\right)^2 R_L$$

where I_{max} and $I_{1/2}$ are the values of current at $\omega t = 0$ and $\omega t = \pi/3$.

5-37 (a) For FET devices show that in (5-108), $a_n = 0$ for $n \geq 3$.
(b) Derive (5-112).

5-38 For distortion calculations, suppose we choose four points instead of five (as done in the text). If these four points are $\omega_0 t = 0, \pi/3, \pi/2, \pi$, and the corresponding values of y are: y_m, y_1, y_0, y_{-m}, show that the coefficients of the Fourier series in this case are given by

$$a_0 = \frac{y_m + y_{-m}}{4} - \frac{y_0}{2}$$

$$a_1 = \frac{y_m}{4} - \frac{5}{12}y_{-m} + \frac{2}{3}y_1 - \frac{y_0}{2}$$

$$a_2 = \frac{y_m + y_{-m}}{4} - \frac{y_0}{2}$$

$$a_3 = \frac{y_m}{4} - \frac{y_m}{12} - \frac{2}{3}y_1 + \frac{y_0}{2}$$

5-39 For the push-pull circuit operating in class A, derive the expression for $P_L, P_C,$ and η.
Note that the effective load in this case is $(2n_1/n_2)^2 R_L$.

5-40 For the emitter follower circuit shown in fig P5-40, determine the values of R_{b1} and R_{b2} for a maximum symmetrical output voltage swing. Assume a silicon transistor with $V_{BE} = 0.7$ V and $\beta_0 = 100$.

5-41 If ECAP or CORNAP is available in the computer center of the school, it is suggested that a few problems be assigned for computer-aided analysis and design. For example, the following may be assigned:
(a) Utilize ECAP to analyze the multistage cascaded FET amplifier of fig 5-2b. Use the numerical values of the text example.
(b) Utilize ECAP to analyze the multistage cascaded bipolar transistor amplifier of fig 5-5c. Use the numerical values of the text example.

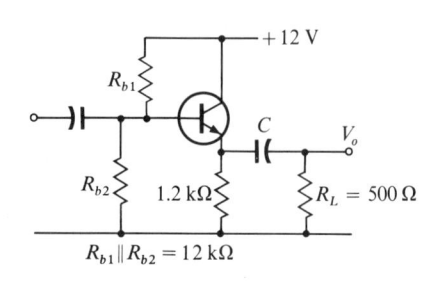

Fig P5-40

Oscillators

Feedback is present in all electronic circuits, intentionally or otherwise. In general, a system with feedback is one in which signal transmission exists in some manner from the output to the input. The nonunilateral nature of transistors and vacuum triodes is an example of unintentional internal feedback. External feedback is used deliberately in many systems to accrue some benefits. One of the many such examples in electronic circuits is the stabilization of the operating point in transistor amplifiers, as discussed in Chapter 3. In fact, sensitivity requirements necessitate the use of feedback, and most amplifiers are thus of the feedback type.

It should be mentioned at the outset that the conventional methods of analysis discussed in Chapter 2 are applicable to all circuits, regardless of the presence and nature of feedback. An approximate method of analysis which focuses attention on feedback is very desirable, however, from the design viewpoint and considerable simplicity is thereby often achieved. The design can then be further refined and improved, if necessary, by more detailed analysis, preferably computer-aided. It should be mentioned, however, that no one analysis method is the best for all situations.

In this chapter we consider the use of feedback in electronic circuits. We discuss some of the advantages resulting from the use of negative feedback, and the potential disadvantage of feedback, which is the introduction of the problem of instability. For amplifiers, instability is, of course, to be avoided at all costs. Stability criteria are also briefly discussed. The discussion of feedback amplifiers will, of necessity, be introductory, since feedback control theory is a vast topic and cannot be adequately covered in one or two chapters. In the analysis and design of systems with multiloop feedback, the signal-flow graph can be of considerable aid. Since we do not discuss multiloop feedback amplifiers in this text, however, the signal-flow graph is discussed briefly in Appendix C.

Instability is deliberately used in some electronic circuits to perform certain circuit functions. Instability in such cases forms the basis of operations, as in oscillators. In this chapter we also discuss, briefly, the linear analysis of sinusoidal feedback oscillators.

6-1 Basic Feedback Concepts and Definitions

An *idealized* model of a single-loop feedback amplifier is shown in fig 6-1. The input and the output signals may be either current or

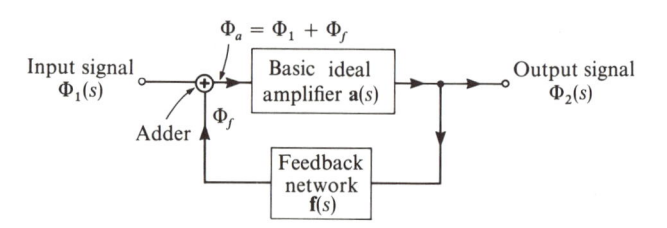

Fig 6-1 Schematic of a single-loop feedback amplifier

voltage quantities. The idealizations are that the signals are transmitted only in the direction of the arrows. We now define some terminologies which are used in the rest of this chapter.

The feedback loop is defined as the closed path consisting of the forward path, namely, the amplifier $a(s)$, and the feedback path $f(s)$. Both $a(s)$ and $f(s)$ are rational functions for lumped networks.

6-1 Basic Feedback Concepts and Definitions

Open-loop transfer function: If the feedback loop is open at the feedback path, i.e., $\Phi_f = 0$ in fig 6-1, the forward *amplifier gain function*, $\mathbf{a}(s)$, is defined as the open-loop transfer function, i.e.,

$$\mathbf{a}(s) = \frac{\Phi_2}{\Phi_1}\bigg|_{\Phi_f=0} = \frac{\Phi_2}{\Phi_a} \qquad (6\text{-}1)$$

Note that we use the symbol Φ instead of a V or I so as to have generality in our subsequent discussions, i.e., Φ_2/Φ_1 may be the ratio of two currents or two voltages (in the transformed domain) depending upon the circuit configurations. Also note that from fig 6-1, Φ_a is the adder output,

$$\Phi_a = \Phi_1 + \Phi_f \qquad (6\text{-}2)$$

The *feedback function* $\mathbf{f}(s)$ is defined as

$$\mathbf{f}(s) = \frac{\Phi_f(s)}{\Phi_2(s)} \qquad (6\text{-}3)$$

From (6-1), (6-2), and (6-3) we get

$$\frac{\Phi_2}{\mathbf{a}(s)} = \Phi_1 + \Phi_2 \mathbf{f}(s) \qquad (6\text{-}4)$$

or

$$A \equiv \frac{\Phi_2(s)}{\Phi_1(s)} = \frac{\mathbf{a}(s)}{1 - \mathbf{a}(s)\mathbf{f}(s)} \qquad (6\text{-}5)$$

Equation (6-5) is very important, and is usually referred to as the *basic feedback* equation. Many of the properties of feedback can be deduced from (6-5). In (6-5), A is referred to as the *closed-loop gain function*. Note that in fig 6-1 we are assuming positive feedback in the choice of sign conventions, but that the feedback is negative if either $\mathbf{a}(s)$ or $\mathbf{f}(s)$ is inverting (i.e., has a negative sign at $s = 0$). The amplifiers discussed in this chapter are of the negative-feedback type because, as we shall see, negative feedback has many advantages.

The *return ratio* or the *loop-transmission* function is defined as

$$T(s) = -\mathbf{a}(s)\mathbf{f}(s) \qquad (6\text{-}6)$$

6/Feedback Amplifiers and Oscillators

The *return difference* $F(s)$, also sometimes referred to as the amount of feedback, is defined by

$$F(s) = 1 + T(s) = 1 - \mathbf{a}(s)\mathbf{f}(s) \qquad (6\text{-}7)$$

Equation (6-7) is also very important, since the roots of $F(s) = 0$ determine the natural frequencies of the system. Note that $F(s)$ is the denominator of $A(s)$. The equation

$$F(s) = 1 - \mathbf{a}(s)\mathbf{f}(s) = 0 \qquad (6\text{-}8)$$

is usually called the *characteristic equation* of the system. Equation (6-8) is used for stability analysis of the system. For a stable system, the roots of $F(s)$ must be in the left half-plane.

The magnitude of $F(j\omega)$, which is usually expressed in dB, namely

$$20 \log |F(j\omega)| = 20 \log |1 - \mathbf{a}(j\omega)\mathbf{f}(j\omega)|$$

determines the amount of feedback. Note that $\mathbf{a}(s)\mathbf{f}(s)$, in general, is a complex quantity. If $|1 - \mathbf{a}(j\omega)\mathbf{f}(j\omega)|$ is greater than unity, it is called *negative* (or *degenerative*) feedback and, if $|1 - \mathbf{a}(j\omega)\mathbf{f}(j\omega)|$ is smaller than unity, it is called *positive* (or *regenerative*) feedback. An unstable single-loop feedback[1] amplifier may exhibit both negative and positive feedback, where the latter is the cause of instability, as discussed later in this chapter.

6-2 Feedback Configurations and Classifications

In the single-loop feedback configuration shown in fig 6-1, the amplifier circuit $\mathbf{a}(s)$ is an active two-port, while the feedback network $\mathbf{f}(s)$ is usually a passive two-port network. The *idealization* of the assumed unilateral nature of $\mathbf{a}(s)$ and $\mathbf{f}(s)$, as depicted in fig 6-1, and thus the equations in Section 6-1 are never exactly met in practice, since amplifiers are never exactly unilateral and passive RLMC circuits are always bilateral. We must, therefore, make some reasonable assumptions in order approximately to achieve the situation in fig 6-1. Fortunately, the assumptions in fig 6-1 can be relaxed and satisfactorily met in many practical problems.

[1] By a single-loop feedback system we mean a system in which the loop transmission is zero if any of the active devices is dead (i.e., $g_m = 0$).

6-2 Feedback Configurations and Classifications

The two basic assumptions we make are the following:
(1) The forward transmission through the feedback network is negligible in comparison with the forward transmission through the amplifier.
(2) The reverse transmission through the amplifier is negligible in comparison with the reverse transmission through the feedback circuit.

The above approximations are very reasonable and often valid when the feedback path is through a passive circuit, which is almost always the case because of sensitivity reasons. Since the signal is amplified via the forward amplifier circuit, and attenuated via the passive feedback circuit, the first assumption is thus quite valid. The second assumption is usually valid, since the amplifiers used in $\mathbf{a}(s)$ are approximately unilateral, while $\mathbf{f}(s)$ is bilateral, hence reverse transmission through the $\mathbf{a}$ circuit is negligible in comparison with that through the $\mathbf{f}$ circuit. If we further assume that the loading effects of the feedback network on the amplifier circuit are negligible, considerable simplification results in the initial design. The latter assumption is not necessary, but is often made when the feedback parameter values are unknown.

Since we are usually dealing with two-port networks let us consider how the feedback connections can be made. The four basic interconnections are shown in fig 6-2. These are the shunt-shunt, series-series, series-shunt, and shunt-series feedback configurations.

We shall briefly consider each case, from the two-port theory viewpoint, in the following. We associate the source and the load immittances with those of the amplifier circuit, since they must be there whether feedback is applied or not. The interconnected two-port is then considered as an overall two-port, and thus the results of Chapter 2 can be used directly in these cases.

The Shunt-Shunt Configuration

A composite two-port containing two parallel connected two-ports is shown in fig 6-2a. Since, the parallel connection, the y-parameters are the most convenient to use, we shall make this choice in our analysis. The short-circuit admittance parameters of the amplifier and feedback circuits are designated by y_{ij}^a and y_{ij}^f respectively. We assume that the direct addition of the y-parameters in the interconnection is valid, thus $y_{ij}^T = y_{ij}^a + y_{ij}^f$.

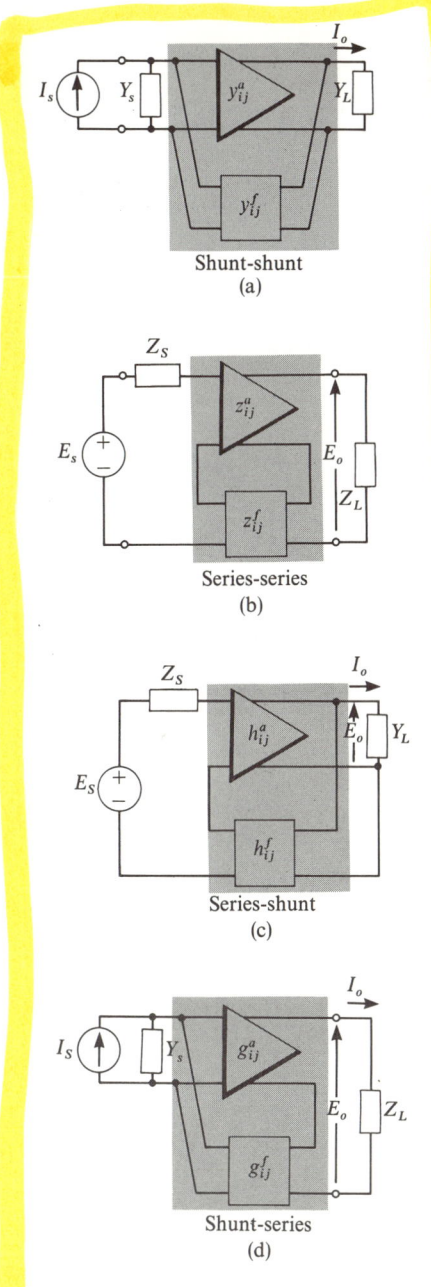

Fig 6-2 Basic feedback connections: (a) Shunt-shunt; (b) Series-series; (c) Series-shunt; (d) Shunt-series

From the two-port results in Section 2-2, or table 2-2, the current-gain function is

$$\frac{I_2}{I_1} = \frac{y_{21} Y_L}{(y_{11} + Y_s)(y_{22} + Y_L) - y_{12} y_{21}} \quad (6\text{-}9)$$

In this case, however, we have the y-parameters of an interconnected two-port. Thus

$$\frac{\Phi_2}{\Phi_1} = \frac{I_o}{I_s} = \frac{-I_2}{I_1} = \frac{-y_{21}^T Y_L}{(y_{11}^T + Y_s)(y_{22}^T + Y_L) - y_{12}^T y_{21}^T} \quad (6\text{-}10)$$

where $y_{ij}^T = y_{ij}^a + y_{ij}^f$. In order to express (6-10) in the form of the basic feedback equation (6-5), we rewrite (6-10) as

$$A_I = \frac{I_o}{I_s} = \frac{-y_{21}^T Y_L/(y_{11}^T + Y_s)(y_{22}^T + Y_L)}{1 - [y_{12}^T y_{21}^T/(y_{11}^T + Y_s)(y_{22}^T + Y_L)]} \quad (6\text{-}11a)$$

In (6-11a), no approximations are made. We can rewrite (6-11a) as

$$\frac{I_o}{I_s} = \frac{-y_{21}^T Y_L/(y_{11}^T + Y_s)(y_{22}^T + Y_L)}{1 - [(-y_{12}^T/Y_L)(-y_{21}^T Y_L)/(y_{11}^T + Y_s)(y_{22}^T + Y_L)]} \quad (6\text{-}11b)$$

From a comparison of (6-11b) and (6-5), we can at once identify

$$\mathbf{a}_i(s) = -y_{21}^T Y_L/(y_{11}^T + Y_s)(y_{22}^T + Y_L) \quad (6\text{-}12)$$

$$\mathbf{f}(s) = -y_{12}^T/Y_L \quad (6\text{-}13)$$

and

$$T(s) = -\mathbf{a}_i(s)\mathbf{f}(s) = -\frac{y_{12}^T y_{21}^T}{(y_{11}^T + Y_s)(y_{22}^T + Y_L)} \quad (6\text{-}14)$$

From (6-12) and (6-13), it is seen that both $\mathbf{a}(s)$ and $\mathbf{f}(s)$ depend on the y-parameters of the individual two-ports. This is inconvenient and does not correspond to the idealized situation depicted in fig 6-1. In order to *approximate* the idealized situation with as few assumptions as possible we reexamine (6-11b) and incorporate the two basic assumptions, $y_{21}^a \gg y_{21}^f$ and $y_{12}^a \ll y_{12}^f$. These approximations are quite reasonable and can provide us considerable simplicity in the analysis and design as we shall soon see. Hence the approximate expression corresponding to (6-11b) is

$$A_I = \frac{I_o}{I_s} \approx \frac{-y_{21}^a Y_L/(y_{11}^T + Y_s)(y_{22}^T + Y_L)}{1 - (-y_{12}^f/Y_L)[-y_{21}^a Y_L/(y_{11}^T + Y_s)(y_{22}^T + Y_L)]} \tag{6-15}$$

Our identification of approximate **a**(s) and **f**(s) corresponding to (6-12) and (6-13) now becomes

$$\mathbf{a}_i(s) = \left.\frac{I_o}{I_s}\right|_{f=0} \approx -\frac{y_{21}^a Y_L}{(y_{11}^T + Y_s)(y_{22}^T + Y_L)} \tag{6-16}$$

$$\mathbf{f}(s) = \frac{I_{fb}}{I_o} \approx -\frac{y_{12}^f}{Y_L} \tag{6-17}$$

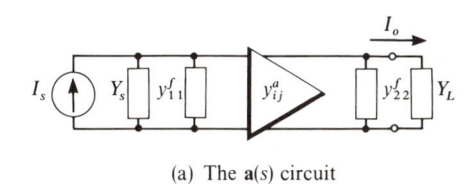

Note that (6-16) depends upon the feedback circuit, since $y_{11}^T = y_{11}^a + y_{11}^f$ and similarly $y_{22}^T = y_{22}^a + y_{22}^f$. This is different from the *ideal* forward path equation **a**(s) in (6-1) which is independent of **f**(s) because of the idealized assumptions. The forward and feedback circuits corresponding to (6-16) and (6-17) are shown in fig 6-3. Note that fig 6-3a includes y_{11}^f and y_{22}^f because in (6-16) y_{11}^T and y_{22}^T are involved. The feedback function in (6-17) does not depend on the parameters of **a**(s). Also note that in fig 6-3b

$$-y_{12}^f = \left.\frac{I_{fb}}{E_o}\right|_{V_1=0} = \frac{I_{fb}}{I_o Z_L}$$

Fig 6-3 (a) **a**(s) circuit of a shunt-shunt connection; (b) **f**(s) circuit of a shunt-shunt connection

If the values of the elements in the feedback network are known, such as in the analysis problem, one simply adds y_{11}^f and y_{22}^f with those of y_{11}^a and y_{22}^a as shown in fig 6-3a. This must be done, since (6-16) involves y_{11}^T and y_{22}^T. In a design problem, however, the feedback network values must usually be determined. In such cases, we make the *initial assumption* of $y_{11}^f \ll y_{11}^a$ and $y_{22}^f \ll y_{22}^a$ in the *preliminary design*. Then we reexamine our approximations and iterate, using the determined values for the feedback circuit, if necessary.

The input admittance of fig 6-2a, using the results of Chapter 2 for the overall two-port, is

$$Y_{in} = (y_{11}^T + Y_s) - \frac{y_{12}^T y_{21}^T}{y_{22}^T + Y_L} = (y_{11}^T + Y_s) - \frac{y_{12}^T y_{21}^T}{y_{22}^T + Y_L}\left(\frac{y_{11}^T + Y_s}{y_{11}^T + Y_s}\right)$$

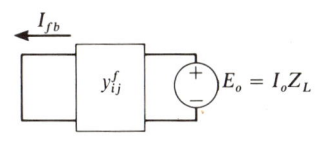

$$\tag{6-18}$$

We can rewrite (6-18) as

$$Y_{in} = (y_{11}^T + Y_s)\left[1 - \frac{y_{12}^T y_{21}^T}{(y_{22}^T + Y_L)(y_{11}^T + Y_s)}\right] \quad (6\text{-}19)$$

From (6-19) and (6-14) we obtain

$$Y_{in} = (y_{11}^T + Y_s)[1 + T(s)] \quad (6\text{-}20a)$$

Similarly the output admittance is

$$Y_{out} = (y_{22}^T + Y_L)[1 + T(s)] \quad (6\text{-}20b)$$

From (6-20a) and (6-20b) it is seen that shunt-shunt feedback increases the low-frequency input and output admittances (since $T(0) > 0$), i.e., the input and output impedances are decreased.

EXAMPLE

Consider the simple single-stage shunt-feedback amplifier, employing negative feedback as shown in fig 6-4 (exclusive of biasing circuitry).

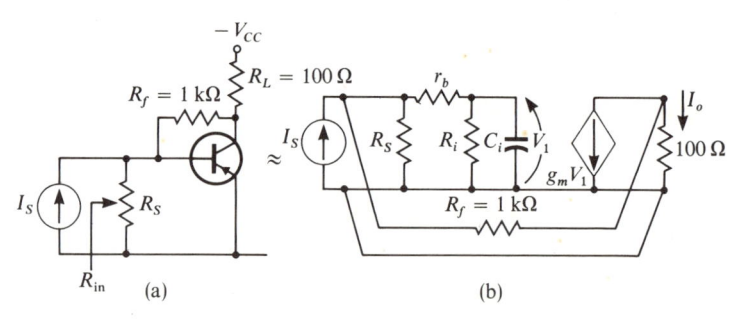

Fig 6-4 (a) Shunt-feedback amplifier for the example; (b) Equivalent circuit of (a)

This circuit has already been analyzed by conventional techniques in Section 4-13. We shall determine the gain and bandwidth of the circuit from the feedback viewpoint.

The transistor parameters are repeated in the following for convenience:

$$\omega_T = 2\pi(4 \times 10^8) \text{ rad/sec}, \quad \beta_0 = 50,$$

$$C_c = 5 \text{ pF}, \quad r_b = 100 \, \Omega, \quad \text{and} \quad r_e = 5 \, \Omega \quad (6\text{-}21)$$

6-2 Feedback Configurations and Classifications

From the above, and noting that $R_L = 100\,\Omega$, we determine the following:

$$R_i = \beta_o r_e = 50(5) = 250\,\Omega$$

$$C_i = \frac{D}{r_e \omega_T} = \frac{1 + R_L C_c \omega_T}{r_e \omega_T} = \frac{2.25}{5(2.51 \times 10^9)} \approx 180\,\text{pF}$$

We use the unilateral approximate model for the transistor as shown in fig 6-4b (see also fig 4-28b). Recall the limitations and caution required in the use of this model.

The **a** and the **f** circuits are shown in fig 6-5. Usually it is simpler to find the expression of **a**(s) directly from the circuit rather than by using (6-16). From fig 6-5a, letting $R'_s = R_s \| R_f$, we have

$$\mathbf{a}_i(s) = \frac{I_o}{I_s} = \left[-\frac{1}{r_e C_i} \frac{R'_s/(R'_s + r_b)}{s + (1/R_i C_i)(R'_s + r_b + R_i)/(R'_s + r_b)} \right] \left(\frac{R_f}{R_L + R_f} \right) \quad (6\text{-}22)$$

Fig 6-5 (a) **a**(s) circuit for fig 6-4; (b) **f**(s) circuit for fig 6-4

We rewrite (6-22) for convenience as

$$\mathbf{a}_i(s) = -\frac{a_o p_o}{s + p_o} \quad (6\text{-}23)$$

where

$$a_o = \beta_o \frac{R'_s}{R'_s + r_b + R_i} \left(\frac{R_f}{R_L + R_f} \right) \approx \beta_o \left(\frac{R_f}{R_f + r_b + R_i} \right) \quad (6\text{-}24\text{a})$$

$$\text{if } R_s \gg R_f \gg R_L$$

For the example where $R_s \to \infty$, and $R_f = 1\,\text{k}\Omega$, we get

$$a_o = 50 \left(\frac{1\,\text{k}\Omega}{1\,\text{k}\Omega + 0.35\,\text{k}\Omega} \right) = 37$$

$$p_o = \frac{1}{R_i C_i} \left(\frac{R'_s + r_b + R_i}{R'_s + r_b} \right) \approx \frac{1}{R_i C_i} \left(\frac{R_f + r_b + R_i}{R_f + r_b} \right) \quad (6\text{-}24\text{b})$$

$$= \frac{1}{(250)(1.82 \times 10^{-10})} \left(\frac{1.35\,\text{k}\Omega}{1.1\,\text{k}\Omega} \right) = 2.70 \times 10^7 \text{ rad/sec}$$

Also **f**(s) from (6-17) and fig 6-5b is

$$\mathbf{f} = \frac{I_f}{I_o} = -\frac{y^f_{12}}{Y_L} = \frac{G^f}{G_L} \equiv f_o \quad (6\text{-}25)$$

$$= \frac{100}{1000} = 0.1$$

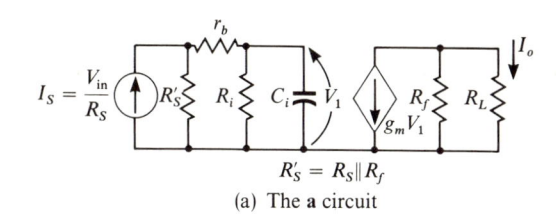

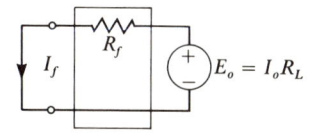

The open-loop gain and bandwidth are

$$a_o = 37, \quad p_o = 2.7 \times 10^7 \text{ rad/sec}$$

The closed-loop gain function using (6-5) is

$$A_I = \frac{\mathbf{a}_i}{1 - \mathbf{a}_i \mathbf{f}} = -\frac{a_o p_o/(s + p_o)}{1 - [-a_o p_o/(s + p_o)] f_o} = -\frac{a_o p_o}{s + p_o(1 + a_o f_o)} \tag{6-26}$$

Note that the midband loop gain is

$$T(0) = -\mathbf{a}_i(s)\mathbf{f}(s)|_{s=0} = a_o f_o = 3.7$$

The closed-loop gain and bandwidth are

$$A_I(0) = -\frac{a_o}{1 + a_o f_o} = -\frac{37}{1 + 3.7} \approx -7.9$$

$$f_{3\text{ dB}} = \frac{p_o}{2\pi}(1 + a_o f_o) = \frac{(2.7 \times 10^7)(4.7)}{2\pi} \approx 21 \text{ MHz}$$

Note also that the approximate midband input impedance with feedback, in shunt with the current source, is

$$(R_{\text{in}})_f = Y_{\text{in}}^{-1} \simeq \frac{1/(y_{11}^T + Y_s)}{1 + a_o f_o} \simeq \frac{(r_b + R_i)\|R_f\|R_s}{1 + a_o f_o}$$

$$= \frac{260}{4.7} \simeq 56 \text{ }\Omega.$$

For the output impedance calculation, recall that the unilateral model is not correct at high frequencies. We have denoted the low-frequency output impedance of a *CE* stage by r_o, which is usually large and can be ignored. Thus the output impedance with feedback, including the load impedance, from (6-20b) is

$$(R_{\text{out}})_f = Y_o^{-1} = \frac{1/(y_{22}^T + Y_L)}{1 + a_o f_o} \approx \frac{r_o\|R_f\|R_L}{1 + a_o f_o} \approx \frac{91}{4.7} \simeq 19 \text{ }\Omega$$

Compare these values of gain and bandwidth with those obtained from the involved expressions (4-98). Note that for impedance calculations, the use of the Miller theorem (see Problem 1-31) can also provide a simple means of finding the input and output impedances, provided the voltage gain can be readily determined or is known.

The reader can appreciate the advantage of the feedback approach further if he were to design the circuit. Say that either the gain or the bandwidth for fig 6-4a is specified, and that we want to determine the value of R_f. In this situation the use of the expressions in (4-98) is hopeless. Using the feedback approach, however, the initial value of R_f is readily obtained as follows. Suppose that we wish to design the circuit in fig 6-4a for an approximate bandwidth of 10 MHz. The termination and the transistor parameters are assumed to be the same as before. We ignore initially the effect of R_f on the **a** circuit. Thus the open-loop gain and bandwidth are

$$a_o \simeq \beta_0 = 50, \quad \text{and} \quad p_o = \frac{1}{R_i C_i} = 2.2 \times 10^7 \text{ rad/sec}$$

The specification calls for a bandwidth $\omega_{3\,dB} = 2\pi \times 10$ MHz. We determine the amount of feedback from

$$\omega_{3\,dB} = p_o(1 + a_o f_o) \to 1 + a_o f_o = \frac{2\pi \times 10^7}{2.2 \times 10^7} = 2.85$$

Hence

$$f_o = \frac{1.85}{a_o} = \frac{1.85}{50} = 3.7 \times 10^{-2} = \frac{R_L}{R_f}$$

or

$$R_f = \frac{R_L}{f_o} = \frac{100}{3.7 \times 10^{-2}} = 2.7 \text{ k}\Omega$$

Now we examine the approximation we have used, i.e., $y_{11}^f \ll y_{11}^a$, and $y_{22}^f \ll y_{22}^a$ or, specifically, $R_f \gg r_b + R_i$ and $R_f \gg R_L$, which are clearly satisfied in this case. There is no real need for iteration, i.e., including a shunt resistor of value 2.7 kΩ at the input and output of the **a** circuit and recalculating R_f. The overall gain of this circuit is

$$A_I \approx \frac{-a_0}{1 + a_o f_o} = \frac{-50}{2.85} \approx -17.5$$

The Series-Series Configuration

In fig 6-2b, the interconnections are in series and, therefore, the z-parameters are the most convenient choice to use for the analysis. Assuming that the interconnection permits the direct addition of the respective z-parameters of each two-port, the overall transfer function is given by

$$\frac{\Phi_2}{\Phi_1} = \frac{E_o}{E_s} = \frac{z_{21}^T Z_L}{(z_{11}^T + Z_s)(z_{22}^T + Z_L) - z_{12}^T z_{21}^T} \qquad (6\text{-}27)$$

where $z_{ij}^T = z_{ij}^a + z_{ij}^f$. Equation (6-27) may be rewritten in the form of (6-5) as

$$A_V = \frac{E_o}{E_s} = \frac{-z_{21}^T Z_L/(z_{11}^T + Z_s)(z_{22}^T + Z_L)}{1 - [z_{21}^T Z_L(z_{12}^T/Z_L)/(z_{11}^T + Z_s)(z_{22}^T + Z_L)]} \qquad (6\text{-}28)$$

In (6-28), no approximations are made, and the exact return ratio function for this case is

$$T(s) = \frac{-z_{12}^T z_{21}^T}{(z_{11}^T + Z_s)(z_{22}^T + Z_L)} \qquad (6\text{-}29)$$

For the same reasons as in the previous case, we incorporate the two assumptions $z_{21}^a \gg z_{21}^f$ and $z_{12}^a \ll z_{12}^f$ in (6-28), which enables us to write (6-28) as

$$A_V = \frac{E_o}{E_s} = \frac{z_{21}^a Z_L/(z_{11}^T + Z_s)(z_{22}^T + Z_L)}{1 - (z_{12}^f/Z_L)[z_{21}^a Z_L/(z_{11}^T + Z_s)(z_{22}^T + Z_L)]} \qquad (6\text{-}30)$$

We can now identify separate **a**(s) and **f**(s) circuits, where the **f** circuit does not depend on the **a** circuit. The identifications are

$$\mathbf{a}_v(s) = \left.\frac{E_o}{E_s}\right|_{f=0} \approx \frac{z_{21}^a Z_L}{(z_{11}^T + Z_s)(z_{22}^T + Z_L)} \qquad (6\text{-}31)$$

$$\mathbf{f}(s) = \frac{E_{fb}}{E_o} \approx \frac{z_{12}^f}{Z_L} \qquad (6\text{-}32)$$

The **a**(s) and the **f**(s) circuits corresponding to (6-31) and (6-32) are shown in fig 6-6.

6-2 Feedback Configurations and Classifications

(a) The a circuit

(b) The f circuit

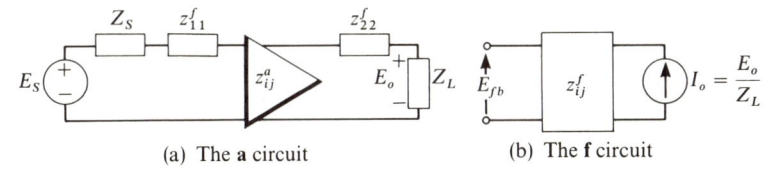

Fig 6-6 (a) a(s) circuit of a series-series connection; (b) f(s) circuit of a series-series connection (for convenience the directions of I_o and E_{fb} are changed)

Note that the **a** circuit, unlike the idealized situation, includes z^f_{11} and z^f_{22}. As in the previous case, this addition does not pose a problem in the analysis. In the design, this effect may be ignored initially, and then included as an iteration step if necessary.

The input impedance of fig 6-2b, using the result of Chapter 2, for the overall two-port is

$$Z_{in} = (z^T_{11} + Z_s) - \frac{z^T_{12} z^T_{21}}{z^T_{22} + Z_L}$$

$$= (z^T_{11} + Z_s)\left[1 - \frac{z^T_{12} z^T_{21}}{(z^T_{22} + Z_L)(z^T_{11} + Z_s)}\right] \quad (6\text{-}33)$$

From (6-33) and (6-29) we have

$$Z_{in} = (z^T_{11} + Z_s)[1 + T(s)] \quad (6\text{-}34)$$

Similarly

$$Z_o = (z^T_{22} + Z_L)[1 + T(s)] \quad (6\text{-}35)$$

From (6-34) and (6-35), it is seen that the series-series feedback connection increases the low-frequency input and output impedances since $T(0) > 0$.

EXAMPLE

Consider the single-stage feedback amplifier employing series negative feedback shown in fig 6-7 (exclusive of biasing circuitry). We shall determine the overall gain and the bandwidth of the circuit from the feedback viewpoint. Assume the following transistor parameters

$$\omega_T = 10^8 \text{ rad/sec}, \quad \beta_0 = 200,$$
$$C_c = 5 \text{ pF}, \quad r_b = 100\,\Omega, \quad r_e = 5\,\Omega \quad (6\text{-}36)$$

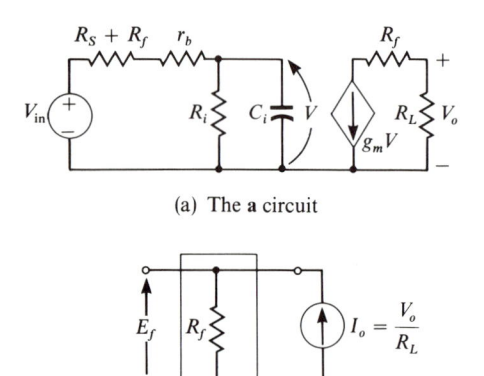

Fig 6-8 (a) **a**(s) circuit for fig 6-7; (b) **f**(s) circuit for fig 6-7

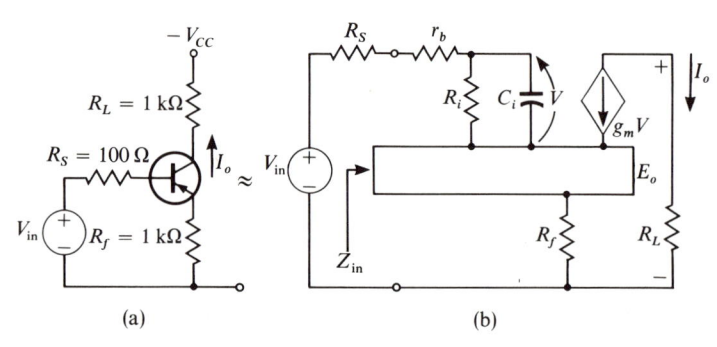

Fig 6-7 (a) Series-feedback amplifier for the example; (b) Equivalent circuit of (a)

We use the unilateral approximate circuit model for the transistor, as shown in fig 6-7b. Note that the effective load resistor of the stage is not small, hence some error is introduced in the bandwidth calculation by the unilateral method. We use this simple model, however, in order to illustrate and emphasize the feedback technique more clearly. The **a** and the **f** circuits for this case are shown in fig 6-8. From fig 6-8a, we have

$$\mathbf{a}_v(s) = \frac{V_o}{V_{in}} = -\frac{g_m R_L}{C_i(R_f + R_s + r_b)} \bigg/ \left(s + \frac{R_i + R_s + R_f + r_b}{R_i(R_f + R_s + r_b)C_i} \right) \quad (6\text{-}37)$$

For convenience, we rewrite (6-37) as

$$\mathbf{a}_v(s) = -a_o \frac{p_o}{s + p_o} \quad (6\text{-}38)$$

Where

$$a_o = \frac{g_m R_L R_i}{R_i + R_s + R_f + r_b} \quad (6\text{-}39)$$

$$p_o = \frac{R_i + R_s + R_f + r_b}{R_i(R_f + R_s + r_b)C_i} \quad (6\text{-}39b)$$

For the circuit element values of this example

$$R_i = \beta_o r_e = 1 \text{ k}\Omega, \quad C_i = \frac{1 + (R_L + R_f)C_c \omega_T}{r_e \omega_T} = 4 \times 10^{-9} \text{ F}$$

From (6-39a) and (6-39b), we obtain $a_o = 91$, $p_o = 7.9 \times 10^5$ rad/sec.

Also, **f**(s) from (6-32) and fig 6-8b is

$$f(s) = \frac{z_{12}^f}{Z_L} = \frac{R_f}{R_L} = f_o \qquad (6\text{-}39c)$$

$$= 1\,\text{k}\Omega/1\,\text{k}\Omega = 1$$

The closed-loop gain function using (6-5) is

$$A_V = \frac{\mathbf{a}_v}{1 - \mathbf{a}_v \mathbf{f}} = -\frac{a_o p_o/(s + p_o)}{1 - [-a_o p_o/(s + p_o)]f_o} = -\frac{a_o p_o}{s + p_o(1 + a_o f_o)} \qquad (6\text{-}40)$$

Hence, the midband loop transmission gain is

$$T(0) = -a(0)f(0) = a_o f_o = 91$$

Thus the closed-loop gain and bandwidth are

$$A_v(0) = -\frac{a_o}{1 + a_o f_o} = -\frac{91}{92} = -0.99$$

$$\omega_{3\,\text{dB}} = p_o(1 + a_o f_o) = (7.9 \times 10^5)(92) \approx 7.3 \times 10^6 \text{ rad/sec}$$

Note that in the above example the open-loop gain is not just the gain with R_f short-circuited. Notice also the increase in bandwidth and the decrease in gain as a result of feedback.

The midband input impedance with feedback, in series with the voltage source, is

$$(R_{\text{in}})_f = (r_b + R_i + R_f + R_s)(1 + a_o f_o)$$

$$= (r_b + R_i + R_f + R_s)\left(1 + \frac{g_m R_i R_f}{r_b + R_i + R_f + R_s}\right)$$

$$(R_{\text{in}})_f \doteq R_s + r_b + R_i + (\beta_0 + 1)R_f \simeq 200\,\text{k}\Omega$$

The output impedance is also increased by the amount of feedback and is hence very large, i.e., $(R_{\text{out}})_f \approx (1 + a_o f_o)(R_L + r_o + R_f)$, where r_o is the output impedance of the CE transistor. Note that fig 6-7 is not valid for output impedance calculations at high frequencies, and that the hybrid π model must be used for such calculations.

The Series-Shunt and the Shunt-Series Configurations

The series-shunt connection is shown in fig 6-2c. For this configuration, since the input is in series and the output in shunt, the h-parameters are the most convenient to use for analysis in terms of the two-port parameters. The overall gain function in this case is given by

$$A_V = \frac{E_o}{E_s} = \frac{h_{21}^T}{(h_{11}^T + Z_s)(h_{22}^T + Y_L) - h_{12}^T h_{21}^T} \tag{6-41}$$

In order to express (6-41) in the form of (6-5) for identification purposes, the two basic assumptions to be used are $h_{21}^a \gg h_{21}^f$ and $h_{12}^a \ll h_{12}^f$. Note that the returned feedback signal must be dimensionally the same as the input signal, since they are added directly. Hence, without going into the details, we rewrite (6-41) in the form of (6-5) and identify

$$\mathbf{a}_v(s) = \frac{E_o}{E_s}\bigg|_{f=0} \approx \frac{h_{21}^a}{(h_{11}^T + Z_s)(h_{22}^T + Y_L)} \tag{6-42}$$

$$\mathbf{f}(s) = \frac{E_{fb}}{E_o} \approx h_{12}^f \tag{6-43}$$

$$T(s) = -\mathbf{a}_v \mathbf{f} \approx -\frac{h_{21}^a h_{12}^f}{(h_{11}^T + Z_s)(h_{22}^T + Y_L)} \tag{6-44}$$

Note again that in these identifications $\mathbf{a}_v(s)$ includes h_{11}^f and h_{22}^f, since $h_{11}^T = h_{11}^a + h_{11}^f$, etc., while $\mathbf{f}(s)$ depends only on the feedback network.

Similarly for the shunt-series connection, because the input signal is in shunt and the output signal in series, the g-parameters are the most convenient to use for analysis. For this case, the gain function is:

$$A_I(s) = \frac{I_o}{I_s} = \frac{-g_{21}^T}{(g_{11}^T + Y_s)(g_{22}^T + Z_L) - g_{12}^T g_{21}^T} \tag{6-45}$$

The use of the two assumptions $g_{21}^a \gg g_{21}^f$ and $g_{12}^a \ll g_{12}^f$ enables us to rewrite (6-45) in the form of (6-5) and make the following identifications:

$$\mathbf{a}_i(s) = \frac{I_o}{I_s}\bigg|_{f=0} \approx \frac{-g_{21}^a}{(g_{11}^T + Y_s)(g_{22}^T + Z_L)} \tag{6-46}$$

6-2 Feedback Configurations and Classifications

$$\mathbf{f}(s) = \frac{I_{fb}}{I_o} \approx -g_{12}^f \tag{6-47}$$

$$T(s) = -\mathbf{a}_i \mathbf{f} \approx -\frac{g_{21}^a g_{12}^f}{(g_{11}^T + Y_s)(g_{22}^T + Z_L)} \tag{6-48}$$

Observe that $\mathbf{a}_i(s)$ includes g_{11}^f and g_{22}^f, since $g_{11}^T = g_{11}^a + g_{11}^f$, etc., and $\mathbf{f}(s)$ depends on the feedback network alone.

It should be noted that when h_{ij}^a and g_{ij}^a are three terminal two-port networks, the input and output ports of the overall two-port cannot have common ground unless a transformer is used and thus such circuits are impractical. Usually modified versions of the series-shunt and shunt-series configurations where the input and output ports have a common ground are used. Such circuits are discussed in Section 6-4. In analysis of such configurations it is best to work directly in terms of the circuits by drawing $\mathbf{a}(s)$ and $\mathbf{f}(s)$ and then obtaining the corresponding equations.

For the series-shunt configuration, the input impedance increases while the output impedance decreases. In the shunt-series connection, the opposite is true, i.e., the input impedance decreases while the output impedance increases. These can be shown exactly in the same manner as we did for the shunt-shunt and the series-series configurations. For convenience the various properties of the four basic connections are compared in Table 6-1.

TABLE 6-1 *Properties of the Basic Feedback Connections*

Configuration	Two-port parameters	Input impedance	Output impedance
shunt-shunt	$[y_{ij}]$	decreases	decreases
series-series	$[z_{ij}]$	increases	increases
series-shunt	$[h_{ij}]$	increases	decreases
shunt-series	$[g_{ij}]$	decreases	increases

In all cases the bandwidth increases and the gain decreases. We shall show in the next section that feedback also improves desensitivity and reduces nonlinear distortion in all of the above connections.

In literature one also finds such classifications as *voltage feedback* and *current feedback*. In a voltage feedback the signal fedback, which may be a current or a voltage, is proportional to the *output voltage*; in other words, if E_o is equal to zero, $\mathbf{f}$ is zero. In a current feedback, the signal fedback is proportional to the *output current*, i.e., if I_o is zero, $\mathbf{f}$ is zero. For example, the circuits in figs 6-2a and 6-2c are also called voltage-shunt and voltage-series feedback configurations, respectively. The circuits in figs 6-2b and 6-2d are also called current-series and current-shunt feedback configurations, respectively.

It should be pointed out that *not* every feedback connection can be classified as being in one of the above four categories. Furthermore, feedback analysis technique is not always the simplest method. In fact, in some feedback circuits, the conventional analysis techniques may prove to be simpler.

It should be emphasized that $\mathbf{a}(s)$ need not be a single-stage amplifier. In fact, more often than not, $\mathbf{a}(s)$ is more than one stage. Usually two or three stages of amplification are employed in order to meet the sensitivity requirements, etc. The design-oriented analysis of such amplifiers is discussed in Section 6-4. Note that all feedback amplifiers discussed in this chapter employ negative feedback.

6-3 Advantages and Disadvantages of Negative Feedback

Before we go further into the analysis and design of feedback amplifiers, let us examine some of the merits and demerits of negative feedback. Some of these we have already discovered in Section 6-2.

Advantages

Improvement in Bandwidth

In addition to the increase in bandwidth (already discovered in Section 6-2), desirable transmission characteristics, such as maximally flat magnitude or linear phase response, can be achieved in the design. This is discussed later in Section 6-4.

Control over Impedance

The impedances of the circuit, e.g., the input and output impedances, can be increased or reduced by proper use of feedback. This property has also been demonstrated in Section 6-2.

Improvement in Sensitivity

The transfer function variation, as a result of changes in the active device (due to aging, replacement, etc.), is reduced by feedback. The

6-3 Advantages and Disadvantages of Negative Feedback

improvement in sensitivity is one of the most attractive features of negative feedback.

To show this, consider the relation between fractional (actually differential) change of $A(s)$ with feedback to the fractional change without feedback, as obtained by differentiating (6-5). We consider the midband values in the following for simplicity, so that A and $\mathbf{a}$ are not functions of s. Differentiation of (6-5) yields

$$\left|\frac{dA}{A}\right| = \frac{1}{(1-af)}\left|\frac{da}{a}\right| \qquad (6\text{-}49)$$

In negative feedback $(1 - af) > 1$, thus the closed-loop variation $|dA/A|$ is smaller than that of the open-loop da/a. For example, in an amplifier with 26 dB of negative feedback, $F = (1 - af) = 20$, a 1-per cent change in gain without feedback is reduced to a 0.05-per cent change with feedback. Note that for $|af| \gg 1$, from (6-5), $A \approx -1/f$.

The sensitivity of $A(s)$ with respect to $\mathbf{a}$ is defined as

$$S_{\mathbf{a}}^A = \frac{dA/A}{d\mathbf{a}/\mathbf{a}} \qquad (6\text{-}50)$$

Note that, in general, sensitivity is a function of s. The topic of sensitivity in active networks is discussed in detail in Section 7-16, and will not be pursued here. From (6-49) and (6-50) at midband we have $|S_{\mathbf{a}}^A| = |1/(1 - af)|$. The smaller the value of $S_{\mathbf{a}}^A$ the better the closed-loop gain sensitivity.

For large changes, which are often the case, one must use differences rather than the differentials, in which case, from (6-5)

$$\Delta A \equiv A_2 - A_1 = \frac{a_2}{1 - a_2 f} - \frac{a_1}{1 - a_1 f}$$

Multiply both sides by $1/A_1$ to get

$$\frac{\Delta A}{A_1} = \frac{a_2}{1 - a_2 f}\frac{1 - a_1 f}{a_1} - 1 = \frac{(a_2 - a_1)/a_1}{1 - a_2 f} \equiv \frac{\Delta a/a_1}{1 - a_2 f}$$

Thus, the fractional change from the original gain level is

$$\left|\frac{\Delta A}{A_1}\right| = \left|\frac{\Delta a/a_1}{1 - a_2 f}\right| = \left|\frac{\Delta a/a_1}{1 - (a_1 + \Delta a)f}\right| \qquad (6\text{-}51)$$

For example, if $a_1 = 10^4$ and $f = -10^{-2}$, then $A_1 = 10^2$. If a_1 changes by 100 per cent, the change in A_1 will be less than 1 per cent.

Reduction of Nonlinear Distortion

At large input signal levels, because of the nonlinear characteristics of the active device, distortion will appear in the output waveform. We assume that the nonlinear distortion is small enough so that the amplifier can be represented by a distortionless amplifier and a distortion generator v_d, as shown in fig 6-9. In other words, we assume

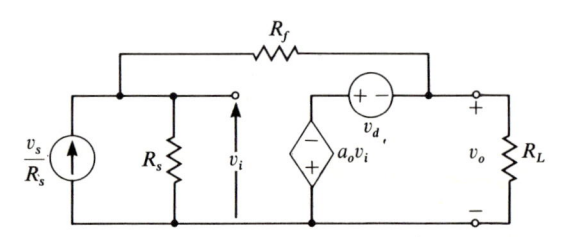

Fig 6-9 Feedback amplifier with distortion generator explicitly shown

approximately linear operation so that the use of the superposition principle is valid. For the sake of illustration, the amplifiers are represented by ideal amplifiers and we assume $R_f \gg R_s$, so that $v_i \approx v_s$. Now, in the absence of feedback

$$v_o \approx -(a_o v_s + v_d) \tag{6-52}$$

With feedback applied in fig 6-9, we note that $v_o/v_s = -a_o/(1 + a_o f_o)$, when $v_d = 0$. Similarly, the gain with feedback due to v_d alone is $v_o/v_d = (1/a_o)(v_o/v_s)$. Hence

$$v_o \simeq -\left(\frac{a_o}{1 + a_o f_o} v_s + \frac{v_d}{1 + a_o f_o} \right) \tag{6-53}$$

where $f_o = R_L/R_f$. From (6-53) it is noted that the effect of the disturbing signal has been reduced by a factor of $1/(1 + a_o f_o)$. Note, however, that the signal-to-noise ratio has not been improved and is the same as in (6-52) and (6-53). The advantage of feedback is that the signal level can be increased relative to the distortion signal without overdriving the amplifier. The linearization effect of feedback is best illustrated by the transfer characteristic of an amplifier with and without feedback, as shown in fig 6-10. Note that signals with amplitudes much larger than v_m can yield linear amplification with feedback, while nonlinear amplification would result if there were no feedback. Note

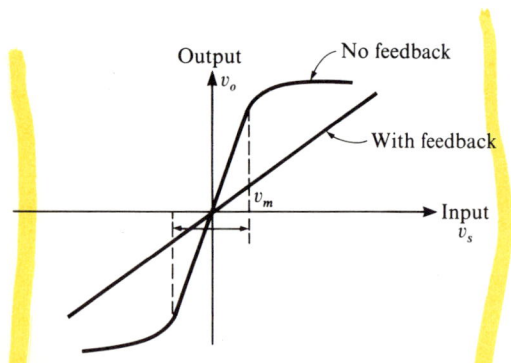

Fig 6-10 Linearization effect of negative feedback on the transfer characteristic of the amplifier

6-3 Advantages and Disadvantages of Negative Feedback

again that the price paid is the reduction of gain. Observe that if one can increase the signal without affecting the disturbance signal, then merely by increasing the signal level, without using feedback, we can swamp out the distortion and improve the signal-to-noise ratio. Feedback is useful when the increase in signal level increases the distortion.

Note that if the distortion appears at the output of the amplifier, the effect of the disturbing signal can be considerably reduced. For example, from fig 6-11, where v_d is the distortion signal and a_1 provides the main

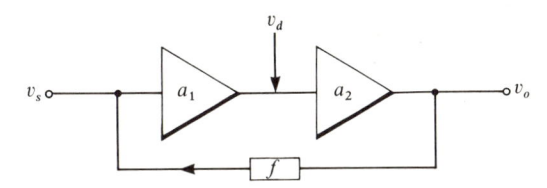

Fig 6-11 Feedback amplifier with the disturbing signal not appearing at the input

amplification, we have

$$v_o = v_d \frac{a_2}{1 - a_1 a_2 f} + v_s \frac{a_1 a_2}{1 - a_1 a_2 f} \qquad (6\text{-}54)$$

For $a_1 a_2 f \gg 1$, (6-54) reduces to

$$v_o \approx -v_s \frac{1}{f} - \frac{v_d}{a_1 f} \qquad (6\text{-}55)$$

If $|a_1 f| \gg 1$, the effect of the disturbing signal is reduced considerably. Thus nonlinear distortion, since it occurs mainly in the output stage, is reduced by feedback. The noise signal at the input stage, however, is the most bothersome and feedback cannot help in improving the signal-to-noise ratio.

Disadvantages of Negative Feedback

From the previous two sections it is quite clear that one of the disadvantages of negative feedback is the reduction of the overall gain of the amplifier. This is not a costly price to pay for the benefits derived. Further, since active devices are often inexpensive, the use of additional stages with feedback can readily take care of the specified closed-loop gain. Another disadvantage is the introduction of the problem of oscillation into the system. This is possible since in general $\mathbf{a}(s)$ and $\mathbf{f}(s)$ are frequency-dependent and, if the loop transmission phase shift and attenuation characteristics are not properly controlled, the system may become unstable. Most often in the analysis and design we ignore the nondominant poles of the active device. As we demonstrated in Chapter 5, nondominant poles introduce phase shift. These were not crucial in cascaded RC-coupled amplifiers, but could be very crucial in a feedback amplifier, in that their inclusion may show the possibility and cause of oscillation. One must be particularly careful if the design is based on the unilateral model for the transistors. For stability analysis, the unilateral model would not normally be used, and thus one must use another model such as the hybrid π circuit.

We next consider design-oriented analysis of some practical feedback amplifiers. Stability and related problems are postponed for Section 6-6.

6-4 Design-Oriented Analysis of Some Feedback Amplifier Circuits

We noted earlier that the conventional analysis methods discussed in Chapter 2 can be applied to feedback systems regardless of the type of feedback connections. There is still another technique, however, which can be used for the analysis of linear systems. This method is the signal-flow graph. The signal-flow graph method is very useful for the analysis of complicated multiloop feedback configurations. An introduction to this topic is given in Appendix C. The techniques developed therein are very useful for reducing the signal-flow graph of a complex system to a simple form. Since we shall subsequently limit ourselves to the single-loop feedback system, the signal-flow graph is not included in the text of this chapter.

In this section, we consider some representative feedback amplifier circuits and present a design-oriented analysis from the feedback viewpoint. We introduce approximations, which are very helpful for design purposes. The initial design based on these methods may be further improved by iteration and/or by a computer-aided analysis of the configuration.

There are a large number of feedback amplifiers using different feedback configurations. Some of these were encountered in Section 5-7. Amplifiers in general are of the feedback type, and appear in both

6-4 Design-Oriented Analysis of Some Feedback Amplifier Circuits

discrete and integrated-circuit form. A complete discussion of all of the various feedback amplifier circuits could easily occupy a volume by itself. We shall, therefore, limit ourselves in the following to consideration of a few basic types of feedback amplifiers. We shall then discuss them in some detail, including their frequency responses, in order to clarify further the feedback techniques.

The Shunt-Series Feedback (or Current-Feedback) Pair

Consider the shunt-series feedback pair circuit shown in fig 6-12, in which biasing and coupling circuitry have been omitted for simplicity. We can do this only if we are interested in merely the midband and high-frequency responses. Note that, in some cases the effect of the biasing resistors may not be entirely negligible and should, therefore, be included. The effect of the biasing resistors on the gain and frequency response of an amplifier stage was discussed in Chapters 4 and 5.

We next introduce the simplified approximate transistor models shown in fig 6-13, in which we have assumed that $R_f \gg R_e$. This approximation, as we shall show, is valid if a gain much larger than unity is

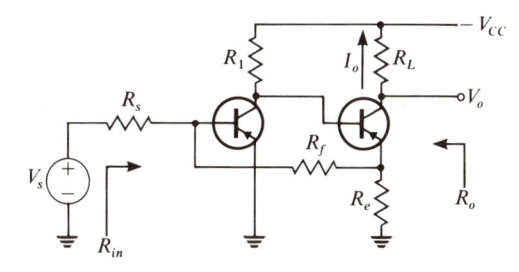

Fig 6-12 Shunt-series feedback pair

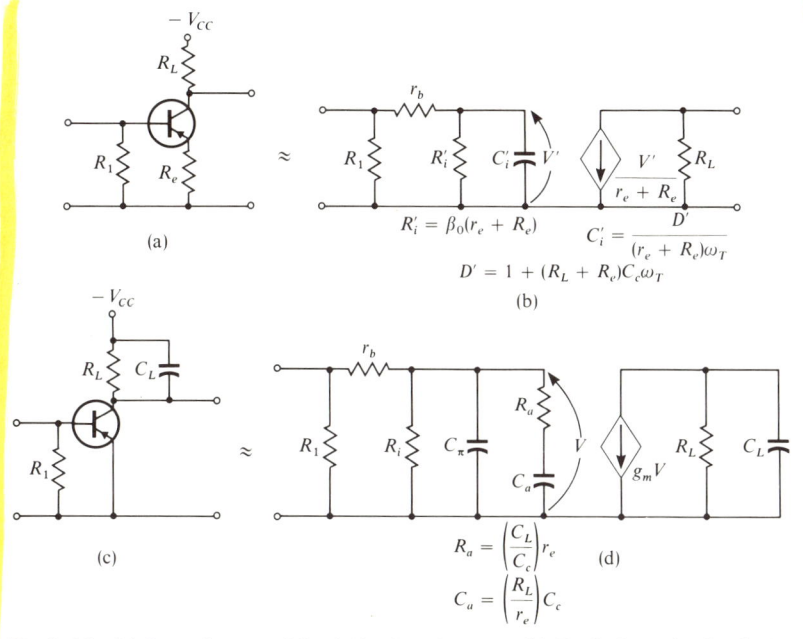

Fig 6-13 (a) Second stage of fig 6-12 when $R_f \to \infty$; (b) Equivalent circuit of (a); (c) First stage of fig 6-12 with the load considered as a shunt RC; (d) Equivalent circuit of (c)

350 6/Feedback Amplifiers and Oscillators

desired. Note that these models are good only for low-pass 3-dB frequency calculations and not for stability prediction.

In the circuit in fig 6-12, we first replace the voltage source and series R_s by a current source and shunt R_s. For open-loop calculations the leading effect of R_f on R_e is negligible, since $R_f \| R_e \approx R_e$, and the effect of R_f on R_s is included in R_p for the **a** circuit, as shown in fig 6-14.

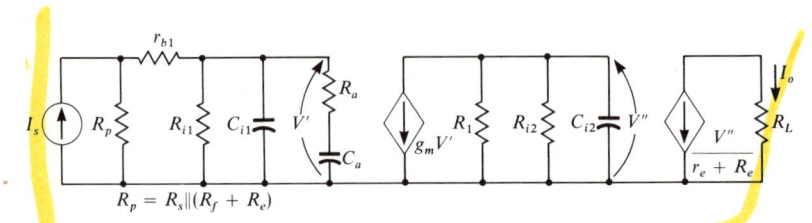

Fig 6-14 **a**(s) circuit for fig 6-12

The second stage transistor is modeled in fig 6-13a (see also fig 4-36). The first transistor stage can be modeled as in fig 6-13b, since R_1 in fig 6-17 is large and the effect of r_{b1} is negligible (because $r_{b1} \ll R_1, R_{i2}$), hence the first stage sees approximately a shunt RC load. In fact, r_b of the first stage can, in this case, also be ignored since $R_p, R_{i1} \gg r_{b1}$; we retain it here to demonstrate this fact.

The approximate forward amplifier circuit **a**(s) is then seen to be as shown in fig 6-14. The **a** function is obtained from the analysis of the circuit in fig 6-14. We have

$$\mathbf{a}_i = \frac{I_o}{I_s} = \left(\frac{I_o}{V''}\right)\left(\frac{V''}{V'}\right)\left(\frac{V'}{I_s}\right) \tag{6-56}$$

Solving for the various ratios in the circuit, and noting a pole-zero cancellation in the first stage model, we obtain

$$\mathbf{a}_i(s) = \frac{I_o}{I_s} = \frac{1}{(R_e + r_{e2})C_{i2}r_{e1}C_{i1}(1 + \delta)}$$

$$\times \left\{ \frac{1}{s^2 + s\{p_{i1} + p_{i2} + C_{c1}/(C_{i1}C_{i2}r_{e1}) + 1/[R_pC_{i1}(1+\delta)]\} + p_{i1}p_{i2}\{1 + R_{i1}/[R_p(1+\delta)]\}} \right\} \tag{6-57}$$

6-4 Design-Oriented Analysis of Some Feedback Amplifier Circuits

where the subscripts 1 and 2 stand for the first and second stage parameters, and

$$C_{i1} = \frac{1 + r_{e1}C_{c1}\omega_{T1}}{r_{e1}\omega_{T1}} \qquad C_{i2} = \frac{1 + (R_L + R_e)C_{c2}\omega_{T2}}{(r_{e2} + R_e)\omega_{T2}}$$

$$R_{i1} = \beta_{01}r_{e1} \qquad R_{i2} = \beta_{02}(r_{e2} + R_e)$$

$$p_{i1} = \frac{1}{R_{i1}C_{i1}} \qquad p_{i2} = \frac{R_1 + R_{i2}}{R_1 R_{i2} C_{i2}} \qquad (6\text{-}58)$$

$$R_p = \frac{(R_e + R_f)R_s}{R_e + R_f + R_s} \approx \frac{R_f R_s}{R_f + R_s} \quad (\text{for } R_f \gg R_e) \qquad \delta = \frac{r_{b1}}{R_p}$$

Note that if $r_{b1} \ll R_p$, $\delta \ll 1$ and its effect can be neglected without any significant error.

We can rewrite (6-57), for convenience, as

$$\mathbf{a}_i(s) = \frac{a_o p_1 p_2}{(s + p_1)(s + p_2)} \qquad (6\text{-}59)$$

where p_1 and p_2 are the poles of the forward $\mathbf{a}(s)$ circuit, namely the roots of the denominator in (6-57). Note that the low-frequency gain from (6-57) is

$$a_o = \mathbf{a}_i(s)\bigg|_{s=0} = \frac{1}{(R_e + r_e)C_{i2}r_{e1}C_{i1}}\left[\frac{1}{p_{i1}p_{i2}[(1 + \delta) + R_{i1}/R_p]}\right] \qquad (6\text{-}60a)$$

Substitution in (6-60a) of the expressions for $p_{i1}, p_{i2}, R_{i1}, R_{i2}$ from (6-58) yields

$$a_o = \frac{\beta_{01}\beta_{02}R_1 R_p}{(R_1 + R_{i2})(R_p + r_{b1} + R_{i1})}$$

$$= \left(\frac{\beta_{01}}{1 + (r_{b1} + R_{i1})/R_p}\right)\left(\frac{\beta_{02}}{1 + R_{i2}/R_1}\right) \qquad (6\text{-}60b)$$

The feedback circuit corresponding to (6-47) is shown in fig 6-15, where

$$\mathbf{f} = \frac{I_{fb}}{I_o} = -\frac{1}{\alpha_{02}}\frac{R_e}{(R_e + R_f)} \equiv -f_o \qquad (6\text{-}61)$$

The closed-loop current-gain function is

$$A_I = \frac{\mathbf{a}_i}{1 - \mathbf{a}_i \mathbf{f}} = \frac{a_o p_1 p_2}{s^2 + s(p_1 + p_2) + p_1 p_2 (1 + a_o f_o)} \qquad (6\text{-}62)$$

We shall consider now a numerical example.

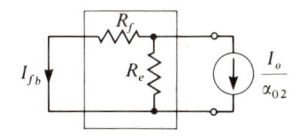

Fig 6-15 f(s) circuit for fig 6-12

EXAMPLE

Consider the following individual transistor parameters at $V_{CE} = -6$ V, $I_C = -5$ mA:

$$f_T = 450 \text{ MHz} \qquad C_c = 2 \text{ pF}$$
$$\beta_0 = 50 \qquad r_b = 75 \text{ }\Omega$$

Let $R_s = R_f = 1$ kΩ, $R_1 = 2.75$ kΩ, $R_L = 500$ Ω, $R_e = 50$ Ω. We then calculate the following expressions:

$$\omega_{T1} = \omega_{T2} = 2\pi f_T = 2\pi(4.5 \times 10^8) = 2.83 \times 10^9 \text{ rad/sec}$$

$$r_{e1} = r_{e2} = \frac{0.025}{5 \times 10^{-3}} = 5 \text{ }\Omega$$

$$C_{i1} = \frac{1 + r_{e1}C_{c1}\omega_{T1}}{r_{e1}\omega_{T1}} = \frac{1 + 0.0283}{1.42 \times 10^{10}} = 72.6 \text{ pF}$$

$$C_{i2} = \frac{1 + (R_L + R_e)C_{c2}\omega_{T2}}{(R_e + r_{e2})\omega_{T2}} = \frac{1 + 3.14}{1.56 \times 10^9} = 26.5 \text{ pF}$$

$$R_{i1} = \beta_{01}r_{e1} = 50(5) = 250 \text{ }\Omega$$

$$R_{i2} = \beta_{02}(r_{e2} + R_e) = 50(55) = 2.75 \text{ k}\Omega$$

$$p_{i1} = \frac{1}{R_{i1}C_{i1}} = \frac{1}{250(72.6 \times 10^{-12})} = 5.50 \times 10^7 \text{ rad/sec}$$

$$p_{i2} = \frac{R_{L1} + R_{i2}}{R_{L1}R_{i2}C_{i2}} = \frac{2(2.75 \times 10^6)}{(2.75 \times 10^6)^2(26.5 \times 10^{-12})} = 2.74 \times 10^7$$

$$R_p = \frac{(R_e + R_f)R_s}{R_e + R_f + R_s} \approx \frac{R_f R_s}{R_f + R_s} = \frac{(10^3 \times 10^3)}{2 \times 10^3} = 500 \text{ }\Omega$$

$$\delta = \frac{r_b}{R_p} = \frac{75}{500} = 0.150$$

The open-loop pole locations are the roots of the denominator of $\mathbf{a}_i(s)$. Substituting these values in (6-57) we get

$$s^2 + (3.413 \times 10^8)s + 0.200 \times 10^{16} = 0$$
$$[s + (0.66 \times 10^7)][s + (33.45 \times 10^7)] = 0$$

6-4 Design-Oriented Analysis of Some Feedback Amplifier Circuits

The open-loop low-frequency gain a_o from (6-60b) is

$$a_o = 50\left(\frac{1}{1 + 0.65}\right)50\left(\frac{1}{1 + 1}\right) = 760$$

Thus the open-loop gain function is

$$\mathbf{a}_i(s) = \frac{760}{[1 + (s/0.66 \times 10^7)][1 + (s/33.4 \times 10^7)]} = \frac{a_o}{(1 + s/p_1)(1 + s/p_2)}$$

Note that the farthest pole, although very large, is still much smaller than ω_T, so that bandwidth calculation on the basis of the dominant pole is still valid. It should be noted, however, that in feedback systems the nondominant poles of the open-loop gain function, in general, *cannot* be ignored. For example, in this case $p_2 = 50p_1$. If we ignore p_2 altogether and use a large amount of feedback, different results will be obtained for the closed-loop gain function.[2] This is one of the reasons why the use of the hybrid π model is preferable in feedback amplifier analysis. The unilateral model is to be used for a very rough calculation of the 3-dB frequency, but not for the prediction of stability.

We may normalize the frequency by a factor of 0.66×10^7 for convenience. This normalization reduces the above equation to the following:

$$\mathbf{a}_i(s_n) = \frac{760}{(1 + s_n)(1 + s_n/50)}$$

The closed-loop gain function is now determined to be

$$A_I(s_n) = \frac{\mathbf{a}_i}{1 - \mathbf{a}_i f} = \frac{a_o p_{1n} p_{2n}}{s_n^2 + s_n(p_{1n} + p_{2n}) + p_{1n}p_{2n}(1 + a_o f_o)}$$

---

[2] The closed-loop gain function, $A_I(s)$, if the nondominant pole of the open-loop gain function is ignored, i.e.,

$$\mathbf{a}_i(s) \approx \frac{760}{1 + s/(0.66 \times 10^7)}$$

would always be a one-pole gain function. For example if $f_o = 0.0476$ it is

$$A_I(s) = \frac{20}{s + 2.47 \times 10^7}$$

This result is quite different from the closed-loop gain function as obtained in the text by including the nondominant open-loop pole.

Hence

$$A_I(s_n) = \frac{(760)(50)}{s_n^2 + s_n(51) + 50(1 + 760f_0)}$$

Since

$$f_0 = \frac{R_e}{R_e + R_f} = \frac{50}{50 + 1000} = 0.0476 \text{ (in the above example)}$$

$$A_I(s_n) = \frac{3.8 \times 10^4}{s_n^2 + 51s + 1850} \approx \frac{3.8 \times 10^4}{(s_n + 25.7 + j35)(s_n + 25.7 - j35)}$$

Note that since $|a_0 f_0| \gg 1$,

$$A_I(0) = \frac{a_0}{1 + a_0 f_0} \approx \frac{1}{f_0} = \frac{R_e + R_f}{R_e} \approx \frac{R_f}{R_e} = 20$$

Note that for $A_I(0) \gg 1$, $R_f \gg R_e$. The pole-zero plots of $\mathbf{a}_i(s_n)$ and $A_I(s_n)$ are shown in fig 6-16. Note that as f_0 is varied, the roots of $A_I(s_n)$ lie on the locus shown by the heavy lines. For $f_0 = 0.0476$, the value of this example, the closed-loop poles are complex, with an approximate $\psi = 53$ degrees with the real axis. Actually, because of the nondominant poles, not included in this analysis (because the unilateral instead of hybrid π model is used), the loci will bend toward the $j\omega$-axis (as will be discussed later). The closed-loop magnitude response corresponding to the pole positions in fig 6-16 will exhibit some frequency peaking, since $\psi > 45$ degrees and the 3-dB bandwidth of the amplifier is approximately $f_{3\,dB} \simeq 45$ MHz. For f_0 smaller than this value, the closed-loop pole angle from the real axis will be smaller. In fact, for a particular design value of f_0, the poles of $A_I(s)$ can be designed to lie on the 45 degree radial line and thus achieve Butterworth two-pole response. For $f_0 = 0.0476$, and by using a shunt C_f across R_f, we can also achieve an approximate maximally flat magnitude response (see Problem 6-9).

For the circuit in fig 6-12, the input resistance is low because of shunt feedback and the output impedance is high because of series feedback. Specifically, the input impedance is similar to that in the shunt-feedback case, namely

$$(R_{in})_f = \frac{R_s \| (r_b + R_i) \| R_f}{1 + a_0 f_0} = \frac{198}{37} \simeq 5.4 \, \Omega,$$

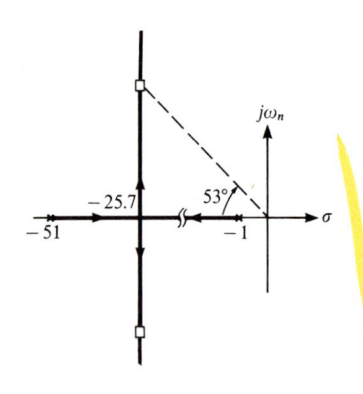

Fig 6-16 Locus of the closed-loop poles of the example: × poles of $A_I(s_n)$ for $f_0 = 0$ (hence poles of $\mathbf{a}_i(s_n)$); □ poles of $A_I(s_n)$ for $f_0 = 0.0476$

6-4 Design-Oriented Analysis of Some Feedback Amplifier Circuits

which is quite low. The above result can also be checked by using the Miller theorem (see Problem 1-31). The output impedance is similar to that in the series-feedback case. If the output impedance of the CE stage is r_o, the output impedance with feedback would be $(R_o)_f = (r_o + R_L + R_e)(1 + a_0 f_0)$, which is quite high. Note that the model in fig 6-14 is not good for high-frequency output impedance calculations. The circuit of fig 6-12, because of its low input impedance and high output impedance, is very attractive for cascading and stagger tuning to obtain multistage amplifiers with a large gain-bandwidth product, since the interaction between the stages can be negligible. Initial design ignoring the interaction will not be too far off from the actual case. The shunt-series feedback pair can also be used to realize approximately a current-controlled current source element (see Problem 6-27).

In addition to obtaining a high gain-bandwidth product, note the reduced sensitivity in the midband gain as a result of feedback. If the transistor betas (β_0) vary by as much as 50 per cent, their effect on the overall gain is still negligible, since $A_I(0) \approx R_f/R_e$ for $a_0 f_0 \gg 1$.

The Series-Shunt (or Voltage Feedback) Pair

The so-called series-shunt feedback pair is shown in fig 6-17, exclusive of biasing and coupling circuitry. We shall analyze the mid-high frequency response of the circuit. For the circuit of fig 6-17, with $R_f \gg R_e, R_L$ the approximate open-loop forward circuit model is shown in fig 6-18. Note that we have assumed R_{i2} to be small so that the first-stage load is approximately resistive as the model shows. Otherwise we would use the model in fig 6-13b for the first stage.

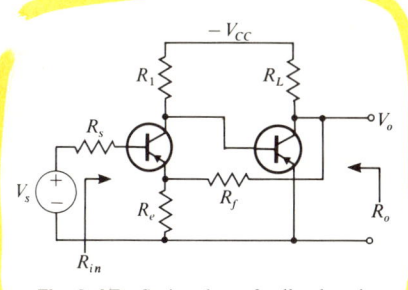

Fig 6-17 Series-shunt feedback pair

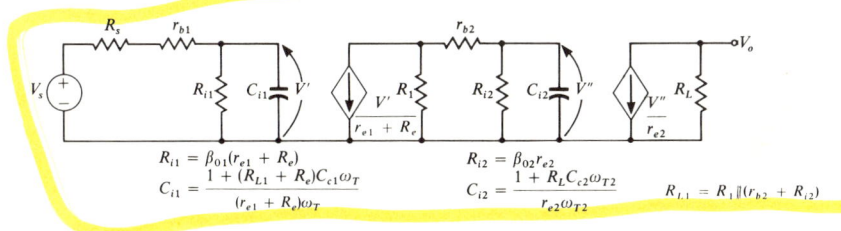

Fig 6-18 a(s) circuit for fig 6-17

$R_{i1} = \beta_{01}(r_{e1} + R_e)$
$C_{i1} = \dfrac{1 + (R_{L1} + R_e)C_{c1}\omega_T}{(r_{e1} + R_e)\omega_T}$

$R_{i2} = \beta_{02} r_{e2}$
$C_{i2} = \dfrac{1 + R_L C_{c2}\omega_{T2}}{r_{e2}\omega_{T2}}$

$R_{L1} = R_1 \| (r_{b2} + R_{i2})$

From the circuit in fig 6-18, we determine the open-loop gain function

$$\mathbf{a}_v = \frac{V_o}{V_s} = \left(\frac{V_o}{V''}\right)\left(\frac{V''}{V'}\right)\left(\frac{V'}{V_s}\right) \qquad (6\text{-}63)$$

Solving for the various quantities, we can write the final expression in the form

$$\mathbf{a}_v = \frac{V_o}{V_s} = \frac{a_o p_1 p_2}{(s + p_1)(s + p_2)} \quad (6\text{-}64)$$

where

$$a_o = \beta_{01}\beta_{02}\left(\frac{R_1}{R_{i1} + R_s + r_{b1}}\right)\left(\frac{R_L}{R_{i2} + R_1 + r_{b2}}\right) \quad (6\text{-}65)$$

$$p_1 = \frac{1}{R_{i1}C_{i1}}\left(\frac{R_{i1} + R_s + r_{b1}}{R_s + r_{b1}}\right) \quad (6\text{-}66a)$$

$$p_2 = \frac{1}{R_{i2}C_{i2}}\left(\frac{R_{i1} + R_1 + r_{b2}}{R_1 + r_{b2}}\right) \quad (6\text{-}66b)$$

The feedback circuit corresponding to (6-43) is shown in fig 6-19:

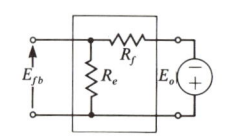

Fig 6-19 f(s) circuit for fig 6-17

$$\mathbf{f}(s) = \frac{E_{fb}}{E_o} = -\frac{R_e}{R_e + R_f} = -f_o \quad (6\text{-}67)$$

The closed-loop gain function $A_V(s)$ is

$$A_V(s) = \frac{\mathbf{a}_v}{1 - \mathbf{a}_v\mathbf{f}} = \frac{a_o p_1 p_2}{s^2 + s(p_1 + p_2) + p_1 p_2(1 + a_o f_o)} \quad (6\text{-}68)$$

Note that if $|af| \gg 1$,

$$A_V(s) \simeq -\frac{1}{\mathbf{f}} \approx \frac{R_f + R_e}{R_e} \quad (6\text{-}69)$$

For A_V to be much larger than one, we use $R_f \gg R_e$, the assumption we have already used. To illustrate the above equations, we again consider a numerical example:

EXAMPLE

Let the individual transistor parameters be approximately equal for simplicity. At $V_{CE} = -5$ V, $I_C = -5$ mA, we assume the individual transistor parameters to be

$$\omega_T = 2\pi(175 \text{ MHz}) = 1.1 \times 10^9 \text{ rad/sec}, \quad r_b = 75 \,\Omega,$$

$$\beta_0 = 50, \quad C_c = 5 \text{ pF},$$

$$R_s = 25 \,\Omega, \quad R_e = R_L = 50 \,\Omega, \quad \text{and} \quad R_1 = 3.9 \text{k}\Omega$$

6-4 Design-Oriented Analysis of Some Feedback Amplifier Circuits

From (6-65), (6-66a), (6-66b) and the transistor parameters we calculate $a_0 = 40.6$, $p_1 = 21.9 \times 10^7$ rad/sec, and $p_2 = 1.8 \times 10^7$ rad/sec. For convenience, we normalize the first pole to unity (i.e., $\Omega_o = 1.8 \times 10^7$), and then, from (6-64) and the numerical results

$$\mathbf{a}_v(s_n) = \frac{(40.6)(12.2)}{(s_n + 1)(s_n + 12.2)}$$

Note again that the farthest pole is much smaller than ω_T, hence bandwidth calculations on the basis of the dominant pole are reasonable. The feedback function, from (6-67), is $\mathbf{f}(s) = -f_0$. Hence, the closed-loop gain function for this example, from (6-68) and above, is

$$A_V(s_n) = \frac{(40.6)(12.2)}{s_n^2 + 13.2s_n + 12.2(1 + 40.6 f_o)}$$

For $R_f = 285\,\Omega$ (i.e., $f_o = 0.15$) we have an interesting design since, for this value of f_o, the denominator becomes

$$s_n^2 + 13.2 s_n + 87 = s_n^2 + a s_n + b$$

Note that in the above

$$(13.2)^2 \approx 2(87), \text{ i.e., } a^2 = 2b$$

Hence the poles are at a 45-degree radial angle, and the transmission shape due to the dominant poles is a two-pole Butterworth response. The normalized 3-dB radian frequency from Section 2-8 is equal to $\sqrt{b} = \sqrt{87} = 9.32$, or the actual bandwidth $f_{3\,\text{dB}} = [9.32(1.8 \times 10^7)]/2\pi \approx 27$ MHz. The amount of feedback $F(0) = 1 + a_o f_o = 7.1$. The closed-loop low-frequency gain is $A_V(0) = 5.7$. The amount of feedback is low because of the low value of R_L, and hence a low value of a_o. The low-frequency input impedance, as obtained in the series feedback case, is $(R_{in})_f \approx (R_s + R_i + r_b)(1 + a_o f_o) = 2.5\,\text{k}\Omega$. The output impedance is determined as in the shunt feedback case, namely

$$(R_o)_f \approx [(r_o \| R_L \| R_f)/(1 + a_o f_o)] \approx 6\,\Omega$$

The series-shunt circuit can also be used to realize approximately a VCVS element (see Problem 6-26).

Shunt-Shunt Feedback Triple

As a final example of feedback analysis, we consider the commonly used feedback amplifier configuration shown in fig 6-20. For the sake of simplicity, we have again omitted the biasing and coupling circuitry.

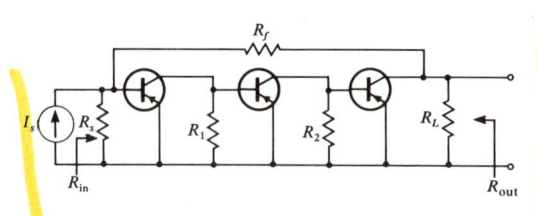

Fig 6-20 A shunt-shunt feedback triple

This circuit belongs to the shunt-shunt configuration (see fig 6-2a). The **a** and **f** circuit models corresponding to fig 6-3a and fig 6-3b are shown in fig 6-21a and fig 6-21b, respectively. Note that we again use

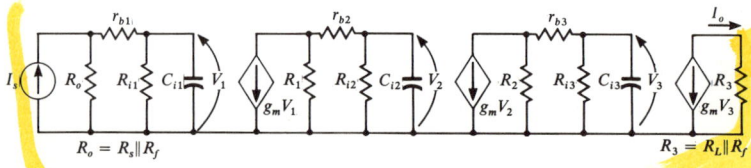

Fig 6-21 (a) **a**(s) circuit for fig 6-20;

Fig 6-21 (b) **f**(s) circuit for fig 6-20

the "unilateral" model approximation to get an approximate result in a simple manner. The preliminary analysis or design may have to be checked by using the hybrid π model, if such a detailed analysis is warranted.

From fig 6-21 the open-loop gain function, which is simply that of three cascaded CE stages, can be readily obtained. Since

$$\mathbf{a}_i(s) = \frac{I_o}{I_s} = \left(\frac{I_o}{V_3}\right)\left(\frac{V_3}{V_2}\right)\left(\frac{V_2}{V_1}\right)\left(\frac{V_1}{I_s}\right) \tag{6-70}$$

the resulting expression can be written in the form

$$\mathbf{a}_i(s) = \frac{I_o}{I_s} = -\frac{a_o p_1 p_2 p_3}{(s+p_1)(s+p_2)(s+p_3)} \tag{6-71}$$

6-4 Design-Oriented Analysis of Some Feedback Amplifier Circuits

where

$$p_1 = \frac{1}{R_{i1}C_{i1}} \frac{R_0 + r_{b1} + R_{i1}}{(R_0 + r_{b1})} \qquad (6\text{-}72a)$$

$$p_2 = \frac{1}{R_{i2}C_{i2}} \frac{R_1 + r_{b2} + R_{i2}}{R_1 + r_{b2}} \qquad (6\text{-}72b)$$

$$p_3 = \frac{1}{R_{i3}C_{i3}} \frac{R_2 + r_{b3} + R_{i3}}{R_2 + r_{b3}} \qquad (6\text{-}72c)$$

$$a_o = \beta_{01}\left(\frac{R_0}{R_0 + r_{b1} + R_{i1}}\right)\beta_{02}\left(\frac{R_1}{R_1 + r_{b2} + R_{i2}}\right)\beta_{03}\left(\frac{R_2}{R_2 + r_{b3} + R_{i3}}\right) \qquad (6\text{-}73)$$

The feedback factor $\mathbf{f}(s) = -y^f_{12}R_L = R_L/R_f = f_o$. Hence the closed-loop gain function is

$$A_I(s) = \frac{\mathbf{a}_i}{1 - \mathbf{a}_i \mathbf{f}} = \frac{-a_o p_1 p_2 p_3}{(s + p_1)(s + p_2)(s + p_3) + p_1 p_2 p_3 a_o f_o} \qquad (6\text{-}74)$$

EXAMPLE

Consider a numerical example for fig 6-20. For simplicity, we assume identical parameters for each of the transistors. Let the individual transistor parameters be

$$\omega_T = 10^9 \text{ rad/sec} \qquad r_b = 50 \, \Omega$$

$$\beta_0 = 50 \qquad C_c = 5 \text{ pF}$$

$$r_e = 5 \, \Omega$$

$$R_f \gg R_s, R_L, \qquad R_s = R_1 = R_2 = 100 \, \Omega, \qquad R_L = 75 \, \Omega$$

The value of R_f will be chosen later for a flat magnitude response. We proceed to find p_3 first because of the D factors. For the third stage

$$R_{i3} = 50(5) = 250 \, \Omega, \qquad R_3 = R_L \| R_f \approx 75 \, \Omega$$

$$C_{i3} = \frac{1 + R_3 C_c \omega_T}{r_e \omega_T} = \frac{1 + 0.375}{5(10^9)} = 275 \text{ pF}$$

From (6-72c) we obtain

$$p_3 = \frac{1}{R_{i3}C_{i3}}\left(\frac{R_2 + r_b + R_i}{R_2 + r_b}\right)$$

$$= \frac{1}{(250)(2.75 \times 10^{-10})}\left(\frac{100 + 50 + 250}{100 + 50}\right) = 3.87 \times 10^7$$

Similarly, $R_{L2} = R_2 \| R_{i3} \approx 75\,\Omega$, $R_{L1} = R_1 \| R_{i2} \approx 75\,\Omega$, and since $R_s = R_1 = R_2 = 100\,\Omega$ and $R_{L1} = R_{L2} = R_3 = 75\,\Omega$ we obtain

$$p_3 = p_2 = p_1 = 3.87 \times 10^7 \text{ rad/sec}$$

The open-loop gain from (6-73) is

$$a_o = \left[(50)\left(\frac{100}{100 + 50 + 250}\right)\right]^3 = (12.5)^3 = 1950$$

We normalize frequency by a factor of 3.87×10^7, hence

$$\mathbf{a}_i(s_n) = \frac{-a_0}{(s_n + 1)^3} = \frac{-1.95 \times 10^3}{(s_n + 1)^3}$$

$$\mathbf{f}(s) = f_o$$

Therefore

$$A_I(s_n) = \frac{-1.95 \times 10^3}{s_n^3 + 3s_n^2 + 3s_n + 1 + a_o f_o}$$

For $f_o = 1/2a_o$, i.e., $R_f = 285$ kΩ, $T_o = a_o f_o = \frac{1}{2}$, it can readily be shown (see Problem 6-41) that the magnitude response of $|A(j\omega)|$ will be as flat as possible without peaking, and that the 3-dB radian frequency for this case is $(\omega_{3\,\text{dB}})_n = 0.88$ while the actual 3-dB frequency is $f_{3\,\text{dB}} = [0.88(3.87 \times 10^7)]/2\pi = 5.43$ MHz. The closed-loop gain is $A_I(0) = \frac{2}{3}a_0 = 1300 = 62.28$ dB. Note that for $f_o \geq 8/a_o$ the two-complex poles of the amplifier will lie in the right half-plane and the system will be unstable.[3] The position of the natural frequencies as the feedback factor is varied can be conveniently examined by root-locus technique as discussed in the next section.

[3] One of the reasons why the unilateral model cannot be used for stability analysis of feedback amplifiers is because the phase-shift contributions due to the neglected nondominant poles are not included in the analysis. If this phase-shift contribution is included in the analysis, the system will be unstable for f_o smaller than $8/a_0$. A reasonably accurate model for such purposes is the hybrid π model since the nondominant poles are included in circuit model.

In most feedback systems, one of the main items of interest is the use of a large amount of feedback in order to obtain desensitivity, as discussed earlier. By using a large value of f_o, i.e., a smaller valued feedback resistor R_f, we can increase $T(0)$. For $T_o \geq 8$, however, the amplifier will become unstable. In order to increase T_o without causing instability, a frequency-compensation scheme is usually used. One approach is to broad-band one stage by a large factor. This will increase one of the poles and decrease the open-loop gain a_o, and thus T_o can be increased without causing instability. Sometimes narrow banding may be used and/or a zero is introduced in the feedback network, e.g., shunting R_f by C_f (see Refs 2 and 3).

So far we have discussed exclusively the midband and high-frequency responses of feedback amplifiers. A word or two for the low-frequency end is also in order. The effect of negative feedback is also to reduce the low-frequency bandedge. In some cases the coupling and bias circuitry capacitors in conjunction with feedback which greatly affects the low-frequency response of a feedback amplifier, can cause instability. This instability occurs at very low frequencies and can be readily observed. To remedy this problem, simply reduce the value of the capacitance in either the coupling circuit or the bypass circuit. This has the effect of pushing a low-frequency pole of the open loop gain function to the left on the σ-axis and thus preventing oscillation. This effect can also be seen readily by employing the root-locus technique (see Problem 6-11).

6-5 The Root-Locus Technique

The root-locus technique is a very simple concept that provides a great amount of information to the designer of feedback systems. Basically, it provides a visual picture of the effects of feedback on the closed-loop poles. The root-locus is a plot of the roots of the characteristic equation $(1 + T(s) = 0)$ as a function of the low-frequency gain of the loop transmission function. It can be used to predict the stability of a system if a realistic model for the device is used in the loop transmission calculation. Let us consider a couple of simple examples.

First, consider a simple single-stage feedback amplifier for which the normalized open-loop gain function is

$$\mathbf{a}(s) = -\frac{a_o}{s+1} \qquad a_o > 0 \qquad (6\text{-}75\text{a})$$

Let the feedback factor

$$\mathbf{f}(s) = f_o \qquad f_o \geq 0 \qquad (6\text{-}75b)$$

The closed-loop gain function is

$$A(s) = \frac{\mathbf{a}}{1 - \mathbf{af}} = -\frac{a_o}{s + 1 + a_o f_o} = -\frac{a_o}{s + 1 + T_o} \qquad (6\text{-}76)$$

Now as T_o is increased from zero, the closed-loop pole, or the natural frequency, moves along the negative real axis, as shown by the heavy line in fig 6-22. The heavy line is the *locus* of the *roots* of $A(s)$, hence the root-locus plot. From the root-locus plot it is evident that the system cannot become unstable, since the pole can never go into the right half-plane no matter what the value of T_o.

As a second simple example, consider

$$\mathbf{a}(s) = -\frac{a_o}{(s+1)(s+3)} \qquad a_o > 0 \qquad (6\text{-}77a)$$

$$\mathbf{f}(s) = f_o \qquad f_o \geq 0 \qquad (6\text{-}77b)$$

The closed-loop gain function is

$$A(s) = \frac{\mathbf{a}}{1 - \mathbf{af}} = \frac{-a_o}{s^2 + 4s + 3 + T_o} \qquad (6\text{-}78)$$

Now, as T_o is increased from zero, the characteristic roots will move on the locus shown in fig 6-23. For $T_o = 0$, $s_{1,2} = -1, -3$, for $T_o = 1$ there are double roots at $s_{1,2} = -2$, and for $T_o > 1$, the roots will be complex at $s_{1,2} = -2 \pm j\omega_0$, e.g., for $T_o = 5$ the characteristic roots will be on the 45 degree radial line ($s_{1,2} = -2 \pm j2$) or the closed-loop gain function is a maximally flat two-pole function.

Root-Locus Construction

Now let us consider the general case where the loop transmission function has k zeros and m poles. We rewrite the characteristic equation as

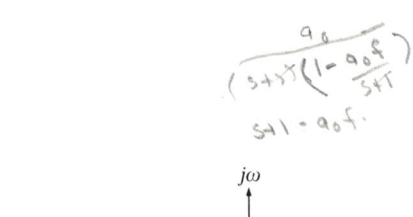

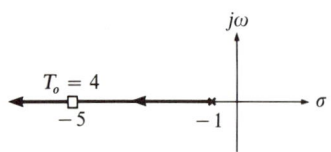

$$1 + T(s) = 1 + KG(s) = 0 \qquad (6\text{-}79a)$$

Fig 6-22 Root-locus for equation (6-76)

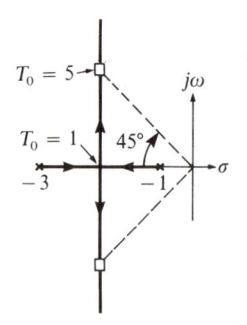

Fig 6-23 Root-locus for equation (6-78)

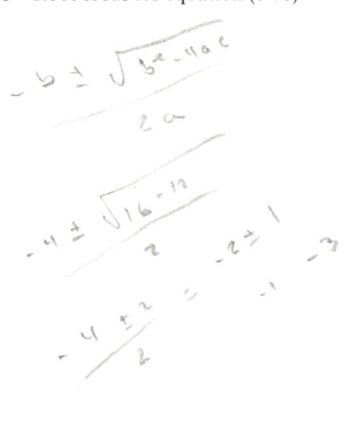

or

$$1 + K \frac{\prod_{i=1}^{k}\left(1 + \frac{s}{z_i}\right)}{\prod_{j=1}^{m}\left(1 + \frac{s}{p_j}\right)} = 0 \quad m \geq k \quad \text{(6-79b)}$$

The root-locus constitutes all s-plane points at which (6-79) is satisfied. Note that for negative feedback $K > 0$, while for positive feedback $K < 0$. From (6-79b), we can write

$$KG(s) = -1 = e^{\pm jn\pi} \quad \text{(6-80)}$$
$$(n \text{ is an odd integer})$$

From (6-80), a set of rules is subsequently derived which facilitate the root-locus plot. These rules, which must all be satisfied, are guidelines in sketching root-locus plots. The root-locus sketch is a useful guide for design purposes. It should be noted, however, that if k and m are large numbers, the use of a computer is the best approach.

From (6-80), we impose the following simultaneous conditions on the magnitude and phase angle of the complex quantity $KG(s)$:

$$|KG(s)| = 1 \quad \text{(6-81a)}$$

$$\arg G(s) = \pm n\pi \quad \text{(6-81b)}$$
$$(n \text{ is an odd integer})$$

From (6-81) we can deduce the following rules that enable us readily to sketch the loci:

(1) Loci start from the poles of $G(s)$ and terminate at the zeros of $G(s)$. This can be seen readily from (6-81a); since $KG(s) = 1$, the value of K must be zero at the poles of $G(s)$ and K must be infinite at the zeros of $G(s)$:

(2) The root-locus is symmetrical about the real axis. This rule is obvious from the fact that the polynomial in (6-79a) has real coefficients and thus the roots, if complex, must occur in complex conjugated pairs.

(3) The number of separate loci equals the number of poles or zeros of $G(s)$, whichever number is larger. From (6-79b) this number is equal to m. This rule is obvious from the fact that the loci start from the poles of $G(s)$ and terminate at the zeros of $G(s)$.

(4) (a) The loci which go toward infinity approach asymptotic lines whose directions can be found by

$$\theta_i = \pm \frac{n\pi}{m - k} \quad \text{(6-82a)}$$
$$(n \text{ is an odd integer})$$

where $(m - k)$ is the difference between the finite number of poles and zeros of $G(s)$. If $m = k$ there is no asymptotic line.

(b) The asymptotes intersect on the real axis at σ_o, which is given by

$$\sigma_o = \frac{\sum_{}^{m}\text{poles} - \sum_{}^{k}\text{zeros}}{m - k} \quad (6\text{-}82\text{b})$$

To prove rules (6-82a) and (6-82b), rewrite (6-79a) as follows:

$$KG(s) = K\frac{s^k + b_1 s^{k-1} + \cdots + b_o}{s^m + a_1 s^{m-1} + \cdots + a_o} \quad (6\text{-}83)$$

By simple division we obtain

$$KG(s) = \frac{K}{s^{m-k} - (b_1 - a_1)s^{m-k-1} + \cdots} \quad (6\text{-}84)$$

But since $KG(s) = -1$ and $s \to \infty$ for asymptotic behavior, we can ignore the terms beyond s^{m-k-1} in the denominator; we can thus write (6-84) as

$$s^{m-k} - (b_1 - a_1)s^{m-k-1} \simeq -K \quad (6\text{-}85)$$

Let $m - k \equiv r$; then (6-85) may be rewritten as

$$s^r - (b_1 - a_1)s^{r-1} \simeq Ke^{jn\pi} \quad (6\text{-}86)$$
$$(n \text{ is an odd integer})$$

Factor s^r in (6-86) to get

$$s^r[1 - (b_1 - a_1)s^{-1}] \simeq Ke^{jn\pi} \quad (6\text{-}87)$$

Take the rth root of (6-87)

$$s[1 - (b_1 - a_1)s^{-1}]^{1/r} \simeq K^{1/r} e^{j(n\pi/r)} \quad (6\text{-}88)$$

Expand (6-88) in a power series of $1/s$ and retain only the first term (since $s \to \infty$)

$$s\left[1 - \frac{b_1 - a_1}{rs}\right] \simeq K^{1/r} e^{j(n\pi/r)}$$

or

$$s - \frac{b_1 - a_1}{r} \simeq K^{1/r}\left[\cos\frac{n\pi}{r} + j\sin\frac{n\pi}{r}\right] \qquad (6\text{-}89)$$

Substituting $s = \sigma + j\omega$ in (6-89) we have

$$\sigma + j\omega \simeq -\frac{a_1 - b_1}{r} + K^{1/r}\left[\cos\frac{n\pi}{r} + j\sin\frac{n\pi}{r}\right] \qquad (6\text{-}90)$$

Equate the real and imaginary parts of (6-90), and eliminate $K^{1/r}$ to obtain

$$\omega = \left(\tan\frac{n\pi}{r}\right)\left(\sigma + \frac{a_1 - b_1}{r}\right) \qquad (6\text{-}91)$$

Equation (6-91) is the equation of a straight line of the form $\omega = m(\sigma - \sigma_o)$, which has the slope $m = \tan n\pi/r$ and the intercept $\sigma_o = -(a_1 - b_1)/r$. But $-a_1$ and $-b_1$ are the sums of the roots of the denominator and numerator polynomials in (6-83), respectively, hence the poles and zeros respectively, of $G(s)$. Thus the asymptotes intersect at $(\sum \text{poles} - \sum \text{zeros})/(m - k)$ which is a real number, and the direction of the asymptotic lines is given by $\pm n\pi/m - k$, as asserted. Recall that for $m = k$ there are no asymptotic loci.

For example, let $G(s) = 1/[(s + 1)(s + 2)(s + 4)]$. The pole-zero plot is shown in fig 6-24. The asymptotes intersect at

$$\sigma_0 = \frac{\sum \text{poles} - \sum \text{zeros}}{m - k} = \frac{(-1 - 2 - 4) - 0}{3 - 0} = -2.3$$

The direction of the asymptotes is given by

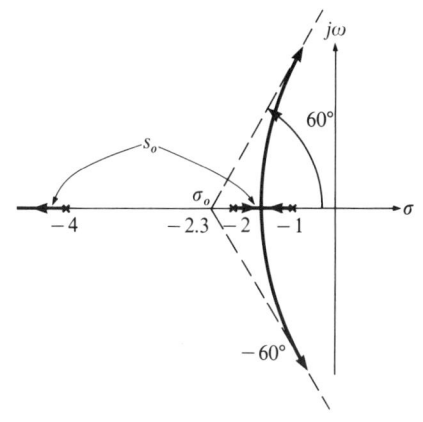

$$\theta = \pm\frac{n\pi}{m - k} = \pm\frac{n\pi}{3} = \pm 60 \text{ degrees},\ 180 \text{ degrees}$$

Fig 6-24 Root-locus of the example

as shown in fig 6-24.

(5) The loci on the section of the real axis are to the left of an *odd* number of critical frequencies (poles or zeros) of $G(s)$. This can be readily shown as follows: Since Arg $G(s) = \pm 180$ degrees, loci can exist only if the point in question (s_o), as shown in fig 6-24, has an odd number of critical frequencies at the right. Note that the argument from complex pole or zero pairs is zero along the real axis. Critical frequencies to the left of (s_o) also have zero contribution.

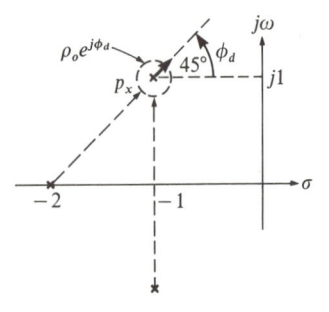

Fig 6-25 Departure-angle calculation of the root-locus example

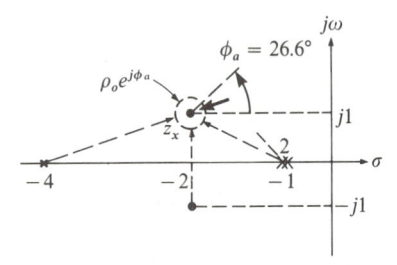

Fig 6-26 Arrival-angle calculation of the root-locus example

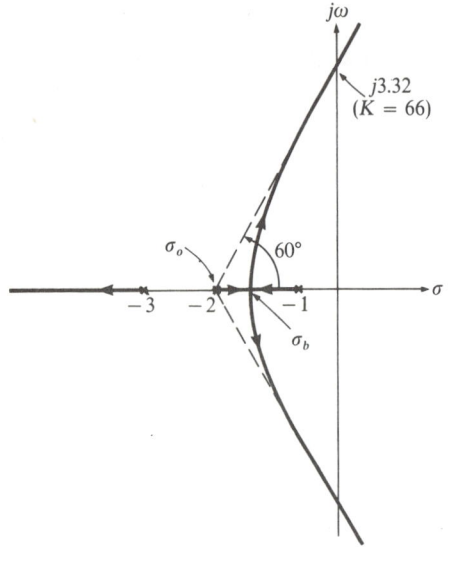

Fig 6-27 Breakaway point and the complete root-locus of example 1

(6) The angle of departure from a pole, p_x, and the angle of arrival at a zero, z_x, are given by

(a) Departure angle

$$\phi_d = \sum_{j=1}^{k} \arg(p_x - z_j) - \sum_{\substack{i=1 \\ i \neq x}}^{m} \arg(p_x - p_i) \mp 180 \text{ degrees} \quad (6\text{-}92a)$$

(b) Arrival angle

$$\phi_a = \sum_{i=1}^{m} \arg(z_x - p_i) - \sum_{\substack{j=1 \\ j \neq x}}^{n} \arg(z_x - z_j) \pm 180 \text{ degrees} \quad (6\text{-}92b)$$

This rule can be verified if one considers the point s_o in the neighborhood of the pole p_x, or the zero z_x, within a circle of very small radius ρ_o, as shown in fig 6-25, and then determine its angle from the other poles and zeros.

For example, let $G(s) = 1/[(s+2)(s^2+2s+2)]$; the pole-zero plot is shown in fig 6-25. From (6-92a)

$$\phi_d = 0 - (90 \text{ degrees} + 45 \text{ degrees}) \mp 180 \text{ degrees} = 45 \text{ degrees}$$

Next consider $G(s) = (s^2 + 4s + 5)/[(s+1)^2(s+4)]$; the pole-zero plot is shown in fig 6-26. From (6-92b)

$$\phi_a = 2(135) \text{ degrees} + 26.6 \text{ degrees} - 90 \text{ degrees} \pm 180 \text{ degrees}$$

$$= 26.6 \text{ degrees}$$

(7) The breakaway point, s_b (usually the one on the real axis is of interest), is computed from setting

$$\frac{d}{ds}G(s) = 0 \quad \text{or} \quad \frac{d}{ds}\left[\frac{1}{G(s)}\right] = 0$$

For example, consider

$$G(s) = \frac{1}{(s+1)(s+2)(s+3)}$$

$$\frac{d}{ds}\left[\frac{1}{G(s)}\right] = \frac{d}{ds}[(s+1)(s+2)(s+3)] = 0$$

or

$$3s^2 + 12s + 11 = 0 \rightarrow s_b = -1.425 \text{ or } -2.575$$

Since the breakaway point must lie between 1 and 2, the result is $s_b = \sigma_b = -1.425$, as shown in fig 6-27.

To prove the above assertion, we assume that at the breakaway points, $1 + KG(s) = 0$ has only roots with a multiplicity of two at $s = s_b$. If this is so, then the derivative of $1 + KG(s) = 0$ is also zero

TABLE 6-2   Rules for the Construction of the Root-Loci

1. *Loci end points*
   The loci start from the poles and terminate at the zeros of $G(s)$
2. *Conjugate values*
   The root loci are symmetrical about the real axis
3. *Number of loci*
   The number of separate loci equals the number of poles or zeros of $G(s)$, whichever number is larger. It is equal to m for (6-79)
4. *Asymptotes of loci*
   a. The loci for large values of s (i.e., near infinity) approach asymptotic lines whose directions are given by
   $$\theta_i = \pm \frac{n\pi}{m - k} \qquad m > k \ (n \text{ is an odd integer})$$
   b. The asymptotes intersect on the real axis at σ_0, which is determined from
   $$\sigma_0 = \frac{\sum_{}^{m} \text{poles} - \sum_{}^{k} \text{zeros}}{m - k} \qquad m > k$$
   where $m - k$ is the difference between the number of finite poles and zeros of $G(s)$
5. *Loci on real axis*
   The parts of the real axis which comprise sections of the loci are to the left of an odd number of critical frequencies (poles or zeros) of $G(s)$
6. *Angles of departure and arrival*
   a. The angle of departure from a pole, p_x, is given by
   $$\phi_d = \sum_{j=1}^{k} \arg(p_x - z_j) - \sum_{\substack{i=1 \\ i \neq x}}^{m} \arg(p_x - p_i) \mp 180 \text{ degrees}$$
   b. The angle of arrival at a zero, z_x, is given by
   $$\phi_a = \sum_{i=1}^{m} \arg(z_x - p_i) - \sum_{\substack{j=1 \\ j \neq x}}^{k} \arg(z_x - z_j) \pm 180 \text{ degrees}$$
7. *Breakaway point, s_b*
   The breakaway point(s) are determined by solving for the roots of:
   $$\frac{d}{ds}[G(s)]^{-1} = 0$$
8. *Intersection with imaginary axis*
   The intersection of the loci with the imaginary axis is usually determined by the Routh–Hurwitz test

at the breakaway point. Hence

$$\frac{d}{ds}[1 + KG(s_b)] = 0 \rightarrow \frac{dG}{ds}(s_b) = 0$$

The roots of $\frac{d}{ds}G(s) = 0$ are the same as those of $\frac{d}{ds}[G(s)]^{-1} = 0$.

Note that the breakaway points need not be on the real axis.

(8) The intersection of the root-loci with the imaginary axis is usually determined by the Routh–Hurwitz test (discussed in Section 6-6). For example, consider the characteristic equation given by

$$1 + \frac{K}{s(s+1)(s+2)} = 0$$

or

$$s^3 + 3s^2 + 2s + K = 0$$

From the Routh–Hurwitz test, the imaginary axis intersection is at $\omega = \pm j\sqrt{2}$ for a value of $K = 6$. In other words, the system will be unstable for $K \geq 6$.

The above rules provide an aid for sketching the loci; they do not give the exact plot. For convenience, the rules are summarized in table 6-2. The construction of the root-loci for low-order systems can be readily sketched by paper-and-pencil methods. It should be emphasized, however, that for high-order systems, resorting to a computer solution is the best choice.

Illustrative Examples

EXAMPLE 1

Let

$$T(s) = \frac{K}{(s+1)(s+2)(s+3)} = -\mathbf{a}\mathbf{f}(s)$$

Using the rules, we obtain the following: The asymptotes intersect at $\sigma_o = 2.0$, the angles of the asymptotes are ± 60 degrees, 180 degrees, and the breakaway point is at $\sigma_b = -1.42$. The imaginary axis intersection is at $\omega = \pm j3.32$, with the value of $K = 66$ (as discussed in Section 6-6). The root-locus is shown in fig 6-27.

EXAMPLE 2

Let

$$\mathbf{a}(s) = -\frac{5a_o}{(s+1)(s+5)} \qquad a_o > 0$$

$$\mathbf{f}(s) = \frac{f_o}{10}(s+10) \qquad f_o \geq 0$$

The loop transmission function is

$$T(s) = -\mathbf{af}(s) = \frac{a_o f_o}{2} \frac{(s+10)}{(s+1)(s+5)}$$

the pole-zero plot and the root-locus are shown in fig 6-28. Note that in this case, the closed-loop system has only two poles and no finite transmission zeros, because

$$A(s) = \frac{\mathbf{a}}{1-\mathbf{af}} = \frac{-5a_o/(s+1)(s+5)}{1+(a_o f_o/2)(s+10)/[(s+1)(s+5)]}$$

$$= \frac{-5a_o}{(s+1)(s+5) + (a_o f_o/2)(s+10)}$$

$$= \frac{-5a_o}{s^2 + s(6 + a_o f_o/2) + 5(1 + a_o f_o)}$$

If there were no zeros in $\mathbf{f}(s)$, e.g., if $\mathbf{f} = f_o$, then the root-locus would be as shown by the dotted line in fig 6-28. The zeros of $\mathbf{f}(s)$ do not produce zeros in $A(s)$, hence these are called *phantom zeros* in the literature. The use of phantom zeros in feedback amplifiers can improve the performance of the amplifiers (Ref 3).

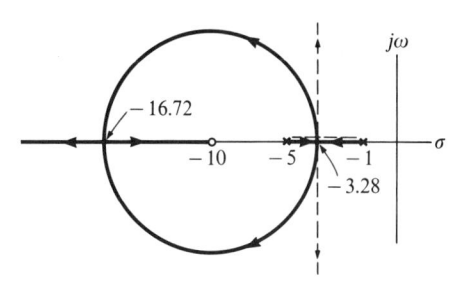

Fig 6-28 Root-locus of example 2

EXAMPLE 3

Consider finally the amplifier configuration of fig 6-20 with one modification; namely, that a shunt capacitor C_f be connected across R_f. The assumptions that $Z_f \gg R_{\text{in}}, R_L$ are still retained. Let the amplifier poles be nonidentical. For this case

$$\mathbf{a}(s) \approx -\frac{a_o}{(1+s/p_1)(1+s/p_2)(1+s/p_3)} \qquad (6\text{-}93)$$

$$\mathbf{f}(s) \approx f_o\left(1 + \frac{s}{z}\right) \qquad (6\text{-}94)$$

From (6-93) and (6-94), the loop transmission function is

$$T(s) \equiv KG(s) = a_o f_o \frac{(1 + s/z)}{(1 + s/p_1)(1 + s/p_2)(1 + s/p_3)} \quad (6\text{-}95)$$

The root-locus for (6-95) will depend on the values of z, p_1, p_2, and p_3. If $p_1 = z$, then the root-locus will be of the form in fig 6-23. But be careful to notice that $A(s)$ will *not be* of the same form, as in (6-78), since a cancellation occurs between the forward pole and the feedback zero (phantom zero). For this case (i.e., $p_1 = z$), $A(s)$ is given by

$$A(s) = \frac{a}{1 - af} = \frac{a_o}{(1 + s/p_1)[(1 + s/p_2)(1 + s/p_3) + a_o f_o]} \quad (6\text{-}96)$$

The above point will be clear if we consider a somewhat general case. Let $p_1 = 1$, $p_2 = 2$, $p_3 = 3$, and $z = 5$. The root-locus is shown in fig 6-29. For a certain value of T_o, the closed-loop poles are shown by rectangles. Note that $A(s)$ has no zeros even though the loop transmission $T(s)$ has a zero. If $z_1 = \infty$, the root-locus is as shown in fig 6-27. Note that the presence of the phantom zero prevents the loci from bending toward the $j\omega$-axis, and hence avoids instability.

Simple design methods based on the root-locus techniques and the discussion in Section 6-3 should be evident to the reader. A few simple design problems are given to the reader as exercises. For a more detailed discussion of feedback amplifier design, the reader should refer to the author's advanced text (Ref 3). Again, it should be noted that the simplified preliminary design may have to be checked and improved by using computer-aided analysis and the hybrid π transistor models.

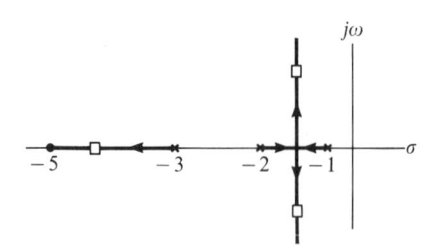

Fig 6-29 Root-locus of example 3

6-6 Stability Considerations in Linear Feedback Systems

All of the advantages discussed in the previous section can be realized only if the feedback amplifier is stable. By stability we mean that the roots of the characteristic equation are in the left half-plane not including the $j\omega$ axis. Note that we are considering rational functions, in which case the characteristic roots are roots of the polynomials. The stability of an amplifier must be ascertained at *all frequencies*, and not just over the passband frequency range of the

6-6 Stability Considerations in Linear Feedback Systems

amplifier. In other words, the effect of nondominant poles and excess phase shift should be taken into consideration if the stability of an amplifier is to be guaranteed. This is why the unilateral model is not good for stability calculations.

For example, consider the series-shunt and the shunt-series feedback pairs discussed in Section 6-4. In each case there are only two dominant poles; and the root-locus is of the form of fig 6-23. From the root-locus we see that the roots never enter the right half-plane for any amount of feedback. We know, however, that we have neglected the nondominant poles of the amplifier. If we insert only one nondominant pole per stage, the root-locus will be modified as shown by the solid line in fig 6-30, which clearly shows the possibility of instability for large amounts of feedback. Note that the root-locus can be used to predict the stability of a system if we have complete accurate expressions for the loop transmission functions. To obtain the loop transmission function for stability analysis, we must at least use the hybrid π model for the transistors. This is one of the drawbacks of the analytical procedure in testing for stability. The drawback is that we must have an accurate realistic model for the device and thus an accurate loop transmission function to begin with. This can be involved and tedious to obtain. We assume in the following analytic procedure that $T(s)$ represents a realistic loop transmission function of the system.

Let us consider the characteristic equation of a single-loop feedback amplifier (6-8), rewritten for convenience in the following:

$$F(s) \equiv 1 + T(s) = 0 \qquad (6\text{-}97)$$

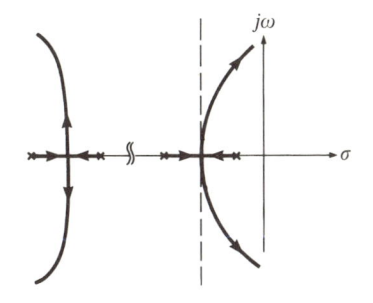

Fig 6-30 Root-locus when two nondominant poles are included with the two dominant poles of the loop-transmission function

To determine the stability of a lumped system, we examine the roots of (6-97). If we know the values of all the roots of $F(s) = 0$, and at least one turns out to be in the right half-plane, or if any of the coefficients of the characteristic polynomial is negative or missing, we immediately dismiss the circuit as unstable. Otherwise we use any one of the following two well known tests. These tests are given in the following without proofs. For proofs, the reader is referred to the literature (Ref 4).

Routh–Hurwitz Test

The Routh–Hurwitz test is an analytic procedure to check the existence of any root or roots in the right half s-plane. In general, the characteristic equation of a lumped linear system can be written as

$$F(s) = a_0 s^n + a_1 s^{n-1} + \cdots + a_{n-1} s + a_n = 0 \qquad (6\text{-}98)$$

We rearrange (6-98) in two lines and compute the triangular arrays as follows:

$$\begin{array}{c|cccc}
s^n & a_0 & a_2 & a_4 & a_6 & \cdots \\
s^{n-1} & a_1 & a_3 & a_5 & a_7 & \cdots \\
s^{n-2} & b_0 & b_2 & b_4 & b_6 & \cdots \\
\cdots & c_1 & c_3 & c_5 & & \\
\cdots & d_0 & & & & \\
s^0 & \uparrow & & & &
\end{array}$$

where

$$b_0 = \frac{a_1 a_2 - a_0 a_3}{a_1} \qquad b_2 = \frac{a_1 a_4 - a_0 a_5}{a_1} \qquad b_4 = \frac{a_1 a_6 - a_0 a_7}{a_1}$$

$$c_1 = \frac{b_0 a_3 - a_1 b_2}{b_0} \qquad c_3 = \frac{b_0 a_5 - a_1 b_4}{b_0} \qquad (6\text{-}99)$$

$$d_0 = \frac{c_1 b_2 - b_0 c_3}{c_1} \qquad \text{etc.}$$

To simplify numerical calculations, we can multiply or divide a row by any convenient positive number without altering the results. The test is simply examining the elements of the entire column (b_0, c_1, d_0, etc.) shown by the arrow. *If any of the numbers is either a zero or negative, the system is unstable*, i.e., there are roots either on the $j\omega$ axis and/or in the right half-plane. The $j\omega$-axis crossover frequencies, if any, are determined by solving for the roots of the row polynomial previous to the row which has entirely zero element values. The corresponding polynomial is always an even polynomial.

EXAMPLE

Let us consider the system in (6-74) for which the three poles are identical and normalized to unity. The characteristic equation is given by

$$s^3 + 3s^2 + 3s + (1 + T_0) = 0 \qquad T_0 > 0 \qquad (6\text{-}100)$$

6-6 Stability Considerations in Linear Feedback Systems

We form the array

$$
\begin{array}{c|ccc}
s^3 & 1 & 3 & \leftrightarrow s^3 + 3s \\
s^2 & 3 & 1 + T_0 & \leftrightarrow 3s^2 + (1 + T_0) \\
\hline
s^1 & (8 - T_0)/3 & 0 & \\
s^0 & (1 + T_0) & &
\end{array}
$$

For $T_0 \geq 8$, the element in the row s^1 becomes zero or negative. Hence the amplifier is unstable for $T_0 \geq 8$. This result was seen in the root-locus plot. Note that for $T_0 = 8$, the entire row corresponding to s^1 has zero element values, hence the $j\omega$-axis crossover is determined from the previous row polynomial, namely the polynomial corresponding to s^2 with $T_0 = 8$, i.e., the roots of $3s^2 + 9 = 0$.

EXAMPLE

Let the loop transmission function of a feedback amplifier, which includes the nondominant poles, be as follows:

$$T(s) = \frac{T_0}{(s + 1)(s + 5)(s + 20)(s + 30)}$$

We would like to use Routh's test to determine the minimum value of T_0 for which the system becomes unstable. The characteristic equation is given by

$$s^4 + 56s^3 + 905s^2 + 3850s + (3000 + T_0) = 0$$

We form the array

$$
\begin{array}{c|ccc}
s^4 & 1 & 905 & 3000 + T_0 \\
s^3 & 56 & 3850 & 0 \\
\hline
s^2 & 836 & 3000 + T_0 & \\
s^1 & \dfrac{30.5 \times 10^5 - 56T_0}{836} & &
\end{array}
$$

Setting $30.5 \times 10^5 - 56T_0 = 0$ yields the desired value of T_0, namely, for $T_0 \geq 5.45 \times 10^4$, the system will be unstable. Note that if the nondominant poles are ignored, the possibility of instability would not be apparent.

The Nyquist Criterion and the Bode Plots

The Nyquist criterion is essentially a graphical test based on the steady-state response of the loop transmission function. It is an extremely useful and practical method, since measured data can be used to predict the stability of the amplifier prior to closing the feedback loop. In applying the Nyquist criterion, one therefore need not worry whether the circuit model represents a realistic situation, if measured data are used.

We shall assume in the following that the open-loop function **a**(s) and the feedback factor **f**(s) are stable to begin with. Thus, the closed-loop gain function can have poles in the right half-plane (RHP) only if $1 + T(s)$ has zeros in the RHP. Hence, we have to find whether there is any value of s with a nonnegative real part at which $T(s) = -1$. To determine this situation, we plot a Nyquist diagram, i.e., plot $T(j\omega)$ for $-\infty < \omega < \infty$. Figure 6-31 shows the plots of $T(j\omega) = T_0/(j\omega + 1)^3$.

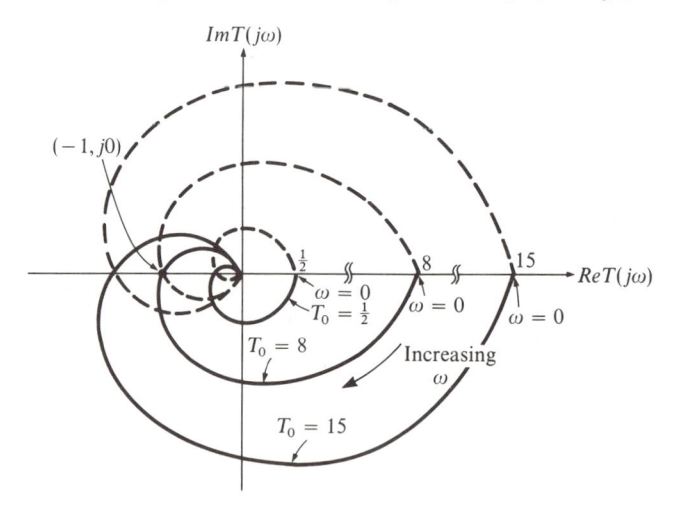

Fig 6-31 Nyquist plot of $T(s) = T_0/(s + 1)^3$

for $T_0 = \frac{1}{2}, 8$, and 15. The dashed curves are mirror images of the solid curves and correspond to $T(j\omega)$ for $-\infty < \omega \leq 0$. It can be shown (Ref 4) that the Nyquist diagram maps the right half s-plane into the interior of the contour in the T-plane. If there are any roots of $1 + T(s)$ in the RHP, the T-plane contour will enclose the point $(-1, j0)$, which is usually referred to as the critical point. Hence, encirclement of the critical point is the criterion of instability. Thus, to use the Nyquist criterion, we plot the imaginary part of $T(j\omega_1)$ along the y-axis and the real part of $T(j\omega_1)$ along the x-axis or, alternatively, we may use a polar plot and locate the complex point by its magnitude

6-6 Stability Considerations in Linear Feedback Systems

and phase for various values of ω_1, where $0 \leq \omega_1 \leq \infty$. We thus determine the curve by locating all the points of $T(j\omega)$, where ω starts from zero and goes to infinity. *The Nyquist criterion states that the amplifier is stable if this curve does not enclose the point $(-1, j0)$, and unstable if the curve encloses this point.*

EXAMPLE

Consider the feedback system with three identical poles where the loop transmission function is given by

$$T(j\omega) = -a(s)f(s)\bigg|_{s=j\omega} = \frac{T_0}{(s+1)^3}\bigg|_{s=j\omega} = \frac{T_0}{(1+j\omega)^3}$$

We next examine the stability of the system by the Nyquist plot for the three cases: $T_0 = \frac{1}{2}$, $T_0 = 8$, and $T_0 = 15$. The plots of Im $T(j\omega)$ vs $Re T(j\omega)$ for $0 \leq \omega \leq \infty$ for the three cases are shown in fig 6-31. The dashed curves correspond to $-\infty \leq \omega \leq 0$.

In a more complicated Nyquist plot, the enclosure of the point $(-1, j0)$ may not be obvious. In such a case, use the following rule: Assume you are walking on the curve in the direction of increasing ω, where $0 \leq \omega \leq \infty$. If, in this walk, the "flag-post" which is located on the point $(-1, j0)$ is not on the path and is always to your left, the system is stable. As an example, consider the Nyquist plot of a feedback system shown in fig 6-32. The system in fig 6-32 is called *conditionally stable*. A conditionally stable system is one in which instability occurs for a decrease of the loop transmission gain.

Two quantities of interest in feedback system design are the *gain margin* and the *phase margin*. The gain margin, G_M, is defined as the value of $[|T(j\omega)|]^{-1}$ in dB at the frequency at which arg $T(j\omega) = 180$ degrees. The radian frequency ω_p at which arg $T(j\omega) = 180$ degrees is called the *phase crossover* frequency. For a stable system $G_M > 0$ dB. The phase margin is defined as 180 degrees plus arg $T(j\omega)$. The radian frequency ω_g at which $|T(j\omega)| = 1$ is called the *gain crossover* frequency. For a minimum phase network (i.e., a network with no zeros in the RHP), which is almost always used in feedback amplifiers for reasons of stability, $\phi_M > 0$ for a stable system. Usually, in a feedback amplifier design, gain and phase margins of at least 10 dB and 60 degrees, respectively, are provided. The gain and phase margins in a Nyquist plot are illustrated in fig 6-33.

Analytically, the gain and phase margins are determined as follows: For example, if $T(s) = 4/(1+s)^3$; from $T(j\omega_g) = 1$, we obtain

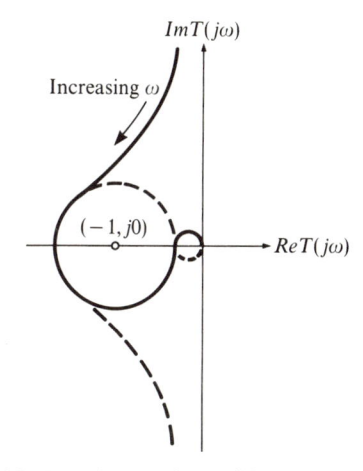

Fig 6-32 Nyquist plot of a conditionally stable system

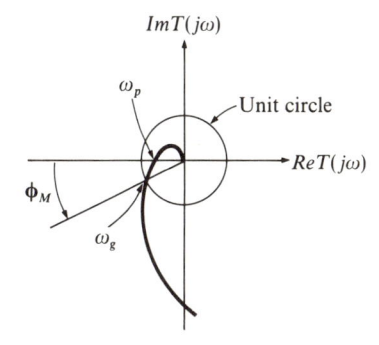

Fig 6-33 Gain and phase margin illustration in a Nyquist plot

$4/|1 + j\omega_g|^3 = 1$, or $\omega_g = 1.23$. Also from $\arg T(j\omega_p) = 180$ degrees, we obtain $3 \tan^{-1} \omega_p = 180$ degrees, or $\omega_p = \sqrt{3}$. Hence

$$G_M = 20 \log \left[\frac{4}{|1 + j\sqrt{3}|^3} \right]^{-1} = 20 \log 2 = 6 \text{ dB}$$

$$\phi_M = 180 \text{ degrees} + \arg T(j\omega_g) = 180 \text{ degrees} - 3(51 \text{ degrees})$$
$$= 27 \text{ degrees}$$

The purpose of using the gain and phase margins is to make allowances for deviation in the system and to obtain a good transient response. For example, in cases where the closed-loop gain function magnitude responses include frequency peaking, the amplifier transient response will not be good (i.e., will exhibit large overshoot) and the amplifier will not be suitable for linear pulse amplification.

The Bode plot (see Section 2-5) can also be used to examine the stability of a system. All the information that can be obtained in the Bode plot is contained in the Nyquist plot. The critical point $(-1, j0)$ corresponds to 0 dB and 180 degrees phase. The gain and phase margins, as illustrated in a Bode plot, are shown in fig 6-34. Finally, the gain and phase margins may also be obtained from a plot of $|T(j\omega)|$ vs $\arg T(j\omega)$.

Since, in the Nyquist diagram and Bode plot, we examine $T(s)$ for $s = j\omega$ only, i.e., steady-state response, transient response performance is completely covered because we have set $\sigma = 0$ at the outset. In a root-locus we examine $T(s)$ for $s = \sigma + j\omega$, hence the transient response information may also be readily obtained from the root-locus plots.

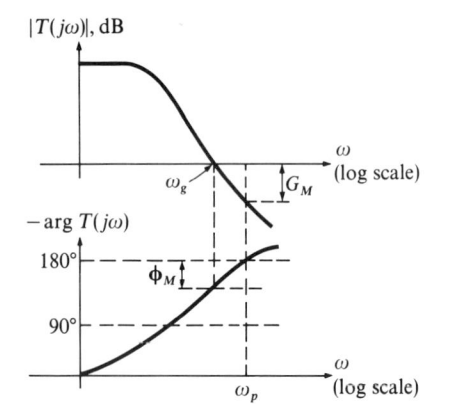

Fig 6-34 Gain and phase margin illustration in a Bode plot

6-7 Operational Amplifiers and Applications

An operational amplifier is a special type of very high-gain amplifier that is very useful in electronic circuits. Operational amplifiers are readily available in integrated-circuit and discrete-circuit forms and are used in analog computers to perform mathematical operations and computations. The operational amplifier is also the work horse in linear integrated circuit signal processing. They are also used as small-signal amplifiers and in the design of active RC filters, as discussed in Chapter 8.

The schematic diagram and the equivalent circuit of an operational amplifier is shown in fig 6-35. The amplifier is a very-high voltage gain $(A_v \geq 10^4)$ direct-coupled ac amplifier in the form of a differential

6-7 Operational Amplifiers and Applications

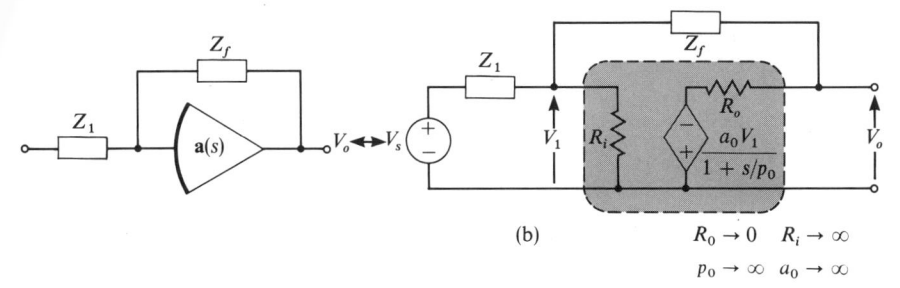

Fig 6-35 (a) A very high-gain amplifying block—basic operational amplifier circuit; (b) Circuit model of (a)

amplifier, with a bandwidth such that the gain magnitude will be unity at approximately 10 MHz, or somewhat higher (depending on the type, e.g., see Appendix D). We represent the amplifier by a single block, though in fact it consists of several amplifying stages. For stability reasons, the amplifier block is intentionally frequency-compensated, either externally or internally, in order to have a dominant one-pole rolloff. In practical amplifiers, R_i is of the order of 500 kΩ and R_o of the order of 50 Ω. For simplicity and purposes of illustration, in the following we let $|s| \ll p_o$, $R_i \to \infty$, and $R_o \to 0$, where p_o is the pole of the operational amplifier. From fig 6-35b, under the said assumption, we obtain

$$\frac{V_o}{V_s} = -\left(\frac{Z_f}{Z_1}\right)\frac{1}{1 + (1/a_o)(1 + Z_f/Z_1)} \qquad (6\text{-}101)$$

For a_o very large, i.e., $a_o \to \infty$, (6-101) reduces to

$$\frac{V_o}{V_s} \approx -\frac{Z_f}{Z_1} \qquad (6\text{-}102)$$

Note that for V_o finite, the condition of $a_o \to \infty$ requires that the V_1 in fig 6-35b be zero. These conditions may be conveniently represented by a conceptual circuit with a "virtual" ground, as shown in fig 6-36. By virtual ground we mean a short circuit through which no current flows. From fig 6-36, since $V_o = -I_1 Z_f$ and $V_s = I_1 Z_1$, one obtains simply and directly (6-102). Equation (6-102) is very important

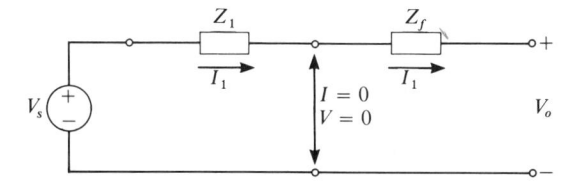

Fig 6-36 Basic operational amplifier circuit with virtual ground

in circuit calculations using operational amplifiers. From (6-102) we can readily deduce the various mathematical operations in which operational amplifiers are employed. These are shown in fig 6-37 and the operations are as follows.

Inverter (or Sign Changer)

If $R_f = R_1$ in fig 6-37a

$$\frac{V_o}{V_s} = -\frac{R_f}{R_1} = -1 \qquad (6\text{-}103)$$

Scale Changer

If $R_f = KR_1$, then in fig 6-37a

$$\frac{V_o}{V_s} = -\frac{R_f}{R_1} = -K \qquad (6\text{-}104)$$

Adder

For fig 6-37b we have

$$V_o = -\left[\frac{R_f}{R_1}V_1 + \frac{R_f}{R_2}V_2 + \cdots + \frac{R_f}{R_n}V_n\right] \qquad (6\text{-}105)$$

Integrator

For fig 6-37c and (6-102) we have

$$V_o = \left(-\frac{1}{R_1 C_f}\right)\left(\frac{1}{s}\right)V_s, \quad \text{or} \quad v_o = -\frac{1}{R_1 C_f}\int v_s\, dt \qquad (6\text{-}106)$$

The initial condition for the capacitor can be supplied by having a battery of Q_o/C_f in shunt with C_f (as shown dotted in fig 6-37c) and opening the switch at $t = t_0$. Because of the virtual ground at the input of the operational amplifier, the voltage across the capacitor of an integrator equals the output voltage.

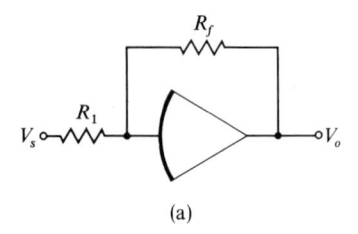

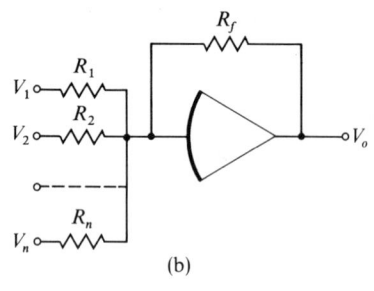

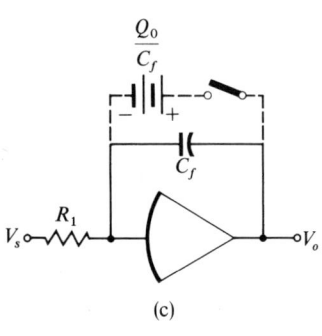

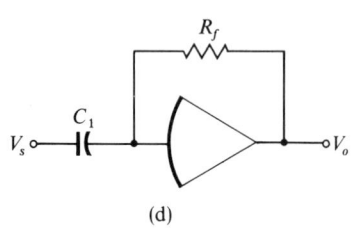

Fig 6-37 Mathematical operations with operational amplifier: (a) Inverter ($R_f = R_1$) and scale changer ($R_f = KR_1$); (b) Adder; (c) Integrator (initial condition simulated by the dotted line); (d) Differentiator

Differentiator

For fig 6-37d and (6-102) we have

$$V_o = -R_f C_1 s V_s \quad \text{or} \quad v_o = -R_f C_1 \frac{dv_s}{dt} \qquad (6\text{-}107)$$

Multiplication

There are several ways of realizing multiplication. One such method is based on the identity

$$xy = \tfrac{1}{4}[(x+y)^2 - (x-y)^2]$$

Thus the multiplication process is changed to the operation of summing and squaring. The squaring operation is performed by diode function generators, which provide a piecewise-linear synthesis of a square-law curve (see Chapter 9). In fact, other nonlinear functions, such as log x, etc., are also obtained by the use of diode function generators. Another approach is to use multipliers directly. Monolithic wideband analog multipliers are commercially available. When used with an operational amplifier, the multiplier may provide multiplication, division, squaring, square-rooting, and other operations (see *Motorola Semiconductors*: linear four-quadrant multiplier integrated circuits MC1595L and MC1495L).

Applications

These simple components can be arranged to solve for any transfer function of the form

$$\frac{V_o(s)}{V_s(s)} = \frac{N(s)}{D(s)} \qquad (6\text{-}108)$$

Or in the time domain we may solve the differential equation of the form

$$\frac{d^n v}{dt^n} + K_{n-1} \frac{d^{n-1} v}{dt^{n-1}} + \cdots + K_1 \frac{dv}{dt} + K_o v$$

$$= C_m \frac{d^m v_s}{dt^m} + C_{m-1} \frac{d^{m-1} v_s}{dt^{m-1}} + \cdots + C_o v_s \qquad (6\text{-}109)$$

with the appropriate initial conditions.

6/Feedback Amplifiers and Oscillators

EXAMPLE

Let us consider programming an analog computer to solve the second-order differential equation

$$\frac{d^2x}{dt^2} + K_1 \frac{dx}{dt} + K_0 x + f(t) = 0 \qquad (6\text{-}110a)$$

We rewrite (6-110a) as follows:

$$\frac{d^2x}{dt^2} = -K_1 \frac{dx}{dt} - K_0 x - f(t) \qquad (6\text{-}110b)$$

From (6-110b) we consider the highest order term, d^2x/dt^2, and imagine it is available. We next start drawing the circuit and use various outputs to generate d^2x/dt^2 as the output, which is the sum of three components in this case. This is shown in fig 6-38. Now one integration by an

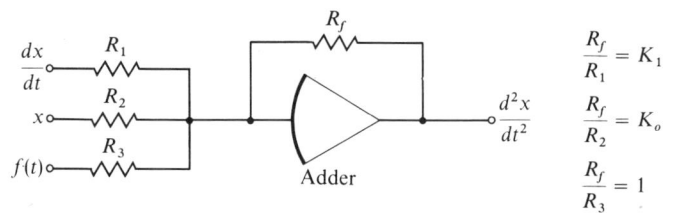

Fig 6-38 Operational amplifier simulation of equation (6-110b)

operational amplifier of d^2x/dt^2 yields $-dx/dt$, another integration yields x. We provide one inverter for the sign change of dx/dt. The overall loop is then as shown in fig 6-39, where the batteries provide

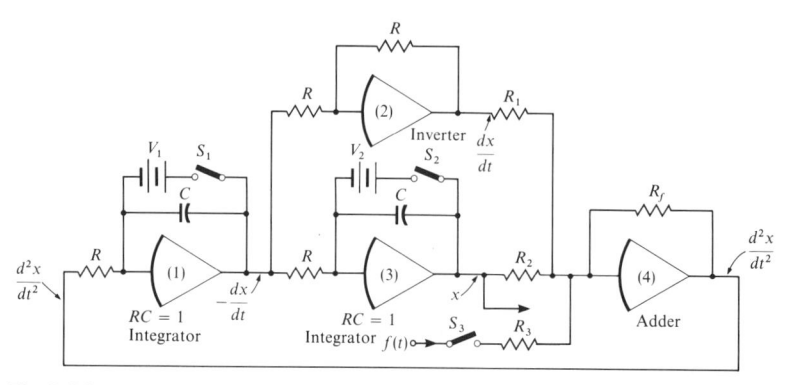

Fig 6-39 Analog computer setup for solving equation (6-110a) including arbitrary initial conditions

the initial conditions. Note that the operational amplifiers (1) and (3) are used as integrators, (2) is used as an inverter, and (4) is used as an adder. The solution is obtained by simultaneously opening switches S_1 and S_2, closing switch S_3, and observing the waveform at terminal x. Note that in this problem we could have used differentiators to effect a solution. For many practical reasons, however, integrators are usually preferable to differentiators in analog computer applications. The preference follows from the following reasons:

(a) It is more convenient to introduce initial conditions in an integrator.

(b) The gain of an integrator decreases with frequency while that of a differentiator increases with frequency. The former is therefore less prone to instability and saturation.

(c) In a differentiator, a sudden change of input waveform may overload and saturate the amplifier.

EXAMPLE

Analog simulation for simultaneous differential equations in the state variable formulation is as follows. Consider for example

$$\frac{dx_1}{dt} = a_{11}x_1 + a_{12}x_2$$

$$\frac{dx_2}{dt} = a_{21}x_1 + a_{22}x_2$$

The simulation is shown schematically in fig 6-40. (For a higher-order system, the extension is obvious.) For a given set of initial conditions and values of a_{ij}, the actual simulation is left as an exercise in Problem 6-31. Note that the integrator represented by $1/s$, for convenience, is identical to the one shown in fig 6-37c.

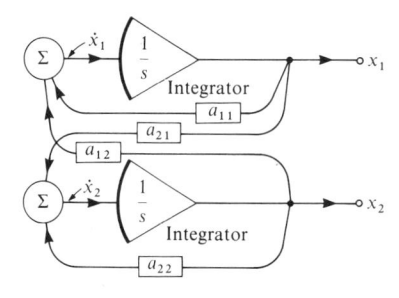

Fig 6-40 Analog computer setup for second-order differential equations in state variable form

Magnitude and Time Scaling in Analog Computers

In the solution of a given problem, the "real" time scale may be either too fast or too slow for the computer, and the magnitude scale may be too large, so that saturation and, hence, nonlinear effects will occur. In order to avoid these two problems, we use magnitude and time scaling in an analog computer, which is akin to the circuit normalization with which we are familiar.

Consider the following differential equation of the problem

$$A\frac{d^2x}{dt^2} + B\frac{dx}{dt} + Cx(t) = f(t) \qquad (6\text{-}111)$$

We wish to use both magnitude and time scaling in order to make (6-111) convenient in terms of the machine time and the linear dynamic range of the amplifiers in the computer. The procedure is quite simple. We let

$$\tau = \alpha t \qquad (6\text{-}112a)$$

and

$$y = \beta x \qquad (6\text{-}112b)$$

where τ is the "machine" time, t is the "real" time, y is the machine "variable" in volts, and x is the real variable of the problem. The constants α and β are time and magnitude coefficients. Note that α is dimensionless, while β may have dimensions. Substitution of (6-112a) and (6-112b) into (6-111) yields

$$\alpha^2 A \frac{d^2y}{d\tau^2} + \alpha B \frac{dy}{d\tau} + Cy = \beta f\left(\frac{\tau}{\alpha}\right) \qquad (6\text{-}113)$$

Note that the effect of magnitude scaling is a change in the magnitude of the forcing function $f(t)$, while the effect of time scaling is a change in the coefficients of the differential equation. By appropriate choices of α and β, we solve (6-113) using analog computers and then obtain the real time and the variables from (6-112a) and (6-112b).

6-8 Practical Considerations in Operational Amplifiers

Operational amplifiers using either differential input, differential output, or both are also available (see Appendix D). These amplifiers are invariably designed in the form of differential amplifiers, in order to achieve low dc drift and direct-coupled ac amplification.

The equivalent circuit of a differential-input, differential-output, operational amplifier is shown in fig 6-41. Usually, however, the differential-input, single-output, operational amplifier is used in practice. The controlled-source model and circuit representation of such an amplifier for the *ideal* case is shown in fig 6-42. (See also Section 8-7.) The output voltage responds only to the voltage differential at the input terminals. The output voltage is zero when the voltage difference between the input terminals is zero. This is referred to as a

6-8 Practical Considerations in Operational Amplifiers

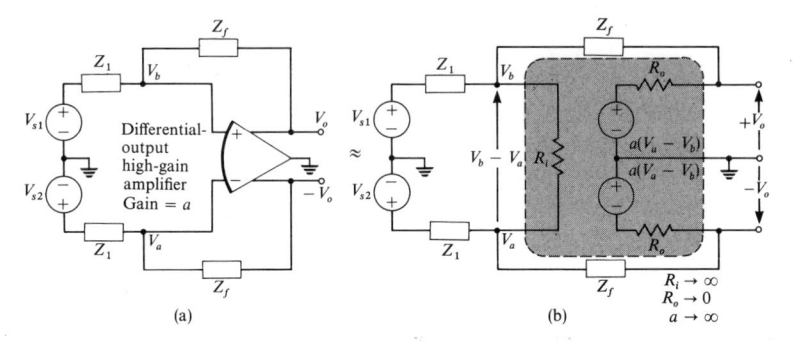

Fig 6-41 (a) Schematic of a differential-input, differential-output operational amplifier; (b) Circuit model for (a)

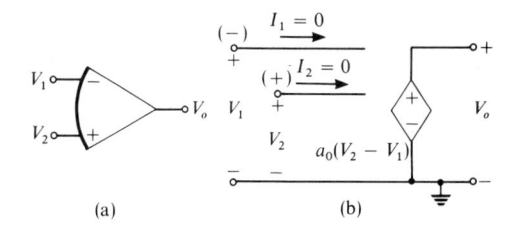

Fig 6-42 (a) Schematic of a differential-input, single-ended output operational amplifier; (b) Circuit model for (a)

zero "offset" property. The open-loop gain of such an ideal operational amplifier is infinity (i.e., $a_0 \to \infty$). Furthermore, $V_o \to 0$ when $(V_2 - V_1) \to 0$.

When a differential-input, single-output, operational amplifier is used to perform the mathematical operations listed above, e.g., differentiation, integration, addition, or inversion, simply ground the positive input terminal of the amplifier. For subtraction and noninverting scaling operation, use both of the input terminals, as shown in fig 6-43.

In an actual operational amplifier there are restrictions imposed on the input and output signal levels. The common mode signal, although negligibly small in a well designed amplifier, is present due to imperfect impedance matching, etc. Finally, the offset voltage is not identically zero. Thus, the fundamental equation of an actual operational amplifier may be written as

$$V_o = a(V_2 - V_1) + \varepsilon(V_1 + V_2) + V_{os} \qquad (6\text{-}114)$$

where a is the open-loop gain and is frequency dependent. The second term in (6-114) is due to the common mode signal, and limits the resolution of a differential amplifier. The third term is the offset voltage, which is usually specified by the manufacturer (see Appendix D). The offset voltage is determined by grounding both inputs. Usually external adjustment is provided to make the offset zero. The frequency dependence of the gain is an important practical consideration. Usually the frequency dependence of the open-loop gain can be approximated by

$$a(s) = \frac{a_o}{(1 + s/p_1)(1 + s/p_2)(1 + s/p_3)}$$

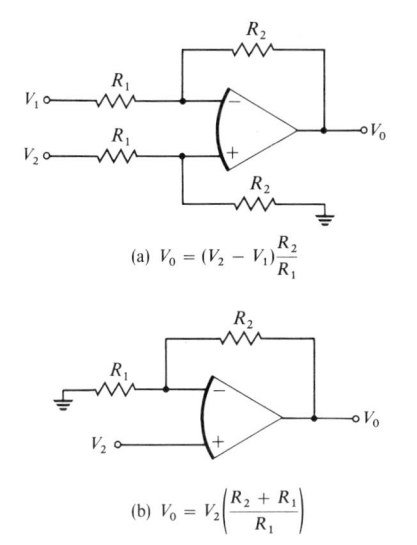

(a) $V_o = (V_2 - V_1)\dfrac{R_2}{R_1}$

(b) $V_o = V_2\left(\dfrac{R_2 + R_1}{R_1}\right)$

Fig 6-43 (a) The output signal when inputs are applied at both terminals; (b) A noninverting gain block (for $R_2 = 0$, it is a unity gain block)

384 **6/Feedback Amplifiers and Oscillators**

Thus, when feedback is used around an operational amplifier that is not frequency compensated, instability is imminent due to the high value of a_o. For stability reasons, external frequency compensation is usually used to make the frequency dependence a one-pole function (i.e., 6 dB/octave rolloff). The frequency compensation is in the form of external lead or lag RC networks. The types of compensation to be used are also suggested by the manufacturers, as is seen in their specification sheets (see Appendix D: μ702C or M1520 operational amplifiers). Internally compensated operational amplifiers are also available commercially.

The frequency-compensated operational amplifier with a one-pole gain function has a limited gain-bandwidth product. The gain-bandwidth product of a typical frequency-compensated operational amplifier (such as listed in Appendix D) is around 10 MHz. Thus, when an operational amplifier is used at very high gain, the pole must be taken into consideration. This can be done simply by using the gain function $a_o/(1 + s/p_o)$ instead of a_o for the compensated operational amplifier.

For low-frequency applications, the operational amplifier provides a very useful functional block in linear circuits. The very high input impedance and very low output impedance of the operational amplifier enables the designer to use these blocks with negligible interaction. Operational amplifiers are extensively used in integrated circuits for active filters. This topic is discussed in Chapter 8.

6-9 Linear Analysis of Sinusoidal Oscillators

A sinusoidal oscillator circuit is a circuit configuration that delivers a sinusoidal output waveform without an externally applied input signal. In other words, such systems supply their input from their own output without external excitation. From the feedback viewpoint, the above condition may be expressed as positive feedback with unity loop transmission, namely

$$T(j\omega) = 1\angle 0° \qquad (6\text{-}115)$$

The condition in (6-115) is also referred to as the Barkhausen condition. In other words, instability is the necessary condition for oscillation. If the system is stable, the steady-state output will be zero in the absence of an input. If the system is unstable, the output would theoretically indefinitely build up exponentially with time, as will be predicted by a linear analysis. In an actual circuit, however, this cannot occur since, due to nonlinearities of the active device, saturation occurs, which limits the amplitude of the output. Thus, a linear analysis is not strictly valid, and is insufficient to characterize completely the operation of an oscillator. Linear analysis may be used in a preliminary design, however, to establish approximately the condition of oscillation and to determine the frequency of oscillation.

The final design is to be refined experimentally and/or by using more accurate analytic techniques. In order to ensure oscillation, we often use $|T(j\omega)|$ slightly larger than unity so that the environmental effect or replacement of the device does not cause $|T(j\omega)|$ to be less than unity. If $|T(j\omega)|$ is much larger than unity, the waveform will be very much distorted. The amplitude of oscillation in all cases will be limited by the nonlinearity of the circuit.

Note that the amplitude of oscillation cannot be predicted by a linear analysis, since the natural frequencies of a linear system do not depend on the input signal.

6-10 RC Oscillators

RC oscillators are commonly used for audio frequencies, i.e., several hertz (Hz) to several kilohertz (kHz). The simplest practical oscillator circuit at low frequencies is the so-called "phase-shift oscillator," shown schematically in fig 6-44a. The amplifier gain A, which has a

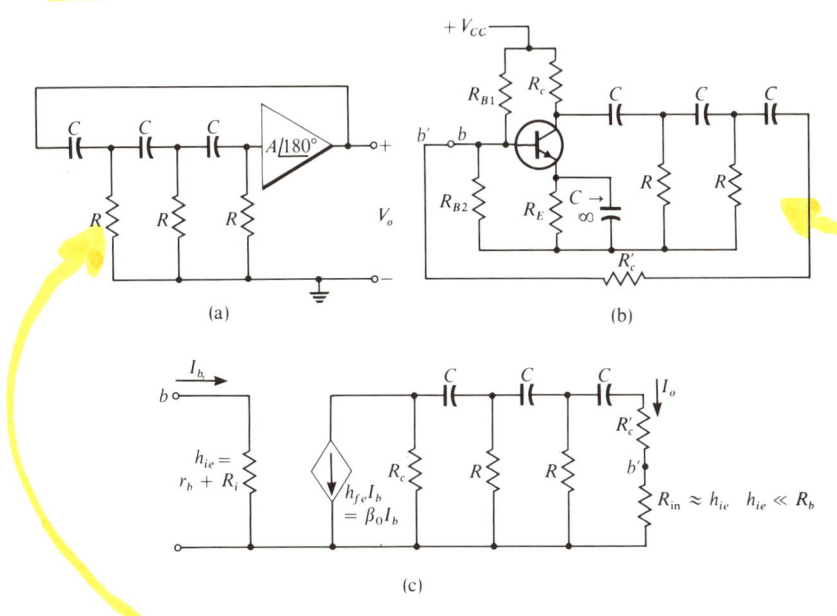

Fig 6-44 (a) General *RC* phase-shift oscillator; (b) Transistor *RC* phase-shift oscillator; (c) Circuit model for (b)

phase shift of 180 degrees, may be realized by a bipolar transistor, FET, or vacuum tube.

The operation of the circuit is roughly as follows: Each RC section provides roughly a 60 degree phase shift, and with the 180 degree phase shift of the amplifier making arg $T(j\omega) = 0$. For example, if a bipolar transistor is utilized, the circuit is as shown in fig 6-44b. The oscillation condition may be determined by using feedback techniques and setting $T(j\omega) = 1$, or by loop or nodal analysis, setting the determinant equal to zero. Since RC oscillators are designed for low frequencies, the device model need only be the low-frequency equivalent circuit. The circuit model of fig 6-44b using the h-parameters is shown in fig 6-44c. Note that in fig 6-44c R_c could include the shunting effect of the large resistor value of h_{oe}^{-1}. We assume that biasing resistors R_E and $R_b = R_{B1} \| R_{B2}$ have negligible effect on the performance of the circuit so that they can be ignored in the following analysis.

The loop transmission function is readily evaluated by opening the loop at point b and injecting a current I_b in the base. In order not to change the loading effect, an impedance equal to the input impedance of the amplifier is placed from b' to ground as in fig 6-44c. The loop transmission function is then given by I_o/I_b. From fig 6-44c, either by nodal analysis or loop analysis, we obtain

$$T(s) = -\frac{I_o}{I_b} = \frac{-h_{fe}}{3 + R/R_c + (1/s)(4/RC + 6/R_cC) + (1/s^2)(1/R^2C^2 + 5/RR_cC^2) + (1/s^3)(1/R^2R_cC^3)}$$

(6-116)

The oscillation condition is $T(j\omega) = 1\angle 0°$. Thus, from (6-116), the imaginary part of the denominator must be zero:

$$-j\left[\frac{1}{\omega}\left(\frac{4}{RC} + \frac{6}{R_cC}\right) - \frac{1}{\omega^3}\left(\frac{1}{R^2R_cC^3}\right)\right] = 0 \qquad (6\text{-}117a)$$

$$\frac{1}{\omega^2 RR_cC^2} = 4 + 6\left(\frac{R}{R_c}\right) \qquad (6\text{-}117b)$$

Similarly, setting $|T(j\omega)| = 1$ and using (6-117a) yields

$$\frac{-h_{fe}}{3 + R/R_c - (1/\omega^2)(1/R^2C^2 + 5/RR_cC^2)} = 1 \qquad (6\text{-}118)$$

Substituting (6-117b) into (6-118) and simplifying we get

$$h_{fe} = 23 + \frac{29}{(R_c/R)} + 4\left(\frac{R_c}{R}\right) \qquad (6\text{-}119a)$$

6-10 RC Oscillators

The expression on the right of (6-119) is minimum when $R_c/R = 2.7$. Since $|T(j\omega)| \geq 1$, we must have

$$h_{fe} \geq 44.5 \qquad (6\text{-}119b)$$

Thus, if the transistor h_{fe}, i.e., β_0, is less than 44.5, the circuit will not oscillate. If (6-119) is satisfied, the circuit will oscillate and the frequency of oscillation will be given by (6-117), namely

$$f = \frac{1}{2\pi RC} \frac{1}{\sqrt{6 + 4R_c/R}} \qquad (6\text{-}120)$$

In practice, usually transistors with a β_0 much larger than 44.5 are used and a variable resistor is used for R_c. The variable resistor is adjusted such that $|T(j\omega)|$ is slightly larger than unity in order to eliminate waveform distortion.

Another RC oscillator circuit is shown in fig 6-45a. The RC circuit is used to provide attenuation as well as phase adjustment. The transistorized version of fig 6-45a is shown in fig 6-45b. Note that the

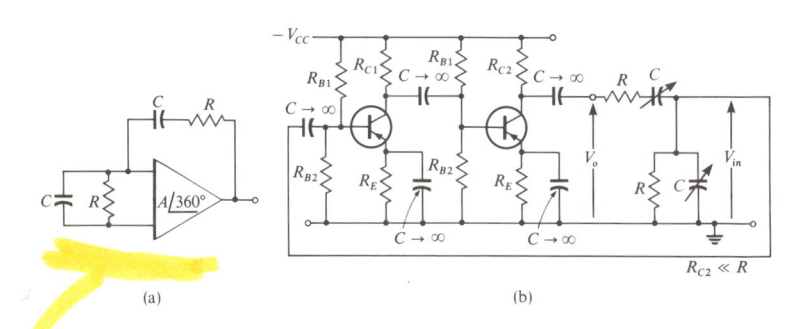

Fig 6-45 (a) General Wien-bridge oscillator circuit; (b) Transistor Wien-bridge oscillator

RC circuit is in a bridge form, and is usually referred to as the *Wien-bridge oscillator*. The signal V_o will be in phase with V_{in} when $X_c = R$. Usually potentiometers are used for R_E in order to vary the amplifier gain to the appropriate value for oscillation and to eliminate distortion. Detailed analysis of this circuit is left as an exercise in Problem 6-35.

6-11 LC Oscillators

For radio-frequency oscillators, i.e., in the range beyond 100 kHz up through the megahertz (MHz) range, LC oscillators are usually used because of the relatively smaller sizes of the reactive elements. There are a number of such circuits; some of the classical ones are shown in fig 6-46. The inductance labeled RF choke is a large inductance having a very high impedance at the frequency of oscillation. The labeled circuit elements are the principal frequency-controlling elements. The biasing circuit-element values are such that they have negligible effect on the operation of the circuit.

The condition of oscillation and the computation of the frequency of oscillation can be made from the feedback viewpoint using the interconnected two-port network theory approach, or by setting the circuit determinant, from loop or nodal analysis, equal to zero. We shall illustrate in the following the analysis of the Colpitts circuit. The analysis of the Hartley oscillator and the tuned-collector oscillator is left as an exercise for the reader.

Since the operating frequency of LC oscillators is usually in the megahertz region, the reactive elements of the circuit models for the active device, in general, cannot be ignored. The small-signal model of the Colpitts oscillator is shown in fig 6-47. For simplicity, we have ignored r_b. Note that $C_i \approx (1 + R_1 C_c \omega_T)/r_e \omega_T$.

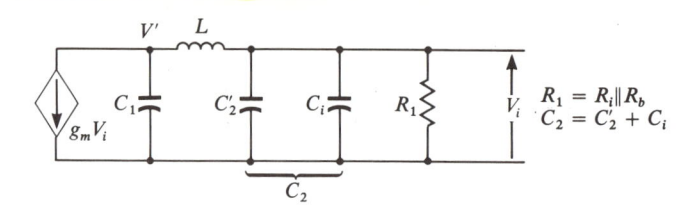

Fig 6-47 Circuit model for Colpitts oscillator

We can now determine the return ratio function and set it equal to unity as in the previous section (see Problem 6-38). For the sake of illustration, however, we shall use an alternative method. We write the nodal equation for the circuit and set the determinant equal to zero. The nodal equations rearranged are (note that $C_2 = C_2' + C_i$)

$$V_i\left(g_m - \frac{1}{sL}\right) + V'\left(\frac{1}{sL} + sC_1\right) = 0$$

$$V_i\left(sC_2 + G_1 + \frac{1}{sL}\right) + V'\left(-\frac{1}{sL}\right) = 0 \quad (6\text{-}121)$$

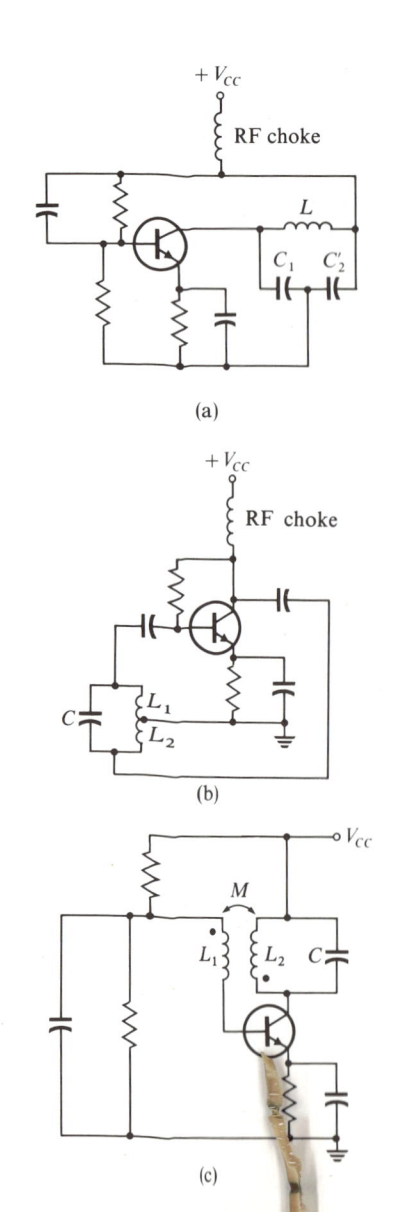

Fig 6-46 (a) Colpitts oscillator; (b) Hartley oscillator; (c) Tuned oscillator

We set the determinant of (6-121) equal to zero,

$$\left(g_m - \frac{1}{sL}\right)\left(-\frac{1}{sL}\right) - \left(\frac{1}{sL} + sC_1\right)\left(sC_2 + G_1 + \frac{1}{sL}\right) = 0 \quad (6\text{-}122)$$

Rearranging (6-122) and substituting for $s = j\omega$ we get

$$\omega^4(L^2C_1C_2) - \omega^2 L(C_1 + C_2) + j[\omega L(G_1 + g_m) - \omega^3 L^2 C_1 G_1] = 0$$
$$(6\text{-}123)$$

For the determinant, Δ, to be zero, both real ($Re\Delta$) and imaginary ($Im\Delta$) parts must be zero, hence

$$Re\Delta = 0 \Rightarrow \omega^2 = \frac{C_1 + C_2}{LC_1C_2} \quad (6\text{-}124a)$$

$$Im\Delta = 0 \Rightarrow \omega^2 LC_1 G_1 = G_1 + g_m \quad (6\text{-}124b)$$

Dividing (6-124a) by (6-124b) yields

$$1 + g_m R_1 = 1 + \frac{C_1}{C_2} \quad (6\text{-}125)$$

If $R_b \gg R_i$ then $R_1 \approx R_i = \beta_0 r_e = \beta_0/g_m$. Hence, for oscillation to take place, from (6-125)

$$\beta_0 \approx \frac{C_1}{C_2} \quad (6\text{-}126)$$

Note that β_0 slightly larger than C_1/C_2 is to be used, since the losses in the reactive elements and r_b are ignored. The frequency of oscillation from (6-124a) is

$$f = \frac{1}{2\pi}\sqrt{\frac{C_1 + C_2}{LC_1C_2}} \quad (6\text{-}127)$$

6-12 Crystal Oscillators

For very good frequency stability, crystals are often used as the sole controlling element for the frequency of oscillation. The symbol and the circuit model of a vibrating piezoelectric crystal are shown in figs 6-48a and 6-48b, respectively. The Q of a crystal can be as high as several hundred thousand. Typical values for a 10-MHz quartz crystal are: $Q = 1.5 \times 10^5$, $C_0/C = 300$, $L = 12$ mH, $C_0 = 10^{-11}$ F, and $R = 5\,\Omega$. In different cut quartz crystals usually $C_0 \approx 10^{-11}$ F, $C = 10^{-13}$ F to 10^{-14} F, L is determined from $L = 1/C\omega_0^2$ where ω_0 is the resonant frequency of the crystal, R is determined from $R = \omega_0 L/Q$, and Q is in the range of 10^4 to 5×10^5 (see table 7-1). A variety of crystal oscillator circuits is possible. The Colpitts-derived crystal oscillator is shown in fig 6-49.

Finally, it is to be noted that a negative resistance, such as that exhibited by the tunnel diode, can also be used to construct a high frequency oscillator. The analysis of one such circuit is given as an exercise in Problem 6-43. The topic of nonlinear oscillators is discussed in Chapter 9.

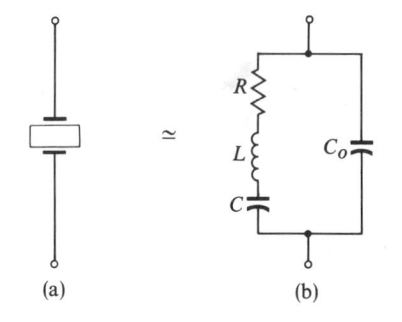

Fig 6-48 (a) Graphic symbol for a piezoelectric crystal; (b) Equivalent circuit of (a)

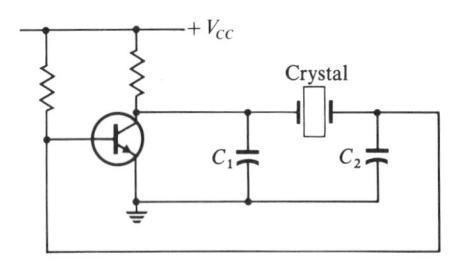

Fig 6-49 Colpitts crystal oscillator

REFERENCES AND SUGGESTED READING

1. Thornton, R. D., et al., *Multistage Transistor Circuits*, New York: Wiley, 1965, Volume 5, Chapters 3 and 4.
2. Pederson, D. O., *Electronic Circuits*, New York: McGraw-Hill, 1965 (Prelim. ed.) Chapters 10 and 11.
3. Ghausi, M. S., *Principles and Design of Linear Active Circuits*, New York: McGraw-Hill, 1965, Chapter 16.
4. Kuo, B., *Automatic Control Systems*, Englewood Cliffs: Prentice-Hall, 1962, Chapter 8 (for root-locus techniques and stability tests).
5. Comer, D. J., *Introduction to Semiconductor Design*, Reading, Mass: Addison-Wesley, 1968, Chapter 8.
6. Glasford, G. M., *Linear Analysis of Electronic Circuits*, Reading, Mass: Addison-Wesley, 1965, Chapter 5.

PROBLEMS

6-1 For the single-stage, shunt-feedback amplifier shown in fig 6-4a, the transistor parameters are as in (6-36). Design the circuit (i.e., find R_f) for a midband current gain of 10. What is the 3-dB bandwidth of the amplifier?

6-2 In the single-stage, shunt-feedback amplifier shown in fig 6-4a, assume that β_0 can vary by ± 50 per cent from its nominal value in (6-36). Design the circuit such that the effect of this variation in the midband gain is within 5 per cent. What is the midband gain of the circuit?

Problems

6-3 For the single-stage, series-feedback amplifier shown in fig 6-7, assume that the transistor parameters are as in (6-36). If $R_s = 100\,\Omega$, and $R_L = 1\,\text{k}\Omega$, determine the value of R_f such that the midband voltage gain of the amplifier is $A_v = -10$. What is the 3-dB bandwidth of the amplifier in this case?

6-4 A single-stage FET amplifier is shown in fig P6-4. Determine the value of R_f for a midband voltage gain of 10.

6-5 For the shunt-shunt, triple-feedback amplifier shown in fig P6-5, show the $\mathbf{a}(s)$ and $\mathbf{f}(s)$ circuits if $y_{11}^f \ll y_{11}^a$ and $y_{22}^f \ll G_L$. Write the expressions for the midband $\mathbf{a}_i$, $\mathbf{f}$, and A_I.

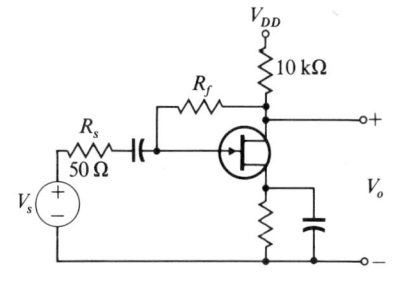

Fig P6-4

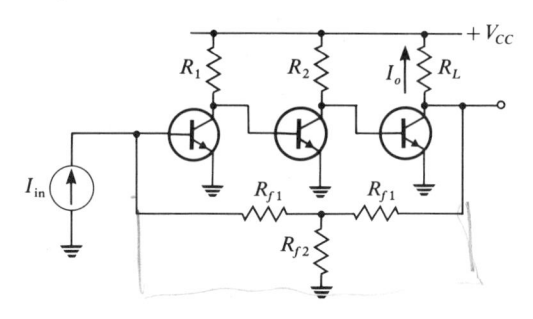

Fig P6-5

6-6 In fig P6-5 the circuit element values are as follows: $R_1 = R_2 = R_L = 50\,\Omega$, $R_{f1} = 100\,\text{k}\Omega$, and $R_{f2} = \infty$. The transistor parameters are as in (6-21). Determine the dominant pole-zero location of the closed-loop amplifier. If the circuit is stable (for simplicity, use the unilateral model), what is the gain and the approximate 3-dB bandwidth of the amplifier?

6-7 For the series-series, triple-feedback amplifier shown in fig P6-7 show the $\mathbf{a}_v(s)$ and $\mathbf{f}(s)$ circuits. Determine the expression for the midband $\mathbf{a}_v$, $\mathbf{f}$ and A_V.

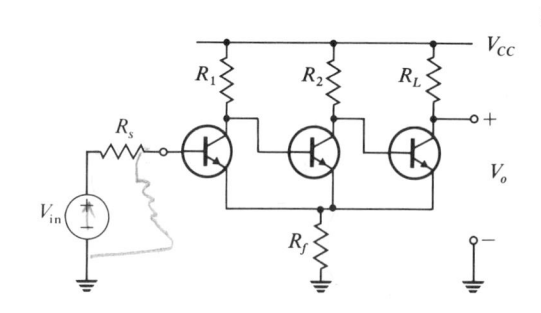

Fig P6-7

6-8 In fig P6-7 the circuit element values are as follows: $R_1 = R_2 = R_s = 50\,\Omega$, $R_f = 100\,\Omega$, $R_L = 5\,\text{k}\Omega$. The transistor parameters are as in (6-21). Determine the dominant pole-zero location of the closed-loop amplifier. If the circuit is stable (use the unilateral model for simplicity), what is the gain and the approximate 3-dB bandwidth of the amplifier?

6-9 For the shunt-series example of the text (fig 6-12), consider the following:
(a) A shunt capacitor C_f is used across R_f. Determine the value of C_f such that the magnitude response is approximately maximally flat. What is the 3-dB bandwidth for this case?
(b) If $C_f = 0$ and R_f is varied, determine the value of R_f for a maximally flat magnitude response. What is the 3-dB bandwidth for this case?
(c) Compare the above two cases from the gain-sensitivity viewpoint.

392 6/Feedback Amplifiers and Oscillators

6-10 This problem illustrates the effectiveness of overall feedback as compared to the local feedback case in regards to sensitivity. Assume no interaction between the blocks in fig P6-10 and use differential sensitivity. Show that for an equal overall gain in (a) and (b):

$$S_a^{A_2} = \left(\frac{1}{1 - af_1}\right) S_a^{A_1}$$

Comment on the above result.

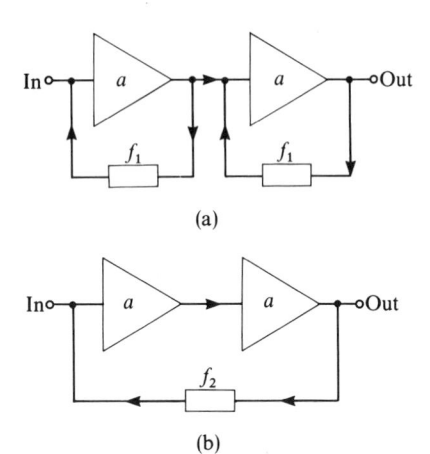

Fig P6-10

6-11 The low-frequency loop transmission function of an ac coupled feedback amplifier is given by

$$T(s) = \frac{T_0 s^3 (s + 5)}{(s + 10)(s + 15)(s + 20)(s + 2)}$$

Draw the root-locus and comment on the stability of the system.

6-12 For a feedback system, the following are known:

$$\mathbf{a}(s) = \frac{5 \times 10^4}{(s/10^7 + 1)(s/10^7 + 5)}$$

$$\mathbf{f}(s) = -f_o$$

Determine the values of f_o for each of the following cases:
(a) For a phase margin of 45 degrees.
(b) For $A(s)$ to have a monotonically decreasing magnitude response with maximum bandwidth. What can be said about the gain and phase margins?

6-13 A pentode feedback amplifier is shown in fig P6-13. Determine the expressions for $\mathbf{a}_v(s)$, $\mathbf{f}(s)$, and $A_V(s)$. For what value of f_0 will the circuit enter the instability region? If R_f is chosen such that $f_0 = 1/20$ of this value, determine the midband voltage gain and the 3-dB bandwidth.

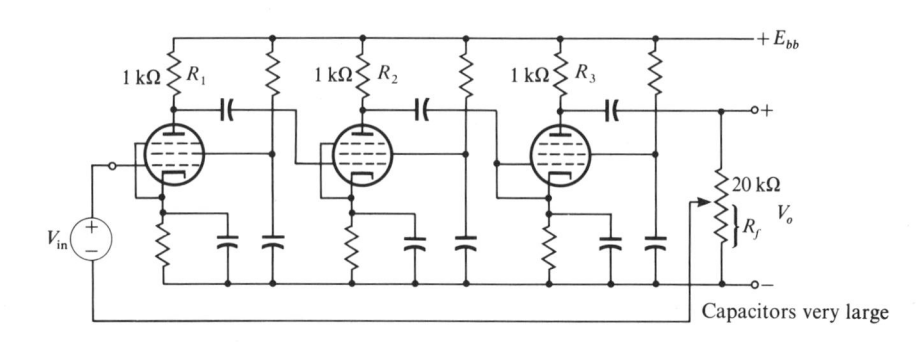

Fig P6-13

6-14 For the FET feedback amplifier shown in fig P6-14, determine $\mathbf{a}(s)$, $\mathbf{f}(s)$, $A_V(s)$, and determine *roughly* whether the circuit is stable (i.e., use the unilateral model). If it is stable, what is the overall voltage gain and bandwidth of the circuit?

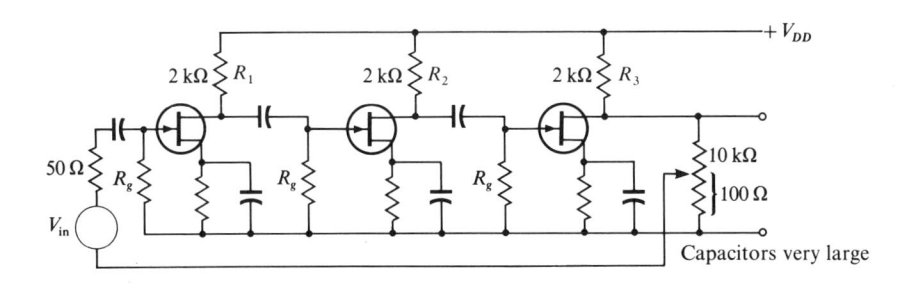

Fig P6-14

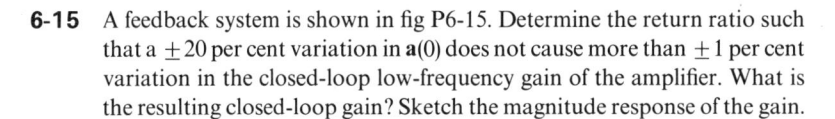

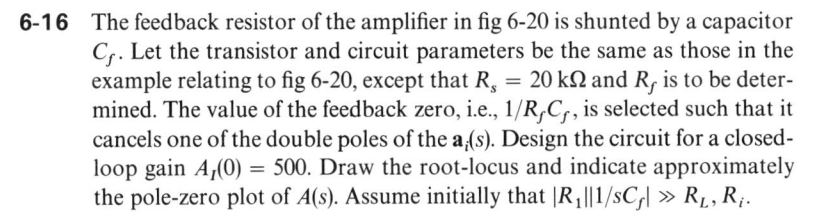

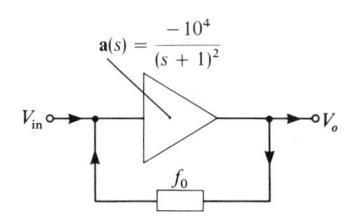

Fig P6-15

$$a(s) = \frac{-10^4}{(s+1)^2}$$

6-15 A feedback system is shown in fig P6-15. Determine the return ratio such that a ± 20 per cent variation in $\mathbf{a}(0)$ does not cause more than ± 1 per cent variation in the closed-loop low-frequency gain of the amplifier. What is the resulting closed-loop gain? Sketch the magnitude response of the gain.

6-16 The feedback resistor of the amplifier in fig 6-20 is shunted by a capacitor C_f. Let the transistor and circuit parameters be the same as those in the example relating to fig 6-20, except that $R_s = 20\ \text{k}\Omega$ and R_f is to be determined. The value of the feedback zero, i.e., $1/R_f C_f$, is selected such that it cancels one of the double poles of the $\mathbf{a}_i(s)$. Design the circuit for a closed-loop gain $A_I(0) = 500$. Draw the root-locus and indicate approximately the pole-zero plot of $A(s)$. Assume initially that $|R_1\|1/sC_f| \gg R_L, R_i$.

6-17 Sketch the root-locus for the following loop transmission functions and show the values of the key points:

(a) $$T(s) = \frac{T_0 s^2}{(s+1)(s+2)}$$

(b) $$T(s) = \frac{6T_0(s^2+1)}{(s+1)(s+2)(s+3)}$$

6-18 Repeat Problem 6-17 for the following cases:

(a) $$T(s) = \frac{T_0(s+1)(s+3)}{(s^2+2s+2)(s+5)}$$

(b) $$T(s) = \frac{T_0(s+1)}{(s+2)^2(s+3)^2}$$

6/Feedback Amplifiers and Oscillators

6-19 Determine the stability of the following characteristic equations:
(a) $s^4 + 4s^3 + 6s^2 + 4s + K = 0$
(b) $s^3 + a_2 s^2 + a_1 s + a_0 = 0$

6-20 Estimate the roots of the following polynomial by the root-locus technique
$$s^4 + 10s^3 + 45s^2 + 105s + 105 = 0$$
(Hint: The intersection of any two root-loci as obtained from the above locates the roots.)

6-21 (a) For the feedback system shown in fig P6-21a, the following are given: $a_0 = -200$, $f_1 = 0.1$. The closed loop gain function $A \equiv V_o/V_i$.

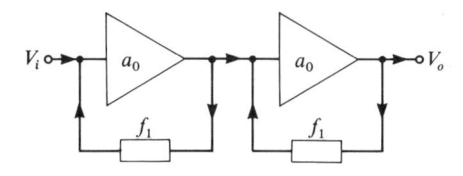

Fig P6-21a

Find the value of A and dA/A for $da_0/a_0 = 0.05$.
(b) The feedback system shown in fig P6-21b is used with the following data given: $a_0 = -200$, $A = -100$. Find the values of f_2 and dA/A for $da_0/a_0 = 0.05$:

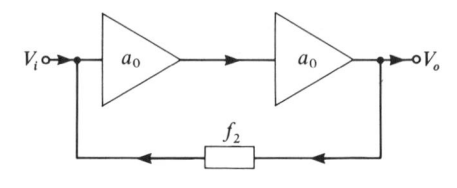

Fig P6-21b

(c) If $a(s) = -200/(1 + s/10^6)$ in each case above, determine the corresponding closed-loop pole-zero locations.

6-22 In a certain feedback system, the loop transmission function is given by
$$T(s) = \frac{K}{(s+1)^2(s+3)^2}$$

(a) Sketch the root-locus showing clearly the asymptotes and the departure angles.
(b) For what value of K will the system be unstable?
(c) For K equal to one-half of the value in (b), what are the gain and phase margins?

6-23 In a certain feedback amplifier, the following data are given:

$$a(s) = 5 \times 10^{14} \frac{s^2}{(s+1)(s+2)(s+10^6)(s+2 \times 10^6)}$$

$$f(s) = -0.002$$

Find the midband closed-loop gain and the upper and lower 3-dB cutoff frequencies of the closed-loop gain function.

6-24 For the four-stage, shunt-series feedback circuit shown in fig P6-24, exclusive of biasing and coupling circuitry, determine the following:
(a) $a_i(0)$, $f(0)$, $A_I(0)$. Let $\beta_o = 50$, $r_e = 10\,\Omega$, and $r_b = 0$.
(b) Find the approximate midband input impedance.

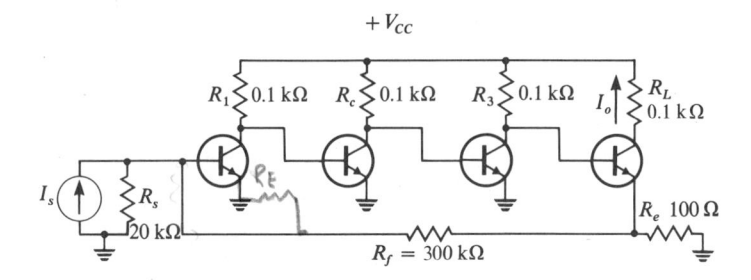

Fig P6-24

6-25 For the four-stage, series-shunt feedback circuit shown in fig P6-25, exclusive of biasing and coupling circuitry, determine the following:
(a) $a_v(0)$, $f(0)$, $A_V(0)$.
(b) Find the approximate midband input impedance.

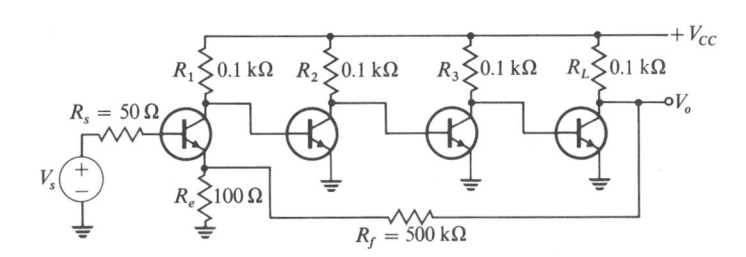

Fig P6-25

The transistor parameters are as given in Problem 6-24.

6/Feedback Amplifiers and Oscillators

6-26 Show that the series-shunt feedback circuit shown in fig 6-17 realizes approximately a voltage-controlled voltage source (VCVS) element. (Hint: Determine the midband g-parameters.) Assume $r_b = 0$, $\alpha_0 \to 1$, $r_c \to \infty$.

6-27 Show that the shunt-series feedback pair shown in fig 6-12 realizes approximately a current-controlled current source (CCCS) element. (Hint: Determine the midband h-parameters.) Assume $r_b = 0$, $\alpha_0 \to 1$, $r_c \to \infty$.

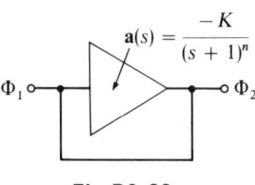

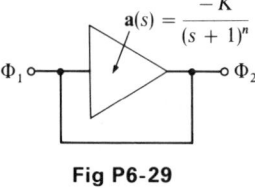

6-28 Analyze the circuit of fig 6-12, but use only a resistive load for the modeling of the first stage. Compare your results with those in the text.

6-29 For the feedback system shown in fig P6-29, determine the maximum value of K as a function of n for which the system is stable.

Fig P6-29: $a(s) = \dfrac{-K}{(s+1)^n}$

6-30 (a) The following measured data were obtained from a single loop negative feedback amplifier:

Frequency (MHz)	0.01	0.1	1
$\|a\|/\arg a(j\omega)$:	$1000/{-5°}$	$800/{-20°}$	$200/{-80°}$
	5	10	50
	$90/{-120°}$	$40/{-200°}$	$1/{-260°}$

What is the value of the resistive feedback, f_0, that can be applied around this amplifier and provide a gain margin of 10 dB? What is the corresponding phase margin? What is the corresponding desensitivity? For what values of f_0 will the feedback amplifier become unstable?

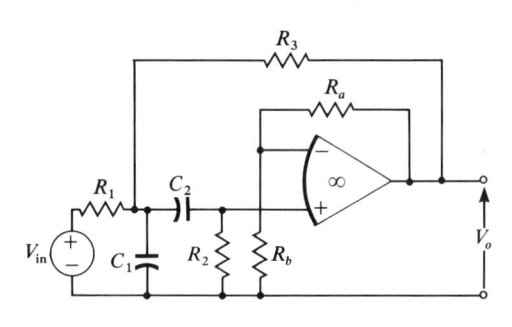

6-31 Show the analog computer setup for the solution of the following system of differential equations and show the values of the circuit elements.

$$\dot{x}_1 = -5x_1 + x_2, \qquad \dot{x}_2 = 3x_1 - 3x_2$$

with

$$x_1(0) = 4, \qquad x_2(0) = 8$$

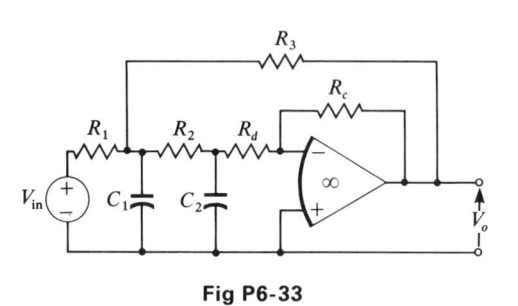

6-32 For the circuit shown in fig P6-32, determine the voltage gain function $A_v(s)$. If $R_1 C_1 = R_2 C_2 = 10^{-5}$ sec, $R_3 = 10$ kΩ. $R_1 = R_2 = 1$ kΩ. Draw the root-locus of the poles of V_o/V_{in} as the ratio (R_a/R_b) is varied from 0 to 10^4. Can this circuit become unstable in this range?

6-33 For the circuit shown in fig P6-33, determine the expression for the voltage-gain function $A_v(s)$. Design the circuit for a voltage gain of 10 and a two-pole Butterworth response with $\omega_{3\,dB} = 10^4$ rad/sec. Let $R_1 = R_2, R_3 = R_d = 10$ kΩ, and $R_c = 900$ kΩ.

Fig P6-32

Fig P6-33

6-34 A differential-input, differential-ouput operational amplifier has external resistors connected as shown. You may assume $R_{in} = \infty$, $R_0 = 0$ for the operational amplifier.

(a) Show that

$$V_o = \frac{[1 + (R_d/R_c)]V_2 - (R_d/R_c)[(1 + R_b/R_a)]V_1}{1 + 2(R_b/R_a) + R_bR_d/R_aR_c + (1/a_0)(1 + R_b/R_a)(1 + R_d/R_c)}$$

(b) If $a_0 \to \infty$, $R_d/R_c = R_b/R_a = K_0$, what are the expressions for K_1 and K_2.

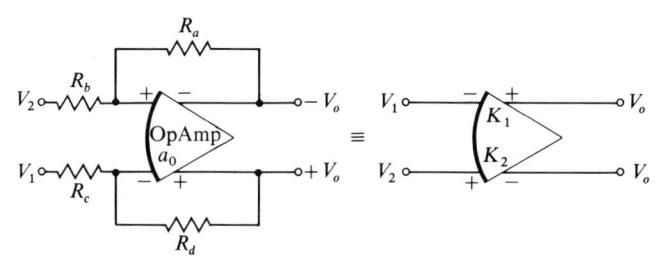

Fig P6-34

6-35 Determine the oscillation condition and the frequency of oscillation for the Wien-bridge oscillator in fig 6-45b.

6-36 For the FET circuit shown, exclusive of biasing circuitry, in fig P6-36 Z_1, Z_2, and Z_3 are purely reactive elements. Set up the expression for the condition of oscillation. What type of reactive elements, i.e., L and/or C, would result in an oscillator? Use the low-frequency model for the FET.

6-37 The circuit model of a tuned-collector oscillator (for $f < f_\beta$) is shown in fig P6-37. Derive the expression for the oscillation condition and the equation which determines the frequency of oscillation.

6-38 For the Colpitts oscillator in fig 6-47, use the method of setting $|T(j\omega)| = 1\angle 0°$ to derive (6-126) and (6-127).

6-39 The circuit model for a Hartley oscillator (for $f < f_\beta$) is shown in fig P6-39, where r_b is neglected. Derive the expression for the oscillation condition and the equation which determines the frequency of oscillation.

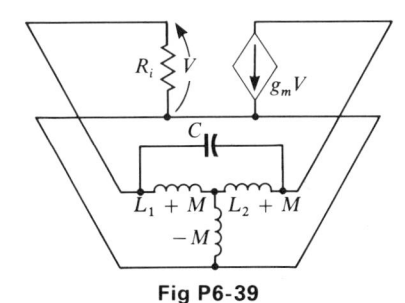

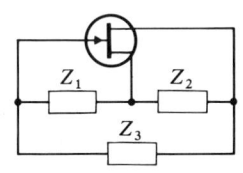

Fig P6-36

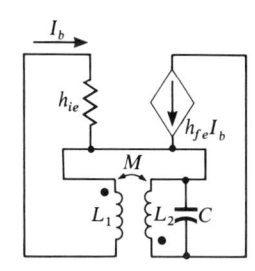

Fig P6-37

Fig P6-39

6/Feedback Amplifiers and Oscillators

6-40 For the MOSFET oscillator shown in fig P6-40, determine the condition for oscillation and the frequency of oscillation. Use the low-frequency model for the FET.

6-41 The closed-loop gain function of a feedback system is given by

$$A(s) = \frac{-a_0}{s^3 + 3s^2 + 3s + 1 + a_0 f_0} \qquad a_0 f_0 > 0$$

(a) Show that for $a_0 f_0 = \frac{1}{2}$, the magnitude response is flat.
(b) Show that the 3-dB bandwidth (for $a_0 f_0 = \frac{1}{2}$) is equal to 0.88.

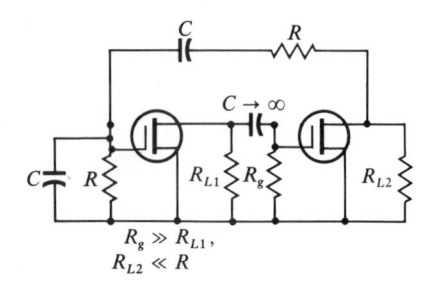

Fig P6-40

$R_g \gg R_{L1}$,
$R_{L2} \ll R$

6-42 The loop transmission of a certain feedback amplifier is given by

$$T(s) = -K \frac{s^2(s + 2)}{(s + 1)^2(s + 5)(s + 10^6)^2}$$

Note that the low and high frequencies are included.
(a) Sketch the root-locus.
(b) For what value of K will the circuit become unstable?
(c) If the low-frequency pole at 5 can be varied without affecting the other parameters in $T(s)$, what is the smallest value of this pole so that the system is stable and the high-frequency closed-loop response is a Butterworth two-pole.

6-43 For the circuit shown in fig P6-43, R_n is a negative resistance. Determine the expressions for the oscillation condition and the frequency of oscillation for the two forms of Z_1.

(a)   (b)

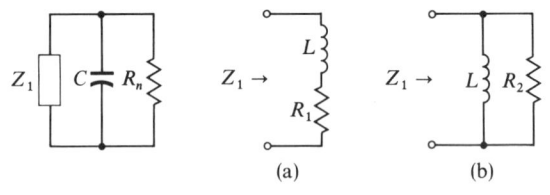

Fig P6-43

*6-44 Utilize ECAP to analyze the shunt-series feedback pair in fig 6-12. Use the hybrid π model for the transistor and the numerical values of the text example.

*6-45 Analyze the series-shunt feedback pair of fig 6-17 by ECAP. Use the hybrid π model and the numerical values of the text example.

*6-46 Analyze the shunt-shunt feedback triple of fig 6-20 by ECAP. Use the hybrid π model and the numerical values of the text example.

---

* The problems marked by an asterisk are to be assigned if ECAP program is available in the Computation Center. These may be replaced by CORNAP or any other available packaged program.

Design Examples of Bandpass Amplifiers

7-1 Introduction

In this chapter we consider the analysis and design of bandpass amplifiers. Bandpass amplifiers are also commonly referred to as tuned amplifiers. These amplifiers find many applications in telecommunications, such as in radio, radar, and television receivers, to name a few. To illustrate the requirements on the bandpass amplifier, consider the block diagram of a simple communication system, as shown in fig 7-1. The input signal is a message wave $v_m(t)$ which can be expressed in terms of its Fourier series or Fourier integral and is assumed to consist of a

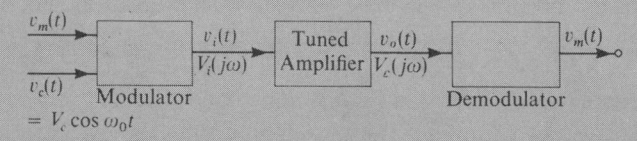

Fig 7-1 Simple amplitude-modulated system

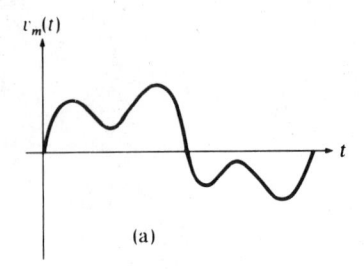

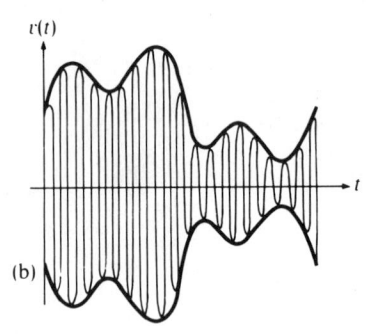

Fig 7-2 (a) Message signal; (b) Amplitude-modulated signal

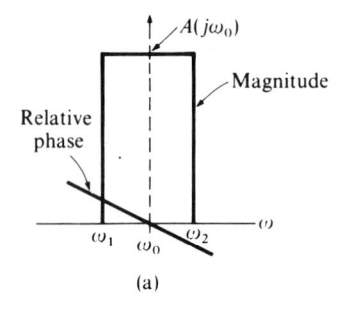

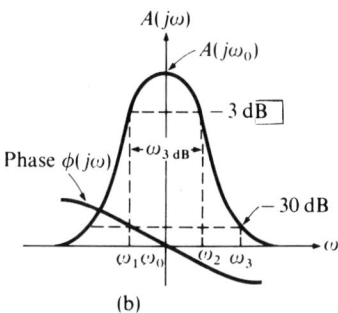

Fig 7-3 (a) Ideal bandpass characteristics; (b) Actual bandpass characteristics

band of low frequencies in the vicinity of dc frequency. The modulator translates this band of frequencies to a new band of frequencies centered around the carrier frequency ω_c without disturbing the relative position of various frequency components of the signal. For example, for an amplitude-modulated (AM) signal

$$v(t) = [V_c + v_m(t)] \cos \omega_c t \tag{7-1}$$

where V_c is the amplitude of the carrier signal. The magnitude of $v_m(t)$ is always adjusted to be less than V_c, hence the envelope of the modulated signal never drops to zero. The message signal and the modulated waveform are shown in fig 7-2. To illustrate, consider a special case of a single frequency, i.e., $v_m(t) = V_m \cos(\omega_m t + \theta_m)$. In this case, from (7-1)

$$\begin{aligned} v(t) &= [V_c + V_m \cos(\omega_m t + \theta_m)] \cos \omega_c t \\ &= V_c[1 + m \cos(\omega_m t + \theta_m)] \cos \omega_c t \end{aligned} \tag{7-2}$$

where $m = V_m/V_c$ is called the modulation index and is always less than unity, and $\omega_c \gg \omega_m$. We may rewrite (7-2), using the trigonometric identity of the product of two cosines, as

$$\begin{aligned} v(t) = V_c \cos \omega_c t &+ \frac{m}{2} V_c \cos[(\omega_c + \omega_m)t + \theta_m] \\ &+ \frac{m}{2} V_c \cos[(\omega_c - \omega_m)t - \theta_m] \end{aligned} \tag{7-3}$$

Note that (7-3) consists of three frequency components, the carrier frequency and two side wave frequencies, located symmetrically on either side of the carrier frequency.

In the general case, there are many signal frequency components and hence many pairs of side waves. These are called the upper and the lower sidebands of the AM wave. The carrier frequency is usually much higher than the highest frequency in the modulating signal. The demodulator in fig 7-1 converts the signal information back to its original form. The purpose of the bandpass amplifier is to perform a filtering function in addition to frequency-selective amplification. The magnitude and phase characteristics of the ideal bandpass amplifier are shown in fig 7-3a. The corresponding characteristics for the realizable tuned-amplifier are shown in fig 7-3b. The bandwidth $\omega_{3\,\text{dB}} = \omega_2 - \omega_1$ in a narrow-band amplifier is much smaller than the center radian frequency ω_0. In other words, for a narrow-band situation, $\omega_{3\,\text{dB}} \leq 0.1\,\omega_0$, and for this case considerable simplification is achieved in the design as demonstrated in later sections of this chapter.

The effectiveness of the filtering is measured by the "skirt" of the magnitude response in fig 7-3b. Quantitatively, this measure is given by *selectivity*, which is given either by -60-dB to -6-dB frequency ratio or -30-dB to -3-dB ratio ω_3/ω_2, as shown in fig 7-3b.

The design of a bandpass amplifier, as we shall soon see, has some aspects common to those of lowpass amplifier design, especially in the narrow-band case. There are certain features of the design in a bandpass amplifier that are quite different from those of lowpass amplifiers. The chief problem in bandpass amplifier design is the possibility of oscillation, due to the reactive load and the internal feedback of the device, e.g., the C_c of a CE transistor or the C_{gd} of a common-source FET. Since C_c plays an important, undesirable role in the design, and exhibits the potential instability of the device (see Section 2-3), the unilateral model cannot, in general, be used for an inductive load. For narrow-band tuned amplifiers the transistor two-port parameters are measured at the center frequency and then the design is based on the approximation that these parameters do not change in the passband of the amplifier. Since the passband extends at most to ± 5 per cent of the center frequency, in a narrow-band situation, this approximation is usually valid. For a wide-band design, mismatching technique must be used, as illustrated in Sections 7-12 and 7-13.

7-2 Single-Tuned Interstage

One of the simplest and most commonly used bandpass amplifier interstages is the so-called single-tuned amplifier. The simplest single-tuned circuit is shown in fig 7-4. This circuit has a single resonant circuit and provides a good vehicle for introducing the analysis and design of tuned amplifiers. The voltage-gain function of the circuit is given by

$$A_v = \frac{V_o}{V_i} = -\frac{g_m}{G + sC_T + 1/sL} \quad (7\text{-}4a)$$

$$= -\frac{g_m}{C_T} \frac{s}{s^2 + (G/C_T)s + 1/LC_T} \quad (7\text{-}4b)$$

where C_T is the total interstage capacitance and $G = 1/R$. Equation (7-4) is also sometimes referred to as the resonator expression.

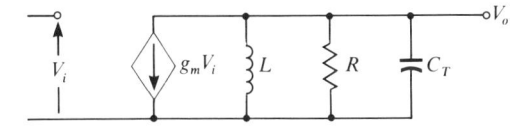

Fig 7-4 Single-tuned circuit

The magnitude of the gain function is

$$|A_v(j\omega)| = \frac{g_m}{C_T} \frac{\omega}{\sqrt{[(G/C_T)\omega]^2 + (1/LC_T - \omega^2)^2}} \qquad (7\text{-}5)$$

The magnitude is maximum at the frequency where the denominator of (7-5) is minimum. This frequency is commonly called the center or the resonant frequency and is denoted by ω_0. Hence

$$\omega_0 = \frac{1}{\sqrt{LC_T}} \qquad (7\text{-}6)$$

Note that at ω_0 the gain is

$$A_v(j\omega_0) = -\frac{g_m}{G} = -g_m R \qquad (7\text{-}7)$$

The frequencies at which the gain function is down by 3 dB from the center frequency (also called half-power frequencies) are determined from (7-5) and (7-7), that is,

$$\left(-\frac{g_m}{G}\right)^2 \frac{1}{2} = \left(-\frac{g_m}{C_T}\right)^2 \frac{\omega^2}{[(G/C_T)\omega]^2 + (\omega_0^2 - \omega^2)^2} \qquad (7\text{-}8)$$

The two solutions of (7-8) are

$$\omega_2 = \frac{G}{2C_T} + \sqrt{\omega_0^2 + G^2/4C_T^2} \qquad (7\text{-}9a)$$

$$\omega_1 = -\frac{G}{2C_T} + \sqrt{\omega_0^2 + G^2/4C_T^2} \qquad (7\text{-}9b)$$

Note that at ω_0, the phase shift of A_v is zero, while at $\omega_{2,1}$ the relative phase shifts are $\mp 45°$, respectively. The -3-dB bandwidth of the circuit is defined as the difference between the -3-dB bandedge frequencies, namely

$$\omega_{3\,\text{dB}} \equiv \omega_2 - \omega_1 = \frac{G}{C_T} \qquad (7\text{-}10)$$

Also, from (7-9) and (7-10), we can solve for ω_0 in terms of the bandedge frequencies. The result is

$$\omega_0 = \sqrt{\omega_1 \omega_2} \quad (7\text{-}11)$$

Most often we deal with *narrow-band* tuned amplifiers, where $\omega_{3\,\mathrm{dB}} \leq 0.1\omega_0$. For such cases, from (7-9) and (7-10), we obtain

$$\omega_0 \approx \frac{\omega_1 + \omega_2}{2} \qquad \omega_{3\,\mathrm{dB}} \ll \omega_0 \quad (7\text{-}12)$$

The pole-zero plot of (7-4) for a narrowband tuned amplifier is shown in fig 7-5.

A useful term in tuned amplifiers is the so-called Q of the circuit. The Q of a single-tuned circuit is defined as

$$Q = \frac{\omega_0}{\omega_{3\,\mathrm{dB}}} \quad (7\text{-}13)$$

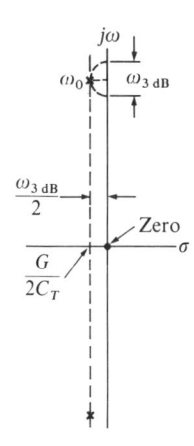

Fig 7-5 Pole-zero plot for a narrow-band single-tuned stage ($\omega_0 \gg \omega_{3\,\mathrm{dB}}$)

From (7-6), (7-10), and (7-13) we have

$$Q = \omega_0 R C_T = \frac{R}{L\omega_0} \quad (7\text{-}14)$$

Note that a narrow-band case ($\omega_{3\,\mathrm{dB}} \ll \omega_0$) means a high Q (i.e., $Q \gg 1$).

Substitution of (7-6) and (7-10) into (7-4) yields a useful form for the single-tuned amplifier:

$$A_v(s) = -\frac{g_m}{C_T} \frac{s}{s^2 + \omega_{3\,\mathrm{dB}} s + \omega_0^2} \quad (7\text{-}15a)$$

Or in a normalized form

$$A_v(s_n) = -\left(\frac{g_m R}{Q}\right) \frac{s_n}{s_n^2 + (1/Q)s_n + 1} \quad (7\text{-}15b)$$

where $s_n = s/\omega_0$.

Note that (7-4a) can also be expressed as

$$A_v(j\omega) = \frac{-g_m R}{1 + j\omega R C_T - j(R/\omega L)}$$

$$= -\frac{-g_m R}{1 + j\omega_0 R C_T(\omega/\omega_0 - \omega_0/\omega)} \quad (7\text{-}16a)$$

Substitution of (7-7) and (7-14) into (7-16a) yields

$$\frac{A_v(j\omega)}{A_v(j\omega_0)} = \frac{1}{1 + jQ(\omega/\omega_0 - \omega_0/\omega)} \qquad (7\text{-}16b)$$

A plot of the magnitude and phase of (7-16b) for two values of Q is shown in fig 7-6. We may express the response in the passband of the narrow-band amplifier by letting $\omega/\omega_0 = 1 + \Delta$ with $\Delta \ll 1$.

Hence (7-16b) reduces to

$$\frac{A_v(j\omega)}{A_v(j\omega_0)} \approx \frac{1}{1 + j2Q\Delta} \qquad (7\text{-}16c)$$

The response characteristic of (7-16c) is essentially the same as in fig 7-6. The curve of fig 7-6 is commonly called the *resonance curve*.

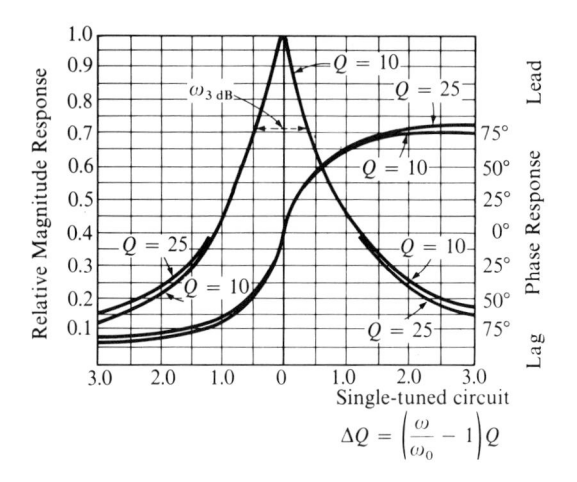

Fig 7-6 Universal resonance curve for a single-tuned stage

The gain-bandwidth product (GBP) of the single-tuned stage from (7-7) and (7-10) is given by

$$\text{GBP} = |A_v(j\omega_0)|\omega_{3\,\text{dB}} = \frac{g_m}{C_T} \qquad (7\text{-}17)$$

We shall illustrate the above expressions by two examples in the following.

EXAMPLE 1

Consider the resistive-loaded single-tuned transistor stage shown, exclusive of biasing circuitry, in fig 7-7a. For simplicity, we assume that $r_b = 0$ and that the coil loss and the parasitic capacitance of the inductor are included in R_1 and C_a, respectively (this will be discussed later).

Since the load is resistive, we can use the simplified unilateral model which incorporates the Miller effect, as shown in fig 7-7b. Note that from (4-70) $C_i = (1 + R_L C_c \omega_T)/r_e \omega_T$. The circuit is redrawn where all the capacitances and conductances are combined.

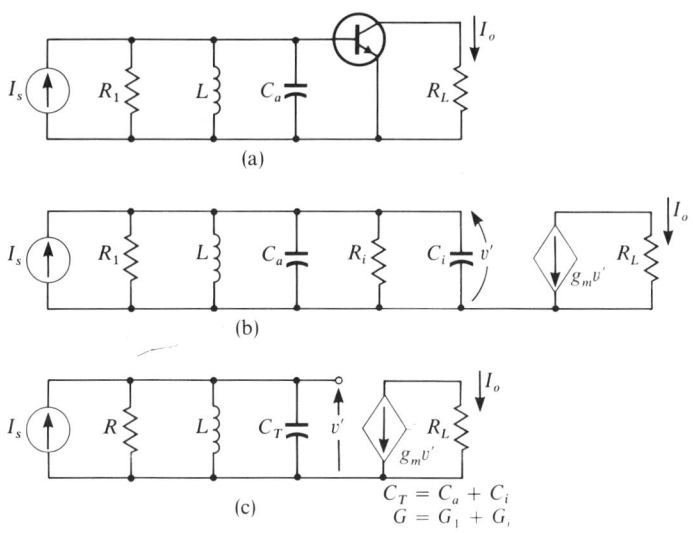

Fig 7-7 (a) Simple transistor single-tuned stage of example; (b) Approximate equivalent circuit of (a); (c) Simplified form of (b)

$C_T = C_a + C_i$
$G = G_1 + G_i$

The current gain function of the amplifier is

$$A_i = \frac{I_o}{I_s} = \frac{I_o}{v'} \cdot \frac{v'}{I_s}$$

$$= \frac{-g_m}{G + sC_T + 1/sL} = \frac{-g_m R}{1 + sRC_T + R/sL}$$

The above equation is of identical form to (7-4), hence the midband gain and the 3-dB bandwidth are given by

$$A_i(j\omega_0) = -g_m R \qquad \omega_{3\,dB} = \frac{1}{RC_T}$$

To illustrate the above numerically, consider a transistor with the following parameters:

$$f_T = 200 \text{ MHz} \qquad \beta_0 = 100$$
$$r_e = 20\,\Omega \qquad C_c = 5 \text{ pF}$$

The other circuit element values are

$$R_L = 200\,\Omega, \qquad R_1 = 2 \text{ k}\Omega, \qquad L = 10\,\mu H, \qquad C_a = 2000 \text{ pF}$$

From fig 7-7c, we find

$$R = R_1 \| R_i = (2\|2) \text{ k}\Omega = 1 \text{ k}\Omega$$
$$C_T = C_a + C_i = (2000 + 90)\,\text{pF} = 2.09 \times 10^{-9}\,F$$

The resonant frequency, from (7-6), is

$$f_0 = \frac{\omega_0}{2\pi} = \frac{1}{2\pi}[\sqrt{(10^{-5})(2.09 \times 10^{-9})}]^{-1} = 1.1 \text{ MHz}$$

The midband gain, the 3-dB bandwidth, and the Q of the circuit are

$$A_i(j\omega_0) = -50, \qquad f_{3\,dB} = 76 \text{ kHz}, \qquad Q = 14.5$$

The analysis and design of one single-tuned stage employing an FET or a vacuum tube are quite similar to the above, except that the parallel *RLC* circuit is formed at the output circuit for these devices.

The design of a single-tuned stage for a specified center frequency, f_0, and a bandwidth, $f_{3\,dB}$, follows directly from (7-6) and (7-10), respectively.

EXAMPLE

Consider a simple design example, using an FET (in fig 7-8a) with the following parameters:

$$g_m = 4 \times 10^{-3} \text{ mho}, \; C_{gs} = 40 \text{ pF} \qquad C_{gd} = C_{ds} = 5 \text{ pF}, \; r_d = 20 \text{ k}\Omega$$

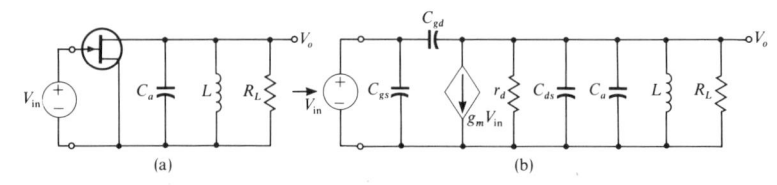

Fig 7-8 (a) FET single-tuned stage of design example; (b) Equivalent circuit of (a)

Note that the FET has the shunt-tuned circuit at its output, since the device is excited by a voltage source.

The design specifications for the stage, shown exclusive of biasing circuitry, in fig 7-8, are

$$\omega_0 = 2\pi(10^6) \text{ Hz}, \qquad \omega_{3\text{ dB}} = 2\pi(10^4) \text{ Hz}$$

and a midband voltage gain $A_v(j\omega_0) = -10$.
From the gain requirements, we have

$$A_v(j\omega_0) = -g_m(R_L \| r_d) = -10 \Rightarrow R_L = 2.9 \text{ k}\Omega$$

From the bandwidth specification

$$\omega_{3\text{ dB}} = \frac{G}{C_T} = \frac{(G_L + g_d)}{C_a + C_{ds} + C_{gd}} = 2\pi(10^4) \text{ Hz}$$

Substitution of the values in the above and solving for C_a yields

$$C_a = \frac{(G_L + g_d)}{2\pi(10^4)} - C_{ds} - C_{gd} \approx 0.0064 \; \mu F$$

From the resonant frequency requirement we have

$$\frac{1}{LC_T} = \omega_0 \Rightarrow L = \frac{1}{C_T \omega_0^2} = \frac{1}{6.4 \times 10^{-9}[2\pi(10^6)]^2} = 3.97 \; \mu H$$

In this case, the circuit values are practical and the initial design is acceptable.

Often in the design of tuned amplifiers, the value of the inductor may not be practical. In other words, the inductor may not have a high Q, and the use of transformers may be necessary. For example, consider the inclusion of the inductor coil losses, as shown by a series resistance, r_s, in fig 7-9a. For convenience, the circuit may be transformed into a parallel circuit, as shown in fig 7-9b, where the equivalent parallel resistance, r_p, is readily determined from

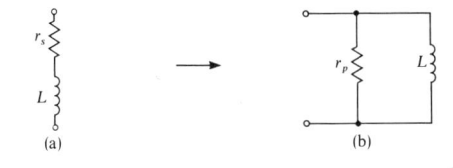

Fig 7-9 (a) Coil, including the loss; (b) Equivalent model for a high-Q coil

$$Y(j\omega) = \frac{1}{r_s + j\omega L} = \frac{r_s - j\omega L}{r_s^2 + \omega^2 L^2} \qquad (7\text{-}18a)$$

For a high Q coil

$$Q_L = \frac{\omega L}{r_s} \gg 1$$

and (7-18a) can be written approximately as

$$Y(j\omega) \simeq \frac{r_s}{\omega^2 L^2} - j\frac{1}{\omega L} = r_p^{-1} + \frac{1}{j\omega L} \qquad (7\text{-}18b)$$

where

$$r_p = \frac{\omega^2 L^2}{r_s} = \omega L Q_L \approx \omega_0 L Q_L \qquad (7\text{-}18c)$$

Thus, if the coil Q is not high enough, the value of r_p may be of the same order of magnitude as R in fig 7-7c or fig 7-8b, and the circuit Q will be lowered (i.e., $\omega_{3\,dB}$ will be increased). Thus, the effect of the coil loss, or coil Q_L cannot, in general, be ignored.

7-3 Impedance Transformation and Transformer Coupling

The circuit-element values (L's and C's) can be made practical by using suitable impedance-transforming networks. There are many types of impedance-transforming networks, the choice of which depends on the frequency range and the impedance levels. One such network which is particularly attractive in tuned amplifiers is the autotransformer, shown in fig 7-10a.[1] The coils are wound on a ferrite core and a coefficient of coupling very close to unity can be achieved. The voltage-current relationship of the ideal autotransformer is given by

$$-\frac{I_2}{I_1} = \frac{V_1}{V_2} = n > 1 \qquad (7\text{-}19)$$

The model for the autotransformer, neglecting the losses, is shown in fig 7-10b, where L_m is the total magnetizing inductance characterized on the primary side, and n is the turns ratio. Note also that when the transformer is used, the voltage gain is reduced by a factor of $1/n$, and the current gain is increased by a factor of n in a stepdown transformer. For this reason, one usually considers power gain in tuned amplifiers, which would be unaffected in a lossless transformer. Often a double-tapped inductance or a combination of tapped primary interstage

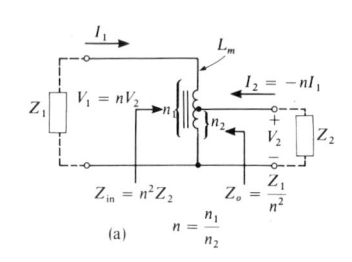

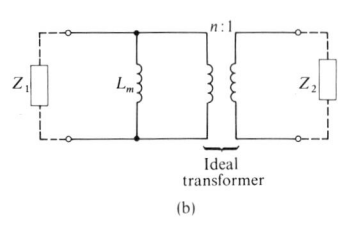

Fig 7-10 (a) Autotransformer with impedance-transforming relations; (b) Equivalent model of (a) (losses are ignored)

---
[1] Impedance transformation can also be obtained with tapped capacitors for high-frequency tuned amplifiers (see Problem 7-17).

7-4 Transistor Single-Tuned Amplifier with Tuned Circuit at the Output

transformer is used in tuned amplifier design. These are shown in fig 7-11a and fig 7-11b, respectively. Their use is illustrated in the design examples later in this chapter.

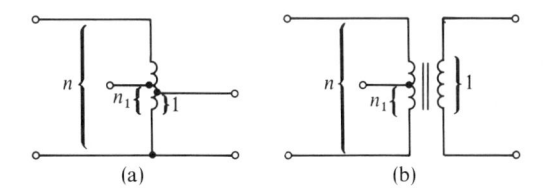

Fig 7-11 (a) Double-tapped autotransformer; (b) Double-tapped transformer

7-4 Transistor Single-Tuned Amplifier with Tuned Circuit at the Output

A single stage is seldom used in a tuned amplifier. For the sake of clarity of discussion of some basic ideas which will be encountered in multistage design, however, we consider briefly a single-stage single-tuned bipolar transistor where the tuned circuit is at the output. An interior stage of a triode or FET with a tuned-circuit load may also be handled in a manner similar to that of the following treatment. The reason that the design of such stages can be different is because of the inductive nature of the load just under the resonant frequency. For the inductive load, the Miller approximation can lead to erroneous results. In other words, if the tuned circuit is at the output, as shown in fig 7-12a, we cannot use the unilateral model, since the output

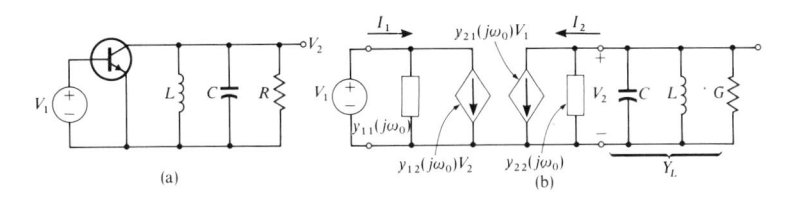

Fig 7-12 (a) Transistor stage with tuned circuit at the output; (b) Equivalent circuit of (a) at the center frequency

impedance of the transistor is *very high* and can be inductive or capacitive near the resonant frequency. In fact, for such a case, it is possible for the circuit to oscillate. This fact makes the problem of a tuned-amplifier design quite different when compared to the lowpass case. The instability problem of a tuned stage is discussed in Section 7-8.

In the majority of applications, we deal with the narrow-band tuned amplifiers, i.e., $\omega_{3\,dB} \leq 0.1\omega_0$. This condition yields considerable simplification in the analysis and design. At the center frequency, we can represent the transistor by its *y*-parameters. This characterization at ω_0 is very accurate and can include the parasitic and other effects.

7/Analysis and Design Examples of Bandpass Amplifiers

The small-signal model of the tuned amplifier is shown in fig 7-12b. In the neighborhood of the center frequency, i.e., for $|\omega - \omega_0| \leq 0.05\omega_0$, the two-port parameters are approximately the same as those measured at ω_0, and can be expressed approximately as[2]

$$y_{11} \approx g_{11} + j\omega b_{11} \qquad y_{21} \approx g_{21} + j\omega b_{21}$$
$$y_{12} \approx g_{12} + j\omega b_{12} \qquad y_{22} \approx g_{22} + j\omega b_{22} \qquad (7\text{-}20)$$

The voltage gain of a linear two-port with load admittance $Y_L = G + j\omega C + 1/j\omega L$, from table 2-2, is given by

$$A_v(j\omega) = \frac{V_2}{V_1} = -\frac{y_{21}}{y_{22} + Y_L} = -\frac{y_{21}}{y_{22} + G + j\omega C + 1/j\omega L}$$
$$= \frac{-y_{21}}{g_{22} + G + j(\omega C + b_{22}) - j(1/\omega L)} \qquad (7\text{-}21)$$

Note that, in bandwidth calculations, the effect of the real and the imaginary parts of y_{22} (especially the latter) is very important since we deal with the difference of two nearly equal quantities near resonance. However, y_{21} is not involved in the resonant part and since for a narrow band, y_{21} is fairly constant in the frequency range of interest, we can approximate it by a constant over the bandwidth. Hence we can write (7-21) as

$$A_v(j\omega) = -\frac{|y_{21}(j\omega_0)|}{g_T + j\omega C_T + 1/j\omega L} = -\frac{g_m}{g_T + j\omega C_T + 1/j\omega L} \qquad (7\text{-}22)$$

where $g_T = g_{22} + G$, $C_T = C + b_{22}/\omega_0$, and $|y_{21}(j\omega_0)| \equiv g_m$.

A comparison of (7-22) and (7-4a) shows immediately that the expressions are identical in the passband and thus the same design equations can be used.

EXAMPLE

Consider the circuit shown in fig 7-12b. The transistor y-parameters measured at the desired center frequency, $\omega_0 = 2\pi(10 \text{ MHz})$, are given

---

[2] Equation (7-20) is really the Taylor series expansion of $y_{ij}(\omega)$ near ω_0, with the approximation that the linear terms are retained.

7-4 Transistor Single-Tuned Amplifier with Tuned Circuit at the Output

as follows (all in mhos):

$$y_{11} = y_{ie} = (2.0 + j0.5) \times 10^{-3}$$
$$y_{12} = y_{re} = -(1.0 + j5.0) \times 10^{-5}$$
$$y_{21} = y_{fe} = (20 - j5.0) \times 10^{-3}$$
$$y_{22} = y_{oe} = (2.0 + j4.0) \times 10^{-5}$$

(7-23)

The design requires a bandwidth of $\omega_{3\,dB} = 2\pi\,(100\text{ kHz})$ (i.e., $\pm 50\text{ kHz}$ centered at $f_0 = 10\text{ MHz}$). The stage voltage gain at ω_0 is to be approximately -50.

From the gain specifications at ω_0, we have

$$A_v(j\omega_0) = -|y_{21}|/g_T = -50$$

or

$$\frac{21 \times 10^{-3}}{2 \times 10^{-5} + G_L} = 50 \rightarrow R_L = \frac{1}{G_L} = 2.56\text{ k}\Omega$$

From the bandwidth specification and (7-10), we have

$$\frac{g_T}{C_T} = \frac{g_{22} + G}{b_{22}/\omega_0 + C} = 2\pi(10^5)$$

or

$$\frac{2 \times 10^{-5} + 3.9 \times 10^{-4}}{0.64 \times 10^{-12} + C} = 2\pi(10^5) \rightarrow C = 652\text{ pF}$$

The value of L is determined from the specified ω_0 and (7-6):

$$\frac{1}{LC_T} = \omega_0^2 \rightarrow \frac{1}{L(6.5 \times 10^{-10})} = [2\pi(10^7)]^2 \rightarrow L = 0.385\,\mu\text{H}$$

The element values are acceptable as an initial design. In practice, inductors with a low value of inductance have a small equivalent shunt resistor ($r_p \approx \omega_0 L Q_L$) which will have the effect of increasing the bandwidth, thus reducing the circuit Q. A more desirable value for the inductor is obtained if an autotransformer with $n = 3$ is used. In this case

$$L_m = n^2 L = 9(0.385 \, \mu H) = 3.47 \, \mu H$$

$$C_a = \frac{C}{n^2} = \frac{652}{9} = 72.5 \, pF$$

The designed circuit, exclusive of biasing circuitry, is shown in fig 7-13. In passing, we remark again that the above design method can be carried over identically to any other *nonunilateral* active device.

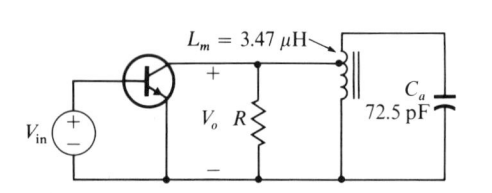

Fig 7-13 Tuned-amplifier circuit of the design example

7-5 The Narrow-Band Approximation

Narrow-band tuned amplifiers play a very important role in telecommunication systems. They are used widely in the transmitter and receiver systems using modulated signals. This is fortunate, since considerable simplification can be achieved in the analysis and design of narrow-band tuned amplifiers. Because of the tremendous usefulness of the narrow-band approximation, we consider a brief discussion of this topic in this section.

The admittance function of a single-tuned bandpass amplifier (fig 7-4) is given by

$$Y(s) = G + sC + \frac{1}{sL} = G + C\left(s + \frac{\omega_0^2}{s}\right) \quad (7\text{-}24)$$

where

$$\omega_0^2 = 1/LC$$

Now consider the admittance of the single-pole lowpass case shown in fig 7-14. For the lowpass case, we use the p-complex frequency plane ($p = x + jy$) so as not to confuse the lowpass and bandpass complex frequency variable planes. The admittance of fig 7-14 is

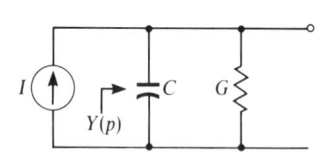

$$Y(p) = G + Cp \quad (7\text{-}25)$$

Fig 7-14 A lowpass admittance with a single-pole network function

From (7-24) and (7-25), we introduce a *lowpass-to-bandpass transformation*:[3]

$$p = s + \frac{\omega_0^2}{s} \quad (7\text{-}26)$$

---

[3] Other transformations which are useful in other types of filters are the following:

lowpass-to-highpass $\quad p = \dfrac{1}{s}$

lowpass-to-band elimination $\quad p = \dfrac{s}{s^2 + \omega_0^2}$

7-5 The Narrow-Band Approximation

or, solving for s in terms of p from (7-26), we have

$$s = \frac{p}{2} \pm \sqrt{\left(\frac{p}{2}\right)^2 - \omega_0^2} \qquad (7\text{-}27)$$

For a narrow-band case, $|p/2| \ll \omega_0$, and (7-27) reduces to

$$s \approx \frac{p}{2} \pm j\omega_0 \qquad (7\text{-}28)$$

In other words, the bandpass pole-zero location is the pole-zero location of the lowpass case scaled by a factor of $\frac{1}{2}$ and translated around new origins at $\pm j\omega_0$. For example, a single-pole lowpass pole location, under the narrow-band frequency transformation, has the bandpass pole locations shown in fig 7-15, which are those of a single-tuned circuit. In fact, for the frequency response calculations near the

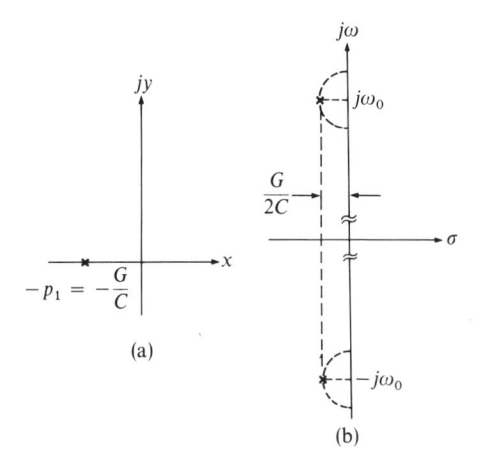

Fig 7-15 (a) Lowpass pole location in the p-plane; (b) Bandpass pole locations in the s-plane

center frequency, in a narrow-band case, the zeros and the conjugate poles can also be ignored, as seen in the following:

Consider the pole-zero locations of a single-tuned, narrow-band, bandpass amplifier shown in fig 7-16. The gain function corresponding to this pole-zero location is given by

$$\frac{V_o}{V_i} = -\frac{g_m}{C} \frac{s}{(s + s_1)(s + s_1^*)} \qquad (7\text{-}29)$$

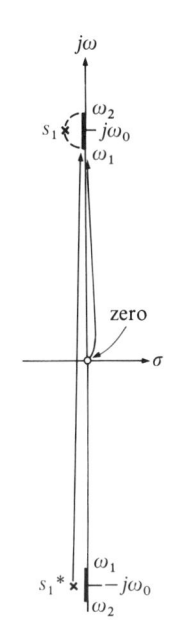

Fig 7-16 Pole-zero locations of a narrow-band single-tuned stage

The magnitude response is

$$\left|\frac{V_0}{V_i}(j\omega)\right| = \frac{g_m}{C} \frac{|j\omega|}{|j\omega + s_1||j\omega + s_1^*|} \quad (7\text{-}30)$$

For the frequencies near $j\omega_0$ and in the range $\omega_1 \leq \omega \leq \omega_2$, actually in the range $|\omega| > |\omega_1|$, the following magnitude expression (see fig 7-16) is approximately constant:

$$\frac{|j\omega|}{|j\omega + s_1^*|} \approx \frac{1}{2} \quad \omega \geq \omega_1 \quad (7\text{-}31)$$

Note that for the narrow-band approximation, i.e., $|\omega_2 - \omega_1| \leq 0.1\omega_0$, the error in (7-31) is less than 5 per cent. Therefore, from (7-31) and (7-30) we can write

$$\left|\frac{V_0}{V_i}(j\omega)\right| \approx \frac{g_m}{2C_T} \frac{1}{|j\omega + s_1|} \quad |\omega - \omega_0| \leq 0.05\omega_0$$

or

$$\frac{V_0}{V_i} \approx -\frac{g_m}{2C_T} \frac{1}{(s + s_1)} \quad (7\text{-}32)$$

Therefore, in frequency-response calculations, in the bandpass region for narrow-band tuned amplifier, the zeros at the origin (or real axis) and the conjugate poles, and the poles on the real axis can be ignored.

7-6 Synchronous-Tuning and Stagger-Tuning

Synchronous-tuning and stagger-tuning were discussed in Section 5-8 in conjunction with lowpass amplifiers. In a multistage tuned amplifier, if the individual stages are unilateral or designed to have negligible interaction, and if each stage has the same center frequency and bandwidth, then the system is said to be synchronously tuned. For such a case, the pole-zero pattern of synchronously tuned single-tuned stages is shown in fig 7-17a. The overall bandwidth of the n-stage is determined, as in the lowpass case, to be

$$(\omega_{3\,\text{dB}})_{\text{overall}} = S_n(\omega_{3\,\text{dB}})_i = \sqrt{2^{1/n} - 1}\,(\omega_{3\,\text{dB}})_i \quad (7\text{-}33)$$

where S_n is the shrinkage factor and $(\omega_{3\,\text{dB}})_i$ is the bandwidth of the individual stage. The overall response and the response of individual

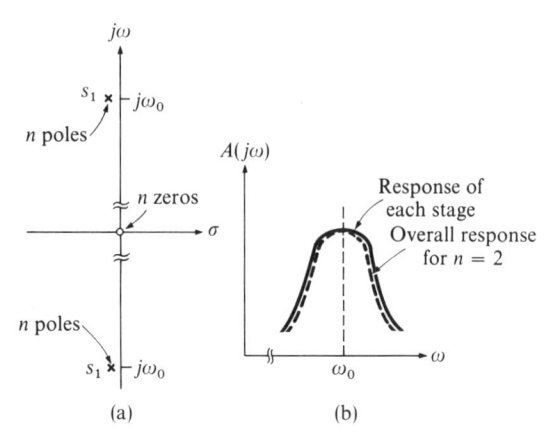

Fig 7-17 (a) Pole-zero locations of n identical single-tuned synchronously tuned design; (b) Magnitude response of the individual single-tuned stage and that of a two-stage synchronously tuned design

7-6 Synchronous-Tuning and Stagger-Tuning

stages for $n = 2$ are shown in fig 7-17b. Synchronously tuned design can also be used for double-tuned stages. This will be discussed later. In this case, of course, the shrinkage factor will be different.

The shrinkage factor can be avoided by the use of *stagger-tuning*. In stagger-tuning, the individual noninteracting stages do not have the same pole-zero locations, but rather are arranged in such a manner that the overall response has the desired response shape. For example, for a narrow-band case, a cascade of two noninteracting single-tuned stages can be designed to have maximally flat magnitude transmission characteristics. The pole-zero pattern for the bandpass is readily obtained from the lowpass maximally flat pole locations and the narrow-band approximation. The pole-zero pattern is shown in fig 7-18. Thus, the individual single-tuned circuits can be designed to realize a zero and a pair of the complex poles from the pattern shown in fig 7-18a. The individual tuned-circuit response and the overall response are shown in fig 7-18b. When many noninteracting stages are used, the poles for a stagger-tuned design may be designed to be on a circle corresponding to the Butterworth angles (see Section 2-8), in order to get a smooth magnitude response and maximum bandwidth. Whenever a better selectivity is required, stagger-tuning is often employed.

Illustrative design examples for synchronously tuned and stagger-tuned amplifiers using a cascode transistor amplifier are discussed in Section 7-10. Often nonunilateral transistor stages are designed to have negligible interaction by deliberately mismatching the stages. This approach is discussed in Section 7-12.

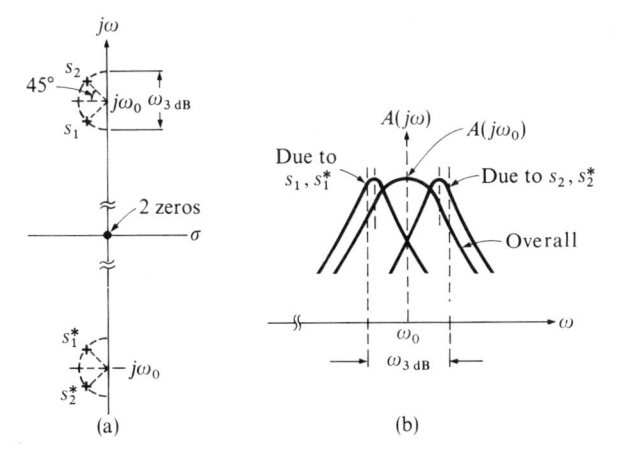

Fig 7-18 (a) Pole-zero plot of a two-stage stagger-tuned maximally flat magnitude design using two single-tuned stages; (b) Magnitude response of the individual single-tuned stages and of the overall stagger-tuned design

7-7 Double-Tuned Interstage

A double-tuned interstage provides a better gain-bandwidth product and selectivity in tuned-amplifier design than the single-tuned interstage. An inductively coupled double-tuned interstage is shown in fig 7-19. This circuit could also correspond to the transistor circuit

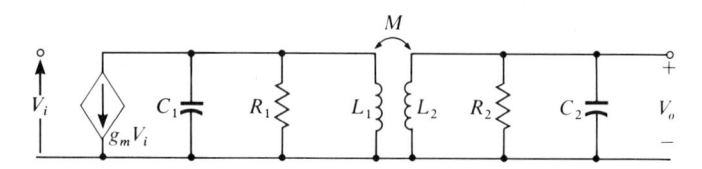

Fig 7-19 Inductively-coupled double-tuned stage

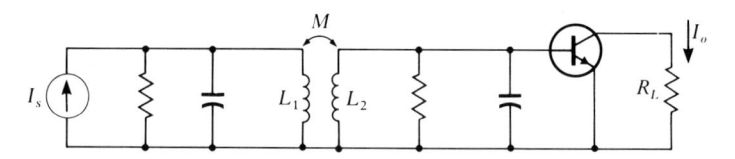

Fig 7-20 Inductively-coupled double-tuned transistor stage

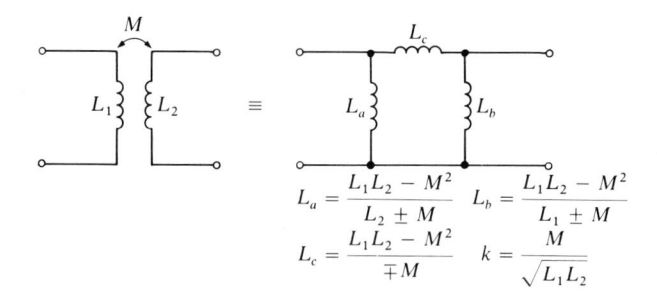

$$L_a = \frac{L_1 L_2 - M^2}{L_2 \pm M} \quad L_b = \frac{L_1 L_2 - M^2}{L_1 \pm M}$$

$$L_c = \frac{L_1 L_2 - M^2}{\mp M} \quad k = \frac{M}{\sqrt{L_1 L_2}}$$

Fig 7-21 A transformer with its equivalent circuit (losses are ignored)

shown in fig 7-20, if r_b is neglected and the Miller approximation is used. The transformer can be replaced by its equivalent circuit, shown in fig 7-21, where k is the coefficient of coupling and M is the mutual inductance. The circuit to be analyzed is shown in fig 7-22. The nodal

7-7 Double-Tuned Interstage

circuit equations are

$$V_1 \left[G_1 + sC_1 + \frac{1}{sL_c} + \frac{1}{sL_a} \right] - \frac{V_2}{sL_c} = -g_m V_{in} \quad (7\text{-}34a)$$

$$-V_1 \left(\frac{1}{sL_c} \right) + V_2 \left[G_2 + sC_2 + \frac{1}{sL_c} + \frac{1}{sL_b} \right] = 0 \quad (7\text{-}34b)$$

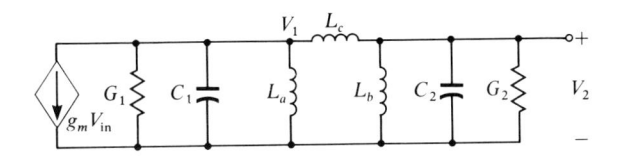

Fig 7-22 Circuit of fig 7-19 where equivalent circuit of fig 7-21 is used for analysis

From (7-34), the voltage gain function, after some manipulations, can be written as

$$A_V = \frac{V_2}{V_{in}} = -\frac{g_m k \omega_1 \omega_2}{(1-k^2)\sqrt{C_1 C_2}} \frac{s}{s^4 + a_3 s^3 + a_2 s^2 + a_1 s + a_0} \quad (7\text{-}35)$$

where

$$a_3 = \frac{\omega_1}{Q_1} + \frac{\omega_2}{Q_2} \qquad a_2 = \frac{\omega_1 \omega_2}{Q_1 Q_2} + \frac{\omega_1^2 + \omega_2^2}{1-k^2}$$

$$a_1 = \frac{\omega_1^2 \omega_2}{Q_2(1-k^2)} \left(1 + \frac{\omega_2 Q_2}{\omega_1 Q_1} \right) \qquad a_0 = \frac{\omega_1^2 \omega_2^2}{1-k^2} \quad (7\text{-}36)$$

and we have defined:

$$\omega_1 = \frac{1}{\sqrt{L_1 C_1}} \qquad \omega_2 = \frac{1}{\sqrt{L_2 C_2}}$$

$$Q_1 = \frac{R_1}{\omega_1 L_1} \qquad Q_2 = \frac{R_2}{\omega_2 L_2} \quad (7\text{-}37)$$

The gain function (7-35) for the narrow-band case can also be written as

$$A_V(s) = -\frac{g_m k \omega_1 \omega_2}{\sqrt{C_1 C_2}(1-k^2)} \frac{s}{(s+s_1)(s+s_2)(s+s_1^*)(s+s_2^*)} \quad (7\text{-}38)$$

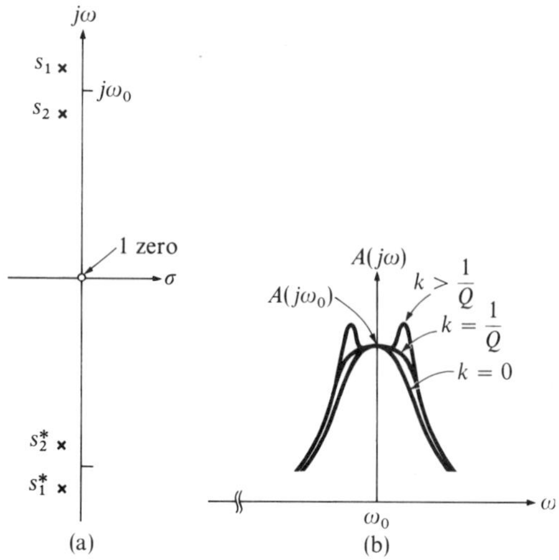

The pole-zero pattern of (7-38) is shown in fig 7-22 where s_1 and s_2 can be designed for, say a maximally flat magnitude, a linear phase, or $s_1 = s_2$. If we let $\omega_1 = \omega_2 = \omega_0$, $Q_1 = Q_2 = Q \gg 1$, and $k^2 \ll 1$, the approximate pole positions of (7-38) can be found:

$$s_1, s_1^* = -\frac{\omega_0}{2Q} \pm j\omega_0\left(1 - \frac{k}{2}\right) \qquad (7\text{-}39a)$$

$$s_2, s_2^* = -\frac{\omega_0}{2Q} \pm j\omega_0\left(1 + \frac{k}{2}\right) \qquad (7\text{-}39b)$$

Hence, for a maximally flat magnitude, the poles are on the 45° radial line around $j\omega_0$ as origin, i.e., $k = 1/Q$. For a linear phase, the poles are on the 30° radial line, i.e., $k = 1/\sqrt{3}Q$; for $k = 0$, the poles are superimposed, i.e., $s_1 = s_2$, which is similar to two synchronously tuned single-tuned stages. The magnitude responses for $k = 0$, $k = 1/Q$, and $k > 1/Q$ are shown in fig 7-23. Detailed response calculations in the vicinity of the center frequency can easily be made by the narrow-band approximation. For the narrow-band case, the approximation to (7-38) for $\omega_1 = \omega_2 = \omega_0$, $Q_1 = Q_2 = Q \gg 1$, and $k^2 \ll 1$ allows us to ignore the zero and the conjugate poles:

$$A_V(s) \approx \frac{-g_m k \omega_0}{4\sqrt{C_1 C_2}(s + s_1)(s + s_2)} \qquad (7\text{-}40)$$

For the maximally flat magnitude case, also called critical coupling, i.e., $k_c = 1/Q$, the gain-bandwidth product, GBP, for the circuit is determined as follows. The bandwidth for this case is equal to twice the radius of the circle centered around $j\omega_0$ and passing through s_1, s_2 which are at the 45° radial lines, namely $2[\sqrt{2}(\omega_0/2Q)] = \sqrt{2}(\omega_0/Q)$. Hence, from (7-40) the gain-bandwidth product is

$$\text{GBP} = |A_V(j\omega_0)|\frac{\omega_0\sqrt{2}}{Q} = \frac{g_m k \omega_0}{4\sqrt{C_1 C_2}(k\omega_0/\sqrt{2})^2}\sqrt{2}\omega_0 k = \frac{g_m}{\sqrt{2 C_1 C_2}}$$

$$(7\text{-}41)$$

For a single-tuned interstage, i.e., using a single inductance in fig 7-4 the gain-bandwidth product from (7-17) is

$$(\text{GBP})_s = \frac{g_m}{C_T} = \frac{g_m}{C_1 + C_2} \qquad (7\text{-}42)$$

Fig 7-23 (a) Pole-zero pattern of circuit in fig 7-22; (b) Magnitude response for various values of coefficient of coupling. For critical coupling (maximally flat magnitude response) $k_c = 1/Q$

where the subscript s denotes single-tuned case. The ratio of GBP is

$$\frac{\text{GBP}}{(\text{GBP})_s} = \frac{C_1 + C_2}{\sqrt{2C_1 C_2}} \qquad (7\text{-}43)$$

From (7-43) the GBP of double-tuned interstage is better by a factor of $\sqrt{2}$ (for $C_1 = C_2$) than the single-tuned interstage. If the primary and secondary Q's are unequal, with either $Q_1 \gg Q_2$ or $Q_2 \gg Q_1$, then for $C_1 = C_2$ it can be shown that the GBP of double-tuned is better by a factor of 2.0 than the single-tuned case. The drawback of this type of double-tuned stage (fig 7-19) is that in practice it is difficult to vary the mutual inductance.

Other possibilities for coupling as a double-tuned circuit also exist. These are self-inductance coupling, such as using three inductors directly in a connection, as in fig 7-22, capacitive coupling (Problem 7-8), resistive coupling, called the tuned-feedback pair (Problem 7-9), and the cascode circuit (Section 7-10). Since C_m can be easily varied in the capacitive-coupling circuit, it is used more often in tuned amplifiers. We shall next consider an example for the mutually-coupled doubled-tuned interstage discussed in this section.

EXAMPLE

In the double-tuned interstage shown in fig 7-19 the resistances are $R_1 = 10 \text{ k}\Omega$, $R_2 = 2 \text{ k}\Omega$. The circuit is to provide a maximally flat magnitude response centered at $f_0 = (10^7 \text{ Hz})$ with a bandwidth $f_{3\,\text{dB}} = (10^5 \text{ Hz})$. The pentode parameters are $g_m = 5 \times 10^{-3}$ mho, $C_i = C_o = 5$ pF. The interstage gain and the values of the circuit parameters are to be obtained.

From the relations developed, $k_c = 1/Q$ and $\omega_{3\,\text{dB}} = \sqrt{2}(\omega_0/Q)$. Hence

$$Q = \frac{\sqrt{2}\omega_0}{\omega_{3\,\text{dB}}} = \frac{\sqrt{2}(2\pi \times 10^7)}{2\pi(10^5)} = \sqrt{2}(100)$$

$$k_c = \frac{1}{Q} = 0.707 \times 10^{-2}$$

$$\frac{G_1}{C_1} = \frac{G_2}{C_2} = \frac{\omega_0}{Q} = \frac{\omega_{3\,\text{dB}}}{\sqrt{2}}$$

Thus

$$\frac{1}{(10)^4 C_1} = \frac{2\pi(10^5)}{\sqrt{2}} \rightarrow C_1 = 225 \text{ pF}$$

$$\frac{1}{2(10^3)C_2} = \frac{2\pi(10^5)}{\sqrt{2}} \rightarrow C_2 = 0.0011 \ \mu\text{F}$$

Hence, to C_i and C_o, we must add 220 pF and 0.001 μF capacitors, respectively. From

$$\omega_0^2 = \frac{1}{L_1 C_1} \to L_1 = \frac{1}{225 \times 10^{-12}[2\pi(10^7)]^2} = 1.1 \ \mu H$$

$$\omega_0^2 = \frac{1}{L_2 C_2} \to L_2 = \frac{1}{1125 \times 10^{-12}[2\pi(10^7)]^2} = 0.22 \ \mu H$$

$$M = k\sqrt{L_1 L_2} = 0.707 \times 10^{-2}\sqrt{(0.22)(1.1) \times 10^{-12}} = 0.0034 \ \mu H$$

The voltage gain of the circuit, at center frequency, from (7-41) is

$$|A_V(j\omega_0)| = \frac{g_m}{\sqrt{2C_1 C_2}} \frac{1}{\omega_{3\,dB}} = \frac{5 \times 10^{-3}}{\sqrt{2(225)(1125) \times 10^{-24}}} \frac{1}{2\pi(10^5)} = 11.2$$

7-8 Oscillation Possibilities in Tuned Amplifiers

In most tuned-amplifier designs, we wish to maximize the power gain. In Section 2-4, however, we showed that such a maximization can be achieved only if the device is inherently stable. Unfortunately many transistors, triodes, and FET's are potentially unstable in their useful frequency range of operations. For example, consider the transistor parameters given in (7-23) repeated for convenience: at $f_0 = 10$ MHz, the parameters are

$y_{11} = (2.0 + j0.5) \times 10^{-3}$ mho $\qquad y_{12} = -(1.0 + j5.0) \times 10^{-5}$ mho

$y_{21} = (20 - j5.0) \times 10^{-3}$ mho $\qquad y_{22} = (2.0 + j4.0) \times 10^{-5}$ mho

Using the potential instability criterion (2-61), we have

$$Re \ y_{11} = 2.0 \times 10^{-3} > 0 \qquad Re \ y_{22} = 2 \times 10^{-5} > 0$$

where Re stands for the real part of

$$Re \ y_{11} \ Re \ y_{22} > \tfrac{1}{2}|y_{12}y_{21}|(1 + \cos \theta) \qquad (7\text{-}44)$$

where

$$\theta = \tan^{-1} \frac{Im \ y_{12}y_{21}}{Re \ y_{12}y_{21}}$$

$$Re \ y_{11} \ Re \ y_{22} = (2 \times 10^{-3})(2 \times 10^{-5}) = 4 \times 10^{-8}$$

$$\tfrac{1}{2}|y_{12}y_{21}|(1 + \cos \theta) = \tfrac{1}{2}(106 \times 10^{-8})[1 + \cos(-114°)] = 30 \times 10^{-8}$$

7-9 Maximum Frequency of Oscillation of a Bipolar Transistor

Since the inequality in (7-44) is not satisfied, the device is potentially unstable. Thus, we can not maximize the power gain, since instability arises under this condition. Note that, for an ideal unilateral two-port device, $y_{12} = 0$ and (7-44) is always satisfied.

The possibility of oscillation can be further demonstrated from the transistor circuit shown in fig 7-24. Since the load of a tuned transistor stage is inductive for frequencies below the center frequency, we consider the shunt RL load. For $\omega C_c \ll g_m, |Y_L|$ the input admittance of the circuit is (see 4-65)

$$Y_{in} \approx j\omega C_c(1 + g_m Z_L) = j\omega C_c\left(1 + \frac{g_m}{G - j(1/\omega L)}\right)$$

$$= -\frac{g_m C_c}{L(G^2 + 1/\omega^2 L^2)} + j\left[\omega C_c + \frac{g_m C_c G}{G^2 + 1/\omega^2 L^2}\right] \quad (7\text{-}45)$$

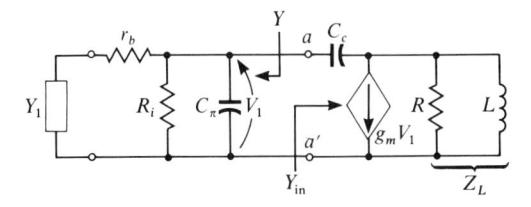

Fig 7-24 The Miller effect in a transistor stage for an inductive load

The appearance of the negative real part in (7-45) is the potential cause of oscillation. In fact, from fig 7-24 and equation (7-45), if

$$\text{Re } Y < |\text{Re } Y_{in}| = \frac{g_m C_c}{L(G^2 + 1/\omega^2 L^2)} \quad (7\text{-}46)$$

the circuit will be unstable. This fact makes the design with triodes, FET's, and bipolar transistors difficult and requires particular attention on the part of the designer.

7-9 Maximum Frequency of Oscillation of a Bipolar Transistor

Power amplification and oscillation in a circuit at any frequency can be achieved only if the device is active for that frequency. For two-port linear networks, it can be shown (Ref 4) that a necessary condition for passivity is the following:

$$U \equiv \frac{|y_{21} - y_{12}|^2}{4(g_{11}g_{22} - g_{12}g_{21})} \leq 1 \quad (7\text{-}47)$$

where $g_{11} = \text{Re } y_{11}$, etc. In other words, if $U > 1$, the two-port is active. It is very important to know beyond just what frequency the transistor ceases to be active. To find this frequency limitation, we set

$U = 1$ in (7-47) and determine the maximum radian frequency ω_m. Consider the hybrid π transistor model shown in fig 7-25a. For very high frequencies (in the neighborhood of ω_T), $1/\omega C_\pi \ll R_i$ and the transistor model is approximately as shown in fig 7-25b. For the circuit in fig 7-25b, if we assume $C_\pi \gg C_c$ and $r_b \gg r_e$, the y-parameters at very high frequencies are

$$y_{11} \approx \frac{1}{r_b} \qquad y_{12} \approx 0$$
$$y_{21} = -j\frac{\omega_T}{\omega r_b} \qquad y_{22} = C_c(\omega_T + j\omega) \tag{7-48}$$

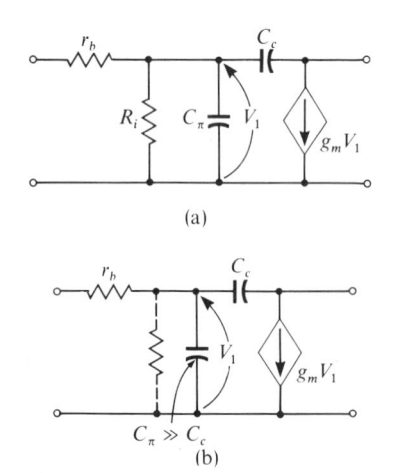

Fig 7-25 (a) Hybrid π model of a transistor; (b) The simplified model of (a) at very-high frequencies

Substitution of (7-48) into (7-47) yields the maximum frequency of activity, i.e., a frequency beyond which the transistor is passive. Hence, from (7-47) and (7-48)

$$\frac{(\omega_T/\omega_m r_b)^2}{4(1/r_b)(\omega_T C_c)} = 1$$

we obtain

$$\omega_m = \sqrt{\frac{\omega_T}{4r_b C_c}} \tag{7-49}$$

Notice that ω_m can be smaller or larger than ω_T; ω_m is called the *maximum frequency of oscillation* since, for frequencies larger than ω_m, the transistor cannot be made to oscillate because the device will be passive for $\omega \geq \omega_m$. Since the transistor is passive for $\omega \geq \omega_m$, the interchange of terminals cannot affect its passivity or activity; therefore, we conclude that (7-49) is independent of the transistor configuration: CC, CE, or CB. The frequency $f_m = \omega_m/2\pi$ provides a *figure of merit* for the transistor. Note that for amplifiers, as discussed in the previous chapters, a high value of ω_T and low values of r_b and C_c are desirable in order to achieve a high gain-bandwidth product.

As a numerical example for the value of ω_m, consider the following transistor parameters:

$$\begin{aligned} \omega_T &= 2\pi(4 \times 10^8) \text{ rad/sec} & r_e &= 5\,\Omega \\ \beta_0 &= 75, & C_c &= 4\,\text{pF} & r_b &= 50\,\Omega \end{aligned} \tag{7-50}$$

From (7-49) and (7-50) we obtain

$$\omega_m = \sqrt{\frac{\omega_T}{4r_b C_c}} \approx \sqrt{\frac{2\pi(4 \times 10^8)}{4(50)(4 \times 10^{-12})}} = 1.77 \times 10^9 \text{ rad/sec}$$

Note that in this case $\omega_m < \omega_T$. In some other transistors the reverse can be true.

In the design of nonunilateral tuned amplifiers utilizing more than one stage, several approaches are possible. One approach is to use a cascode amplifier where, for example, in a bipolar transistor, the common-base circuit serves to reduce considerably the Miller effect and provides isolation. Another method is to use neutralization in order to cancel the feedback effect due to y_{re}. The final and probably the most practical approach is to use deliberate mismatching in order to avoid oscillation. These methods are discussed in the following sections. Actually the cascode can also be viewed as a mismatched block with the CB stage providing the mismatching.

7-10 The Cascode Tuned Amplifier

We have already shown that when the load for a stage is a tuned circuit or is inductive, be it a bipolar transistor, FET, or a triode device, oscillation is a distinct possibility. One way to avoid this problem is to use a cascode amplifier, as shown in fig 7-26. In this configuration, the small resistive input impedance of the CB stage reduces the Miller effect to the CE stage to make it almost negligible. Further, for frequencies such that $f \ll f_\alpha$, the CB stage provides isolation between the two tuned circuits (see also Section 5-5).

In the following analysis we shall assume, for simplicity, that the transistors are identical, that $f_0 \ll f_\alpha$, so that the internal capacitances of the CB stage can be neglected. We also assume that $r_b = 0$. The approximate equivalent circuit is shown in fig 7-26b. Note that since $h_{ib} = r_e + r_b(1 - \alpha_0) \approx r_e$ and $h_{ib} \ll R$, the contribution of this usually very small resistance to the Miller effect is negligible. Since

$$C_i = \frac{D}{r_e \omega_T} \simeq \frac{1 + h_{ib} C_c \omega_T}{r_e \omega_T} \simeq \frac{1 + r_e C_c \omega_T}{r_e \omega_T} = \frac{1}{r_e \omega_T} + C_c \approx \frac{1}{r_e \omega_T}$$

and $C_c \ll 1/r_e \omega_T$, the effect of the Miller capacitance can be ignored

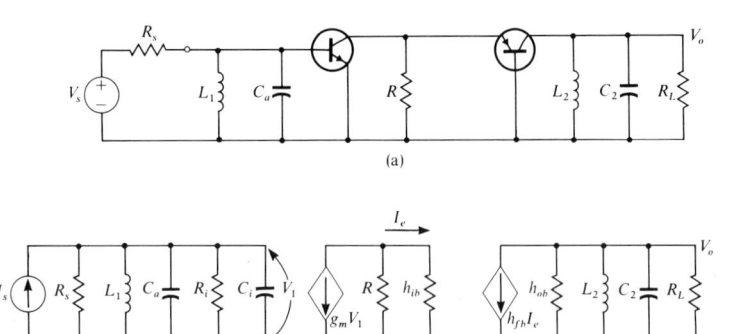

Fig 7-26 (a) Cascode amplifier with two single-tuned stages; (b) Equivalent circuit of (a)

altogether without any significant error. The two noninteracting single-tuned circuits of fig 7-26b can be designed as a synchronously tuned or a stagger-tuned amplifier such as those of the double-tuned stage. We shall illustrate both of these cases in the following:

From the circuit of fig 7-26b, the gain function is given by

$$A_v = \frac{V_0}{V_s} = \left(\frac{V_o}{I_e}\right)\left(\frac{I_e}{V_1}\right)\left(\frac{V_1}{V_s}\right) \tag{7-51}$$

Substitution of the various expressions, as obtained from fig 7-26b, in (7-51) yields

$$\frac{V_o}{V_s} = \left(\frac{-h_{fb}}{G_2 + sC_2 + 1/sL_2}\right)\left(\frac{-g_m R}{R + h_{ib}}\right)\left(\frac{G_s}{G_1 + sC_1 + 1/sL_1}\right) \tag{7-52}$$

where

$$G_1 = G_s + G_i \qquad G_2 = G_L + h_{ob} \simeq G_L$$
$$C_1 = C_a + C_i$$

Equation (7-52) may be rewritten for convenience as

$$\frac{V_o}{V_s} = -\frac{g_m G_s}{C_1 C_2}\left(\frac{s}{s^2 + (G_1/C_1)s + 1/L_1 C_1}\right)\left(\frac{s}{s^2 + (G_2/C_2)s + 1/L_2 C_2}\right) \tag{7-53}$$

where the approximations $R \gg h_{ib}$ and $h_{fb} \approx -\alpha_0 \simeq -1$ are used in the above. We consider first the synchronously tuned design.

Synchronously Tuned Design

In a synchronously tuned design, all the single-tuned stages are identical and tuned to the same resonant frequency f_0. For this case, the pole-zero location is designed as in fig 7-17, where $n = 2$. Thus, for a given center frequency, $f_0 = \omega_0/2\pi$, and a desired bandwidth, $f_{3\,dB} = \omega_{3\,dB}/2\pi$, we design *each* tuned stage to have the center frequency at f_0 and to have a bandwidth at $f_{3\,dB}/0.643$, where 0.643 is the shrinkage factor due to two identical poles. The design equations for (7-53) as a synchronously tuned design are as follows:

$$\frac{G_1}{C_1} = \frac{G_2}{C_2} = \frac{\omega_{3\,dB}}{0.643} \qquad (7\text{-}54)$$

$$\frac{1}{L_1 C_1} = \frac{1}{L_2 C_2} = \omega_0^2 \qquad (7\text{-}55)$$

The voltage gain at the center frequency from (7-53) is

$$[A_v(j\omega_0)]_{\text{sync}} = -\frac{g_m G_s}{G_1 G_2} \qquad (7\text{-}56)$$

If the source and load terminations are also specified, transformers may be needed at the input and output circuits to match the source and load impedances. G_1 and G_2 are determined by the midband gain specification.

Stagger-Tuned Design

In a stagger-tuned design, each circuit resonates at a different frequency. In this case, the pole-zero configuration for a maximally flat magnitude corresponds to fig 7-18. We rewrite (7-53) in a factored form as

$$A_v(s) = -\frac{g_m G_s}{C_1 C_2} \frac{s}{(s + s_1)(s + s_1^*)} \frac{s}{(s + s_2)(s + s_2^*)} \qquad (7\text{-}57)$$

For a maximally flat magnitude design with a specified center frequency f_0 and an overall bandwidth, $f_{3\,dB}$, we assign the pole location as in fig 7-18. Thus

$$s_1, s_1^* = -\frac{\omega_{3\,dB}}{2\sqrt{2}} \pm j\left(\omega_0 - \frac{\omega_{3\,dB}}{2\sqrt{2}}\right) \tag{7-58a}$$

Or the resulting quadratic with roots at s_1, s_1^* for the narrow-band case is given approximately by

$$s^2 + \frac{\omega_{3\,dB}}{\sqrt{2}} s + \omega_0^2\left(1 - \frac{\omega_{3\,dB}}{\sqrt{2}\omega_0}\right) \tag{7-58b}$$

Similarly

$$s_2, s_2^* = -\frac{\omega_{3\,dB}}{2\sqrt{2}} \pm j\left(\omega_0 + \frac{\omega_{3\,dB}}{2\sqrt{2}}\right) \tag{7-59a}$$

The corresponding polynomial is

$$s^2 + \frac{\omega_{3\,dB}}{\sqrt{2}} s + \omega_0^2\left(1 + \frac{\omega_{3\,dB}}{\sqrt{2}\omega_0}\right) \tag{7-59b}$$

Hence, in a design, the corresponding coefficients of (7-58b) and (7-59b) are equated with those of (7-53) and the various quantities are thus determined.

The midband gain $A_v(j\omega_0)$ is readily determined by the use of the narrow-band approximation. From (7-57), using the narrow-band approximation, we have

$$A_v(s) = -\frac{g_m G_s}{C_1 C_2} \frac{1}{2(s+s_1)} \frac{1}{2(s+s_2)} \tag{7-60}$$

Note, from fig 7-18 or from (7-58a) and (7-59a), that

$$|j\omega_0 + s_1| = |j\omega_0 + s_2| = \frac{\omega_{3\,dB}}{2}$$

Hence

$$[A_v(j\omega_0)]_{\text{stagger}} \approx -\frac{g_m G_s}{C_1 C_2}\left(\frac{1}{\omega_{3\,dB}}\right)\left(\frac{1}{\omega_{3\,dB}}\right) \tag{7-61}$$

But

$$\omega_{3\,dB} = \sqrt{2}\frac{G_1}{C_1} = \sqrt{2}\frac{G_2}{C_2} \qquad (7\text{-}62)$$

Therefore

$$[A_v(j\omega_0)]_{stagger} = -\frac{g_m G_s}{2G_1 G_2} \qquad (7\text{-}63)$$

It should be noted that, for the same bandwidth and capacitances, the gain of the stagger-tuned stage is larger by a factor of 2.44, since

$$\frac{[A(j\omega_0)]_{sync}}{[A(j\omega_0)]_{stagger}} = \frac{g_m G_s/(\omega_{3\,dB}/0.64)^2 C_1 C_2}{g_m G_s/2(\omega_{3\,dB}/\sqrt{2})^2 C_1 C_2} = \frac{(0.64)^2}{1} = \frac{1}{2.44} \qquad (7\text{-}64)$$

If in addition to ω_0 and $\omega_{3\,dB}$ specifications, the source and load terminations are also specified, as in the synchronously tuned design, the use of transformers at the input and output may be required. The use of the above expressions in a design example is straightforward (Problems 7-13 and 7-14) and will not be belabored any further.

Note that the cascode has negligible interaction, and thus improved stability and alignability (Section 7-12) result when such blocks are used in a design.

It should be pointed out that the two-transistor block using the emitter-coupled pair can also be used in tuned amplifiers in order to reduce interaction, and thus achieve better stability and alignability properties than the *CE* cascades only. The analysis of the emitter-coupled (*CC-CB* cascade) is similar to that of a cascode, and is left as an exercise in Problem 7-17.

7-11 Neutralization in Narrow-Band Tuned Amplifiers

The principal cause of instability in multistage *CE* tuned amplifiers is the feedback due to C_c in conjunction with the inductive load and source impedances, as discussed in Section 7-8. There are basically several methods to insure stability in narrow-band tuned amplifiers. One method was already discussed in conjunction with cascode amplifiers. Another method is to neutralize the effect of feedback, and yet another method is to sacrifice gain by mismatching the input and the output ports of the device. Of course, a combination of the latter two is also possible. We discuss neutralization in this section and mismatch is discussed in the next section.

Neutralization is a method whereby external circuitry is used and adjusted to approximately cancel y_{12} over the passband of the tuned amplifier. Recall that in a unilateral circuit $y_{12} = 0$ over the entire range of frequencies. In a neutralized circuit $y_{12}(j\omega_0) = 0$ and it is approximately equal to zero in the passband of the narrow-band tuned amplifier. One such neutralized CE stage is shown in fig 7-27, where Y_n is adjusted to cancel the effect of y_{12}. Usually the contribution

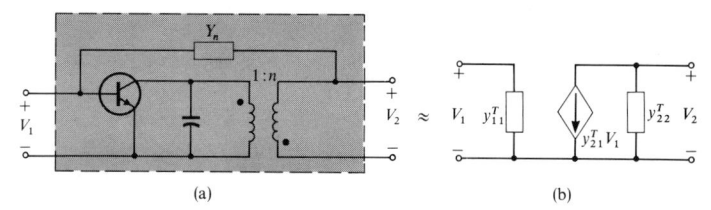

Fig 7-27 (a) Neutralization in a common-emitter amplifier stage; (b) Circuit model of (a) for $y^T_{12} = 0$

to y_{12} is due to C_c, thus Y_n is mainly capacitive. The negative sign applied to Y_n is provided by the polarity of the transformer. For such cases, i.e., $y_{12} \approx (b_{12}/\omega_0)$, we use a capacitance for Y_n, and thus choose the capacitance from the relation

$$nC_n \simeq \frac{b_{12}}{\omega_0} \qquad (7\text{-}65)$$

The overall two-port parameters of the neutralized stage in the passband of the tuned amplifier, in general, are given by

$$\begin{aligned} y^T_{11} &= y_{11} + Y_n & y^T_{12} &= -\frac{y_{12}}{n} - Y_n \approx 0 \\ y^T_{21} &= -\frac{y_{21}}{n} - Y_n & y^T_{22} &= \frac{y_{22}}{n^2} + Y_n \end{aligned} \qquad (7\text{-}66)$$

When $-y_{12}$ is purely capacitive, i.e., $y_{12}(j\omega_0) = 0 - jb_{12}$, then Y_n is capacitive and the real parts of the neutralized stage are the same as that of the original unneutralized stage. From (7-66), however, it should be noted that, in general $-y_{12}$, hence Y_n, may have a conductive component and the real parts would be affected.

For the neutralized stage, the maximum unilateral power gain G_U (for $y_{12}^T = 0$) is given by Problem 2-12:

$$G_U \equiv (G_A)_{\max}\bigg|_{y_{12}^T = 0} = \frac{|y_{21}^T|^2}{4g_{11}^T g_{22}^T} \qquad (7\text{-}67)$$

where g_{11}^T and g_{22}^T are the real parts of y_{11}^T and y_{22}^T, respectively. For the unneutralized stage, G_U provides a measure of the power gain capability of the device.

Neutralization is not the best solution to the stability of the amplifier. Since the transistor parameters are different, even for the same type, and they are functions of temperature and the operating point, effective neutralization cannot be achieved for mass production. For wide-band tuned amplifiers, these problems are further compounded. Hence, the neutralized design is not desirable and is, in general, to be discouraged. The usual practice in solving the stability problem is by mismatching the source and the load ports. In this manner, we achieve stability at the price of gain loss, as discussed in the next section. If the gain requirements are high, a combination of mismatch and neutralization may be used.

7-12 Mismatch in Tuned Amplifiers

By a mismatch, we mean the condition where the source and/or load conductance is deliberately chosen to be much larger than the conductive part of the input and/or output admittance parameters of the two-port. Mismatch achieves stability and reduces the interaction between the stages, as we shall demonstrate in the following. Consider the nonunilateral transistor and the associated input and output load admittances, as shown in fig 7-28. From the potential instability criterion for the overall two-port, we have

$$G_{11}G_{22} > \tfrac{1}{2}|y_{12}y_{21}|(1 + \cos\theta) \qquad (7\text{-}68)$$

Fig 7-28 Schematic of a nonunilateral transistor with source and load admittances at the center frequency

where

$$\theta = \tan^{-1} \frac{Im\ y_{12}y_{21}}{Re\ y_{12}y_{21}}$$

$$G_{11} = g_{11} + G_s \quad \text{and} \quad G_{22} = g_{22} + G_L$$

Note that *Re* and *Im* stand for "the real part of" and "the imaginary part of", respectively. If the two-port is mismatched such that $G_s \gg g_{11}$ and/or $G_L \gg g_{22}$, the inequality in (7-68) can be satisfied, and hence stability can be achieved to any desired margin. In the design of tuned amplifiers, a stability factor (or margin) K is defined, from (7-68), by

$$K = \frac{2G_{11}G_{22}}{|y_{12}y_{21}|(1 + \cos\theta)} \tag{7-69}$$

Usually K is chosen in the range $2 < K \leq 10$, the larger the K, the better the stability, and the lower the power gain of the two-port.

The effect of mismatch on reducing the interaction between the stages can be shown as follows. The power gain for a two-port, as derived in (2-49), is

$$G_P = \frac{P_o}{P_i} = \frac{|y_{21}|^2\ Re\ Y_L}{|y_{22} + Y_L|^2\ Re(Y_{in})} \tag{7-70}$$

where

$$Y_{in} = y_{11} - \frac{y_{12}y_{21}}{y_{22} + Y_L} \tag{7-71}$$

The interaction is exhibited by the dependence of Y_{in} on Y_L. In a unilateral circuit, $y_{12} = 0$ and Y_{in} is independent of Y_L. From (7-71), if $Y_L \gg y_{22}$ and

$$\left|\frac{y_{12}y_{21}}{Y_L}\right| \ll |y_{11}| \tag{7-72}$$

the effect of the load on the input circuit is negligible. We can rewrite (7-72) as

$$\delta = \frac{|y_{12}y_{21}|}{|Y_L y_{11}|} \ll 1 \tag{7-73}$$

Thus, if we keep δ small ($\delta \leq 0.1$ is usually quite acceptable), the variations in Y_L have a negligible effect on Y_{in}. A small value of δ also simplifies the alignment or tuning of the amplifier, and it is a measure of the alignability of the amplifier. Note that δ is exceedingly small for the cascode and the emitter-coupled amplifiers, and thus amplifiers using such amplifying blocks are also readily alignable.

In a mismatched tuned-amplier design, we choose a small δ. This simplifies analysis and design in that we can ignore the interaction between the input and output circuits in the preliminary design. The final design is aligned experimentally to compensate for this approximation. The power gain at the resonant frequency (actually, the resonant frequency is somewhat off from ω_0 due to y_{12}, but it is close enough) from (7-70), using $(g_{22} + G_L) = y_{22} + Y_L$ at resonance, is

$$G_P(\omega_0) = \frac{|y_{21}|^2 G_L}{|g_{22} + G_L|^2 \{g_{11} - [1/(g_{22} + G_L)] \operatorname{Re} y_{12} y_{21}\}} \quad (7\text{-}74)$$

Substitution of the mismatch factor $m \equiv G_L/g_{22}$ in (7-74) yields

$$G_P(\omega_0) = \frac{m|y_{21}|^2}{(1 + m)^2(g_{11}g_{22}) - (1 + m)\operatorname{Re} y_{12} y_{21}} \quad (7\text{-}75)$$

Note that (7-75) is small compared to (7-67), which shows the sacrifice of gain due to mismatch. The mismatch loss L_m is defined by

$$L_m = \frac{G_P(\omega_0)}{G_U(\omega_0)} = \frac{4m}{(1 + m)^2 - [(1 + m)/g_{11}g_{22}]\operatorname{Re} y_{12} y_{21}} \quad (7\text{-}76)$$

If $(1 + m)g_{11}g_{22} \gg \operatorname{Re} y_{12} y_{21}$, (7-76) reduces to

$$L_m = \frac{G_P(\omega_0)}{G_U(\omega_0)} \approx \frac{4m}{(1 + m)^2} \quad (7\text{-}77)$$

The use of the above expressions in a design is best illustrated by the following example.

7-13 A Design Example of a Nonunilateral Tuned Amplifier

In order to illustrate the design of nonunilateral tuned amplifiers, we consider an example for the mismatch method discussed in Section

7-12. Specifically, we use a large mismatch in order to reduce considerably the effect of the interaction and achieve an alignable tuned amplifier. We then ignore the interaction in the preliminary design, and the final amplifier is aligned experimentally to compensate for this approximation.

Let the transistor parameters, measured at $f = 10$ MHz, be given as in (7-23):

$$y_{11} = (2.0 + j0.5) \times 10^{-3} \text{ mho} \qquad y_{12} = -(1.0 + j5.0) \times 10^{-5} \text{ mho}$$

$$y_{21} = (20 - j5.0) \times 10^{-3} \text{ mho} \qquad y_{22} = (2.0 + j4.0) \times 10^{-5} \text{ mho}$$

The specifications are as follows: The overall bandwidth is to be $f_{3\,\text{dB}} = 100$ kHz centered at $f_0 = 10$ MHz. The source and load impedances are $R_s = R_L = 100\,\Omega$. The power gain at the center frequency is desired to be $G_P(j\omega_0) \approx 4000$, i.e., 36 dB. To have a small noise figure, the input impedance, as seen by the first stage transistor, is required to be 500 ohms.

The transistor at $f = 10$ MHz is potentially unstable [see (7-44)] and we must use mismatch if neutralization is not used. We *estimate* the power gain capability of a stage from the unilateral power gain, namely

$$G_U = \frac{|y_{21}|^2}{4g_{11}g_{22}} = \frac{425 \times 10^{-6}}{4(2 \times 10^{-3})(2 \times 10^{-5})} = 26.6 \times 10^2 = 34.2 \text{ dB}$$

The power gain of the stage is lower than the unilateral gain by the amount of the mismatch loss, as given in (7-77).

It would appear that two stages would provide the required gain. It turns out, as will be shown soon, that for good alignability i.e., $\delta = 0.05$, $m = 500$, and from (7-77) $L_m = -21$ dB. If we allow about 1 dB loss for each transformer, it is seen that for the specifications of this problem we must use at least three stages. We thus proceed with the design for three stages. We also choose to use single-tuned stages for simplicity. If better selectivity is required, we may use double-tuned stages. For three stages of tuned amplifiers, we can use, at most, four single-tuned circuits: one at the input, one at the output, and two at the intermediate stages. In some problems, we may not need a tuned circuit at the input. Since the selectivity is not specified, we have a choice between the synchronously tuned design and the staggered-tuned design. We choose the former. For the four single-tuned synchronously tuned designs, the individual stage bandwidth $(\omega_{3\,\text{dB}})_1$ and the overall

7-13 A Design Example of a Nonunilateral Tuned Amplifier

bandwidth are related by

$$(\omega_{3dB})_1 = \frac{(\omega_{3dB})_{overall}}{S_n} \frac{(\omega_{3dB})_{overall}}{\sqrt{2^{1/4} - 1}}$$

Hence

$$(\omega_{3dB})_1 = \frac{(\omega_{3dB})_{overall}}{0.44} = 2.28[2\pi(10^5)]\,\text{rad/sec}$$

We choose $\delta \leq 0.1$ and thus ignore the interaction of the input and output circuits in the preliminary design.

The circuit for the last stage, as shown in fig 7-29a, reduces to that shown in fig 7-29b. The output circuit in fig 7-29b can now be designed

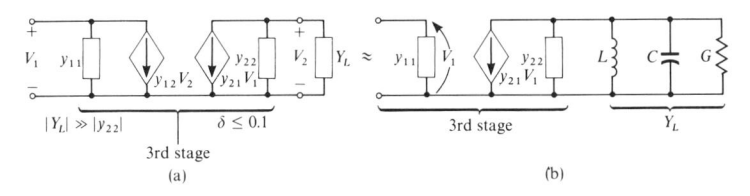

Fig 7-29 (a) Nonunilateral two-port amplifier stage with a tuned-circuit load; (b) Simplified approximate equivalent of (a) if the effect of changes in the load admittance on the input admittance is negligible, i.e., $\delta \leq 0.1$

as discussed in Section 7-4, for $\omega_{3dB} = 2\pi(2.28 \times 10^5)\,\text{rad/sec}$ and $\omega_0 = 2\pi(10^7)\,\text{rad/sec}$. From

$$\delta = \frac{|y_{12}y_{21}|}{|y_{11}Y_L|} \leq 0.1$$

we obtain

$$|Y_L| \geq \frac{106 \times 10^{-8}}{0.1(2.06 \times 10^{-3})} \geq 5 \times 10^{-3}$$

Since, at resonance, the reactances cancel out, we can choose ($G_L = 5 \times 10^{-3}$). However, the load resistance specified $R_L = 100\,\Omega$. Since $G_L = 1/R_L = 10^{-2}$ is not too far from the requirements of G_L by δ, we choose $G_L = 10^{-2}$ and thus eliminate the need of an impedance-transforming network. This choice also yields $\delta = 0.05$, which results

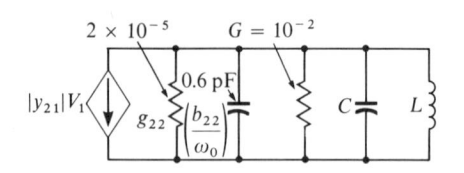

Fig 7-30 Last (output) stage single-tuned circuit

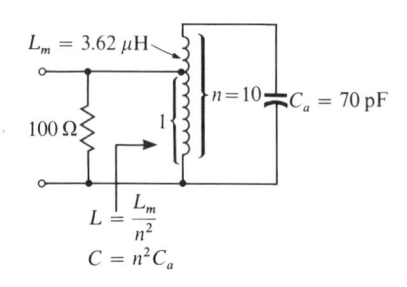

Fig 7-31 Impedance-transforming network

in reduced interaction and better alignability. Of course, the price paid for this high mismatch is reduced power gain, which in some cases may necessitate an additional stage. For this case, $m = G_L/g_{22} = 500$ and L_m, from (7-77), is equal to -21 dB. If a gain of 12 dB/stage is achieved, three stages would be sufficient for the specifications of this problem. Since $G_P = 34.2$ dB $- 21$ dB ≈ 13 dB/stage, we can achieve the design with good alignability using the three-stage, four single-tuned circuit design.

The output single-tuned circuit is shown again in fig 7-30. From the relation

$$\frac{G_t}{C_t} = (\omega_{3\,\text{dB}})_1 \qquad (7\text{-}78)$$

we determine C_t, namely

$$C_t = \frac{g_{22} + G}{2\pi(2.28 \times 10^5)} \simeq \frac{10^{-2}}{1.43 \times 10^6} = 7000 \text{ pF}$$

$$C = C_t - \frac{b_{22}}{\omega_0} \approx 7000 \text{ pF}$$

$$L = \frac{1}{\omega_0^2 C_t} = \frac{1}{[2\pi(10^7)]^2 7.0 \times 10^{-9}} = 0.0362 \text{ }\mu\text{H}$$

Since the element values are not practical (low L and high C), we use an autotransformer, as shown in fig 7-31. For example, if we choose $n = 10$, the total magnetizing inductance L_m and the actual capacitance C_a are

$$L_m = n^2 L = 100(0.0362 \text{ }\mu\text{H}) = 3.62 \text{ }\mu\text{H}$$

$$C_a = \frac{C}{n^2} = \frac{7000}{100} \text{ pF} = 70 \text{ pF}$$

The last intermediate stage is designed next. Because δ is intentionally chosen to be very small ($\delta \simeq 0.05$ in this case) we can ignore the effect of the interaction. The interstage circuit is shown in fig 7-32, and we use again the same δ factor, that is, we choose the load of the second stage to be 100 ohms. Since the input impedance of stage three, ignoring the interaction, is a shunt RC with the resistance equal to $g_{11} = 2 \times 10^{-3}$ mhos and a capacitance equal to $b_{11}/\omega_0 = 7.97$ pF, the transformer turns ratio required is

$$g_{11} = \frac{G}{n_1^2} \rightarrow n_1^2 = \frac{G}{g_{11}} = \frac{10^{-2}}{2 \times 10^{-3}} = 5$$

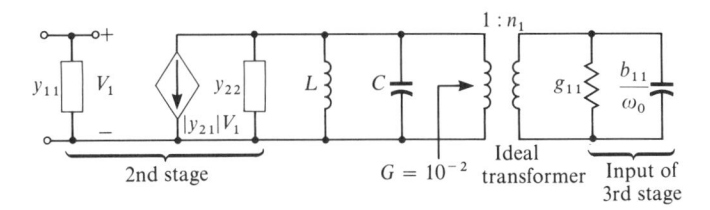

7-13 A Design Example of a Nonunilateral Tuned Amplifier

Fig 7-32 Second stage and its load circuit for δ very small ($\delta = 0.05$).

The values of C and L are determined as in the last stage. From (7-78), we determine C_t and

$$C = C_t - \frac{b_{22}}{\omega_0} - n^2\left(\frac{b_{11}}{\omega_0}\right) \approx 7000 - 40 = 6960 \text{ pF}$$

$$L = \frac{1}{\omega_0^2 C_t} = 0.0362 \text{ } \mu\text{H}$$

Again we use an autotransformer to make these values practical. Note that we must add an additional tap in the coil for matching g_{11} to $G = 10^{-2}$, as shown in fig 7-33. The first intermediate stage design is identical to the one shown in fig 7-33 and needs no further discussion. The input stage, ignoring the interaction, is shown in fig 7-34. The

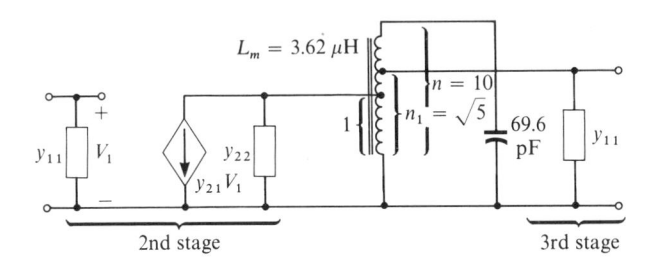

Fig 7-33 Circuit of fig 7-32 with transformer replaced by a double-tapped autotransformer

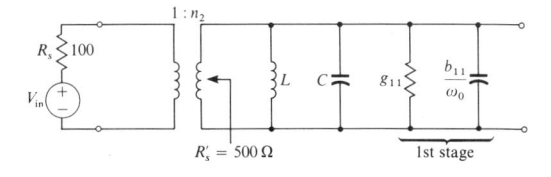

Fig 7-34 Input stage if the load effect on the input admittance is negligible

transformer turns ratio required to get the desired 500 Ω source facing the first transistor stage is

$$R'_s = n_2^2 R_s \rightarrow n_2^2 = \frac{500}{100} = 5$$

From (7-78), we have

$$C_t = \frac{g_{11} + G'_s}{2\pi(2.28 \times 10^5)} = \frac{2 \times 10^{-3} + 2 \times 10^{-3}}{2\pi(2.28 \times 10^5)} = 2795 \text{ pF}$$

$$C = C_t - \frac{b_{11}}{\omega_0} = 2795 - 7.9 = 2787 \text{ pF}$$

$$L = \frac{1}{\omega_0^2 C_t} = \frac{1}{[2\pi(10^7)]^2 2.79 \times 10^{-9}} = 0.0915 \text{ }\mu\text{H}$$

Again we can use an impedance-transforming network, with $n = 10$, which gives

$$L_m = n^2 L = 9.15 \text{ }\mu\text{H}, \qquad C_a = 27.9 \text{ pF}$$

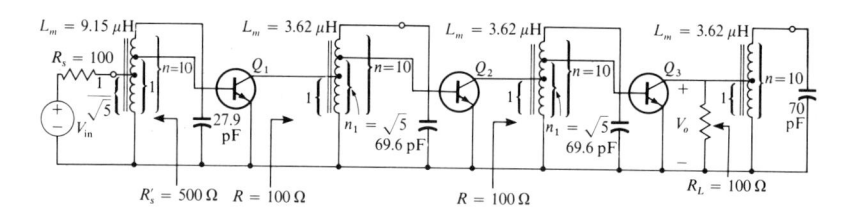

Fig 7-35 Circuit of the design example, exclusive of biasing circuitry

The complete circuit, exclusive of biasing circuitry, is shown in fig 7-35. The power gain of the individual stages of the circuit, from (7-75), is

$$G_{P1}(\omega_0) = \frac{m|y_{21}|^2}{(1 + m)^2 g_{11} g_{22} - (1 + m) \text{Re}[y_{12} y_{21}]}$$

$$= \frac{500(425) \times 10^{-6}}{(1 + 500)^2 2 \times 10^{-3} 2 \times 10^{-5} - (1 + 500)(-9.2 \times 10^{-8})}$$

$$= 21 \text{ or } 13.2 \text{ dB}$$

7-14 Crystal Bandpass Filters

The total power gain, in dB, is

$$G_P = G_{P1} + G_{P2} + G_{P3} = 39.6 \text{ dB}$$

If we allow a total loss of 3.6 dB for the transformers we still would have a gain of 36 dB, which meets the specification of this problem. The final important step is the experimental alignment of the complete amplifier to compensate for the approximation of ignoring the interaction. Since $\delta \approx 0.05$ in this example, the interaction is very small and alignment is not a difficult problem.

Note that in this example, the specified G_s is matched to the input conductance of the first transistor stage, hence $G_P = G_T$. In general, this may not be so and, as you recall from (2-58) and (2-60), $G_P \geq G_T$. Finally, it should be pointed out that mismatch design must be used in wide-band tuned amplifiers, since neutralization cannot be effective over a wide band of frequencies. For a wide-band tuned amplifier, the transistor parameters may vary appreciably from those at the center frequency, and hence the design is not a simple problem.

7-14 Crystal Bandpass Filters

When extremely high Q and excellent stability are required in narrow-band bandpass filters, the use of crystal filters is practical. In crystal filters, narrow-band has a stricter interpretation, roughly a bandwidth which is less than 0.4 per cent of the center frequency. The use of quartz crystals as a component for frequency control of oscillators was discussed briefly in Section 6-12. In this section, we consider the use of crystals in very high-Q bandpass filters.

The electrical equivalent circuit of a crystal is shown in fig 7-36. Crystals in the range of 10 kHz to 100 MHz are commercially available. The shunt capacitance C_o is of the order of 5 pF, while L_m ranges from several thousand henries to millihenries in the above frequency range, respectively. The resonant frequency of the crystal is $f_0 = 1/2\pi\sqrt{L_m C_m}$ and $Q = \omega_0 L_m/R_m$. For practical LC circuits, a Q of 100 or better can be achieved, while for crystal filters a Q larger than 20,000 can easily be achieved. Table 7-1 lists the parameters of several different type cut crystals. The various cuts, such as AT, NT, etc., are described in crystal literature and need not concern us here. Stability of frequency with both age and temperature is of the order of 0.001 per cent (i.e., 10 parts per million)

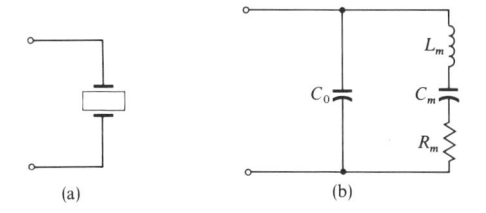

Fig 7-36 (a) Crystal graphic symbol; (b) Circuit model for (a)

7/Analysis and Design Examples of Bandpass Amplifiers

TABLE 7-1 *Typical Characteristics of Precision Crystals*[a]

Characteristic	10 kHz XY'	100 kHz NT	1000 kHz AT	10 MHz AT	100 MHz AT
Q	100,000	120,000	$2\text{–}5 \times 10^6$	145,000	110,000
Inductance (L_m)	7900 h	365 h	1.56 h	11.6 mH	7.0 mH
Shunt capacitance (C_0)	20 pF	8 pF	10 pF	7 pF	7 pF
Capacitance ratio (C_0/C_m)	800	1000	592	275	8700
Series resistance (R_m)	5000 Ω	2000 Ω	2–5 Ω	5 Ω	40 Ω
Aging f vs t	0.5 ppm/wk	0.3 ppm/wk	0.35 ppm/wk	0.35 ppm/wk	0.7 ppm/wk

[a] From *Electro-Technology*, June 1969, page 49.

In crystals, usually C_o and C_o/C_m, in addition to nominal f_0, are specified in a design. The value of f_0 can then be trimmed with a high degree of precision, by adjusting L_m. Since Q is very high, it can be assumed infinite in the initial analysis and design. Because of the very high value of Q, the use of crystals puts a severe limitation on achievable bandwidth. As will be shown later, if the design calls for a pole pair with a Q of less than $2(C_0/C_m)$, shunt inductors will be required across the crystal to tune out part of C_0.

When crystals are used in bandpass filtering for good attenuation characteristics, bridge circuits must generally be used in order to balance out the parasitic capacitance (shunt capacitance C_0). The lattice structure (see fig P5-25) is one of the most general two-port

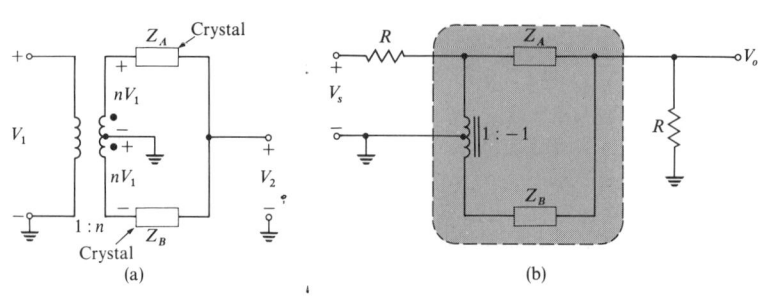

Fig 7-37 (a) Semilattice structure; (b) The semilattice of (a) shown in simplified form for $n = 1$ and including terminations

7-14 Crystal Bandpass Filters

structures. Because of the absence of common ground, however, the semilattice structure shown in fig 7-37a is more practical.

Consider the semilattice structure in fig 7-37a. The open-circuit impedance parameters are

$$[z_{ij}] = \frac{1}{4n^2}\begin{bmatrix} Z_B + Z_A & n(Z_B - Z_A) \\ n(Z_B - Z_A) & n^2(Z_B + Z_A) \end{bmatrix} \quad (7\text{-}7a)$$

Equation (7-7a), for $n = 1$, is identical to the z-parameters of the lattice (Problem 5-25) except for the factor of $\frac{1}{2}$. Note that, because the z-parameters are reciprocal and symmetrical, for $n = 1$ (i.e., $z_{12} = z_{21}$ and $z_{11} = z_{22}$), we could interchange the input and output without any effect on the circuit. Also observe that, in the structure shown in fig 7-37, the tapped transformer can be replaced by a differential output operational amplifier if so desired (Problem 7-29).

In the sequel, we shall consider $n = 1$ without any loss of generality. The transfer voltage ratio of the terminated lattice, shown in simplified form in fig 7-37b, is given by

$$\frac{V_o}{V_s}(s) = \frac{\frac{1}{2}(Z_B - Z_A)}{Z_B + Z_A + Z_A Z_B/2R + 2R} \quad (7\text{-}80)$$

where Z_A and Z_B represent the impedances of the crystals in branches A and B, respectively. These impedances will include shunt inductors if the design requirements call for Q which is less than $2C_0/C_1$ of the crystal, as we shall soon see. In the following, it will be assumed that crystals A and B have equal shunt capacitance and equal capacitance ratios. This is not a limitation, since a trimmer can always be placed across the crystal with the lowest shunt capacitance to equalize them. The assumption, however, simplifies the design, since the transmission zeros of (7-80) under this condition will be at the origin. The poles of the transfer voltage ratio are found from (7-80),

$$Z_A Z_B + 2R(Z_A + Z_B) + 4R^2 = 0 \quad (7\text{-}81)$$

which can be factored as

$$(Z_A + 2R)(Z_B + 2R) = 0 \quad (7\text{-}81b)$$

Therefore, we will consider the roots of each factor. For convenience, we drop the subscripts, A and B, and consider the roots of

$$Z_1 + 2R = 0 \quad (7\text{-}82)$$

where, from fig 7-36b, the impedance Z_1 is given by

$$Z_1 = \frac{s^2 + (\omega_x/Q_x)s + \omega_x^2}{sC_o[s^2 + (\omega_x/Q_x)s + \omega_x^2(1 + C_m/C_o)]} \quad (7\text{-}83)$$

where

$$\omega_x^2 = \frac{1}{L_m C_m} \quad \text{and} \quad Q_x = \omega_x \frac{L_m}{R_m}$$

The value of Q_x is typically of the order 10^5, and if the Q's of the desired poles are much less than this value, Q_x can be assumed to be infinite, i.e., R_m ignored with negligible error. Equation (7-82) then becomes

$$s^3 + \frac{1}{2RC_o}s^2 + \omega_x^2\left(1 + \frac{C_m}{C_o}\right)s + \frac{\omega_x^2}{2RC_o} = 0 \quad (7\text{-}84)$$

One pole of (7-84) must be real and, therefore, nondominant in a narrow-band response. The remaining pair of poles will be the desired complex conjugate pairs lying close to the $j\omega$ axis with a loaded Q_L, which is a function of termination resistors R. We can rewrite (7-84) as

$$(s + p_o)\left(s^2 + \frac{\omega_0}{Q_L}s + \omega_0^2\right) = 0 \quad (7\text{-}85)$$

By equating coefficients of (7-84) and (7-85), we obtain

$$\omega_x^2 = \omega_0^2\left\{1 - \frac{1}{2Q_L^2(1+\alpha)} - \frac{\alpha}{2(1+\alpha)} \times \left[1 - \sqrt{1 - \frac{2}{\alpha Q_L^2}\left(1 + \frac{2}{\alpha}\right) + \frac{1}{\alpha^2 Q_L^4}}\right]\right\} = \omega_0^2(1 - \varepsilon) \quad (7\text{-}86)$$

$$R = \frac{Q_L}{2\omega_0 C_o}\left(1 - \frac{\omega_x^2}{\omega_0^2}\right) = \frac{Q_L \varepsilon}{2\omega_0 C_o} \quad (7\text{-}87)$$

where ε is determined from (7-86) and

$$\alpha = \left(\frac{C_o}{C_m}\right)^{-1} \quad (7\text{-}88)$$

7-14 Crystal Bandpass Filters

For the above equations, we make the following observations:
(a) ω_x is less than ω_0, namely, $\omega_x \approx \omega_0(1 - \frac{1}{2}\varepsilon)$ since $\varepsilon \ll 1$
(b) the ratio of ω_0 to ω_x is a function of α and Q_L
(c) there exists a minimum value of Q_L, Q_{min}, for which the sign of the term under the radical in (7-86) is positive.

The minimum value Q_{min} is determined from (7-86),

$$1 - \frac{2}{\alpha Q_{min}^2}\left(1 + \frac{2}{\alpha}\right) + \frac{1}{\alpha^2 Q_{min}^4} = 0 \qquad (7\text{-}89a)$$

Since $\alpha = (C_0/C_m)^{-1}$ is typically much less than unity, we have

$$Q_{min} \approx \frac{2}{\alpha} = \frac{2C_0}{C_m} \qquad (7\text{-}89b)$$

Hence, for $Q < 2C_0/C_m$, the shunt capacitance C_0 of the crystal must be tuned out by a shunt inductance. Also note that the value of R determined from (7-87) will determine the roots of both:

$$Z_A + 2R = 0 \quad \text{and} \quad Z_B + 2R = 0$$

Therefore

$$\frac{Q_A}{2\omega_{0A}C_{0A}}\left[1 - \left(\frac{\omega_x}{\omega_{0A}}\right)^2\right] = \frac{Q_B}{2\omega_{0B}C_{0B}}\left[1 - \left(\frac{\omega_x}{\omega_{0B}}\right)^2\right] \qquad (7\text{-}90)$$

Equation (7-90) is approximately satisfied if the real parts of both pairs of complex frequencies are identical. This situation generally occurs in most of the narrow-band design techniques. Two examples will illustrate the design procedure.

EXAMPLE 1

It is desired to design a second-order Butterworth filter centered at $f_0 = 10$ MHz and a bandwidth $f_{3\,dB} = 10$ kHz. The parameters of the crystal at 10 MHz are given as $C_0/C_m = 250$, $C_0 = 3.2$ pF, $Q \approx 2 \times 10^5$.

From the specifications, the poles ought to be located as in fig 7-38a (only the upper half s-plane is shown). Thus we have

$$\omega_{0A} = 2\pi\left(10^7 + 0.707 \times \frac{10^4}{2}\right) = 2\pi(10{,}003{,}536) \text{ rad/sec}$$

$$\omega_{0B} = 2\pi\left(10^7 - 0.707 \times \frac{10^4}{2}\right) = 2\pi(9{,}996{,}466) \text{ rad/sec}$$

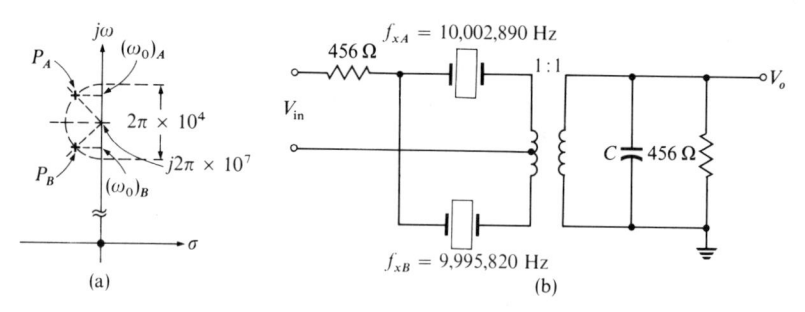

442   7/Analysis and Design Examples of Bandpass Amplifiers

Fig 7-38 (a) Pole-zero plot of the design example; (b) Circuit of the narrow-band design example

Hence

$$Q_L = \frac{\omega_{0B}}{2\,\mathrm{Re}\,(P_B)} \approx \frac{\omega_{0A}}{2\,\mathrm{Re}\,(P_A)} = 1414$$

Note that Q_L is much smaller than the Q of the crystal ($Q \approx 2 \times 10^5$). The assumption of ignoring the Q of the crystal in the analysis is valid. Knowing C_o and C_o/C_m of the crystal (which are specified) and the desired Q_L, from (7-86) and (7-87) we find

$$f_0 - f_x \approx \tfrac{1}{2}\varepsilon f_0 = 646 \text{ Hz}$$

$$R \approx \frac{Q_L \varepsilon}{2\omega_0 C_o} = 456 \text{ }\Omega$$

Thus, the resonant frequencies of the crystals are at

$$f_{xA} = f_{0A} - 646 = 10{,}002{,}890 \text{ Hz}$$
$$f_{xB} = f_{0B} - 646 = 9{,}995{,}820 \text{ Hz}$$

The above accuracy in f_x can be achieved in practice for crystals. The circuit realization is shown in fig 7-38b. Note that the input and output of the semilattice are interchanged for convenience and the capacitor C has been added to resonate with the inductance of the nonideal transformer at the center frequency $f_0 = 10$ MHz, so as to cancel it out.

EXAMPLE 2

It is desired to design a second-order Butterworth filter with center frequency at $f_0 = 1$ MHz, and a bandwidth $f_{3\,\mathrm{dB}} = 2$ kHz. Use the following crystal, which is typical, at 1 MHz:

$$\frac{C_o}{C_m} = 500, \qquad C_o = 3 \text{ pF}, \qquad Q \approx 10^6$$

As before, we determine f_{0A}, f_{0B} and Q_L:

$$\left. \begin{array}{l} f_{0A} = 1{,}000{,}707 \\ f_{0B} = 999{,}294 \end{array} \right\} \quad \text{and} \quad Q_L = 707$$

Using a crystal with a capacitance ratio of 500 (i.e., $Q_{\min} = 1000$), a Q_L of 707 cannot be obtained, as was discussed earlier. This class of filters is referred to as "wide-band" in crystal filter literature. A Q_L of 707 can be obtained, however, if an inductance L_0 is placed across the crystal to resonate with the capacitor C_o. It will be assumed that this inductor has a series resistance R_0. By equating the coefficients as before, the corresponding equations for the resonant frequency of the crystal and for the termination resistance are found to be

$$\omega_x^2 = \omega_0^2 \left\{ 1 - \frac{1}{2Q_L^2} \frac{1-\gamma}{1+\alpha-\gamma} + \frac{\alpha}{2(1+\alpha-\gamma)} \times \right.$$

$$\left. \left[-1 + \sqrt{1 - \frac{2}{\alpha Q_L^2}(1-\gamma) + \left(\frac{1}{Q_L^4} - \frac{4}{Q_L^2} \right) \left(\frac{1-\gamma}{\alpha} \right)^2 } \right] \right\} \quad (7\text{-}91)$$

and

$$R = \frac{1}{2C_o\omega_0^2} \frac{\omega_0/Q_L + (R_o/L_o)[\omega_x^2/\omega_0^2 - 1]}{\eta - 1 + 1/Q^2 - R_o/\omega_0 L_o Q_L + (\omega_x^2/\omega_0^2)(1+\alpha-\eta)} \quad (7\text{-}92)$$

where

$$\eta = \frac{1}{L_o C_o \omega_0^2} \quad \text{and} \quad \gamma = \frac{\eta}{1 - R_o/\omega_0 L_o Q_L + R_o^2/\omega_0^2 L_o^2} \quad (7\text{-}93)$$

Note that if $\gamma = 1$ then (7-91) becomes $\omega_x^2 = \omega_0^2$.

Returning to the example, let us arbitrarily choose a 10 mH inductor which has a 10 Ω series resistance. From (7-93), we have

$$\eta = 0.8444 \qquad \gamma \approx \eta = 0.8444$$

From (7-91) and (7-92), we obtain

$$f_0 - f_x = 79 \text{ Hz} \quad \text{and} \quad R = 19 \text{ k}\Omega$$

The circuit realization is shown in fig 7-39. The capacitor C is added for the reasons stated in example 1. Finally, note that an arbitrary load resistor can be accommodated since the turns ratio of the transformer need not be unity to begin with.

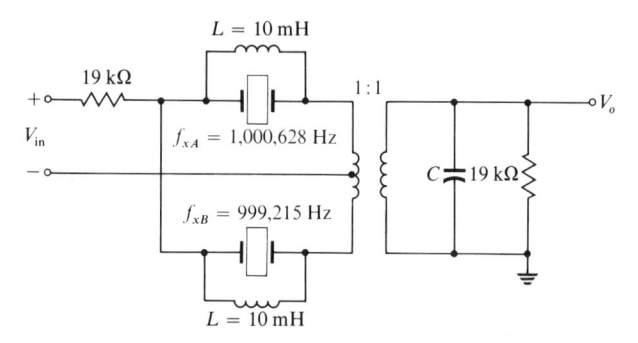

Fig 7-39 Circuit of the wideband design example

A general active crystal configuration which can be used to design any order filter is shown in fig 7-40a. The section with one crystal realizes one pole, while those with two crystals realize a pole-pair around $j\omega_0$ as shown in fig 7-40b. (Only the portion of the s-plane around $j\omega_0$ in the upper half plane is shown.) The buffer amplifiers are readily implemented by operational amplifiers, and gain can also be achieved if so desired at ω_0, provided the gain-bandwidth product of the compensated operational amplifier is larger than f_0. If gain requirements are unimportant, simple emitter followers can be used for the buffer stages.

Ceramic resonators, which exhibit properties similar to those of quartz crystals, and which are compatible with hybrid integrated circuits, are also finding wide applications in *IF* amplifiers, *RF* tuned amplifiers, and other communication networks.

7-15 Tuned Amplifiers Using Active-*RC* Circuits

In the preceding sections we considered the use of coils or transformers in the tuned circuits. This is not so desirable in integrated

7-15 Tuned Amplifiers Using Active-RC Circuits

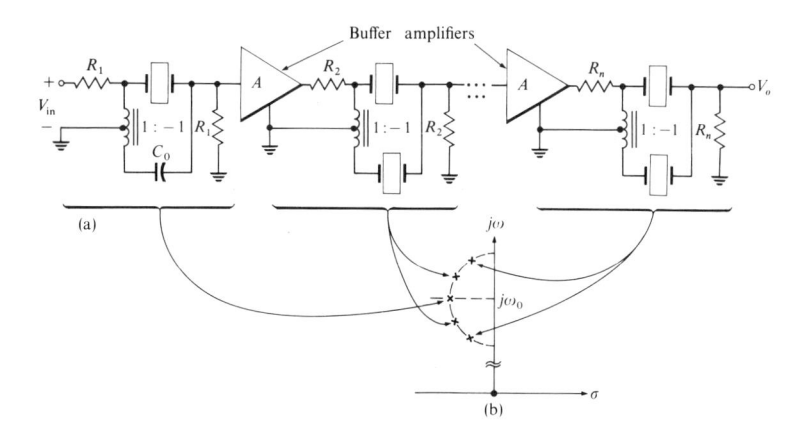

Fig 7-40 (a) Configuration for a general nth order active crystal filter; (b) Pole-zero plot of the circuit

circuits, and cannot be used in monolithic integrated circuits. Even in hybrid integrated circuits, the coils will not be suitable in size at low frequencies. In this section, we discuss briefly methods of realizing tuned circuits with amplifiers and *RC* circuits. The main drawback of active-*RC* tuned amplifiers is that a very high value of Q at high frequencies cannot be achieved that has good stability and sensitivity and, at the same time, is suitable for assembly-line production. The sensitivity topic will be discussed in the next section.

The basic idea in the design of active-*RC* tuned amplifiers is the use of feedback in the circuit either to modify the characteristics of the active device or the entire circuit, to obtain the desirable transmission characteristics. As an example, consider the schematic representation of the single-loop feedback amplifier intended as an active-*RC* tuned amplifier, shown in fig 7-41, where **a**(s) is a high-gain wide-band amplifier; **f**(s) is an *RC* two-port or an active-*RC* two-port. The *RC* two-port may be a null network such as the Twin-Tee circuit shown in fig 7-42.

Consider the use of the Twin-Tee network in the feedback path. The short-circuit reverse transfer admittance of the Twin-Tee circuit of fig 7-42, as obtained by straightforward analysis, is

$$y_{12} = -\frac{C_2}{2} \frac{[s^2 + (8R_2/R_1)/\tau^2]}{(s + 2/\tau)} \quad (7\text{-}94)$$

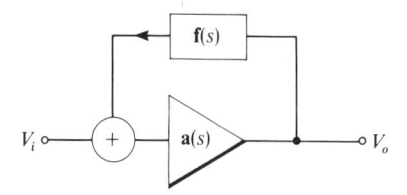

Fig 7-41 Schematic of an active *RC* tuned-amplifier

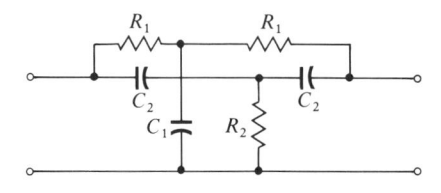

Fig 7-42 Twin-tee null-producing network

7/Analysis and Design Examples of Bandpass Amplifiers

where $\tau \equiv R_1 C_1$ and the constraint $R_1 C_1 = 4 R_2 C_2$ has been used in (7-94). Since the zeros of y_{12} from (7-94) are on the $j\omega$-axis, we have a null at

$$\omega_0 = \pm j \frac{1}{\tau} \sqrt{8 R_2/R_1} = \pm j \frac{\sqrt{8 R_2/R_1}}{R_1 C_1} \quad (7\text{-}95)$$

From the basic feedback equation (6-5), provided the system is stable, we have

$$A = \frac{a}{1 - af} \approx -\frac{1}{f} \text{ for } |af| \gg 1 \quad (7\text{-}96)$$

From (6-14), for the shunt-shunt configuration $\mathbf{f} = -R_L y_{12}$, hence we obtain the responses shown in figs 7-43a, b. It is to be noted that both ω_0 and Q of the circuit depend upon the gain of the amplifier, $\mathbf{a}$, as can be readily verified by a root-locus sketch.

In Chapter 8, various techniques and many circuits capable of realizing tuned-circuit characteristics are given. We consider only one such circuit in this section. The simplest one is shown in fig 7-44.

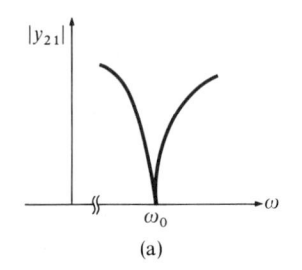

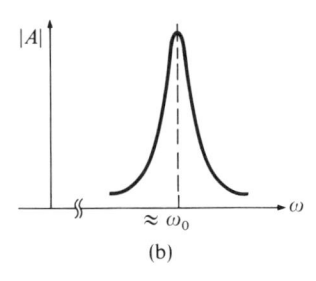

Fig 7-43 (a) Magnitude response of the $\mathbf{f}(s)$ circuit; (b) Magnitude response of the closed-loop gain function when the amount of feedback is very large

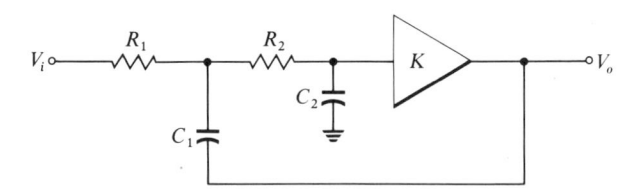

Fig 7-44 Simple active RC bandpass network

The voltage gain of the circuit in fig 7-44, assuming for the moment that K is an ideal voltage-controlled voltage source (such as would be obtained by a noninverting operational amplifier, see fig 8-31), is given by

$$A_v = \frac{V_o}{V_i} = \frac{K}{R_1 R_2 C_1 C_2 s^2 + [R_1 C_1 (1 - K) + (R_1 + R_2) C_2] s + 1} \quad (7\text{-}97)$$

For a high Q (i.e., narrow-band case), the presence or absence of the zero has no effect on the response in the passband of the tuned amplifier as long as it is far removed from the dominant poles. Hence the poles of the circuit can be designed for the desired center frequency and Q by the following relations [compare with (7-15b)]:

$$R_1 R_2 C_1 C_2 = \frac{1}{\omega_0^2} \qquad (7\text{-}98)$$

$$R_1 C_1 (1 - K) + (R_1 + R_2) C_2 = \frac{1}{\omega_0 Q} \qquad (7\text{-}99)$$

EXAMPLE

Consider the design of fig 7-44 for $f_0 = 1\,\text{kHz}$ and $Q = 25$. From (7-98), set $R_1 C_1 = R_2 C_2 = \tau$ and

$$\tau^2 = \frac{1}{\omega_0^2} \rightarrow \tau = \frac{1}{\omega_0} = 1.59 \times 10^{-4}$$

Among the many possible choices, one may select $R_1 = R_2 = 10\,\text{k}\Omega$. Hence $C_1 = C_2 = 0.0159\,\mu\text{F}$.

From (7-99), we determine the value of K,

$$\tau(1 - K) + 2\tau = \frac{1}{\omega_0 Q}$$

or

$$K = 3 - \frac{1}{\omega_0 Q \tau} = 3 - \frac{1}{Q} = 2.96$$

The voltage gain of this design at the center frequency is

$$|A_v(j\omega_0)| = KQ = 2.96(25) = 74$$

Note that, since $1/Q = 3 - K$ in this design, a difference between two almost equal quantities, the value of Q is quite dependent on the gain of the amplifier and must be designed and adjusted to the desired accuracy. Note also that if the gain of this amplifier is equal to or larger than 3, the circuit will be unstable. One method to guarantee stability in this design is to use the operational amplifier K in the unity-gain mode rather than any other low value. Unity gain in an operational amplifier is achieved as shown in fig 7-45 (see also fig 6-43). For this special case, the circuit of fig 7-44 is always stable, since the loop gain

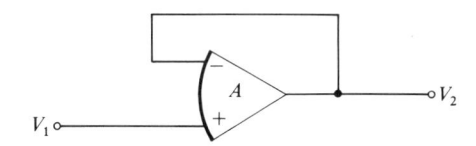

Fig 7-45 Noninverting unity-gain block as obtained from an operational amplifier ($K = V_2/V_1 \approx 1$ for $A \rightarrow \infty$)

can never exceed unity. The above design example under this condition is as follows:

For $K = 1$, and if we let $R_1 = R_2 = R = 50\ \text{k}\Omega$, from (7-99) we obtain

$$2RC_2 = \frac{1}{\omega_0 Q} \rightarrow C_2 = \frac{1}{2\omega_0 QR} = 63.7\ \text{pF}$$

From (7-98) we obtain $C_1 = 2Q/\omega_0 R = 0.159\ \mu\text{F}$. The voltage gain at the center frequency is $|A_v(j\omega_0)| = KQ = 25$. The inclusion of the input and the output impedance and the frequency dependence of the operational amplifier as a one-pole will show that the maximum value of Q is quite limited. This is generally true for almost all active RC networks, especially where very high-Q, low-sensitivity, frequency-selective amplification at high frequencies is desired.

7-16 Sensitivity in Active-*RC* Circuits[4]

The transfer function of a network changes from its nominal designed values due to changes in the active and/or passive elements. The change in circuit element values may be caused by a variety of reasons. For example, in an active network the active element parameters, such as β_0 of a transistor, change because of environmental temperature change and/or transistor replacement.

The passive and active circuit elements have tolerance variation, etc. In an active network, these variations may be large enough to cause instability. Therefore, it is very important to have a knowledge about the degree of dependence of one quantity upon the other, hence, the sensitivity of a system. We have already briefly encountered this topic in Section 6-3, where it was shown that a proper use of feedback can reduce the sensitivity of the circuit. We shall examine this topic a little further in this section. We shall restrict ourselves in the sequel to variations in the single element only. However, it should be pointed out that, in any physical system, the characteristics change due to variation of many elements in the circuit and, therefore, a complete analysis must consider multiparameter sensitivity. The latter topic is beyond the scope of this book. A sensitivity measure which depends on variations of ω_0 and Q, however, is briefly discussed.

---

[4] This section may be combined with the material in Chapter 8.

7-16 Sensitivity in Active-RC Circuits

If $A(s, x)$ is the gain function and x any parameter, *classical sensitivity* is defined as [see also (6-50)]

$$S_x^A(s, x) = \frac{dA/A}{dx/x} = \frac{d(\ln A)}{d(\ln x)} \qquad (7\text{-}100)$$

In (7-100), it is assumed that the variations are incremental. If the variations are not small, a logical extension of the above definition is

$$S_x^A(s, x) = \frac{\Delta A/A}{\Delta x/x} \qquad (7\text{-}101)$$

Note that S_x^A is really the linear term of the Taylor series expansion for $\Delta A/A$. Both of these definitions have the physical interpretation of percentage change in A due to percentage change in x. If the variations of x do not cause any change in A, the system has the ideal zero sensitivity. In general, $A(s)$ is the ratio of two polynomials, i.e.,

$$A(s, x) = \frac{N(s, x)}{D(s, x)} \qquad (7\text{-}102)$$

From (7-100) and (7-102), we have

$$S_x^A = \frac{dA}{dx}\left(\frac{x}{A}\right) = \frac{DN' - ND'}{D^2}\left(x\frac{D}{N}\right)$$

$$= x\left(\frac{N'}{N} - \frac{D'}{D}\right) \qquad (7\text{-}103)$$

where

$$N' = \frac{\partial N}{\partial x} \quad \text{and} \quad D' = \frac{\partial D}{\partial x}$$

EXAMPLE

Consider the single-tuned circuit shown in fig 7-4. The gain function from (7-15) is

$$A(s) = \left(-\frac{g_m}{C}\right)\frac{s}{s^2 + s(\omega_0/Q) + \omega_0^2} \qquad (7\text{-}104)$$

The sensitivity of A due to variations in g_m from (7-100) and (7-104) is given by

$$S_{g_m}^A = 1$$

Thus, from (7-101) a 1 per cent change in g_m causes a 1 per cent change in A. Similarly, the sensitivity of A due to variations in ω_0 is found from (7-100), namely

$$S_{\omega_0}^A = \omega_0 \left(-\frac{2\omega_0 + s/Q}{s^2 + s(\omega_0/Q) + \omega_0^2} \right) \quad (7\text{-}105)$$

The sensitivity of A due to changes in ω_0 at steady-stage (i.e., $s = j\omega$) and specifically at $\omega = \omega_0$ is

$$S_{\omega_0}^A(s = j\omega_0) = -1 + j2Q \quad (7\text{-}106)$$

We shall now show that the real part of S_x^A corresponds to the normalized magnitude sensitivity, while the imaginary part of S_x^A corresponds to the phase sensitivity. To do this, for $s = j\omega$ we may write

$$A(j\omega) = |A(j\omega)| e^{j\phi(\omega)} \quad (7\text{-}107)$$

where $\phi(\omega)$ is $\arg A(j\omega)$. Then, from (7-107)

$$\ln A(j\omega) = \ln |A(j\omega)| + j\phi(\omega) \quad (7\text{-}108)$$

From (7-108) and (7-100)

$$S_x^A = \frac{d[\ln A(j\omega)]}{d(\ln x)} = \frac{d \ln |A(j\omega)|}{dx/x} + j \frac{d\phi(\omega)}{dx/x} \quad (7\text{-}109)$$

Since x is usually a real quantity, we have, from the above

$$\text{Magnitude sensitivity} = Re\, S_x^A \quad (7\text{-}110\text{a})$$
$$\text{Phase sensitivity} = Im\, S_x^A \quad (7\text{-}110\text{b})$$

where Re and Im designate the real and the imaginary parts of the term, respectively. Note that, in (7-109), $Re\, S_{\omega_0}^A$ corresponds to the normalized change in magnitude of $A(j\omega)$ and $Im\, S_{\omega_0}^A$ corresponds to the change in the phase function.

For example, in (7-97) and Example 1, a 1 per cent change in ω_0 produces approximately a 1 per cent change in the magnitude of the gain at the center frequency, and a change of 0.5 radians in the phase function (since $Q = 25$).

7-16 Sensitivity in Active-RC Circuits

There are also other useful definitions of sensitivity. One such is the *root sensitivity*, which is defined as

$$S_x^{s_i} = \left.\frac{ds_i}{dx/x}\right|_{s=s_i} \qquad (7\text{-}111)$$

where s_i is the root of the polynomial (pole or zero) when x takes the nominal value. The root sensitivity is a measure of the change in the pole and zero of a network function.

For example, in (7-97) and the design values, the roots are at

$$s_1, s_1^* \simeq -2\pi(20 \pm j1000)$$

The root sensitivity due to changes in the amplifier gain K, as determined from (7-97), namely, differentiating

$$s_1 = -\frac{(3-K)}{2\tau} + j\sqrt{\frac{1}{\tau^2} - \left(\frac{3-K}{2\tau}\right)^2}$$

with respect to K and then letting $K = 2.96$, yields

$$\frac{ds_1}{dK} = \frac{1}{2\tau}\left[1 + j\frac{(+\frac{3}{2} - K/2)}{\sqrt{[-(3-K/2)]^2 + 1}}\right]_{K=2.96}$$

$$\approx \frac{1}{2\tau}(1 + j0.02)$$

Hence

$$S_K^{s_1} = K\frac{ds_1}{dK} = \frac{1}{\tau}(1.48 + j0.0296) \qquad (7\text{-}112)$$

In other words, if K changes by 10 per cent, i.e., $\Delta K/K = 0.1$, then the pole s_1 is shifted approximately by $\Delta s_1 = (0.148 + j0.00296)/\tau$. Another useful sensitivity criterion for a tuned circuit is the *Q-sensitivity*. This sensitivity, as we might expect, is related to the pole sensitivity of the single-tuned network function. In fact, it can be shown (Problem 7-24) that

$$S_x^Q \equiv \frac{dQ/Q}{dx/x} = -2Q\,\text{Im}\left[\frac{S_x^{s_i}}{s_i}\right] \simeq -\frac{2Q}{\omega_0}\left[\text{Re}\,S_x^{s_i} + \frac{\text{Im}\,S_x^{s_i}}{2Q}\right] \qquad (7\text{-}113)$$

where s_i are the poles of the high-Q, single-tuned circuit. Thus, in order to reduce S_x^Q, the variation of the real and the imaginary part of the pole-sensitivities must be made very small. In most active network design, i.e., those that involve a subtraction to realize high Q, it is the variation of the real part of the pole that causes the high sensitivity, since this is movement of the pole directly toward the $j\omega$-axis.

The sensitivity of Q due to variations of the gain of the active device is larger than $2Q - 1$ when Q is obtained as differences of terms (Problem 7-23). To see this specifically, consider the example corresponding to fig 7-44 where

$$Q = \frac{1}{3 - K}$$

From (7-100) and the above, we have

$$S_K^Q = \frac{K}{3 - K} = KQ = 74 \quad \text{for} \quad Q = 25 \quad (7\text{-}114)$$

Thus, when the design equation for Q depends on the differences of terms, active RC networks suffer considerably in sensitivity, since an extremely tight tolerance on the active device will be required and the sensitivity will be high for a high-Q design.

Finally we mention another measure which has been found to be very useful for assessing the sensitivity of high-Q active-RC circuits. This sensitivity measure is obtained by expanding the gain function into a Taylor series with respect to ω_0 and Q. Assuming that the percentage changes in ω_0, Q, and $\omega_{3\,dB}$ are small, i.e., $\Delta\omega_0/\omega_0$, $\Delta Q/Q$, and $Q\Delta\omega_0/\omega_0$ are small enough so that the terms beyond the linear term can be ignored, we obtain

$$\frac{\Delta A(s)}{A(s)} \approx S_{\omega_0}^A(s)\frac{\Delta\omega_0}{\omega_0} + S_Q^A(s)\frac{\Delta Q}{Q} \quad (7\text{-}115)$$

where

$$A(s) = \frac{K}{s^2 + s(\omega_0/Q) + \omega_0^2} \quad Q \gg 1 \quad (7\text{-}116)$$

From (7-116) and (7-103) we obtain

$$S_{\omega_0}^A(j\omega_0) = 1 + j2Q, \quad S_Q^A(j\omega_0) = 1 \quad (7\text{-}117)$$

The sensitivity measure for a high Q' circuit is then defined as

$$|S| = \left|\frac{\Delta A(j\omega_0)}{A(j\omega_0)}\right| = \sqrt{\left(2Q\frac{\Delta\omega_0}{\omega_0}\right)^2 + \left(\frac{\Delta Q}{Q}\right)^2} \quad (7\text{-}118)$$

Note that, in (7-118), the fractional change in ω_0 is weighted by Q, and this is meaningful since, for a high-Q circuit, the bandwidth is a very small fraction of the center frequency ω_0. The variation of center frequency must be kept very tight. The measure in (7-118) has been found to be a valid basis of comparison for the various high-Q active RC realizations, especially when the nonideal nature of the active element is taken into account. The measure indicates that the limiting factor on high Q active RC networks is not the Q-sensitivity but rather the center frequency sensitivity which is always limited by RC tracking and manufacturing tolerances. That is why crystal tuned-amplifiers are used for very high Q application.

Low-sensitivity, high-Q design, which does not depend on differences of terms, is discussed in the next chapter (Section 8-8). In these circuits, the various sensitivities are compatible with those of RLC-tuned circuits. The topic of active filters is discussed in some detail in the next chapter and will not be pursued further here.

REFERENCES AND SUGGESTED READING

1. Searle, C. L., et al., *Elementary Circuit Properties of Transistors*, New York: Wiley, 1964, Vol. 3, Chapter 8.
2. Thornton, R. D., et al., *Multistage Transistor Circuits*, New York: Wiley, 1965, Vol. 5, Chapter 7.
3. Gibbons, J. F., *Semiconductor Electronics*, New York: McGraw-Hill, 1966, Chapter 18.
4. Ghausi, M. S., *Principles and Design of Linear Active Circuits*, New York: McGraw-Hill, 1965, Chapter 15.
5. Schilling, D., and Belove, C., *Electronic Circuits: Discrete and Integrated*, New York, McGraw-Hill, 1968, Chapter 14.
6. Linvill, J. and Gibbons, J. F., *Transistors and Active Circuits*, New York: McGraw-Hill, 1961, Chapter 18.
7. Huelsman, L., *Theory and Design of Active RC Circuits*, New York: McGraw-Hill, 1968. For Active RC filters and sensitivity.

PROBLEMS

7-1 For the single-stage FET tuned-amplifier shown in fig P7-1, determine the expression for the voltage gain function.

Determine the gain, the bandwidth, and the resonant frequency if the FET parameters are:

$g_m = 4 \times 10^{-3}$ mho, $r_d = 20$ kΩ, $C_{iss} = 8$ pF, $C_{rss} = 2$ pF, $L = 10$ μH with a Q of 200 at ω_0, $C_a = 500$ pF and includes the stray capacitance of the inductor, $R_L = 1$ kΩ.

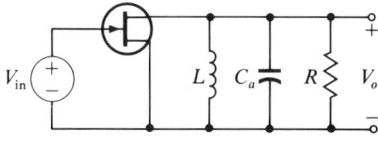

Fig P7-1

454 7/Analysis and Design Examples of Bandpass Amplifiers

7-2 In the example of fig 7-7a, use the Miller effect and include r_b. Determine the expression for the gain function. Find the pole-zero location of the transfer function if $r_b = 50\,\Omega$.
Determine the midband gain and the bandwidth.

7-3 Find the midband gain and the bandwidth for Problem 7-2 under the following cases:
(a) $f_0 \gg 1/R_i C_i$, i.e., neglect R_i as compared to the shunt C_i.
(b) $f_0 \ll 1/R_i C_i$, i.e., neglect C_i as compared to the shunt R_i.

7-4 Design the single-stage bipolar transistor amplifier of fig 7-7a for the following specifications:

$$\omega_0 = 2\pi(500\,\text{kHz}), \qquad Q = 50, \qquad A_i = -20$$

i.e., determine R_1, L, C. The load resistance $R_L = 1\,\text{k}\Omega$ and the transistor parameters are the same as in the example 1.

7-5 A single-stage bipolar transistor has the tuned circuit at the output, as in fig 7-12. The transistor parameters at the resonant frequency $f_0 = 10\,\text{MHz}$ are given by (7-23). Design the circuit for a bandwidth $f_{3\,\text{dB}} = 500\,\text{kHz}$ and a midband voltage gain of -100.

7-6 A tuned amplifier has been designed such that the pole-zero pattern of the voltage gain function is as shown in fig P7-6. The midband voltage gain $A_v(j\omega_0) = 10^3$. Determine the 3-dB bandwidth of the amplifier using the narrow band approximation.

7-7 Analyze the transistor double-tuned amplifier circuit shown in fig 7-20. Assume $r_b = 0$ and use the Miller approximation. Derive the design equations for the critically-coupled case and obtain the simplified gain function for the narrow-band approximation.

7-8 Repeat Problem 7-7 for the capacitively coupled double-tuned amplifier circuit shown in fig P7-8.

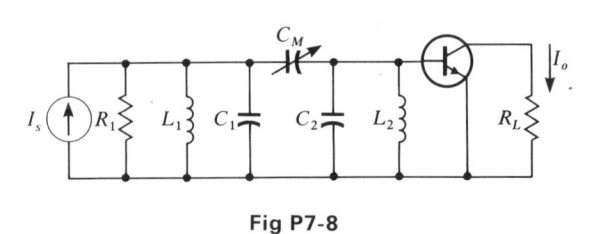

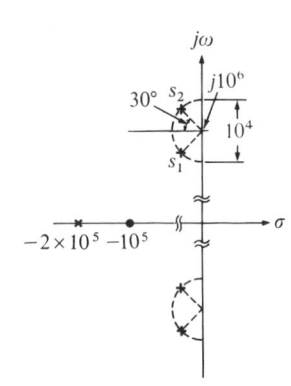

Fig P7-6

Fig P7-8

7-9 For the circuit shown in fig P7-9, the pentodes are assumed to be identical with the following parameters:

$$g_m = 5 \times 10^{-3}\,\text{mho}, \qquad C_o = C_i = 8\,\text{pF}$$

Also given:

$$G_1 = G_2 = (1 \text{ k}\Omega)^{-1}$$
$$L_1 = L_2 = 10 \text{ }\mu\text{H}$$
$$C_1 = 48 \text{ pF}, \quad C_2 = 56 \text{ pF}$$

(a) If $G_f = 0$ determine the midband voltage gain, the center frequency, and the 3-dB bandwidth.

(b) For what value(s) of G_f is the response similar to the critically coupled double-tuned circuit?

Hint: use the low pass case (see Problem 5-17).

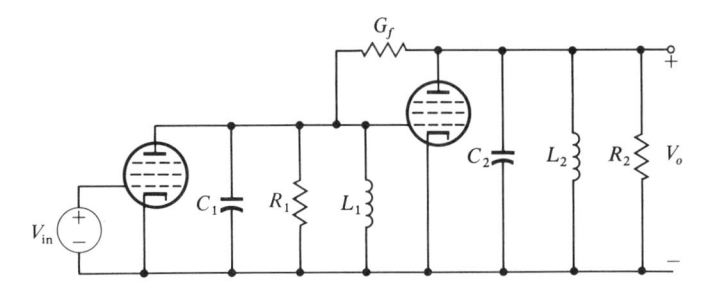

Fig P7-9

7-10 The circuit shown in fig P7-10 is to resonate at $f_o = 20$ MHz and with $f_{3\text{ dB}} = 1$ MHz. The transistor parameters are as follows:

$$\beta_o = 50 \qquad f_T = 500 \text{ MHz}$$
$$C_c = 2 \text{ pF} \qquad r_e = 10 \text{ }\Omega, \quad \text{(assume that } r_b = 0\text{)}$$

Determine the values of L_m, n, n_1 and C to achieve the above specifications in addition to realizing the maximum midband current gain. What is the resulting midband current gain?

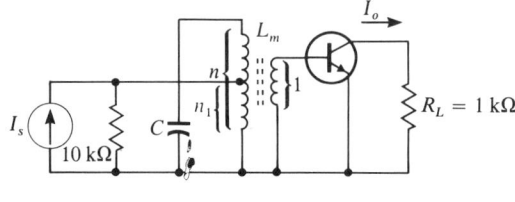

Fig P7-10

7-11 For the nonunilateral active two-port shown in fig P7-11, the following specifications are given: The two-port parameters at $\omega_0 = 2\pi(10 \text{ MHz})$ are given in (7-23). The source and load terminations are $R_S = R_L = 100 \, \Omega$.

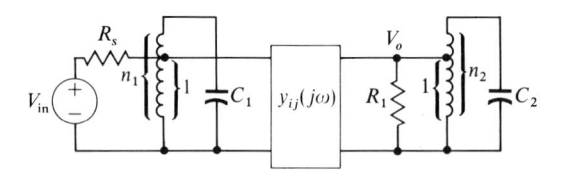

Fig P7-11

The design is to be synchronously tuned for an overall bandwidth $f_{3\text{ dB}} = 500 \text{ kHz}$. Determine n_1, C_1, n_2, C_2 if coils with $L_{m1} = L_{m2} = 5 \, \mu\text{H}$ are used. (Hint: Note that the $y_{12}V_2$ generator can be ignored.)

7-12 The y-parameters of a device measured at $f = 1$ MHz are given as follows:

$$y_{11} \approx (1 + j2) \times 10^{-3} \text{ mho} \qquad y_{12} \approx -j(10^{-5}) \text{ mho}$$

$$y_{21} \approx 4 \times 10^{-2} \text{ mho} \qquad y_{22} \approx (2 + j10) \times 10^{-6} \text{ mho}$$

Is the device absolutely stable? What value of shunt conductance added at the input or the output will make the overall two-pole absolutely stable?

7-13 For the cascode amplifier shown in fig 7-26, the transistor parameters are given in (7-50) with $r_b = 0$ and $R_s = 100 \, \Omega$. Determine the element values for a synchronously tuned design and the following specifications:

$$f_0 = 10 \text{ MHz}, \qquad f_{3\text{ dB}} = 1 \text{ MHz}, \qquad A_v(j\omega_0) = -100$$

7-14 Repeat Problem 7-13, but with a stagger-tuned design, corresponding to the critically coupled double-tuned case.

7-15 A common-base tuned circuit is shown in fig P7-15. The center frequency

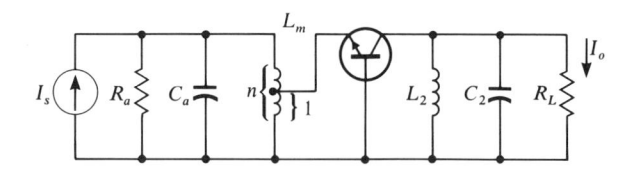

Fig P7-15

$f_0 \ll f_\alpha$, so that the internal capacitance of the CB stage may be omitted (as in fig 7-26). Derive the design equations for a synchronously tuned narrowband design where ω_0 and $\omega_{3\,dB}$ are specified.

7-16 Repeat Problem (7-15) for a stagger-tuned design with a maximally flat magnitude response characteristic.

7-17 In the emitter-coupled tuned-amplifier circuit shown in fig P7-17, tapped capacitors are used for impedance transformation.
(a) Consider the CB load alone and derive the impedance transforming equation. Assume $C_2 \gg C_1$.
(b) If $R_L = R_E = 1\,k\Omega$, $R_s = 50\,\Omega$, $L = 2\,\mu H$, and the transistor parameters at the operating point ($I_E = 2\,mA$, $V_{CE} = 5\,V$) are

$$\beta_0 = 100, \qquad f_T = 400\,MHz, \qquad r_b = 50\,\Omega, \qquad C_c = 2\,pF$$

design the circuit (i.e. find C_1 and C_2) for $f_0 = 50\,MHz$ and $Q = 20$. Make reasonable approximations. (See Problem 5-11.) What is the voltage gain at the resonant frequency?

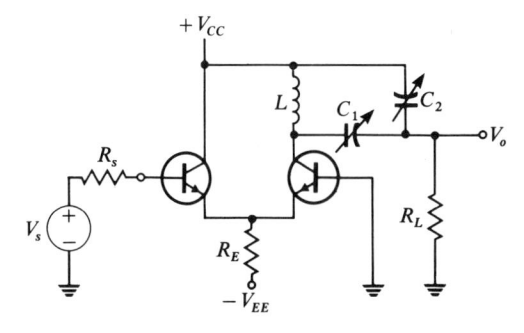

Fig P7-17

7-18 Use the design method given in Section 7-13 for the following specification:

$$f_0 = 10\,MHz, \qquad f_{3\,dB} = 1\,MHz, \qquad R_s = R_L = 1\,k\Omega$$

The power gain required is $G_P \geq 25\,dB$. Use the minimum number of transistors and the maximum number of single-tuned stages and $\delta = 0.1$. Use the same impedance-matching transformer, if possible.

7-19 Repeat Problem 7-18, but with a stagger-tuned design (see Section 2-8), for a three-pole maximally flat magnitude pole location.

7-20 For the active RC network shown in fig P7-20, the amplifier is an ideal VCVS element and $K < 0$. Show that the voltage gain function is given by:

$$A_V = \frac{V_0}{V_{in}} = -\frac{sR_2C_1K}{s^2(1+K)(R_1R_2C_1C_2) + s(R_1C_2 + R_2C_1 + R_1C_1) + 1}$$

Design the circuit for a center frequency $f_0 = 1\,kHz$, $f_{3\,dB} = 200\,Hz$. Let $R_1C_1 = R_2C_2$, choose $R_1 = R_2 = 1\,k\Omega$ and suitable element values.

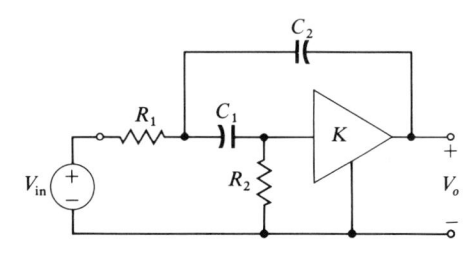

Fig P7-20

7-21 For the circuit shown in fig P7-21, the amplifier is an ideal VCVS with $K > 0$. Show that the transfer function is given by

$$A_v = \frac{V_o}{V_i} = \frac{sKG_1/C_2}{s^2 + s(1/C_1C_2)[G_1(C_1+C_2) + G_2(C_1+C_2-KC_1)} $$
$$+ G_3C_1] + [G_3(G_1+G_2)]/C_1C_2$$

If $R_1 = R_2 = R_3 = 1\ k\Omega$, $C_1 = C_2$, design the circuit for $f_0 = 1\ kHz$ and $f_{3\ dB} = 200\ Hz$. For what value of K will the circuit begin to oscillate?

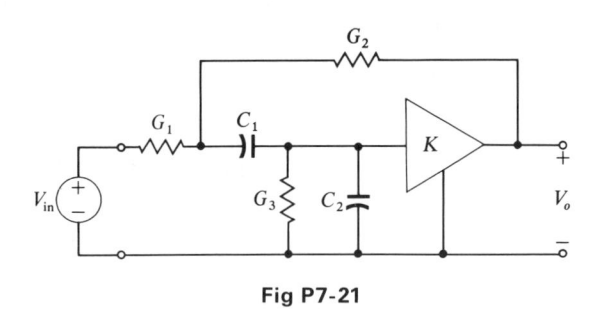

Fig P7-21

7-22 The circuit shown in fig P6-32 is to be designed for the following specifications:

$$\omega_0 = 2\pi(1\ kHz), \qquad \omega_{3\ dB} = 2\pi(100\ Hz)$$

Design the circuit using $R_1 = R_2 = 1\ k\Omega$, $C_1 = C_2 = 0.2\ \mu F$.

7-23 The natural frequencies of an active network, as obtained via differences of terms to realize a bandpass filter, are expressed by

$$(s^2 + as + \omega_0^2) - ks$$

The quadratic term in the parentheses is usually due to an RC network, hence its roots are real and distinct. Show that

$$S_k^Q > 2Q - 1$$

7-24 For a high-Q single-tuned circuit, show that

$$S_x^Q = -2Q\ Im\left[\frac{S_x^{s_1}}{s_1}\right]$$

where $s_1 = -\omega_{3\ dB}/2 + j\omega_0$ and x is any network parameter. Note that $Q = \omega_0/\omega_{3\ dB}$.

7-25 For the single-tuned circuit shown in fig 7-4, determine

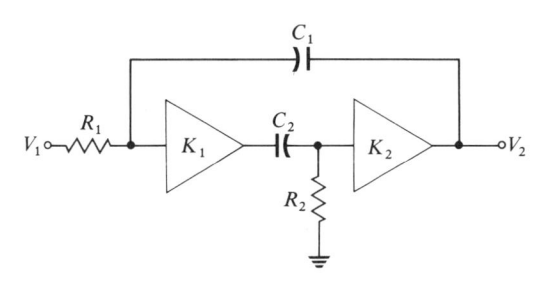

$$S_G^Q, \quad S_L^Q \quad \text{and} \quad S_C^Q$$

Fig P7-26

7-26 Consider the active RC tuned network shown in fig P7-26. The amplifiers are ideal $VCVS$ elements.
(a) Show that the voltage transfer function is given by

$$T_V = \frac{V_2}{V_1} = \frac{K_1 K_2 R_2 C_2 s}{(1 - K_1 K_2) R_1 R_2 C_1 C_2 s^2 + (R_1 C_1 + R_2 C_2) s + 1}$$

(b) If

$$R_1 = R_2 = R$$
$$C_1 = C_2 = C$$
$$K_1 = -K_2 = |K|$$

design the circuit for a center frequency $f_0 = 1$ kHz and a bandwidth $f_{3\,dB} = 50$ Hz. If you have a choice in the element values use $R = 1$ kΩ.

7-27 For Problem (7-20), if $R_1 = R_2 = 1$, $C_1 = C_2 = 1/3Q$, and $K = 9Q^2 - 1$,
(a) Show that

$$S_K^Q = \frac{1}{2} - \frac{1}{18Q^2}$$

(b) Show that $|S_x^Q| \approx \frac{1}{6}$, where x is any of the elements R_1, R_2, C_1, or C_2.

7-28 For the circuit in Problem (7-26) if

$$R_1 = R_2 = 1, \quad C_1 = C_2 = \frac{1}{2Q} \quad \text{and} \quad K_1 = -K_2 = \sqrt{4Q^2 - 1}$$

show that

(a) $\qquad |S_x^Q| \leq \frac{1}{2}$

(b) $\qquad |S_x^{\omega_0}| = \frac{1}{2}$

where x is any of the elements R_1, R_2, C_1, C_2, K_1, or K_2.

7-29 Show an equivalent structure to fig 7-37b, but replacing the tapped transformer by a differential output operational amplifier.

7-30 Show that the voltage ratio of the terminated lattice (fig 7-37b) for any n and different terminations is given by

$$\frac{V_2}{V_1} = \left[\frac{n}{1 + n^2(R_s/R_L)}\right]\frac{Z_B - Z_A}{Z_B + Z_A + (Z_A Z_B/R_L)[1/(1 + n^2 R_s/R_L)] + [4n^2 R_s/(1 + n^2 R_s/R_L)]}$$

Note that for $n = 1$ and $R_s = R_L$ the above reduces to (7-80).

7-31 Derive (7-86).

7-32 Design a three-pole Butterworth-response tuned-filter with $f_0 = 10$ MHz and $f_{3\,dB} = 1$ kHz. Use the crystal parameters given in table 7-1. Show the complete circuit including the buffer amplifier.

7-33 Design a second-order Butterworth filter with center frequency at $f_0 = 1$ MHz and a bandwidth of $f_{3\,dB} = 5$ kHz. Use the crystal parameters in table 7-1. Also specify the turns ratio of the transformer if R_L is specified to be 500 ohms.

8
Active *RC* Filters

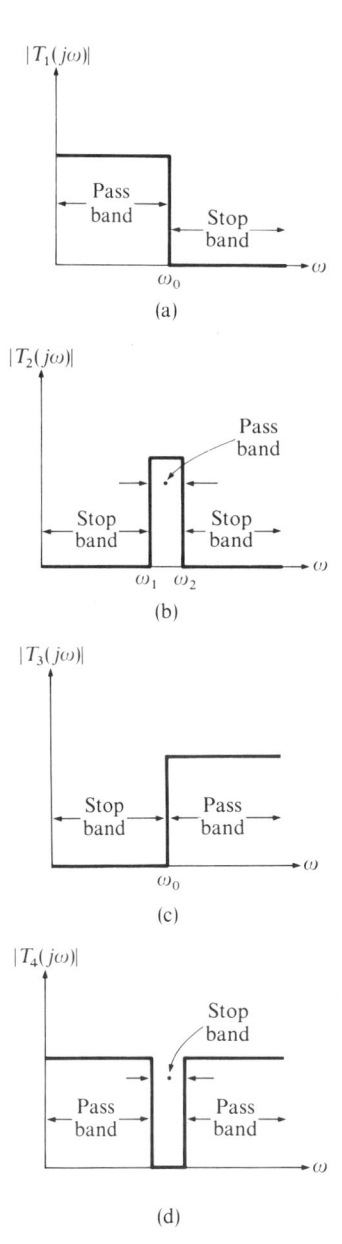

Fig 8-1 The various forms of ideal filter characteristics: (a) Lowpass filter characteristic; (b) Bandpass filter characteristic; (c) Highpass filter characteristic; (d) Band-elimination filter characteristic

Our goal in this chapter is to introduce the design of active *RC* filters. The treatment that follows will necessarily be at an elementary level, since we assume no circuit synthesis background on the part of the reader.

Filters are widely used in electronic circuits for a variety of reasons. They may take any one of the following forms: low-pass, band-pass, high-pass, and band-elimination filter. The idealized magnitude responses of such filters are shown in fig 8-1. Of course, such responses cannot be realized exactly by any physical system. However, they can be approximated to a satisfactory degree in most practical applications. The general problem of approximation is a vast topic that cannot be adequately covered in one chapter of an introductory text. We shall consider in this chapter, however, only the simple techniques, in order to familiarize the reader with the design aspects of such filters. For example, the Butterworth transfer function to approximate the low-pass response of fig 8-1a was already discussed briefly in Chapter 2. Bandpass amplifiers were discussed in Chapter 7, where frequency-selective *amplification* was desired. In passive filters utilizing mainly *LC* elements, the specification may require no loss in the passband and a certain amount of rejection in the stopband. There exists a large body of literature on this subject. We will focus our attention only on active *RC* filters. These filters are used in telephone communication and other electronic circuitry. In fact for frequencies below 1 kHz, active *RC* filters are predominantly used because of size, performance, and cost. For frequencies over 500 kHz, either crystal filters (for very high Q) or passive *LC* filters are used. We shall assume in the following discussion that the specified transfer function, which approximates a desired filter response, is given and we want to realize it via active *RC* circuits. The transfer function $T(s)$ of the two-port network shown in fig 8-2 may be in any one of the following forms:

$$A_I = \frac{I_2}{I_1}(s) \qquad A_V = \frac{V_2}{V_1}(s)$$
$$Y_{21} = \frac{I_2}{V_1}(s) \qquad Z_{21} = \frac{V_2}{I_1}(s) \tag{8-1}$$

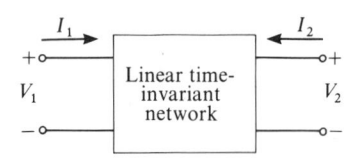

Fig 8-2 Schematic of a two-port network

Any one of the above may be written in the general form

$$T(s) = \frac{N(s)}{D(s)} = \frac{s^m + a_{m-1}s^{m-1} + \cdots + a_1 s + a_0}{s^n + b_{n-1}s^{n-1} + \cdots + b_1 s + b_0} \tag{8-2}$$

where $m \leq n$ for practical reasons and the coefficients b_i are all non-negative due to stability. Also, the orders of m and n are usually low due to sensitivity reasons. A high-order transfer function is usually broken into a cascade of low-order simple transfer functions (see Section 8-3).

Because of integrated-circuit (*IC*) technology, much attention has been drawn to the study of active *RC* filters. Practical values of inductors are unavailable in monolithic *IC*'s. In hybrid *IC*'s, the size and weight of the discrete inductance for low-frequency filters make it unsuitable for miniaturization. Furthermore, inductors are lossy, and in many practical circuit applications their magnetic fields may create problems. Thus, there are advantages to be gained by eliminating inductors in the design. However, as will be seen later, due to sensitivity and stability problems, active *RC* filters are not presently practical for mass production beyond frequencies in the 100 kHz range. It should also be pointed out that by discarding inductors, we do not discard any of the properties provided by the inductor. Thus, active *RC* circuits, that is circuits which consist of resistors, capacitors, and some form of active elements, can realize any network function that can be realized if inductors are used. The active element may take any of the following forms: a controlled source, a negative-impedance converter, an operational amplifier, or a gyrator used in a passive mode. We shall define these active elements in this chapter, show their realizations, and demonstrate their applications in realizing various types of filters. It is to be noted that the various forms above can be realized by operational amplifiers alone, or they can be viewed as controlled sources. For convenience, however, and to familiarize the reader with the various forms of active elements, we shall discuss them separately. The topic of sensitivity in active RC networks is discussed in Section 7-16 and the reader is urged to review that section in conjunction with the topics discussed in this chapter.

Since distributed RC networks arise naturally in the fabrication of integrated circuits, we also briefly discuss active filters with distributed RC circuits. It will be shown that simple and practical filters can be realized using such elements.

8-1 RC Driving-Point Functions

Before we embark on the design of active RC filters and transfer function synthesis, it is expedient to discuss briefly the synthesis of RC driving-point functions. A knowledge of these functions is essential to the solution of most active RC synthesis problems.

At any two terminals of a network, such as shown in fig 8-3, recall that the driving-point functions are defined as follows:

$$\text{Driving-point impedance function: } Z(s) = \frac{V}{I} \quad (8\text{-}3a)$$

$$\text{Driving-point admittance function: } Y(s) = \frac{I}{V} \quad (8\text{-}3b)$$

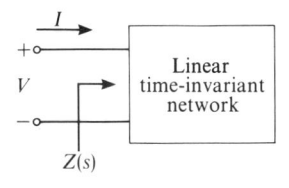

Fig 8-3 Schematic of a one-port network

If we restrict the network to being RC, then naturally there are restrictions on the poles and zeros of $Z(s)$ and $Y(s)$.

The necessary and sufficient conditions for $Z(s)$ to be an RC driving-point impedance functions can be shown to be (Refs 1 and 2) the following:

(a) All poles and zeros are on the negative real axis and are simple, i.e., of multiplicity one. The poles and zeros alternate with each other.

8/Active RC Filters

The critical frequency (pole or a zero) nearest to or at the origin is a pole while the nearest to or at infinity is a zero. For example, the following functions are *not* RC driving-point impedance functions:

$$Z_a = \frac{(s+3)^2}{(s+2)(s+4)}, \quad Z_b = \frac{(s+1)(s+2)}{s(s+3)}, \quad Z_c = \frac{s(s+2)}{(s+1)(s+3)}$$

Since Z_a has a zero of multiplicity 2, in Z_b the poles and zeros do not alternate while in Z_c the critical frequency nearest infinity is a pole and nearest the origin is a zero. The following are RC driving-point impedance functions:

$$Z_d = \frac{(s+1)}{s(s+2)}, \quad Z_e = \frac{(s+2)(s+5)}{(s+1)(s+4)}$$

(b) The alternative necessary and sufficient conditions may be stated as: all poles are simple and restricted on the negative real axis, the residue of $Z(s)$ at each pole is real and positive, and at infinity $Z(s)$ is a finite nonnegative constant. Therefore, on this basis the RC driving-point function can be expressed as

$$Z_{RC}(s) = K_\infty + \frac{K_o}{s} + \sum_{i=1}^{n} \frac{K_i}{s + \sigma_i} \tag{8-4}$$

where σ_i and the residues K_∞, K_o, and K_i are nonnegative quantities. The realization or synthesis of each term in (8-4) is shown in fig 8-4. Thus, any general RC driving-point function can be realized in the form shown in fig 8-5. This form of synthesis, which is based on partial fraction expansion, is referred to as the Foster first canonical form. The word canonical means a realization with the minimum number of circuit elements. For example, let

$$Z(s) = 1 + \frac{2}{s} + \frac{1}{s+1} + \frac{2}{s+3}$$

A realization of the above is shown in fig 8-6.

The necessary and sufficient conditions for $Y(s)$ to be an RC driving-point admittance function are either of the following:

(a) All poles and zeros are simple and on the negative real axis. The poles and zeros alternate with each other. The highest critical frequency is a pole and the lowest critical frequency is a zero.

(b) All poles are simple and on the negative real axis. No pole is at the origin. The residue of $Y(s)/s$ at each pole is real and positive. At

Fig 8-4 Circuit realizations of equation (8-4): (a) $Z(s) = R$; (b) $Z(s) = 1/sC$; (c) $Z(s) = (1/C_i)/(s + 1/R_iC_i)$

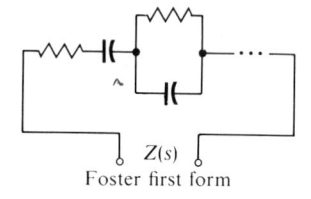

Fig 8-5 General form of RC driving-point impedance realization—Foster's first canonical form

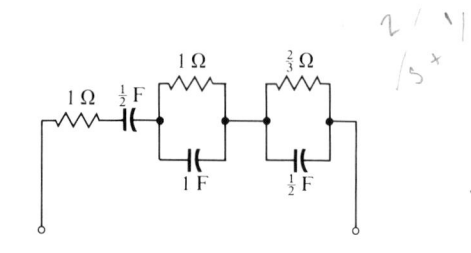

Fig 8-6 Circuit realization of the example

8-1 RC Driving-Point Functions

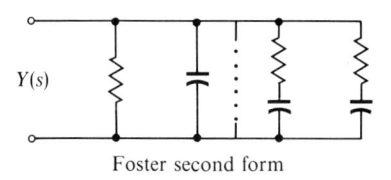

infinity $Y(s)/s$ is a finite nonnegative constant. Thus on the basis of (b), an RC admittance function can be expressed as

$$\frac{1}{s}[Y_{RC}(s)] = H_\infty + \frac{H_o}{s} + \sum_{j=1}^{n} \frac{H_j}{s + \sigma_j} \qquad (8\text{-}5)$$

where σ_j, H_∞, H_o, and H_j are nonnegative quantities. From (8-5) we may rewrite:

$$Y_{RC}(s) = H_\infty s + H_o + \sum_{j=1}^{n} \frac{H_j s}{s + \sigma_j} \qquad (8\text{-}6)$$

The realization of each term in (8-6) is shown in fig 8-7. Hence, any general RC driving-point function can be realized in the form of fig. 8-8.

Fig 8-8 General form of RC driving-point admittance realization—Foster's second canonical form

Fig 8-7 Circuit realizations of equation (8-6): (a) $Y(s) = G = R^{-1}$; (b) $Y(s) = sC$; (c) $Y(s) = \{(1/R_j)s/[s + 1/R_jC_j]\}$

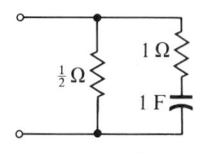

In figure 8-7:
(a) $G = H_o$, $Y(s) = G = \frac{1}{R}$
(b) $C = H_\infty$, $Y(s) = sC$
(c) $R_j = \frac{1}{H_j}$, $C_j = \frac{H_j}{\sigma_j}$, $Y(s) = \frac{(1/R_j)s}{s + 1/R_jC_j}$, $(\sigma_j = 1/R_jC_j)$

Fig 8-9 Circuit realization of the example

This form is usually referred to as the Foster second canonical form. As an example, let $Y(s) = 2 + s/(s + 1)$. The realization is shown in fig 8-9.

Two other canonical realizations based on the continued fraction expansion of $Z(s)$ or $Y(s)$ are the Cauer forms shown in fig 8-10.

To obtain the Cauer forms one expresses $Z(s)$ or $Y(s)$ in the following forms:

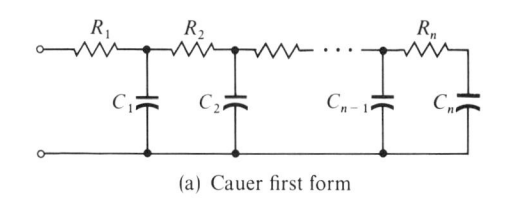

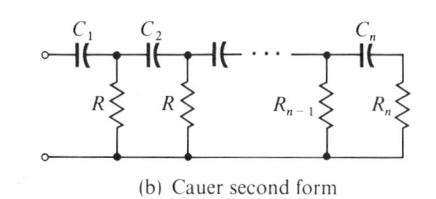

$$Z_{RC}(s) = R_1 + \cfrac{1}{C_1 s + \cfrac{1}{R_2 + \cfrac{1}{C_2 s + \cfrac{1}{\ddots + \cfrac{1}{C_n s}}}}} \qquad (8\text{-}7)$$

(a) Cauer first form

(b) Cauer second form

Fig 8-10 Alternate forms of RC driving-point realizations: (a) Cauer's first canonical form; (b) Cauer's second canonical form

8/Active RC Filters

For the second Cauer form, we have

$$Z_{RC}(s) = \cfrac{1}{C_1 s} + \cfrac{1}{\cfrac{1}{R_1} + \cfrac{1}{\cfrac{1}{C_2 s} + \cfrac{1}{\cfrac{1}{R_2} + \cfrac{1}{\ddots \cfrac{1}{R_n}}}}} \quad (8\text{-}8)$$

In the Cauer form of synthesis, one need *not* factor either the numerator or the denominator polynomials. Simply start the continued division as follows.

To obtain Z_{RC} in Cauer first form, arrange the numerator and denominator polynomials in descending order, perform the continued division, and assign the quotients of the division as R and C in the ladder network. If $Z(\infty)$ is zero, there is no initial series resistance in fig 8-10a and the division starts with $Y = 1/Z$.

To obtain Z_{RC} in Cauer second form, arrange the numerator and the denominator polynomials in ascending order, perform the continued division, and assign the quotients of the division as C and R in the ladder network. If $Z(0)$ is not infinite, the initial series capacitance in fig 8-10b is missing, then start the division with $Y = 1/Z$.

EXAMPLE

Let

$$Z_{RC} = \frac{s+2}{s^2 + 4s + 3}$$

We shall obtain the Cauer first form. Since $Z(\infty)$ is zero, the first resistor in fig 8-10a is $R_1 = 0$, hence we start the division with

$$Y = \frac{s^2 + 4s + 3}{s+2}$$

8-1 RC Driving-Point Functions

$$\begin{array}{r}
\overset{\displaystyle \overset{C}{\searrow}\,\frac{1}{s}}{s^2+4s+3}\\
s+2\,\overline{)\,s^2+4s+3\,}\\
\underline{s^2+2s} \quad \overset{R}{\leftarrow}\tfrac{1}{2}\\
2s+3\,\overline{)\,s+2\,}\\
\underline{s+\tfrac{3}{2}} \quad \overset{C}{\leftarrow}4s\\
\tfrac{1}{2}\,\overline{)\,2s+3\,}\\
\underline{2s} \quad \overset{R}{\leftarrow}\tfrac{1}{6}\\
3\,\overline{)\,\tfrac{1}{2}\,}\\
\underline{\tfrac{1}{2}}\\
0
\end{array}$$

The circuit realization is shown in fig 8-11.

To obtain the Cauer second form, we arrange the numerator and the denominator in ascending series form

$$Z_{RC} = \frac{2+s}{3+4s+s^2}$$

Now, since $Z_{RC}(0)$ is not infinity, the first series capacitor in fig 8-10b is $C_1 = 0$, hence we start the division with $Y = (3 + 4s + s^2)/(2 + s)$:

$$\begin{array}{r}
\overset{\displaystyle \overset{G}{\leftarrow}\,\tfrac{3}{2}}{3+4s+s^2}\\
2+s\,\overline{)\,3+4s+s^2\,}\\
\underline{3+\tfrac{3}{2}s}\quad \overset{1/C}{\leftarrow}\tfrac{4}{5}\,1/s\\
\tfrac{5}{2}s+s^2\,\overline{)\,2+s\,}\\
\underline{2+\tfrac{4}{5}s}\quad \overset{G}{\leftarrow}\tfrac{25}{2}\\
\tfrac{1}{5}s\,\overline{)\,\tfrac{5}{2}s+s^2\,}\\
\underline{\tfrac{5}{2}s}\quad \overset{1/C}{\leftarrow}\tfrac{1}{5}\,1/s\\
s^2\,\overline{)\,\tfrac{1}{5}s\,}\\
\underline{\tfrac{1}{5}s}\\
0
\end{array}$$

The circuit realization is shown in fig. 8-12. As a final comment, note that if an *RC* impedance function is multiplied by s, another *RC* admittance function is obtained.

We have now established the synthesis of one-port *RC* networks. All four types of realizations are possible for any given *RC*-realizable impedance or admittance functions. This knowledge will be utilized in the following where the synthesis of a general transfer function will be reduced to that of one-port *RC* synthesis. We shall now discuss the various transfer function realization methods.

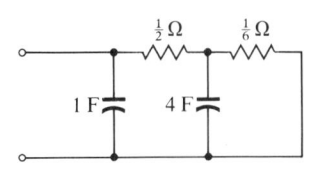

Fig 8-11 Circuit realization of the example

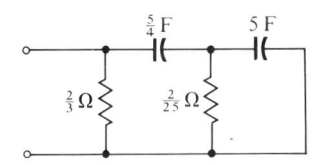

Fig 8-12 Circuit realization of the example

8-2 Controlled-Source Realizations ($RC: -RC$ Decomposition)

The various forms of the idealized controlled-source representation were introduced in Chapter 1. Some of the practical transistor circuit realizations of these elements were given in the previous chapters. For example, an operational amplifier can be used to realize a voltage-controlled voltage source network element (see fig 8-31). We shall now make use of some of these elements in the synthesis of general transfer functions. Since in the methods of this section the polynomials are decomposed into RC and $-RC$ driving-point functions, we shall refer to it in the subsequent sections as the $RC: -RC$ decomposition techniques.

Realization of Voltage Ratio Functions

The voltage ratio function may be specified in the general form

$$\frac{V_o}{V_i} = \frac{N(s)}{D(s)} = \frac{a_k s^k + a_{k-1} s^{k-1} + \cdots + a_1 s + a_0}{b_r s^r + b_{r-1} s^{r-1} + \cdots + b_1 s + b_0} \quad k \leq r \quad (8\text{-}9)$$

where $N(s)$ and $D(s)$ are arbitrary polynomials with real coefficients. We wish to realize (8-9) by active RC networks.

Consider the circuit shown in fig 8-13. The amplifiers are assumed to be ideal voltage-controlled voltage source elements (VCVS). The inverting or noninverting nature of these amplifiers is indicated by the signs associated with the gain, K, of these elements.

The analysis of this circuit yields the following transfer function (see Problem 8-4):

$$A_V = \frac{V_o}{V_{in}} = \frac{(Y_A - K_1 Y_B)}{Y_A + Y_B + Y_1 - (K_2 - 1)Y_2} \quad (8\text{-}10)$$

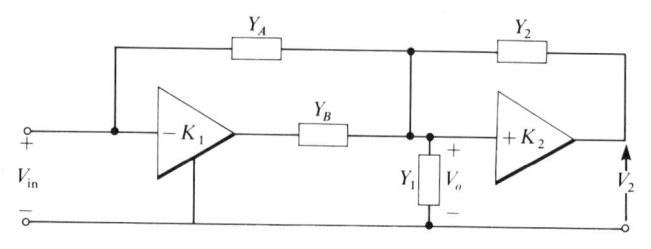

Fig 8-13 Active RC configuration for the realization of general voltage-ratio transfer functions

In (8-10), without any loss of generality, we can set $K_2 - 1 = K_1 = 1$. Hence, (8-10) reduces to

$$\frac{V_o}{V_{in}} = \frac{(Y_A - Y_B)}{[Y_A + Y_B + Y_1 - Y_2]} \qquad (8\text{-}11)$$

The specified voltage ratio function (8-9) may be written as

$$\frac{V_o}{V_{in}} = \frac{N(s)}{D(s)} = \frac{N(s)/Q(s)}{D(s)/Q(s)} \qquad (8\text{-}12)$$

with

$$Q(s) = \prod_{i=1}^{n-1} (s + \sigma_i) \qquad (8\text{-}13)$$

where σ_i are real, positive and distinct and $n - 1$ is the degree of $Q(s)$ in (8-13) ($n \geq r \geq k$). Under these conditions $N(s)/Q(s)$ and $D(s)/Q(s)$ are $RC: -RC$ decomposable, i.e., we can *always* express the numerator and the denominator of (8-12) as differences of two RC admittance functions. All that is left then is to equate the corresponding numerators and denominators of (8-12) and (8-11), and synthesize the RC driving-point functions. Before we go any further, we shall digress briefly to prove this statement, since it will appear time and again in active RC synthesis.

Polynomial Decomposition in Active *RC* Synthesis

Let $N(s)$ be a specified polynomial with arbitrary real coefficients, and let $Q(s)$ be a polynomial with distinct negative real roots. Then $N(s)/Q(s)$ can be expressed as the difference of two RC immittance functions. That is

$$\frac{N(s)}{Q(s)} = Z_{RC}^a - Z_{RC}^b \text{ for degree of } Q(s) \geq \text{degree of } N(s) \qquad (8\text{-}14)$$

$$\frac{N(s)}{Q(s)} = Y_{RC}^c - Y_{RC}^d \text{ for degree of } Q(s) \geq \text{degree of } N(s) - 1$$

$$(8\text{-}15)$$

To prove the above theorem, consider (8-14) first. Since the coefficients of $N(s)$ and $Q(s)$ are real, and $Q(s)$ has only distinct negative real roots, we obtain the partial fraction expansion of the form

$$\frac{N(s)}{Q(s)} = H_\infty + \frac{H_o}{s} + \sum_{i=1}^{} \frac{K_i^+}{s + \sigma_i} - \sum_{j=1}^{} \frac{K_j^-}{s + \sigma_j} \tag{8-16}$$

where the residues, K's, are all real and nonnegative. H_∞ and/or H_o may be either positive or negative. If positive, we designate by K_∞^+, K_o^+, if negative, we designate these by K_∞^-, K_o^-. We then regroup (8-16) as

$$\frac{N(s)}{Q(s)} = \underbrace{\left(K_\infty^+ + \frac{K_o^+}{s} + \sum_i \frac{K_i^+}{s + \sigma_i} \right)}_{Z_{RC}^a} - \underbrace{\left(K_\infty^- + \frac{K_o^-}{s} + \sum_j \frac{K_j^-}{s + \sigma_j} \right)}_{Z_{RC}^b}$$
$$\tag{8-17}$$

Since the individual groups inside the parentheses are RC networks, as in (8-4), we have proved the theorem.

To prove (8-15), we make a partial fraction expansion of $[N(s)/sQ(s)]$ to obtain

$$\frac{N(s)}{sQ(s)} = Z_{RC}^A - Z_{RC}^B \tag{8-18}$$

or

$$\frac{N(s)}{Q(s)} = sZ_{RC}^A - sZ_{RC}^B = Y_{RC}^c - Y_{RC}^d \tag{8-19}$$

Note that since the choice of σ_i in (8-13) and the degree of $Q(s)$ are arbitrary, the decomposition is *not* unique. This arbitrariness can be utilized in sensitivity minimization, which we shall not cover here. To obtain a realization with fewer elements, however, we choose $n = r$.

EXAMPLE

Consider

$$\frac{N(s)}{Q(s)} = \frac{(s-1)(s^2+1)}{(s+1)(s+2)}$$

We obtain the partial expansion of $[N(s)/sQ(s)]$, since the degree of $Q(s)$ = degree of $N(s) - 1$. Note that this is not a proper function, thus we can write

$$\frac{N(s)}{sQ(s)} = \frac{(s-1)(s^2+1)}{s(s+1)(s+2)} = 1 - \frac{4s^2+s+1}{s(s+1)(s+2)}$$

$$= 1 - \left[\frac{K_1}{s} + \frac{K_2}{s+1} + \frac{K_3}{s+2}\right]$$

$$K_1 = \left.\frac{4s^2+s+1}{(s+1)(s+2)}\right|_{s=0} = \frac{1}{2}$$

$$K_2 = \left.\frac{4s^2+s+1}{s(s+2)}\right|_{s=-1} = -4$$

$$K_3 = \left.\frac{4s^2+s+1}{s(s+1)}\right|_{s=-2} = \frac{15}{2}$$

hence

$$\frac{N(s)}{sQ(s)} = \left(1 + \frac{4}{s+1}\right) - \left(\frac{\frac{1}{2}}{s} + \frac{\frac{15}{2}}{s+2}\right)$$

or

$$\frac{N(s)}{Q(s)} = \left(s + \frac{4s}{s+1}\right) - \left(\frac{1}{2} + \frac{15s/2}{s+2}\right) = Y_{RC}^c - Y_{RC}^d$$

where

$$Y_{RC}^c = s + \frac{4s}{s+1}, \qquad Y_{RC}^d = \frac{1}{2} + \frac{15s/2}{s+2}$$

We now return to the voltage ratio synthesis, namely (8-12). Hence, according to the theorem (8-12) can be written as

$$\frac{N(s)/Q(s)}{D(s)/Q(s)} = \frac{Y_A - Y_B}{Y_C - Y_D} \qquad (8\text{-}20)$$

From (8-11) and (8-20), we make the following identifications, which will always guarantee a realization using RC driving-point admittance functions:

$$\begin{aligned} Y_1 &= Y_C \\ Y_2 &= Y_D + Y_A + Y_B \end{aligned} \qquad (8\text{-}21)$$

8/Active RC Filters

Thus, the synthesis is complete. Note that by subtracting common terms from Y_1 and Y_2, we do not change the results, as is seen in (8-20), hence by doing so a simpler circuit realization is achieved.

EXAMPLE

Consider the realization of an all-pass function given by

$$A_V = \frac{N(s)}{D(s)} = \frac{s^2 - \sqrt{2}s + 1}{s^2 + \sqrt{2}s + 1}$$

We arbitrarily choose $Q(s) = s + 1$. Since the degree of Q is one less than the numerator or denominator, we expand $[N(s)/sQ(s)]$ according to (8-18) and (8-19) to get $Y_A - Y_B$. Omitting the details we have

$$A_V = \frac{N/Q}{D/Q} = \frac{(s^2 - \sqrt{2}s + 1)/(s+1)}{(s^2 + \sqrt{2}s + 1)/(s+1)} = \frac{(s+1) - [(2 + \sqrt{2})s/(s+1)]}{(s+1) - [(2 - \sqrt{2})s/(s+1)]}$$

Thus from (8-20), (8-21), and (8-11) we have

$$Y_A = s + 1 \qquad Y_1 = s + 1$$

$$Y_B = \frac{3.4142s}{s+1} \qquad Y_2 = s + 1 + \frac{4s}{s+1}$$

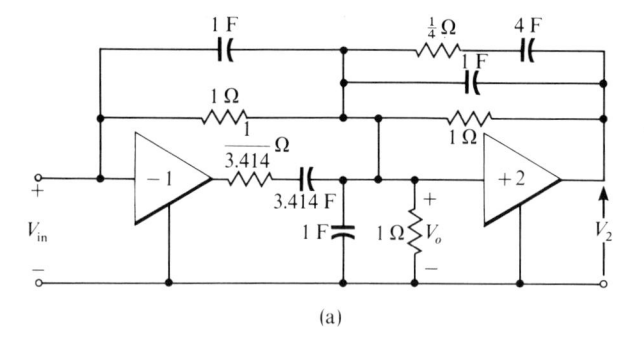

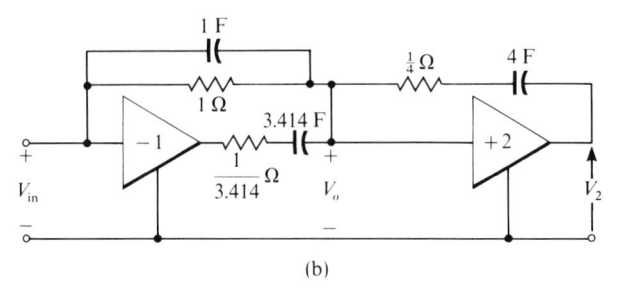

Fig 8-14 (a) Circuit realization of the design example; (b) Alternate simple form of circuit realization for the design example

The realization is shown in fig 8-14a. Note that in this case we could set $Y_1 = 0$, since Y_1 and Y_2 have the common terms $(s + 1)$. This choice leads to the following:

$$Y_2 = \frac{4s}{s+1} \quad \text{and} \quad Y_1 = 0$$

Of course, Y_A and Y_B are the same as before. In this case, the circuit is simpler than that of fig 8-14a, as shown in fig 8-14b.

It should be noted that we can add an arbitrary value of G_1 to Y_C and Y_D in (8-20), hence to Y_1 and Y_2 without affecting the results. This addition can account for the input impedance of K_2, or we can obtain a realization of $Y_{21} = I_o/V_{in}$, where $I_o = V_o G_1$. Note also, that if the output is taken at V_2 the voltage-ratio function is merely multiplied by 2. This choice is preferable for cascading purposes.

Realization of Current Ratio Functions

If the transfer function specified is a current-ratio function of the general form in (8-2), the circuit shown in fig 8-15 can be used. The active elements are assumed to be ideal voltage-controlled voltage sources. The analysis of the circuit of fig 8-15 yields

$$A_I = \frac{I_o}{I_{in}} = \frac{K_2(Y_A - K_1 Y_B)}{Y_1 - (K_2 - 1)Y_2} \tag{8-22}$$

Again we can set $K_2 - 1 = K_1 = 1$, without any loss of generality. Thus (8-22) reduces to

$$A_I = 2\frac{Y_A - Y_B}{Y_1 - Y_2} \tag{8-23}$$

Thus given any $A_I = [N(s)/D(s)]$, we introduce $Q(s)$ of (8-13) and use the decomposition theorem to express

$$A_I = \frac{N/Q}{D/Q} = 2\frac{(Y_A - Y_B)}{Y_1 - Y_2} \tag{8-24}$$

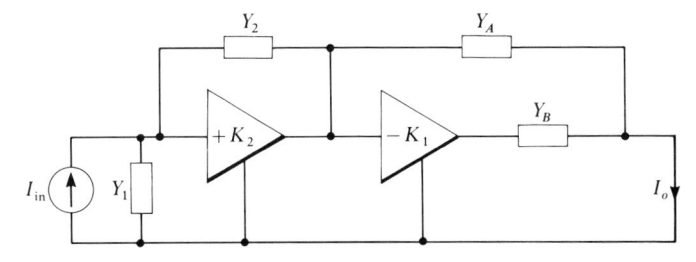

Fig 8-15 Active RC configuration for the realization of general current-ratio transfer functions

8/Active RC Filters

Comparison of (8-24) and (8-23) yields a one-to-one identification, and thus the synthesis is complete.

EXAMPLE

Consider the realization of

$$A_I = \frac{s^2 - \sqrt{2}s + 1}{s^2 + \sqrt{2}s + 1}$$

We arbitrarily introduce $Q(s) = s + 1$ and, as in the previous example, obtain

$$A_I = \frac{(s^2 - \sqrt{2}s + 1)/(s + 1)}{(s^2 + \sqrt{2}s + 1)/(s + 1)} = \frac{(s + 1) - 3.4142s/(s + 1)}{(s + 1) - 0.5858s/(s + 1)}$$

Thus

$$Y_A = (s + 1)/2 \qquad Y_1 = s + 1$$

$$Y_B = \frac{1.7071s}{s + 1} \qquad Y_2 = \frac{0.5858s}{s + 1}$$

The circuit realization is shown in fig 8-16. Again, note that the source conductance, if specified, can always be realized by adding G_1 to Y_1 and Y_2 in (8-24). The above example already has a conductance of unity, which can accomodate the source termination. It should be noted that all methods based on $RC: -RC$ decomposition suffer from high Q sensitivity, i.e., the sensitivity of Q due to variations of the gain of the active devices is larger than $2Q$ (see Section 7-16). In fact the lowest Q sensitivity obtainable is $2Q$ (see Problem 7-23). Thus, such methods should not be used for the realization of high Q selective filters.

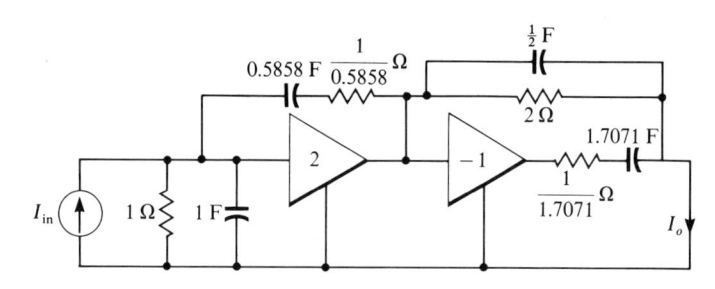

Fig 8-16 Circuit realization of the design example

8-3 Realization by Cascade of Simple Second-Order Networks

Practical *RC* filters are usually designed as a cascade of simple second-order networks, because of sensitivity and stability reasons. Thus, if we have a realization of a general biquadratic function, namely, the ratio of two second-order polynomials

$$T(s) = \frac{a_2 s^2 + a_1 s + a_0}{b_2 s^2 + b_1 s + b_0} \qquad (8\text{-}25)$$

then any general transfer function of the form (8-2) can be realized by cascaded noninteracting blocks with transfer functions given by (8-25). In (8-25) the denominator coefficients are either *all* negative or all nonnegative constants for stability reasons.

There is a large number of circuit configurations, using controlled sources and *RC* elements, capable of realizing various second-order transfer functions. We shall only consider a few simple circuits that utilize only one voltage-controlled voltage source with either an inverting or noninverting property, i.e., $K > 0$ or $K < 0$, and no more than three *R*'s and three *C*'s. (These circuits are usually referred to as the Sallen and Key circuits, see Ref 4.) Note that since active elements with very high input impedance and very low output impedance are readily available, in the form of the operational amplifier, the interaction is negligible and no buffer stages are required. The methods of this section, because of sensitivity problems, are useful only for low and moderate *Q* designs. High-*Q*, low-sensitivity methods based on biquadratic transfer functions and utilizing more than one amplifying block are discussed in Section 8-8.

Low-Pass Section of the Form $H_1/(s^2 + \alpha s + \beta)$

The circuits shown in fig 8-17a and fig 8-17b are two such circuits that provide a lowpass, two-pole transfer function of the form

$$A_v(s) = \frac{H_1}{s^2 + \alpha s + \beta} \qquad (8\text{-}26)$$

The voltage transfer ratio function of the network in fig 8-17a as obtained by analysis is given by

$$A_v = \frac{V_o}{V_i} = \frac{K/R_1 C_1 R_2 C_2}{[s^2 + s(1/R_1 C_1 + 1/R_2 C_1 + 1/R_2 C_2 - K/R_2 C_2) + 1/R_1 C_1 R_2 C_2]} \qquad (8\text{-}27)$$

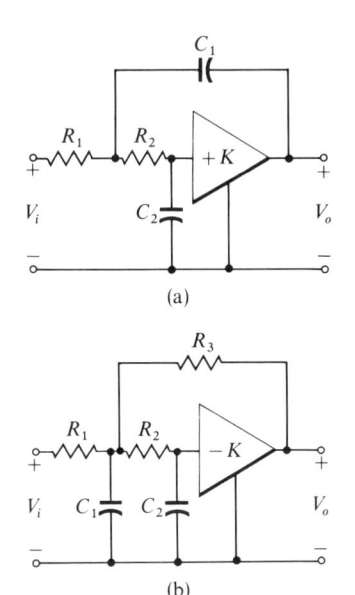

Fig 8-17 Circuit configurations which realize lowpass transfer functions of the form $A_v(s) = H_1/(s^2 + \alpha s + \beta)$: (a) Canonical configuration using a noninverting gain block; (b) Configuration using an inverting gain block and one more resistor than (a)

8/Active RC Filters

Equation (8-27) may be rewritten as

$$A_v = \frac{V_o}{V_i} = \frac{K/\tau_1\tau_2}{[s^2 + s(1/\tau_1 + 1/\tau_2 - K/\tau_2 + 1/R_2C_1) + 1/\tau_1\tau_2]} \quad (8\text{-}28)$$

where $\tau_1 = R_1C_1$ and $\tau_2 = R_2C_2$. From (8-28) and (8-27), by matching coefficients we obtain

$$H_1 = \frac{K}{\tau_1\tau_2} \qquad \beta = 1/\tau_1\tau_2$$

$$\alpha = \frac{1}{\tau_1} + \frac{1}{\tau_2}(1 - K) + \frac{1}{R_2C_1}, \quad (8\text{-}29)$$

There are five design variables, 2 R's, 2 C's, and the gain K. Since the number of unknowns is larger than the number of equations, two variables may be preselected to have convenient values, and the others are then determined from (8-29). Note that the source resistance specification can be incorporated in R_1. If $|K| \leq 1$ is used, the network is absolutely stable. A unity gain amplifier is readily obtained in several ways. An operational amplifier with unity gain is obtained by making $R_2 = 0$ in fig 8-31b.

DESIGN EXAMPLE

Let $V_o/V_i = 10/(s^2 + \sqrt{2}s + 1)$ with a 3-dB bandwidth at $\omega = 2\pi(10^3)$ rad/sec. The source resistance is given, $R_s = 100\,\Omega$. Let us choose $R_1 = 1$ (normalized), which may be assigned to the source resistance, and let $C_2 = 1$. Then from the specification, and (8-29), we obtain

$$K = 10, \qquad R_2 = 3.21 \quad \text{and} \quad C_1 = 0.311$$

The actual element values corresponding to the normalization factors $R_0 = 100$, $\Omega_0 = 2\pi \times 10^3$ are

$$R_1 = 100\,\Omega \qquad R_2 = 321\,\Omega \qquad K = 10$$

$$C_1 = 0.50\,\mu\text{F} \qquad C_2 = 1.59\,\mu\text{F}$$

Note that if the two poles of the specified transfer function are real, a ladder RC network with two R's and two C's can be used. For cascading purposes, however, buffer amplifiers will be required.

8-3 Realization by Cascade of Simple Second-Order Networks

Band-Pass Section of the Form $H_2 s/(s^2 + \gamma s + \xi)$

The circuits shown in fig 8-18a and fig 8-18b may be used to realize a bandpass transfer function of the form:

$$A_v(s) = \frac{H_2 s}{s^2 + \gamma s + \xi} \tag{8-30}$$

The voltage transfer function of the circuit in fig 8-18a is given by Problem 8-10:

$$A_v = \frac{V_o}{V_i} = \frac{-[K/(1+K)R_1 C_1]s}{\{s^2 + s[(R_2 C_2 + R_1 C_1 + R_1 C_2)/(1+K)R_1 C_1 R_2 C_2] + 1/(1+K)R_1 C_1 R_2 C_2\}} \tag{8-31}$$

Equation (8-31) may be rewritten as

$$A_v = \frac{-[K/(1+K)\tau_1]s}{s^2 + s[(\tau_1 + \tau_2 + R_1 C_2)/(1+K)\tau_1 \tau_2] + 1/(1+K)\tau_1 \tau_2} \tag{8-32}$$

where $\tau_1 = R_1 C_1$ and $\tau_2 = R_2 C_2$. The design equations are obtained by matching coefficients, that is

$$H_2 = -\frac{K}{(1+K)\tau_1} \qquad \xi = \frac{1}{(1+K)\tau_1 \tau_2}$$

$$\gamma = \frac{\tau_1 + \tau_2 + R_1 C_2}{(1+K)\tau_1 \tau_2} \tag{8-33}$$

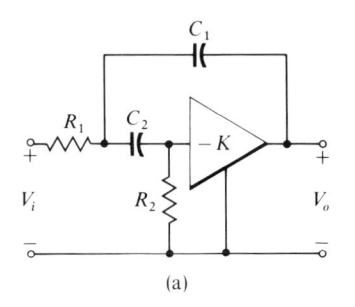

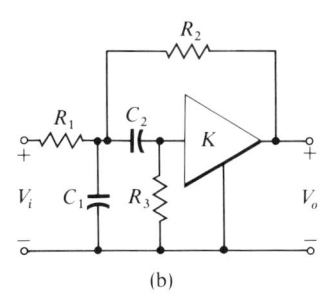

Fig 8-18 Circuit configurations which realize band-pass transfer functions of the form $A_v(s) = H_2 s/(s^2 + \gamma s + \xi)$: (a) Canonical configuration using an inverting gain block; (b) Configuration using a noninverting gain block and one more resistor than (a)

The design procedure is similar to the above, and will not be repeated. It should be noted, however, that the Q of these circuits is quite limited due to sensitivity reasons. The practical Q's realized are usually below 25.

High-Pass Section of the Form $H_3 s^2/(s^2 + \eta s + \delta)$

The circuits shown in fig 8-19a and fig 8-19b are two such circuits that provide transfer functions of the high-pass form, that is

$$A_v(s) = \frac{H_3 s^2}{s^2 + \eta s + \delta} \tag{8-34}$$

The voltage transfer function of fig 8-19a is given by

$$A_v = \frac{V_o}{V_i} = \frac{Ks^2}{s^2 + s[1/R_1C_1 - K/R_1C_1 + 1/R_2C_2 + 1/R_2C_1] + 1/R_1R_2C_1C_2} \quad (8\text{-}35)$$

$$A_v = \frac{Ks^2}{s^2 + s[(1-K)/\tau_1 + 1/\tau_2 + 1/R_2C_1] + 1/\tau_1\tau_2} \quad (8\text{-}36)$$

where $\tau_1 = R_1C_1$, and $\tau_2 = R_2C_2$. The design equations as obtained by matching the coefficients are

$$H_3 = K \qquad \delta = \frac{1}{\tau_1\tau_2}$$
$$\eta = \frac{1-K}{\tau_1} + \frac{1}{\tau_2} + \frac{1}{R_2C_1} \quad (8\text{-}37)$$

The design procedure is quite similar to the previous cases. Note that all of the configurations discussed in this section, namely figs 8-17a, 8-18a, and 8-19a, use the minimum number of R and C elements, i.e., 2 R's and 2 C's for a quadratic denominator realization, and are therefore canonic. The configurations in figs 8-17b, 8-18b, and 8-19b use one more element than the canonic realization. Finally, it should be noted that the ideal controlled sources can be realized very closely by operational amplifiers as discussed in Section 8-7. For example, the VCVS is realized by the circuit shown in fig 8-31b and fig 8-31c.

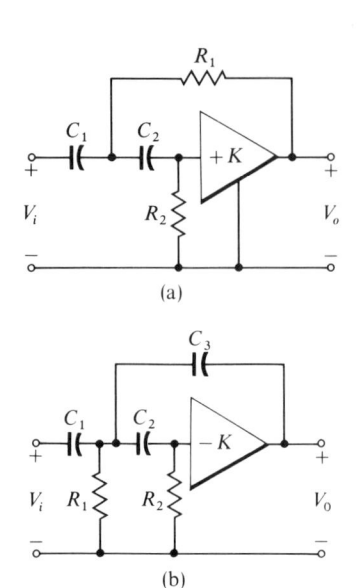

Fig 8-19 Circuit configurations which realize highpass transfer functions of the form $A_v(s) = H_3s^2/(s^2 + \eta s + \delta)$: (a) Canonical configuration using a noninverting gain block; (b) Configuration using an inverting gain block and one more capacitor than (a)

8-4 The Negative-Impedance Converter as a Network Element

An ideal negative-impedance converter (abbreviated NIC) is an active two-port network that converts an impedance termination, Z_L, at one port into a negative impedance at the other port, proportional to Z_L. Consider a linear two-port network as shown in fig 8-20. The terminated input and output immittances of the two-port in terms of, say, the h-parameters are

$$Z_i = h_{11} - \frac{h_{12}h_{21}}{h_{22} + Y_L} \quad (8\text{-}38a)$$

8-4 The Negative-Impedance Converter as a Network Element

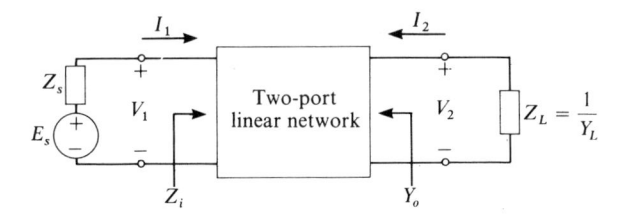

Fig 8-20 Linear two-port network with source and load terminations

$$Y_o = h_{22} - \frac{h_{12}h_{21}}{h_{11} + Z_s} \qquad (8\text{-}38b)$$

Necessary and sufficient conditions for the active two-port to be an ideal NIC are

$$h_{11} = h_{22} = 0 \qquad (8\text{-}39a)$$

$$h_{12}h_{21} = K \qquad (8\text{-}39b)$$

where K is an arbitrary real positive number, usually referred to as the conversion factor. Thus, the condition of NIC may be satisfied in a variety of methods and in particular, we may choose $h_{12} = h_{21} = \pm\sqrt{K}$.

The *ABCD* parameters of the NIC circuit from (8-39) and table 2-1 are

$$\begin{bmatrix} A & B \\ C & D \end{bmatrix} = \begin{bmatrix} \pm\sqrt{K} & 0 \\ 0 & \mp 1/\sqrt{K} \end{bmatrix} \qquad (8\text{-}40)$$

Alternatively, we may write the input impedance of fig 8-20 directly in terms of the *ABCD* parameters. The input impedance $Z(s)$ is given by table 2-2:

$$Z_i = \frac{AZ_L + B}{CZ_L + D} \qquad (8\text{-}41)$$

Now in (8-41) if $B = C = 0$ and $A/D = -K$, the input impedance is

$$Z_i = -KZ_L \qquad (8\text{-}42)$$

Usually an NIC with unity conversion factor (also referred to as a unity-gain NIC) is considered, i.e., $K = 1$. For this particular choice we have:

$$h_{12} = h_{21} = \begin{cases} 1 \leftrightarrow A = 1 \quad D = -1 & (8\text{-}43a) \\ -1 \leftrightarrow A = -1 \quad D = 1 & (8\text{-}43b) \end{cases}$$

$$(h_{11} = h_{22} = 0) \qquad B = C = 0$$

For case (8-43a), the voltage at the output is not inverted, while the actual current has an inversion in sign (note that I_2 is defined as flowing into the output port), i.e.,

$$V_1 = V_2 \qquad I_1 = I_2 \tag{8-44}$$

The converter in this case is referred to as of the current inversion type, or INIC for short. Thus, for a unity gain current inversion type we have

$$\left. \begin{array}{ll} h_{12} = h_{21} = 1 & A = 1 \quad D = -1 \\ h_{11} = h_{22} = 0 & B = C = 0 \end{array} \right\} \text{INIC} \tag{8-45}$$

For case (8-43b), the same is true only for the voltage, i.e.,

$$V_1 = -V_2 \qquad I_1 = -I_2 \tag{8-46}$$

The converter in this case is referred to as of the voltage inversion type or VNIC. Note that the same current entering port 1 leaves port 2, hence the actual current direction is not changed. Thus for a unity gain voltage inversion type we have

$$\left. \begin{array}{ll} h_{12} = h_{21} = -1 & A = -1, \quad D = 1 \\ h_{11} = h_{22} = 0 & B = C = 0 \end{array} \right\} \text{VNIC} \tag{8-47}$$

There are many transistor circuit realizations of both types of NIC's in the literature. Two practical circuits, which realize the NIC action, are shown in fig 8-21 and fig 8-22. For the sake of clarity, only the major components of the circuit are shown, exclusive of biasing circuitry.

The analysis of the midband h-parameters of INIC in fig 8-21 with $\alpha_0 \to 1$ yields

$$\begin{array}{ll} h_{11} \simeq 0 & h_{22} \approx 0 \\ h_{12} \approx 1 & h_{21} \approx \dfrac{R_1}{R_2} \end{array} \tag{8-48}$$

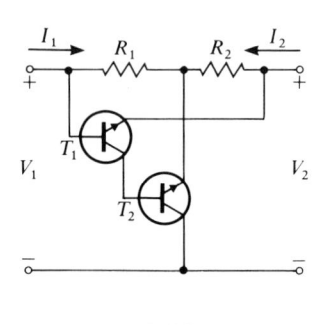

Fig 8-21 Circuit realization of current-inversion type negative-impedance converter (*INIC*)

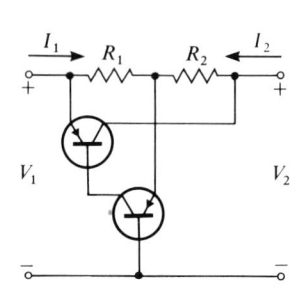

Fig 8-22 Circuit realization of voltage-inversion type negative-impedance converter (*VNIC*)

8-4 The Negative-Impedance Converter as a Network Element

Similarly for the VNIC in fig 8-22, the h-parameters as $\alpha_0 \to 1$ are given by

$$h_{11} \approx 0 \qquad h_{22} \approx 0$$
$$h_{12} \approx -\frac{R_1}{R_2} \qquad h_{21} \approx -1 \tag{8-49}$$

More accurate expressions than in above may be obtained by using the low frequency Tee model of the transistors in the analysis. However, α_0 very close to unity can be achieved if Darlington compound transistors are used instead of a single transistor. Operational amplifiers may also be used to realize NIC circuits, as shown later in fig 8-33.

From the analysis of any practical NIC circuit, it is obvious that h_{11} and h_{22} are not identically equal to zero, albeit they are very small. The NIC can be designed to achieve this ideal condition. Consider for example, fig 8-23. The inner box is a NIC circuit with h_{11} and h_{22} very small, but not equal to zero. Let $h_{11} = -Z_a$ and $h_{22} = -Y_b$.

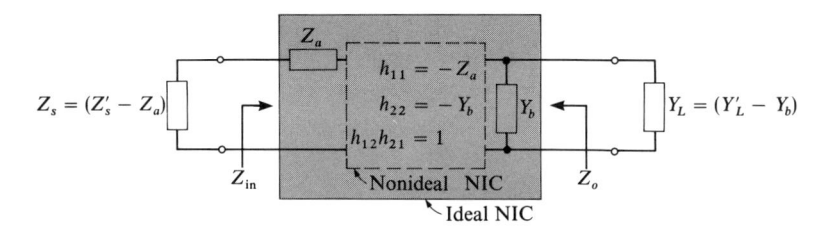

Fig 8-23 Compensated nonideal *NIC* which exhibits an ideal *NIC* characteristic

(If $h_{11} = Z_a$ and $h_{22} = Y_b$ in this case, these may be put outside the box, and the two-port within the box will be an ideal NIC.) If we associate part of the source and load immittances artificially with the nonideal NIC, it is immediately seen that the outer box represents an ideal NIC. In other words, the NIC has a very interesting property of compensating for its own nonidealness. For a given load and source immittance, parts of these are used for self-correction to obtain the ideal NIC action. For example, if the low-frequency h-parameters of a practical VNIC circuit are given by

$$[h_{ij}] = \begin{bmatrix} -10\,\Omega & -1 \\ -1 & -10^{-5}\,\text{mho} \end{bmatrix}$$

and if the terminations are $Z_s = 100\,\Omega$ and $Z_L = 1\,\text{k}\Omega$, the impedances seen at the input and the output of the NIC are

$$Z_{in} = -(1 + 0.01)\,\text{k}\Omega = -1.01\,\text{k}\Omega$$
$$Z_o = -(100 - 10)\,\Omega = -90\,\Omega$$

In other words, the circuit behaves as an ideal NIC with $Z_s = 90\,\Omega$ and $Z_L = 1010\,\Omega$.

Since NIC's are potentially unstable active two-ports (see Section 2-3), care should be exercised in designing their terminations. It has one port as a short-circuit stable side, and the other port as an open-circuit stable side. So in a design, the short-circuit stable side must see a low impedance termination, while the open-circuit stable side should face a high impedance termination.

8-5 Transfer Function Synthesis Using Negative-Impedance Converter

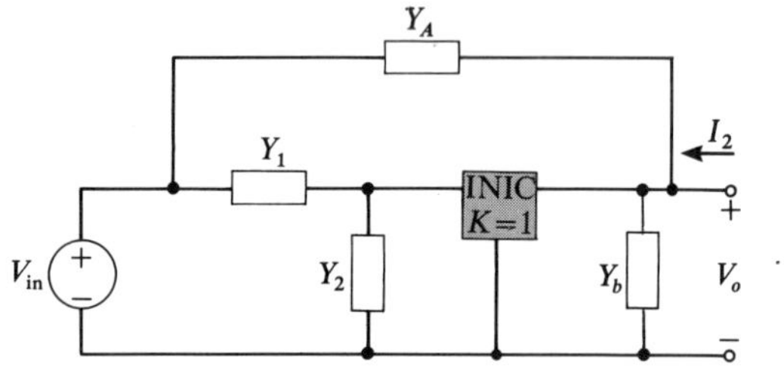

There are many useful synthesis techniques using NIC and RC networks. In the following section, as previously, we consider only simple techniques which require one-port RC synthesis in the realization of general transfer functions.

A circuit configuration which can be used to realize any general voltage ratio transfer function of the form (8-9) is shown in fig 8-24. Analysis of the circuit in fig 8-24 yields (see Problem 8-14) the following voltage ratio transfer function:

Fig 8-24 Active RC configuration, using $INIC$ and passive RC one-port networks, for the realization of general voltage-ratio transfer functions

$$A_V = \left.\frac{V_o}{V_{in}}\right|_{I_2=0} = \frac{Y_A - Y_1}{Y_A - Y_1 + Y_B - Y_2} \qquad (8\text{-}50)$$

Given any voltage ratio transfer function $A_V = [N(s)/D(s)]$, we may rewrite

$$A_V = \frac{N(s)}{D(s)} = \frac{N(s)}{N(s) + D(s) - N(s)} = \frac{N/Q}{N/Q + (D-N)/Q} \qquad (8\text{-}51)$$

in order that the numerator and part of the denominator are identical, as in (8-50). Now we introduce a polynomial $Q(s)$ with distinct negative

8-5 Transfer Function Synthesis Using Negative-Impedance Converter

real roots as in (8-13), namely

$$Q(s) = \prod_{i=1}^{n-1} (s + \sigma_i) \qquad \sigma_i > 0 \qquad (8\text{-}52)$$

where n is equal to or greater than the degree of $N(s)$ or $D(s)$, whichever is larger. Partial fraction expansion of N/Q and $D - N/Q$ yields the individual driving-point RC admittances, namely

$$\frac{N(s)}{Q(s)} = Y_A - Y_1 \qquad (8\text{-}53)$$

$$\frac{D - N}{Q} = Y_B - Y_2 \qquad (8\text{-}54)$$

where the individual Y's are guaranteed to be RC driving-point admittances. Thus, the synthesis is complete by the realization of admittances in (8-53) and (8-54). Note also, that since an admittance can be added to Y_B and Y_2 without changing any expression in (8-50), we can, therefore, specify any termination desired. Thus, if we add a conductance G to Y_B and Y_2, the same configuration can also be used to synthesize $Y_{21} = I_o/V_{in}$.

EXAMPLE

Consider the realization of the transfer function

$$A_V = \frac{s^2 - s + 1}{s^2 + s + 1}$$

We introduce arbitrarily $Q(s) = s + 1$ and rewrite the above as

$$A_V = \frac{(s^2 - s + 1)/(s + 1)}{(s^2 - s + 1)/(s + 1) + 2s/(s + 1)}$$

Hence

$$\frac{N(s)}{Q(s)} = \frac{s^2 - s + 1}{s + 1} = s + 1 - \frac{3s}{s + 1}$$

or

$$Y_A = s + 1 \qquad Y_1 = \frac{3s}{s + 1}$$

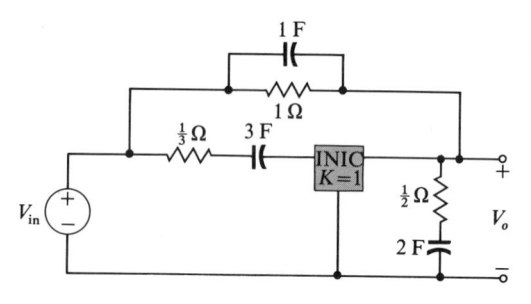

Fig 8-25 Circuit realization of the design example for specified $A_V(s)$

Similarly,

$$\frac{D(s) - N(s)}{Q(s)} = \frac{2s}{s+1}$$

or

$$Y_2 = 0 \qquad Y_B = \frac{2s}{s+1}$$

The realization is shown in fig 8-25. If we have the termination also specified to be 1 Ω, the above synthesis can be readily modified as follows. Add a conductance of 1 mho to Y_B and Y_2, so that

$$Y_B = 1 + \frac{2s}{s+1} \qquad \text{and} \qquad Y_2 = 1$$

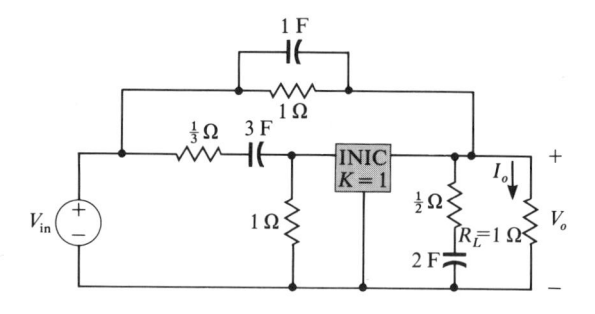

Fig 8-26 Circuit realization of the design example for specified $Y_{21}(s)$

The realization in this case is as shown in fig 8-26. Note that the above circuit also realizes

$$Y_{21} = \frac{I_o}{V_{in}} = \frac{s^2 - s + 1}{s^2 + s + 1}$$

since $V_o = I_o G_L = I_o$. Again the reader is cautioned about the poor sensitivity of this approach, since the synthesis is based on RC–RC decomposition.

8-6 The Gyrator as a Network Element

An ideal gyrator is a passive, nonreciprocal two-port network characterized by the following two-port z-parameters:

$$[z_{ij}] = \begin{bmatrix} 0 & R_o \\ -R_o & 0 \end{bmatrix} \text{ ohms} \quad [y_{ij}] = \begin{bmatrix} 0 & -G_o \\ G_o & 0 \end{bmatrix} \text{ mhos} \quad (8\text{-}55)$$

where $R_o = G_o^{-1}$ is a real positive quantity referred to as the gyration resistance (conductance). Symbolically, the gyrator is represented as shown in fig 8-27a. The equivalent circuit model is shown in fig 8-27b, with $G_1 = G_2 = G_o$. If $G_1 \neq G_2$, the gyrator is an active element, and is usually referred to as an active gyrator (see Problem 8-27). Necessary and sufficient conditions for a two-port to be a gyrator, in terms of the z-parameters, are

$$z_{11} = z_{22} = 0 \quad z_{12}z_{21} = K_1 < 0 \quad (8\text{-}56a)$$

or alternatively in terms of ABCD parameters:

$$A = D = 0 \quad \frac{B}{C} = K_2 > 0 \quad (8\text{-}56b)$$

One of the most important properties of a gyrator is that an inductor can be simulated, simply by terminating a gyrator with a capacitor. This is readily seen as follows. Consider the circuit in fig 8-28. Using (8-55) we have

$$Z_{in} = z_{11} - \frac{z_{12}z_{21}}{z_{22} + Z_L} = \frac{R_o^2}{1/sC} = (R_o^2 C)s \quad (8\text{-}57)$$

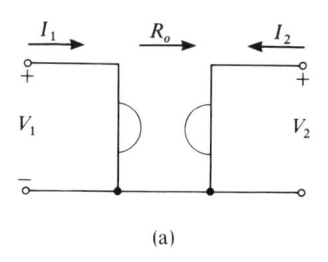

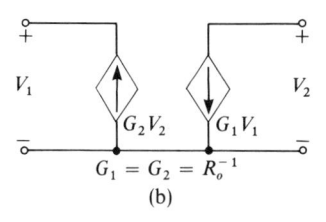

Fig 8-27 (a) Symbol for an ideal gyrator with voltage polarity and current reference directions; (b) Circuit model of an ideal gyrator $G_1 = G_2 = R_o^{-1}$

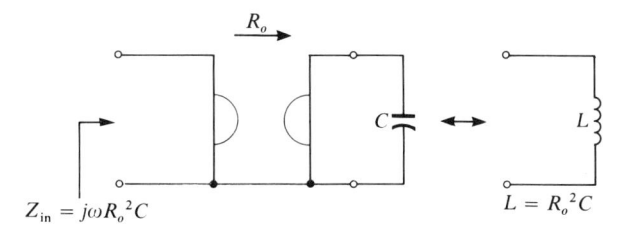

$Z_{in} = j\omega R_o^2 C$

$L = R_o^2 C$

Fig 8-28 Capacitively terminated ideal gyrator which simulates an inductor $L = R_o^2 C$

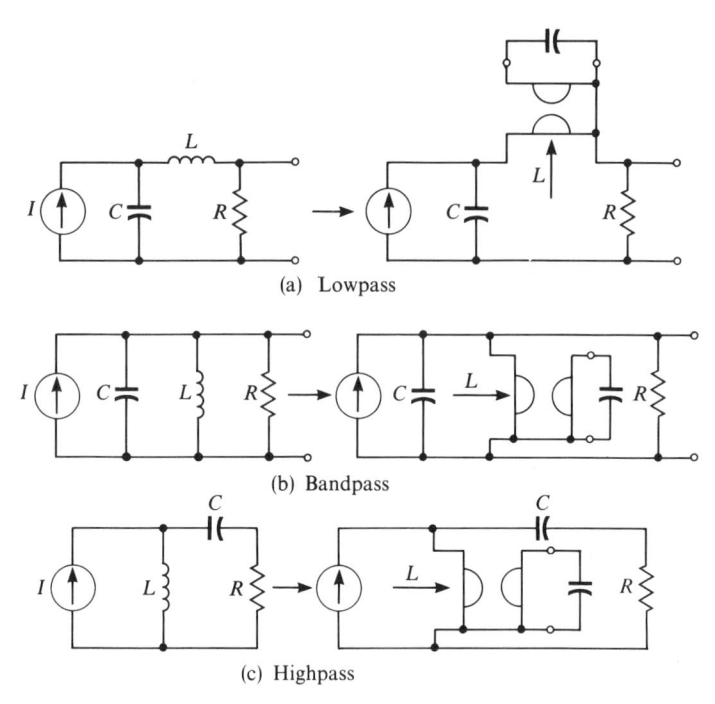

Fig 8-29 Various filtering circuits and those with the inductor replaced by a capacitively terminated gyrator: (a) Lowpass circuits; (b) Bandpass circuits; (c) Highpass circuits

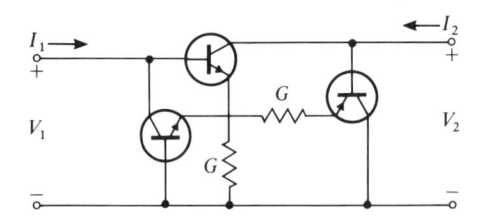

Fig 8-30 Circuit realization of the gyrator

Thus, the impedance Z_{in} is the same as would be obtained from an inductor with $L = R_o^2 C$. This property immediately suggests that an RLC filter can be transformed into an inductorless filter merely by replacing the inductor by a capacitively terminated gyrator. For example, the lowpass, bandpass, and highpass RLC filters can be replaced by the RC-gyrator networks as shown in figs 8-29a, b, c, respectively. A floating inductance can be realized by two grounded gyrators and a capacitor (see Problem 8-18). Thus the vast literature for passive LC filters can be used directly, by replacing the inductor by gyrator-capacitor networks. Since the active elements are used in the passive mode for an ideal gyrator, the circuit realization has low sensitivity. The performance of an inductive simulated active RC filter depends heavily on the quality and performance of the gyrator (see Problem 8-28).

There are many transistor circuit realizations of gyrators in the literature. One such circuit is shown in fig 8-30. For the sake of clarity, the major components of the circuit are shown exclusive of biasing circuitry. The analysis of the circuit in fig 8-30 for $\alpha \to 1$ yields the following y-parameters:

$$y_{11} \simeq 0 \qquad y_{22} \approx 0$$
$$y_{12} \simeq -G \qquad y_{21} \simeq G$$

For a more accurate expression in the above, the low-frequency Tee model of the transistor may be used, and the resulting circuit can then be analyzed.

Similar to the controlled source realization, second-order network functions can be realized by gyrator-RC networks. The analysis and design of two such circuits are given as exercises in Problems 8-16 and 8-17.

8-7 Operational Amplifier Realizations

The operational amplifier has already been discussed in Section 6-7. It was noted then that ideally, the gain of an operational amplifier is frequency independent and infinite, its input impedance is infinite, and its output impedance is zero.[1] However, in practice, the gain is usually in the range of 80 dB to 100 dB and is frequency dependent, its input impedance is about 100 kilohms, and the output impedance is about 100 ohms. Usually external frequency compensation is used to make the frequency dependence of the gain have a one-pole response for stability reasons. These considerations are important in practical circuit design. In order to focus our attention on the basic techniques, however, we shall assume ideal operational amplifiers in the following discussion.

A differential input operational amplifier is shown symbolically in fig 8-31a. The minus terminal is called the *inverting terminal* while the positive terminal is called the *noninverting terminal*. If the operational amplifier is to be used with a single-ended input, such as used in analog computation, the plus terminal is grounded. Operational amplifiers can be obtained in integrated circuit form, and are readily available from a large number of manufacturers in flat packages or in the same TO cans as are discrete transistors. Their use in active RC filters is, therefore, quite attractive. In fact, operational amplifiers

---

[1] The precise definition of an *ideal* operational amplifier is (see fig 8-31a)
$$V_o = A(V_2 - V_1), \qquad I_1 = I_2 = 0, \qquad A \to \infty, \qquad V_o \to 0 \text{ when } (V_2 - V_1) \to 0$$
where A is a constant independent of frequency, temperature, and input levels.

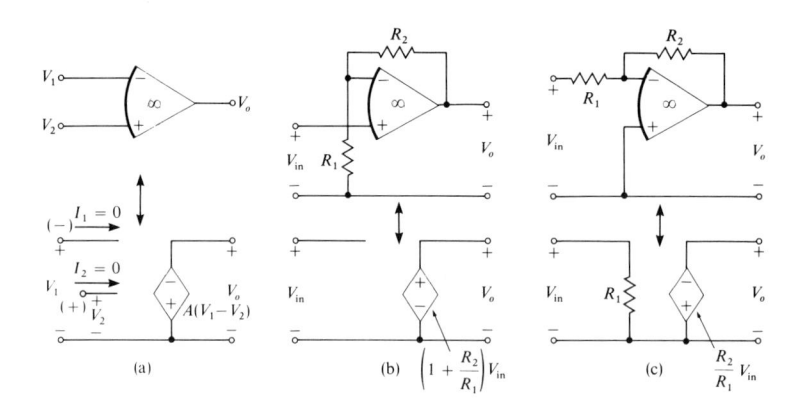

Fig 8-31 (a) Ideal differential-input operational amplifier and its circuit model; (b) Operational amplifier connection for a noninverting gain block, with gain $K = (R_1 + R_2)/R_1$, and its circuit model; (c) Operational amplifier connection for an inverting gain block, with gain $K = -R_2/R_1$, and its circuit model

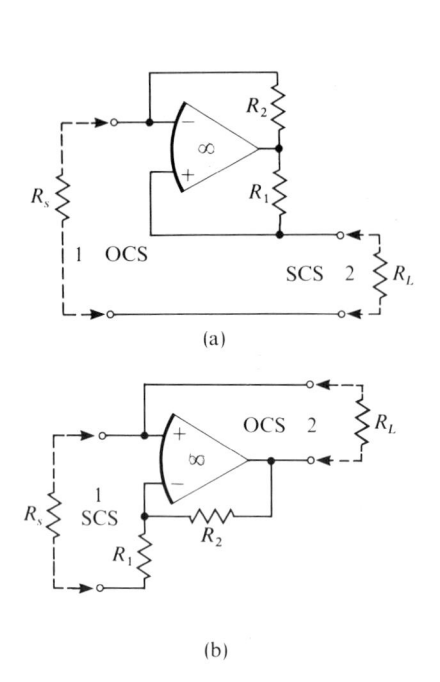

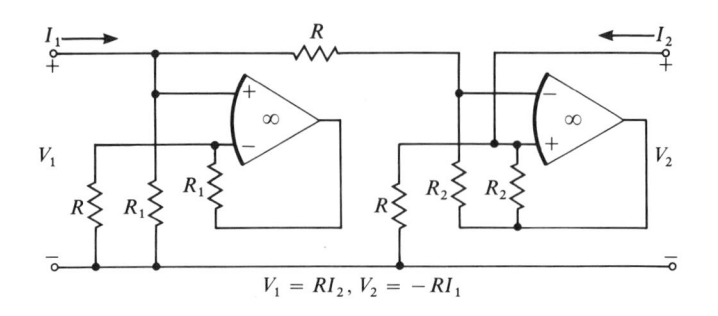

Fig 8-32 (a) *INIC* realization with operational amplifier; (b) *VNIC* realization with operational amplifier

can be used to realize the controlled source, NIC, and gyrators. The circuits shown in fig 8-31b and fig 8-31c are realizations of voltage-controlled voltage sources with K positive and K negative, respectively. Hence, all the methods discussed in Section 8-3 can utilize operational amplifiers. The INIC and VNIC realizations are shown in fig 8-32a and fig 8-32b, respectively. Since NIC's are potentially unstable devices, the short-circuit stable side (SCS) and the open-circuit stable side (OCS) are also indicated therein. In other words, the SCS should be terminated by a small load resistor, while the OCS should be terminated by a large resistor for stability reasons. A gyrator realization with operational amplifiers is shown in fig 8-33. Hence, the NIC and

Fig 8-33 Gyrator realization with operational amplifiers

gyrator realization techniques can also be viewed as operational amplifier realizations. We shall consider, however, transfer function synthesis directly in terms of the operational amplifier. We will consider only the use of a single operational amplifier in this section. Other general techniques utilizing two or more operational amplifiers are also available in the literature (Ref 4 and Ref 5). Practical low sensitivity high-Q realizations using more than one operational amplifier are discussed in the next section.

General Realization

Consider the circuit configuration shown in fig 8-34. The voltage transfer function of the circuit as obtained by analysis is (see Problem 8-20)

$$\frac{V_2}{V_1} = \frac{Y_1(Y_3 + Y_4 + Y_6) - Y_3(Y_1 + Y_2 + Y_5)}{Y_6(Y_1 + Y_2 + Y_5) - Y_5(Y_3 + Y_4 + Y_6)} \qquad (8\text{-}58)$$

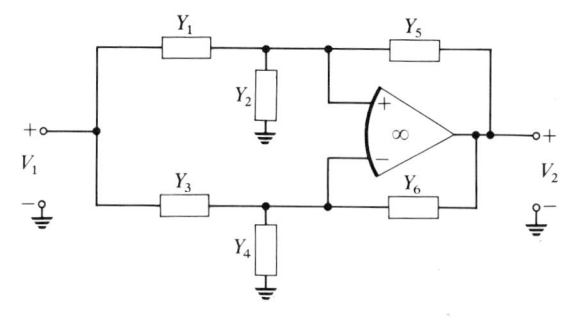

Fig 8-34 Active RC configuration using a single operational amplifier and RC one-ports to realize any general voltage-ratio transfer function

For convenience we choose

$$Y_1 + Y_2 + Y_5 = Y_3 + Y_4 + Y_6 \qquad (8\text{-}59)$$

Using (8-59) in (8-58), we obtain the simple expression

$$A_V(s) = \frac{V_2}{V_1} = \frac{Y_1 - Y_3}{Y_6 - Y_5} \qquad (8\text{-}60)$$

The desired transfer function may be written as

$$\frac{V_2}{V_1} = \frac{N(s)/Q(s)}{D(s)/Q(s)}$$

where $Q(s)$ is chosen as in (8-13), with a degree one less than the degree of $N(s)$ or $D(s)$, whichever is larger. The method of Section 8-2 may now be used so that

$$\frac{N}{Q} = Y_1 - Y_3 \qquad (8\text{-}61)$$

$$\frac{D}{Q} = Y_6 - Y_5 \qquad (8\text{-}62)$$

Having obtained Y_1, Y_3, Y_5, and Y_6, we determine the others from the following relation:

$$Y_2 = Y_3 + Y_6 \qquad (8\text{-}63)$$

$$Y_4 = Y_1 + Y_5 \qquad (8\text{-}64)$$

The synthesis is thus complete. Because of the cancellation of terms and differences of admittances involved in the realizations, this method is not practical. This method also falls into the *RC–RC* decomposition category, and thus suffers from poor sensitivity at high Q realization. However, the method provides a good illustration of a general synthesis technique.

It is to be noted that, in most of the above realizations, the differences of admittances are involved in the design (i.e., *RC–RC* decomposition). As was shown in Section 7-16, this leads to a very high Q sensitivity (larger than $2Q$). The circuits are, therefore, not practical for a high Q pole pair realization.

The methods of Section 8-3 can also be used in the realization of transfer functions, where the VCVS are realized by operational amplifiers as in fig 8-31. The methods of Section 8-3 are practical and useful for low Q pole pairs. A different method, which avoids the differences of admittance in the design, is presented in the following section.

8-8 Low-Sensitivity High-Q Transfer-Function Realization

A circuit configuration utilizing the operational amplifier, which has the attractive feature of very low sensitivity, is shown in fig 8-35.

8-8 Low-Sensitivity High-Q Transfer-Function Realization

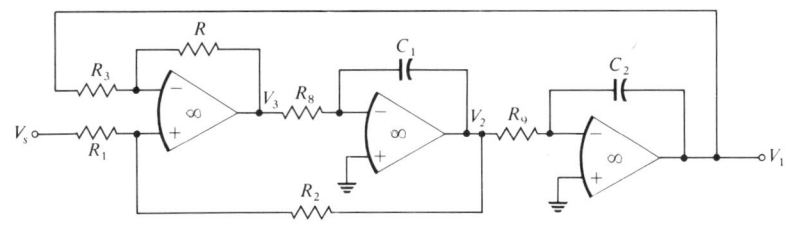

Fig 8-35 Active RC configuration which can be used for high-Q low-sensitivity second-order transfer function (low value of center frequency)

The analysis of the circuit of fig 8-35 yields (see Problem 8-22) the following transfer functions of the lowpass, bandpass and highpass forms, respectively:

$$A_l = \frac{V_1}{V_s} = \frac{R_2(R + R_3)/R_3(R_1 + R_2)}{s^2 C_1 C_2 R_8 R_9 + s[R_1(R + R_3)C_2 R_9/R_3(R_1 + R_2)] + R/R_3} \quad (8\text{-}65)$$

$$A_b = \frac{V_2}{V_s} = \frac{-[R_2(R + R_3)/R_3(R_1 + R_2)]R_9 C_2 s}{s^2 C_1 C_2 R_8 R_9 + s[R_1(R + R_3)C_2 R_9/R_3(R_1 + R_2)] + R/R_3} \quad (8\text{-}66)$$

$$A_h = \frac{V_3}{V_1} = \frac{[R_2(R + R_3)/R_3(R_1 + R_2)]C_1 C_2 R_8 R_9 s^2}{s^2 C_1 C_2 R_8 R_9 + s[R_1(R + R_3)C_2 R_9/R_3(R_1 + R_2)] + R/R_3} \quad (8\text{-}67)$$

A number of the circuit values can be chosen arbitrarily, and thus for convenience we can let

$$R_1 = R_3 = R = 1, \quad R_8 C_1 = R_9 C_2 = \tau \quad (8\text{-}68)$$

The use of (8-68) in (8-65), (8-66), and (8-67) simplifies the resulting transfer functions considerably. Also, observe that the output terminal in each case has zero output impedance and it thus can be cascaded directly with other circuits. Note that for a high Q pole pair, the resonant frequency from the denominator of the above transfer functions is given by

$$\omega_0 = \sqrt{\frac{R}{R_8 C_1 R_9 C_2 R_3}} = \frac{1}{\tau} \quad (8\text{-}69a)$$

$$Q = \left[\frac{R_1 + R_2}{R_1(R + R_3)}\right]\sqrt{\frac{RR_3 R_8 C_1}{R_9 C_2}} = \left[\frac{1 + R_2}{2}\right] \quad (8\text{-}69b)$$

In (8-69b), since a high Q does not depend on differences of terms, the sensitivity can be expected to be low. In fact, the Q sensitivity from (7-113) due to variations of any R's or C's in (8-69b) can be shown (see Problem 8-29) to be less than unity. Hence, the sensitivity of this circuit is of the same order of magnitude as those of the RLC filter. Finally, it is to be noted that the operational amplifiers are used at high gain, and that the poles of the operational amplifiers are neglected in the analysis for fig 8-35. Hence very high Q cannot be realized at high frequencies (i.e., $f_0 \geq 10 \text{ kHz}$). This is because of the limitation on the gain-bandwidth product of the operational amplifier when frequency compensation with a one pole rolloff is used.

A network configuration which utilizes the operational amplifiers at low gain and the Q of the network does not depend on the pole of the operational amplifier (if a compensated one pole rolloff operational amplifier is used) is shown in fig 8-36. This configuration can be used to realize high Q ($Q \geq 100$) with similar sensitivity as in fig 8-35, but at a much higher center-frequency.

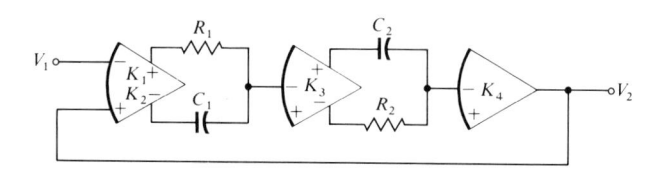

Fig 8-36 Active RC configuration which can be used for high-Q low-sensitivity second-order transfer function (moderate and low values of center frequency)

The differential-input, differential-output operational amplifier with gain A can be used to obtain gains K_1 and K_2 as in Problem 6-34. The transfer function of the circuit, as obtained by analysis (see Problem 8-30), is given by

$$A_v(s) = \frac{V_2}{V_1}(s) = \frac{K_1 K_3 K_4 (1 - Ts)^2}{\{(1 + K_2 K_3 K_4)T^2 s^2 + [2T(1 - K_2 K_3 K_4)]s + 1 + K_2 K_3 K_4\}} \quad (8\text{-}70)$$

where $T = R_1 C_1 = R_2 C_2$. For high Q the design equations are

$$\omega_0 = \frac{1}{T} \quad (8\text{-}71\text{a})$$

$$Q = \frac{1 + K_2 K_3 K_4}{2(1 - K_2 K_3 K_4)} \quad (8\text{-}71\text{b})$$

8-8 Low-Sensitivity High-Q Transfer-Function Realization

Because of the difference of terms, it may appear that the sensitivity of the circuit is not good. This is not the case for the following reasons. If we make K_3 and K_4 unity gain amplifiers (see fig 7-45), the design equation for Q becomes

$$\frac{1}{Q} = \frac{2(1 - K_2)}{1 + K_2} \qquad (8\text{-}72\text{a})$$

If we choose $K_2 = 1 - 1/Q$, then (8-72a) becomes

$$\frac{2(1 - K_2)}{1 + K_2} = \frac{2(1/Q)}{2 - 1/Q} \approx \frac{1}{Q}\left(1 + \frac{1}{2Q}\right) \approx \frac{1}{Q} \qquad (8\text{-}72\text{b})$$

The realization of K_1 and K_2 is given in fig P6-34. If $R_d/R_c = R_b/R_a = 1/Q$, then, from Problem 6-34

$$K_2 = \frac{1 + 1/Q}{1 + 2/Q + 1/Q^2} = \frac{1}{1 + 1/Q} \approx 1 - \frac{1}{Q}$$

which is the desired result. Also note that since $|K_2 K_3 K_4| < 1$, the network is absolutely stable, regardless of how much excess phase shift is associated with the amplifiers.

When K_2 is determined by (8-72), the value of K_1 is fixed. From Problem 6-34

$$K_1 = \frac{1/Q(1 + 1/Q)}{(1 + 1/Q)^2} \approx \frac{1}{Q}$$

Now, if we normalize T to unity, (8-70) becomes

$$A_v(s) = \frac{1}{Q} \frac{(1 - s)^2}{2[s^2 + (1/Q)s + 1]} \qquad (8\text{-}73\text{a})$$

so that

$$|A_v(j1)| = 1 \qquad (8\text{-}73\text{b})$$

Finally, observe that ω_0 can be tuned by varying T, which has negligible effect on Q, and Q can be varied by tuning $K_2 K_3 K_4$, which has negligible effect on ω_0. Thus, the tuning for ω_0 and Q is relatively independent, and the circuit has an excellent tunability property.

We end the discussion on high Q active RC networks by pointing out the fact, from Section 7-16, that in a filter with high Q the frequency stability is more important than the Q stability (7-118). Thus the RC product tracking and tolerance will always be a limiting factor on the maximum value of Q and one must, therefore, use a crystal filter for very high Q tuned amplifiers.

8-9 Distributed *RC* Circuit as a Network Element

Integrated circuits are usually designed on the basis of lumped resistors, capacitors, and transistors. However, the fabrication of monolithic diffused resistors intrinsically involves distributed effects. Thus, in some cases one may have to consider the distributed nature of the elements. In fact, as will be shown in the next sections, distributed elements can be used to advantage. One of the most commonly used is the distributed RC network, usually denoted by $\overline{RC}$, shown symbolically in fig 8-37a. This network is obtained physically by the structure shown in fig 8-37b, where the uppermost thin sheet is the resistive layer, the sandwiched layer is the dielectric sheet, and the bottom layer is the ground plane. In semiconductors, this functional block is readily obtainable in the form of thin films, as discussed in Appendix B. The parameters of the structure in fig 8-37 are usually given as R_o and C_o, where R_o is the resistance per square, and C_o is the capacitance per unit area. The total R and C of the structures in fig 8-37 and fig 8-38 are obtained from

$$R = R_o \frac{d}{w} \quad \text{and} \quad C = C_o \, dw \qquad (8\text{-}74)$$

where d is the length and w is the width of the network. Note that for the same materials, i.e., same R_o and C_o, we can obtain different R's and

Fig 8-37 (a) Distributed-*RC* circuit graphic symbol; (b) Physical realization of a distributed-*RC* circuit

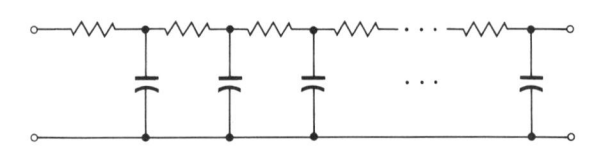

Fig 8-38 Iterative lumped-*RC* structure to simulate a distributed-*RC* network

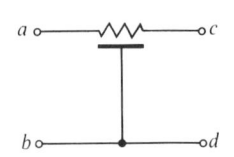

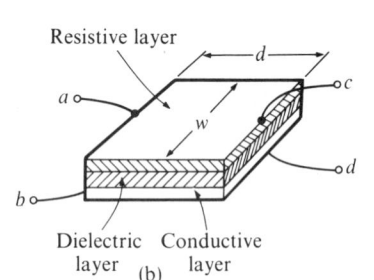

8-9 Distributed RC Circuit as a Network Element

different C's by merely changing the length, or width. The product of $RC = R_o C_o d^2$ depends on the material and the length, but is independent of the width.

To simulate such a structure exactly in lumped form would require an infinite number of RC sections, as shown in fig 8-38. The distributed structure is equivalent to an RC transmission line, and can be analyzed as such. For example, an elemental lumped section approximation of an $\overline{RC}$ network is shown in fig 8-39. The equilibrium equations in the transformed s-plane, for zero initial conditions, are

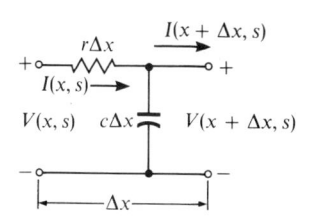

Fig 8-39 Elemental section of fig 8-38

$$V(x, s) - r\Delta x I(x, s) \approx V(x + \Delta x, s) \quad (8\text{-}75a)$$

$$I(x, s) - c\Delta x s V(x + \Delta x, s) \approx I(x + \Delta x, s) \quad (8\text{-}75b)$$

where r and c are the resistance and capacitance per unit length, respectively.

From (8-75) we obtain

$$\frac{V(x + \Delta x, s) - V(x, s)}{\Delta x} \approx -rI(x, s) \quad (8\text{-}76a)$$

$$\frac{I(x + \Delta x, s) - I(x, s)}{\Delta x} \approx -csV(x + \Delta x, s) \quad (8\text{-}76b)$$

In the limit as $\Delta x \to 0$, from (8-76) we obtain

$$\frac{dV(x, s)}{dx} = -rI(x, s) \quad (8\text{-}77a)$$

$$\frac{dI(x, s)}{dx} = -cs V(x, s) \quad (8\text{-}77b)$$

Equations (8-77a) and (8-77b) are usually referred to as the Telegrapher's equations for an RC line.

The two-port parameters of the network can now be obtained as follows. Differentiate (8-77a) with respect to x and obtain

$$\frac{d^2V}{dx^2} + r\frac{dI}{dx} = 0 \quad (8\text{-}78)$$

Substitution of (8-78) into (8-77b) yields

$$\frac{d^2V}{dx^2} - rcs\, V = 0 \quad (8\text{-}79)$$

Similarly

$$\frac{d^2 I}{dx^2} - rcs\, I = 0 \qquad (8\text{-}80)$$

The solutions of (8-79) and (8-80) are

$$V(x, s) = A_1 \cosh \sqrt{rcs}\, x + A_2 \sinh \sqrt{rcs}\, x \qquad (8\text{-}81a)$$

$$I(x, s) = B_1 \cosh \sqrt{rcs}\, x + B_2 \sinh \sqrt{rcs}\, x \qquad (8\text{-}81b)$$

The constants are found as follows: From (8-81) at $x = 0$

$$A_1 = V(0, s), \qquad B_1 = I(0, s)$$

Differentiating (8-81) with respect to x and substituting (8-77) into the resulting expression at $x = 0$ yields

$$A_2 = -\sqrt{\frac{r}{cs}} I(s, 0), \qquad B_2 = -\sqrt{\frac{cs}{r}} V(s, 0)$$

For a network of length d, the voltage and current at the output ($x = d$) in terms of the voltage and current at the input ($x = 0$) from (8-81) are given by the following matrix:

$$\begin{bmatrix} V(d, s) \\ -I(d, s) \end{bmatrix} = \begin{bmatrix} \cosh \sqrt{rcs}\, d & \sqrt{r/cs} \sinh \sqrt{rcs}\, d \\ \sqrt{cs/r} \sinh \sqrt{rcs}\, d & \cosh \sqrt{rcs}\, d \end{bmatrix} \begin{bmatrix} V(0, s) \\ -I(0, s) \end{bmatrix}$$

$$(8\text{-}82)$$

The above is immediately recognized to be the $\mathcal{ABCD}$ parameter description of a linear two-port network. From (8-82) and table 2-1, one can find the open-circuit impedance parameters of the line, namely

$$[z_{ij}] = \sqrt{\frac{r}{cs}} \begin{bmatrix} \coth d\sqrt{rcs} & \operatorname{csch} d\sqrt{rcs} \\ \operatorname{csch} d\sqrt{rcs} & \coth d\sqrt{rcs} \end{bmatrix} \qquad (8\text{-}83)$$

Note that unlike a lumped network, the impedance parameters of a distributed network are transcendental functions. The poles and zeros of the impedance parameters are found as follows:

$$z_{11} = z_{22} = \sqrt{\frac{r}{cs}} \coth d\sqrt{rcs} = \sqrt{\frac{r}{cs}} \frac{\cosh \sqrt{\tau s}}{\sinh \sqrt{\tau s}} \qquad (8\text{-}84a)$$

8-9 Distributed RC Circuit as a Network Element

The transcendental functions may be expressed in infinite power series or infinite product expansion form. The infinite product expansion is (Ref 7)

$$z_{11} = z_{22} = \frac{1}{cds} \frac{\prod_{n=1}^{\infty}\left[1 + \frac{4\tau s}{(2n-1)^2\pi^2}\right]}{\prod_{n=1}^{\infty}\left(1 + \frac{\tau s}{n^2\pi^2}\right)} \tag{8-84b}$$

Similarly,

$$z_{12} = z_{21} = \sqrt{\frac{r}{cs}}\operatorname{csch} d\sqrt{rcs} = \frac{1}{cds}\frac{1}{\prod_{n=1}^{\infty}\left(1 + \frac{\tau s}{n^2\pi^2}\right)} \tag{8-85}$$

where $\tau = rcd^2 = RC$, $R = rd$, and $C = cd$.

From (8-85) it is seen that there are an infinite number of poles and zeros. This is one of the basic properties of any distributed network. For convenience of analysis from z-parameters in (8-83), an RC line may be represented by the Tee model, as shown in fig 8-40, where

$$\mathscr{R} = \frac{R}{\sqrt{sRC}\sinh\sqrt{RCs}} \quad \text{and} \quad \mathscr{P} = \cosh\sqrt{sRC} \tag{8-86}$$

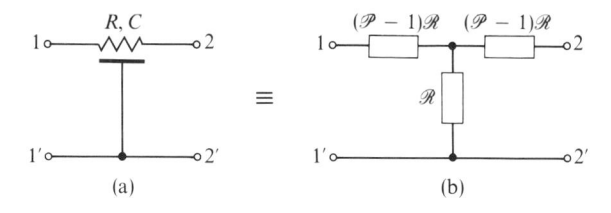

Fig 8-40 (a) Distributed-RC circuit; (b) T-model of the circuit in the $\mathscr{P}$-plane

The open-circuit voltage transfer function of the RC line is given by

$$A_V = \frac{E_o}{E_i} = \frac{z_{21}}{z_{11}} = \operatorname{sech}\sqrt{s\tau} = \frac{1}{\prod_{n=1}^{\infty}\left[1 + \frac{4\tau s}{(2n-1)^2\pi^2}\right]} \tag{8-87}$$

The magnitude and phase plot of (8-87) are shown in fig 8-41. Note that the phase shift steadily increases for increasing frequency, and becomes unbounded as $\omega \to \infty$.

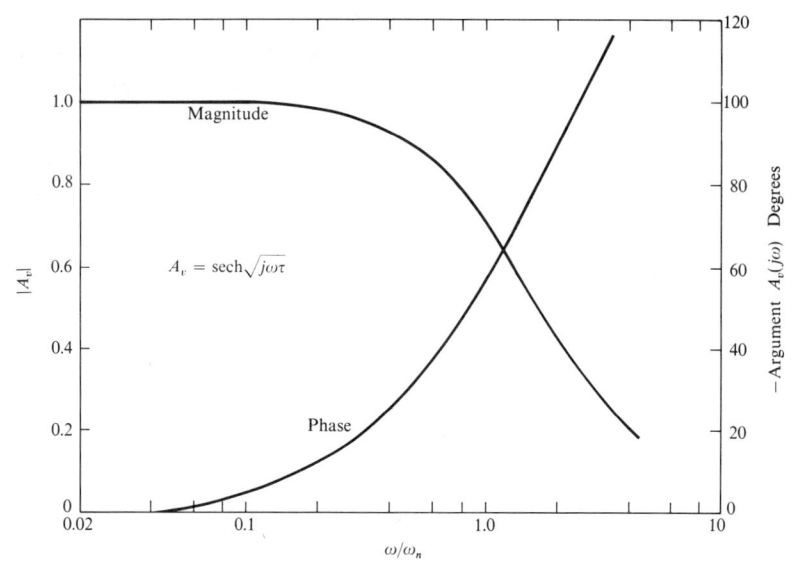

Fig 8-41 Magnitude and phase response of equation (8-87) normalized to $\omega_n = 2.43/\tau$

8-10 Applications of Distributed *RC* Networks

Distributed *RC* networks may be used advantageously in a variety of applications. Representative examples are the following:

$\overline{RC}$ Phase-Shift Oscillator

The great amount of phase shift which is obtained from an $\overline{RC}$ network can be utilized effectively in designing a phase-shift oscillator. The $\overline{RC}$ phase-shift oscillator, exclusive of biasing circuitry, is shown in fig 8-42. Compare fig 8-42 with the conventional *RC* phase-shift oscillator shown in fig. 6-43.

Fig 8-42 Simple $\overline{RC}$ phase-shift oscillator

$\overline{RC}$ Notch Filter

By adding a lumped series resistor or lumped shunt capacitor in an $\overline{RC}$ network, as shown in fig 8-43a and fig 8-43b, respectively, we can design a notch filter. The notch filter has a null output at certain discrete frequencies. A notch filter could also be obtained from a parallel ladder, bridged-Tee, or a twin-Tee network, as was shown in fig 7-42. The null frequencies of the $\overline{RC}$ notch filter can be determined as follows. Using the *z*-parameters in fig 8-43a, because of the series connection (*y*-parameters for fig 8-43b), we write the open-circuit voltage ratio of the overall two-port network

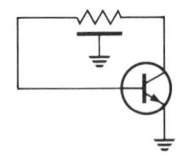

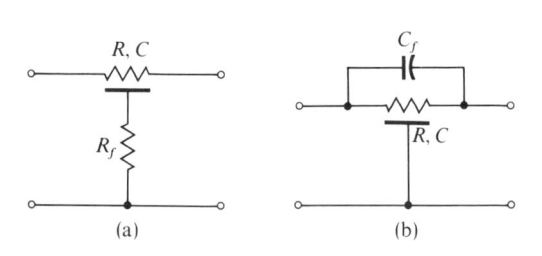

Fig 8-43 Two forms of $\overline{RC}$ notch filters

8-11 Active-Distributed RC Low-Pass Filters

$$A_V = \frac{V_0}{V_i} = \frac{z_{21}^T}{z_{11}^T} = \frac{\sqrt{R/Cs}\,\text{csch}\,\sqrt{RCs} + R_f}{\sqrt{R/Cs}\,\text{coth}\,\sqrt{RCs} + R_f} \quad (8\text{-}88)$$

Equation (8-88) may be rearranged in the following convenient form:

$$A_V = \frac{1 + K\theta \sinh \theta}{\cosh \theta + K\theta \sinh \theta} \quad (8\text{-}89)$$

where $\theta = \sqrt{s\tau}$, $\tau = RC = rcd^2$, and $K = R_f/R$. Similar expressions result for fig 8-43b where $K = C_f/C$.

The null frequencies are the particular frequencies at which the numerator is zero. The null frequencies are found from the roots of the equation

$$1 + K\theta \sinh \theta = 0 \quad (8\text{-}90)$$

There are an infinite number of roots of (8-90); the first and the second smallest roots, which may be verified by direct substitution, are $\omega_{01} = 11.19/\tau$ and $\omega_{02} = 149.3/\tau$. The values of K at these frequencies are 5.61×10^{-2} and 2.9×10^{-5} respectively (Ref 7).

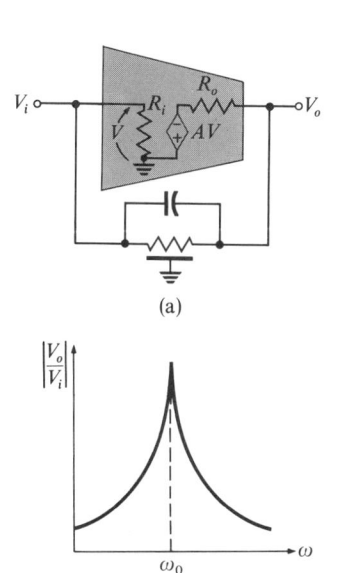

Narrow-Band Tuned Amplifier

Another interesting application of the $\overline{RC}$ notch filter is that it can be used in conjunction with a high-gain amplifier to obtain a narrow-band tuned amplifier. The application of this idea for lumped networks was discussed in Section 7-15. A basic network for the $\overline{RC}$ tuned amplifier is shown in fig 8-44. For a high gain, the response approaches the reciprocal of the notch filter characteristics. Since for $C_f > 5.6 \times 10^{-2} C$ the zero of the notch filter will lie in the right half s-plane, a very high Q tuned amplifier is susceptible to oscillation. Therefore, the circuit is not practical for very high Q's.

Fig 8-44 $\overline{RC}$-active tuned amplifier

8-11 Active-Distributed RC Low-Pass Filters

Distributed RC networks, in conjunction with active devices, can also be effectively used for the design of low-pass filters. Even though the network functions of distributed networks have an infinite number of poles, using the dominant-pole concept, an all pole low-pass filter can be very economically designed. Consider the basic circuit configuration shown in fig 8-45. From the circuit in fig 8-45, assuming an ideal voltage-controlled voltage source for A, and using the Tee model for the $\overline{RC}$ network (fig 8-40b), we obtain (see Problem 8-23) the following transfer function:

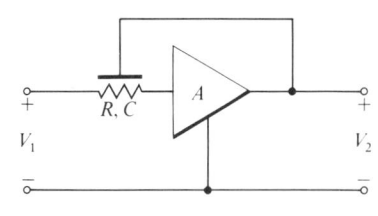

Fig 8-45 Active-$\overline{RC}$ configuration with transfer function which can be approximated by a pair of dominant poles

8/Active RC Filters

$$A_v = \frac{V_2}{V_1} = \frac{A/(1-A)}{\mathscr{P} + A/(1-A)} \tag{8-91a}$$

Substitution of the expression for $\mathscr{P}$ from (8-86) in (8-91a) yields

$$A_v = \frac{A/(1-A)}{\cosh\sqrt{sRC} + A/(1-A)} \tag{8-91b}$$

The denominator of A_v in (8-91) has an infinite number of roots. The dominant-pole location of the transfer function for various values A, namely in the range $0.333 \leq A \leq 0.921$, and normalized $\overline{RC}$ time constant $RC = 1$, is shown in fig 8-46. From fig 8-46, it is readily seen that the dominant poles of the transfer function may be expressed in the form

$$A_v = \frac{V_2}{V_1} \approx \frac{H}{s^2 + a_1 s + a_0} \tag{8-91c}$$

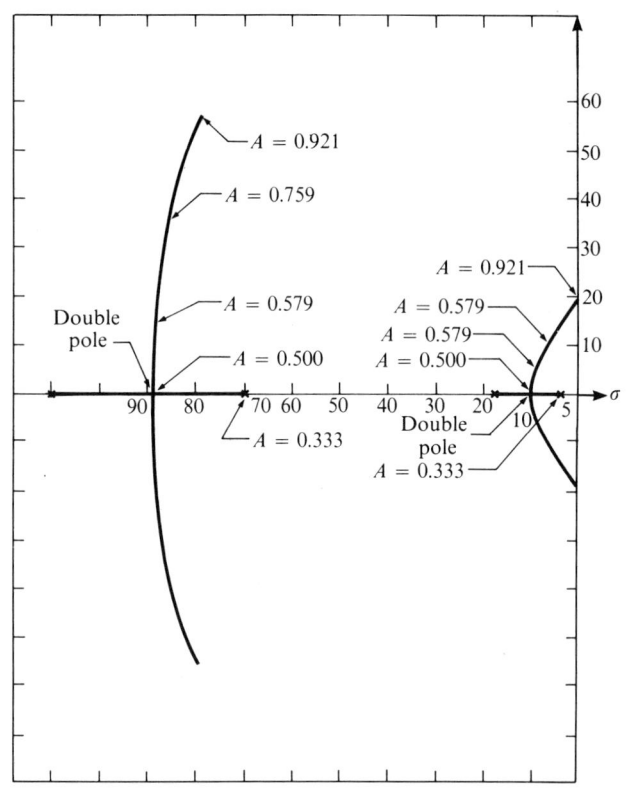

Fig 8-46 First two pole pairs of the voltage-ratio transfer function of fig 8-45 for different values of the gain A

8-11 Active-Distributed RC Low-Pass Filters

where a complex dominant pole pair is obtained for $0.50 < A < 0.92$. Note that the second pole pair is farther, by at least a factor of 9, so that *magnitude response calculation* using (8-91) will give reasonably accurate results. For a single-pole transfer function, the circuit realization in fig 8-45 could be used; however, it is most convenient to use the simpler configurations shown in fig 8-47a and fig 8-47b. Note that the circuit of fig 8-47b was already analyzed, and it was found that the second pole is nine times larger than the dominant pole [see (8-87)]. The error by using (8-91) will be appreciable in phase calculations, but this can be handled simply as discussed later in this section. The amplifier A, which is assumed to be a voltage-controlled voltage source (VCVS), can be realized by operational amplifiers. Compensated operational amplifiers designed for low-gain operation are nearly ideal, and can approximate a VCVS satisfactorily in most applications. As we shall see in the next section, for purposes of design to be described in this section, an emitter follower can also do the job. However, note that the pole positions are very sensitive to the amplifier gain variations. Because of the isolation property of the amplifier, blocks of the form of fig 8-45 can be cascaded with negligible interaction, and thus the design of an all-pole voltage ratio function can be accomplished by a configuration consisting of cascades of the circuits of fig 8-45 and fig 8-47.

For example, a fifth-order Butterworth filter, Thomson filter, or any other five-pole, low-pass filter can be written as

$$T(s) = \frac{V_o}{V_{in}} = \frac{K}{(s + p_1)(s^2 + a_1 s + a_0)(s^2 + b_1 s + b_0)} \quad (8\text{-}92)$$

If we are considering a five-pole Butterworth filter, the two quadratic factors will have angles of 36 degrees and 72 degrees from the negative real axis with radii of p_1. The gain function in (8-92) may be written as a product of three terms:

$$\frac{V_1}{V_{in}} = \frac{K_1}{s + p_1} \quad (8\text{-}93\text{a})$$

$$\frac{V_2}{V_1} = \frac{K_2}{s^2 + a_1 s + a_0} \quad (8\text{-}93\text{b})$$

$$\frac{V_o}{V_2} = \frac{K_3}{s^2 + b_1 s + b_0} \quad (8\text{-}93\text{c})$$

where the gain-function of (8-93a) is realized by the circuit shown in fig 8-47a or fig 8-47b, and (8-93b) and (8-93c) are realized by blocks

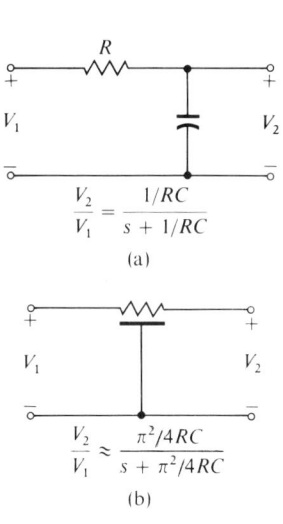

Fig 8-47 Circuits for realization of a single real pole (a) Lumped-RC section; (b) $\overline{RC}$ section with dominant single real pole

$$\frac{V_2}{V_1} = \frac{1/RC}{s + 1/RC}$$
(a)

$$\frac{V_2}{V_1} \approx \frac{\pi^2/4RC}{s + \pi^2/4RC}$$
(b)

of the form shown in fig 8-45. The specification of a dc gain level can also be achieved by adding an additional wideband amplifier. It should be noted that if there are nondominant poles in the given $T(s)$, as a first approximation, the nondominant poles of $T(s)$ should be dropped at the outset.

Design Charts

In order to facilitate the design of active $\overline{RC}$ networks by the method just described, an expanded design chart, which yields the dominant poles of fig 8-46, is given in fig 8-48. The next nearest poles are an order of magnitude larger and are, therefore, ignored in magnitude calculations. The phase error can be severe, and its inclusion in the approximation will be described shortly.

The values of A, R, and C of fig 8-45, which approximate the magnitude of a *complex pole pair*, are determined as follows. We express the desired complex pole pairs in the normalized form as

$$T(s) = \frac{K}{s^2 + a_1 s + a_0} = \frac{K'}{(s/\sqrt{a_0})^2 + (a_1/\sqrt{a_0})(s/\sqrt{a_0}) + 1} \qquad (8\text{-}94a)$$

$$= \frac{K'}{s_n^2 + 2\cos\phi\, s_n + 1} \qquad (8\text{-}94b)$$

where a_1 and a_0 are known, thus $\phi = \cos^{-1}(a_1/2\sqrt{a_0})$ is given. For a given ϕ, the values of A and ρ ($\rho = RC\Omega_0$) are read from the chart of fig 8-48. The values of ρ and Ω_0 ($=\sqrt{a_0}$), and the fact that the chart is for $(RC) = 1$, determine the actual RC product of the distributed network. Thus for a convenient choice of value of either R or C, the other parameter is determined. Note that a real pole of order 2,

$$T(s) = \frac{K}{(s_n + 1)^2}$$

can be realized using fig 8-48 with $A = 0.50$ and $\rho = 9.87 = \pi^2$. As an example of the use of the chart[2] consider the following.

[2] The analytic solution for A and RC is obtained by identifying the given roots in (8-94) with the dominant roots in (8-91b), which are

$$s_0, s_0^* = \frac{1}{RC}(\lambda^2 - \pi^2 \pm j2\pi\lambda)$$

where $\lambda = \ln(-B - \sqrt{B^2 - 1})$, or alternatively, $B = -\cosh\lambda$. The results are

$$\lambda = \pi\left(\frac{\cos\phi - 1}{\sin\phi}\right), \qquad RC = \frac{\lambda^2 + \pi^2}{\sqrt{a_0}}, \qquad A = \frac{B}{B-1}, \qquad \rho = \lambda^2 + \pi^2$$

8-11 Active-Distributed RC Low-Pass Filters

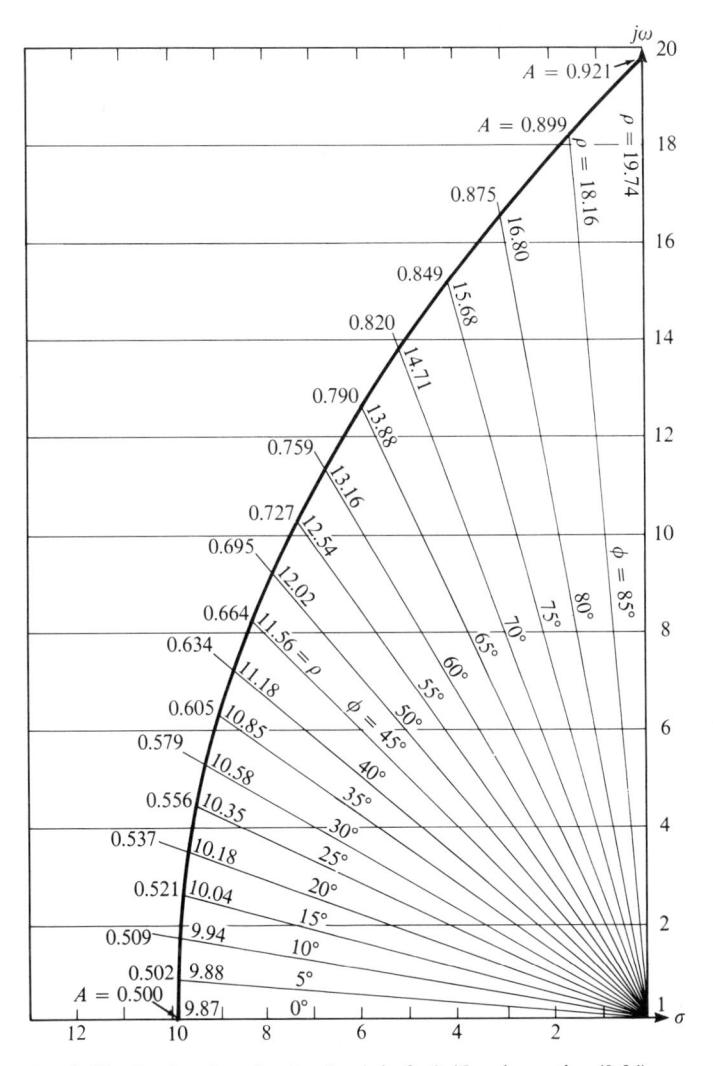

Fig 8-48 Design chart for the circuit in fig 8-45 and equation (8-94)

EXAMPLE 1

Consider the design of an active $\overline{RC}$ network with a magnitude response $|V_2/V_1|$, corresponding to a two-pole Butterworth filter with a 3-dB cutoff frequency, $f_{3\,dB} = 20$ kHz. The desired lumped transfer function may be written as

$$T(s) = \frac{V_2}{V_1}(s) = \frac{K}{s_n^2 + \sqrt{2}s_n + 1}$$

8/Active RC Filters

where

$$s_n = \frac{s}{\Omega_0} = \frac{s}{2\pi(2 \times 10^4)} \quad \text{and} \quad \phi = 45°$$

Corresponding to $\phi = 45$ degrees, we read $A = 0.664$ and $\rho = 11.56$. Thus the actual RC product of the distributed network is given by

$$RC = \frac{\rho}{\Omega_0} = \frac{11.56}{1.25 \times 10^5} = 92.4 \ \mu\text{sec}$$

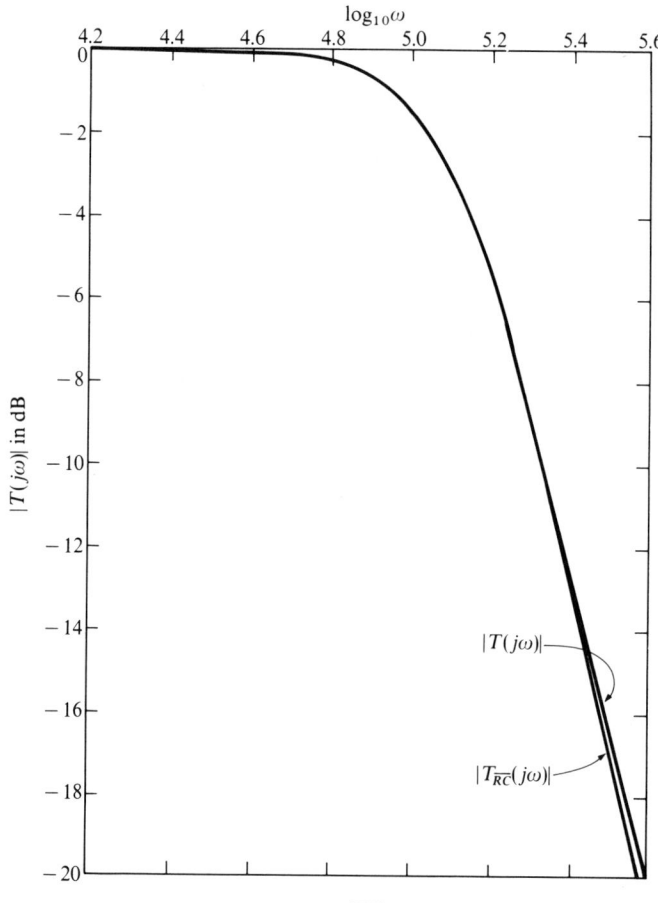

Fig 8-49 Comparison of $|T(s_n)|$ with its $\overline{RC}$ approximation for the design example. ($s_n = j\omega/\Omega_0$ and $\Omega_0 = 1.257 \times 10^5$ rad/sec)

For any convenient value of R, the value of C is thus determined. If the values of R_o and C_o of the material used in the distributed $\overline{RC}$ network are given from (8-74), we have the length $d = \sqrt{RC/R_oC_o}$. Any convenient value of R, say 10 kΩ, determines the width w and the total capacitance C of the distributed network. For this example we may choose $R = 10$ kΩ, then $C = 9.24 \times 10^{-9}$ F. The magnitude response of the distributed and lumped network function is shown in fig 8-49.

EXAMPLE 2

An active $\overline{RC}$ network is to be designed such that the magnitude of its voltage gain function approximates the magnitude of the following lumped network voltage ratio function, $T(s)$, within the passband of the filter:

$$T(s) = \frac{K}{(s + 0.494 \times 10^7)(s^2 + 0.494 \times 10^7 s + 0.994 \times 10^{14})}$$

(This is a three-pole Chebyshev function with 1-dB ripple, see Ref 2.) We rewrite $T(s) = T_1(s)T_2(s)$, where

$$T_1(s) = \frac{K_1}{s + 0.494 \times 10^7}$$

and

$$T_2(s) = \frac{K_2}{s^2 + 0.494 \times 10^7 s + 0.994 \times 10^{14}}$$

The first factor, $T_1(s)$, may be realized either by the circuit shown in fig 8-47a or fig 8-47b, and thus the RC product of the section, say $(RC)_1$, is determined. If fig 8-47a is used, $RC = 0.203$ μsec, and if fig 8-47b is used, $(RC)_1 = (\pi^2/4)/0.494 \times 10^7 = 5.0$ μsec. For the second section, namely $T_2(s)$, $\Omega_0 = \sqrt{0.994 \times 10^{14}} = 0.998 \times 10^7$, $\phi = 75.7$ degrees. Corresponding to $\phi = 75.7$ degrees we read from fig 8-48, $A = 0.85$, and $\rho = 15.83$. Hence $(RC)_2 = \rho/\Omega_0 = 15.83/0.998 \times 10^7 = 1.58$ μsec. Hence, convenient choices of the values of R determine the values of C. Note that in this design method and configuration, a very close tolerance on the gain of the amplifiers is required. However, since the gain of the amplifier is less than unity, one can achieve a close tolerance in the design.

EXAMPLE 3

Consider the approximate realization of a four-pole Thomson (maximally flat delay) filter within the passband of the filter. The transfer function of a normalized four-pole Thomson filter from table 2-3 is

$$T(s) = \frac{K}{s^4 + 10s^3 + 45s^2 + 105s + 105}$$

$$= \frac{K_1}{(s^2 + 5.792s + 9.140)} \times \frac{K_2}{(s^2 + 4.208s + 11.488)}$$

From the above we have $\phi_1 = 16.67$ degrees, $\Omega_{01} = 3.023$, $\phi_2 = 51.53$ degrees, and $\Omega_{02} = 3.389$. The corresponding normalized RC products are

$$(RC)_1 = \frac{\rho_1}{\Omega_{01}} = \frac{10.07}{3.023} = 3.335$$

and

$$(RC)_2 = \frac{\rho_2}{\Omega_{02}} = \frac{12.17}{3.389} = 3.593$$

The amplifier gains for each section are $A_1 = 0.525$, and $A_2 = 0.706$. The circuit realization is shown in fig 8-50.

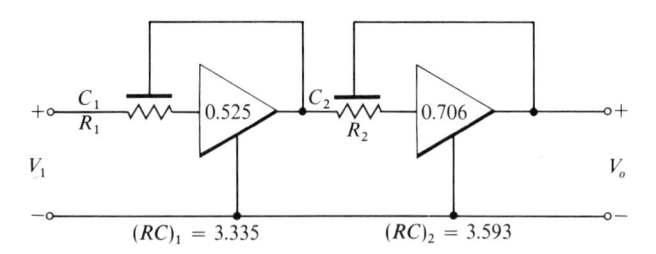

Fig 8-50 Active-$\overline{RC}$ realization of the design example

Phase Error Considerations

In our study of the dominant poles in Section 4-10, we discovered that ignoring the nondominant poles causes considerably more error in the phase response of low-pass functions than in the magnitude function. We now consider the phase difference between the lumped network function and the active $\overline{RC}$ network function.

8-11 Active-Distributed RC Low-Pass Filters

The phase difference between (8-91b), the active $\overline{RC}$ network, and that of the lumped network function in (8-91c) (fig 8-45) has been found by computation to be approximately a linear function of frequency. In other words,

$$\arg T_{\overline{RC}}(\omega) - \arg T(\omega) \simeq -\tau_n \omega \qquad (8\text{-}95)$$

where $\arg T_{\overline{RC}}$ and $\arg T$ denote the argument of the actual active $\overline{RC}$ network function of fig 8-45 and the argument of the lumped network function approximation, respectively. (Note that $T(s)$ instead of $A_v(s)$ is used for convenience.) The delay or dead time τ_n is a constant expressed in seconds, which depends on ϕ and Ω_0. In other words, the approximate transfer function of the circuit of fig 8-45 can be written as

$$T_{\overline{RC}}(s) \approx T(s)e^{-\tau_n s} \qquad (8\text{-}96)$$

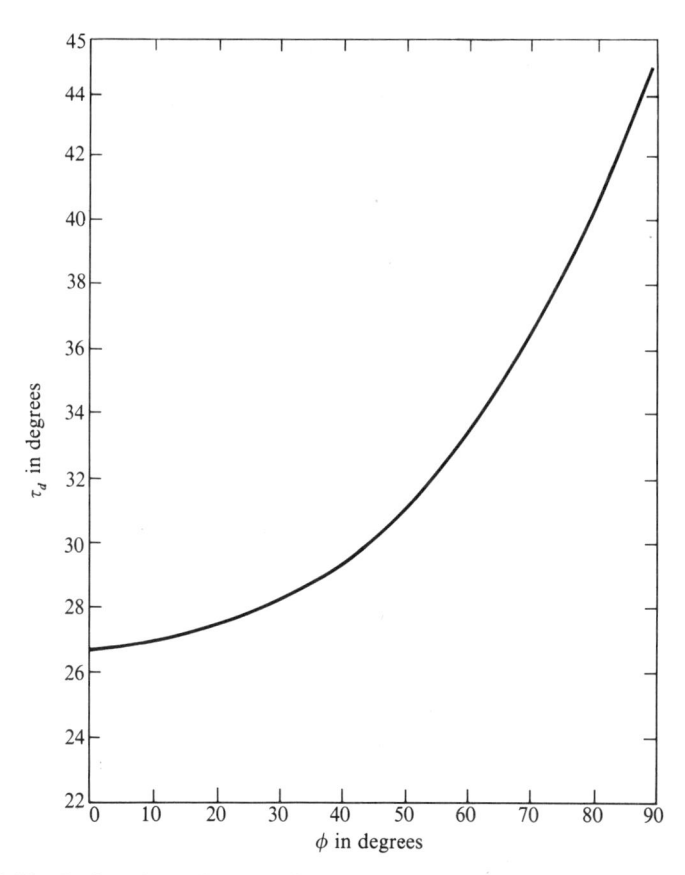

Fig 8-51 Design chart of τ_d vs ϕ for the circuit in fig 8-45 and equation (8-96). $T(s_n) = 1/(s_n^2 + 2\cos\phi s_n + 1)$, $s_n = j\omega/\Omega_0$

From (8-96), it is seen that the transient response of the active $\overline{RC}$ network will be the same as that of $T(s)$ delayed by τ_n seconds.

A design chart for finding $\tau_d \equiv \Omega_0 \tau_n$, given ϕ, is given in fig 8-51, where both ϕ and τ_d are expressed in degrees. For convenience, table 8-1 is included, which may be more convenient to use than the charts in fig 8-48 and fig 8-51.[3] For the two-pole Butterworth filter (example 1) considered above, $\phi = 45$ degrees and $\tau_d = 30.3$ degrees ($\tau_n = 4.2$ μsec). The phase error between (8-96) and the actual phase, as computed from fig 8-45 with the aid of a computer, is shown in fig 8-52.

If fig 8-47b is used for a real pole, the value of $\tau_d \approx 26.757/2$ degrees and $\tau_n \approx 0.23/\Omega_0$ sec. (Note the factor of $\frac{1}{2}$ since the transfer function of fig 8-47b has one dominant pole given by (8-87) while in Table 8-1 for $\phi = 0$ the poles are real of order 2.) The time constant is determined by the pole location from $RC = (\pi^2/4\,\Omega_0)$. This type of approximation has also been used to approximate the transcendental function, $\alpha(s)$, of a transistor by a one-pole function with an excess phase factor [Eq (4-86)].

For the four-pole maximally flat delay filter realization, the delay due to each section as obtained from fig 8-51, or Table 8-1, is

For $\phi_1 = 16.67°$

$$\tau_1 = 27.2° \qquad \tau_{n1} = \frac{\tau_1(1/57.3) \text{ radians}}{\Omega_{01}} = 0.157 \text{ sec}$$
$$\Omega_{01} = 3.023$$

For $\phi_2 = 51.53°$

$$\tau_2 = 31.5° \qquad \tau_{n2} = \frac{\tau_2(1/57.3)}{\Omega_{02}} = 0.162 \text{ sec}$$
$$\Omega_{02} = 3.389$$

The total delay is $\tau_{nt} = \tau_{n1} + \tau_{n2} = 0.32$ sec.

The transient response of the active-$\overline{RC}$ filter is expected to be approximately the same as that of a lumped four-pole Thomson filter delayed by 0.32 sec. The actual step response of the Thomson four-pole

---

[3] This chart was obtained with (8-96) evaluated at ω_c of $T(s)$ which is

$$\omega_c = \sqrt{1 - 2\cos\phi^2 + 2(\cos\phi^4 - \cos\phi^2 + 0.5)^{1/2}}$$

τ_d may be found analytically as follows. If $T(s)$ is given by (8-94) and if we evaluate (8-96) at $\omega = 0$ we obtain

$$\tau_d = -\frac{a_1}{a_0} + \frac{\rho(1-A)}{2} \quad \text{radians where} \quad \rho = RC\Omega_0$$

The values obtained in Table 8-1 are approximately equal to those obtained using the design chart. Tables 8-1 and the response curves were obtained by Dr. Vincent Bello.

TABLE 8-1 *Design Table for Active $\overline{RC}$ Network in Figure 8-45*[a]

ϕ in deg	λ	B	$\rho = RC\Omega_n$	τ_d in deg
0	0.0000	−1.0000	9.8696	26.7572
2	−0.0548	−1.0015	9.8726	26.7638
4	−0.1097	−1.0060	9.8816	26.7834
6	−0.1646	−1.0136	9.8967	26.8165
8	−0.2197	−1.0242	9.9179	26.8627
10	−0.2749	−1.0380	9.9451	26.9220
12	−0.3302	−1.0550	9.9786	26.9951
14	−0.3857	−1.0753	10.0184	27.0816
16	−0.4415	−1.0991	10.0645	27.1818
18	−0.4976	−1.1264	10.1172	27.2957
20	−0.5539	−1.1574	10.1765	27.4240
22	−0.6107	−1.1923	10.2425	27.5662
24	−0.6678	−1.2314	10.3155	27.7233
26	−0.7253	−1.2748	10.3956	27.8950
28	−0.7833	−1.3228	10.4831	28.0816
30	−0.8418	−1.3757	10.5782	28.2836
32	−0.9008	−1.4339	10.6811	28.5014
34	−0.9605	−1.4978	10.7921	28.7353
36	−1.0208	−1.5678	10.9116	28.9858
38	−1.0817	−1.6444	11.0398	29.2532
40	−1.1434	−1.7281	11.1771	29.5381
42	−1.2059	−1.8197	11.3239	29.8410
44	−1.2693	−1.9197	11.4807	30.1625
46	−1.3335	−2.0290	11.6479	30.5032
48	−1.3987	−2.1485	11.8260	30.8639
50	−1.4649	−2.2792	12.0157	31.2455
52	−1.5323	−2.4223	12.2174	31.6486
54	−1.6007	−2.5792	12.4319	32.0742
56	−1.6704	−2.7513	12.6599	32.5234
58	−1.7414	−2.9403	12.9021	32.9970
60	−1.8138	−3.1484	13.1595	33.4964
62	−1.8877	−3.3777	13.4328	34.0226
64	−1.9631	−3.6308	13.7233	34.5769
66	−2.0402	−3.9110	14.0319	35.1608
68	−2.1190	−4.2216	14.3599	35.7758
70	−2.1998	−4.5669	14.7086	36.4232
72	−2.2825	−4.9516	15.0794	37.1048
74	−2.3674	−5.3814	15.4740	37.8224
76	−2.4545	−5.8631	15.8941	38.5779
78	−2.5440	−6.4046	16.3416	39.3729
80	−2.6361	−7.0152	16.8186	40.2101
82	−2.7309	−7.7063	17.3277	41.0913
84	−2.8287	−8.4913	17.8711	42.0189
86	−2.9296	−9.3866	18.4520	42.9953
88	−3.0338	−10.4121	19.0735	44.0230
90	−3.1416	−11.5919	19.7392	45.1049

[a] The table is listed for $\Omega_n = 1$. The value of RC for any other value of Ω_0 is determined from ρ/Ω_0. For example, for $\Omega_0 = 10^5$ rad/sec and $\phi = 30$ degrees, $RC = 10.578 \times 10^{-5}$ sec, and $\tau_n = [\tau_d(1/57.3)]/\Omega_0 = 4.94$ μsec.

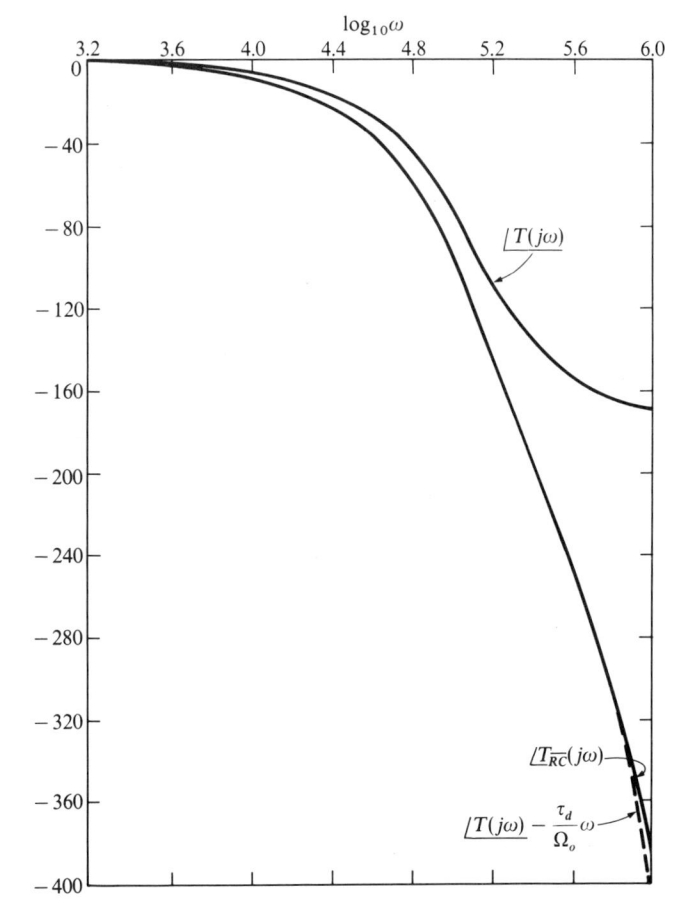

Fig 8-52 Comparison of arg $T_{\overline{RC}}(j\omega)$ for fig 8-49 with arg $T(j\omega) - (\tau_d/\Omega_0)\omega$, where $T(s_n) = [1/(s_n^2 + 1.414s_n + 1)]$, $s_n = j\omega/\Omega_0$, $\Omega_0 = 1.257 \times 10^5$ rad/sec and $\tau_d = 30.3$ degrees

filter and of the active-RC realization, as obtained by computer aided analysis, is shown in fig 8-53. The actual delay time is 0.32 sec. The results substantiate the conclusions.

Finally, it should be noted that the phase difference between a lumped network and its distributed approximation is nearly a linear function of ω, as seen in (8-96). Thus, if the lumped network function has a linear phase characteristic, its distributed approximation based on magnitude response, as is done in this section, will also have a linear phase response. In fact, the linearity of the phase vs frequency in the distributed realization is further improved. This fact is readily seen by comparing the delay (the derivative of the phase) of both cases,

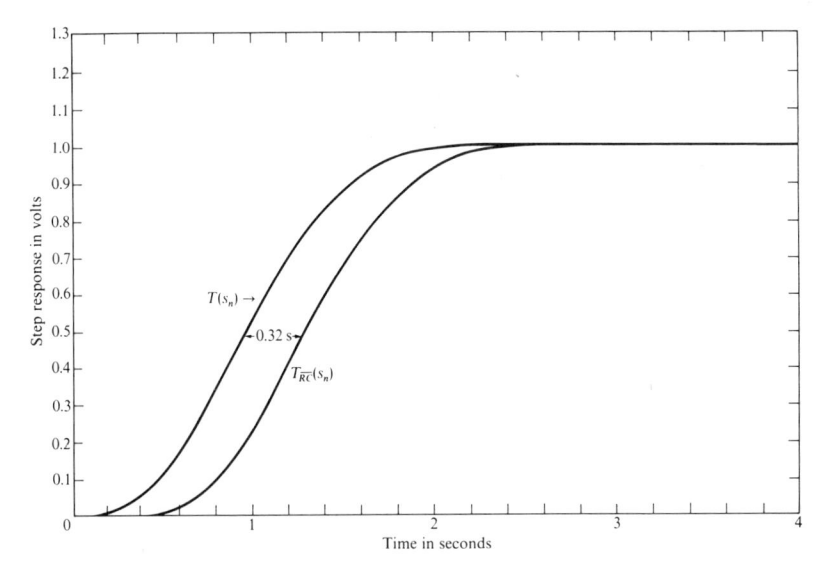

Fig 8-53 Normalized step response of a lumped four-pole Thomson filter $T(s_n)$ and that of an active-$\overline{RC}$ dominant four-pole approximation (the approximation is based on the magnitude response only)

as shown in fig 8-54. Note that the active-$\overline{RC}$ realization has a flat delay over a wider frequency range than that of the lumped Thomson filter.

A single $\overline{RC}$ network with a voltage-controlled voltage source can also be used to realize approximately a bandpass transfer function of the form obtained from a single tuned circuit (see Problem 8-34). The design equations are similar, and will not be discussed any further here. The reader is no doubt aware that these circuits are quite sensitive to gain variations, and the pole location will deviate considerably if the amplifier gain changes.

8-12 Summary

The purpose of this chapter has been to introduce the reader to the techniques and design of active RC filters. The discussion, and its scope, has been purposely kept at an elementary level requiring practically no previous background from the reader. It is assumed, however, that the reader can start with the proper transfer function, which would be suitable for his desired filter. Part of this information has been supplied in Sections 2-7 and 2-8 in conjunction with low-pass Butterworth and Thomson filters. Bandpass filtering has been discussed in Chapter 7. Highpass filter transfer functions are obtained by the low-pass to high-pass transformation, namely $s = 1/p$, while the band elimination filters are obtained by using another suitable transformation (see Section 7-5, footnote 3).

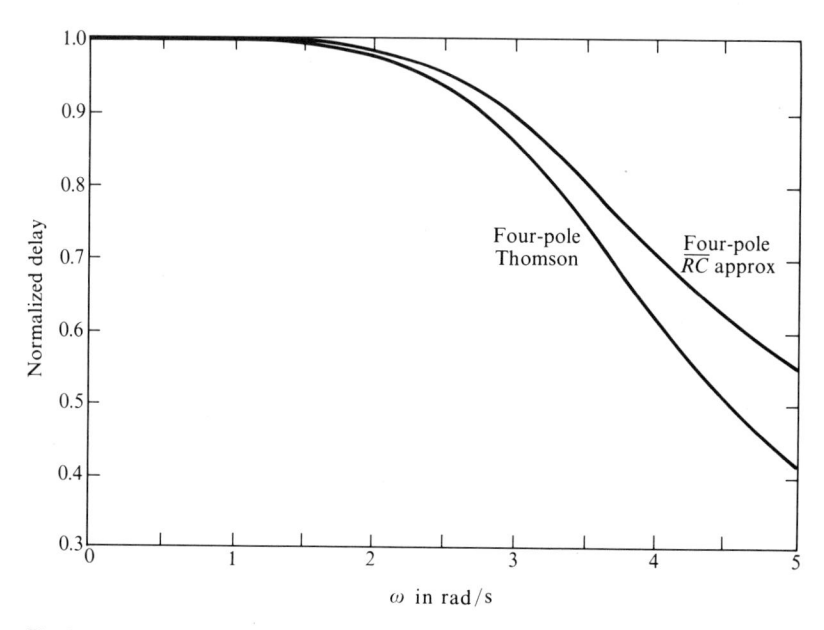

Fig 8-54 Normalized delay characteristics of a lumped four-pole Thomson filter $T(s_n)$ and that of an active-RC dominant four-pole approximation (the approximation is based on the magnitude response only)

We have categorized the active RC synthesis techniques by controlled source, negative impedance, gyrator, and operational-amplifier realizations. This has been done merely to familiarize the reader with the various forms of network elements used in active filters. One could argue that all the realization techniques are some form of the controlled sources or operational amplifier realizations. The basic ingredients are active elements, RC networks, and some form of feedback.

All the realizations based on the $RC: - RC$ decomposition suffer from the sensitivity viewpoint. These realizations cannot be practically used for Q's higher than 25. For a very high Q pole pair at low frequencies, the low sensitivity circuit of fig 8-35 may be used. However, for filters with very high Q and $f_0 \geq 1$ kHz, other circuits are to be used. For example, for frequencies below 500 kHz, the circuit of fig 8-36 may be used for a very high Q realization. For frequencies beyond 500 kHz, either crystals or passive LC filters are more practical. The LC filters are relatively cheap, and at high frequencies can be used in hybrid integrated circuits, but the Q is limited. For very high Q at high frequencies, the use of crystals in hybrid IC's is practical. In fact, even at moderate frequencies the RC tolerances limit the high value of Q in an active RC network. Thus by very high Q in active RC we mean Q's on the order of 100. For Q's larger than these values crystal filters are to be used (see Section 7-14). The use of crystals in high Q filtering was briefly introduced in Section 7-14. Active RC filters dominate at low frequencies, but face a strong challenge by other methods at very high Q and high f_0.

As a final note, we mention another class of entirely different filters. These filters are known as *digital filters*. A digital filter consists of digital circuits. It is a device, or a computational process, which performs numerical filtering. The input and output are in numerical form and compatible with digital computer data processing. Digital filters are becoming more and more important, and LSI integrated circuit technology has made the realization of such filters practically feasible. The discussion of this topic is beyond the scope of this text, and interested readers are referred to the literature on this subject (Ref 8).

REFERENCES AND SUGGESTED READING

1. Kuo, F., *Network Analysis and Synthesis*, New York: Wiley, 1966. Chapters 11 and 12.
2. Van Valkenberg, M., *Introduction to Modern Network Synthesis*, New York: Wiley, 1960. Chapter 6.
3. Kuh, E. S., and Pederson, D. O., *Principles of Circuit Synthesis*, New York: McGraw-Hill, 1959.
4. Huelsman, L. P., *Theory and Design of Active RC Circuits*, New York: McGraw-Hill, 1968.
5. Mitra, S. K., *Analysis and Synthesis of Linear Active Networks*, New York: Wiley, 1968.
6. Burr-Brown Research Corp, *Handbook of Operational Amplifier Applications*, Tucson, Arizona: 1963. Also, *Handbook of Operational RC Networks*, 1964.
7. Ghausi, M. S., and Kelly, J. J., *Introduction to Distributed Parameter Networks: With Application to Integrated Circuits*, New York: Holt, Rinehart & Winston, 1968.
8. Huelsman, L. P., Editor, *Active Filters: Lumped, Distributed, Integrated, Digital, and Parametric*, New York: McGraw-Hill, 1970. Chapters 2, 3, and 5.

PROBLEMS

8-1 The following functions are given:

(a) $\dfrac{s^2 + 5s + 4}{s^3 + 7s^2 + 10s}$ (b) $\dfrac{s^2 + 3s + 2}{s^3 + 3s + 5}$

(c) $\dfrac{s^2 + 4s + 4}{s^2 + s}$ (d) $\dfrac{s^2 + s + 1}{s^2 + 5s + 6}$

Identify the *RC* driving point immittance function(s). For the *RC* driving point function(s) show one realization.

8-2 Repeat (Problem 8-1) for the following:

(a) $\dfrac{s(s + 1)}{(s + 2)(s + 3)}$ (b) $\dfrac{s(s + 2)(s + 4)}{(s + 1)(s + 3)}$

(c) $\dfrac{(s^2 + 2)(s + 3)}{(s + 4)(s + 1)}$ (d) $\dfrac{s^2(s + 1)}{(s + 2)(s + 3)}$

8-3 The following functions are given:

(a) $\dfrac{s^2 + \sqrt{2}s + 1}{(s + 1)}$ (b) $\dfrac{s^2 + 3s + 3}{s(s + 2)}$

Use the polynomial decomposition theorem, and show the two *RC* driving-point function realizations in each case.

8-4 Derive (8-10), and design the circuit for the following specifications: A Butterworth two-pole filter with $f_{3\,dB} = 1\,\text{kHz}$, $A_V(0) = 1$, and $R_L = 1$ kilohm.

8-5 In the configuration of fig 8-13, design the circuit to meet the following bandpass specifications for V_o/V_{in}:

$\omega_0 = 2\pi(1\,\text{kHz})$, $\omega_{3\,dB} = 2\pi(100\,\text{Hz})$,
$A(j\omega_0) = 10$, $R_L = 1\,\text{k}\Omega$

8-6 Design the circuit of fig 8-13 to realize

$$Y_{21} = \dfrac{I_o}{V_{in}} = \dfrac{K}{s^2 + \sqrt{2}s + 1}$$

8-7 Derive (8-22), and design the circuit for the following specification: A Thomson two-pole filter with $f_{3\,dB} = 1\,\text{kHz}$ and $A_i(0) = 10$. What modification is to be made if we are also given that $R_s = 1$ kilohm?

8-8 Derive (8-27) and show a circuit realization for

$$\dfrac{V_o}{V_{in}} = \dfrac{K}{s^3 + 2s^2 + 2s + 1}$$

8-9 Derive the expression for the transfer function for the circuit of fig 8-17b.

8-10 Derive (8-31), and design the circuit for $\omega_0 = 2\pi(100 \text{ Hz})$, $\omega_{3\text{ dB}} = 2\pi(10 \text{ Hz})$, and $A_v(j\omega_0) = -50$. The source resistance is given to be 1 kilohm.

8-11 Derive the expression for the transfer function for the circuit in fig 8-18b.

8-12 Derive (8-35), and design the circuit such that

$$T_v = \frac{As^2}{s^2 + \sqrt{2} \times 10^5 s + 10^{10}}$$

where A is a constant. Is there any restriction on A?

8-13 The following driving-point function is given:

$$Y(s) = \frac{s^2 + 2s + 2}{(s+1)(s+3)}$$

Show a circuit realization using R's, C's, and an NIC.

8-14 Derive (8-50), and design the circuit to have a three-pole Butterworth response with $\omega_{3\text{ dB}} = 10^5$ rad/sec and $A_v(0) = 100$. The load resistance is given as $R_L = 1$ kilohm.

8-15 The active RC configuration due to Linvill is shown in fig P8-15. Show that

$$z_{21} = \left.\frac{V_2}{I_1}\right|_{I_2=0} = \frac{\pm\sqrt{Kz_{21}^a z_{21}^b}}{z_{22}^a - Kz_{11}^b}$$

where the plus and minus signs are for VNIC and INIC, respectively.

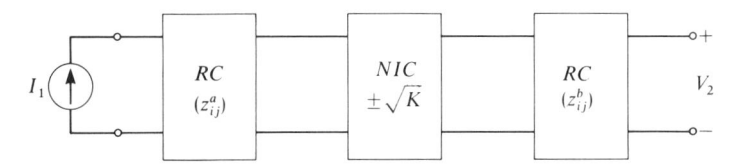

Fig P8-15

8-16 For the gyrator-RC network shown in fig P8-16, show that

(a) $$\frac{V_2}{V_1} = \frac{G_o G_1/C_1 C_2}{s^2 + s(G_1/C_1 + G_2/C_2) + (G_1 G_2 + G_o^2)/C_1 C_2}$$

(b) Design the circuit for the following specifications: A Butterworth filter with $\omega_{3\text{ dB}} = 2\pi(1 \text{ kHz})$, $R_L = R_S = 1$ kilohm, and $G_o = 10^{-3}$ mho.

516   8/Active *RC* Filters

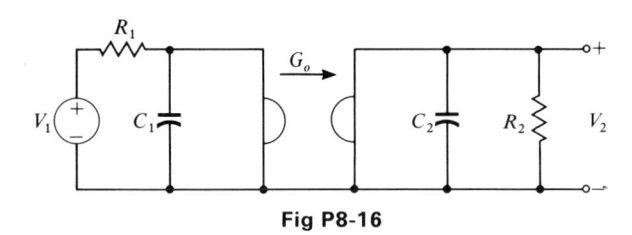

Fig P8-16

8-17 For the gyrator *RC* network shown in fig P8-17, show that

(a) $$\frac{V_2}{V_1} = \frac{sG_o/C_2}{s^2 + s(G_1/C_1 + G_2/C_2) + (G_1 G_2 + G_o^2)/C_1 C_2}$$

(b) Design the circuit for the following specifications, $\omega_0 = 2\pi$ (1 kHz), $\omega_{3\,\text{dB}} = 2\pi(100)$ Hz, $R_1 = R_2 = 2\,\text{k}\Omega$, and $G_0 = 10^{-2}$ mho.

Fig P8-17

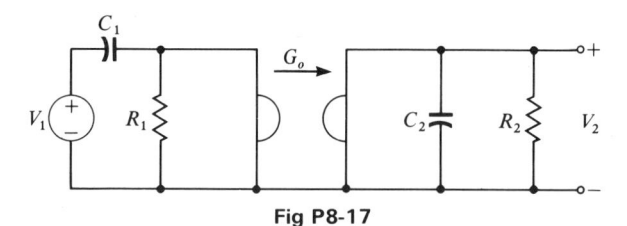

8-18 A floating inductance can be replaced by two grounded gyrators, if $G_{o1} = G_{o2} = G_o$, and a capacitor as shown in fig P8-18. Verify this assertion. Is the conclusion still valid if $G_{o1} \neq G_{o2}$?

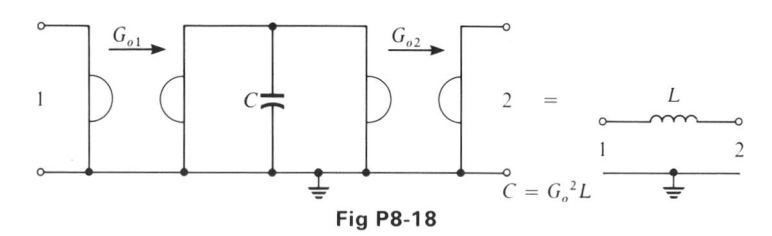

$C = G_o^2 L$

Fig P8-18

8-19 Derive the transfer function of the gyrator circuit shown in fig P8-19.

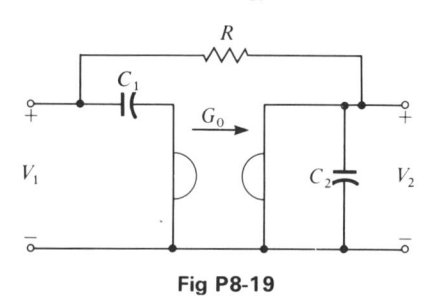

Fig P8-19

8-20 Derive (8-60), and design the circuit for a Butterworth three-pole response characteristic with $\omega_{3\,dB} = 10^6$ rad/sec and $A_V(0) = 10$.

8-21 The operational amplifier circuit shown in fig P8-21 can be used to realize any general transfer function.

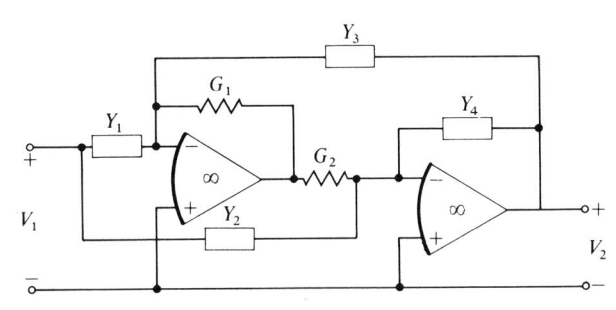

Fig P8-21

(a) Show that

$$\frac{V_2}{V_1} = \frac{Y_1 G_2 - Y_2 G_1}{G_1 Y_4 - Y_3 G_2}$$

(b) Design the circuit to realize the following transfer function:

$$\frac{V_2}{V_1} = \frac{s^2 - s + 1}{s^2 + s + 1}$$

8-22 Derive (8-65) in fig 8-35. Assume ideal operational amplifiers.

8-23 Derive (8-91).

8-24 Design an active-distributed RC filter with a three-pole Butterworth response with $\omega_{3\,dB} = 10^6$ rad/sec. Use only one amplifier.

8-25 What is the additional delay time of the circuit of Problem 8-24 as compared with that of a lumped three-pole Butterworth filter?

8-26 Show the circuit component values of an active-distributed RC filter for a four-pole Butterworth filter. The desired cutoff frequency is $\omega_{3\,dB} = 2\pi\,(10^6)$ rad/sec.

8-27 A gyrator has the following admittance matrix:

$$[y] = \begin{bmatrix} 0 & -G_1 \\ G_2 & 0 \end{bmatrix}$$

Show that the gyrator is not passive if $G_1 \neq G_2$. (Hint: Examine the power entering at each port, see Problem 2-13).

8-28 In practical gyrators, y_{12} and y_{21} are not identically zero. Furthermore, G_1 and G_2 are not identically equal. Thus, it can be accurately represented at low frequencies by the following y-matrix:

$$[y] = \begin{bmatrix} g_a & -G_1 \\ G_2 & g_b \end{bmatrix}$$

If the gyrator is terminated at port 2 by a capacitance C, determine the value of Leq and Req as seen from port 1. What is the Q if the inductor at $\omega = 10^5$ rad/sec, if $C = 0.1\ \mu F$, $g_a = g_b = 10^{-5}$ mho, and $G_1 = G_2 = 10^{-3}$ mho.

8-29 For the circuit of fig 8-35, show that the Q sensitivity with respect to any R's or C's is less than unity.

8-30 (a) Derive (8-70).
(b) Derive the Q-sensitivity in terms of K_1 and K_2 assuming that K_3 and K_4 are unity.

8-31 If the compensated operational amplifiers in fig 8-36 have finite gain and bandwidth, namely A_0 and P_0, with a one-pole roll-off, show that the inclusion of finite gain and bandwidth for operational amplifiers in the expressions for Q and ω_0 are given by

$$\frac{1}{\hat{\omega}_0^2} \simeq T^2 + \frac{3T}{A_0 P_0}$$

$$\frac{1}{\hat{Q}} \simeq \left(1 - K_2 K_3 K_4 + \frac{3}{A_0}\right)$$

(Hint: let $\hat{K}_i = K_i/(1 + 1/A_0 + s/A_0 P_0)$ in the expression for the transfer function).

8-32 The circuit shown in fig P8-32 is an actual realization of a very high Q response. The operational amplifier used is the Motorola MC1520 (see Appendix D).
(a) Determine K_1, K_2, K_3, and K_4 if $R_5 = 0$.
(b) Determine the values of Q and ω_0.

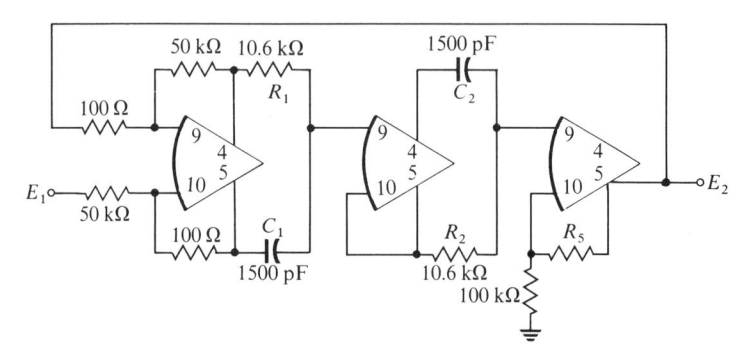

Fig P8-32

(c) If the operational amplifier has the pole at $P_0 = 2\pi(5\text{ kHz})$ and a gain bandwidth product $A_0 P_0 = 2\pi(10\text{ MHz})$, with a one-pole roll-off, determine their effect on the values of Q and ω_0. What value of R_5 would result in a Q of 500? (Hint: Use the results of Problem 8-31.)

8-33 Derive the expression for Q and ω_0 of fig 8-35 if the operational amplifier has a finite gain A_0, and a bandwidth P_0 with a one-pole roll off.

8-34 Consider the active $-\overline{RC}$ network shown in fig P8-34. The amplifier is assumed to be an ideal VCVS.

(a) Use the Tee model for $\overline{RC}$ (see fig 8-40b) and show that

$$A_v = \frac{V_o}{V_{\text{in}}} = K \frac{\cosh\sqrt{sRC} - 1}{\cosh\sqrt{sRC} - K}$$

(b) The dominant pole-zero of the above can be approximated by

$$A_v = K_1 \frac{s}{s^2 + a_1 s + a_0}$$

where the relations between a_1, a_0, K, and RC are the same as in the text. Design the circuit for $Q = 10$ and $\omega_{3\text{ dB}} = 10^3$ rad/sec.

Fig P8-34

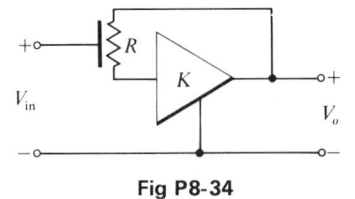

9
Elementary Analysis of Nonlinear Electronic Circuits

9-1 Introduction

We have already seen in Chapter 3 that all electronic devices exhibit *nonlinear* behavior and that one cannot always use a linear model for large signal analyses. Only for small-signal applications, where the signal swings are small such that the operation is in the linear region of the characteristics of the device, could we use linear models to design and predict the circuit behavior. This has been the case for the topics discussed in Chapters 4 to 8, except for power amplifiers, where small-signal conditions do not apply and graphical analysis was used.

In the analysis and design of truly nonlinear circuits we have several methods of attack. The simplest is the *piecewise-linear* analysis, where the nonlinear characteristic of the device is approximated by several linear segments. Then, for each segment, a linear analysis is used. The method is quite general; however, caution is necessary when some segments display negative resistance. Some computer programs,

9-2 Piecewise-Linear Approximation of Nonlinear Resistive Circuits

namely ECAP (Section 1-13), require piecewise-linear approximation for nonlinear functions. A general approach for solving nonlinear differential equations by a computer is through numerical techniques. The other approach is to solve the nonlinear equations, if possible, using any of several available analytical and graphical methods. Nonlinear analysis is quite a vast topic and a comprehensive treatment of it is beyond the scope of this text. We, therefore, restrict ourselves merely to a brief and elementary analysis of nonlinear circuits. Monotonic resistive circuits are treated first in Sections 9-2 to 9-6 as a basis of nonlinear circuit analysis. These are followed by limited analytical methods for treatment of nonlinear circuits with reactances and a nonlinear reactance.

9-2 Modeling of Nonlinear Resistive Circuits by Piecewise-Linear Approximation

The static characteristics of electronic devices over any arbitrarily large region of operation are quite nonlinear. Electronic devices are often subjected to large-signal operations in digital circuitry. For example, a transistor is driven from saturation to cutoff or vice versa. The nonlinear characteristics can be adequately approximated by a nonlinear model which consists of a *piecewise-linear* (PWL) model. For example, the characteristics of an actual nonideal diode including the reverse breakdown region can be approximated by the piecewise-linear characteristics shown in fig 9-1. The approximation of the characteristic of a tunnel diode is shown in fig 9-2. Note that piecewise-linear approximation consists of *distinct regions, break points,* and *resistances,* i.e., the slopes of the v-i characteristic in the various regions or segments. Usually, the larger the number of linear segments, the better the approximation, and the more complicated the resulting model.

By means of ideal diodes, resistors, and batteries, almost any given dc v-i characteristics can be approximated by PWL segments. Such an approximation is usually referred to as diode function generation. The graphic symbol and the v-i characteristics of an *ideal diode* are shown in fig 3-3, and they are shown again for convenience in fig 9-3. The model acts as a short circuit when a forward voltage is applied and as an open circuit when a reverse voltage is applied. A forward bias corresponds to a positive potential at the arrow (p-terminal or anode) side. With no externally applied bias the *break point,* which is defined here as the

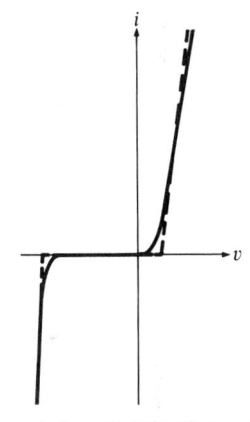

Fig 9-1 The v-i characteristic of an actual junction diode (reverse region not to scale) and its piecewise-linear approximation (dotted lines)

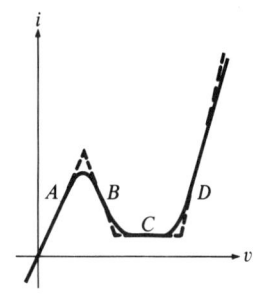

Fig 9-2 The v-i characteristic of a tunnel diode and its piecewise-linear approximation. The four linear segments are labeled as A, B, C and D

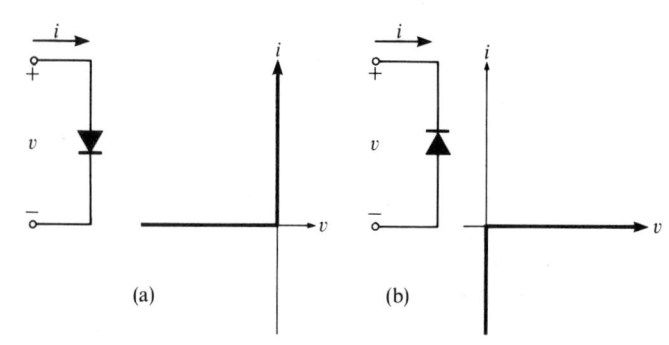

522 9/Elementary Analysis of Nonlinear Electronic Circuits

Fig 9-3 (a) An Ideal diode with its v-i characteristic for the reference voltage and current shown; (b) An Ideal diode with its v-i characteristic for the reference voltage and current shown

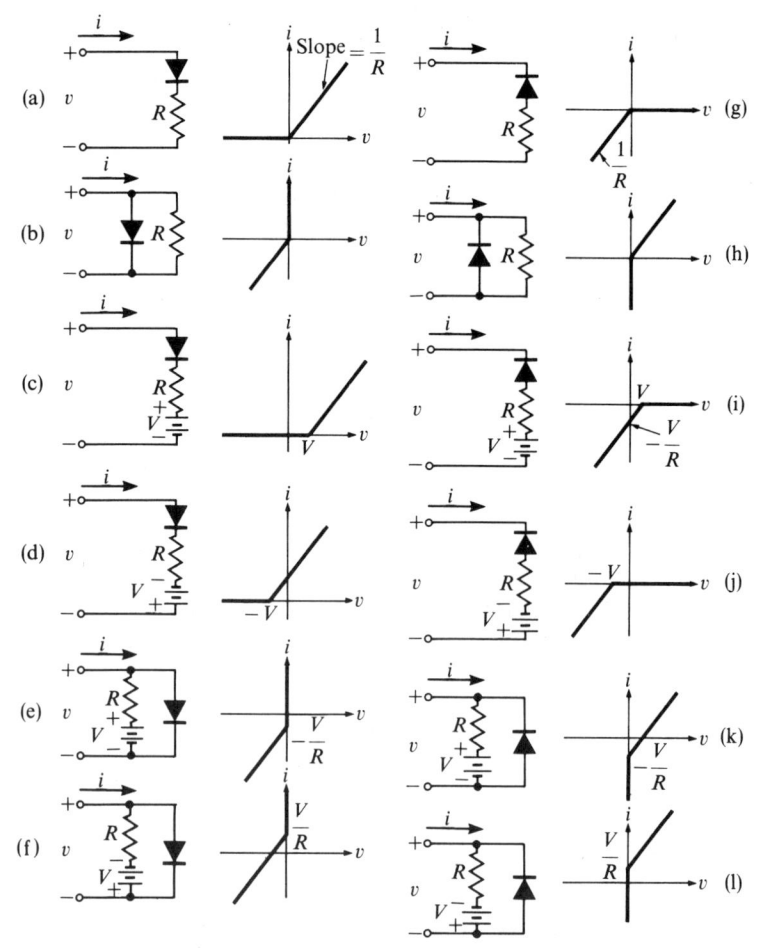

Fig 9-4 Various piecewise-linear building blocks

9-2 Piecewise-Linear Approximation of Nonlinear Resistive Circuits

point of intersection of two line segments, is at the origin. By adding a battery in series with the diode, the break point may be moved to left or right, depending on the battery polarity.

The characteristics of elementary building blocks of piecewise linear models are shown in fig 9-4. Note that the slope of the v-i characteristics is equal to $1/R$ in all cases and the break point shift is provided by the battery $\pm V$. Note further that the voltage in series with a resistance can also be represented by a current generator, $I_o(= V/R)$, in shunt with the resistor R.

Once a given v-i characteristic is approximated by a "suitable" number of linear segments, the building blocks can be utilized in successive steps to synthesize the model. The technique is illustrated by two simple examples.

EXAMPLE 1

Consider the curve shown in fig 9-5. The curve is approximated by three segments, as shown by the dashed lines on the same figure. The model is then built up using the building blocks of fig 9-4. For example, in this case, proceeding from left to right, segments A and B are obtained from fig 9-4d, as shown in fig 9-6a. The C segment has a slope (conductance) larger than the B segment, hence the conductance must be increased for $v \geq V_2$, which can be done by adding a shunt conductance. The segment realization is readily accomplished by connecting the building block of fig 9-4c in shunt with the previous circuit. The resulting model is shown in fig 9-6b. The v-i characteristics can be readily checked. For $v < -V_1$, D_1 and D_2 are open circuit and $i = 0$. For $-V_1 < v < V_2$, D_1 is short circuit, D_2 open circuit, hence $i = i_1$ and the v-i corresponds to that of segment B. For $v > V_2$, both diodes are short circuits, hence $i = i_1 + i_2$, and the Thévenin equivalent circuit corresponds to that of segment C.

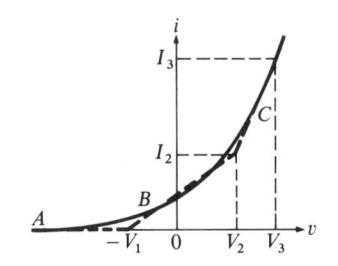

Fig 9-5 Nonlinear v-i characteristic with its piecewise-linear approximation for the example

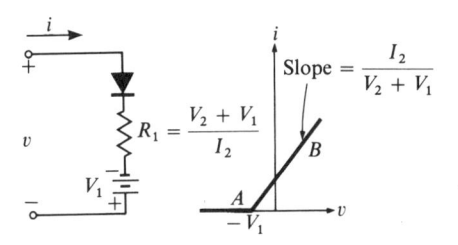

Fig 9-6 (a) Realization of piecewise-linear segments A and B

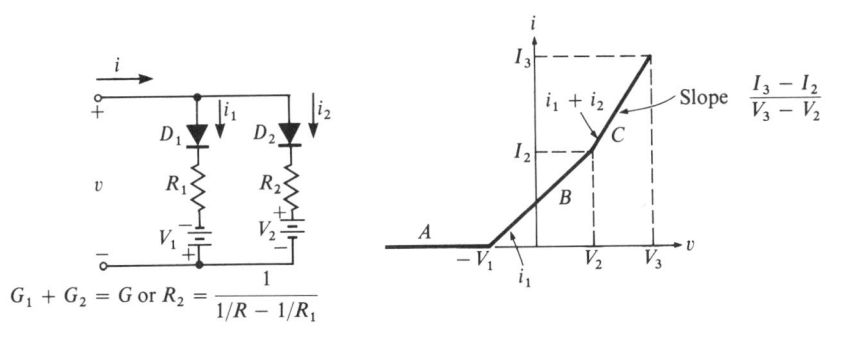

$G_1 + G_2 = G$ or $R_2 = \dfrac{1}{1/R - 1/R_1}$

Fig 9-6 (b) Realization of piecewise-linear seyments A, B, and C

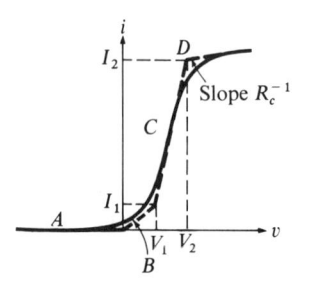

Fig 9-7 Nonlinear v-i characteristic with its piecewise-linear approximation for the example

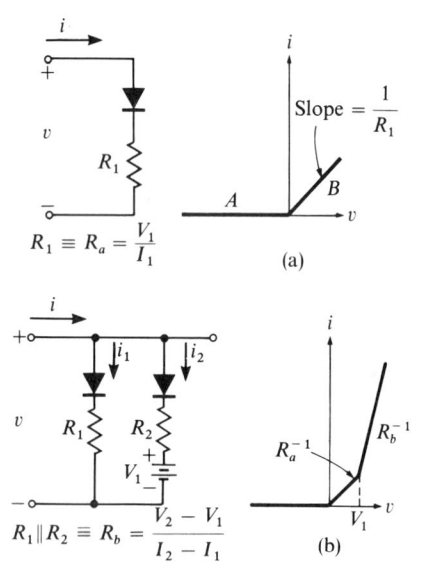

Fig 9-8 (a) Realization of piecewise-linear segments A and B; (b) Realization of piecewise-linear segments A, B, and C

EXAMPLE 2

As a second example of obtaining a piecewise-linear model, consider the curve shown in fig 9-7. The piecewise-linear approximation is also shown therein by the dashed lines. Proceeding as in the previous example, segments A and B are realized by the building block of fig 9-4a, as shown in fig 9-8a. The realization of segment C, which has a slope larger than that of B, was already discussed. The realization of segments A, B, C is shown in fig 9-8b. The D segment has a smaller slope than the C segment, hence the resistance must be increased for $v > V_2$, which can be done simply by adding a series resistance. The segment realization is accomplished by connecting the building block of fig 9-4l in series with the previous circuit. The realization for the entire A, B, C, D segments is shown in fig 9-8c.

So far we have considered monotonically increasing v-i relationships, where the resulting resistors in the synthesis are all positive. If we allow negative resistors, the above idea can be extended in a straightforward manner to the approximation of any nonlinear dc v-i relationship. As an example, consider Problem 9-8c, the synthesis of fig 9-2, where a negative resistance will appear in the piecewise-linear model.

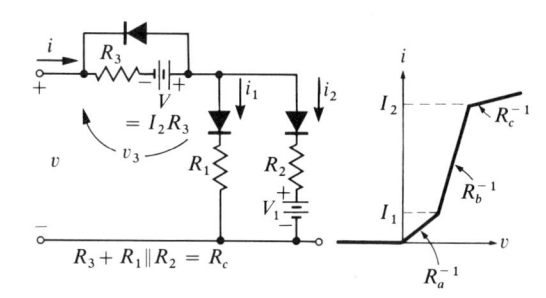

Fig 9-8 (c) Realization of piecewise-linear segments A, B, C, and D

9-3 Synthesis of Piecewise-Linear Functions

The v-i characteristic of a one-port linear resistive network, when approximated by the PWL method, can be synthesized as a driving-point function. If the nonlinear device is a general resistive two-port, however, the synthesis of driving points at the input and output as well as transfer functions is required to characterize the device completely. We shall now consider the synthesis of driving-point and transfer functions.

9-3 Synthesis of Piecewise-Linear Functions

(a) Driving-Point Functions

Any piecewise-linear driving-point function can be synthesized using one diode per break point. The simplest v-i characteristics to synthesize by positive resistors, diodes, and batteries are those which are monotonically increasing, and are either concave or convex and can be synthesized by the diode networks of the Foster or Cauer forms (see Section 8-1).

A typical concave v-i characteristic is shown in fig 9-9a. The canonical realizations in the Foster and Cauer forms are shown in figs 9-9b and

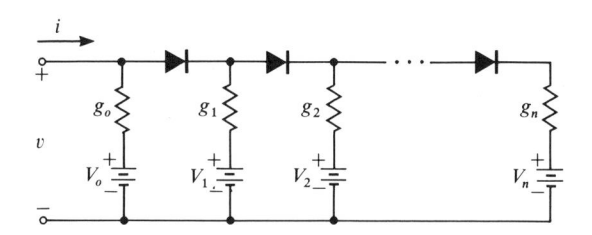

Fig 9-9 (c) General Cauer form canonical realization of (a)

9-9c, respectively. For the convex v-i characteristics shown in fig 9-10a, the canonical realizations are shown in figs 9-10b and 9-10c. It should be clear that the networks of figs 9-9b and 9-9c realize concave v-i characteristics, while those of figs 9-10b and 9-10c realize convex v-i characteristics. Thus, any general driving-point function can be synthesized directly in the above manner by partitioning the v-i

$$r_o = R_o$$
$$r_n = R_n - R_{n-1} \quad n = 1, 2, 3, \ldots$$

(b)

(c)

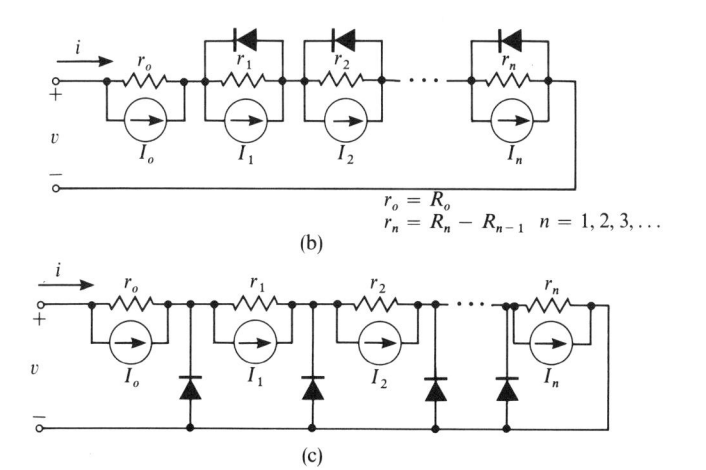

Fig 9-10 (b) General Foster form canonical realization of (a); (c) General Cauer form canonical realization of (a)

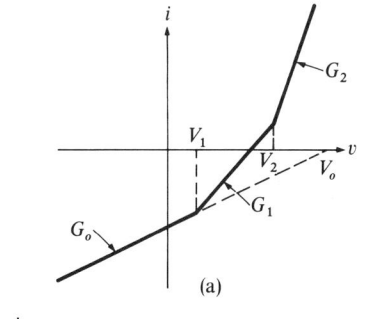

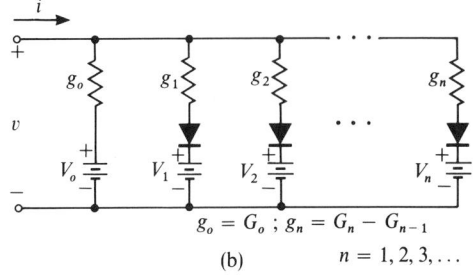

Fig 9-9 (a) Monotonically increasing concave piecewise-linear v-i characteristic; (b) General Foster form canonical realization of (a);

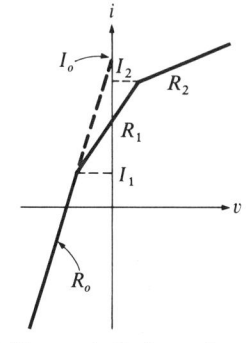

$$g_o = G_o \, ; \, g_n = G_n - G_{n-1}$$
$$n = 1, 2, 3, \ldots$$

Fig 9-10 (a) Monotonically increasing convex piecewise-linear v-i characteristic

characteristics into concave and convex parts and then using the above networks to achieve canonical realizations. As an example, consider the v-i characteristics shown in fig 9-11a. The v-i characteristic is made up of one concave and one convex part. One possible canonical network realization is shown in fig 9-11b. The resistors will all be positive if the v-i characteristic, such as in fig 9-11a, is a monotonically increasing function of the voltage. This is the class we shall be most concerned with this section. It should be noted that in the above figures of this section g_i, r_i, G_i, and R_i are *not* reciprocals of each other.

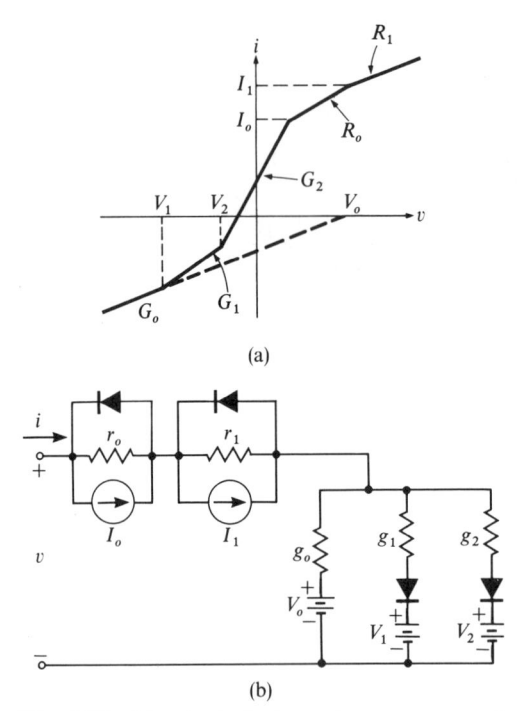

Fig 9-11 Monotonically increasing concave-convex piecewise-linear v-i characteristic; (b) Canonical realization of (a)

(b) Transfer Function

A piecewise-linear voltage *transfer function* may be synthesized by several methods. We shall discuss only two methods of PWL v_o vs v_i synthesis. Both methods reduce to that of synthesizing driving-point functions.

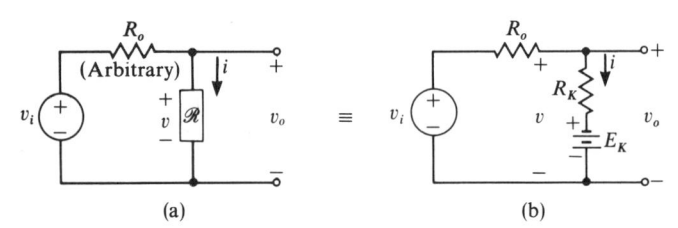

Fig 9-12 (a) Voltage-divider configuration for the synthesis of monotonically increasing transfer function with slope less than unity; (b) Piecewise-linear model of (a) for the Kth segment

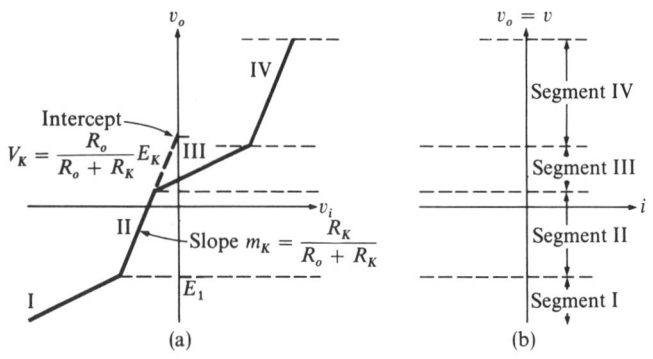

Fig 9-13 (a) Monotonically increasing piecewise-linear transfer function characteristic with slope of each segment less than unity; (b) Alignment of i-v plane with the v_i-v_o plane to obtain the piecewise-linear driving-point function

9-3 Synthesis of Piecewise-Linear Functions

The first method consists of a voltage divider configuration, as shown in fig 9-12. The linear resistance R_o is arbitrary and the arbitrariness can be utilized to accommodate the source resistance. Our problem is to reduce the transfer synthesis to that of a driving point, i.e., to specify an i-v curve for $\mathscr{R}$ given v_o vs v_i curve. If the plot of v_o vs v_i has a positive slope less than or equal to unity in the v_i-v_o plane, as we shall see, the driving-point plot to be synthesized is always monotonically increasing.

Since the transfer function to be synthesized is given as a piecewise-linear function, we can treat the network as a linear network at each of the K different segments. For example, the transfer plot of fig 9-13 has four different linear segments. The equation for v_o vs v_i corresponding to each segment, from fig 9-13, is

$$v_o = \frac{R_K}{R_o + R_K} v_i + \frac{R_o}{R_o + R_K} E_K \qquad (9\text{-}1)$$

Note that (9-1) is the equation of a straight line with slope, m_K, equal to $R_K/(R_o + R_K)$, and intercept V_K equal to $E_K R_o/(R_o + R_K)$, and for $K = 2$ it is shown on fig 9-13a. Now, corresponding to each segment in the prescribed v_i-v_o plane, the slope and the intercept are known, hence R_K and E_K are determined. Known values of R_K and E_K enable us directly to find v vs i for each segment, and thus the piecewise-linear driving-point is specified. We are already familiar with the synthesis of the driving-point function and so the synthesis is complete.

Note further that the slope is equal to $R_K/(R_o + R_K)$, thus, if the resistances are all to be positive valued, the slope of v_o vs v_i must be less than or equal to unity. Only if the slope is less than or equal to unity will the driving-point function will be monotonic. Also note that the breakpoint voltages of v and v_o are identical since $v_o = v$, and for this reason we align the i-axis of the i-v plane with the v_i axis of v_i-v_o plane.

EXAMPLE

Suppose that the v_o-v_i function to be synthesized is given as in fig 9-14a. We choose R_o arbitrarily, say $R_o = 1 \text{ k}\Omega$. The values of R_K and V_K for each segment are as follows:

segment I: $R_\text{I} = 0$, $V_\text{I} = -1$ volt,

segment II: $R_\text{II} = 2 \text{ k}\Omega$, $V_\text{II} = -\frac{1}{3}$ volt,

segment III: $R_\text{III} = \infty$, $V_\text{III} = -1$ volt,

The v-i plot, according to the above segment descriptions, is shown in

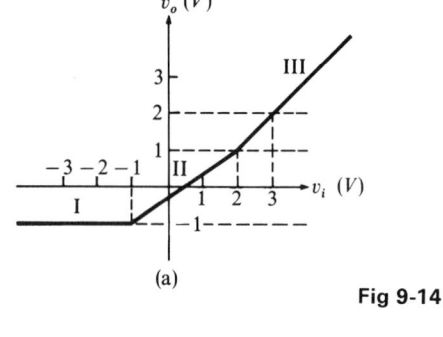

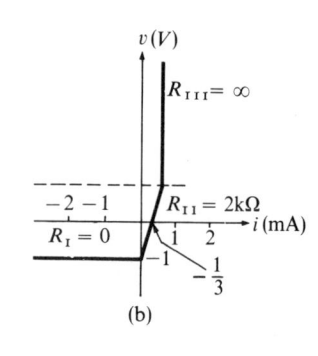

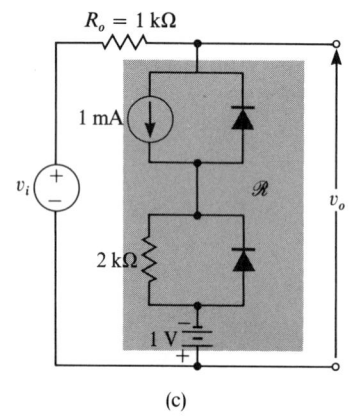

528 9/Elementary Analysis of Nonlinear Electronic Circuits

Fig 9-14

Fig 9-14 (a) Piecewise-linear transfer function for the example; (b) *i-v* characteristic obtained from (a); (c) Realization of the circuit which has the transfer function of (a)

fig 9-14b. The driving-point realization has already been discussed and will not be repeated. Hence, the diode-resistive network realization of fig 9-14b is put in the place of one-port $\mathscr{R}$ and the transfer function realization is complete, as shown in fig 9-14c.

A piecewise-linear voltage-transfer function may also be synthesized by using the operational amplifier configuration, as shown in fig 9-15. This method leads to a monotonic driving-point function as long as the transfer function is monotonic. Thus this method is *not* restricted to a transfer slope less than unity. The transfer-function synthesis is as follows: From the circuit in fig 9-15, the input of the operational amplifier is at virtual ground, $v_{ab} \simeq 0 = i_a$. Let the *v-i* characteristic of the nonlinear resistor $\mathscr{N}$ be

$$v = f(i) \tag{9-2}$$

From the circuit in fig 9-15, we have

$$v_o = v \quad \text{and} \quad v_i = R_o i_1 \tag{9-3}$$

But $i = -i_1$, so that, from (9-2) and (9-3), we obtain

$$v = f(-i_1) = f\left(-\frac{v_i}{R_o}\right) \tag{9-4}$$

Since R_o is a constant and arbitrary, the value of R_o can be used to change the scale of the transfer function. Thus, as long as the transfer function is monotonic, the driving-point function will be monotonic. Note that if we use $R_o = 1\,\Omega$ and add a unity-gain inverting amplifier so that v_i is changed to $-v_i$, then from (9-4), the transfer-function synthesis is identical to that of the driving-point synthesis.

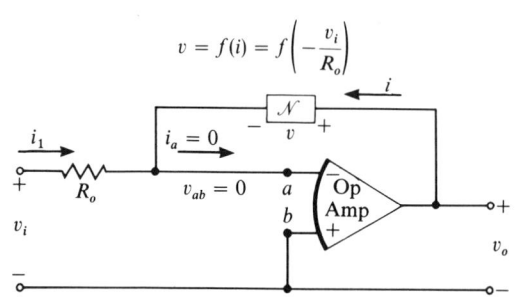

Fig 9-15 Operational-amplifier configuration for the synthesis of a monotonically increasing transfer function

9-4 Piecewise-Linear Models of Some Electronic Devices

An alternate version of the transfer-function synthesis where a linear feedback resistor is used in conjunction with a nonlinear resistor is shown in fig P9-13. The method is quite similar to above and is left as an exercise in Problem 9-12.

Finally, it should be mentioned that there are other methods available for synthesizing nonmonotonic driving-point and transfer functions. For this topic as well as synthesis of jointly prescribed driving-point and transfer functions, interested readers are referred to the literature (Ref 4).

9-4 Piecewise-Linear Models of Some Electronic Devices

We now consider briefly the piecewise-linear models of some electronic devices. Although we consider only a few typical cases, the method is general and can be applied to obtain a piecewise-linear model to approximate the current-voltage relationships of any device. Consider first the dc characteristics of a *pnp* transistor in the common-base configuration, as shown in fig 3-15. The piecewise-linear approximation of the characteristics is shown in fig 9-16b and c. The piecewise-linear model which approximates these characteristics is seen to be as shown in fig 9-17. Notice that if the *PWL* approximation of the characteristics has finite nonzero slopes, the model will have resistances to account for these. More accurate models are given in Refs 1 and 4.

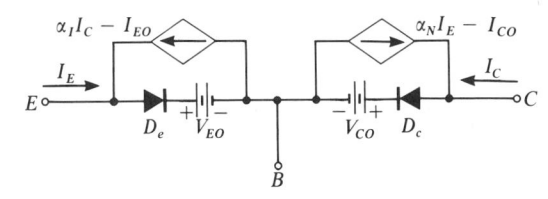

Fig 9-17 Large-signal piecewise-linear model of a common-base transistor

A piecewise-linear approximation of the *v-i* characteristics of a *pnp* transistor in the common-emitter configuration is shown in fig 9-18b and c. The *PWL* model for these characteristics is shown in fig 9-19a. It should be noted that if the *PWL* approximation is more detailed, the model will naturally be further involved. In general, simple models are preferable, if the accuracy as a result of such simplifications is

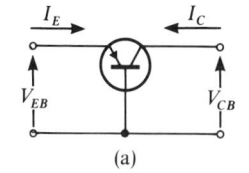

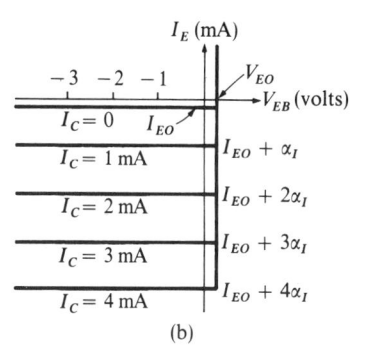

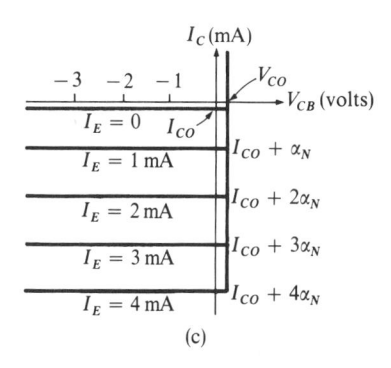

Fig 9-16 Common-base transistor with its input and output piecewise-linear *v-i* characteristics. (a) Common-base transistor with voltage polarity and current reference directions; (b) Input piecewise-linear *v-i* characteristics; (c) Output piecewise-linear *v-i* characteristics

530   9/Elementary Analysis of Nonlinear Electronic Circuits

satisfactory. Typically, the resistances R_B and R_I are small and are often omitted from the model. If these approximations are made, the two diodes can then be combined into one and the simplified model of fig 9-19b results.

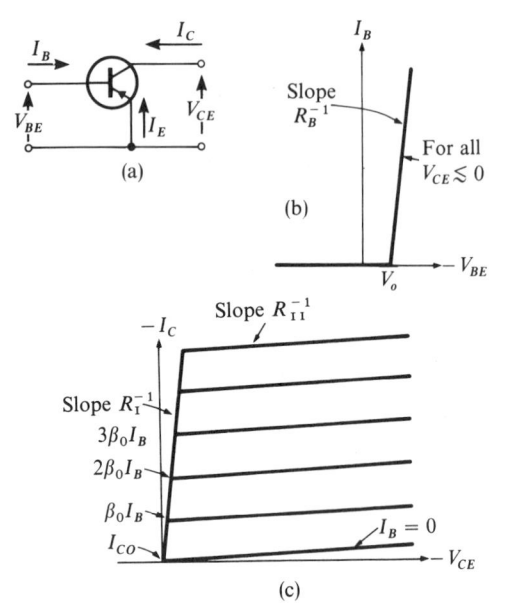

Fig 9-18 Common-emitter transistor with its input and output piecewise-linear v-i characteristics: (a) common-emitter transistor with voltage polarity and current reference directions; (b) Input piecewise-linear v-i characteristics; (c) Output piecewise-linear v-i characteristics

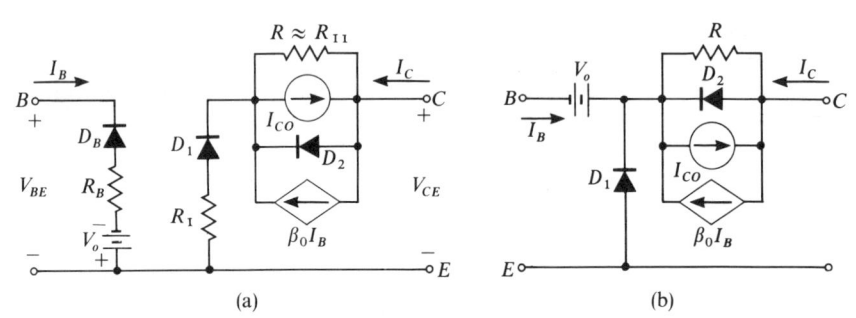

Fig 9-19 (a) Large-signal piecewise-linear model of a common-emitter transistor; (b) Simplified large-signal model of a common-emitter transistor

The *PWL* characteristics of a vacuum-tube triode are shown in fig 9-20b and c. The piecewise-linear model of the triode corresponding to these characteristics is shown in fig 9-21. Diode D_g at the grid circuit approximates the average transfer characteristics. Diode D_p in the plate circuit does not permit e_b ever to become negative.

A piecewise-linear approximation of the characteristics of an n-channel JFET, for regions below breakdown operation, is shown in fig 9-22b and c. The model corresponding to the *PWL* characteristics is shown in fig 9-23. Note that for V_{DS} positive and small, the slope is equal to the parallel combination of g-g_d and g_d, hence the slope is g; and for V_{DS} large, the slope is equal to g_d. Since the pentode characteristics are similar to those of the FET, the model of fig 9-23 also represents the piecewise-linear model of pentodes, with appropriate change in the notation and symbols.

It should be noted that for ac calculations, the independent generators may be ignored in the models, provided that operation is restricted to within a single *PWL* segment at a time. Note further that the small-signal ac models are readily obtained from the piecewise-linear models. At the operating points, the diodes are either on or off, hence either short-circuited or open-circuited, e.g., for the triode case, D_g and D_s are off and D_p is on. Observe that the element values of the linear and piecewise-linear models may be quite different since, in the linear model, the values of the elements are based on the slopes *at* the operating point, while in the piecewise-linear model, the slopes are determined on the basis of *average* values over a large region of operation.

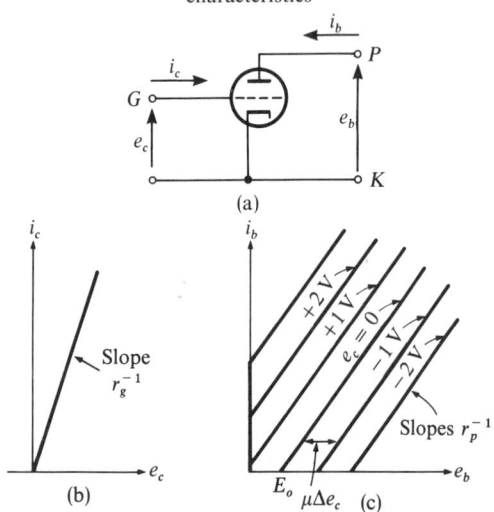

Fig 9-20 Vacuum triode with its input and output piecewise-linear v-i characteristics: (a) Vacuum triode with voltage polarity and current reference directions; (b) Input piecewise-linear v-i characteristics; (c) Output piecewise-linear v-i characteristics

9-5 Piecewise-Linear Analysis

The dc analysis of circuits containing nonlinear elements may often be reduced to a piecewise-linear analysis. In one approach, the nonlinear device is approximated by a piecewise-linear model which consists of diodes, resistors, batteries, and/or current sources. Thus the problem reduces to that of analysis of the diode-resistor-source circuits which, in turn, reduces to that of determining the breakpoints and the slopes of the various segments. A more general approach, the iterative piecewise-linear model, is described in Ref 4. The advantage of this method over the former approach is that it can handle two-terminal resistors characterized by arbitrary piecewise-linear v-i curves, thus avoiding the necessity of constructing models using ideal diodes.

There are essentially two approaches for *PWL* analysis. One is the method of *assuming the diode states* and the other is the so-called *break point* method.

In the first method, the circuit is analyzed by assuming all possible combinations of states of the ideal diodes and then finding the corresponding v-i relationships for the combination. The method has a severe drawback for hand computation since, for an n diode circuit, there are 2^n combinations to be analyzed. Of course the constraints of a network may eliminate some states from consideration. The method, however, can be implemented via computer-aided analysis.

In the second method, the break points are determined and, since the model is piecewise-linear, each breakpoint is connected by a linear segment. The slope at the outermost breakpoints is determined separately. The maximum possible number of breakpoints for a network with n diodes is $(2^n - 1)$. This method is useful since, at the breakpoint, both the current and the voltage of the ideal diodes are simultaneously zero. The application of these two constraints reduces the n diode circuit to one containing $(n - 1)$ diodes. Again, the analysis

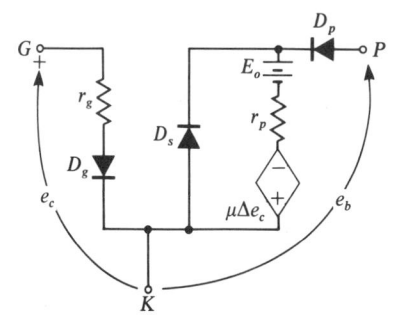

Fig 9-21 Large-signal piecewise-linear model for a vacuum triode

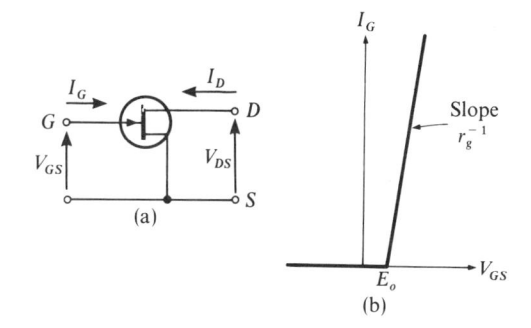

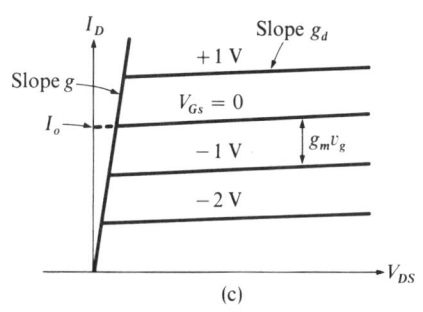

Fig 9-22 Junction field-effect transistor with its input and output piecewise-linear v-i characteristics: (a) *JFET* with voltage polarity and current reference characteristics; (b) Input piecewise-linear v-i characteristics; (c) Output piecewise-linear v-i characteristics

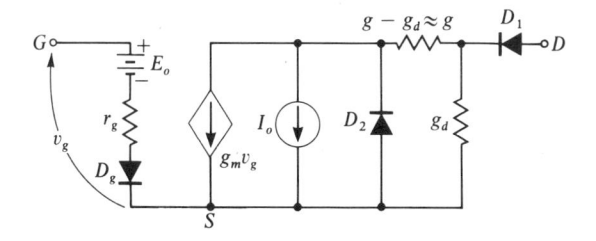

Fig 9-23 Large-signal piecewise-linear model for a junction field-effect transistor

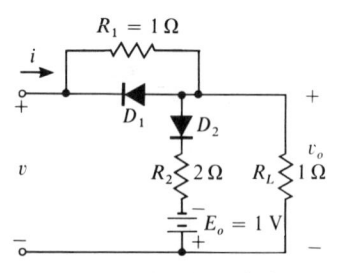

Fig 9-24 Circuit for the analysis example

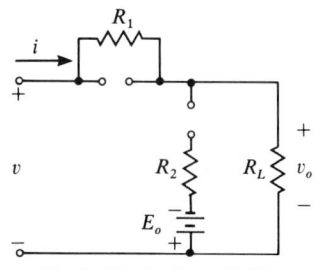

Fig 9-25 (a) State [0, 0]

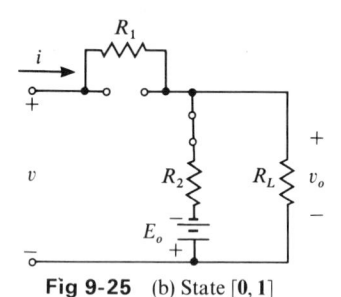

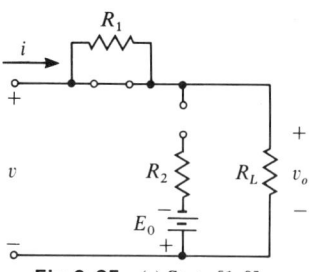

Fig 9-25 (b) State [0, 1]

Fig 9-25 (c) State [1, 0]

of circuits containing a large number of diodes by this method is quite cumbersome and tedious. This approach is also best implemented by a computer solution. In fact, user-oriented computer programs are available for *PWL* analysis of electronic circuits.[1]

When the number of diodes in the circuit is very small, the two methods of analysis mentioned above are simple and straightforward. The following examples illustrate the methods.

EXAMPLE OF ASSUMED STATES

Consider the circuit shown in fig 9-24, which employs two diodes. The normalized values of the resistors and the battery are also shown therein. The i-v driving-point characteristic and the $v - v_o$ transfer characteristic are to be determined by the method of assumed states.

There are two diodes and thus $2^2 = 4$ possible combinations of states. The four possibilities are listed in table 9-1. Each state may be identified

TABLE 9-1 Possible States of the Circuit of Fig 9-24

Diode 1	Diode 2	State	
Off	Off	0	0
Off	On	0	1
On	Off	1	0
On	On	1	1

for convenience simply by **1** and **0** which denote the on and off states, respectively. The circuit for each assumed state is shown in fig 9-25. For each state, the i-v and v-v_o characteristic is readily determined, as listed below:

For [**0,0**] state:

$$v = i(R_1 + R_L) \tag{9-5}$$

$$v_o = \frac{R_L}{R_L + R_1} v \tag{9-6}$$

[1] For example, see "Multivalued Electronic Circuit Analysis Program, MECA. This program can be used to find operating points, driving-point characteristics, and transfer characteristics of resistive nonlinear networks containing linear resistors, dc and ac independent sources, linear controlled sources, and two-terminal nonlinear resistors characterized by piecewise linear v-i curves. For details and the reference to the manual see Ref 4.

For **[0,1]** state:

$$v = i\left[\frac{R_L(R_1 + R_2) + R_1R_2}{R_L + R_2}\right] - E_o\left(\frac{R_L}{R_L + R_2}\right) \quad (9\text{-}7)$$

$$v_o = v\left[\frac{R_LR_2}{(R_1 + R_2)R_L + R_1R_2}\right] - E_o\frac{R_LR_1}{(R_1 + R_2)R_L + R_1R_2} \quad (9\text{-}8)$$

For **[1,0]** state:

$$v = iR_L \quad (9\text{-}9)$$

$$v = v_o \quad (9\text{-}10)$$

For **[1,1]** state:

$$v = \frac{R_LR_2}{R_L + R_2}i - E_o\frac{R_L}{R_L + R_2} \quad (9\text{-}11)$$

$$v = v_o \quad (9\text{-}12)$$

For the assumed values of R's and E_o, as given in fig 9-24, the driving-point and transfer characteristics corresponding to (9-5), (9-7), (9-9), (9-11) and (9-6), (9-8), (9-10), (9-12) are shown in figs 9-26 and 9-27 by the dashed lines. The problem now is to determine which of the four possible states exists for particular values of the input voltage. Once the state is determined, then the appropriate portion of the piecewise-linear segment can be chosen.

To find the particular state for the values of input, we vary v from a very large negative value to a very large positive value and note the changes in the states of the diodes. For this example, when v is very large and negative, from the circuit of fig 9-24, we see that D_1 is on and D_2 off, hence the state is **1, 0**. When v is very large and positive, D_1 is off and D_2 on, hence the circuit is in the **0, 1** state. From fig 9-26, it is seen that the change of state from **1, 0** to **0, 1** may occur either through **1, 1** or directly. However, for $v = 0$, i.e., the input terminals of fig 9-24 are short-circuited, D_1 is off, D_2 on, hence the **0, 1** state, and thus the change of state, occur through **1, 1**, as shown by the heavy line in fig 9-26. The transfer function is shown by the heavy line in fig 9-27. Note that, for $-\frac{1}{3} > v > -1$, the circuit is in the **1, 1** state.

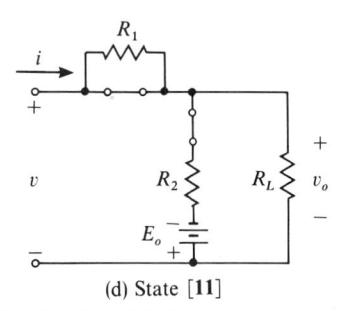

(d) State [11]

Fig 9-25 Circuit models for the various possible states

EXAMPLE OF BREAKPOINT ANALYSIS

In this method of analysis, the breakpoints are first determined. Recall that, at the breakpoints, the current through and the voltage across the diode are simultaneously zero. We then analyze the circuit

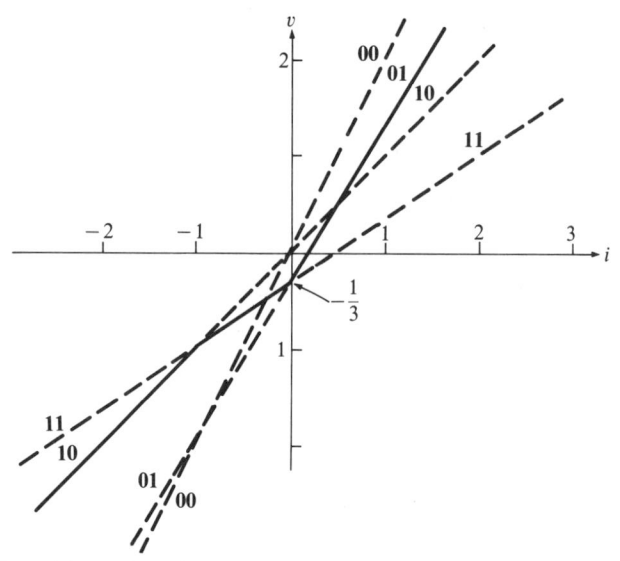

Fig 9-26 Driving-point characteristics of the various states (dotted) and of the circuit in fig 9-24

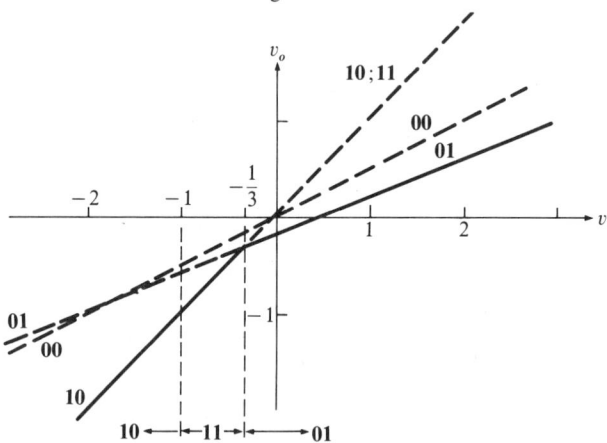

Fig 9-27 Transfer-function characteristics of the various states (dotted) and of the circuit in fig 9-24

with $n - 1$ diodes, by beginning at the internal terminal pair of the diode at the break point.

Consider the driving-point characteristic of fig 9-24 which is to be determined by the break point method. The circuits, when D_1 and D_2 are individually at break points, are shown in figs 9-28a and 9-28b, respectively. The coordinate of the break point in the v-i characteristic when D_1 is at break point is determined from fig 9-28a. From the circuit, it is seen that $V_1 = R_1 i$, but since $V_1 = 0$, hence $i = 0$. Now we determine

9-5 Piecewise-Linear Analysis

the state of D_2, which can be determined by assuming a state and then checking it for consistency. We first assume D_2 is off, in which case there is no current through D_2. For this to be true, the voltage across the diode must be such that it is reverse-biased. But from the circuit of fig 9-28a, the voltage across R_L is zero and the anode of diode D_2 is at a positive potential with respect to the other side, hence it is forward-biased and, therefore, a contradiction. Now, for D_2 on,

$$v = -E_o \frac{R_L}{R_L + R_2} \quad (9\text{-}13)$$

Thus the coordinates of the break point (D_2 on) for the assumed parameter values, in fig 9-24, are

$$i = 0, \quad v = -\tfrac{1}{3}\text{ V} \quad (9\text{-}14)$$

Similarly, for D_2 at break point, from the circuit of fig 9-28b, we determine the coordinate of the next break point. For $V_2 = 0$, and the current through it equal to zero, the voltage across R_L is $v_L = -E_o$ and thus

$$i = -\frac{E_o}{R_L} \quad (9\text{-}15)$$

To determine the state of D_1, assume it is off. For this assumed state, the voltage across D_1 is $(R_1/R_L)E_o$ with the positive polarity on the anode side, but this implies a forward bias, hence D_1 cannot be off. For D_1 on, we thus have

$$v = -E_o \quad (9\text{-}16)$$

For the given parameter values, the coordinate of the break point (D_1 on) is

$$i = -1 \text{ A}, \quad v = -1 \text{ V} \quad (9\text{-}17)$$

These two break points are connected by a linear segment, as shown in fig 9-29. The slopes at the outermost break points are easily determined as follows. For v very large and negative, from the circuit of fig 9-24, we see that D_1 is on and D_2 is off, hence the slope is R_L. For v very large and positive, D_1 is off and D_2 is on, hence, from the circuit of fig 9-24, under these conditions the slope is $[R_L(R_1 + R_2) + R_1 R_2]/(R_L + R_2)$. The complete v-i characteristic is shown in fig 9-29.

Note that both methods of analysis have significant drawbacks for a large number of diodes. The methods, for a larger number of diodes, are practical only if a digital computer is utilized in the solution of the

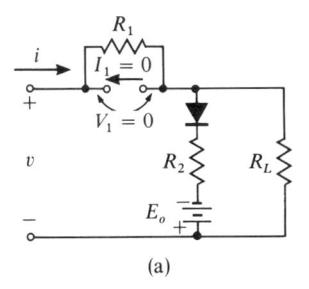

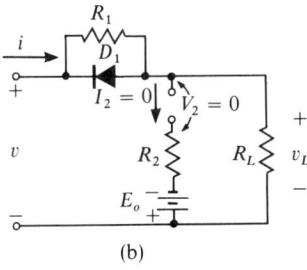

Fig 9-28 (a) Circuit of fig 9-24 for D_1 at breakpoint; (b) Circuit of fig 9-24 for D_2 at breakpoint

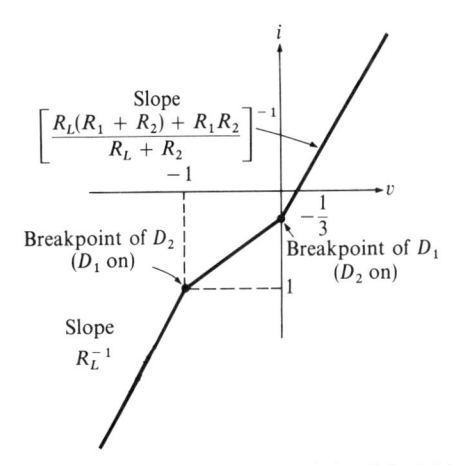

Fig 9-29 Driving-point characteristic of fig 9-24 by the breakpoint method (compare with fig 9-26)

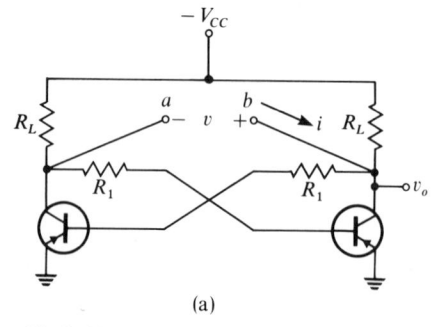

Fig 9-30 (a) A transistor flip-flop circuit

problem. In fact, this becomes mandatory in most circuits where PWL models are used for the devices.

EXAMPLE OF A TRANSISTOR BISTABLE MULTIVIBRATOR

Consider the symmetrical circuit shown in fig 9-30. This is a bistable circuit usually referred to as a flip-flop circuit. The operation of the circuit is discussed in detail in the next chapter. In this example, we

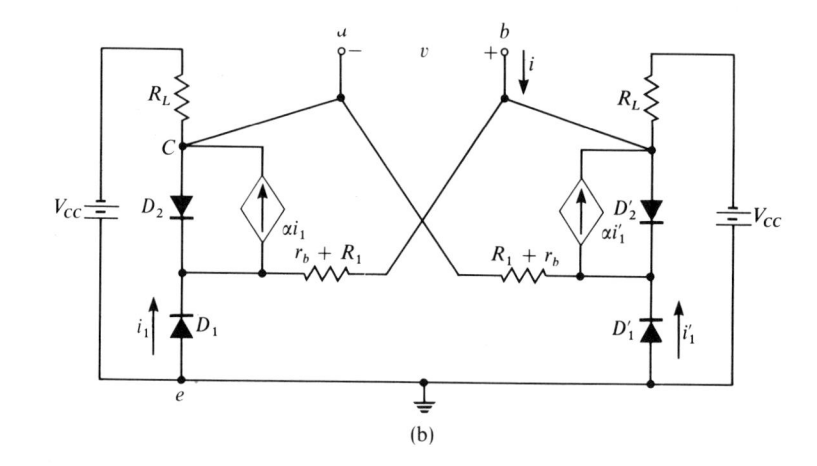

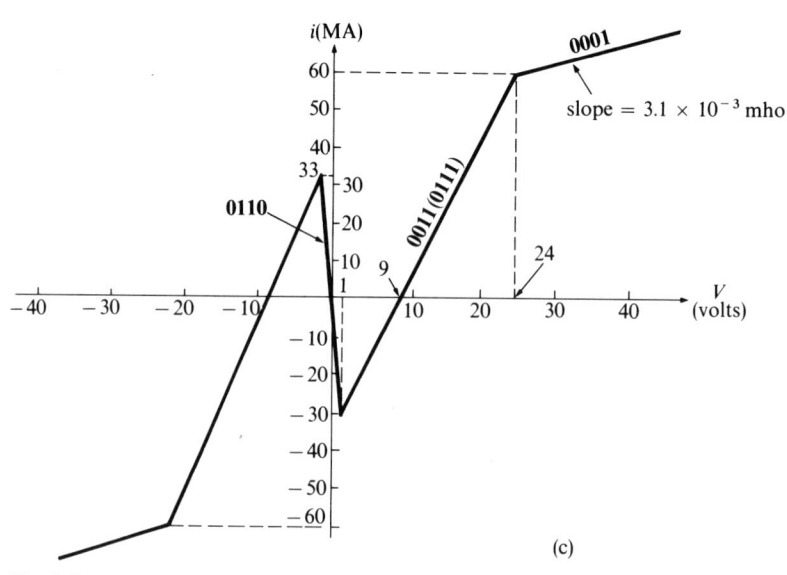

Fig 9-30 (b) Piecewise-linear model of (a); (c) Driving-point characteristic of (b) at the port shown

would like to obtain the driving-point characteristics at the terminals, *ab*, as shown in fig 9-30a, by the PWL method.

First of all, we can make use of symmetry in simplifying our work. Note that for identical transistors and resistors there is odd symmetry, i.e.,

$$i(v) = -i(-v) \qquad (9\text{-}18)$$

Using the breakpoint analysis discussed earlier, the *PWL* driving-point characteristic may be determined. The circuit model utilizing ideal diodes is shown in fig 9-30b. As a numerical example,

$$V_{CC} = 12 \text{ V}, \quad R_L = 300 \text{ }\Omega, \quad R_1 = 900 \text{ }\Omega, \quad r_b = 100 \text{ }\Omega, \quad \alpha_0 = 0.98$$

The *v-i* characteristic is shown in fig 9-30c. The details of the calculations (which are quite tedious to say the least) are left as an exercise in Problem 9-34. The tediousness of the calculations may be avoided if a computer-aided analysis program (such as MECA, for example) is used. Note that the *v-i* relationship exhibits a negative-resistance characteristic.

9-6 Negative-Resistance Characteristics

In the analysis and design of nonlinear circuits, negative-resistance characteristics play an important role. There are two general forms of *v-i* curves which are basic in negative-resistance characteristics. These are shown in fig 9-31a and b. Note that, for convenience, these are shown symmetrically with respect to the origin; of course, this is not necessary.

The curve in fig 9-31a is known as the *voltage-controlled* type (also called *N*-type curve), since i is a single-valued function of v; in other words, the current is uniquely controlled by the voltage. In general, the *v-i* curve crosses the axis at zero voltage only once. At the points of inflection $di/dv = 0$; hence, the driving-point exhibits zero differential admittance, i.e., poles of the impedance. It is also sometimes referred to as *short-circuit stable* (SCS) (see Section 1-11). Practical devices which exhibit *v-i* characteristics of this type are tunnel diodes (see fig 3-9) and flip-flop circuits (see fig 9-30c).

The curve in fig 9-31b is known as the *current-controlled* type (also called *S*-type curve), since v is a single-valued function of i, and the

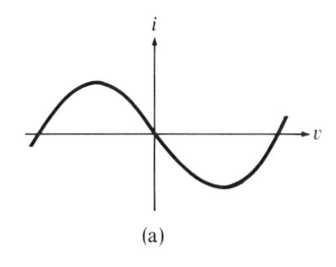

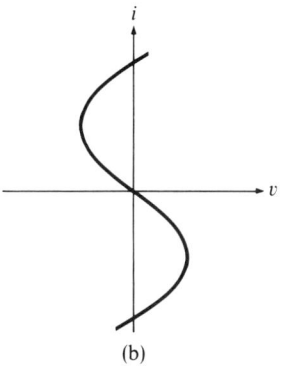

Fig 9-31 (a) Voltage-controlled type nonlinear *v-i* characteristic (*n*-type curve); (b) Current-controlled type nonlinear *v-i* characteristic (*s*-type curve)

voltage is uniquely determined by the current. At the points of inflection, $di/dv \to \infty$; hence the driving point exhibits infinite differential admittance, i.e., poles of the admittance. This type of characteristic is also sometimes referred to as *open-circuit stable* (OCS). Practical devices which exhibit the *v-i* characteristic of this type are unijunction transistors (see fig 3-49) and four-layer diodes.

Stable and Unstable Operating Points

We shall next consider the question of operating points in circuits with negative-resistance characteristics. As an example, consider the tunnel diode in the circuit shown in fig 9-32a. The load line is shown in fig 9-32b, where it is assumed that $1/R$ is *less* than the magnitude of the negative conductance of the tunnel diode. From fig 9-32b, it is seen that there are *three* intersection points with the load line. However, at any given time, there is only one unique value of the quiescent Q point. The quiescent point, as we shall presently show, is only at A or C, depending on the past history of the circuit. To see this, consider the following experiment. Let the resistance R be a variable resistor. Begin with R very large, so that the current is extremely small and the load line intersection is only at one point; hence the operating point is at A_1, as shown in fig 9-32c. Then as R is decreased, the operating point will vary successively to A_2, A_3, and A_4. Note that, for the third value, the operating point will be at A_3 in spite of the fact that there are three points of intersection with the load line. At the fifth value, the operating point will be at C_5, and so on. Now, if R is increased, the operating point moves from C_6 to C_5 to C_4 to C_2, etc. Thus, depending on the past history of the circuit, the operating point will be in either of the positive-slope portions of the curve. These points are referred to as the stable points. The above arguments may also be verified by keeping R in fig 9-32b constant and making E a variable supply. Again we notice that the operating point cannot be in the negative-resistance portion of the *v-i* characteristics of the tunnel diode as long as $1/R$ is less than the magnitude of the slope of the negative-resistance region of the device.

Access to the unstable point (i.e., in the negative-resistance region) is possible, of course, and may be arrived at by a suitable choice of E and a resistance R such that $1/R$ is *larger* than the magnitude of the slope of the negative-resistance region. However, if a change in E causes the load line to intersect with either of the positive-slope regions of the curve, the operating point will again be stabilized in either of the positive-slope regions.

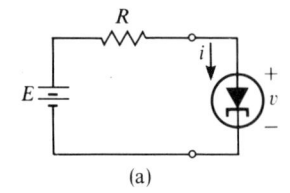

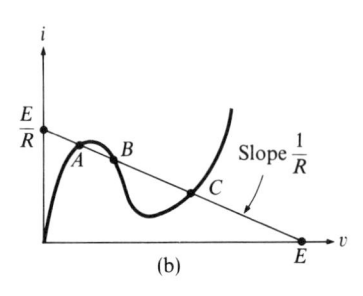

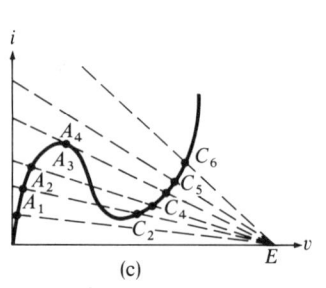

Fig 9-32 (a) Simple tunnel-diode dc circuit; (b) *v-i* characteristic of the tunnel diode with load line and possible operating points; (c) Operating point locations as R is varied from a very large value to decreasing value

In the next section as well as in the next chapter, it is shown how one of the stable points, in a negative-resistance-characteristic circuit, may be favored by using an energy-storage element such as a capacitor or an inductor, to provide *monostable* operation, or how neither of the points may be favored, to give *astable* operation, in which case the circuit will freely switch its own states without any external source. *Bistable* operation, i.e., switching from one state to the other, both stable, may be achieved by an external signal source such as a trigger pulse.

9-7 Relaxation Oscillators

The application of a current-controlled negative-resistance characteristic in a relaxation oscillator is best illustrated by a simple example. Consider the circuit shown in fig 9-33a, where the v-i characteristic is shown as a piecewise-linear approximation in fig 9-33b. It is assumed that the capacitance is initially uncharged and the switch is closed at $t = 0$. The current and voltage waveforms $i_c(t)$ and $v_c(t)$ are determined as follows:

We first note that, for $t \geq 0$ from fig 9-33a, $i_c = i$ and $v_c = v$. Also, since the capacitor voltage cannot be changed instantaneously by finite current, the operating point at $t = 0$ is at A_0. The operation is in the stable region I. The current goes exponentially to zero with time constant $R_{\rm I}C$ and initial value of A_0, while the voltage rises from zero to a target value of V_1 with a time constant $R_{\rm I}C$ ($R_{\rm I}$ is the reciprocal of the slope for region I) as shown in fig 9-33c and d.

When the current reaches the value at A_2, any further change in region I is not allowed since the $i = f(v)$ relation will not be satisfied. Hence the operation must then go along either region III or region II. However, when the capacitor voltage begins to decrease, the current cannot be negative, therefore, the operating point switches instantaneously from point A_2 to B_1. At point B_1, the voltage decreases to a target value V_2 along state II, i.e., exponentially, with a time constant $R_{\rm II}C$. The operating point moves from B_1 to B_2. From B_2, the transition occurs to A_1, similar to that described from A_2 to B_1. The capacitor voltage begins to increase again at A_1 and the process continues along the $A_1 A_2 B_1 B_2$ locus. The repetitive time intervals, as shown in fig 9-33c and 9-33d, may be readily calculated as

$$\tau_2 = R_{\rm I} C \ln \frac{i_{A1}}{i_{A2}} \qquad (9\text{-}19)$$

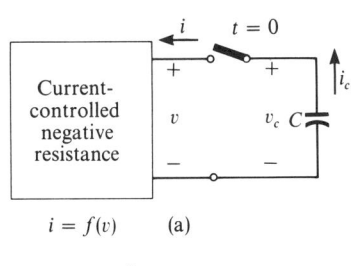

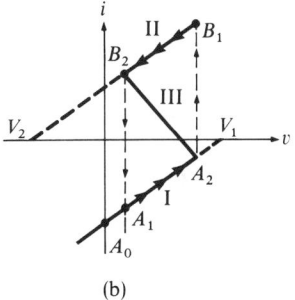

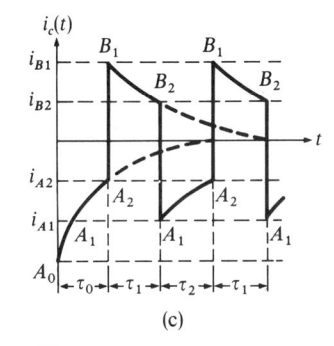

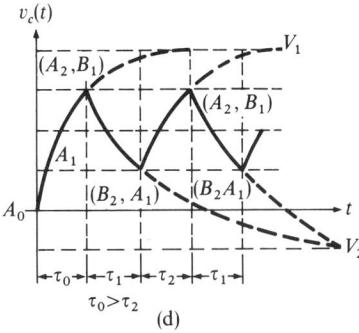

Fig 9-33 (a) Schematic of a basic astable relaxation oscillator; (b) Piecewise-linear v-i characteristic and operating point locus; (c) Waveform of current through the capacitor; (d) Waveform of voltage across the capacitor

$$\tau_1 = R_{\text{II}} C \ln \frac{i_{B1}}{i_{B2}} \qquad (9\text{-}20)$$

The above example illustrates an astable operation where the circuit has no stable states. In this case, the exponential charge or discharge of the capacitor cannot reach completion before a transition occurs.

Monostable operation may be achieved by modifying the circuit slightly such that $i_c = 0$ at either of the stable regions. In other words, the exponential charge or discharge of a capacitor C can reach completion (i.e., $i_c = 0$) before a state transition occurs. If a constant-current source is added to fig 9-33a as shown in fig 9-34a, we obtain a monostable relaxation oscillator. In this case, any change in the current away from the operating point, Q, results in a polarity change for dv_c/dt such that the operation returns to the stable equilibrium point Q.

The above simple example illustrates the basic concepts involved in analyzing relaxation oscillators. The application of a voltage-controlled negative resistance, shown in fig 9-31a, in a relaxation oscillator is similar to the one described in this section and will not be repeated. Problems 9-18 and 9-21 show the black-box viewpoint of the three regenerative types of circuits, also called multivibrators. Practical circuits, utilizing transistors, are discussed in the next chapter.

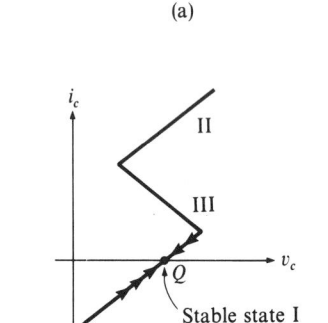

Fig 9-34 (a) Astable relaxation oscillator of fig 9-33a changed to a monostable oscillator by the addition of a current source; (b) Modified piecewise-linear v-i characteristic and stable operating point

9-8 Nonlinear Analysis Techniques

In the analysis and design of electronic circuits, the nonlinearities may be small and can be ignored, and thus we can get an approximate solution by linear analysis. If the nonlinearity is large, often piecewise-linear analysis may be used. In some cases, however, we must face the fundamental nonlinearity of the problem. For example, the system may be operated in the neighborhood of a highly nonlinear region where piecewise-linear analysis may not be a good approximation to use. By a nonlinear system, we mean a system which is described by a nonlinear differential equation. In essence, the problem is to solve a nonlinear differential equation. Unfortunately, in the solution of nonlinear differential equations, the highly developed theories of linear systems cannot be applied. The difficulty is that the superposition principle is invalid and cannot be applied in nonlinear systems. The frequencies present in the output of a nonlinear system, in general, are not the same as those of the input. In other words, harmonics and/or subharmonic frequencies can be generated in such a system. The stability of the system may depend on the amplitude of excitation and/or initial conditions in the system.

There is no single method of attack which is generally applicable to the analysis and design of all nonlinear systems. Unfortunately, nonlinear differential equations generally cannot be solved exactly or in terms of known tabulated mathematical functions. In the approximate methods of solution often several approaches must be used by the designer in order to gain adequate familiarity and information about the system. Numerical techniques, using computers, are often used in practice. In the sequel, we shall consider mainly the solution of first- and second-order systems. A vast majority of electronic oscillators fall in the second-order system category.

9-9 First-Order System, Isocline Method

Consider the general form of a first-order equation

$$\frac{dx}{dt} = f(x, t) \tag{9-21}$$

where $f(x, t)$ is a continuous and single-valued function. The *isocline* method is a graphical method of solution for equations of the form of (9-21).

In the isocline method (isocline means same slope), specific numerical values are assigned to the slope. In other words, we let $m = dx/dt$, and thus (9-21) reduces to

$$m = f(x, t) \tag{9-22}$$

From (9-22) the x-t plane is filled with sufficient isoclines (short straight lines having slopes corresponding to the values of m) by assigning different values to m. Then a solution curve starting from the given initial condition is drawn such that it has the same slopes as the isoclines. The solution curve is really made up from just short straight line segments drawn through the isoclines. Since $f(x, t)$, hence m, is single-valued, no two solution curves can cross one another. If the initial condition is specified, the solution curve crossing the initial condition (x_0, t_0) in the x-t plane is the desired solution. An example illustrates the procedure.

EXAMPLE

Let the solution of the following equation

$$\frac{dx}{dt} = \tfrac{1}{4}x^2 + t$$

be determined using the isocline method. Consider the initial condition as $[t = 0, x = 0]$.

For this equation, the isoclines are parabolas

$$m = \tfrac{1}{4}x^2 + t$$

Isocline curves for $m = 0, \pm 4$ and 8 are shown in fig 9-35. Note that, since m is the slope in the t-x plane, the scales of x and t are the same.

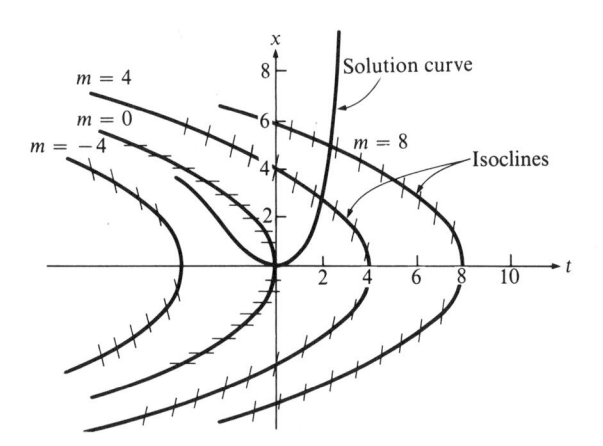

Fig 9-35 Isoclines and the solution curve for the example

The short straight lines with the slope characterizing the isoclines are aids in the approximate graphical construction of a solution curve. The desired solution curve is then drawn for the given initial condition with the same slope of the isoclines at each point of the solution curve. We start from $t = 0$, $x = 0$, with a zero slope (since at $t = 0$, $x = 0$, $m = 0$). For $t > 0$ in the neighborhood of the initial point, we may draw isoclines with slope $+2$, $+1$ (not shown), etc. The more isoclines that are drawn, the more accurate the solution curve that can be obtained.

9-10 Second-Order System, Phase-Plane Diagram

Certain second-order equations that can be reduced to a single first-order equation can be solved by the isocline method. The general form of these equations is[2]

$$\frac{d^2x}{dt^2} + f\left(\frac{dx}{dt}, x\right) = 0 \qquad (9\text{-}23)$$

Equation (9-23) may be reduced to a first-order as follows:

Let $y = \dfrac{dx}{dt}$ so that

$$\frac{d^2x}{dt^2} = \frac{dy}{dt} = \frac{dy}{dx}\frac{dx}{dt} = \left(\frac{dy}{dx}\right)(y) \qquad (9\text{-}24)$$

From (9-23) and (9-24) we obtain

$$\frac{dy}{dx} = -\frac{f(y, x)}{y} \qquad (9\text{-}25)$$

For various values of slopes dy/dx, the solution curves corresponding to (9-25), also known as the *trajectories*, can now be plotted in the y-x plane. The trajectory which is a solution to (9-25) describes how y varies vs x with time as a parameter. The y-x plane diagram is commonly referred to as the *phase-plane* diagram. Note that the phase-plane diagram involves time implicitly and one must go one integration step further to determine the desired waveform, i.e., to plot x vs t. This step will be illustrated shortly.

---

[2] Alternatively the equations may be given as:

$$\frac{dx}{dt} = P(x, y) \qquad \frac{dy}{dt} = Q(x, y)$$

From the above, $dy/dx = [P(x, y)/Q(x, y)]$. The points at which both $P(x, y) = 0$ and $Q(x, y) = 0$ are called singularities or equilibrium states.

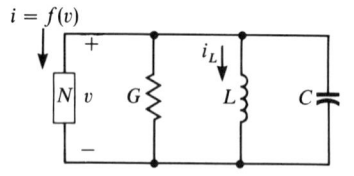

Fig 9-36 Second-order nonlinear oscillator circuit

Self-Oscillating System

Consider the circuit shown in fig 9-36. The equilibrium equation of the circuit is given by

$$C\frac{dv}{dt} + [Gv + f(v)] + \frac{1}{L}\int_0^t v\, dt = 0 \quad (9\text{-}26)$$

Multiply (9-26) by L and differentiate with respect to time once, to get

$$LC\frac{d^2v}{dt^2} + L[Gv' + f'(v)] + v = 0 \quad (9\text{-}27)$$

Equation (9-27) is of the form of (9-23) with v corresponding to x.

Before we begin the analysis of the nonlinear circuit, let us consider the special case where $f(v) = -Kv$, i.e., a linear negative resistance for N in fig 9-36. In this case, (9-27) reduces to a linear second-order equation:

$$LC\frac{d^2v}{dt^2} + L(G\text{-}K)\frac{dv}{dt} + v = 0 \quad (9\text{-}28)$$

The oscillatory solution of (9-28) may be subdivided into three cases, depending on the values of $G\text{-}K$:

(a) If $G\text{-}K > 0$, the solution is an exponentially damped sinusoid.
(b) If $G\text{-}K < 0$, the solution is an exponentially increasing sinusoid, and thus unstable.
(c) If $G\text{-}K = 0$, the solution is a sinusoid.

The phase-plane diagram (also referred to as a phase-portrait) of (9-28) is obtained by letting $y = dv/dt$ to get the equation corresponding to (9-25),

$$\frac{dy}{dv} = -\left[\frac{(G\text{-}K)y/C + v/LC}{y}\right] = m \quad (9\text{-}29)$$

Before we discuss the nonlinear case, it is instructive to consider case (c), the *harmonic oscillator*, i.e., $G\text{-}K = 0$. For this special case, (9-28) becomes

$$\frac{d^2v}{dt^2} + \omega_0^2 v = 0 \quad (9\text{-}30)$$

9-10 Second-Order System, Phase-Plane Diagram

where

$$\omega_0^2 = \frac{1}{LC}$$

The solution of (9-30) is known to be

$$v = A \cos(\omega_0 t + \phi) \quad (9\text{-}31)$$

where A and ϕ are constants determined by the initial conditions.

Consider now the graphical solution by the phase-plane method. For convenience, we make a change of variable, $\tau = \omega_0 t$. Thus, from (9-30) by letting $y = dv/d\tau$, we have

$$\frac{dv}{dt} = \left(\frac{dv}{d\tau}\right)\left(\frac{d\tau}{dt}\right) = \omega_0 y \quad (9\text{-}32\text{a})$$

and

$$\frac{d^2v}{dt^2} = \omega_0^2 \frac{d^2v}{d\tau^2} \quad (9\text{-}32\text{b})$$

Hence, using the above, (9-30) may be written as

$$\frac{d^2v}{d\tau^2} = \frac{dy}{d\tau} = \frac{dy}{dv}\left(\frac{dv}{d\tau}\right) = -v \quad (9\text{-}33\text{a})$$

or

$$\frac{dy}{dv} = -\frac{v}{y} = m \quad (9\text{-}33\text{b})$$

The isoclines corresponding to $m = 0, 1, -1, \infty$ are shown in fig 9-37. The closed contours of the isoclines in the phase-plane are solution curves. The closed trajectory of the solution curve indicates that the steady-state oscillation is periodic. For a harmonic oscillator, the closed trajectory is a circle (or an ellipse, depending on the scale of the t axis).

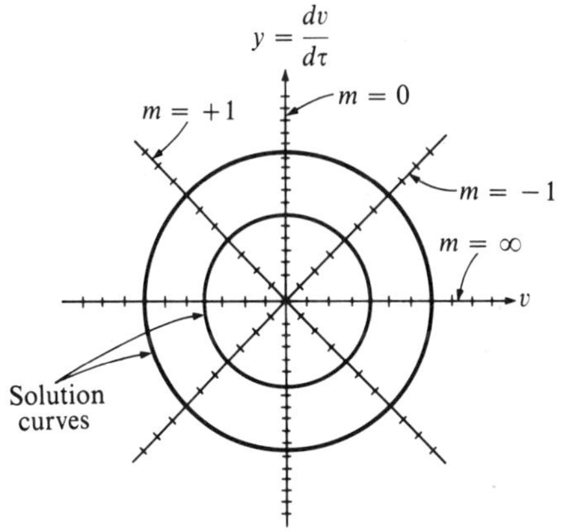

Fig 9-37 Phase-plane diagram showing isoclines and solution curves

Waveform Determination

In order to obtain the waveform, we use the trajectory in the phase-plane wherein the time dependence is implicit (steady-state waveform is determined from the closed trajectory). We first note that increasing

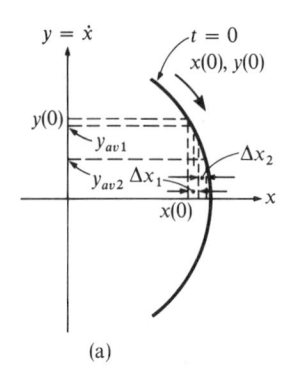

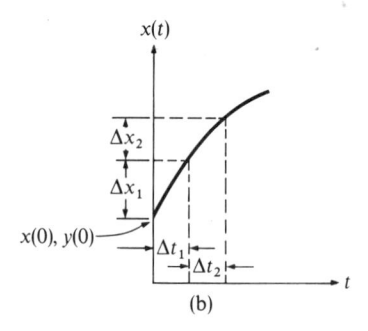

Fig 9-38 (a) Trajectory in the phase-plane for a small increment of time as time increases; (b) Waveform $x(t)$ as obtained from (a) for various increments of time

time always corresponds to clockwise rotation on the trajectory. This is readily seen from the fact that a positive velocity (dx/dt) requires displacement to become more positive with time. Thus, in the upper half of the phase-plane where $dx/dt > 0$, the point must move toward the right, while in the lower half of the phase-plane, the point must move toward the left. This also implies that the trajectory must cross the x-axis perpendicularly.

With the direction along the trajectory established, we shall now see how to determine the waveform. This can be done by numerical integration in several ways. A graphical procedure is as follows:

For small increments Δx and Δt, the average value of y, noting that $y = dx/dt$, is

$$y_{av} \approx \frac{\Delta x}{\Delta t} \tag{9-34}$$

For a small increment Δx, from the phase-plane trajectory (see fig 9-38a and the corresponding value of y_{av}), the value of Δt can be determined from (9-34). Thus starting from the initial conditions [for $t = 0$, $x(0), y(0)$], the increments in t corresponding to the increments in x can be tabulated and the waveform $x(t)$ is then plotted, as in fig 9-38b. Note that the smaller the value of Δx, the more accurate the result, and thus the choice of the value of Δx is a compromise. Usually Δx is chosen such that changes in Δy are also relatively small.

For a fixed increment of time, the following graphical method may be used.

From (9-34)

$$\Delta x = y_{av} \Delta t \tag{9-35}$$

If Δt is small enough so that the changes in x and y are relatively small, we can use the following approximation:

$$y_{av} \approx y(0) + \frac{\Delta y}{2} \tag{9-36}$$

where $y(0)$ is the value of y at the beginning of the increment. Eliminating y_{av} from (9-35) and (9-36), we obtain

$$\Delta y \approx \left(\frac{2}{\Delta t}\right) \Delta x - 2y(0) \tag{9-37}$$

Now, since Δt is fixed, (9-37) represents a straight line in the phase-plane of slope $(2/\Delta t)$ and intercept of $-2y(0)$. The value of Δy and Δx are found by measuring the distance from the original point to the intersection point. The intersection of (9-37) with the solution curve determines the point, and thus the value of Δx at the end of the interval Δt. As in any numerical calculation, the choice of the value of Δt is a compromise. Values of Δt in the order of 0.2 are typical. Of course, the smaller the value of Δt, the better the accuracy. This construction is illustrated in fig 9-39. The intersection of the straight line and the solution curve locates the point $t = t_0 + \Delta t$ and thus Δx_1 is determined. The process

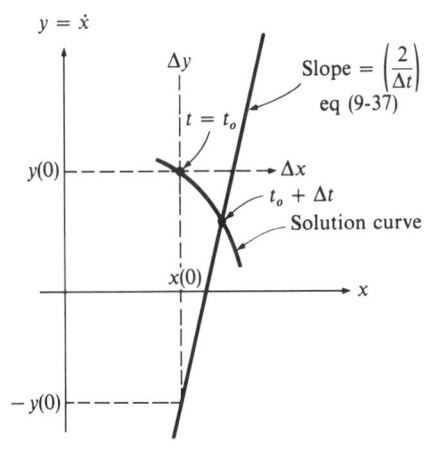

Fig 9-39 Waveform determination from solution curve in the phase-plane for a fixed small-time interval Δt

is repeated starting from $t = t_0 + \Delta t$ to get Δx_2, etc., and thus, the waveforms at the discrete points t_0, $t_0 + \Delta t$, $t_0 + 2\Delta t$, etc., are determined.

As an example, the reader can readily determine (Problem 9-23) the sinusoidal waveform from the closed trajectory shown in fig 9-37. Note that for this case, the amplitude of oscillation depends on the initial state. In nonlinear systems, as we shall soon see, this is generally not the case.

9-11 Lienard's Method

The solution by isocline method requires that much of the plane be filled with small line segments having the slope of a solution curve. This is indeed a tedious process. If only a single solution curve is required, considerable simplification results by using a graphical technique known as the Lienard construction. The method applies to systems whose equations may be written in the form

$$\frac{d^2x}{dt^2} - f\left(\frac{dx}{dt}\right) + x = 0 \qquad (9\text{-}38)$$

Fortunately, many physical systems are described by (9-38). For example, Raleigh's equation[3]

$$\frac{d^2x}{dt^2} - \varepsilon\left[\frac{dx}{dt} - \frac{1}{3}\left(\frac{dx}{dt}\right)^3\right] + x = 0 \qquad (9\text{-}39)$$

is one such example. This equation is often encountered in the study of nonlinear oscillators. By introducing $y = dx/dt$ in (9-38), we obtain the phase-plane form

$$y\frac{dy}{dx} - f(y) + x = 0 \qquad (9\text{-}40)$$

The isocline equations of (9-40) are

$$m = \frac{dy}{dx} = \frac{-x + f(y)}{y} \qquad (9\text{-}41)$$

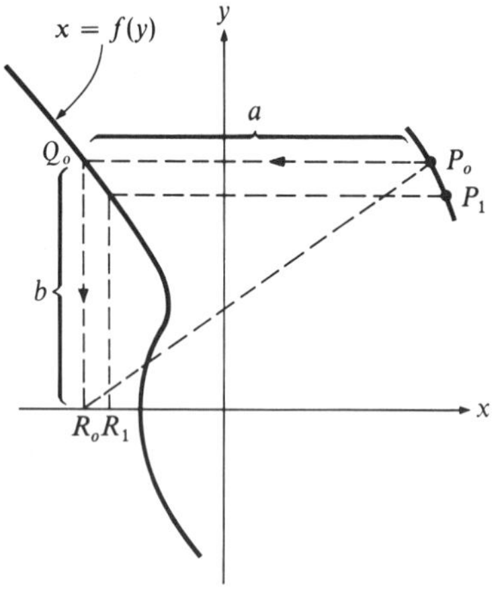

The Lienard construction of (9-41) is as follows (see fig 9-40): First plot the locus of points with $m = 0$, i.e., from (9-41)

$$x = f(y) \qquad (9\text{-}42)$$

We assume that $f(y)$ is single-valued, i.e., for every value of y there is a single value of x, as shown in fig 9-40.

Then, starting from any point $P_o(x, y)$ its slope, i.e., the slope $m = dy/dx$ of the solution trajectory through P_o, is evaluated (constructed) simply by drawing a line from P_o parallel to the x-axis until it intersects the curve $f(y)$. From the point of intersection with $f(y)$ (point Q_o), another line is drawn parallel to the y-axis. The intersection of this line with the x-axis locates the point R_o. Next, swing an arc through P_o with center at R_o. This arc is tangent to the trajectory and has the slope m. This is readily proved by geometry. From fig 9-40, we

Fig 9-40 Illustration of Lienard's method of obtaining solution curve in the phase-plane

---

[3] If $x = dy/dt$ is substituted in Van der Pols' equation (9-50), and each term integrated with respect to t (the constant of integration is set equal to zero), Raleigh's equation is obtained.

have the lengths

$$a = x - f(y) \quad b = y \quad (9\text{-}43)$$

Thus, the slope of the line P_oR_o is equal to b/a. But the arc through P_o is perpendicular to this line, hence its slope m is $-a/b$. From (9-43) we have

$$m = \frac{-x + f(y)}{y} \quad (9\text{-}44)$$

which is identical to (9-41) and thus the arc through P_o is tangent to the true trajectory.

We next extend a short segment of this arc to the neighboring point P_1 and repeat the steps identically to locate R_1. Draw a new arc through P_1, and repeat the process to construct the complete trajectory.

EXAMPLE

Consider (9-39) with $\varepsilon = 3$. The trajectory in the phase-plane and the waveform $x(t)$ is to be determined. The initial conditions are at $t = 0$, $x = 2$, $dx/dt = 2$. For $\varepsilon = 3$, $f(y)$ in (9-40) is

$$f(y) = 3(y - \tfrac{1}{3}y^3) \quad (9\text{-}45)$$

According to the Lienard construction, we obtain the trajectory and the limit cycle shown in fig 9-41a. The closed curve in the phase-plane (fig 9-41a) is called a *limit cycle*. Note that a limit cycle is a closed trajectory (hence the trajectory of a periodic solution) such that *no* trajectory sufficiently near it is also closed. Thus, the closed trajectory in fig 9-37 is not a limit cycle. In other words, a limit cycle is an *isolated* closed trajectory. Every trajectory beginning sufficiently close to a stable limit cycle approaches it as $t \to \infty$. Limit cycles are most important in the theory of nonlinear oscillation.

In fig 9-41a, the time interval before the limit cycle is associated with the transient response. The limit cycle is associated with the steady-state response. The waveforms $y(t)$ and $x(t)$ for a constant increment of time $\Delta t = 0.4$ are shown in fig 9-41b. Note that, for the initial value P_o ($x = 2$, $y = 2$), the transient response is over in approximately 2.4 seconds.

Finally, it is noted that there are other methods which extend the Lienard method to a broader class of systems. One such method is the

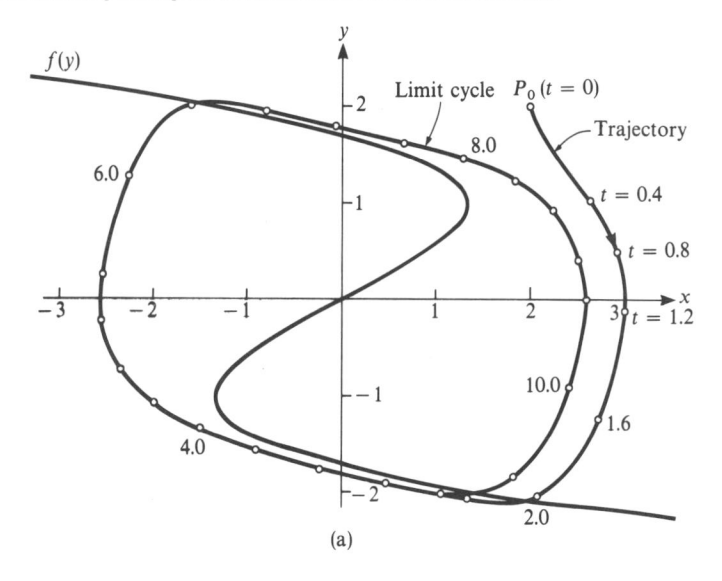

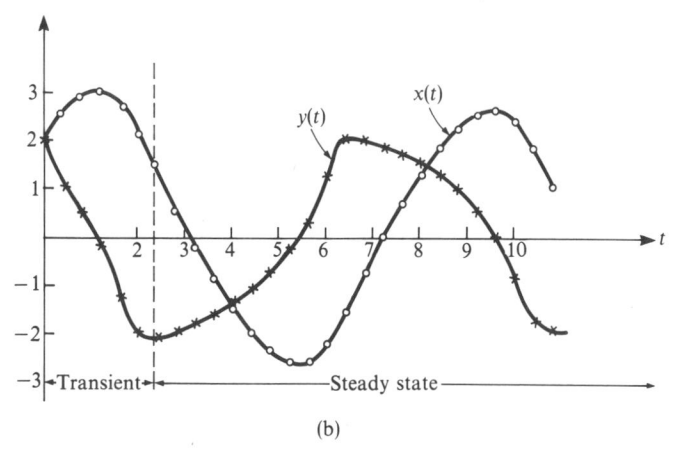

550 9/Elementary Analysis of Nonlinear Electronic Circuits

Fig 9-41 (a) Trajectory and limit cycle for numerical values of circuit in fig 9-36; (b) Waveforms of $x(t)$ and $y(t)$

δ-method which is applicable to the systems described by

$$\frac{d^2x}{dt^2} + f\left(\frac{dx}{dt}, x, t\right) = 0 \tag{9-46}$$

The δ-method is discussed in Ref 5 and will not be covered here.

9-12 Example of a Nonlinear Oscillator: Van der Pols' Equation

Consider the circuit shown in fig 9-42, with a special type of nonlinear device, namely, a voltage-controlled device with the following v-i characteristics:

$$i = f(v) = -\rho v + \gamma v^3 \qquad (9\text{-}47)$$

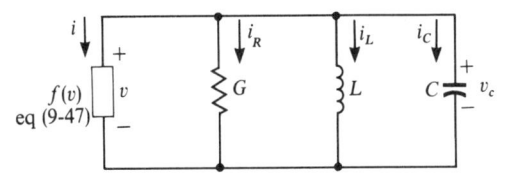

Fig 9-42 Second-order nonlinear oscillator circuit for the example

where ρ and γ are constants. We assume that $\rho > G$. We shall find the limit cycle and the solution $v(t)$ for the circuit.

The differential equation of the circuit is given by (9-27). Substitution of (9-47) into (9-27) yields

$$LC\frac{d^2v}{dt^2} - L[(\rho - G) - 3\gamma v^2]\frac{dv}{dt} + v = 0 \qquad (9\text{-}48)$$

Equation (9-48) may be rewritten as

$$\frac{d^2v}{dt^2} - \frac{(\rho - G)}{C}\left[1 - \left(\frac{3\gamma}{\rho - G}\right)v^2\right]\frac{dv}{dt} + \frac{1}{LC}v = 0 \qquad (9\text{-}49)$$

The phase-plane diagram can now be used. Before doing this, however, we shall use some normalizations, for convenience, to cast (9-49) in the following form:

$$\frac{d^2x}{dt^2} - \varepsilon(1 - x^2)\frac{dx}{dt} + x = 0 \qquad (9\text{-}50)$$

Equation (9-50) is called Van der Pols' equation and is very well known in nonlinear analysis. The nature of its solution for various values of ε has been investigated extensively in literature. By making the following normalizations:

$$t = \sqrt{LC}\tau \quad \text{and} \quad x = \sqrt{\frac{3\gamma}{\rho - G}}\,v \qquad (9\text{-}51)$$

equation (9-49) reduces to (Problem 9-22)

$$\frac{d^2x}{d\tau^2} - (\rho - G)\sqrt{\frac{L}{C}}(1 - x^2)\frac{dx}{d\tau} + x = 0 \qquad (9\text{-}52)$$

Equation (9-52) is identical to (9-50) with

$$\varepsilon = (\rho - G)\sqrt{\frac{L}{C}} \qquad (9\text{-}53)$$

For various values of ε, the phase-plane solution of the previous section can be used. Note that for $\varepsilon = 0$, the solution is a sinusoid, namely, that of a harmonic oscillator.

The phase-plane trajectory and the solution for $L = 1$ H, $C = 1$ F, $\rho = 2$ mho, $G = 1$ mho, $\gamma = \frac{1}{3}$ mho/volt^2, and the arbitrary initial condition $v(0) = 1$ V, $dv/dt(0) = 1$ and $v(0) = 2V, dv/dt(0) = -2$ will be determined. From (9-53), the values correspond to $\varepsilon = 1$.

We rewrite (9-50) as

$$\frac{dy}{dx} = \frac{\varepsilon(1-x^2)y - x}{y} \qquad (9\text{-}54)$$

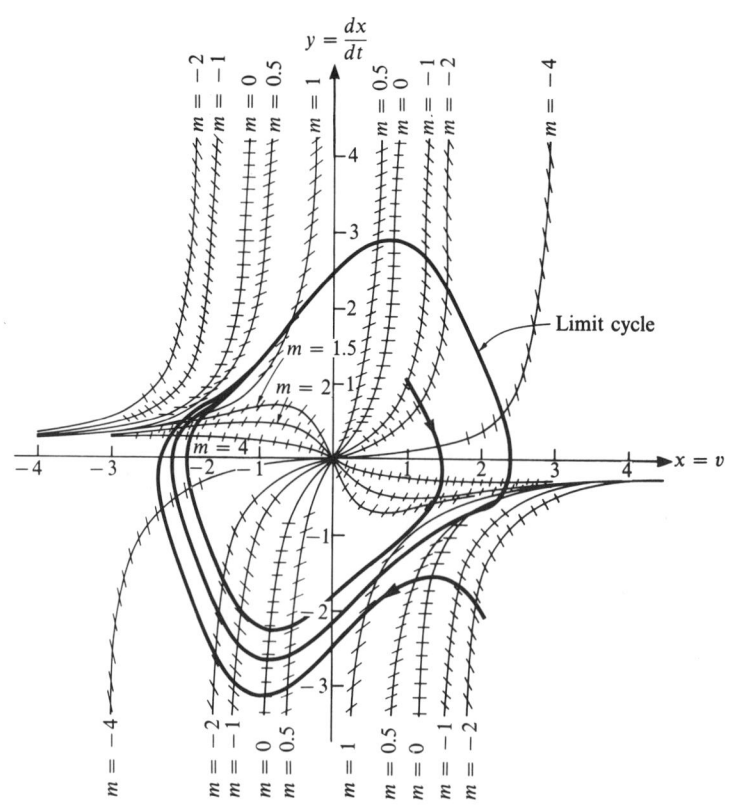

Fig 9-43 Solution curves by the isocline method (for two different initial conditions) and limit cycle for the example

9-13 Numerical Method of Solution for the nth Order Nonlinear System

The isocline curves are then

$$y = \frac{x}{\varepsilon(1 - x^2) - m} \qquad (9\text{-}55)$$

The isoclines and the limit cycle in the phase-plane trajectory are shown in fig 9-43. Note that, for the initial condition ($v \equiv x = 1, y = dx/dt = 1$), which is inside the limit cycle, the solution curve spirals outward. For a different initial condition (e.g., $x = 2, y = -2$), the solution curve spirals inward. In either case, the steady-state limit cycle is attained after the transient elapses. The trajectory and its limit cycle for this problem could be obtained more easily if (9-50) is cast in the form of (9-39), and then the Lienard construction method is used (Problem 9-30).

In the literature, there are many special nonlinear differential equations, such as Duffing's equation, Bernoulli's equation, Riccati's equation, etc., which can be solved by conventional methods (Ref 5). However, instead of considering the special cases, it is more sensible to consider a numerical method of solution via digital computers for the general nth order case, which will include all the above cases.

9-13 Numerical Method of Solution for the nth Order Nonlinear System

The numerical solution of ordinary differential equations is a vast topic and is treated in great detail in specialized texts on numerical methods. We consider only one simple approach in the following discussion. The formulation that follows is basically a truncated-matrix Taylor series.

Higher-order nonlinear differential equations are first put in the *normal form*, as follows:

$$\frac{d[x]}{dt} = [f(x, t)] \qquad (9\text{-}56)$$

where $[x]$ is a column vector, i.e., an $n \times 1$ matrix for an nth-order system. We assume that in a small interval of time $(0, \Delta t)$, $d[x]/dt$ will be approximately constant and that all the necessary conditions are specified as initial conditions at a *single point*:

$$\frac{[x(\Delta t)] - [x(0)]}{\Delta t} \approx \frac{d[x]}{dt} \quad \text{with } [x(0)] \text{ given} \quad (9\text{-}57\text{a})$$

Thus

$$[x(\Delta t)] \approx [x(0)] + \frac{d[x(0)]}{dt}\Delta t \quad (9\text{-}57\text{b})$$

$$= [x(0)] + [f(x(0))]\Delta t \quad (9\text{-}57\text{c})$$

For the next interval $(\Delta t, 2\Delta t)$, we use the same approximation, viz:

$$\frac{d[x(\Delta t)]}{dt} = [f(x(\Delta t))] \quad (9\text{-}58)$$

Hence

$$[x(2\Delta t)] \approx [x(\Delta t)] + \frac{d[x(\Delta t)]}{dt}\Delta t \quad (9\text{-}59\text{a})$$

$$= [x(\Delta t)] + [f(x(\Delta t))]\Delta t \quad (9\text{-}59\text{b})$$

Continuing the above procedure, we get

$$[x((n+1)\Delta t)] \approx [x(n\Delta t)] + [f(x(n\Delta t))]\Delta t \quad n = 0, 1, 2, \ldots, N \quad (9\text{-}60)$$

For good accuracy, the value of Δt should be chosen to be relatively very small. As in any numerical method, the choice of the value of Δt is a compromise. The relative value of Δt depends on the desired accuracy, the number of significant figures carried out in the computation, the values of the constants in the problem, and the computer time. Further discussion of this topic is beyond the scope of this text. We shall illustrate the method, however, by solving the example of Section 9-12 by this method.

EXAMPLE

Consider again the circuit of fig 9-42. We choose the state variables as the inductor currents and the capacitor voltage. The nonlinear device is assumed to have the *v-i* characteristic given in (9-47). The

9-13 Numerical Method of Solution for the nth Order Nonlinear System

equilibrium equations of the circuit are

$$\frac{di_L}{dt} = \frac{v}{L} \qquad i_L(0) = I_0 \qquad (9\text{-}61)$$

$$\frac{dv_c}{dt} = \frac{dv}{dt} = \frac{i_c}{C} \qquad v_c(0) = V_0 \qquad (9\text{-}62)$$

But

$$i_c = -i_L - i_R - i$$
$$= -i_L - Gv - (-\rho v + \gamma v^3) = -i_L + (\rho - G)v - \gamma v^3 \qquad (9\text{-}63)$$

Substitution of (9-63) into (9-62) yields

$$\frac{dv}{dt} = \frac{-i_L}{C} + \frac{(\rho - G)v - \gamma v^3}{C} \qquad (9\text{-}64)$$

Combining (9-61) and (9-64), we obtain the normal form of the equation

$$\frac{d[x]}{dt} = [f(x)] \qquad (9\text{-}65)$$

Namely

$$\frac{d}{dt}\begin{bmatrix} i_L \\ v \end{bmatrix} = \begin{bmatrix} v/L \\ (-1/C)i_L + [(\rho - G)v - \gamma v^3]/C \end{bmatrix} \qquad (9\text{-}66)$$

with the initial state $[x(0)] = \begin{bmatrix} I_0 \\ V_0 \end{bmatrix}$.

Suppose that $L = 1$ H, $C = 1$ F, $\rho = 2$ mho, $G = 1$ mho, $\gamma = 1/3$ mho/volt^2 and $V_0 = -1/3$ V, $I_0 = 1$ A. Arbitrarily, we choose $\Delta t = 0.1$. From (9-66), we have

$$[x(\Delta t)] \approx [x(0)] + [f(x_0)]\Delta t$$

$$= \begin{bmatrix} 1 \\ -1/3 \end{bmatrix} + \begin{bmatrix} -1/3 \\ -1.32 \end{bmatrix} 0.1 = \begin{bmatrix} 0.966 \\ -0.465 \end{bmatrix} \qquad (9\text{-}67)$$

9/Elementary Analysis of Nonlinear Electronic Circuits

We continue the procedure for the next value:

$$[x(2\Delta t)] \approx [x(\Delta t)] + [f(x(\Delta t))]\Delta t$$

$$= \begin{bmatrix} 0.966 \\ -0.465 \end{bmatrix} + \begin{bmatrix} -0.46 \\ -1.40 \end{bmatrix} 0.1 = \begin{bmatrix} 0.920 \\ -0.605 \end{bmatrix} \quad (9\text{-}68)$$

and so on. Since this is a second-order system, the trajectory can be

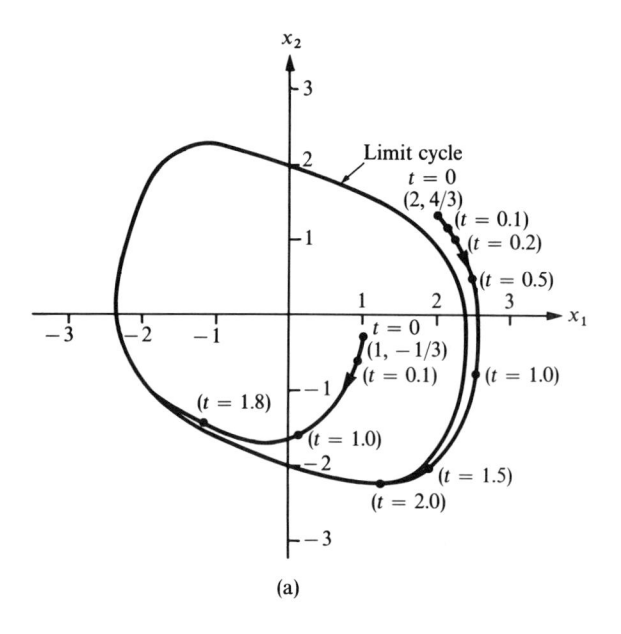

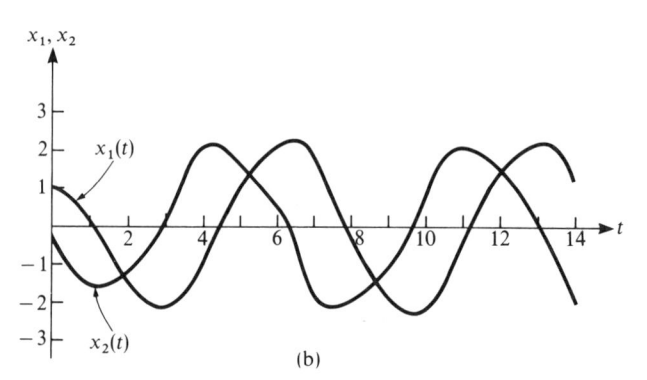

Fig 9-44 (a) Solution curves for two different initial conditions and limit cycle for the example (note that x_2 and x_1 are different quantities from those of y and x in fig 9-43); (b) Waveforms of $x_1(t)$ and $x_2(t)$ for the initial conditions $x_1 = 1.00$, $x_2 = -\frac{1}{3}$

drawn in the phase-plane, i.e., x_1 vs x_2 with t as a parameter. The results, as obtained for two initial conditions, are shown in fig 9-44a.

In this approach, however, the waveform is given directly for each increment of time and no additional computation is necessary. The waveform for $v = x_2$ and $i_L = x_1$, for the initial condition $x_1 = 1$, $x_2 = -\frac{1}{3}$, is shown in fig 9-44b.

9-14 Nonlinear Reactance

Before concluding this chapter, it may be of interest to mention that nonlinear reactances are also very useful in electronic systems. For example, the nonlinear behavior of the p-n junction capacitance (see (A-24)) can be utilized to obtain amplification. Power amplification can be achieved by using a linear time-varying capacitor. Such amplifiers are called *parametric amplifiers* (Ref 8).

As a very simple example of how power amplification can be achieved by a linear time-varying capacitor, consider the circuit of fig 9-45, where $C = C_1(1 - \delta \sin \omega_1 t)^{-1}$ with $\delta \ll 1$, and let $i(t) = I_o \times \cos \omega_0 t$.

The v-i relationship of the variable capacitance is

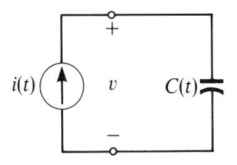

Fig 9-45 Simple circuit with a time-varying capacitor

$$\frac{d}{dt}(Cv) = i \qquad (9\text{-}69)$$

Integration of (9-69) yields

$$Cv = \int i\, dt = I_o \int \cos \omega_0 t\, dt \qquad (9\text{-}70a)$$

$$Cv = \frac{I_o}{\omega_0} \sin \omega_0 t + K \qquad (9\text{-}70b)$$

Let the initial conditions be zero, so that $K = 0$. Solving for v, we have

$$v = \frac{I_o}{\omega_0 C_1}(1 - \delta \sin \omega_1 t) \sin \omega_0 t \qquad (9\text{-}71)$$

The instantaneous power is the product of v and i, hence

$$p = vi = \frac{I_o^2}{C_1 \omega_0}(1 - \delta \sin \omega_1 t) \sin \omega_0 t \cos \omega_0 t$$

$$= \frac{I_o^2}{2C_1\omega_0}(1 - \delta \sin \omega_1 t) \sin 2\omega_0 t \qquad (9\text{-}72)$$

If we select $\omega_1 = 2\omega_0$, the average power is

$$P_{\text{av}} = \frac{1}{2\pi} \int_0^{2\pi} \frac{I_o^2}{2C_1\omega_0}(1 - \delta \sin 2\omega_0 t) \sin 2\omega_0 t\, d(\omega_0 t)$$

$$= \frac{-\delta I_o^2}{2C_1\omega_0}\left[\frac{1}{2\pi}\int_0^{2\pi} \sin^2 2\omega_0 t\, d(\omega_0 t)\right] = -\frac{\delta I_o^2}{2C_1\omega_0}\left(\frac{1}{2}\right)$$

$$= -\frac{\delta I_o^2}{4C_1\omega_0} \qquad (9\text{-}73)$$

Note that a negative average power implies a negative resistance, hence a source. Thus, time-varying capacitors can be utilized to obtain amplification of signals. Also note that we obtain amplification only if $\omega_1 = 2\omega_0$ and if the phase relationship between the source variation and the capacitor variation is correct, namely, δ positive.

Since a time-varying capacitor is difficult to achieve, one uses a nonlinear capacitor. Capacitors used for these purposes are commonly called *varactors* (abbreviated from *var*iable re*actors*). Varactors are usually modeled as time-varying capacitors in the analysis of parametric amplifiers. Problem 9-33 illustrates how this could be done.

Parametric amplifiers have extremely low noise figures and very large bandwidths in the GHz frequency range. Because of these two properties and compactness, they are frequently used as RF and microwave amplifiers in communication systems, such as in satellites. For further information on this interesting topic, the reader is referred to Ref 8 below and to Ref 8 of Chapter 8.

REFERENCES AND SUGGESTED READING

1. Zimmerman, H., and Mason, S., *Electronic Circuit Theory*, New York: Wiley, 1959.
2. Anner, G., *Elementary Nonlinear Electronic Circuits*, Englewood Cliffs: Prentice-Hall, 1967.
3. Pederson, D. O., *Electronic Circuits* (Preliminary Edition), New York: McGraw-Hill, 1965, Chapter 15.
4. Chua, L., *Introduction to Nonlinear Network Theory*, New York: McGraw-Hill, 1969.
5. Cunningham, W. J., *Introduction to Nonlinear Analysis*, New York: McGraw-Hill, 1958.

6. Minorsky, N., *Nonlinear Oscillations*, Princeton, N.J: D. Van Nostrand Co., 1962.
7. Stern, T. E., *Theory of Nonlinear Networks and Systems*, Reading, Mass: Addison-Wesley, 1965.
8. Blackwell, L. A., and Kotzebue, K. L., *Semiconductor-Diode Parametric Amplifiers*, Englewood Cliffs: Prentice-Hall, 1961.

PROBLEMS

9-1 Determine the driving-point (DP) of the circuit shown in fig P9-1. Assume ideal diodes.

9-2 Repeat Problem 9-1 for the circuit in fig P9-2. Assume $R_1 > R_2$.

9-3 Find the DP of the circuit shown in fig P9-3 (use symmetry).

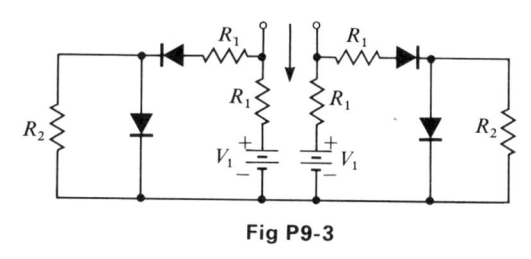

Fig P9-1

Fig P9-2

Fig P9-3

9-4 Show a realization, using diodes, resistors, and batteries for each of the driving-point characteristics shown in fig P9-4(a and b).

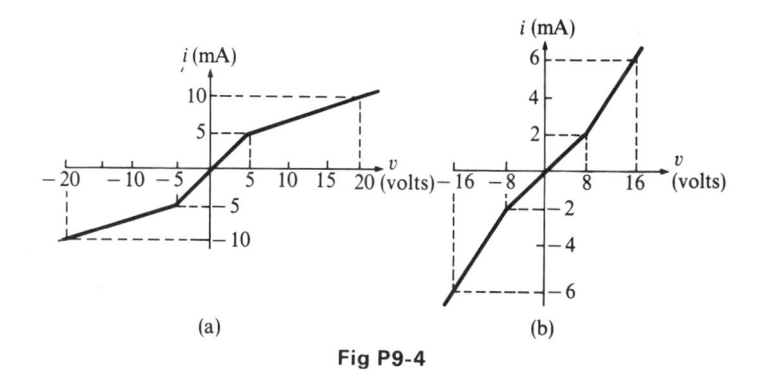

Fig P9-4

9-5 The idealized piecewise-linear v-i characteristic of a Zener diode is shown in fig P9-5.
(a) Show the piecewise-linear model.
(b) If two such diodes with breakdown voltages V_{z1} and V_{z2} are connected back-to-back, show the circuit realization using ideal diodes.

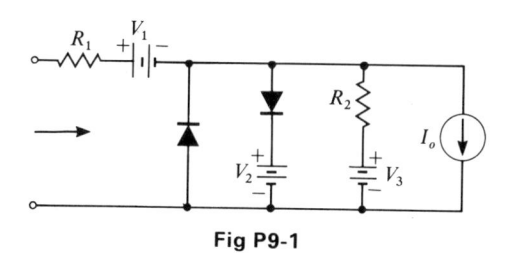

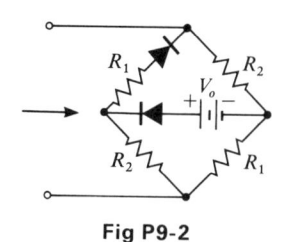

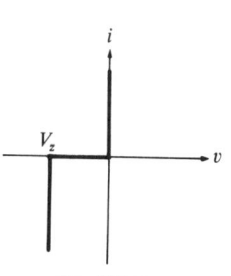

Fig P9-5

9-6 Show the driving-point plot for both parts of fig P9-6.

9-7 Find a realization for each of the DP in fig P9-7(a and b).

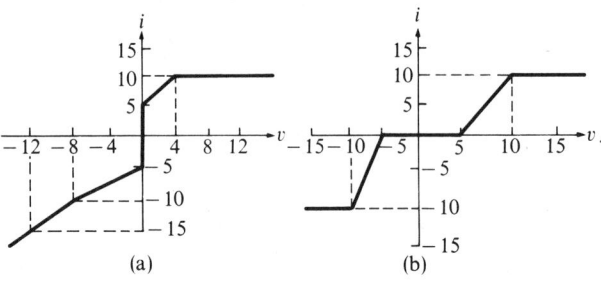

Fig P9-7

9-8 Synthesize the DP characteristics for the following cases. Negative resistors are permissible.
   (a) For fig P9-8a.
   (b) For fig P9-8b.
   (c) Do the above for fig 9-2.

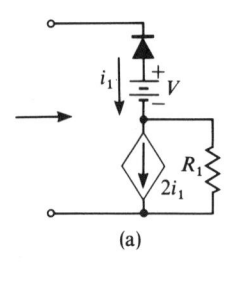

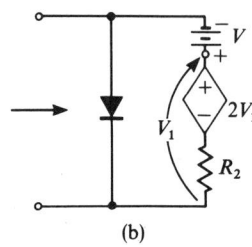

Fig P9-6

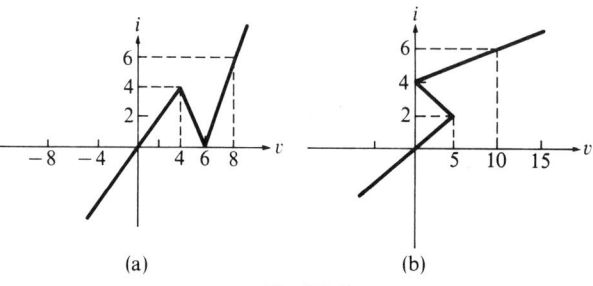

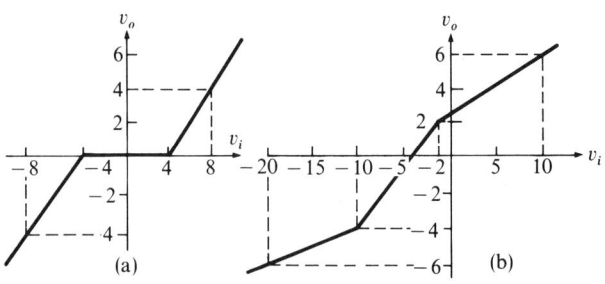

Fig P9-8

9-9 Synthesize the transfer functions in fig P9-9. Do not use operational amplifiers.

9-10 Synthesize the transfer functions in fig P9-10. Do not use operational amplifiers.

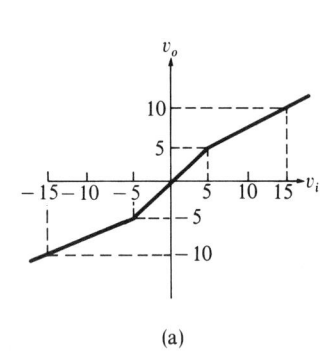

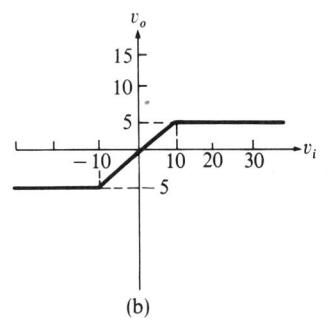

Fig P9-9

Fig P9-10

Problems 561

9-11 Synthesize the transfer functions in fig P9-11. Use positive valued resistors only.

9-12 Synthesize the transfer functions given in fig P9-12, using positive-valued resistors only.

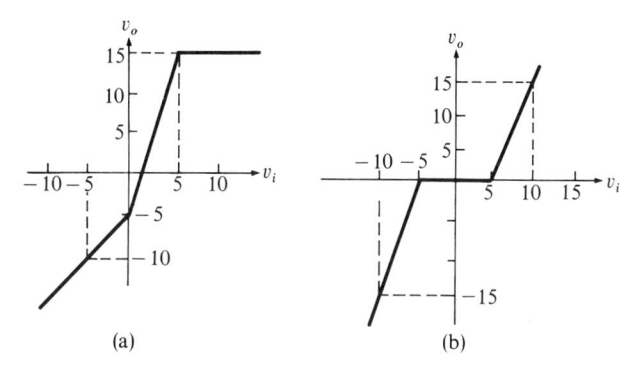

Fig P9-12

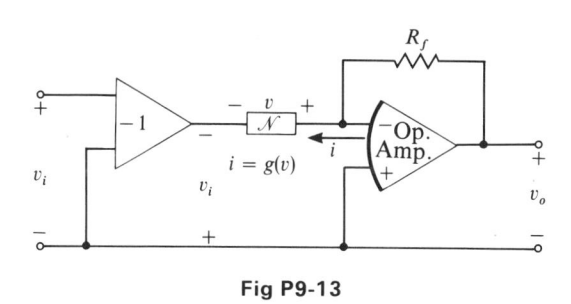

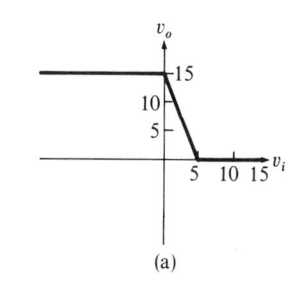

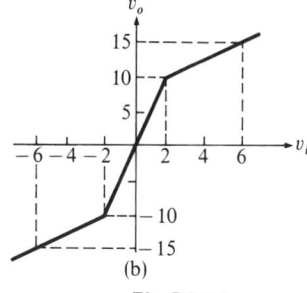

Fig P9-11

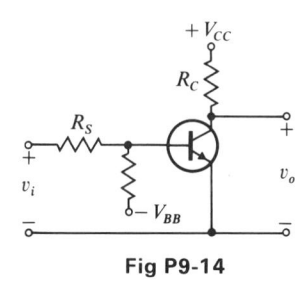

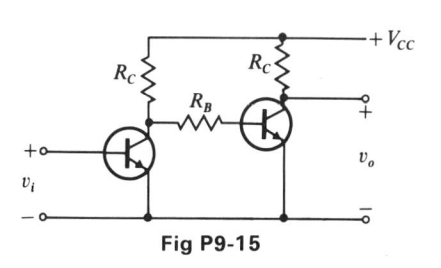

9-13 Show that the following circuit can be utilized in realizing any monotonic transfer function v_o vs v_i.

Fig P9-13

Fig P9-14

Synthesize the transfer function in fig P9-12b.

9-14 Find the transfer function, v_o vs v_i for the circuit shown in fig P9-14. Use model of fig 9-17a. (Note that the transistor here is *npn*.)

9-15 Sketch the transfer function v_o vs v_i of the circuit shown in fig P9-15. Assume that the transistors are open circuit when in cutoff and short circuit when in the saturation mode.

Fig P9-15

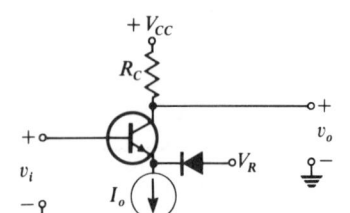

9-16 Find the transfer function v_o vs v_i for the circuit shown in fig P9-16.

9-17 The PWL transfer plot of a Schmitt trigger (fig P9-17a) for certain values of resistors and transistors is given as shown in fig P9-17b. Sketch the output waveform if the input voltage is a sinusoidal signal $v_i = V_m \sin \omega t$ where $V_m > V_2$.

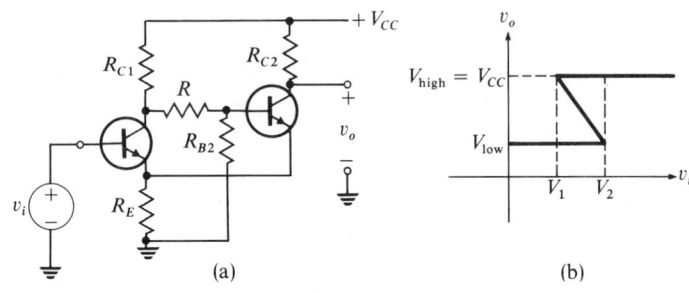

Fig P9-16

Fig P9-17

9-18 The v-i driving point characteristics of negative resistance devices and the quiescent points are shown in fig P9-18a, b, c. Show which of the characteristics are bistable, monostable, or astable state. Why?

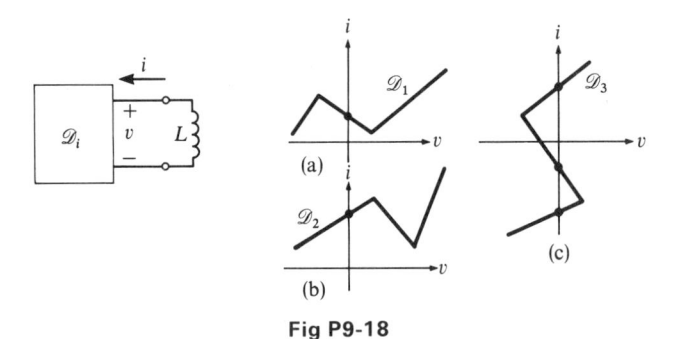

Fig P9-18

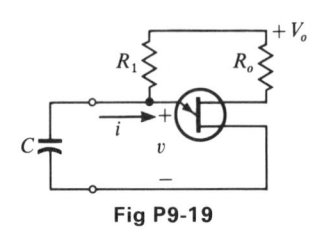

9-19 A unijunction transistor is biased so as to be used as a relaxation oscillator in fig P9-19. The V-I characteristic of the UJT is given in fig 3-49c. Sketch the steady state voltage across the capacitor vs t. Make reasonable approximations.

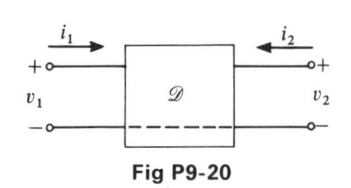

Fig P9-19

9-20 A certain fictitious three terminal device, $\mathscr{D}$, is characterized by the following nonlinear equations:

$$i_1 = 2v_1 + 0.1\, i_2^2$$
$$v_2 = 0.1\, e^{-0.1 v_1} + 0.01\, i_2^2$$

Devise a piecewise linear circuit model for the device in the range $|v_1|, |i_2| \leq 10$ and for $i_1, i_2 \geq 0$.

Fig P9-20

9-21 Repeat Problem 9-18 for the characteristics shown in fig P9-21.

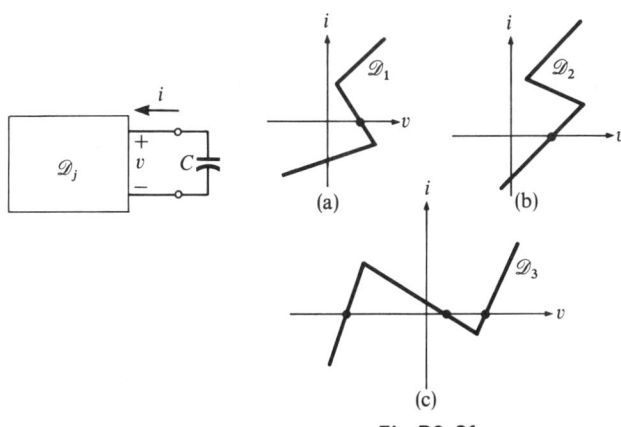

Fig P9-21

9-22 Derive equation (9-52).

9-23 Obtain the waveform $x(t)$ corresponding to the closed trajectory in fig 9-37.

9-24 Show a relaxation circuit using the tunnel diode characteristic in fig 9-2. (Use the piecewise-linear model.)

9-25 Draw the isoclines and the solution curve for the following first-order differential equations:

(a) $$\frac{dx}{dt} = x + 4 + t$$

(b) $$\frac{dx}{dt} = x^2 + 5x + t^2$$

9-26 Draw the phase-plane trajectories for the following second-order differential equations:

$$\frac{d^2x}{dt^2} - 2\left(\frac{dx}{dt}\right)^2 + 4x = 0$$

$$x(0) = 1 \quad \frac{dx}{dt}(0) = 1$$

9-27 Repeat Problem 9-26 for the following:

$$\frac{d^2x}{dt^2} + (x^2 - 2)\left(\frac{dx}{dt}\right)^2 + x\frac{dx}{dt} + x^2 = 0$$

$$x(0) = 0, \quad \frac{dx}{dt}(0) = 0$$

9-28 Obtain the waveform $x(t)$ of Problem 9-26 by using $\Delta t = 0.2$.

9-29 Repeat Problem 9-26 but use Lienard's method.

9-30 Put (9-50) in the form of (9-39) and use Lienard's method to find the trajectory and the limit cycle.

9-31 Write the differential equation of Problem 9-27 in the normal form. Sketch the trajectory by iterative method with $x_1(0) = x_2(0) = 1$ and $\Delta t = 0.2$ sec.

9-32 Solve Equation 9-48 for the values indicated in Section 9-12, but use piecewise linear approximation for (9-47).

9-33 The nonlinearity of a parametric diode capacitance is given by the diode junction capacitance (Appendix A)

$$C = \frac{C_o}{(V_d - V_a)^{1/n}} \qquad n = 2 \text{ or } 3$$

where C_o and V_d are constants.
If $V_a = -V_o + V_1 \sin \omega_1 t$ and $V_1/(V_o + V_d) \ll 1$, show that

$$C \approx \frac{C_1}{1 - \delta \sin \omega_1 t}$$

What are the expressions for C_1 and δ?

9-34 (a) Verify the driving-point characteristic shown in fig 9-30c.
  *(b) If the MECA program is available use computer-aided analysis to verify the DP characteristic of fig 9-30c.

***9-35** Utilize MECA program to solve Problem 9-16. Use the transistor model given in fig 9-19a with $R_B = 10\,\Omega$, $R_I = 25\,\Omega$, $R_{II} = 20\,\text{k}\Omega$, $\beta_0 = 100$, $V_0 = 0.6$ V, and $I_{CO} = 0$.

---

* The problems marked with an asterisk are to be assigned only if MECA program is available. If it is available it is suggested that some of these problems be assigned as small projects.

10
Regenerative, Switching, and Waveshaping Circuits

10-1 Introduction

Switching circuits are very important in electronic systems. These are used extensively in digital computers, and in communications systems. These circuits are highly nonlinear, and small signal models cannot be used for their analysis and design. The signal swings are usually large enough to drive the transistor (discrete or IC) from cutoff to saturation or vice-versa. For the analysis of switching circuits, either large signal nonlinear models for the active devices, or piecewise linear analysis may be used. Both of these approaches can be best implemented with a computer. The latter method is simple, since many of the principles developed in studying linear systems can be applied to nonlinear analysis.

The transistor can be used as a controlled switch in the following manner. Consider the circuit in fig 10-1a. When the input voltage is such that the base-emitter junction is reverse biased, the transistor is cutoff, and no collector current flows (except the leakage current I_{CO}, which we neglect here). Hence, the transistor model for the cutoff case is as shown in fig 10-1b. In this case, v_o is equal to V_{CC}. When the input voltage is such that the base-emitter junction is forward biased, and the base current is large enough (i.e., $I_b > I_{cs}/\beta_0$) so as to drive the transistor into saturation, the transistor model is as shown in fig 10-1c. (A more accurate model which includes $V_{CE(\text{sat})}$ and $V_{BE(\text{sat})}$ is shown in fig 10-1d.[1]) In this case, v_o is equal to $V_{CE(\text{sat})}$ which is very small (of the order of 200 mV). Hence, the transistor operates as a switch. For low-frequency analysis, the models of fig 10-1 are sufficiently accurate. For high-speed switching analysis, however, the capacitance associated with the transistor model must be included (see Section 10-2). For fast (very high f_T) transistors operated in moderate-speed circuits, the turn-on and turn-off time of the transistor may be ignored.

The transition time between the states of the individual device puts a limit on the speed of switching circuits. Generally, this transition time is very small compared to the time intervals designed, or those governed by external circuitry, so that we can ignore it, and obtain tremendous simplification and useful design information. First let us consider the calculation of turn-on and turn-off times in a transistor switch.

10-2 The Transistor Switch: On-Off Time-Interval Calculations

The collector current for a large input voltage pulse ($V_1 > R_s I_c/\beta_0$) is shown in fig 10-2. Note that the turn-on time consists of a delay time and a rise time, i.e.,

$$t_{on} = t_d + t_r \tag{10-1}$$

where t_d is defined to be the time interval between the application of base drive and the 10 per cent value of collector saturation current, and t_r is defined to be the 10 to 90 per cent rise time of the collector

---

[1] For Si, npn:

$$V_{CE(\text{sat})} \approx 0.2 \text{ V}, \quad V_{BE(\text{sat})} \approx 0.7 \text{ V}$$

and for Ge, npn:

$$V_{CE(\text{sat})} \approx 0.1 \text{ V}, \quad V_{BE(\text{sat})} \approx 0.3 \text{ V}$$

for pnp transistors the above voltages are all negative.

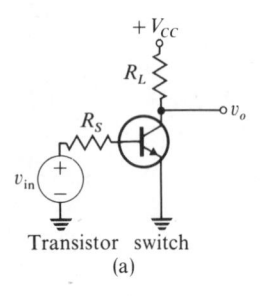

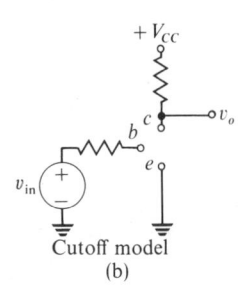

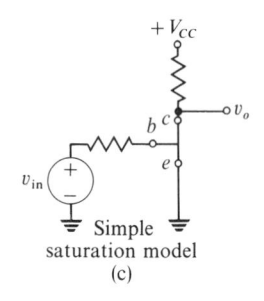

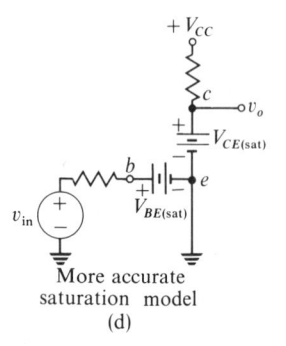

Fig 10-1 (a) Circuit using transistor as a switch; (b) Simplified model when transistor is cut off; (c) Simplified model when transistor is saturated; (d) A more accurate model for (c)

10-2 The Transistor Switch: On-Off Time-Interval Calculations

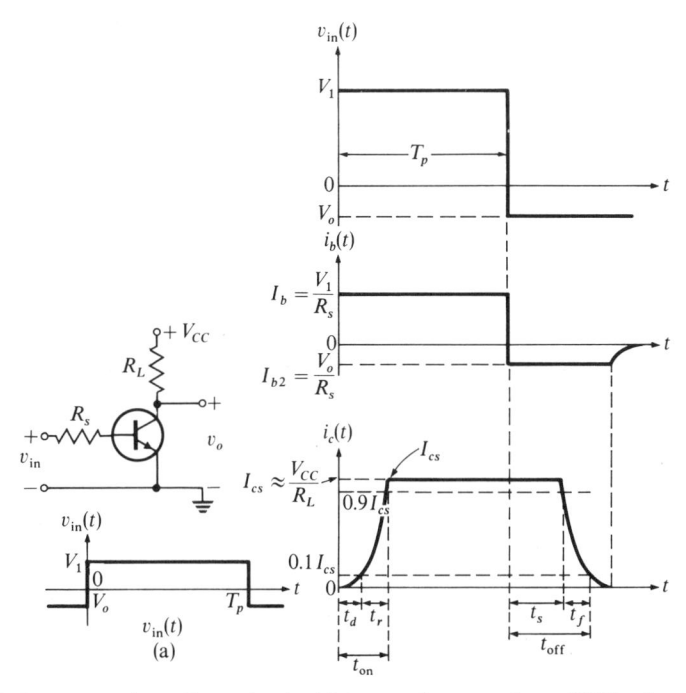

Fig 10-2 (a) Transistor clipper circuit with input voltage waveform; (b) Waveforms of input voltage, base current and collector current

current. The turn-off time consists of a storage time and a fall time, i.e.,

$$t_{off} = t_s + t_f \tag{10-2}$$

where t_s is the storage time, and t_f is defined as the 90 to 10 per cent fall time of the collector current. The storage time is the time interval between the removal of the base drive ($t = T_p$) and the point where i_c just *starts* to decrease toward zero. The various time intervals are calculated as follows.[2]

Delay Time

To calculate t_d, we must find the time required for the emitter-base junction voltage to increase from $-V_o$ to a value which will result in a

---

[2] The dynamic switching of the transistor may also be characterized by the *charge control* model. For the description of the charge control model, the reader is referred to Ref 1.

collector current equal to 0.1 I_{cs}. However, the following approximations are often made for simplicity. We find instead the time interval for v_{be} to increase from $-V_o$ to $V_{BE(sat)} \simeq 0.7$ V for Si(0.3 V for Ge) with a single time constant $\tau_d = R_s C_{ibo}$, where C_{ibo} is the value of the (reverse-biased) emitter-base junction capacitance. The value of C_{ibo} may be specified by the manufacturer. If this value is not available, a rough estimate is $C_{ibo} \simeq 2C_c$ at $-V_o$ (see footnote 3). Usually, t_d is very small and the error in calculating t_d does not have a pronounced effect on t_{on}.

The expression for the emitter-base voltage increasing from $-V_o$ to a target voltage V_1, with a single time constant τ_d, is given by

$$v_{be} = V_1 - (V_o + V_1) e^{-t/\tau_d} \tag{10-3}$$

The delay time, t_d, is found by setting $v_{be} = V_{BE(sat)} \simeq 0.7$ V (for silicon transistors). Hence

$$0.7 = V_1 - (V_o + V_1) e^{-t_d/\tau_d} \tag{10-4a}$$

or

$$t_d = \tau_d \ln\left(\frac{V_1 + V_o}{V_1 - 0.7}\right) \tag{10-4b}$$

Rise Time

In order to calculate t_r, we can use the method of analysis discussed in Chapters 2 and 4, since the transistor is in the active region of operation. In other words, we use the approximate equivalent circuit of the transistor and consider a step input as shown in fig 10-3. An important difference between the model of fig 10-3 and that of fig 4-28

$$\overline{R}_i = \overline{\beta}_0 \overline{r}_e = \frac{\overline{\beta}_0}{g_m} \qquad \overline{C}_i = \frac{D}{r_e \omega_T} = \frac{1 + R_L \overline{C}_c \omega_T}{r_e \omega_T}$$

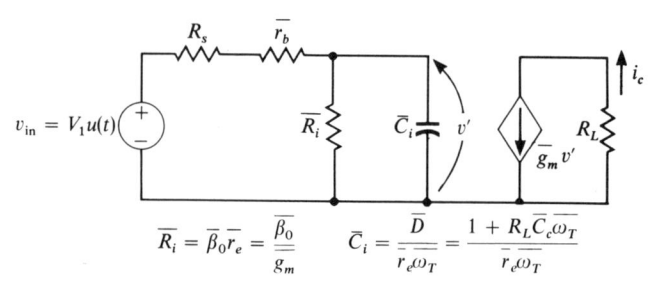

Fig 10-3 Transistor model in the active mode for t_r and t_f calculations

10-2 The Transistor Switch: On-Off Time-Interval Calculations

is that average values are used, indicated by a bar in fig 10-3. This must be done, since the operating point values change considerably, i.e., i_c varies from the initial off state (i.e., zero) to the saturation value $I_{cs} = V_{CC}/R_L$, the collector junction voltage v_{ce} varies from the initial off state at V_{CC} to $V_{CE(\text{sat})}$, which are different by an order of magnitude. For these changes, I_e, β_0, C_c, r_e, and ω_T all change. At a first approximation, we may assume r_b, β_0, and ω_T to be constants. We linearize r_e and C_c. One such approximation, which has been found satisfactory, is that r_e is determined from the midpoint value of I_c, namely, the average of the initial value $I_{ci} = 0$ and the final saturation value $I_{cs} = V_{CC}/R_L$,

$$\overline{g_m} \simeq \frac{q}{kT} \frac{(I_{ci} + I_{cs})}{2} \simeq \frac{q}{kT}\left(\frac{V_{CC}}{2R_L}\right) = \frac{1}{\overline{r_e}} \tag{10-5}$$

and the linearized collector junction capacitance is determined from[3]

$$\overline{C_c} \approx 2C_c \text{ evaluated (or measured) at } V_{CC} \tag{10-6}$$

For a current drive, i.e., $R_s \gg r_b + \overline{R_i}$, we have $i_b = (V_1/R_s)u(t)$. The step response for the circuit of fig 10-3 is a one-pole response. For an overdriven condition ($i_c > V_{CC}/R_L$), the step response is shown in fig 10-4. The expression for the collector current (dashed curve), in terms of the base drive current for the single-pole circuit of fig 10-3, can be written as

$$i_c(t) = I_\infty + (I_o - I_\infty)e^{-t/\tau_1} \tag{10-7}$$

where

$$\tau_1 = \frac{1}{p_1} = \frac{\overline{D}}{\omega_\beta} = \frac{\beta_0(1 + R_L\overline{C_c}\omega_T)}{\omega_T} \tag{10-8a}$$

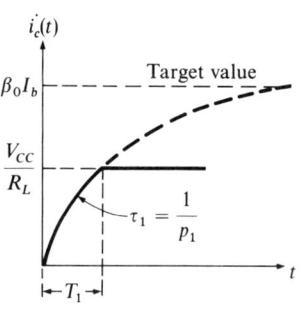

Fig 10-4 Collector-current waveform for rise-time calculation

---

[3] This value is arrived at for the following reason: For a reverse-biased junction, neglecting the electrostatic potential from (A-24), $C_c = K/v^{1/m}$, where $m = 2$ for a step junction, and $m = 3$ for a graded junction.

$$\overline{C_c} = \frac{\Delta Q}{\Delta V} = \frac{Q(V_1) - Q(V_2)}{V_1 - V_2}$$

$$\Delta Q = \int_{V_1}^{V_2} C_c\, dV = K\int_{V_1}^{V_2} V^{-1/m}\, dV = 2K(V_1^{1/2} - V_2^{1/2}) \quad \text{for} \quad m = 2$$

$$\overline{C_c} = 2K\frac{(V_1^{1/2} - V_2^{1/2})}{V_1 - V_2} \approx 2KV_{CC}^{-1/2} \quad \text{for} \quad V_1 = V_{CC}, \quad V_2 \approx 0$$

$$\approx 2C_c$$

and

$$i_c(0) = I_o = 0 \quad \text{and} \quad i_c(\infty) = I_\infty = \beta_0 I_b = \beta_0 \frac{V_1}{R_s} \quad \text{(10-8b)}$$

The actual collector current will, of course, not reach the overdrive value, but will saturate at $I_{cs} = V_{CC}/R_L$. The time required for the collector current to reach from the zero value to the saturated value is calculated from (10-7), and (10-8), that is,

$$I_{cs} = \beta_0 I_b - \beta_0 I_b e^{-T_1/\tau_1} \quad \text{(10-9)}$$

From (10-8a) and (10-9), we obtain

$$T_1 = \tau_1 \ln \frac{\beta_0 I_b}{\beta_0 I_b - I_{cs}} = \frac{\overline{D}}{\omega_\beta} \ln \left(\frac{K}{K-1} \right) \quad \text{(10-10)}$$

where

$$K \equiv \frac{\beta_0 I_b}{I_{cs}} > 1 \quad \text{(10-11)}$$

K is usually called the turn-on overdrive. Usually (10-10) is used as the value for t_r for an overdriven stage, i.e., $t_r \approx T_1$. However, it is a simple matter to find the 10 to 90 per cent rise time t_r. From (10-9)

$$0.1 I_{cs} = \beta_0 I_b - \beta_0 I_b e^{-t_1/\tau_1} \quad \text{(10-12a)}$$

$$0.9 I_{cs} = \beta_0 I_b - \beta_0 I_b e^{-t_2/\tau_1} \quad \text{(10-12b)}$$

From (10-12) and (10-11) we obtain

$$t_r = t_2 - t_1 = \tau_1 \ln \left(\frac{1 - 0.1/K}{1 - 0.9/K} \right) \quad \text{(10-13)}$$

For $K \gg 1$, and using the approximation

$$\ln(1 + X) = X - \frac{X^2}{2} + \cdots \qquad X \ll 1 \quad \text{(10-14)}$$

(10-13) simplifies to

$$t_r \approx \tau_1 \frac{0.8}{K} \quad \text{(10-15)}$$

10-2 The Transistor Switch: On-Off Time-Interval Calculations

Note that, for K large, t_r is much smaller than τ_1. However, a large value of K increases the storage time t_s, as we shall soon see.

Storage Time

The transistor in saturation has a saturation charge of excess minority carriers stored in the base. The transistor, therefore, cannot respond to the trailing edge of the driving pulse until this excess charge has been removed. The lapse of time due to the storage delay and the time constant associated with it is discussed in detail elsewhere (Ref 1). For our purposes, we do not need to develop the exact relationships. We use a dominant storage time constant, τ_s, for calculating the storage time. This time constant, in terms of the normal active and inverse active region parameters, is given by (Ref 1)

$$\tau_s \approx \frac{\omega_\alpha + \omega_I}{\omega_\alpha \omega_I (1 - \alpha_0 \alpha_I)} \tag{10-16a}$$

$$\approx \frac{\beta_I}{\omega_I} \quad \text{for} \quad \omega_\alpha \gg \omega_I, \quad \beta_0, \beta_I \gg 1 \tag{10-16b}$$

where ω_I and α_I are the alpha cutoff frequency and dc alpha for the inverse active region, and α_0 and ω_α are for the normal active region.

Now i_c falls off from a value of I_{c1} to I_{c2}, with a single time constant τ_s. This can be written as

$$i_c(t) = I_{c2} + (I_{c1} - I_{c2}) e^{-t/\tau_s} \tag{10-17}$$

The storage time is the particular value of t, namely t_s, when $i_c = I_{cs}$. Hence, from (10-17)

$$t_s = \tau_s \ln \frac{I_{c1} - I_{c2}}{I_{cs} - I_{c2}} \tag{10-18a}$$

$$= \tau_s \ln \frac{I_{b1} - I_{b2}}{I_{bs} - I_{b2}} \tag{10-18b}$$

where $I_{bs} = I_{cs}/\beta_0 = V_{CC}/\beta_0 R_L$. Equation (10-18) can also be written as

$$t_s = \tau_s \ln \left[\frac{1 + K/M}{1 + 1/M} \right] \tag{10-19}$$

where K is defined by (10-11), and $M \equiv -I_{b2}/(I_{cs}/\beta_0)$ is the turn-off over-drive factor. Note that the value of $I_b = V_1/R_s$ is used for I_{b1}. If both $K, M \gg 1$, then (10-19) simplifies to

$$t_s \approx \tau_s \ln\left(1 + \frac{K}{M}\right) \tag{10-20}$$

Note that a large value of K increases t_s, and that this is the price paid for getting a smaller t_r. Usually τ_s is given by the manufacturer, either directly, or it is found from (10-16) in terms of the parameters of the inverse and normal active region. The value of t_s can also be determined by measuring the storage time for known base-drive conditions from (10-18). Note that (10-18b) assumes positive current I_{b2}. Further, for $I_b < I_{bs}$, the transistor will come out of saturation.

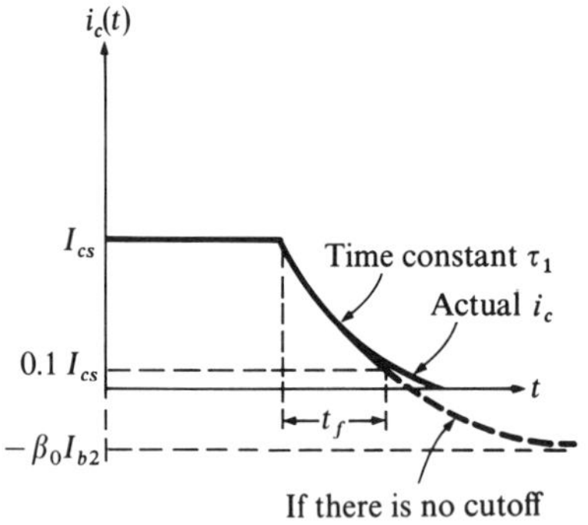

Fall Time

The fall-time calculation is straightforward, since the transistor operates in the active region. It parallels the rise-time calculation. Hence, we write the equation for fall time, for i_c to fall from $0.9 I_{cs}$ to $-\beta_0 I_{b2}$ with a time constant τ_1, as shown in fig 10-5. Note that the negative sign of I_{b2} is included in this case. This can be written as

$$i_c(t) + \beta_0 I_{b2} = (0.9\, I_{cs} + \beta_0 I_{b2})\, e^{-t/\tau_1} \tag{10-21}$$

The fall time is the particular value of t, namely t_f, when $i_c = 0.1\, I_{cs}$. Hence

$$t_f = \tau_1 \ln \frac{0.9 I_{cs} + \beta_0 I_{b2}}{0.1 I_{cs} + \beta_0 I_{b2}} \tag{10-22}$$

Equation (10-22) can also be written as

$$t_f = \tau_1 \ln\left[\frac{1 + 0.9/M}{1 + 0.1/M}\right] \tag{10-23}$$

For large turn-off overdrive, $M \gg 1$, (10-23) simplifies to

$$t_f \approx \tau_1 \frac{0.8}{M} \tag{10-24}$$

Fig 10-5 Collector-current waveform for fall-time calculation

10-2 The Transistor Switch: On-Off Time-Interval Calculations

From (10-24), it is seen that the larger the turn-off drive, the smaller the fall time for the transistor. Compare (10-24) with the rise-time result, (10-15).

Finally, note that if the stage is driven by a voltage source (i.e., R_s is small) rather than a current source, the expression for τ_1, namely (10-8), will be different, and indeed smaller than the current-drive stage. In order to illustrate the derived relations, we shall consider the following example.

EXAMPLE

Consider the pulse response of the circuit shown in fig 10-2a. The circuit parameters are as follows: $V_{CC} = 10$ V, $R_L = 1$ kΩ, $R_s = 10$kΩ. The transistor parameters, which may be assumed constant in the operating range, are: $\beta_0 = 100$, $f_T = 200$ MHz, $r_b = 50$ Ω. We are also given that $C_c = 5$ pF at 10 V, and that in the inverse active region $\omega_I = 10^8$ rad/sec and $\beta_I = 10$. The input step is a pulse voltage of amplitude $+5$ V and a pulse duration of $T_p = 10$ μsec. The turn-on and turn-off times of the transistor switch are to be determined.

We estimate $C_{ibo} \simeq 2C_c \simeq 10$ pF. Hence, $\tau_d = R_s C_{ibo} = 10^{-7}$ sec. From (10-4b) we calculate

$$t_d = 10^{-7} \ln\left(\frac{5+0}{4.3}\right) = 1.5 \times 10^{-8} \text{ sec}$$

We next determine the average values for the active region. I_c changes from an off value of 0 mA to an on value of 10 mA. Hence

$$\overline{I_e} \approx \frac{I_{ci} + I_{cs}}{2} = \frac{0 + 10}{2} = 5 \text{ mA}$$

Hence $\overline{r_e} = \frac{25}{5} = 5$ Ω

$$\overline{C_c} = 2C_c = 2(5 \text{ pF}) = 10 \text{ pF}$$

From (10-8),

$$\tau_1 = \frac{\overline{D}}{\omega_\beta} = \frac{\beta_0(1 + R_L \overline{C_c} \omega_T)}{\omega_T} = 1.08 \text{ μsec}$$

From (10-11) and (10-13), with $K = [100(0.5)/10] = 5$, we have

$$t_r = 1.08 \times 10^{-6} \ln\left[\frac{4.9}{4.1}\right] = 0.19 \text{ μsec}$$

Thus

$$t_{on} = t_d + t_r \simeq 0.21\,\mu\text{sec}$$

From (10-16), since $\beta_0 \gg 1$, $\omega_\alpha > \omega_T \gg \omega_I$, the storage time constant is

$$\tau_s \approx \frac{\beta_I}{\omega_I} = \frac{10}{10^8} = 0.1\,\mu\text{sec}$$

Hence, from (10-18b) with $I_{b2} = 0$ (i.e., $M = 0$)

$$t_s = 0.1 \times 10^{-6} \ln\left[\frac{0.5}{0.1}\right] = 0.16\,\mu\text{sec}$$

Finally, the fall time for zero overdrive,[4] i.e., $I_{b2} = 0$ is found from (10-22)

$$t_f = 1.08 \times 10^{-6} \ln\left(\tfrac{9}{1}\right) = 2.38\,\mu\text{sec}$$

Thus

$$t_{off} = t_s + t_f = 2.54\,\mu\text{sec}$$

The turn-on and turn-off times of the transistor can be reduced by external circuitry in a number of ways. Some of these are discussed in the following. It is to be noted that fast-switching transistors, with t_{off} and t_{on} under 20 nanoseconds, are commonly used in digital computer systems.

10-3 Circuitry to Improve the Switching Time of a Transistor

Some of the simple techniques to reduce t_{on} and t_{off} of a transistor are briefly discussed in the following.

---

[4] Note that if a turn-off overdrive factor of 8, i.e., $M = 8$ is used from (10-24), $t_f \simeq 1.08 \times 10^{-6}(0.8/8) = 0.108\,\mu\text{sec}$, and from (10-20), $t_s \approx 0.1 \times 10^{-6} \ln(1 + 5/8) = 0.05\,\mu\text{sec}$.

10-3 Circuitry to Improve the Switching Time of a Transistor

Speed-up Capacitors

The turn-on time can be improved by reducing t_r. From (10-13), it is seen that t_r can, in turn, be reduced by broadbanding the stage, i.e., reducing τ_1.

From the bandwidth calculations in Chapter 4, it is obvious that when R_s is small, the bandwidth is larger and the stage current gain lower than if R_s is large. In switching circuits it is desirable, however, to drive the stage by a current source, since it is not possible to control the base current accurately when driving it with a voltage source. We shall show in this section how to improve τ_1, hence t_r and t_f, by using a capacitor across the large value of R_s. It should be mentioned at this point that it is also possible to decrease t_r by overdriving the stage. Overdrive increases the storage time, however, as is evident in (10-20), and thus it is undesirable to increase the on overdrive in order to decrease the turn on time. We shall show later how to eliminate the storage time by using nonsaturating circuits.

Let us consider the potential advantage of using a capacitor across R_s, as shown in fig 10-6. The capacitor C_s is usually referred to as the "speed-up" capacitor because, as we shall show, it improves the turn-on and turn-off times. Consider the linear equivalent circuit of the base input of the transistor in the active region shown in fig 10-7a. When $r_b \ll R_s, \bar{R}_i$, the circuit simplifies to that shown in fig 10-7b (similar to a compensated RC attenuator). The transfer function V_{be}/V_s for this circuit is

$$\frac{V_{be}}{V_s} = \frac{Z_2}{Z_1 + Z_2} = \frac{\bar{R}_i/(1 + s\bar{R}_i\bar{C}_i)}{\bar{R}_i/(1 + s\bar{R}_i\bar{C}_i) + R_s/(1 + sR_sC_s)} \quad (10\text{-}25)$$

For perfect compensation, $R_s C_s = \bar{R}_i \bar{C}_i$, or in terms of the transistor parameters, the proper value of C_s is given approximately by

$$C_s \approx \frac{\bar{R}_i \bar{C}_i}{R_s} = \frac{(\beta_0 \bar{r}_e)(\bar{D}/\bar{r}_e \omega_T)}{R_s} = \frac{\bar{D}}{R_s \omega_\beta} \quad (10\text{-}26)$$

Note that, in practice, perfect compensation is not possible. This is mainly due to the actual complicated model of the base-input circuit, and also the fact that r_b is ignored in the simplified analysis in the above. If the value of r_b is included in the analysis, and $C_s \approx \bar{R}_i \bar{C}_i / R_s$, then

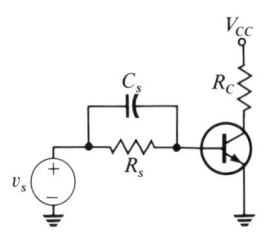

$$\frac{V_{be}}{V_s} = \frac{\bar{R}_i}{s(r_b \bar{R}_i \bar{C}_i) + r_b + \bar{R}_i + R_s} \quad (10\text{-}27)$$

Fig 10-6 (a) Transistor clipper circuit with speed-up capacitor to improve rise and fall times

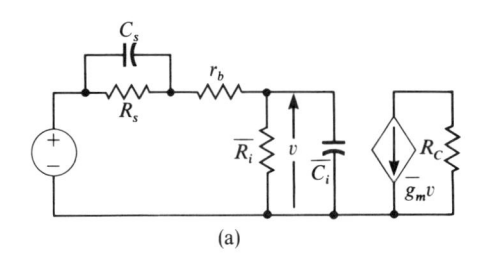

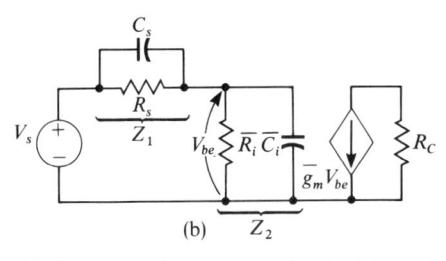

Fig 10-7 (a) Transistor clipper circuit with speed-up capacitor; (b) Input-voltage waveform and collector-current waveform for various values of $R_s C_s$ time constants

The time constant from (10-27) is

$$\tau_{1s} = \frac{r_b \overline{R}_i \overline{C}_i}{r_b + \overline{R}_i + R_s} = \frac{\overline{D}}{\omega_\beta}\left[\frac{1}{1 + (\overline{R}_i + R_s)/r_b}\right] \quad (10\text{-}28)$$

A comparison of (10-28) and (10-8a) shows a decrease in time constant by a factor of $\{1/[1 + (\overline{R}_i + R_s)/r_b]\}$. Since $\overline{R}_i$ and R_s are usually much larger than r_b, it is a significant improvement. From (10-13) and (10-23), it is seen that the speed-up capacitor improves both t_r and t_f significantly, since τ_1 is reduced to τ_{1s}.

In practice, the value of R_s is chosen such that the base current drives the transistor at the edge of saturation, and C_s is then chosen approximately (slightly higher) by (10-26), and then adjusted experimentally. An approximate value of C_s with proper adjustments gives significant improvement in rise and fall times.

Circuits Which Avoid Saturation

The storage time of a transistor switch can be reduced by using transistors with a lower τ_s, and by preventing the transistor from being saturated when it turns on. One such circuit is shown in fig 10-8, where a clamping diode is used to prevent saturation. The diode becomes forward-biased as the transistor enters the saturation region, and thus the excess base current is shunted through the diode into the collector circuit. The diode also stores the charge, but the recovery time of the diode is much faster than τ_s of the transistor.

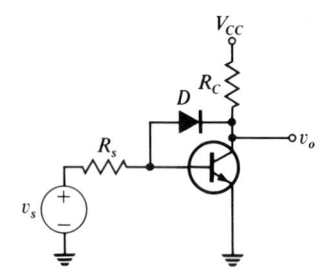

Fig 10-8 Clipper circuit with clamping diode used to reduce the storage time

Current-Mode Circuit (Emitter-Coupled Clipper)

One of the most widely used circuits, in applications where fast switching time is important, is the so-called nonsaturating current-mode switch or the emitter-coupled clipper, as shown in fig 10-9. The circuit is seen to be an overdriven differential amplifier. The emitter-coupled circuit is commonly found in fast integrated-circuit logic, as discussed in the next chapter. The operation of the circuit may be briefly discussed as follows. Consider v_i to be considerably more negative than the reference voltage V_r, so that Q_1 is off and Q_2 is in the active region. As the voltage v_i is increased, at some value Q_1 enters the active region, and both transistors are on. As v_i is further increased in the positive direction, Q_2 eventually cuts off, since the base is at a fixed reference potential. Thus, the output voltage v_o is at a

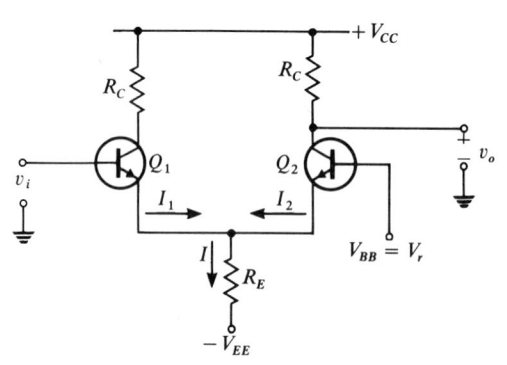

Fig 10-9 Emitter-coupled clipper, also called a current-mode circuit, which prevents the transistors from going into saturation

value ($V_{CC} - IR_c$) for large negative values of v_i, and v_o is equal to V_{CC} for large positive values of v_i. The circuit is called a current-mode switch, because the current is switched from Q_2 to Q_1 with the total emitter current constant, namely $I = (V_r + V_{EE} - V_{BE})/R_E$. Thus, when Q_1 is off (Q_2 on), $I_1 = 0$, $I_2 = I$, and when Q_2 is off (Q_1 on), $I_2 = 0$, $I_1 = I$. Note that neither of the transistors is in saturation over the entire range of operation, and thus storage time is avoided in the switch.

10-4 Classification of Regenerative Switching Circuits

Switching circuits are usually classified into *regenerative* and *nonregenerative* types. In a nonregenerative circuit the value, or level, of the output variable at any time depends only on the value, or level, of the externally applied input drive. Examples of this type are the linear circuits discussed in Chapters 1 through 8, some of the nonlinear circuits discussed in Chapter 9, the transistor switch, the individual logic gate circuits discussed in the next chapter, and the sweep circuits discussed later in this chapter. In each of the above cases, the circuit stays at a single state when no input signal is applied. These circuits possess no "memory."

Regenerative switching circuits are those where the signal at some point in the circuit is determined, in part, by the value, or level, of output variables, as well as by the input variables. In some cases, no input signal is required to cause an output. These circuits exhibit memory, i.e., the output variable depends on the past history, or state, of the circuit. Regenerative circuits possess a common feature: that the driving-point characteristic plot measured across at least one pair of terminals is a *multivalued* function of either the driving voltage or the driving current. In particular, many regenerative circuits possess negative resistance characteristics. This property has already been encountered in Section 9-7.

Regenerative circuits include multivibrators, which are classified in three categories. The classification depends on whether the circuit, in the absence of an external drive, can remain indefinitely in one or both of its stable states. If the circuit can remain in either of the two stable states in the absence of an input signal, it is called a *bistable multivibrator*. A bistable multivibrator is usually referred to as a *flip-flop* and is used as a memory element in digital systems. If the regenerative circuit can remain in only one state in the absence of an input signal

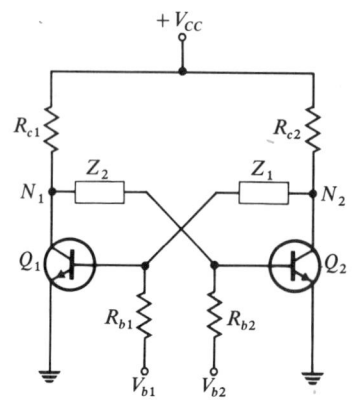

Fig 10-10 General circuit of the basic multivibrator

it is called a *monostable multivibrator*, or a single-shot multivibrator. If it cannot remain permanently in either of the two states, it is called an *astable multivibrator*, or a free-running multivibrator. The latter two forms of multivibrators (MV) are usually used in waveform generation and shaping.

A general form of the basic multivibrator circuit for the bipolar transistor is shown in fig 10-10. Because of the greater speed of bipolar transistors, as compared with field-effect transistors, the former are almost always used in fast switching circuits. The various types or subclasses of multivibrators are obtained by the appropriate choices of Z_1, Z_2, and the polarities of V_{b1} and V_{b2}. The similarities and differences between the three types of multivibrators will subsequently become obvious. Typical input-output waveforms for the three types corresponding to fig 10-10 are shown in fig 10-11. For the bistable case, the output changes from one stable state to another upon the application of an input signal. The circuit will remain in this state until another pulse is applied that will bring it back to its original state. The mono-

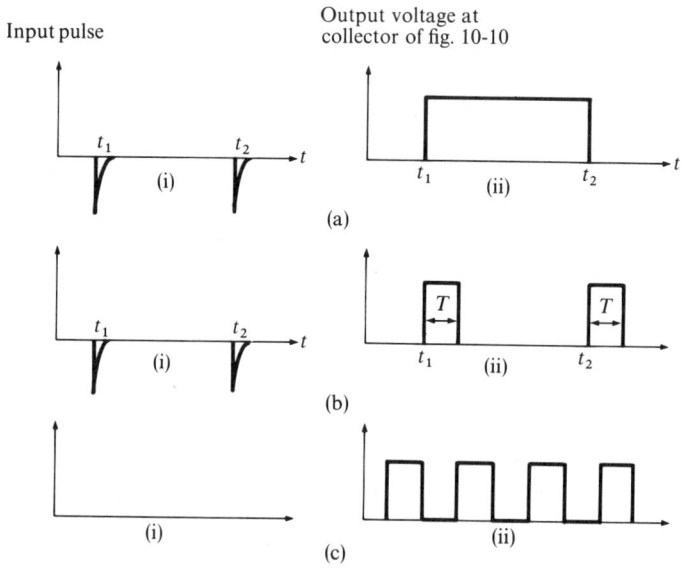

Fig 10-11 (a) Bistable multivibrator waveforms (fig 10-10: if V_{b1} and V_{b2} are both negative, Z_1 and Z_2 are both resistors); (i) Signals applied at different nodes of fig 10-10 (nodes N_1 and N_2); (ii) Output voltage at collector of the on transistor (transistor switching time is ignored); (b) Monostable multivibrator waveforms (fig 10-10: if V_{b1} negative, $Z_1 \to R$ and V_{b2} positive and $Z_2 \to C$); (i) Signal applied at the same node (N_2) of fig 10-10; (ii) Output voltage at node N_2 of fig 10-10 (transistor switching time is ignored); (c) Astable multivibrator waveforms (fig 10-10: if V_{b1} and V_{b2} both positive, Z_1 and Z_2 are both capacitors); (i) No signal applied; (ii) Output voltage at either collector (transistor switching time is ignored)

stable circuit remains in its stable state under no-signal input conditions. When the input signal is applied, it changes its state to the quasi-stable state, but after a time T, governed by the values of the circuit components, it switches back to the original stable state. The astable circuit has no stable state, and the output continually changes between the two quasi-stable states in the absence of any input signal. The length of time spent in each state is determined by certain circuit-element values, as will be discussed shortly.

It should be mentioned that there are many other possible multivibrator circuits which we cannot discuss here. Some of these were already discussed in Section 9-7 and Problems 9-18 and 9-21. We only emphasize the basic transistor circuits and the operation of these circuits, in order to provide general knowledge for the reader. The design of multivibrators is not discussed, since these are available in integrated circuits for a variety of applications and specifications. An example, however, is worked out in the next section, which gives an idea how a design might be performed since the calculations involved are similar.

10-5 Bistable Multivibrators

The basic circuit for a bipolar transistor bistable multivibrator, usually called a flip-flop, is shown in fig 10-12. This circuit is also sometimes referred to as an Eccles-Jordan circuit. Other names such as binary circuit or "toggle" are also occasionally used. This kind of circuit has two stable states, which are ensured by its configuration. To show the conditions for bistability, we redraw the circuit of fig 10-12 in the form of a feedback configuration in fig 10-13, to expose the positive feedback mechanism of the multivibrator. For simplicity, we have assumed identical stages. From fig 10-13, the open loop is a cascade of two identical inverters. The static transfer characteristics of fig 10-13 can now be readily obtained. For v_i negative, Q_1 is cutoff, hence $v_m \approx V_{CC}$, and under this condition Q_2 is saturated so that $v_o = V_{CE(sat)}$. We now increase v_i to become positive, so that Q_1 just enters the active normal region. At the edge of the active region the collector current is very small, so that v_m is still large enough to keep Q_2 in saturation. For v_i larger than a certain value, the collector current of Q_1 becomes large enough so that Q_2 will come out of saturation, and v_o will increase for an increase of v_i. For v_i large enough so that Q_1 saturates, Q_2 reaches cutoff, hence $v_o \approx V_{CC}$. The static transfer

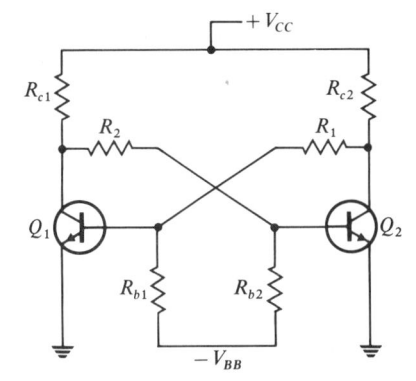

Fig 10-12 Bistable multivibrator; also called a flip-flop

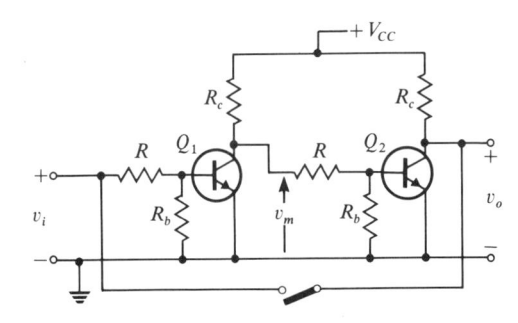

Fig 10-13 Bistable multivibrator redrawn to exhibit the positive feedback mechanism

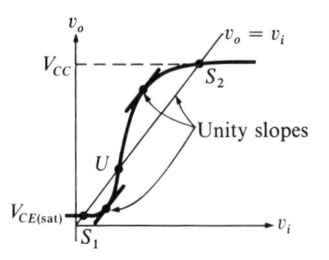

Fig 10-14 Transfer-function characteristic of fig 10-13

characteristic of fig 10-13 is shown in fig 10-14. When the feedback loop is closed (assuming negligible loading effect), we have the constraint that $v_o = v_i$. Hence the possible operating points are S_1, U, and S_2, as shown by the intersection of the $v_o = v_i$ line with the static-transfer characteristic in fig 10-14. But at point U, the loop transmission is larger than unity, and since this is a positive feedback system (v_o is in phase with v_i), the circuit will be unstable. The instability forces one of the amplifiers into saturation to reduce the gain, hence the operating point will move to either one of the two stable points S_1 or S_2. This positive feedback with loop gain larger than unity is an alternative way of stating that a negative resistance v–i characteristic exists. This was seen to be the case in Section 9-6, and Problem 10-3 clearly illustrates the instability of the system.

The regenerative mechanism of the flip-flop may also be demonstrated in the following manner. Consider the circuit of fig 10-13, in which we assume linearized models for both transistors during the regeneration time. This is an idealization which is not strictly true, since the nonlinearities, delays, etc., must be included in a more complete analysis. For our purposes, however, the gross behavior can be exhibited by the following simplified linear analysis. For convenience, we assume identical transistors, $R_1 = R_2 = R$, $R_{c1} = R_{c2} = R_c$, $R_{b1} = R_{b2} = R_b$, and assume[5] that $R \ll R_c, R_b$; $R_c \| R_b \gg \bar{R}_i$, and r_b is negligible. The approximate incremental equivalent circuit for the

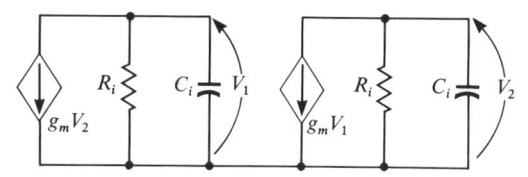

Fig 10-15 Simplified equivalent circuit for fig 10-12

transistors in the active region, with the forementioned assumptions, is shown in fig 10-15. The equilibrium equations of the circuit in the transformed s-plane are

$$V_1 = -g_m V_2 / (G_i + sC_i) \quad (10\text{-}29a)$$

$$V_2 = -g_m V_1 / (G_i + sC_i) \quad (10\text{-}29b)$$

where

$$G_i \simeq \frac{1}{\beta_0 r_e} \quad \text{and} \quad C_i \simeq \frac{1 + R_i C_c \omega_T}{r_e \omega_T} = \frac{D}{r_e \omega_T}$$

---

[5] These assumptions are not necessary, and are made only to simplify the expressions. If r_b, $\bar{R}_i$, and R are included in the analysis, the value of R_{eq} will be somewhat different.

10-5 Bistable Multivibrators

Equation (10-29) may be rewritten as

$$(G_i + sC_i)V_1 + g_m V_2 = 0 \qquad (10\text{-}30a)$$

$$g_m V_1 + (G_i + sC_i)V_2 = 0 \qquad (10\text{-}30b)$$

The natural frequencies of the circuit, from (10-30), are determined by setting the determinant equal to zero, namely

$$\begin{vmatrix} G_i + sC_i & g_m \\ g_m & G_i + sC_i \end{vmatrix} = 0 \qquad (10\text{-}31)$$

From (10-31) we obtain

$$(G_i + sC_i)^2 - g_m^2 = 0 \qquad (10\text{-}32a)$$

or

$$G_i + sC_i = \pm g_m \qquad (10\text{-}32b)$$

The two poles of the network function are

$$s_1 = \frac{g_m - G_i}{C_i} = \frac{\omega_\beta}{D}(\beta_0 - 1) \approx \frac{\omega_T}{D} \qquad (10\text{-}33a)$$

$$s_2 = -\frac{g_m + G_i}{C_i} = -\frac{\omega_\beta}{D}(\beta_0 + 1) \approx -\frac{\omega_T}{D} \qquad (10\text{-}33b)$$

Hence, the currents and the voltages of the network, in the absence of excitation, are of the form

$$i = K_1 e^{t\omega_T/D} + K_2 e^{-t\omega_T/D} \qquad (10\text{-}34)$$

The first term in (10-34) dominates the response, because it increases with time while the second term decays. Hence the switching time depends on ω_T, and thus the regeneration time can be improved by using transistors with high f_T. From (10-34) it is seen that both transistors *cannot* be on simultaneously, since instability will drive one transistor into the cutoff state. Thus one transistor will be off, while the other is on.

Since the transition time from the on to the off state, and vice-versa, depends on the t_{on} and t_{off} of each transistor, circuitry which improves these switching times will also improve the transition time. This was discussed in Section 10-3. A nonsaturating flip-flop, also called an emitter-coupled MV, is shown in fig 10-16. The function of C_E is to smoothen the voltage fluctuation during transients. It can be left out without serious effects. A discrete flip-flop using speed-up capacitors is shown in fig 10-17. The on-off switching times of the transistors put a limit on the maximum frequency at which the bistable MV can be triggered without failing to respond, i.e., to ensure a change of state. The maximum pulse repetition rate is roughly $f_T/10$.

Finally, note that a flip-flop is actually an interconnection of two inverters, as shown in fig 10-18. In digital circuits, the inverter plays an important role of negation in a logic circuit. These topics are discussed in the next chapter.

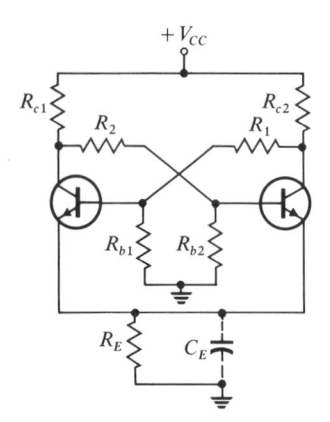

Fig 10-16 Emitter-coupled bistable multivibrator

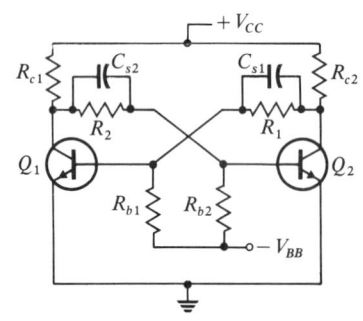

Fig 10-17 Bistable multivibrator using speed-up capacitors for discrete circuits

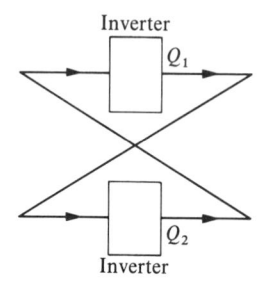

Fig 10-18 (a) Schematic of flip-flop in terms of two inverter interconnections

EXAMPLE

A bistable MV is obtained by interconnecting two simple inverters, as shown in fig 10-17. The circuit parameters are

$$V_{CC} = -V_{BB} = 10 \text{ V}, \quad R_{c1} = R_{c2} = 1.2 \text{ k}\Omega$$
$$R_{b1} = R_{b2} = 39 \text{ k}\Omega \quad R_1 = R_2 = 10 \text{ k}\Omega$$

The transistors are Si, with β_0 larger than 30. Assume that $V_{CE(\text{sat})} \approx 0.2$ V and $V_{BE(\text{sat})} \simeq 0.7$ V, and I_{co} neglible. Compute the stable-state currents and voltages.

In order to analyze the circuit, we assume that Q_1 is off and Q_2 is on and saturated. We then verify the validity of this assumption. Recall that the transistor is saturated whenever $I_c < \beta_0 I_b$.

For Q_1 off and Q_2 saturated, the MV circuit is as shown in fig 10-19. We determine I_{b2} and I_{c2} independently, and then check the saturation condition. Now, for Q_2 on and saturated, from fig 10-19 we have

$$I_{b2} = I_1 - I_2 = \frac{10 - 0.7}{(10 + 1.2) \text{ k}\Omega} - \frac{10 + 0.7}{39 \text{ k}\Omega}$$
$$= (0.83 - 0.26) \text{ mA} = 0.57 \text{ mA}$$

$$I_{c2} = I_3 - I_4 = \frac{10 - 0.3}{1.2 \text{ k}\Omega} - \frac{10 + 0.3}{49 \text{ k}\Omega}$$
$$= (8.1 - 0.21) \text{ mA} = 7.89 \text{ mA}$$

since $I_{c2} < (\beta_0)_{\min} I_{b2}$ (i.e., $7.89 < 17.1$), transistor Q_2 is indeed in

10-5 Bistable Multivibrators

saturation. For Q_2 in saturation, the base voltage of Q_1 is

$$V_{B1} = \frac{10 \text{ k}\Omega}{(10 + 39) \text{ k}\Omega}(-10) + \left(\frac{39}{39 + 10}\right)0.3 = -1.80 \text{ V}$$

This back-bias voltage indeed has Q_1 cutoff, since only about 0 V in the emitter junction of a Si transistor can cutoff the transistor. The collector potential of Q_1, $V_{c1} = 10 - (1.2)0.83 = 9$ V.

Thus the stable-state currents and voltages are as follows:

$$Q_1: I_{b1} = I_{c1} = 0, \quad V_{C1} = 9.0 \text{ V}, \quad V_{B1} = -1.80 \text{ V}$$
$$Q_2: I_{b2} = 0.57 \text{ mA}, \quad I_{c2} = 7.89 \text{ mA}, \quad V_{C2} = 0.3 \text{ V}$$
$$V_{B2} = 0.7 \text{ V}$$

The flip-flop has two stable states. One state is as outlined above. The other state is Q_2 off and Q_1 saturated, and thus the currents and voltages of Q_2 and Q_1 are interchanged. The output swing of the flip-flop is $9 - 0.3 = 8.7$ V.

Triggering Considerations

The state of a flip-flop can be changed by appropriately applying an external trigger pulse to the circuit. Usually the trigger is applied through a diode circuit to turn the on transistor off, or vice versa. In practice, it is advantageous to apply a turn-off trigger to the on transistor rather than a turn-on trigger to the off transistor. This advantage comes from the fact that, in the former case, a smaller trigger is required, and also there is a smaller delay associated with the on transistor. A commonly used circuit for this purpose is shown in fig 10-20. The diodes in this case are referred to as "steering diodes," which direct the trigger pulse to the on transistor. The off transistor collector potential causes the diode to be reverse-biased, and thus the signal does not pass through to the off transistor. Note that positive triggers are ignored on account of the diodes, and that the (npn) transistor flip-flop responds only to negative input pulses. The requirements of the trigger pulse are the following: the pulse must have sufficient amplitude and duration to remove the charge from the on transistor and bring it well enough into the active region so that regeneration takes place. (These are discussed in some detail in Ref 3.) The source impedance of the trigger must be sufficiently high so as not to reduce the loop gain to less than unity. If these conditions are met, the change of flip-flop state can be ensured by the application of a trigger pulse.

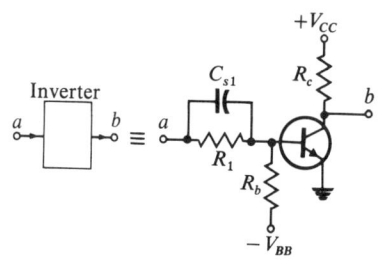

Fig 10-18 (b) Circuit realization of a discrete inverter

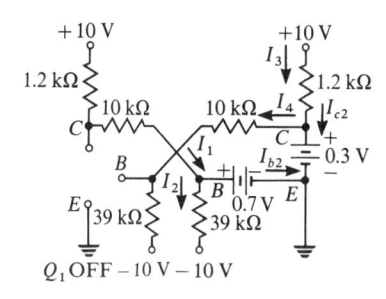

Fig 10-19 Bistable multivibrator of the example

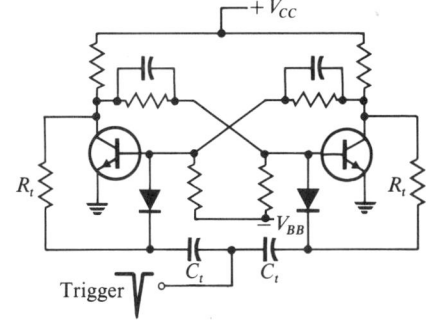

Fig 10-20 Discrete bistable multivibrator with triggering circuitry

10-6 Monostable Multivibrators

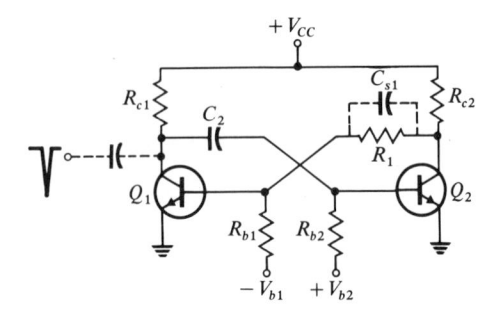

Fig 10-21 Monostable multivibrator circuit

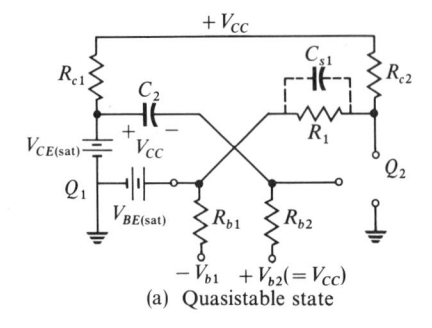

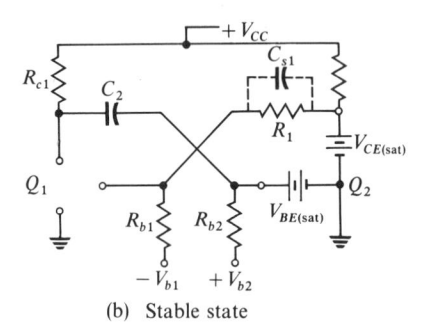

Fig 10-22 (a) Model for the quasi-stable state of the monostable multivibrator; (b) Stable-state model of the monostable multivibrator

(a) Quasistable state

(b) Stable state

The basic circuit of a monostable multivibrator, for the bipolar transistor, is shown in fig 10-21. In most cases R_{b2} is returned to the supply voltage, V_{CC}, in order to avoid requiring another supply potential. Note that the difference between the monostable and bistable multivibrators is that, in the monostable case, the coupling between the collector of Q_1 and the base of Q_2 is capacitive rather than resistive. As we shall see shortly, this capacitor reduces the number of stable states of the circuit from two to one.

From the circuit of fig 10-21, it is readily seen that there is only one stable state, namely Q_2 on and Q_1 off, since under steady-state conditions, Q_2 is biased on through R_{b2}, while Q_1 is off due to a reverse-biased emitter-base junction.

If a negative trigger of sufficient amplitude and duration is applied to the collector of Q_1 (or a positive trigger to the base of Q_1), transistor Q_1 will be turned on, and regeneration will cause a transition in the state of Q_2, and will turn Q_2 off. This state is referred to as the quasi-stable, or metastable, state and cannot last forever. The quasi-stable state for the monostable circuit at an instant after regeneration is shown in fig 10-22a. Note that it is assumed that the stable state has been in effect for a long time before the application of the trigger, so that C_2 has assumed a charge approximately equal to the supply voltage.

The collector and base waveforms of the circuit, which include the saturation voltages, resulting from a single trigger pulse, are shown in fig 10-23. In obtaining these waveforms, the switching times of the transistors, and the effects of r_b and I_{co} are neglected, which are usually valid approximations. The reader is encouraged to verify each waveform shown. The monostable period, which is defined as the time during which Q_2 is in the off state (or, alternatively stated, the time duration in the quasi-stable state), can be calculated as follows. From fig 10-22 and fig 10-23, the charging curve of the base voltage of Q_2 rises from the initial voltage $V_i = -(V_{CC} - V_{BE(sat)} - V_{CE(sat)})$ to a target voltage $+ V_{b2}$, with a time constant $R_{b2}C_2$. The curve is governed by the equation

$$v_{b2}(t) = V_{b2} + (V_i - V_{b2})e^{-t/R_{b2}C_2} \qquad (10\text{-}35)$$

or

$$v_{b2} = V_{b2} - (V_{CC} - V_{BE(sat)} - V_{CE(sat)} + V_{b2})e^{-t/R_{b2}C_2} \qquad (10\text{-}36)$$

10-6 Monostable Multivibrators

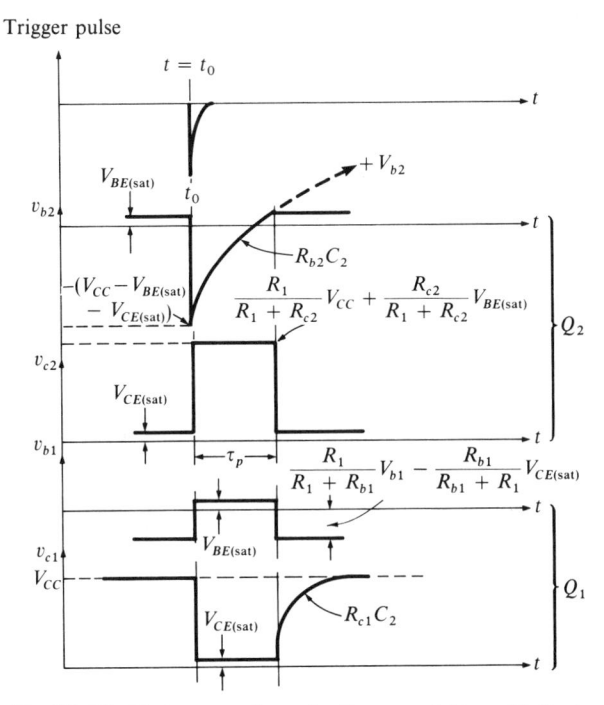

Fig 10-23 Various waveforms for the monostable multivibrator

The time at which v_{b2} equals $V_{BE(\text{sat})}$ (i.e., Q_2 turns on) is then obtained from (10-36), namely

$$V_{BE(\text{sat})} = V_{b2} - (V_{CC} - V_{BE(\text{sat})} - V_{CE(\text{sat})} + V_{b2})e^{-\tau_p/R_{b2}C_2} \quad (10\text{-}37)$$

or

$$\tau_p = R_{b2}C_2 \ln \frac{V_{CC} + V_{b2} - V_{BE(\text{sat})} - V_{CE(\text{sat})}}{V_{b2} - V_{BE(\text{sat})}} \quad (10\text{-}38)$$

Thus, the pulse width may be adjusted by varying R_{b2}, C_2, or V_{b2}. If R_{b2} is returned to the supply voltage, i.e., $V_{b2} = V_{CC}$, and the saturation voltages are neglected, then (10-38) becomes

$$\tau_p = R_{b2}C_2 \ln \frac{2V_{CC}}{V_{CC}} = R_{b2}C_1 \ln 2 = 0.69\, R_{b2}C_2 \quad (10\text{-}39)$$

A one-shot multivibrator requiring a single power source is obtained by utilizing an emitter resistor as shown in fig 10-24. The emitter

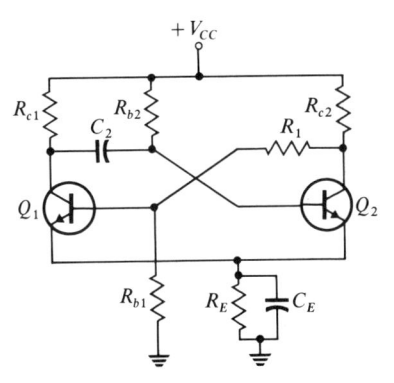

Fig 10-24 Emitter-coupled monostable multivibrator circuit

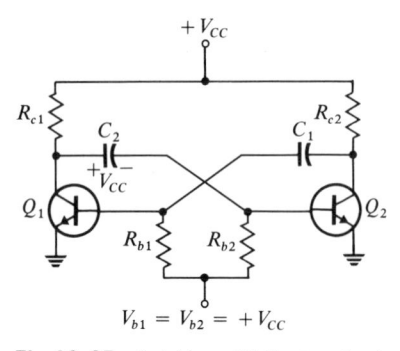

Fig 10-25 Astable multivibrator circuit

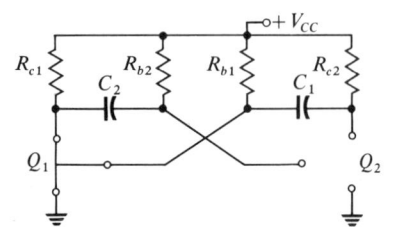

Fig 10-26 Circuit model of fig 10-25 for Q_2 off and Q_1 on (states at $t = t_0^+$)

is often bypassed by a capacitor, as in the flip-flop case. The stable state corresponds to Q_1 off and Q_2 on.

10-7 Astable Multivibrators

The basic circuit of an astable multivibrator, for the bipolar transistor, is shown in fig 10-25. The main difference between the astable and the other two subclasses of multivibrators is that there is no dc coupling between the transistors in the astable case. The loop transmission at dc is thus always less than one. In the astable multivibrator, there is no stable state and both states are quasi-stable. In other words, the circuit changes from one to another state periodically without an applied signal. The circuit, for this reason, also is called a *free-running multivibrator*, since it requires no external trigger to cause the transition of states. The multivibrator output is very much like that of a square-wave generator.

The analysis of the circuit is as follows. First note that R_{b1} and R_{b2} are usually connected to V_{CC}, so that no extra battery is needed. Let us assume that the circuit in fig 10-25 is initially in a state with Q_2 saturated and Q_1 off, i.e., Q_1 is held off by opening the collector or the base circuit of Q_1 and keeping it open until the voltage across capacitor C_2 is approximately $+V_{CC}$ (since one end of C_2 is connected to the base of Q_2, which is on). When the collector or the base circuit is closed, transistor Q_1 enters the active region and regeneration transition occurs, driving Q_1 into saturation and Q_2 off. The equivalent circuit at an instant after regeneration, where Q_1 is on and Q_2 is off, is shown in fig 10-26. The saturation voltages are neglected for simplicity. The base and the collector waveforms are then as shown in fig 10-27. The time, t_1, during which Q_2 is off is now readily determined by writing the charging curve for the base voltage of Q_2, namely v_{b2}. Since $v_{b2}(t) = -V_{CC}$ at $t = t_0^+$, and the target voltage is $+V_{CC}$, with a time constant $R_{b2}C_2$, we write

$$v_{b2}(t) = V_{CC} - 2V_{CC}e^{-t/R_{b2}C_2} \tag{10-40}$$

For $v_{b2}(t) = 0$ at $t = t_1$, the solution of (10-40) is

$$t_1 = R_{b2}C_2 \ln 2 = 0.69\, R_{b2}C_2 \tag{10-41}$$

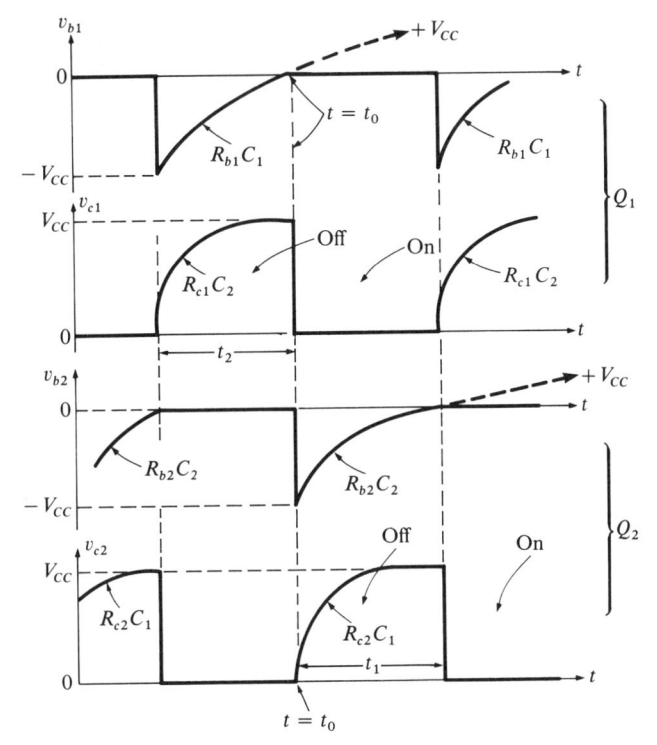

Fig 10-27 Various waveforms for the astable multivibrator

Similarly, the time, t_2, during which Q_1 is off is

$$t_2 = R_{b1}C_1 \ln 2 = 0.69 R_{b1}C_1 \qquad (10\text{-}42)$$

Hence, the total full-cycle period is

$$T = t_1 + t_2 = 0.69(R_{b1}C_1 + R_{b2}C_2) \qquad (10\text{-}43)$$

The repetition frequency is

$$f = \frac{1}{T} = \frac{1}{t_1 + t_2} \qquad (10\text{-}44)$$

It should be noted that the astable circuit of fig 10-25 may not always start when the power supply is turned on. This is so because both transistors Q_1 and Q_2 can be simultaneously on and saturated through the base drive. This case is referred to as the "locked-up" situation. Usually the imbalance in the circuit causes oscillation to start when

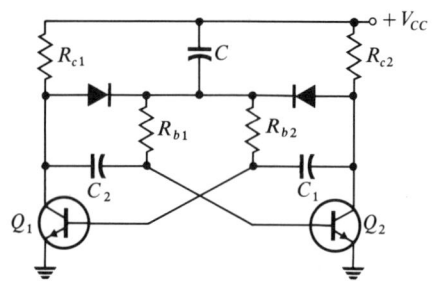

Fig 10-28 Sure-starting astable multivibrator

the power supply is turned on. In some digital circuit applications, however, the sure-starting condition is very important. The circuit shown in fig 10-28 is guaranteed always to start oscillating. The sure-starting results because, when both transistors turn on, the base-drive circuit will discharge C until one of the transistors turns off. One of the diodes will subsequently restore the charge on C.

10-8 Application of Multivibrators

Multivibrators are used in a variety of electronic circuits, such as in computers and communications systems. For example, flip-flops are used as memory elements in shift registers, counters, etc. Monostable multivibrators are used for the generation of well defined pulses, the logic design of pulse delay, variable pulse width, etc. Astable multivibrators are used as square-wave generators, and in pulse synchronization as a clock for binary logic signals. These are but a few applications of regenerative switching circuits. Some of these are discussed in the next chapter. The interested reader is referred to the literature for specifics, and for the details of the various applications (Ref 2).

10-9 Blocking Oscillators

One need not use two transistors in order to get a regenerative switching circuit. All that is required is a negative-resistance characteristic, as discussed in Section 9-6. Alternatively stated, a positive feedback mechanism with a loop gain greater than unity is required. Thus we can replace one transistor with a pulse transformer[6] that is properly phased. Such circuits are known as blocking oscillators. Blocking oscillators may be used for large-power pulse applications. Note that the use of a pulse transformer is necessary, since for a common-emitter stage, phase reversal is needed; the gain is provided by the CE stage. For a common-collector, or a common-base stage, phase reversal is not necessary, but current gain or voltage gain are needed to make the loop gain larger than unity. The pulse transformer has the disadvantage of being bulky and not suitable for integrated circuits. A blocking oscillator has only the monostable and astable forms, and these are discussed in the following.

[6] A pulse transformer is a ferrite-cored transformer designed to handle fast waveforms. For a detailed description and models of pulse transformers, see Ref 2, Chapter 3.

10-10 Monostable Blocking Oscillator

A simple, bipolar transistor monostable blocking oscillator is shown in fig 10-29. The operation of the circuit may be briefly described as follows. In the quiescent state, the transistor is off, since the voltage V_{BB} (which is usually very small) reverse-biases the emitter-base junction. Now, if a negative trigger pulse is applied to the collector of the transistor to lower its potential, the base potential will become positive through the transformer action, and base current will flow. As a result of this, the transistor enters the active region, and the increase in collector current further reduces the collector potential which, in turn, increases the base potential, thus further increasing the base and the collector currents. The result of this positive feedback mechanism, if the loop gain is larger than unity, is to drive the transistor into saturation. The condition for the loop gain to be larger than unity is readily determined by a simplified analysis. We assume that the transistor is in the active region, and thus the approximate circuit model (which ignores r_b) is as shown in fig 10-30. This circuit may be used to find the loop gain,

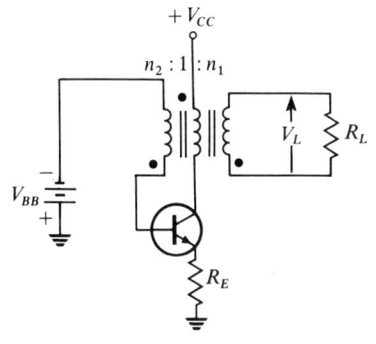

Fig 10-29 Monostable blocking-oscillator circuit

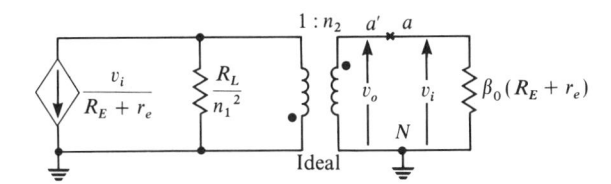

Fig 10-30 Simplified midband circuit model of fig 10-29

i.e., we open the base lead (a-a'), put a voltage v_i across the base to ground terminals (a-N), place the impedance $\beta_0(R_E + r_e)$ across (a'-N), and then determine v_o/v_i. From the circuit

$$\frac{v_o}{v_i} = \left(\frac{n_2}{R_E + r_e}\right)\left[\left(\frac{R_L}{n_1^2}\right) \Big\| \frac{\beta_0(R_E + r_e)}{n_2^2}\right] \qquad (10\text{-}45)$$

For $R_E \gg r_e$, and setting $v_o/v_i \geq 1$ in (10-45) yields

$$R_L \geq \frac{n_1^2}{n_2} \frac{\beta_0 R_E}{\beta_0 - n_2} \qquad (10\text{-}46a)$$

Thus if $\beta_0 > n_2$ and (10-46) is satisfied, the circuit will regenerate. In practice, the value of n_2 is usually in the range $\frac{1}{5} \leq n_2 \leq 1$, and (10-46a) reduces to

$$R_L \geq \frac{n_1^2}{n_2} R_E \tag{10-46b}$$

If the transistor switching time, t_{on}, and the saturation voltages are ignored, a simplified model for the pulse transformer is used, and V_{BB} is assumed to be approximately zero, the equivalent circuit at $t = t_0^+$ is as shown in fig 10-31. Note that the saturation voltages are ignored for simplicity. An analysis of this circuit will result in the waveforms shown in fig 10-32. The pertinent circuit equations are for the outside loop, from KVL,

$$V_{CC} - v - n_2 v = 0$$

or

$$v = \frac{V_{CC}}{n_2 + 1} \tag{10-47}$$

and for the transistor

$$i_c + i_b = -i_e = -\frac{n_2 v}{R_E} = -\left(\frac{n_2}{n_2 + 1}\right)\frac{V_{CC}}{R_E} \tag{10-48}$$

We also have

$$L_m \frac{di_L}{dt} = v = \frac{V_{CC}}{n_2 + 1} \tag{10-49}$$

But at $t = 0$, $i_L = 0$, hence from (10-49) by integration

$$i_L = tv/L_m = [V_{CC}/(n_2 + 1)]t/L_m \tag{10-50}$$

Since the sum of ampere-turns in an ideal transformer is zero, we have a relation between i, i_0, and i_b, namely

$$i - n_2 i_b - n_1 i_0 = 0 \tag{10-51}$$

The collector current is

$$i_c = i + i_L = n_2 i_b + n_1 \left(\frac{n_1 v}{R_L}\right) + \frac{tv}{L_m} \tag{10-52}$$

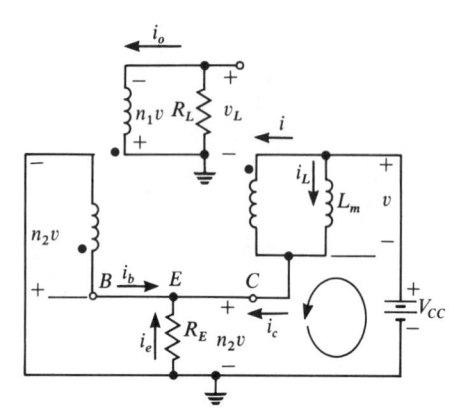

Fig 10-31 Circuit model of fig 10-29 when transistor is in saturation

From (10-47), (10-48), and (10-52) we obtain

$$i_c = \frac{V_{CC}}{(n_2+1)^2}\left(\frac{n_2^2}{R_E} + \frac{n_1^2}{R_L} + \frac{t}{L_m}\right) \qquad (10\text{-}53)$$

and

$$i_b = \frac{V_{CC}}{(n_2+1)^2}\left(\frac{n_2}{R_E} - \frac{n_1^2}{R_L} - \frac{t}{L_m}\right) \qquad (10\text{-}54)$$

The waveforms for i_c, i_b, and i_e are thus as shown in fig 10-32. Note that i_e is constant during the pulse. As long as $i_c < \beta i_b$, the transistor is in the saturation mode. As the collector current increases, however, the base current decreases. Eventually, at a time T_p, we have the situation where $i_c = \beta i_b$, and the transistor enters the active region. This time duration is then obtained by setting $i_c = \beta_0 i_b$, namely from (10-53) and (10-54)

$$\frac{V_{CC}}{(n_2+1)^2}\left(\frac{n_2^2}{R_E} + \frac{n_1^2}{R_L} + \frac{T_p}{L_m}\right) = \beta_0\left[\frac{V_{CC}}{(n_2+1)^2}\left(\frac{n_2}{R_E} - \frac{n_1^2}{R_L} - \frac{T_p}{L_m}\right)\right] \qquad (10\text{-}55)$$

and the result is

$$T_p = \frac{n_2 L_m}{R_E}\left(\frac{\beta_0 - n_2}{\beta_0 + 1}\right) - \frac{n_1^2 L_m}{R_L} \qquad (10\text{-}56)$$

For $\beta_0 \gg 1, n_2$, (10-56) simplifies to

$$T_p \approx \frac{n_2 L_m}{R_E} - \frac{n_1^2 L_m}{R_L} \qquad (10\text{-}57)$$

Note that (10-57) is always positive, because of the regenerative condition in (10-46b).

Once the transistor enters the active region at $t = T_p$, the regenerative action drives the transistor into the cutoff mode, and thus a pulse is formed at the load as shown in fig 10-32d. The polarity of the transformer at the load is arbitrary. Note that the current in the inductors cannot change instantaneously, thus current in the inductors will continue to flow even when the transistor currents are zero. The undershoot is a result of the magnetizing current through the capacitance of the transformer. Since the capacitance is usually small, the

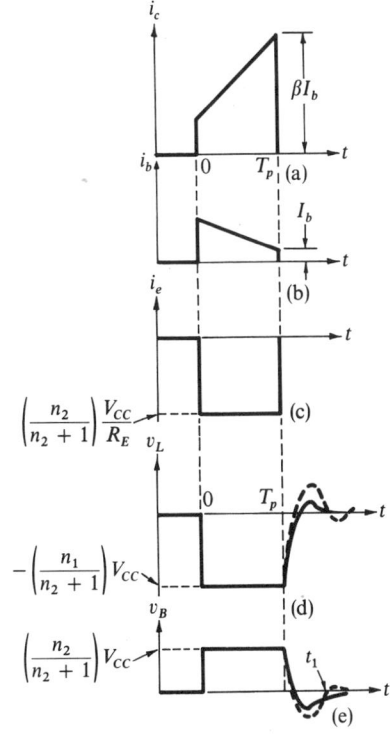

Fig 10-32 Various waveforms of monostable blocking oscillator as obtained from the analysis of fig 10-31

undershoot can be very large and oscillatory in nature (as shown by the dashed curve). This can start regeneration again at $t = t_1$, and thus astable operation will result. Usually R_L is selected to provide sufficient damping, in order to avoid this situation.

10-11 Astable Blocking Oscillator

One form of an astable blocking oscillator is shown in fig 10-33, where an $R_1 C_1$ circuit is added in the emitter lead. The circuit operation is as follows. Let there be a voltage V_o across the capacitor, such that $V_o > V_{BB} - V_{BE}$, so that the transistor is off. Thus, the capacitor discharges exponentially to zero voltage with a time constant $R_1 C_1$. When the voltage v_1 reaches the value of $V_{BB} - V_{BE}$, the transistor enters the active region, drawing base, and collector currents. Regenerative action begins, and the transistor is driven into saturation, as discussed earlier. Thus, the waveform for v_L is similar to fig 10-32 for $t \leq t_p$, as shown in fig 10-34. The transistor is thus on during t_p and off during t_f. These values are determined as follows. To find t_f, we have the discharge equation

$$v_1(t) = V_o e^{-t/R_1 C_1} \tag{10-58}$$

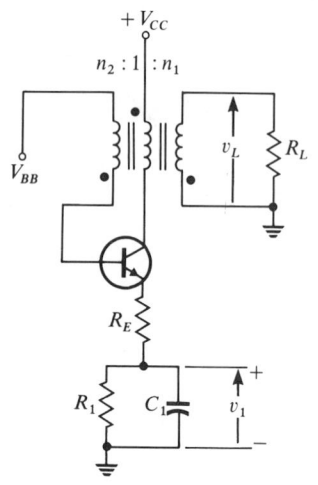

Fig 10-33 Astable blocking-oscillator circuit

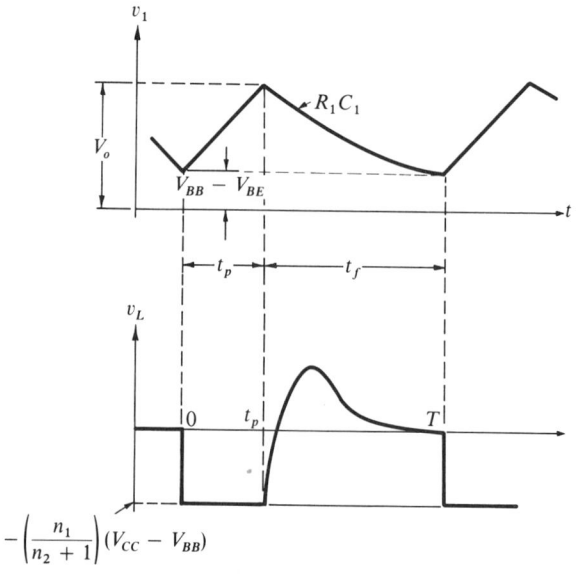

Fig 10-34 Waveforms associated with circuit of fig 10-33

When $t = t_f$, $v_1 = V_{BB} - V_{BE}$, thus

$$t_f = R_1 C_1 \ln \frac{V_o}{V_{BB} - V_{BE}} \tag{10-59}$$

To find the value of V_o and t_p, we use the equivalent circuit of fig 10-31, but we must add to this $R_1 C_1$ and V_{BB}. The expressions are quite involved. After tedious algebraic manipulations, and if we use the following often used approximations,

$$\beta_0 \gg 1, \quad R_1 \gg R_E, \quad \frac{t_p}{R_E C_1} \ll 1$$

so that

$$e^{-t_p/R_E C_1} \simeq 1 - \frac{t_p}{R_E C_1}$$

we obtain (see Problem 10-23)

$$t_p \approx \left(\frac{n_2 L_m}{R_E} - \frac{n_1^2 L_m}{R_L} \right) \left(\frac{1}{1 + n_2 L_m / R_E^2 C_1} \right) \tag{10-60}$$

and

$$V_o \approx V_{BB} + \left(\frac{n_2}{1 + n_2} \right)(V_{CC} - V_{BB}) \frac{t_p}{R_E C_1} \tag{10-61}$$

Note that for C_1 very large, (10-60) reduces to (10-57). The period of the astable blocking oscillator is

$$T = t_p + t_f \tag{10-62}$$

where t_p and t_f are calculated from (10-59), (10-60), and (10-61). Other forms of astable and monostable blocking oscillators are also possible. Interested readers are referred to the literature for a discussion of blocking oscillators and their applications (see Refs 2 and 3).

10-12 Time-Base Generators (Sweep Circuits)

Circuits that provide a linear time scale are called time-base generators. In such circuits, either the output current or the output voltage is a linear function of time over a specified time interval. These circuits are extremely important in electronic systems. For example, a linear time-base voltage is required on the deflection plates of a cathode-ray oscilloscope to sweep the electron beam from left to right across the screen. Similarly, a linear time-base current waveform is required in the deflection coils of a television receiver. Because of the sweep application, the circuits are also sometimes called sweep circuits.

The application of the time-base generator in a cathode-ray oscilloscope is illustrated in fig 10-35. The signal waveform to be displayed is $y(t)$, which is applied at the vertical deflection plates, while the sweep

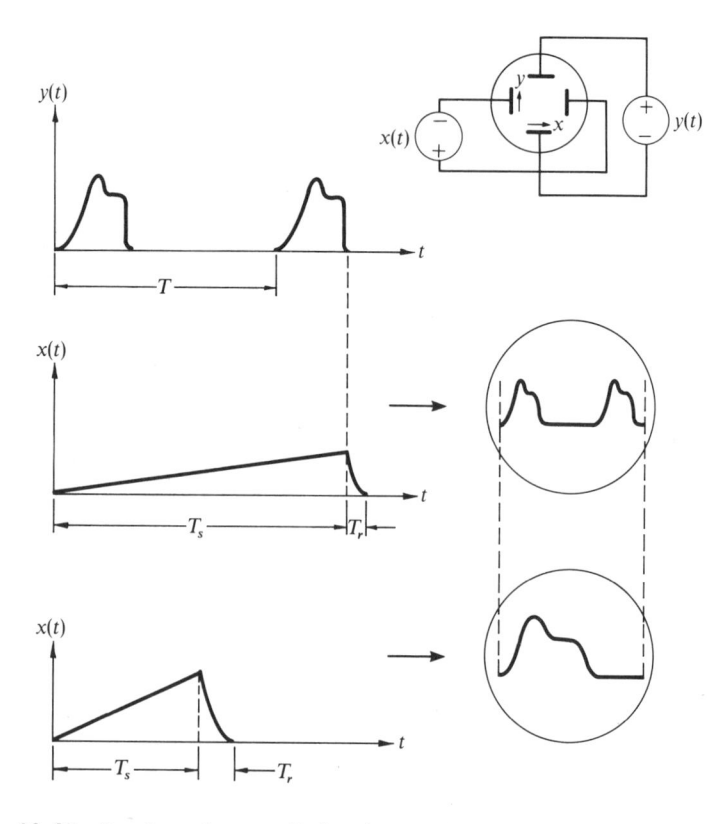

Fig 10-35 Signal waveforms as displayed on a cathode-ray oscilloscope for different sweep times

signal $x(t)$ is applied to the horizontal deflection plates. If $x(t) = kt$, then $y(t) = y(x/k)$. Thus the waveform $y(x)$, displayed on the scope, is of the same form as $y(t)$, except that it is scaled by a constant k. The sweep signal $x(t)$ generally starts from some initial value and increases linearly with time until it reaches some maximum value, after which it returns to the initial value. The time T_s is called the *sweep time*, while the time T_r, required for $x(t)$ to return to the initial value, is called the *restoration* time or *flyback* time. The display for two different values of T_s is shown in fig 10-35. There are basically two types of time-base generators, a free-running time base generator and a triggered time-base generator. In a free-running time-base generator, a periodic sawtooth waveform is generated without the application of any signal. This type of time-base generator is required to display a periodic waveform, and T_s must be larger than the period of the waveform. In a triggered time-base generator, a linear waveform with a prescribed duration of time is generated by the application of a trigger signal. This type of sweep is required to display widely separated narrow-width pulses.

The linearity of a sweep is usually measured in terms of the slope error or the sweep-speed error ε_s. The sweep-speed error is defined by

$$\varepsilon_s = \frac{\text{difference in slope at beginning and end of sweep}}{\text{initial value of slope}} \qquad (10\text{-}63)$$

Note that for an exact linear waveform, $\varepsilon_s = 0$. Thus ε_s expresses deviation from linearity, and we would like to have this as small as possible.

Another term which is encountered in sweep circuits is the sweep speed. If a capacitor C is charged by a constant current I, then $v_c = (I/C)t$. The rate of change of voltage with time is defined as the sweep speed, namely

$$\text{sweep speed} = \frac{I}{C} \qquad (10\text{-}64)$$

There are many methods of generating time-base waveforms (Ref 2), and we shall consider briefly some typical circuits in order to illustrate the basic concepts involved.

10-13 Free-Running Time-Base Generators

Many circuits can be designed to exhibit a negative-resistance driving-point characteristic, as discussed in Chapter 9. These circuits can be utilized to realize approximately free-running time-base

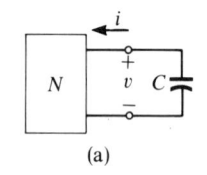

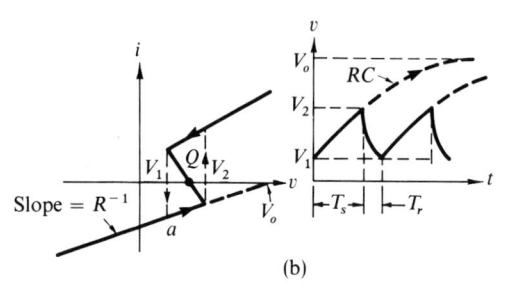

Fig 10-36 Free-running time-base generator circuit with poor sweep linearity: (a) Schematic of the circuit; (b) v-i characteristic of one-port network N and operating point for astable operation; (c) Voltage waveform across the capacitor

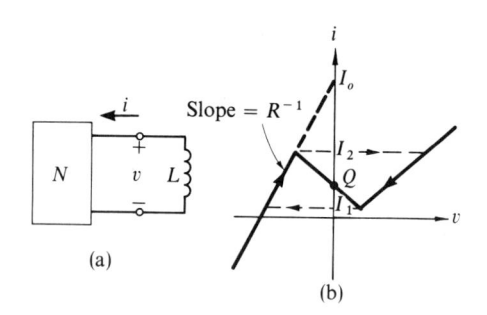

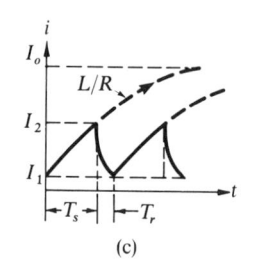

Fig 10-37 Alternate form of free-running time-base generator circuit with poor sweep linearity: (a) Schematic of the circuit; (b) v-i characteristic of one-port network N and operating point for astable operation; (c) Current waveform through the inductor

generators. Two free-running time-base generator circuits are shown in figs 10-36 and 10-37. These circuits generate sawtooth waveforms in the absence of a trigger signal. If the waveforms are approximately linear, the circuits are free-running time-base generators.[7] The operation of these circuits was discussed in Section 9-7. The exponential sweep can approximate a linear time variation as follows. Consider fig 10-36b, and let the initial time be such that the voltage v is at the point a. The voltage across the capacitor will charge towards the target value V_o. The charging curve is given by

$$v(t) = V_1 + (V_0 - V_1)(1 - e^{-t/RC}) \qquad (10\text{-}65a)$$

$$\approx V_1 + V_0(1 - e^{-t/RC}) \quad \text{for} \quad V_0 \gg V_1 \qquad (10\text{-}65b)$$

where R^{-1} is the slope of the linear segment passing through the point a. For $t \leq T_s \ll RC$, we can expand (10-65b) into a series and retain only the linear term as a reasonable approximation, namely

$$v(t) = V_1 + V_0\left[1 - \left(1 - \frac{t}{RC} + \frac{t^2}{(RC)^2} - \cdots\right)\right] \qquad (10\text{-}66a)$$

$$\approx V_1 + V_0\frac{t}{RC} \qquad t \ll RC \qquad (10\text{-}66b)$$

Equation (10-66b) is of the desired form, i.e., linear variation with time. Note, however, that from (10-66b) we have

$$V_2 = v(T_s) = V_1 + \frac{V_o T_s}{RC} \qquad (10\text{-}67)$$

For $T_s \ll RC$, the value of V_2 is only slightly larger than V_1 if V_o is not very large. The sweep error can be readily determined as follows. From (10-66b) and (10-65b)

$$\left.\frac{dv}{dt}\right|_{t=0} = \frac{1}{RC}V_o, \qquad \left.\frac{dv}{dt}\right|_{t=T_s} = \frac{V_o}{RC}e^{-T_s/RC} \qquad (10\text{-}68)$$

Hence, from (10-68) and (10-63), we obtain

$$\varepsilon_s = \frac{(V_o/RC)(1 - e^{-T_s/RC})}{V_o/RC} = 1 - e^{-T_s/RC} \qquad (10\text{-}69)$$

---

[7] Free-running time-base generators can be converted into triggered time-base generators (see Section 9-7).

This linearity improves for $T_s \ll RC$. Under this condition, the sweep speed error is

$$\varepsilon_s \approx \frac{T_s}{RC} \qquad (10\text{-}70)$$

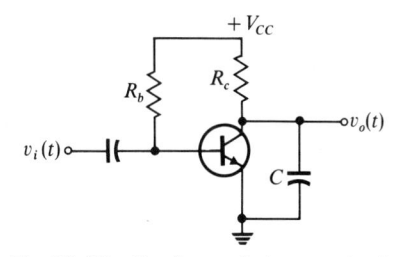

Fig 10-38 Simple practical sweep circuit

10-14 Triggered Voltage Time-Base Generators

The simplest practical sweep circuit is shown in fig 10-38. In this circuit, the transistor is biased on and saturated. When a negative pulse is applied, hence the the name triggered, the transistor turns off and the collector potential rises to the target value of V_{CC} with a time constant $R_c C$, as shown in fig 10-39. The charging curve, ignoring $V_{CE(sat)}$, is then governed by

$$v_o(t) = V_{CC}(1 - e^{-t/R_c C}) \qquad (10\text{-}71)$$

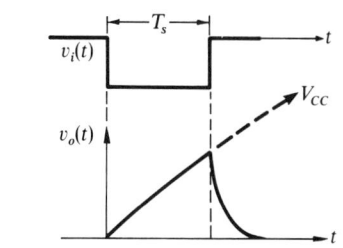

Fig 10-39 Associated waveforms for circuit of fig 10-38

for $t \le T_s \ll R_c C$, we can expand (10-71) into a series as in (10-66), to obtain the approximate expression

$$v_o(t) \approx V_{CC} \frac{t}{R_c C} \qquad (10\text{-}72)$$

Thus (10-72) represents an approximately linear waveform. The slope error as obtained from (10-63) and (10-71) is

$$\varepsilon_s = \frac{v_o(T_s)}{V_{CC}} \approx \frac{T_s}{RC} \quad \text{for} \quad T_s \ll RC \qquad (10\text{-}73)$$

Note that for a small slope error, the output-voltage level must be a very small fraction of V_{CC}, and this is not a desirable feature if a certain output level is desired. This shortcoming can be remedied by the Miller sweep and the bootstrap sweep as discussed in the following.

Miller Sweep Circuit

The operational amplifier used as an integrator (Section 6-7) is one form of the Miller integrator. One need not actually have an operational amplifier. An amplifier in the form of the circuit shown in fig 10-40

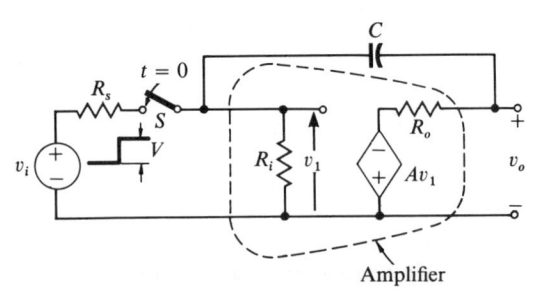

Fig 10-40 Basic form of Miller sweep circuit

would suffice. In fig 10-40, the low-frequency input and output impedances are indicated by R_i and R_o, respectively. The operation of the circuit is as follows. Assume that initially the capacitor is uncharged. If we neglect R_o and *close the switch S* at $t = 0$, v_o will be zero at 0^+ (since the capacitor voltage was zero before the switch was closed and it cannot change instantaneously) and will rise exponentially to the target value $-Av_1$ with a time constant R_eC (where $v_1 = [R_i/(R_i + R_s)]v_i$ and $R_e = R_i \| R_s$). For a given output-voltage level, the linearity of the sweep is considerably improved as we shall now show. The output voltage for a step input of value V, $v_i = Vu(t)$, and $R_eC \ll t$ as in before is

$$v_o(t) \approx A \left(\frac{R_i V}{R_i + R_s} \right) \frac{t}{(R_i \| R_s)C} \qquad (10\text{-}74)$$

The slope error in this case is

$$\varepsilon_s = \frac{v_o(T_s)}{|A|v_1} = \frac{v_o(T_s)}{|A|[R_i/(R_i + R_s)]V} = \frac{v_o(T_s)}{V}\left(\frac{1 + R_s/R_i}{|A|}\right) \qquad (10\text{-}75)$$

Thus, for a given output-voltage level, the slope error can be improved, since $|A|$ can be designed to have a very large value. If an operational amplifier is used, $\varepsilon_s \approx 0$ as we would expect since the integration of a step yields a ramp.

If the value of R_o is included in the analysis, the difference will be only in the starting point, i.e., v_o will not be zero at $t = 0^+$. The time constant and the final value, however, will be the same.

A simple circuit of the Miller sweep is shown in fig 10-41. It should be noted, however, that in practice, more than one stage of amplification is used. Usually the amplifier in fig 10-40 uses emitter followers at the input and the output to increase R_i and reduce R_o, and interior CE stages are used to provide amplification. In order to reduce the flyback time, T_r, a diode is usually used across the capacitor for a rapid discharge.

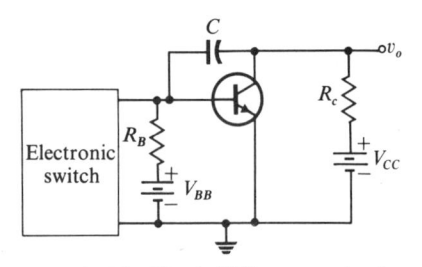

Fig 10-41 Simple Miller sweep circuit

Bootstrap Sweep Circuit

The basic form of a bootstrap circuit is shown in fig 10-42, where the amplifier in this case is usually a unity gain amplifier, as obtained from an emitter follower. A simple circuit of the bootstrap sweep is shown in fig 10-43. The low-frequency input and output impedances are indicated by R_i and R_o, respectively. Usually $R_o \ll R_i$, and the output impedance may be ignored without any noticeable error. Therefore the approximation $R_o \ll R_i$, R_s will be used in the following. The circuit is similar to the Miller integrator, and improved sweep linearity is obtained as follows (fig 10-42). When the *switch S is opened* at $t = 0$, the voltage at the output at $t = 0^+$ is the same as at $t = 0^-$, i.e., there is no jump at $t = 0$, and $v_o = 0$. The voltage across the capacitor rises exponentially with a time constant $R_e C$ (where $R_e = R_s \| R_i$) to a target value

$$v_o(t = \infty) = \frac{AVR_i}{R_i(1 - A) + R_s} \qquad (10\text{-}76)$$

The output voltage for $R_e C \ll t$ is then given by

$$v_o(t) \simeq \frac{AV}{(1 - A) + R_s/R_i} \frac{t}{(R_i \| R_s)C} \qquad (10\text{-}77)$$

The slope error is given by

$$\varepsilon_s = \frac{v_o(T_s)}{AV\{1/[(1-A) + R_s/R_i]\}} = \frac{v_o(T_s)}{V}\left(\frac{1 - A + R_s/R_i}{A}\right) \qquad (10\text{-}78)$$

In a bootstrap circuit, we design the amplifier such that $A \approx 1$. Hence, for a small slope error, we must have R_s/R_i very small, thus the input impedance must be very high. Note that for a given output level, the linearity of the bootstrap sweep is not as good as that of the Miller sweep circuit, see (10-75). Further, in the Miller circuit high input impedance is not critical, since $R_s/R_i \leq 1$.

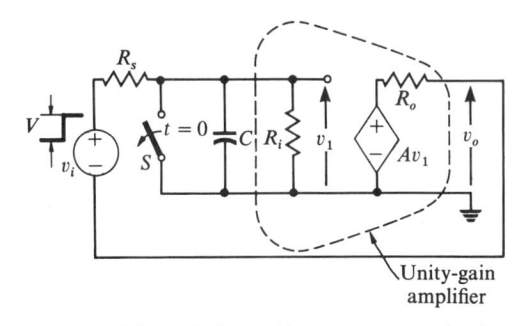

Fig 10-42 Basic form of bootstrap sweep circuit

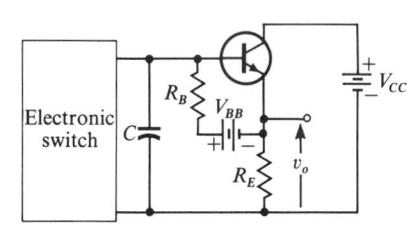

Fig 10-43 Simple bootstrap sweep circuit

10-15 Triggered Current Time-Base Generators

In the preceding section, we considered triggered voltage time-base generators. We next consider briefly triggered current time-base generators, where linear current is caused to flow through an inductor.

This type of time-base generator is most frequently used in radar and television displays.

A simple current time-base generator, or inductive sweep circuit, is shown in fig 10-44a. The associated waveforms are shown in fig 10-44b, and explained as follows. Initially the transistor is off because of the reverse bias, V_1, on the emitter-base junction, thus $v_{CE} = V_{CC}$ and $i_L = 0$. When the switch is turned on, by applying a base drive $v_i = V_2$, current flows through the inductor (diode D does not conduct during the sweep, since it is reverse-biased). The current grows exponentially to a target value of $V_{CC}/(R_L + R_{CS})$, where R_L and R_{CS} are the coil resistance and the transistor saturation resistance, respectively. These are important, and must be included in the analysis, as we shall soon discover. Thus the current growth is governed by the following equation during the sweep:

$$i_L = \frac{V_{CC}}{R_L + R_{CS}}(1 - e^{-[(R_L + R_{CS})/L]t}) \tag{10-79}$$

Thus, if $t \leq T_s \ll L/(R_L + R_{CS})$, (10-79) may be approximated as

$$i_L \approx \frac{V_{CC}}{L} t \tag{10-80}$$

The slope error is found similarly to the voltage sweep case,

$$\varepsilon_s = \frac{i_L(T_s)}{V_{CC}/(R_L + R_{CS})} \approx \frac{T_s(R_L + R_{CS})}{L} \tag{10-81}$$

Linearity improvement can be made in a variety of ways. Some of these are discussed in Ref 2.

In fig 10-44b, note that at the moment when the transistor is turned off, a large voltage spike will appear across the collector junction. This voltage must be kept below the breakdown voltage, so as not to damage the transistor.

10-16 RC Waveform Shaping

In pulse and digital circuits, it is often desired to have signals of nonsinusoidal waveform, e.g., square-wave, ramp, trigger pulse, etc. In this section we consider briefly some simple passive networks that can achieve these purposes.

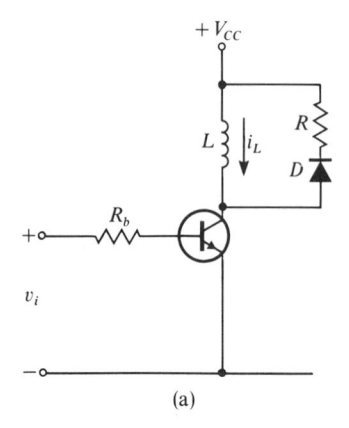

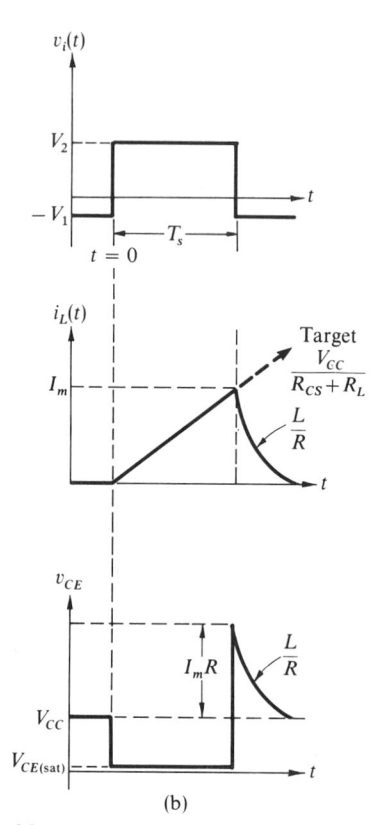

Fig 10-44 (a) Inductive sweep circuit (current time-base generator); (b) Associated waveforms of (a)

Differentiating Circuit

In many instances we may wish to differentiate the input signal in the simplest manner possible. The waveform of the differentiated signal may not be of any importance. For example, in pulse circuitry, trigger pulses may be desired from a square-wave generator. A simple, passive high-pass RC network can do this job very effectively. Consider the high-pass RC circuit shown in fig 10-45. The equilibrium equations of the circuit are

$$e_i(t) = \frac{1}{C} \int i(t)\, dt + i(t)R \qquad (10\text{-}82)$$

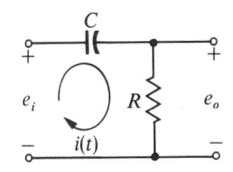

$$e_o(t) = i(t)R \qquad (10\text{-}83)$$

Fig 10-45 RC differentiating circuit

If the time constant of the circuit is very small in comparison with the time required for the input signal to make an appreciable change, the voltage drop across R will be very small in comparison with the drop across C. Thus, for $(1/C)\int i(t)\, dt \gg i(t)R$, (10-82) can be approximated as

$$e_i(t) \approx \frac{1}{C} \int i(t)\, dt \qquad (10\text{-}84)$$

From (10-83) and (10-84), we have

$$e_o(t) \approx RC \frac{de_{i(t)}}{dt} \qquad (10\text{-}85)$$

From (10-85), it is seen that the output is proportional to the derivative of the input, hence the name *differentiator*. An alternative approach may also be used to show this. From the circuit in fig 10-45, we have in the transform domain

$$E_o = \frac{R}{R + 1/sC} E_i = \frac{s}{s + 1/RC} E_i \qquad (10\text{-}86)$$

If the time constant is very small, i.e., $1/RC \gg |s|$, (10-86) can be approximated as

$$E_o = s(RCE_i) \qquad (10\text{-}87)$$

where we know that s corresponds to differentiation in the time domain. Figure 10-46 shows the output for a square-wave input under the

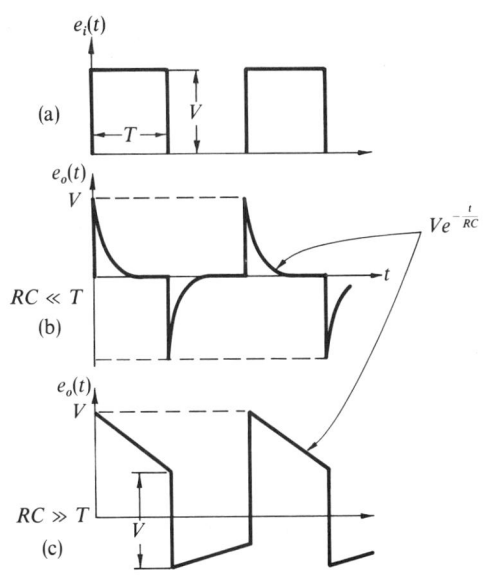

Fig 10-46 Input and output waveforms of fig 10-45: (a) Input waveform; (b) Output when $RC \ll T$ (an approximate differentiator); (c) Output when $RC \gg T$

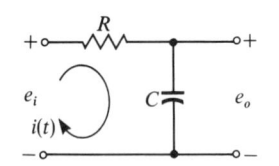

Fig 10-47 *RC* integrating circuit

stated approximation, and also for a case where the time constant is not very small.

Integrating Circuit

A circuit which can approximately integrate the input is the low-pass *RC* circuit shown in fig 10-47. The equilibrium equations of the circuit are

$$e_i(t) = Ri(t) + \frac{1}{C}\int i(t)\,dt \qquad (10\text{-}88)$$

$$e_o(t) = \frac{1}{C}\int i(t)\,dt \qquad (10\text{-}89)$$

If $Ri(t) \ll (1/C)\int i(t)\,dt$, from (10-88) and (10-89), we have

$$e_o(t) \approx \frac{1}{RC}\int e_i(t)\,dt \qquad (10\text{-}90)$$

Thus the output is proportional to the integral of the input. Alternatively, in the transform domain

$$E_o(s) = \frac{1/sC}{R + 1/sC}E_i(s) = \frac{1}{RC}\left(\frac{1}{s + 1/RC}\right)E_i(s) \qquad (10\text{-}91)$$

If $1/RC \ll |s|$, (10-91) can be approximated as

$$E_o(s) \simeq \frac{1}{s}\left(\frac{1}{RC}E_i(s)\right) \qquad (10\text{-}92)$$

where in (10-92), $1/s$ corresponds to integration in the time domain.

The output waveform for a square-wave input is shown in fig 10-48. Note that the simple *RC* integrator circuit is a crude form of the time-base generator discussed in the previous sections.

10-17 Clipping and Clamping Circuits

The nonlinearity of a device, such as a diode, can also be utilized to perform clipping and clamping operations on signal waveforms. In clipping circuits, the energy storage element is not required for the

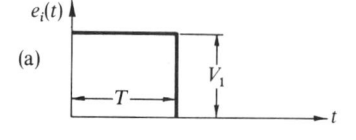

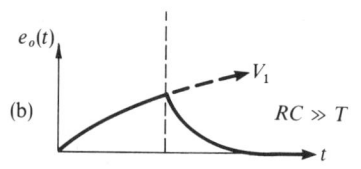

Fig 10-48 (a) Input waveform; (b) Output waveform of fig 10-47 for $RC \gg T$ (an approximate integrator)

10-17 Clipping and Clamping Circuits

operation. We use clipping circuits to select a portion of an arbitrary waveform. These circuits are also referred to as *limiters*, since the amplitude of the output is limited to some particular reference-voltage level. Four simple clipping circuits with their associated waveforms, assuming ideal diodes, are shown in fig 10-49. For example, in fig 10-49a,

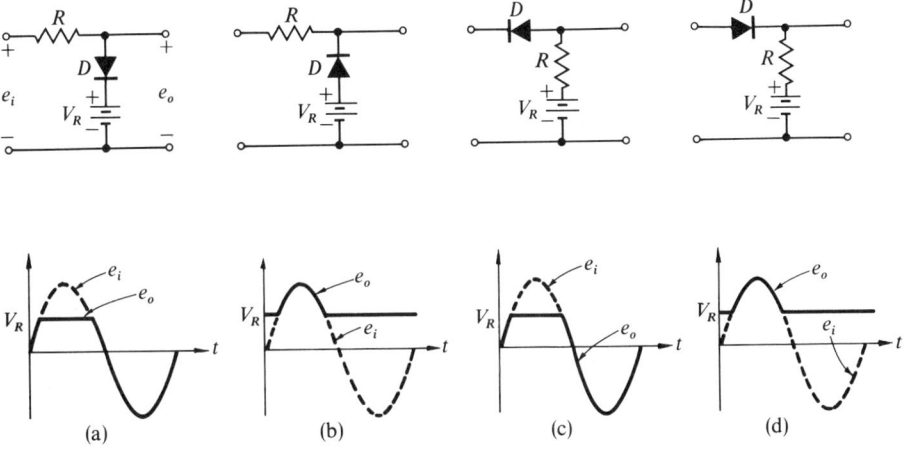

Fig 10-49 Various clipping circuits

if e_i exceeds V_R, then diode D conducts and e_o is limited to V_R. The corresponding waveforms will not be much different if the forward resistance R_f, and reverse resistance R_r, and the threshold voltage of the diode are included in a piecewise-linear analysis. In other words, if $R_f \ll R \ll R_r$, $V_R \gg V_o$, the waveforms will essentially be the same as in fig 10-49. Piecewise-linear analysis may be used to obtain the transfer characteristics, and thus e_o for a given e_i. One simple application of a clipping circuit is to obtain an approximate square wave from a sine-wave input, as shown in fig 10-50. The operation of the circuit, assuming ideal diodes, is described as follows. When $e_i > V_1$ and positive, D_1 is on, D_2 is off, and $e_o = V_1$. For the negative portion, when $|e_i| > |V_2|$, D_2 is on, D_1 is off, and $e_o = V_2$. For all other cases, D_1 and D_2 are off, and thus $e_o = e_i$.

It should also be noted that transistors and vacuum tubes can also be used as clippers. In such applications, the transistor amplifier is driven between saturation and cutoff. A transistor clipper is shown in fig 10-51a, where $R \gg R_i$, and R_i is the input resistance of the transistor. In this circuit, the input signal is large enough to drive the transistor

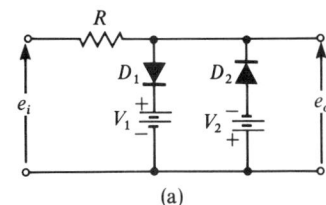

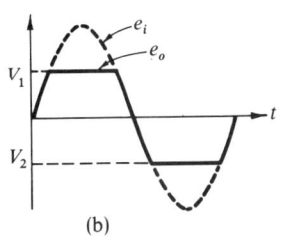

Fig 10-50 Double clipper: (a) Schematic; (b) Associated waveforms

into saturation. The pertinent approximate waveforms are shown in fig 10-51b.

Another function which must often be accomplished in digital circuits is to convert a periodic signal to one which is either unidirectional (i.e., always positive or always negative) or one with a different dc level superimposed on top, without altering its waveform. A circuit which performs this operation is referred to as a *clamping circuit*, or a *dc restorer*. The latter terminology is sometimes used because the clamping circuit reintroduces a dc component to the waveform. In clamping circuits, the capacitors are essential for the operation.

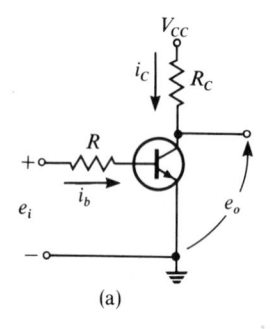

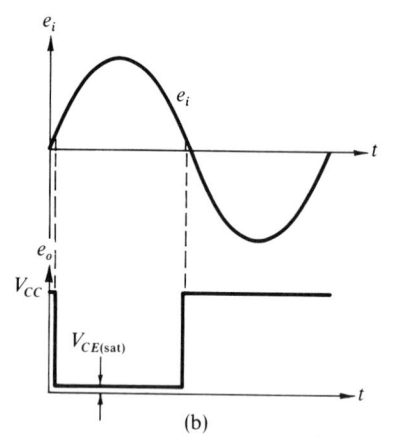

Fig 10-51 (a) Transistor clipper circuit; (b) Associated waveforms

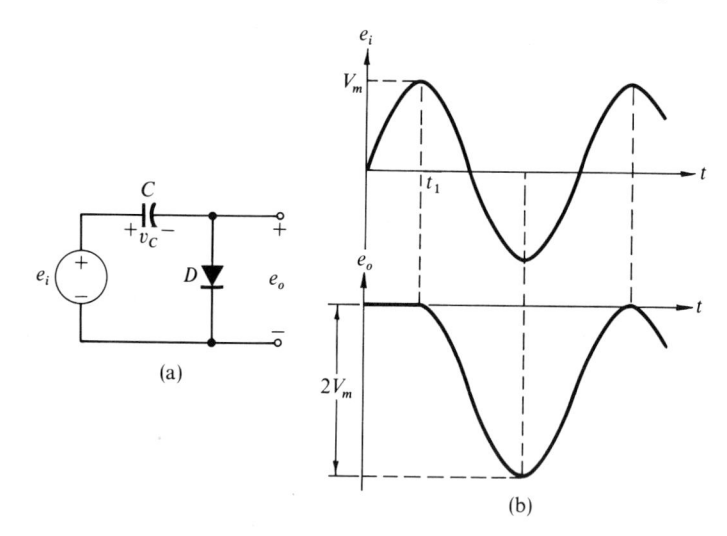

Fig 10-52 (a) Simple clamping circuit; (b) Associated waveforms

A simple clamping circuit is shown in fig 10-52. The operation of the circuit, assuming an ideal diode, is as follows. The capacitor is initially uncharged at $t = 0$. As the input voltage e_i rises from zero to V_m in the first quarter cycle, the voltage across the capacitor is equal to the input voltage with the polarity shown, and $e_o = 0$. At $t = t_1$, the capacitor voltage is equal to V_m. Now, when the signal $e_i(t)$ begins to fall, the voltage across the capacitor cannot follow the input voltage, since it cannot discharge through the ideal diode. Thus, for $t \geq t_1$, the output voltage $e_o = e_i - V_m$. Hence the output signal can never be positive, and we have a positive clamping circuit. Figures 10-53a and 10-53b show positive and negative diode clamping circuits, respectively. In the positive clamping circuit, the top of the pulse train is clamped to zero voltage, while in the negative clamping circuit, the bottom of the pulse train is clamped to zero voltage. The difference in the dc input

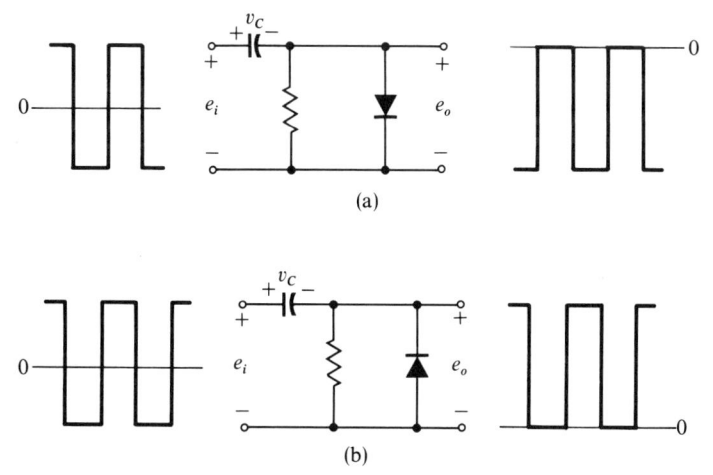

Fig 10-53 (a) Positive clamping circuit with its associated waveforms; (b) Negative clamping circuit with its associated waveforms

and output voltages is taken by the capacitor, with the polarity as shown in the figures. The negative clamping circuit is used in television receivers, where the bottom of the pulse train driving the control electrode of the picture tube is clamped to a fixed level, in order to reproduce the transmitted scene faithfully, i.e., fix the black level correctly. Note that capacitors and diodes can also be used as voltage-multiplier circuits (see Problem 10-27).

Diodes and transistors are also used to perform logic operations. This topic is discussed in some detail in the next chapter.

REFERENCES AND SUGGESTED READING

1. Harris, J., Gray, P. and Searle, C., *Digital Transistor Circuits*, SEEC series, New York: Wiley, 1966, Volume 5, Chapters 1, 2, and 3.
2. Millman, J. and Taub, H., *Pulse, Digital, and Switching Waveforms*, New York: McGraw-Hill, 1965.
3. Chua, L., *Introduction to Nonlinear Network Theory*, New York: McGraw-Hill, 1969.
4. Comer, D. J., *Introduction to Semiconductor Design*, Reading, Mass: Addison-Wesley, 1968, Chapter 9.
5. Pettit, J., and McWhorter, M., *Electronic Switching, Timing, and Pulse Circuits*, New York: McGraw-Hill, second edition, 1970.
6. Strauss, L., *Wave Generation and Shaping*, New York: McGraw-Hill, second edition, 1970.

PROBLEMS

10-1 Consider the circuit shown in fig 10-1. The circuit values are: $R_s = 1 \text{ k}\Omega$, $R_L = 1 \text{ k}\Omega$, and $V_{CC} = +10 \text{ V}$. The input pulse is as shown in fig 10-2a. The one voltage amplitude of the input pulse is 20 V, the pulse duration is 1 μsec and the off voltage amplitude of the input pulse is -10 V. The transistor used is type 2N3011 (see Appendix D). Determine the approximate t_{on} and t_{off} of the transistor switch. What is the effect of capacitive load on the t_{on} and t_{off}?

10-2 Calculate the approximate switching time t_d, t_r, t_s, and t_f of the *pnp* epitaxial planar silicon transistor (type 2N3250), and compare the results with those given in the data sheets (see Appendix D). Use the parameters and circuit values of the test circuits and $\tau_s = 200$ nsec.

10-3 In the bistable multivibrator circuit shown in fig 10-12, assume identical transistors and $R_{C1} = R_{C2} = R_C$, $R_1 = R_2 = R$. For simplicity of calculation, assume $r_b = 0$. If both transistors are assumed to be in the active region of the operation, use the low-frequency hybrid π model to determine the loop transmission of the positive feedback structure. For what values of β_0, in terms of the circuit parameters, is the system unstable?

(*Hint*: From Chapter 6, for instability $|T| > 1$ and $\arg T(j\omega) = 180$ degrees, hence examine $T(0)$.)

10-4 In Problem 10-3, use the unilateral model and do not use the other approximations of the text. Derive the expression for the natural frequencies of the system. Under what conditions are all the natural frequencies in the left half-plane?

10-5 An *n*-channel JFET inverter is shown in fig P10-5b. This circuit has the input signal waveform shown in fig P10-5a, and has a capacitive load. Determine the expressions (and the numerical values) for the turn-on and turn-off times of the FET, and sketch the output voltage waveform $v_o(t)$ and $v_{GS}(t)$. Use a piecewise-linear model and show the t_{on} and t_{off} on the sketch for $v_o(t)$.

(*Hint*: From the output characteristic curve in fig P10-5c for $|V_{GS}| > |V_{PO}|$

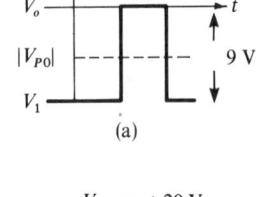

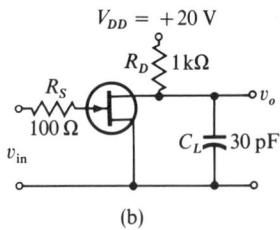

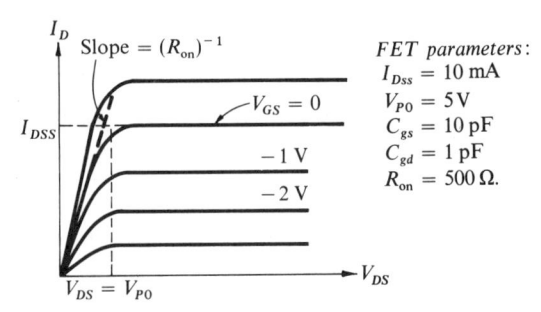

Fig P10-5 (c) Output characteristic

FET parameters:
$I_{DSS} = 10 \text{ mA}$
$V_{PO} = 5 \text{ V}$
$C_{gs} = 10 \text{ pF}$
$C_{gd} = 1 \text{ pF}$
$R_{on} = 500 \Omega$.

(d) OFF model   (e) Saturation-region ON model   (f) Steady state ohmic region ON model

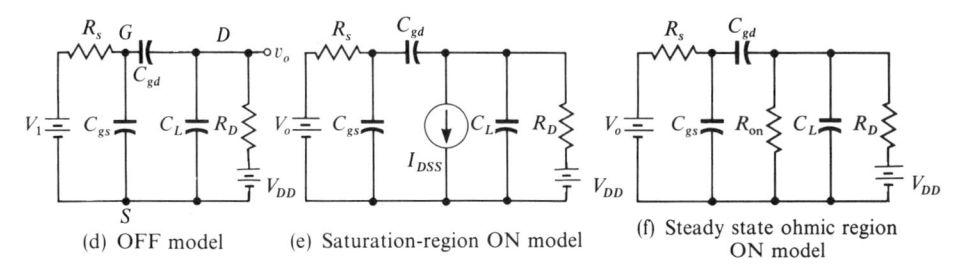

Fig P10-5

in JFET the gate voltage is beyond pinchoff, the FET is off (i.e., the channel is open-circuited), and the model shown in fig P10-5d may be used until $v_{GS} = 0$. During the time when $v_{GS} = 0$ until the time when $V_{DS} = V_{PO}$, the saturation region model of fig P10-5e can be used. From then on, the ohmic region model shown in fig P10-5f is used.)

10-6 Show how you would realize monostable, bistable, and astable multivibrators. The following are available to you: dc current and voltage sources, resistors, a tunnel diode, an inductor, and a capacitor. Use a three-segment piecewise linear model for the tunnel diode and show the operating point. Use as few components as possible.

10-7 Show the FET (depletion-mode) version of the bistable, monostable, and astable multivibrators. Include the biasing circuitry.

10-8 A simple transistor chopper or modulator circuit is shown in fig P10-8. The base drive is sufficiently large such that the transistor in the on mode is deep in saturation. Show the output waveform, ignoring the saturation voltages and the switching transients.

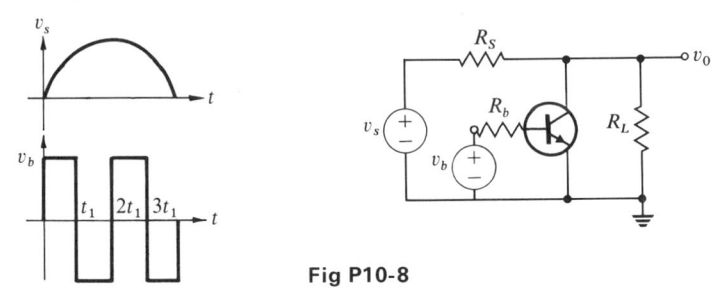

Fig P10-8

10-9 For the common-emitter amplifier circuit shown in fig P10-9, obtain the static transfer characteristic v_o vs v_i.
Use the simplest models for the off, active, and on modes of operation for the transistor.

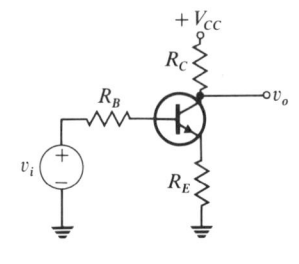

Fig P10-9

10-10 The circuit shown in fig P10-10 is a complementary-symmetry MOSFET multivibrator. Is this a bistable multivibrator? If it is, what are the approximate stable points? What features of this circuit make it attractive in *IC*?

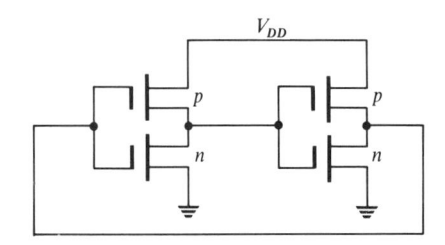

Fig P10-10

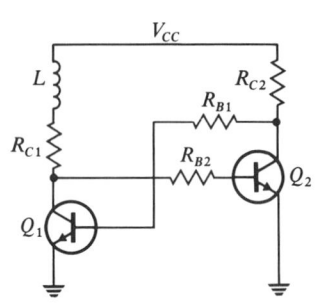

Fig P10-12

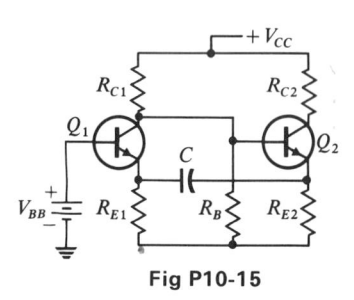

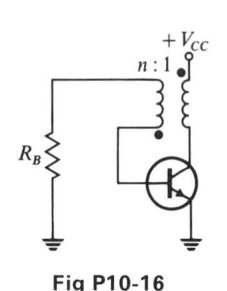

Fig P10-15

Fig P10-16

10-11 Design the free-running MV of fig 10-25 such that it generates a square-wave with a 10 V amplitude and a period of 1 μsec. Let $R_{b1} = R_{b2}$, $C_1 = C_2$, $R_{c1} = R_{c2}$, and assume identical transistors. The on-off times of the transistors are in the nanosecond range and can be ignored.

10-12 An inductively timed monostable multivibrator is shown in fig P10-12.
(a) What is the regenerative condition?
(b) Show that, in the normal state, Q_1 is on and Q_2 is off.
(c) If the circuit is appropriately triggered, show the pertinent waveforms and determine the expression for the pulse width of the MV. Ignore $V_{CE}(\text{sat})$ and $V_{BE}(\text{sat})$, and assume idealized on and off models for the transistors.

10-13 For the emitter-coupled bistable MV (see fig 10-16), the circuit element values are as follows:

$$V_{CC} = +10 \text{ V}, \quad R_{C1} = R_{C2} = 2 \text{ k}\Omega,$$
$$R_{b1} = R_{b2} = 10 \text{ k}\Omega, \quad R_1 = R_2 = 20 \text{ k}\Omega,$$
$$R_E = 500 \text{ }\Omega.$$

Calculate the stable-state currents and voltages if silicon transistors are used with $V_{CE(\text{sat})} = 0.2$ V, $V_{BE(\text{sat})} = 0.7$ V. What is the minimum value of β_0 for which the on transistor is in saturation?

10-14 An emitter-coupled monostable MV is shown in fig 10-24. Sketch the waveforms for the collector and base voltages with respect to ground for Q_2. Determine the expression for the pulsewidth, i.e., the time duration at which Q_2 is off. (Ignore $V_{CE(\text{sat})}$ and $V_{BE(\text{sat})}$ in the analysis.)

10-15 An emitter-coupled astable MV circuit is shown in fig P10-15. Sketch the waveforms for the collector voltages, with respect to ground, for both transistors. Derive the expressions for t_1 (time during which Q_2 is off) and t_2 (time during which Q_1 is off). Ignore $V_{CE(\text{sat})}$ and $V_{BE(\text{sat})}$ in this analysis.

10-16 A simple blocking oscillator circuit is shown in fig P10-16. Ignore the loss and parasitic capacitance of the transformer. Show that the transistor will be driven into saturation if $\beta_0 > n$. Derive the expression for the pulse-width, T_p.

10-17 A neon bulb is often used as an inexpensive relaxation oscillator. One such circuit is shown in fig P10-17. The linearized $v - i$ characteristic of the neon bulb is also shown. E_o and R are chosen to yield the operating point Q, as shown. Sketch the voltage waveform across the capacitance vs time. Derive the expressions for T_s and T_r.

10-18 A four-layer *pnpn* diode is connected in the circuit shown in fig P10-18 so as to function as a saw tooth generator. The linearized $i - v$ characteristic of the diode is also shown.

Sketch the voltage waveform across the capacitor, if E_1 and R_1 are chosen to yield the operating point at Q, as shown. Determine the expression for T_s.

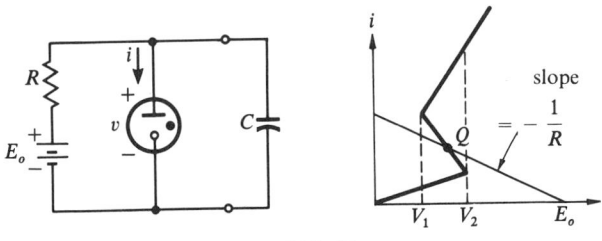

Fig P10-17

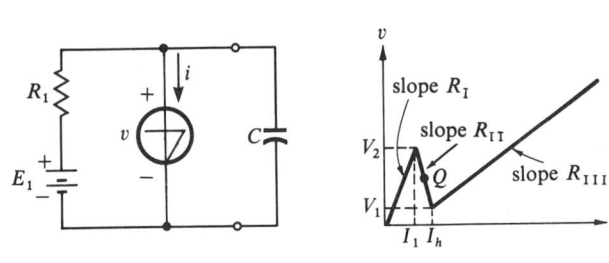

Fig P10-18

10-19 For Problem 10-18, show how E_1 and R_1 may be chosen:
(a) For a monostable operation.
(b) For a bistable operation.
(c) How would you apply a triggering pulse to change the state of the multivibrator?

10-20 For Problem 10-18, a current trigger pulse is applied in shunt with C as shown in fig P10-20.
If the circuit is biased for bistable operation, what should the minimum pulse height be, assuming that (τ) is sufficiently large for triggering, to cause the change of state in MV?

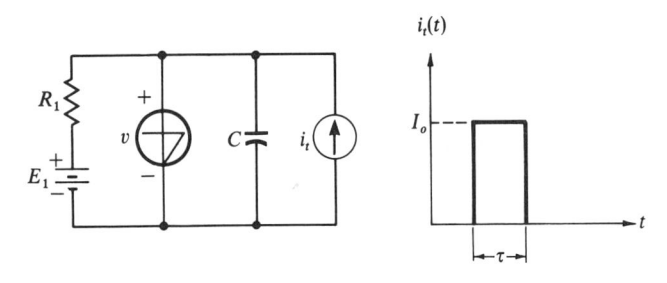

Fig P10-20

10-21 A simplified circuit using a tunnel diode is shown in fig P10-21. The linearized $v - i$ characteristic of the tunnel diode is also shown. For R_1 given, discuss how you would choose the battery E_o; for monostable operation, for bistable operation, for astable operation. If the circuit is biased, show how the trigger pulse may be applied to change the state of the MV.

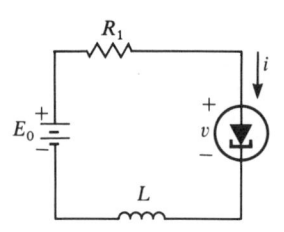

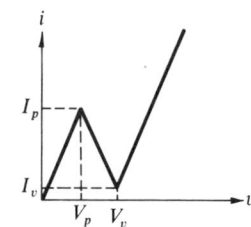

Fig P10-21

10-22 If the circuit in fig P10-21 is biased as a bistable MV, and a voltage trigger pulse is applied in series with L, as shown in fig P10-22, what should the minimum pulse height be (assuming that τ is sufficiently large for triggering) to cause a change in the state of MV?

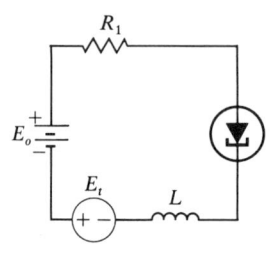

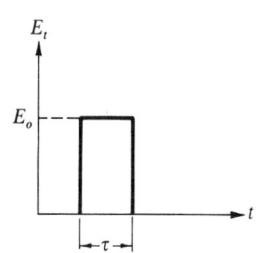

Fig P10-22

10-23 Derive (10-60).

10-24 Show an FET version of the Miller sweep and of the bootstrap sweep circuits. Include the biasing circuitry.

10-25 A Miller sweep, using an operational amplifier, is to be designed for the following specifications:

The peak voltage of the ramp is to be $5\ V$ with a sweep error of approximately 0.1%. The sweep time is $T_s = 1\ \mu sec$. An electronic switch provides a pulse with an amplitude of $1\ V$.

The available operational amplifier specifications are: open loop voltage gain $= 10^5$, $R_i = 100\ k\Omega$, $R_o = 100\ \Omega$. Show your circuit with component values of your design.

Problems 611

10-26 Using a monostable MV with pulse width T_1, an RC differentiator, and a diode, show the circuit diagram for the box to obtain a delayed trigger as shown in fig P10-26.

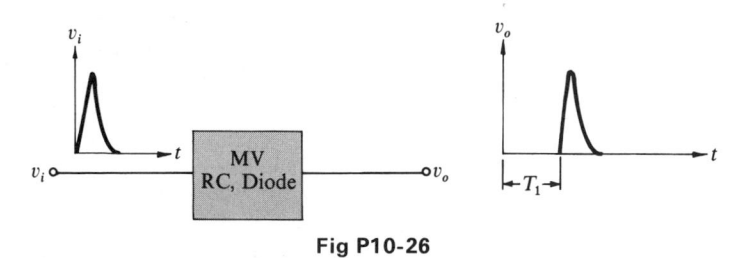

Fig P10-26

10-27 For the voltage multiplier circuit shown in fig P10-27 find V_{o1} and V_{o2}.

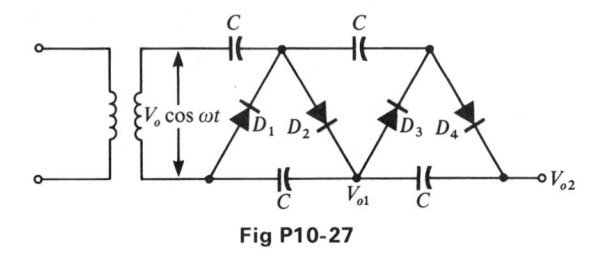

Fig P10-27

10-28 A current clamping circuit is shown in fig P10-28. Assume a two-segment, piecewise-linear model for the diode having the forward and the reverse resistances R_f and R_r, respectively and the breakpoints at the origin. Sketch the output current i_o for $0 \le t \le 3t_1$. Label the various segments and the time constants carefully.

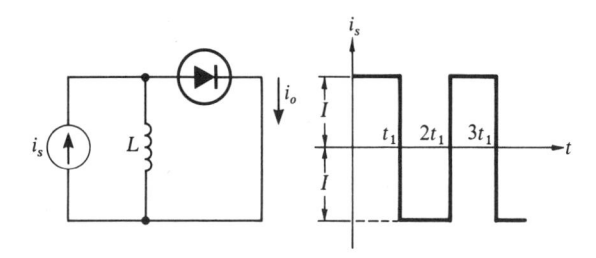

Fig P10-28

11
Logic Circuits and Digital Integrated-Circuit Functional Blocks

11-1 Introduction

The fundamental requirement of electronic circuits used for digital operation is that the electrical variables, current or voltage, that represent information be discrete and two valued, or binary.[1] Each binary digit (abbreviated *bit*) assumes one of two states. Each of the two states of the circuit has a permissible range, but the range is *nonoverlapping*, so that the circuit is always in either one state or the other, but *never* in both or neither of the two states. For example, the circuits may operate in either the saturation or the cutoff state. Of course, nonsaturating circuits are also used but again the operation is always at one or the other of two separate levels. The digital-circuit designer must ensure that the signal is clearly within the range of the two discrete allowed regions and never in the forbidden region. The two discrete

[1] Multivalued discrete circuitry is also possible, but we shall not consider it here.

states are usually designated as *logic* **1** and *logic* **0**, as shown in fig 11-1. In fact, one could have logic **1** corresponding to all voltages $\geq V_2$ and logic **0** corresponding to all voltages $\leq V_1$, with the separation between V_2 and V_1 as large as practicable. Digital circuits thus are tolerant to much wider component variations than are linear circuits. Electronic circuits which perform complex logic functions consist of interconnections of a large number of a few basic elementary digital circuits or iterative functional blocks. Thus, the fundamental characteristics of elementary digital networks are:

(a) Quantization, the capability of the network to produce and preserve the *two* discrete states, usually represented by logic **0** and **1**.
(b) Logic, the capability of the network to perform the basic logic functions.

We shall not undertake a detailed study of the above topics, since the latter is a vast topic and covered in detail in other courses and texts (Ref 1, 2, 4, 6). In this chapter we briefly consider binary system representation of numbers and elementary Boolean algebra. The uses of diodes and transistors to perform the required switching and logic operations in digital circuits are discussed next. The use of these functional blocks to perform counting and shifting functions is also treated.

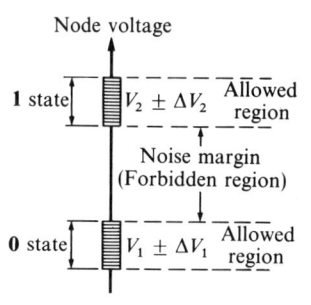

Fig 11-1 Node voltage levels showing allowed and forbidden ranges of values

11-2 Binary System

In a binary system representation of numbers only two numerals, 0 and 1, are allowed. The role played by 0 and 1 in a binary system is the same as that played by 0, 1, 2, ... 9 in the decimal system. In a decimal system we use the base, or *radix*, 10. For example, the digits in the number 1971 (one thousand nine hundred seventy-one) have the following positions and meaning:

$$(1971)_{10} = 1 \times 10^3 + 9 \times 10^2 + 7 \times 10^1 + 1 \times 10^0$$

Note that we use numbers from 0 to 9 multiplied by powers of 10, depending on their positions.

In a binary system we use the base 2, and only the numbers 0 and 1 are used. The interpretation is, otherwise, similar.[2] For example:

$$(1001)_2 = 1 \times 2^3 + 0 \times 2^2 + 0 \times 2^1 + 1 \times 2^0 = (9)_{10}$$

$$(01101)_2 = 0 \times 2^4 + 1 \times 2^3 + 1 \times 2^2 + 0 \times 2^1 + 1 \times 2^0 = (13)_{10}$$

A decimal system of numbers can be converted to a binary system of numbers, as shown in table 11-1. In table 11-1, the given decimal number, D, to be converted to a binary system of numbers, is divided by 2, the quotient C_1 and the remainder R_1 is put as shown. The process is repeated, i.e., divide C_1 by 2 and enter the quotient C_2 and the remainder R_2, until a quotient of 0 is obtained.

TABLE 11-1   Conversion of a Decimal Number to a Binary-Coded Decimal Number

Quotient of Divide by 2	$C_n = 0$	1	...	C_2	C_1	Decimal Number D
Remainder	R_n	R_{n-1}	...	R_2	R_1	Binary Number

The array of number $R_n \cdots R_1$ (which are either 0 or 1) is the binary representation. For example consider the decimal number $(13)_{10}$. The binary representation as obtained from table 11-1, as shown below, is $(1101)_2$.

0	1	3	6	$(13)_{10} = (D)_{10}$
1	1	0	1	Binary number

This can be readily checked since $(1101)_2 = 1 \times 2^3 + 1 \times 2^2 + 0 \times 2^1 + 1 \times 2^0 = (13)_{10}$.

It may be pointed out that the same algorithm holds true for conversion to any other base, i.e., divide repeatedly by the radix r, and continue the procedure.

---

[2] A number system N to any base r can be expressed as

$$(N)_r = a_n r^n + a_{n-1} r^{n-1} + \cdots + a_1 r^1 + a_0 r^0$$

where $n = 1, 2, 3, \ldots$, r = base (e.g., for decimal, $r = 10$; for binary, $r = 2$; for octal, $r = 8$); a_n = numbers with values between 0 and $r - 1$.

A decimal fraction is converted to a binary fraction by repeatedly multiplying the fraction by 2 and saving the integers. As an example, to convert $(0.79)_{10}$ to binary, we have

Decimal Fraction	0.79	2(0.79) = 1.58	2(0.58) = 1.16	2(0.16) = 0.32	2(0.32) = 0.64	2(0.64) = 1.28	etc
Binary Fraction		1	1	0	0	1	

To show how close the binary fraction 0.11001 is to the decimal 0.79

$$(0.11001)_2 = 1 \times 2^{-1} + 1 \times 2^{-2} + 0 \times 2^{-3} + 0 \times 2^{-4} + 1 \times 2^{-5}$$

$$= 0.50 + 0.25 + 0 + 0 + 0.03125 = (0.78125)_{10}$$

Of course, carrying the process farther leads to a more accurate result.

A decimal number which contains an integer and a fraction is converted to binary in two parts separately and then combined. Note that a decimal point (more properly, a radix point) is used to separate a binary integer from a binary fraction in the same manner as is done for decimals.

A simple binary addition is illustrated in the following:

```
        Base 10              Base 2
          19       =       0 1 0 0 1 1
         +38       =       1 0 0 1 1 0
         ───               ─────────────
          47               1 1 0 1 0 1    Sum
           C                         c    Carry
                           ─────────────
                           1 1 0 0 0 1
                                     c
         ────              ─────────────
        (57)₁₀    =       (1 1 1 0 0 1)₂ ← Complete
                                            solution
```

Arithmetic operations (addition, subtraction, multiplication, and division) of binary numbers can be performed by simple operations such as add and shift. For example, a device which accepts two signals representing the augend and the addend digits and produces output signals representing the sum and carry is known as the *half adder* (see Problem 11-26). The output of the half adder may be conveniently expressed by the Boolean functions.

11-3 Boolean Relations

Boolean algebra is naturally suited for the analysis and design of logic, and hence digital circuits with two discrete states. The two discrete states are called the binary or Boolean variables.[3] For these circuits, we arbitrarily consider, for example, for *npn* transistor switches, off as the logic **1** and on as the logic **0** state. Thus, if an *npn* transistor switch is driven between cutoff and saturation, the output voltage at cutoff, $v_c = V_{CC}$, represents logic **1** and the output voltage at saturation, $v_c = V_{CE(sat)}$, represents logic **0**. Since no other states are permitted, the binary variables **0** and **1** are the only possibilities. Thus, the Boolean variables are restricted to *one* of two values at a time, namely, **0** or **1**, i.e., if a variable is not **0**, then it is **1**, and if it is not **1**, then it must be **0**.

The postulates for the complementation, addition, and multiplication of the Boolean variables are listed in table 11-2.

TABLE 11-2 *Boolean Postulates*

The NOT Operation	The OR Operation	The AND Operation
$\bar{\mathbf{1}} = \mathbf{0}$	$\mathbf{0 + 0 = 0}$	$\mathbf{0 \cdot 0 = 0}$
$\bar{\mathbf{0}} = \mathbf{1}$	$\mathbf{0 + 1 = 1}$	$\mathbf{1 \cdot 0 = 0}$
	$\mathbf{1 + 0 = 1}$	$\mathbf{0 \cdot 1 = 0}$
	$\mathbf{1 + 1 = 1}$	$\mathbf{1 \cdot 1 = 1}$

In table 11-2, the bar over the variable indicates complementation, and is read "not" (i.e., $\bar{\mathbf{1}}$ = not **1** = **0**). Bold-face numbers and signs are used to emphasize Boolean relations. Often in the AND operation the sign is omitted, e.g., $A \cdot B = AB$, where A and B are Boolean variables. The postulates listed in table 11-2 are sufficient for all Boolean operations.

11-4 Basic Building Blocks

The various operations, as postulated in table 11-2, are performed by the following building blocks or gates:

---

[3] Statements such as "true" or "false," "yes" or "no," etc., are also binary

11-4 Basic Building Blocks

(a) **The NOT Gate**

The NOT gate, shown symbolically in fig 11-2a, performs the NOT or complementation operation. This is the simplest gate and is realized by an inverter circuit, as will be described shortly. The letter *I* (abbreviation for inverter) or a bar (–), for complementation, is used in place of *N* by some authors. An alternative symbol for the NOT gate is also shown in the figure.

(b) **The OR Gate**

The OR gate, shown symbolically in fig 11-2b, performs the Boolean OR operation. From table 11-2, it is seen that if one or more of the variables is **1**, the OR operation yields **1**. Thus, the OR gate is one which produces a **1** output if at least one of its inputs is **1**. An alternative symbol for the OR gate is also shown in the figure.

(c) **The AND Gate**

The AND gate, shown symbolically in fig 11-2c, performs the Boolean AND operation. From table 11-2, it is seen that the AND operation yields **1** only if each variable is **1**. Thus the AND gate is one which produces a **1** output only if *all* the inputs are **1**. An alternative symbol for the AND gate is also shown in the figure.

(d) **The NOR and NAND Gates**

The NOR (from NOT-OR) gate is logically equivalent to a cascade combination of an OR gate followed by a NOT gate. Symbolic representation of a NOR gate is shown in fig 11-3(a). The NAND (from NOT-AND) gate is logically equivalent to a cascade combination of an AND gate followed by a NOT gate. Symbolic representation of a NAND gate is shown in fig 11-3(b). Note that the NOR and NAND operations are obtained from a combination of the basic OR, NOT, and AND operations. Further, the NOR (or NAND) gate is sufficient to perform all the logic operations performed by OR, AND, and NOT gates (Problems 11-12 and 11-13). Thus the NOR (or NAND) gate

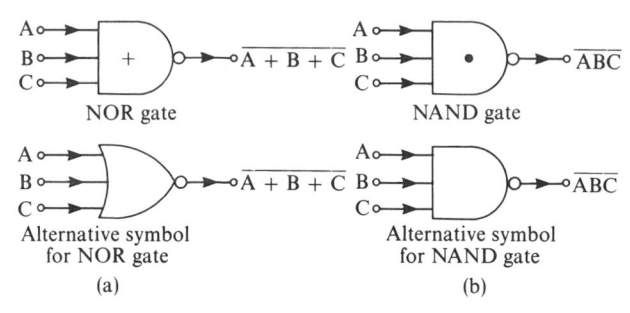

Fig 11-3 (a) Graphic symbol for a *NOR* gate and alternative symbol; (b) Graphic symbol for a *NAND* gate and alternative symbol

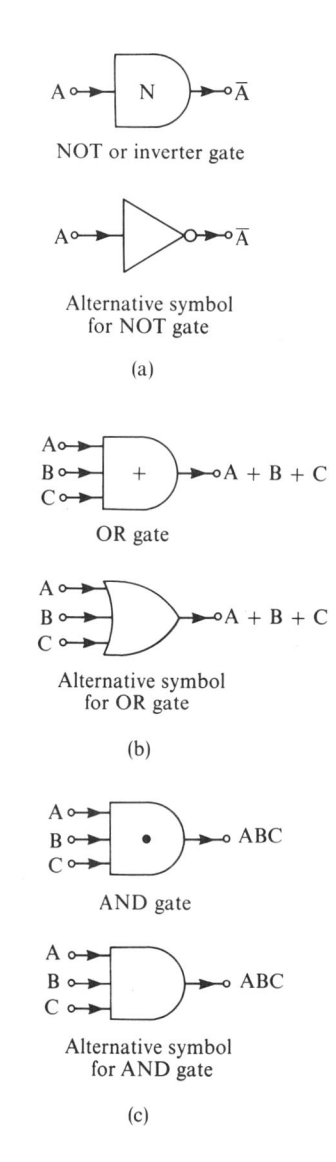

Fig 11-2 (a) Graphic symbol for a *NOT* gate and alternative symbol; (b) Graphic symbol for an *OR* gate and alternative symbol; (c) Graphic symbol for an *AND* gate and alternative symbol

may be viewed as a universal gate. The alternative symbols for the NOR and NAND gates are also shown in the figure. The alternative symbols in all of the above cases are in published standards for digital logic circuit diagrams. The gates with the N, ·, and + sign, however, may be more suggestive of their functions and helpful to the readers of this text.

There is a well known theorem of Boolean algebra known as DeMorgan's theorem which can be used to show the relation between the NOR and the NAND operations. DeMorgan's theorems state

$$\overline{(X + Y)} = \overline{X}\,\overline{Y} \tag{11-1}$$

$$\overline{XY} = \overline{X} + \overline{Y} \tag{11-2}$$

The three gates which perform the basic Boolean operations are, therefore, the OR, AND, and NOT gates. The functions for all possible combinations of three input variables are listed in table 11-3. Such a table is usually called a *truth table*.

TABLE 11-3 *Truth Table for Three Boolean Variables*

Variable Values	OR	AND	NOR	NAND
$A\ B\ C$	$A + B + C$	ABC	$\overline{A + B + C}$	$\overline{ABC}$
0 0 0	0	0	1	1
0 0 1	1	0	0	1
0 1 0	1	0	0	1
0 1 1	1	0	0	1
1 0 0	1	0	0	1
1 0 1	1	0	0	1
1 1 0	1	0	0	1
1 1 1	1	1	0	0

A few other operations, which can be obtained from the basic operations listed earlier, often arise in logic operations so that, for convenience, these are often considered as operations by themselves, even though these are not independent operations. These are the *exclusive OR, inhibitor, coincidence*, and ORNOT operations. The symbols for these operations are shown in fig 11-4. The *exclusive OR* operation, denoted by ⊕, is defined as

$$A \oplus B = \overline{A}B + \overline{B}A \tag{11-3}$$

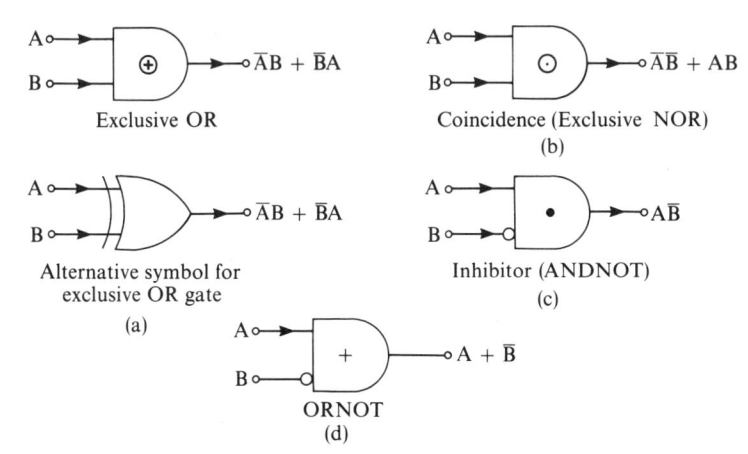

Fig 11-4 (a) Graphic symbol for an exclusive *OR* gate and alternative symbol; (b) Graphic symbol for a coincidence (exclusive *NOR*) gate; (c) Graphic symbol for an inhibitor (*ANDNOT*) gate; (d) Graphic symbol for an *ORNOT* gate

The *coincidence*, also called exclusive NOR operation, denoted by ⊙, is defined as

$$A \odot B = \overline{AB} + AB \qquad (11\text{-}4)$$

From (11-3) and (11-4) we can readily show that

$$\overline{A \oplus B} = A \odot B \qquad (11\text{-}5)$$

The *inhibitor*, also referred to as ANDNOT, denoted by an inhibit sign before the block as shown in fig 11-4(c) is defined as $A\bar{B}$. Note that if there is always a signal present on A, i.e., $A = 1$, then an inhibitor acts as an inverter. The ORNOT is shown in fig 11-4d where the output is $A + \bar{B}$. If $A = 0$ the ORNOT acts as an inverter. The reader is cautioned not to confuse ANDNOT with AND-NOT (also ORNOT with OR-NOT). The AND-NOT is a cascade of an AND gate followed by a NOT gate, hence a NAND operation. The ANDNOT and ORNOT operations are not commonly identified as basic logic gates. They often occur, however, in logic circuitry, and are therefore, included here.

The various logic operations may be graphically visualized by a Venn diagram (see Ref 4). For example, for two variables, the Venn diagram is shown in fig 11-5. The two circular areas represent the elements A and B. The various operations each denoting a subset can be clearly identified. For example the single shaded areas denote the ⊕ operation, according to (11-3). If both A and B are binary, this operation corresponds to $A \neq B$. The double shaded area and the

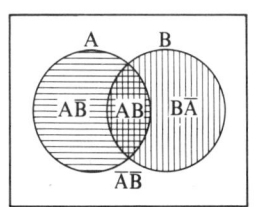

Fig 11-5 Venn diagram for two variables

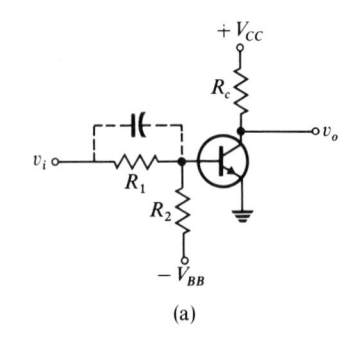

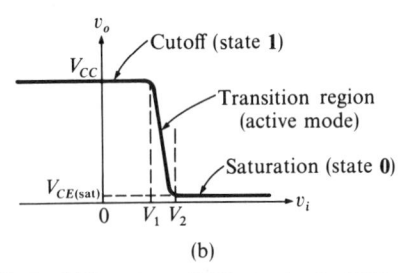

Fig 11-6 (a) Inverter or *NOT* gate circuit; (b) Transfer characteristic of the inverter

area with no shade denote the $\odot$ operation according to (11-4). The relation in (11-5) is thus readily confirmed by the diagrams. If both A and B are binary, (11-5) corresponds to $\overline{(A \neq B)} = (A = B)$.

Next we consider circuit realizations of the various gates.

11-5 Circuit Implementation of the Building Blocks

The complementation operation is performed by the inverter circuit of fig 11-6a. This circuit has already been discussed in Chapter 10. The transfer characteristic of the inverter, which is of interest in logic operations, is shown in fig 11-6b. We shall define the more positive voltage level to be logic **1** and the less positive voltage to be logic **0**. This identification is usually referred to as *positive logic*. The reverse designation is also possible and is called *negative logic*. Actually, a circuit using positive logic can be converted into a circuit using negative logic, performing the same logic function, simply by reversing the polarities of the supply voltages and changing *npn* to *pnp* transistors. Alternatively stated, a circuit performing an AND (or NAND) function with positive logic performs an OR (or NOR) function with negative logic, and vice versa. In integrated circuits, which are universally used for digital systems (because it is the most economical approach due to the batch fabrication of thousands of identical circuits), *npn* transistors are commonly used. This is so because it is more difficult to fabricate fast-switching (high f_T), high-gain *pnp* transistors on an IC chip. Thus, we shall consider positive logic with *npn* transistors throughout the chapter unless otherwise stated.

From fig 11-6b, for positive logic, the two discrete levels are $v_o = V_{CC}$, which corresponds to logic **1**, and $v_o = V_{CE(sat)}$, which corresponds to a logic **0**. If the circuit parameters in fig 11-6a are all known, the transfer characteristic can be completely determined. In digital logic circuits this is seldom necessary, since the input signal will also be binary, i.e., will have values which are either **1** or **0**. Thus, if the input is **1**, the output is **0**, and vice versa. Figure 11-7 shows an idealized representation of the output signal for the given input signal.

The transfer characteristic of an inverter is useful in determining the *noise immunity* of a logic circuit. There are many definitions of noise immunity.

Excessive noise in a system may cause a logic circuit to indicate a false output. Usually the larger the logic swing, i.e., the difference between the voltage levels in state **1** and state **0**, the better the noise

11-6 Diode Logic (DL) 621

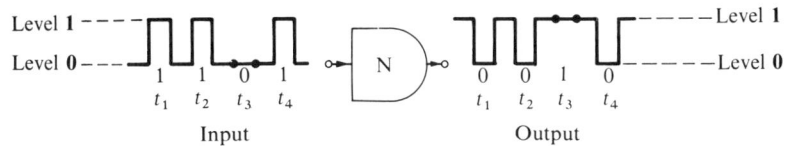

Fig 11-7 *NOT* gate with input and output pulse waveforms

immunity (see fig 11-1). The noise susceptibility of a logic gate is usually determined from the maximum and minimum levels of the transfer characteristics, based on worst-case conditions.

Circuit implementation for logic gates can take a variety of forms. The various schemes which can be used to perform logic operations are as follows:

Diode Logic (DL)
Resistor-Transistor Logic (RTL)
Direct-Coupled Transistor Logic (DCTL)
Diode-Transistor Logic (DTL)
Transistor-Transistor Logic (TTL or T²L)
Emitter-Coupled Logic (ECL), (also called Current Mode Logic CML)

The logic circuits, together with their principles of operation, are discussed in the following sections. It should be noted that of the various schemes for circuit implementation of logic gates, the first two, DL and RTL (the one transistor per gate version), are used in discrete circuits, while the others are used in integrated-circuit technology. We do not plan to discuss the detailed analysis and design of these circuits, because of space limitations as well as rapid changes in the state-of-the-art of circuit design for various gates in digital integrated circuits. Our purpose here is only to familiarize the reader with the different types of logic gates; detailed discussions are given in Refs 1 and 3.

11-6 Diode Logic (DL)

The diode logic circuit (abbreviated DL) is one of the simplest logic circuits. A DL gate which performs the AND logic function is shown in fig 11-8a. A three-input gate is shown for illustrative purposes only. A larger or smaller number of inputs than indicated can be applied. Each input may assume either of two voltage levels, an upper voltage level, denoted by logic **1** or $V(1)$, and a lower level denoted by logic **0** or $V(0)$. The AND function is obtained in the following manner. When

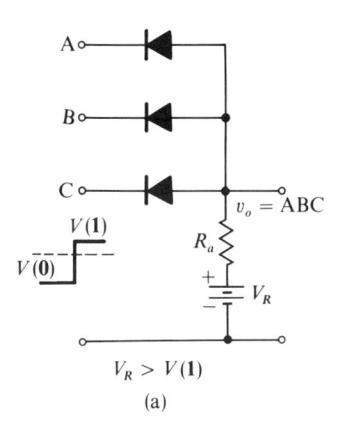

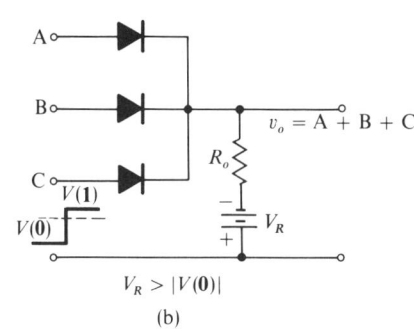

Fig 11-8 (a) Diode logic (*DL*) *AND* gate circuit; (b) Diode logic (*DL*) *OR* gate circuit (V_R is a reference voltage in above)

all the inputs are simultaneously at $V(\mathbf{1})$, all the diodes will conduct and the output voltage will be clamped to $V(\mathbf{1})$.[4] The output impedance will be $(R_f/n) \| R_a$ for $R_s = 0$ (with $n = 3$ in fig 11-8a). Usually $R_a \gg R_f$, where R_f is the forward resistance of the diode, so that the output impedance is very low. Similarly, if all the inputs are simultaneously at $V(\mathbf{0})$, again all the diodes will conduct and the output voltage will be clamped to $V(\mathbf{0})$. However, if there is no coincidence, i.e., if any one or two, but not all of the inputs is at $V(\mathbf{0})$, then the output of the gate will be at $V(\mathbf{0})$, and the other diode(s) is prevented from conducting. In other words the output of the circuit shown in fig 11-8a follows the lowest input voltage. The reader can verify the AND column of the truth table (table 11-3) for this gate.

A DL gate which performs the OR logic function is shown in fig 11-8b. Again, three inputs are shown for illustrative purposes only. The output of this gate follows the highest input voltage.

In other words, the output voltage will be at $V(\mathbf{1})$[5] under all conditions except the case when *all* the inputs are at $V(\mathbf{0})$. When all the inputs are at $V(\mathbf{0})$, the output will be at $V(\mathbf{0})$. This operation corresponds to the OR function, as shown in table 11-3.

The DL gates have the following drawbacks. If several diode logic circuits are connected in cascade to implement a complicated Boolean expression, the logic levels deteriorate because of the voltage drops across the diodes. Hence, level-restoring circuits are needed. This is done by using a one-transistor inverter circuit. This inverter also provides the NOT function, which is often needed in some Boolean function implementations.

Having added a transistor to the logic system, why not design a universal logic circuit anyway? Such circuits, using all transistors, are now being made in integrated form more cheaply than in discrete form. These circuits are taken up next. Except in a few special applications, diode logic has become virtually obsolete.

11-7 Resistor-Transistor Logic (RTL)

The basic resistor-transistor logic (RTL) gate of the discrete circuit is shown in fig 11-9a. The IC version of the RTL gate is shown in fig 11-9b.

---

[4] The actual output voltage level will be $V(\mathbf{1}) + V_o$ where V_o is the breakpoint voltage of the diodes (e.g., ≈ 0.6 V for Si). We shall omit V_o in subsequent discussion for simplicity.

[5] Actually $V(\mathbf{1}) - V_o$, but we shall omit the voltage drop V_o for simplicity.

11-7 Resistor-Transistor Logic (RTL)

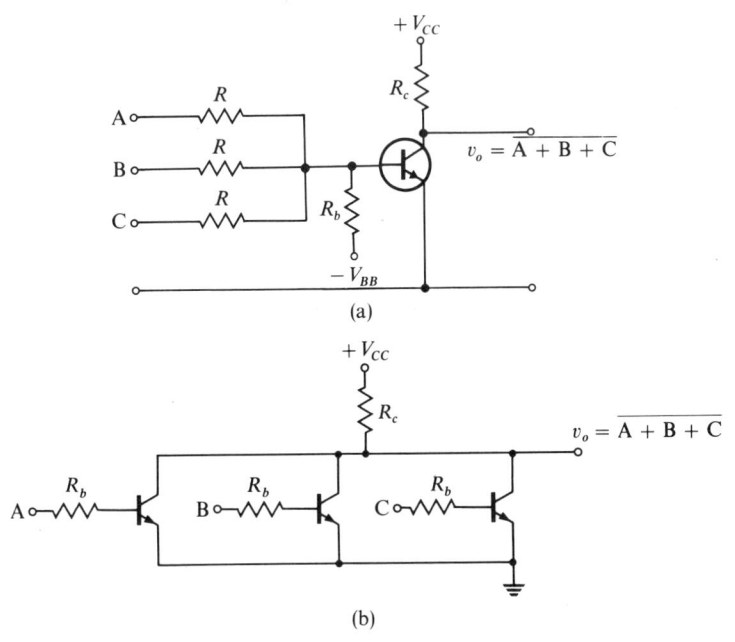

Fig 11-9 (a) Discrete resistor-transistor logic (*RTL*) *NOR* gate circuit (when speed-up capacitors are used across the resistors in discrete circuits, the gate is called *RCTL*); (b) IC resistor-transistor logic (*RTL*) *NOR* gate circuit.

The circuit of fig 11-9b is a modification on the DCTL gate and will be considered in the next section. This circuit, as we shall explain, is a NOR gate.

Since the emitter-base junction is reverse biased by V_{BB}, in the absence of any input, the transistor output is

$$v_o \simeq V_{CC} \equiv V(1)$$

If a signal of appropriate level, i.e., $V(1)$, is applied to at least one of the inputs, the transistor turns on and the output voltage, under saturated conditions, is very low:

$$v_o \simeq V_{CE(sat)} = V(0)$$

The presence or absence of other inputs does not have appreciable effect on the value of v_o. Thus, when any of the inputs is logic **1** (i.e., a voltage of $V(1)$ is applied to any of the input terminals) the output is logic **0** (i.e., $v_o \approx V_{CE(sat)}$). If all the inputs are logic **0** (i.e., $V_A = V_B = V_C \approx V_{CE(sat)}$) then negligible collector current flows and the output is logic **1** (i.e., $v_o = V_{CC}$). If all the inputs are $V(\mathbf{1})$, then the transistor is driven heavily into saturation and $v_o = V(\mathbf{0})$. Thus we see that the gate performs the NOR operation. In order not to have overlap in the states **0** and **1**, we must have the minimum value of $V(\mathbf{1})$ considerably larger than the maximum value of $V(\mathbf{0})$. This would ensure the noise immunity of the gate.

The maximum number of input terminals that can be connected to a gate is usually called the *fan-in* and the maximum number of load gates that can be connected to the output is referred to as the *fan-out*. For example, the logic gate shown schematically in fig 11-10 has a fan-in of 4 and a fan-out of 4. In switching transistors, the output current is usually proportional to the fan-out. A high fan-out requires large collector current. Greater current in a saturated transistor

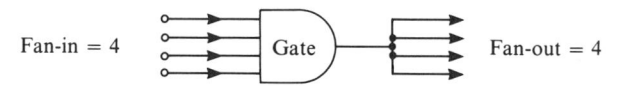

Fig 11-10 A gate with its fan-in and fan-out

reduces the on overdrive, which increases the rise and fall times but decreases the storage time. A large fan-in increases the input capacitance and thus reduces the switching speed of the gate. The sizes of the fan-in and fan-out are important (the larger these numbers, the less the number of circuit elements needed in a digital system), and there is an upper limit for the fan-in and fan-out capabilities of logic circuits. These numbers are dependent upon the types of logic circuits. Certain special circuits can drive more load circuits and, therefore, have higher fan-outs. In order to increase the fan-out capability, usually buffer stages in the form of an emitter follower or a push-pull type output are used in many of the digital logic modules.

The RTL gate has the following disadvantages. The speed is usually slow because of the storage time when driven into saturation. This situation exists in others also, except for ECL, wherein the transistors are not driven into saturation. Temperature change has an effect on the gate, since both I_{CO} and V_{BE} are strongly temperature-dependent; thus, the operating conditions and component values are critical. The speed can be improved by using speed-up capacitors across R in fig 11-9 (but this is not a practical solution). When the capacitors are used, the gate is usually referred to as resistor-capacitor-transistor logic (RCTL). The disadvantage of RCTL is the large space required for capacitors and also the possibility of crosstalk between the input channels. The large area requirement of RCTL makes it unattractive for integrated circuits and it is, therefore, not found in IC form. In fact, the circuit of fig 11-9a is now seldom used in practice and the RTL usually refers to the circuit of fig 11-9b.

11-8 Direct-Coupled Transistor Logic (DCTL)

A direct-coupled transistor logic (DCTL) gate is shown in fig 11-11. The gate performs a NOR operation, as explained in the following. First, notice that the **0** state is $V(0) = V_{CE(sat)}$, since this is the low level for v_o. It occurs when at least one of the transistors, Q_1, Q_2, or Q_3 is saturated. If all the inputs, A, B, and C, are at $V(0)$, transistors Q_1, Q_2, and Q_3 will be essentially off, since such a small voltage cannot turn silicon transistors on. Thus, if there is no load, i.e., no fan-out transistors, v_o will be at V_{CC}. The presence of fan-out transistors, however, which will be driven into saturation clamps the output voltage, v_o, at $V_{BE(sat)}$. Hence when A, B, and C are at $V(0)$, in the presence of fan-out transistors (see fig 11-12a), the output will be at $v_o = V_{BE(sat)} \equiv V(1)$. The difference between the voltage levels in the **0** and **1** states is less than one volt and the gate is thus susceptible to noise. Because the transistors are driven into saturation, the speed is slow, due to storage time. There is also another disadvantage to this gate.

Consider the fan-out transistors Q_1, Q_2, and Q_3 in fig 11-12a. The base current for all the transistors is supplied through V_{CC} and R_L. If the input characteristics of these transistors, which are highly nonlinear, are not identical, one transistor may draw most of the available output current, as shown in fig 11-12a. The other transistors may not draw enough current to be driven into saturation or, in fact, may even not be turned on. This effect is called "current-hogging." To remedy current-hogging, a series resistor, R_B, is usually connected to the base of each transistor, as illustrated in fig 11-12b. The addition of the series resistors makes the current distribution more even, as shown in the v-i characteristics. It should be noted, however, that the larger the value of R_B, the smaller the fan-out capability of the gate (Problem 11-19). Because of the addition of R_B, this gate is referred to as RTL gate. Note that this is the same circuit as shown in fig 11-9b.

The attractive feature of DCTL gate is its simplicity and low standby power. These properties and the fact that transistors are cheap in integrated circuits and quite uniform in performance make the DCTL gate desirable in integrated circuits, if ultimate speed is not a problem. Note further that since the RTL of fig 11-9b has the same features of DCTL but with the additional desirable property of no current hogging

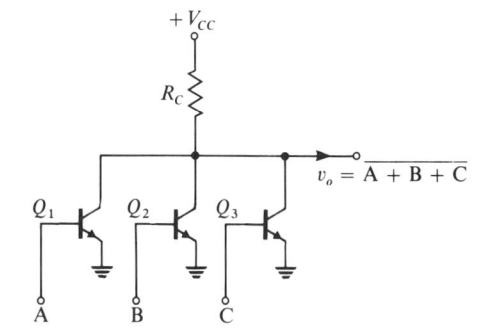

Fig 11-11 Direct-coupled transistor logic (DCTL) NOR gate circuit

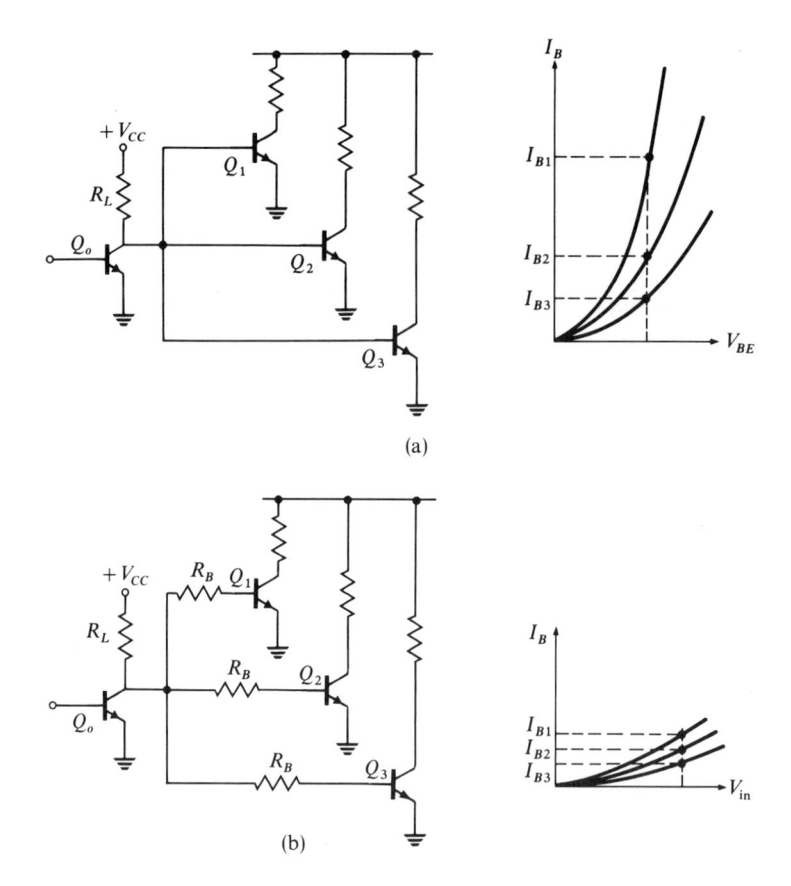

Fig 11-12 (a) *DCTL* gate circuit and input v-i characteristics, showing current-hogging effect; (b) *DCTL* gate circuit with base resistors added and v-i characteristics, showing reduced current-hogging effect

problem it is commonly used in integrated circuits. The circuit of fig 11-9b can, of course, be used in discrete circuit form and speed-up capacitors are then used across the R_B resistors to improve the speed of the gate.

11-9 Diode-Transistor Logic (DTL)

A diode-transistor logic (DTL) gate which performs a NAND operation is shown in fig 11-13. Note that the circuit consists essentially of the DL AND gate of fig 11-8a and a transistor inverter, thus per-

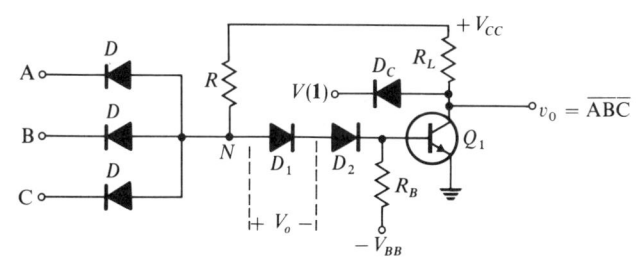

Fig 11-13 Diode-transistor logic (*DTL*) *NAND* gate circuit

forming the NAND operation. In fig 11-13, diodes D_1 and D_2, which are always forward biased, provide an offset voltage (approximately $0.6 + 0.6 = 1.2$ volt) so that the circuit will operate with input and output voltage levels which are compatible. Diode D_C provides clamping so that voltage level $V(1)$ can be low. Furthermore, D_C prevents the succeeding stage from being driven too heavily into saturation and improves the switching speed by reducing the storage time. Voltage level $V(1)$ is thus the clamp voltage, which is usually low (approximately 3 or 4 volts) and $V(0) = V_{CE(\text{sat})}$. The NAND operation is as follows: If at least one of the inputs is at $V(0)$, the potential at node N will be $V(0) + V_o$, which is approximately $(V_{CE(\text{sat})} + 0.6)$ for a Si diode. Because of the voltage drops across the forward-biased diodes (equal to $2V_o$), the transistor will be off, and thus v_o is clamped at $V(1)$. If all the inputs are at $V(1)$, then the potential of node N is $V(1) + V_o$. This potential is larger than $2V_o$ and large enough to drive the transistor into saturation and thus $v_o = V(0)$. Thus, the NAND column of the truth table in table 11-3 is verified.

In a DTL gate, the operation can be made relatively independent of the transistor characteristics. The low logic level of the gate makes it useful in integrated circuits. The coupling and input diodes, however, usually require different characteristics, which is a drawback in integrated circuits.

11-10 Transistor-Transistor Logic (TTL)

Two transistor-transistor logic (TTL or T²L) gates which perform the NAND operation are shown in figs 11-14 and 11-15. The operation of these circuits is similar to the DTL NAND gate, except that the input diodes of the DTL gate are replaced by the transistors in a discrete circuit (fig 11-14), and the multiple-emitter transistor in the integrated-

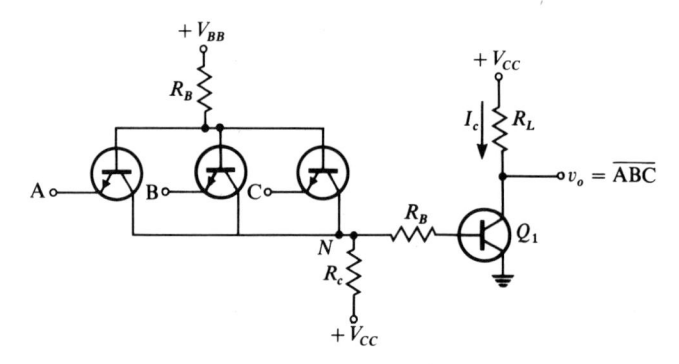

628 11/Logic Circuits and Digital Integrated-Circuit Functional Blocks

Fig 11-14 Transistor-transistor logic (T^2L) $NAND$ gate circuit; discrete version

(a)

(b)

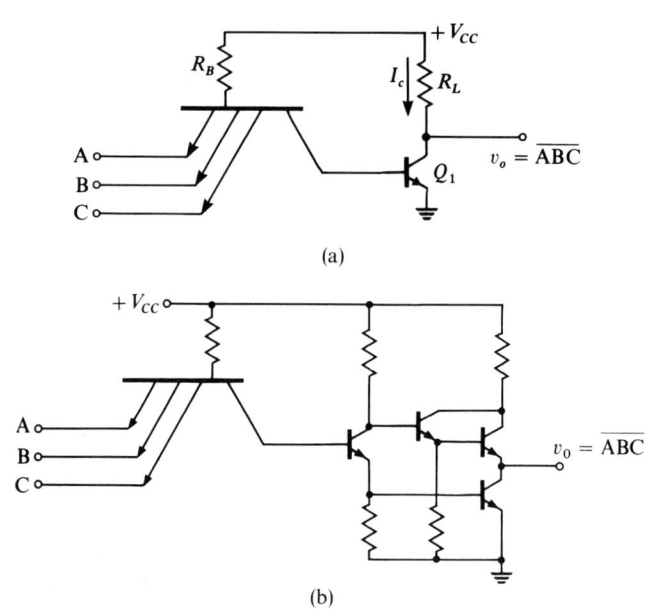

Fig 11-15 (a), (b) Transistor-transistor logic $NAND$ gate circuits; integrated circuit version

circuit version (fig 11-15). The multiple-emitter transistor requires a very small silicon wafer area and is, therefore, attractive in integrated circuits. The use of transistors at the input has eliminated the need for the offset diodes of the DTL gate. When the output stage is driven into saturation, the logic levels are

$$V(0) = V_{CE(\text{sat})}, V(1) = V_{CC} - I_C R_L \qquad (11\text{-}6)$$

The value of R_B is chosen such that, for all loading conditions, the turn-on and turn-off conditions of the input transistors are met. For example, to meet the turn-on conditions of the input transistors, in fig 11-14, we must have $V_{BB} > V(0) + V_{BE}$, and to meet the turn-off conditions, $V_{BB} < V(1) + V_{BE}$. We must also ensure that when Q_1 is on, the low potential at N does not forward-bias any of the input transistors.

In integrated circuit technology the T²L gate, shown in fig 11-15b, is commonly used. The totem-pole output stage provides a great capability for a larger fan-out and improved switching time for the gate.

11-11 Emitter-Coupled Logic (ECL or CML)

The basic form of an emitter-coupled logic (ECL) gate, which is also called current mode logic (CML), is shown in fig 11-16. This gate performs both the OR and NOR logical functions, as will be described.

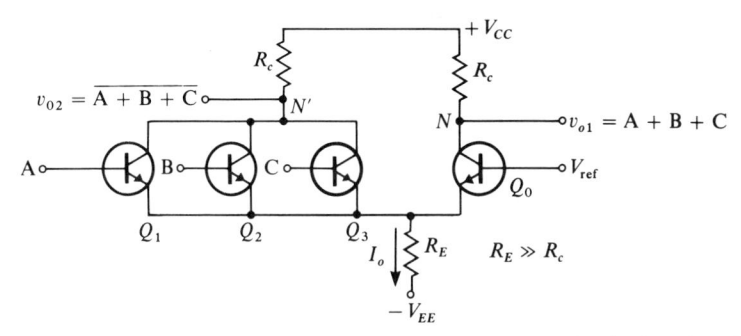

Fig 11-16 Emitter-coupled logic (*ECL*) gate circuit (this connection is also referred to as current mode logic (*CML*))

The switching speed of this circuit is fast because of the current-mode switching (see Section 10-3), which avoids saturation. The voltage $-V_{EE}$ and the emitter resistor R_E are chosen such that they approximate a current source I_o. For this circuit, the output logic levels **1** and **0** are

$$V(\mathbf{1}) = V_{CC} \quad \text{and} \quad V(\mathbf{0}) = V_{CC} - I_o R_c \approx V_{CC}\left(1 - \frac{R_c}{R_E}\right) \quad (11\text{-}7)$$

The reference voltage V_{ref} may be at ground potential if desired. Note that in the approximate expression, (11-7), we have assumed that R_E is very large and that $|V_{EE}| = V_{CC}$. The operation of the circuit has been described in Section 10-3, and is briefly as follows.

If *negative* signals are applied at all the inputs A, B, and C, then Q_1, Q_2 and Q_3 are all off and the current I_o will flow through Q_0. Under this condition $v_{o1} = V(0)$ and $v_{o2} = V(1)$. If a *positive* signal is applied to one or more of the inputs, the corresponding transistor(s) will turn on and the current I_o will no longer flow through Q_0 and Q_0 will be turned off. Thus, $v_{o1} = V(1)$ and $v_{o2} = V(0)$. Note that, for the ECL gate, there is a shift in the voltage levels between the input and the output. The input signal levels vary from some negative to some positive value. The output signal levels, however, vary from $V_{CC}(1 - R_c/R_E)$ to V_{CC}, which are both positive. Thus, level-shifting is necessary when this gate is to be coupled to another such gate.

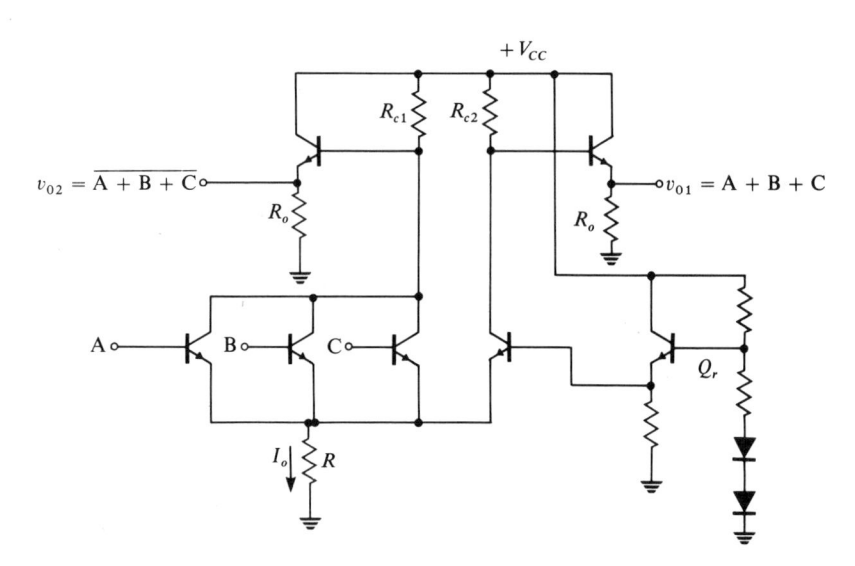

Fig 11-17 Integrated circuit emitter-coupled logic gate (Courtesy Motorola Semiconductor Products Inc.)

In integrated circuits, several modifications are made, and one such circuit is shown in fig 11-17 (see also Appendix D (MC 306)). In this circuit, emitter-followers are used to provide an offset voltage, compensating for the translation of levels discussed above. Transistor Q_r, with the diodes, acts as a base-driver circuit to provide a secondary power supply which is required to establish the reference voltage. This circuit does not require close tolerances and has an extremely large fan-out capability. The circuit configuration is the fastest among all the various types of logic gates. Since small logic swings are required to prevent saturation, however, noise immunity is lower than in other logic types.

11-12 Comparison of Various Logic Gates

In the preceding sections, we have discussed briefly the implementation of logic functions and their principles of operation using six common types of logic circuits: DL, RTL, DCTL, DTL, T^{2}L, and ECL. Among these, the DTL, RTL, T^{2}L, and ECL are common in integrated circuits. A comparison of various logic characteristics and attributes must include speed (propagation delay), power dissipation, noise immunity, fan-in and fan-out capabilities, and cost, among others. In this section, we summarize the salient features of the individual schemes. These are listed for convenience in table 11-4.[6] A complete comparison is very difficult, to say the least, and it should be pointed out that no single logic type is the best for all applications. DTL is popular in both discrete and integrated circuits, while T^{2}L and ECL are commonly used only in integrated circuits.

If low power dissipation is the primary objective the MOSFET gates, discussed in the next section, may be used. MOSFET technology has the additional advantages of very small size and very low cost, but has the main disadvantage of having low speed as compared with bipolar transistors.

In integrated circuits, if the primary requirement is speed the logic circuit to choose is ECL. In fact, this type of gate is very popular in high performance digital computers. If speed and noise immunity are both important, then T^{2}L is a good choice. In digital circuits, the designer is usually interested in speed and power. This is discussed in the next section.

---

[6] For quantitative values of noise immunity, fan-out, power dissipation, and speed, in bipolar transistors, the following values are the best for commercially available integrated-circuit logic gates: propagation delay $\approx$ 1 nsec, power dissipation $\approx$ 10 mW, fan-out $\approx$ 25, noise immunity $\approx$ 1 volt. Further improvements on these numbers are steadily being made. The above values are not necessarily achieved all in one type of gate.

TABLE 11-4 *Comparison of Various Logic Circuit Types*

Logic Circuit Type	Advantages	Disadvantages
DL	Simplicity in discrete circuits.	Not suitable for monolithic IC design.
RTL (fig 11-9a)	Cheap in discrete circuits. Simplicity. Low power dissipation.	Critical component values and operating conditions not suitable for monolithic IC design. Low speed.
DCTL	Suitable for monolithic integrated circuits. Simplicity. Low power dissipation.	Small number for fan-out. Poor noise immunity. Current-hogging problem due to spread in $V_{BE(sat)}$.
DTL	Input diodes provide isolation. Popular in discrete design. Can be designed in monolithic IC's. Good noise immunity. Reasonable speed.	Close-tolerance resistor ratios are required. Coupling and input diodes usually require different characteristics.
TTL	Average-to-good noise immunity. No stringent requirements on components. Commonly used in monolithic IC design. High speed. Low power dissipation. Large fan-out with totem-pole output.	In a multiple-emitter transistor, a high reverse-current gain can cause an emitter to act as a collector under certain conditions.
ECL	Because of nonsaturated operation, has the fastest speed attainable. Emitter-follower output. Large fan-out.	Relatively complex. Poor noise immunity. Some need stabilized low-voltage reference supply. Relatively high power dissipation.
RTL (fig 11-9b)	Same as DCTL, but no current hogging problem.	

11-13 Figure of Merit of a Gate

The product of *power dissipation* and *propagation delay* of a logic circuit is often referred to as a figure of merit of the circuit. The smaller this number the better the gate. This figure of merit plays essentially the same role as does the gain-bandwidth product in amplifiers. The power dissipation of a logic circuit is usually defined as the power the gate will dissipate under 50 per cent duty-cycle conditions. The dissipation should be low to avoid excessive rise in temperature and allow high packing density. In most circuits, the switching speed deteriorates (i.e., increases) with increase in the power dissipation.

The propagation delay time depends upon the switching time of the transistor, namely, the rise time, fall time, and storage time. For a gate, the propagation delay time $\bar{t}_d$ is determined (Ref 3) to be

$$\bar{t}_d = (t_{d1} + t_{d2})/2 \tag{11-8}$$

where t_{d1} and t_{d2} are as indicated in fig 11-18. The smaller the propagation delay, the faster the switching speed of the gate. The figure of merit is usually expressed in units of picojoules (1 pJ = 10^{-12} watt-seconds) where the power dissipation is in milliwatts and the propagation delay in nanoseconds. In a good logic gate, this product is desired to be at a minimum. A typical number for good-quality logic gates in integrated circuits is roughly on the order of 10^{-10} W-sec (i.e., 100 pJ). The highest speed attainable is in the ECL gate with a propagation delay of less than 1 nsec and a power dissipation of about 20 mW. The lowest power dissipation is obtained from MOSFET circuits, with a power dissipation of less than 0.1 mW and a propagation delay of about 100 nsec. The speeds of MOS circuits are steadily being improved. MOS devices are small in size and can have very low dissipation because they can be operated in the enhancement mode. A simple MOSFET integrated circuit inverter is shown in fig 11-19a. In contrast to bipolar transistors, an FET is a voltage-operated device and no steady input current is required to keep it in the on or off state. Since, in integrated circuits, MOSFET's are very economical, usually another MOSFET is used as a nonlinear load instead of R_D, as shown in fig 11-19b. MOSFET integrated circuit NOR and NAND gates are shown in figs 11-20a and 11-20b, respectively. In fig 11-20a, the output v_o is high if all the FET's are off, while in fig 11-20b, the output is low if all the FET's are on. Typical p-channel MOSFET gate characteristics are given in Appendix D.

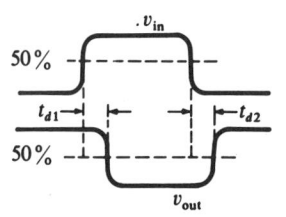

Fig 11-18 Input and output signals of a gate, illustrating propagation delay

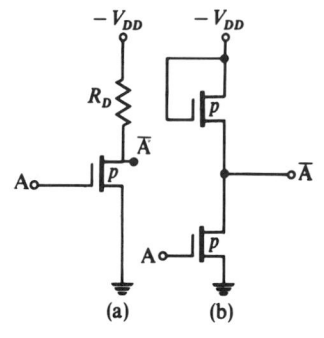

Fig 11-19 MOS gate circuits: (a) MOS inverter with resistive load; (b) MOS inverter with MOS load

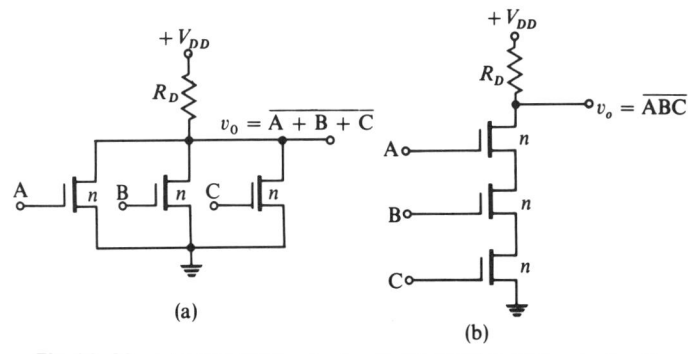

Fig 11-20 (a) MOS-NOR gate circuit; (b) MOS-NAND gate circuit

A logic gate that has extremely low power consumption is shown in fig 11-21a. The inverter is called complementary-symmetry MOSFET logic. The MOSFET's are of the enhancement type. The logic **1** and logic **0** voltages are determined as follows:

A logic **1** signal of $v_i = V_{DD}$ applied to the gates turns the *n*-channel MOSFET on and drives it into the ohmic region. The gate and the source of the *p*-channel MOSFET are at the same potential; hence no channel is induced and negligible current flows through the transistors and thus $v_o \simeq 0$. For the logic **0** (i.e., $v_i \approx 0$) applied to the gates, the *n*-channel has its source and gate at the same potential; hence no channel is induced, and negligible current flows through the MOSFET's. The output voltage is then $v_o = V_{DD}$. Since the power dissipation occurs only in the transient state, the gate has extremely low power dissipation. The NOR and NAND complementary-symmetry MOSFET gates are shown in fig 11-21b and c, respectively. Note that each *n*-channel input requires its own *p*-channel load.

11-14 Memory Circuits: The Flip-Flop (FF)

In many digital systems, in addition to AND, OR, and NOT logic gates, circuits are required which exhibit memory capabilities and can be used as temporary registers. A flip-flop (FF), which was discussed in Section 10-5, does just this, and is, therefore, capable of storing a binary digit (bit); hence, it is sometimes called a *binary*. It should be noted, however, that in most of the present digital computers the main memory function is usually performed by ferrite cores because of their low cost (Ref 5). Large-scale integration (LSI) has made FF's competitive in cost and is changing the picture. For example, the new IBM computers (370 series) use IC memory instead of ferrite cores. The properties and internal circuitry of a FF were covered in Section 10-5. Thus, we shall consider here only the external properties and treat the FF as a memory element as it is utilized in a digital system.

The monolithic integrated-circuit FF is a much more sophisticated functional block. The designers have made liberal use of active components available on the chip and the circuit contains many transistors and diodes in one TO-5 can or flat package. The circuit has many accessible terminals for a variety of logic programming connections. Basically, a flip-flop consists of two interconnected NOT or NOR circuits, as shown in fig 11-22. The outputs are taken from the collectors of the inverters. In integrated circuits, usually either ECL or DCTL

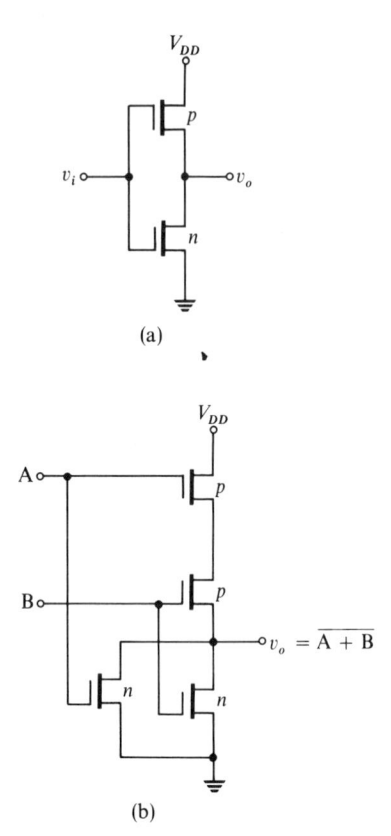

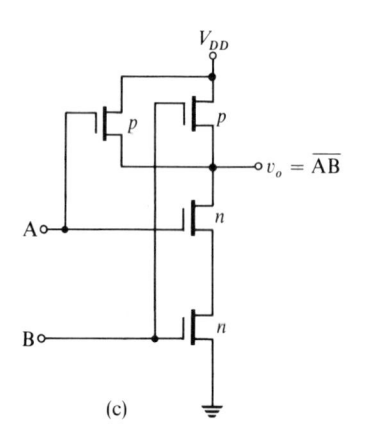

Fig 11-21 (a) Complementary-symmetry *MOSFET* inverter circuit; (b) Complementary-symmetry *MOSFET-NOR* gate circuit; (c) Complementary-symmetry *MOSFET-NAND* gate circuit

11-14 Memory Circuits: The Flip-Flop (FF)

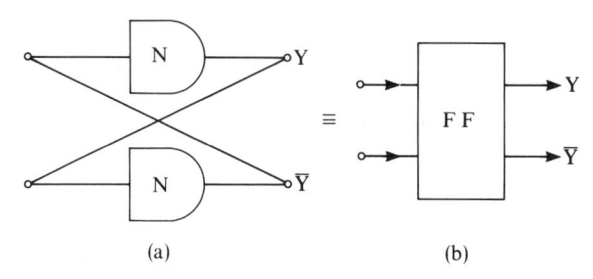

Fig 11-22 Two *NOT* gates interconnected to form a flip-flop (in integrated circuits, *NOR* gates are usually used)

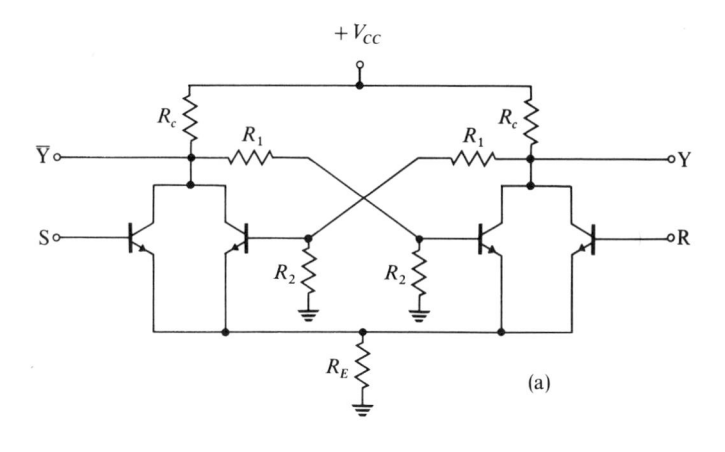

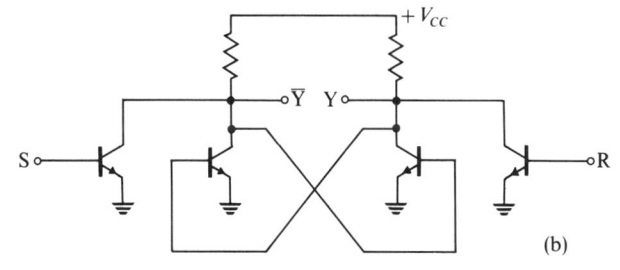

Fig 11-23 (a) Emitter-coupled flip-flop; (b) Direct-coupled flip-flop

inverters are used, as shown in figs 11-23a and 11-23b, respectively. The FF in fig 11-23a has an improved speed (i.e., short switching time) since nonsaturating circuitry is used.

The logic symbol for the RST flip-flop is shown in fig 11-24. The two stable states of the FF correspond to

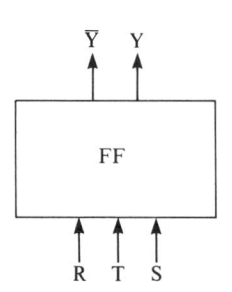

$$Y = 1, \quad \overline{Y} = 0 \quad \text{and} \quad Y = 0, \quad \overline{Y} = 1$$

Fig 11-24 Graphic symbol for an *RTS*-flip-flop

At the input terminals, S corresponds to *set*, i.e., a **1** input to the set terminal puts the FF in state $Y = 1$. If the FF is already in the set state, the input has no effect. A **1** input to the R terminal, which corresponds to *reset*, establishes the FF in the state $Y = 0$. If the FF is already in the reset state, the input signal has no effect. The T terminal corresponds to the *toggle* or *trigger* input which, upon receiving a **1** input, causes a change in the state of the FF (i.e., **0** to **1** or **1** to **0**) and is analogous to the single-input flip-flop operation (see fig 10-20). Other types of FF's are shown in fig 11-25. In the T flip-flop, there is only one input, a

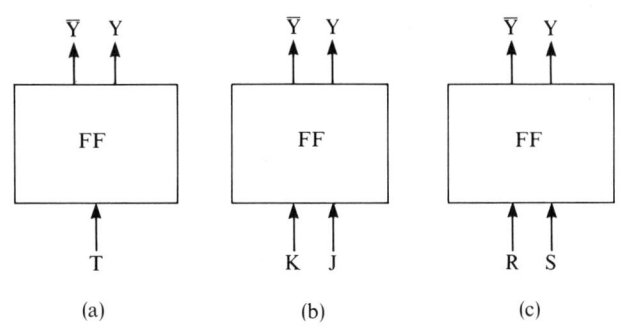

Fig 11-25 (a) T-flip-flop symbol; (b) JK-flip-flop symbol; (c) RS-flip-flop symbol

trigger input (see fig 10-20). The flip-flop complements its state when a **1** input is applied. The JK flip-flop has two inputs. When no input is applied (i.e., **0**), the state of the flip-flop remains unchanged. When a **1** input is applied to J, the flip-flop assumes a state $Y = 1$, and when the input is applied to K, the flip-flop assumes a state $Y = 0$. When inputs are applied simultaneously to both J and K, the flip-flop complements its state. The RS flip-flop has two input terminals, S (set) and R (reset). The operation of the RS flip-flop is similar to that of the JK flip-flop, with S corresponding to J and R to K. The difference between the two, however, is that simultaneous inputs to both terminals of an RS flip-flop are not allowed.

The RST and JK flip-flops are the most common in digital computers. The operation of the RST flip-flop is listed in table 11-5 for convenience. In the table, $t = t_0$ is the initial state and τ is the final state, i.e., sometime after the input has been applied. From table 11-5, note that the T input of the RST flip-flop corresponds to inputs at both J and K of a JK flip-flop. In other words, table 11-5 is identical for a JK flip-flop with terminal J corresponding to terminal S, K corresponding to R, and T corresponding to both J and K (i.e., $T = 1$ corresponds to $J = 1$ and $K = 1$). Some representative applications of the FF's are discussed in the next section.

TABLE 11-5 RST Flip-Flop Operation

$R(t_0)$	$S(t_0)$	$T(t_0)$	$Y(t_0)$	$Y(\tau)$	
0	0	0	0	0	} No change in state
0	0	0	1	1	
1	0	0	0	0	} puts the FF in **0** state
1	0	0	1	0	
0	1	0	0	1	} puts the FF in **1** state
0	1	0	1	1	
0	0	1	0	1	} Complements the FF state
0	0	1	1	0	

11-15 Register Circuits and Shift Registers

Register circuits are used in digital computers to store binary information. Fig 11-26a shows the block diagram of a *register* circuit

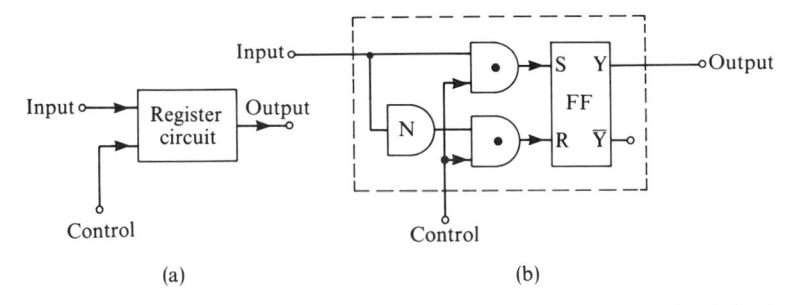

Fig 11-26 Register circuit: (a) Block diagram of a register circuit; (b) Circuit implementation of a register

in which a binary value is to be registered. The control terminal controls the time at which the input bit is stored. An actual circuit for one stage of a register is shown in fig 11-26b, where a FF, two AND gates, and an inverter are used to register a binary digit. The operation of the circuit to enter a bit is as follows (fig 11-26b): The control value is first raised to **1**. If the signal input is **1**, the FF is set ($Y = 1$) due to enabling of the upper AND gate; if the input is **0**, the FF is reset ($Y = 0$) due to enabling of the lower AND gate.

A complete register consists of a large group of individual register circuits, in order to handle very large numbers. For example, a complete register capable of storing four binary digits (i.e., from 0000

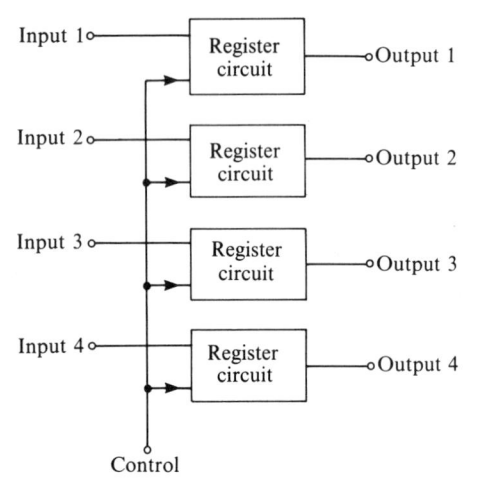

Fig 11-27 Register circuit capable of storing four bits

(zero) up to 1111 (fifteen)) is shown in fig 11-27. Note that a register must contain an individual register circuit for each binary digit, four in this example.

Shift Register

A shift register is a particular type of register which is capable of shifting all the digits toward the right or left, or both, depending upon the particular design of the registers. For example, if a code **10110** is shifted one place to the right it becomes **01011**, and if shifted one place to the left it becomes **01100**. In general, the extreme right-end digit is lost and a zero is shifted into the left end for each digit shift to the right. The opposite is true for each digit shift to the left. The most important application of shift registers is in arithmetic operations, such as in binary multiplication and division. These topics are very well covered in many texts on computer arithmetic (Ref 6) and will not be discussed here.

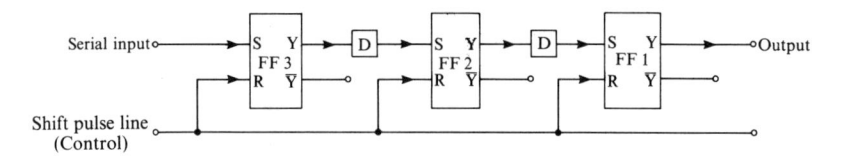

Fig 11-28 Shift-register circuit

A simplified logic diagram of a shift register is shown in fig 11-28. The input information is a series of pulses which are to be stored in the register. The reset line is excited by a train of pulses which are timed to occur midway between the input pulses. The delay section D has a delay time very much smaller than the time interval between pulses, so that the individual FF does not receive trigger signals simultaneously from the shift line and the preceding FF. Since the shift line is connected to R, it always drives the FF to the **0** state. Also, the connection between the FF's is such that a succeeding FF will respond only if the preceding FF goes from **1** to **0**. The pulse due to this transition will drive the succeeding FF to **1**.

Suppose we wish to register the number $3 = (101)_2$. The first input pulse drives FF3 to **1**. The shift pulse resets FF3 and returns its state to **0**. A short time later, the length of which depends on the delay circuit τ, FF2 is driven to state **1** by the pulse from the preceding binary. The shift pulse has been removed by this time. Thus, the first digit has been shifted from FF3 to FF2, and FF3 is cleared to register another pulse. This procedure is repeated until all the digits have been registered. At the time the complete number has been registered, the shift pulse must be stopped.

In integrated circuits, the use of a delay is, in general, not desirable. Almost all integrated circuits have many FF's in each circuit chip, so that if the delay can be replaced by another FF, it is preferable.

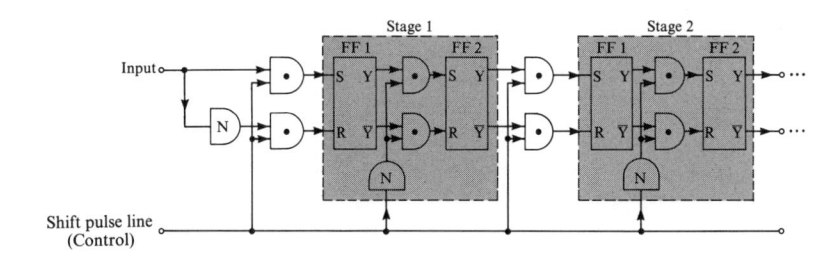

Fig 11-29 Double flip-flop register circuit

A shift register using double FF's is shown in fig 11-29. The operation of the shift register is as follows:

Consider the operation of stage 2. When the shift pulse line is **1**, then one of the AND gates is enabled, the state of stage 1 stored in FF2 of stage 1 is transferred to stage 2 and stored in FF1 of stage 2. When the shift pulse $= 0$, the inhibit input (i.e., inverter output) is **1**, and the state of FF1 of stage 2 is transferred to FF2 of stage 2. Thus, the complete cycle shift pulse, equal to **1** followed by **0**, results in transferring the state of the output of a stage to the output of the next stage. In this way information can be shifted along the register. Stage 1 shows how the information is fed initially into the register. As long as the new input is available just prior to the shift pulse becoming **1**, it does not matter what the relative pulse lengths are.

11-16 Counters

A string of FF's may be connected in several ways to obtain a *counter*. The simplest scheme is to connect the FF's in cascade, as shown in fig 11-30. This is known as a binary counter. The interconnection

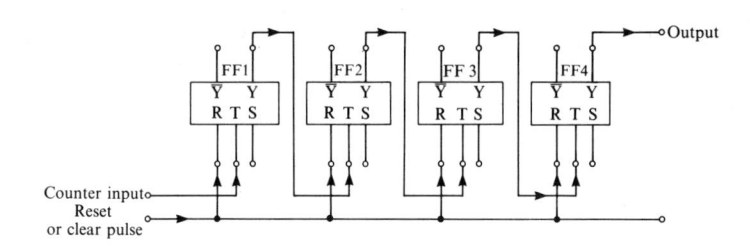

640 11/Logic Circuits and Digital Integrated-Circuit Functional Blocks

Fig 11-30 Schematic of a binary counter

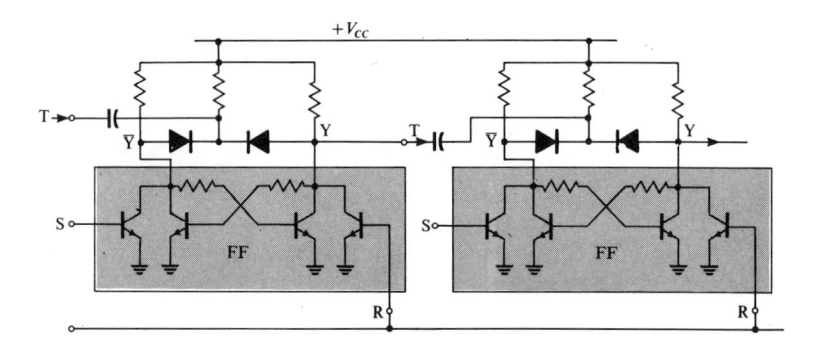

Fig 11-31 Circuit interconnection of FF1 and FF2 in fig 11-30, using *RST* flip-flops

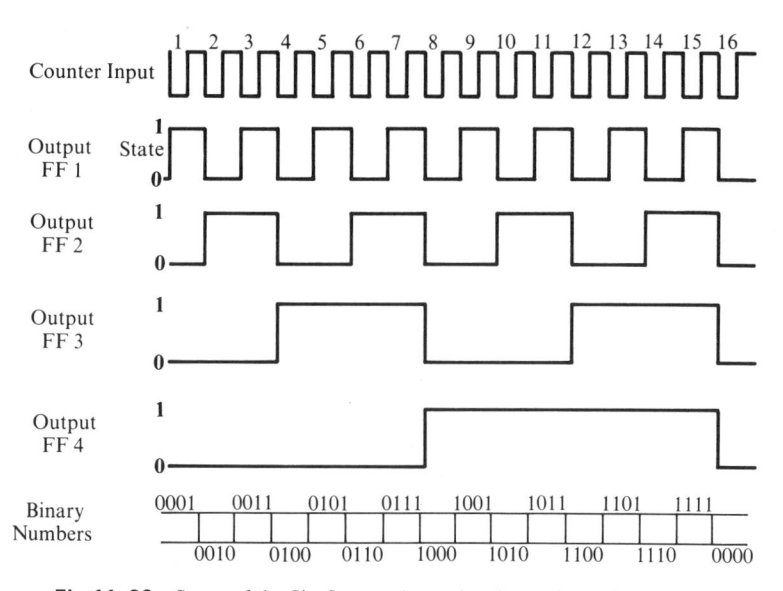

Fig 11-32 States of the flip-flops and associated waveforms for fig 11-30

scheme for the RST flip-flops is shown in fig 11-31. The system shown in fig 11-30 is referred to as a scale of 16 (i.e., 2^4 for 4 FF's). A reset pulse puts the individual FF's in the state $Y = 0$. The state of the FF's for this system as well as the waveforms are shown in fig 11-32. After the sixteenth pulse, the counter returns to its original state and repeats its operation. Thus, an output signal is obtained for each of 16 pulses; hence the 4 FF counter could be used as a digit counter of radix 16. If only 3 FF's are used in the cascade connection, a digit counter of radix 8 (i.e., 2^3) may be obtained.

Often decimal counters, i.e., counters with a scale of 10, are required. To design a counter with some other radix N requires n FF's, where n is the smallest number such that $2^n > N$. Thus, for a decimal counter, we require 4 FF's [$(2^4 > 10)$ and $(2^3 < 10)$]. Since 4 FF's have 16 stable states, we must include techniques to nullify 6 of the states to get a decade counter. There are several ways that this can be done. The methods include feedback connections, pulse advancing, parallel connections, etc.

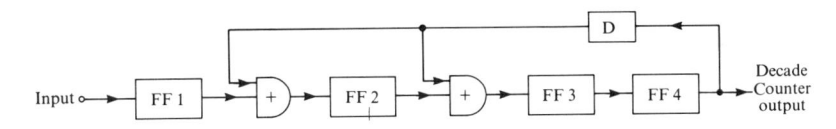

Fig 11-33 Schematic of a decade counter using feedback connection

A decade counter with a feedback connection is shown in fig 11-33. The delay device D is often used to ensure that the FF has changed its state before the inputs via feedback paths arrive. As we shall show presently, the feedback advances the count by $2^4 - 10 = 6$ counts during the first 10 counts. The states of the FF are listed in table 11-6. The counter operates as a normal binary counter for the first 7 pulses. At the eighth pulse, the states of the FF's will momentarily become **1000**. Due to feedback and the OR circuits, however, the states of FF2 and FF3 will be set to **1** so that before the ninth pulse arrives, the states of the FF's are **1110**. Thus, the feedback arrangement has advanced the count by 6, as shown in table 11-6. At the end of the ninth pulse, the FF states are **1111** and, at the tenth pulse, the FF's will return to the original state. For every ten input pulses, then, the decade counter gives one output pulse. In other words, there is a transition from **1** to **0** at the 11th pulse. A complete decimal counter unit is made up of a cascade series of decade counters. Usually auxiliary readout display equipment is also used for direct readout.

TABLE 11-6 *State of the FF's for a Scale of 10 (See fig 11-33)*

No. of input pulses	State of the FF (output Y)			
	FF4	FF3	FF2	FF1
Start or reset pulse	0	0	0	0
1	0	0	0	1
2	0	0	1	0
3	0	0	1	1
4	0	1	0	0
5	0	1	0	1
6	0	1	1	0
7	0	1	1	1
8 advances by 6	1	01	01	0
9	1	1	1	1
10	0	0	0	0

11-17 Summary

A large number of IC manufacturers now market digital integrated circuits. Digital functional blocks of various size, speed, cost, etc., are available in a single package. For example, a complete decade counter, a 10-bit shift register, etc., are available in a small single package. If speed is not a problem, a much larger functional circuit realization on the same chip size is possible by MOS technology. A large and more complete multifunctional block realization of various computing elements is now practical through large scale integration (LSI).

In digital computers, several thousand of a few basic digital integrated circuit functional blocks are used. The design of large and complex computer circuitry is being automated to improve the performance and reliability of the system. Large scale integration and computer-implemented design and testing of electronic circuits and systems may very well revolutionize the electronics industry, and open fascinating and challenging problems and opportunities to engineers.

REFERENCES AND SUGGESTED READING

1. Wickes, W. E., *Logic Design with Integrated Circuits*, New York: Wiley, 1968.
2. Harris, J. N., Gray, P., and Searle, C., *Digital Transistor Circuits*, SEEC Series Vol. 6, New York: Wiley, 1966.

3. Lynn, D., Meyer, C., and Hamilton, D., editors, *Analysis and Design of Integrated Circuits*, New York: McGraw-Hill, 1967.
4. McCluskey, E. J., *Introduction to the Theory of Switching Circuits*, New York: McGraw-Hill, 1965.
5. Lo, A., *Introduction to Digital Electronics*, Reading, Mass: Addison-Wesley, 1967.
6. Chu, Y., *Digital Computer Design Fundamentals*, New York: McGraw-Hill, 1962.
7. Strauss, L., *Wave Generation and Shaping*, New York: McGraw-Hill, second edition, 1970.

PROBLEMS

11-1 Determine the binary numbers corresponding to each of the following numbers in other number systems:
(a) $(1294)_{10}$ (b) $(567)_8$
(c) $(158.640)_{10}$ (d) $(314.36)_8$

11-2 Determine the decimal numbers corresponding to each of the following numbers in other number systems:
(a) $(110101001)_2$ (b) $(1536)_8$
(c) $(1101.01101)_2$ (d) $(172.17)_8$

11-3 Prove each of the following set of theorems with an AND operation which involves only one Boolean variable:
(a) $X \cdot 0 = 0$ (b) $X \cdot 1 = X$
(c) $X \cdot X = X$ (d) $X \cdot \bar{X} = 0$
(Hint: X can take only values of 0 or 1, and use table 11-2)

11-4 Repeat Problem 11-3 for the following OR operations:
(a) $X + 0 = X$ (b) $X + 1 = 1$
(c) $X + X = X$ (d) $X + \bar{X} = 1$

11-5 Make use of Problems 11-3 and 11-4 to prove each of the following simplification theorems for Boolean variables:
(a) $x \cdot y + x \cdot \bar{y} = x$ (b) $x + \bar{x} \cdot y = x + y$

11-6 Repeat Problem 11-5 for the following:
(a) $x + x \cdot y = x$
(b) $x \cdot y + \bar{x} \cdot z + y \cdot z = x \cdot y + \bar{x} \cdot z$

11-7 Prove de Morgan's theorems (11-1) and (11-2) by using the truth table.

11-8 Verify the following identities for exclusive-OR operation:
(a) $A \oplus (A + B) = \bar{A}B$
(b) $A \oplus \bar{A}B = A + B$
(c) $A \oplus (\bar{A} + B) = \overline{AB}$

11-9 Verify the following identities for coincidence operation:
(a) $A \odot \bar{A}B = \overline{A + B}$

(b) $A \odot (\bar{A} + B) = AB$
(c) $A \odot (A + B) = \bar{A}B$

11-10 Are the following relations correct? If not, show why:
(a) $(A \oplus B) + (A \oplus C) \stackrel{?}{=} A \oplus (B + C)$
(b) $(A \oplus B)(A \oplus C) \stackrel{?}{=} A(B \oplus C)$
(c) $(A \odot B) + (A \odot C) \stackrel{?}{=} A \odot (B + C)$
(d) $(A \odot B)(A \odot C) \stackrel{?}{=} A(B \odot C)$

11-11 Simplify each of the following Boolean expressions and draw the logic blocks for each using AND, OR, and NOT gates. All types need not be used.
(a) $(A + B)(\bar{A} + C)(B + C)$
(b) $A \cdot \bar{C} + A \cdot B \cdot C + A \cdot B \cdot (D + \bar{C})$
(c) $A \cdot B + A \cdot B \cdot \bar{C} + \overline{(B + C + \bar{A})}$

11-12 Show that the basic Boolean functions AND, OR, and NOT can be obtained by using NOR gates only. Draw logic block diagrams to show this.

11-13 Show that the basic Boolean functions AND, OR, and NOT can be obtained by using NAND gates only. Draw logic block diagrams to show this.

11-14 Construct a truth table for each of the following Boolean functions:
(a) $A \cdot \bar{B} + B \cdot \bar{C} + C \cdot \bar{A}$
(b) $\bar{A} \cdot \bar{D} + A \cdot C + \bar{B} \cdot \bar{C} \cdot \bar{D} + B \cdot C \cdot D$

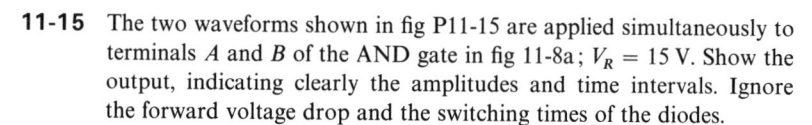

11-15 The two waveforms shown in fig P11-15 are applied simultaneously to terminals A and B of the AND gate in fig 11-8a; $V_R = 15$ V. Show the output, indicating clearly the amplitudes and time intervals. Ignore the forward voltage drop and the switching times of the diodes.

11-16 Repeat Problem 11-15, but apply the waveforms simultaneously to terminals A and B of the OR gate shown in fig 11-8b.

11-17 For the ECL circuit in fig P11-17, if $v_{in}(1) = +2$ V, $v_{in}(0) = -2$ V, what are the logic levels of the output $v_o(1)$ and $v_o(0)$.

11-18 (a) A MOSFET (enhancement type) used as an inverter is shown in fig 11-19a. Sketch the transfer characteristics v_o vs v_i.
(b) Sketch the transfer characteristics for the nonlinear load in fig 11-19b. What are the logic **1** and logic **0** output levels for positive logic?

11-19 For the DCTL NOR gate shown in fig 11-12, series resistors (R_B's) are connected to the base of each fan-out transistor Q_1, Q_2, etc., in order to remedy the current-hogging problem. This gate becomes an RTL NOR gate. You may also assume that the transistors are identical. If these transistors are all to be driven into saturation, show that the fan-out N is limited by

$$N < \beta_0 - \frac{R_B}{R_L}$$

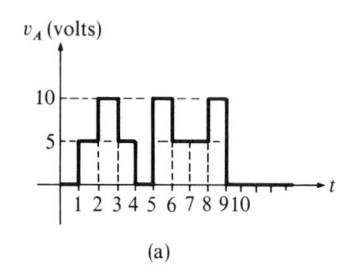

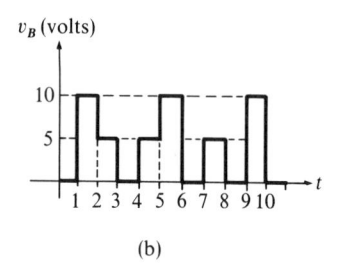

Fig P11-15

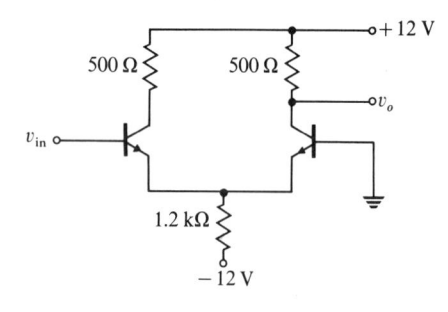

Fig P11-17

11-20 Make a table such as that in table 11-6 for the states of the FF shown in fig P11-20. Show that it is a decade counter.

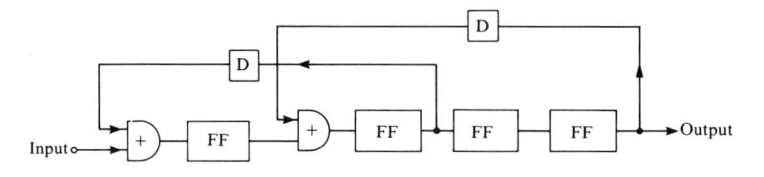

Fig P11-20

11-21 Repeat Problem 11-20 for the counter shown in fig P11-21. Only JK FF's are used.

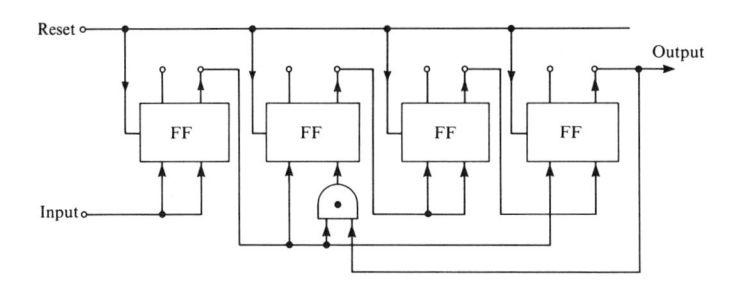

Fig P11-21

11-22 The circuit shown in fig P11-22 performs a pulse shortening function. Describe how the circuit functions and show the output pulse. Assume a delay time $\tau = T_1/10$ for the delay block D.

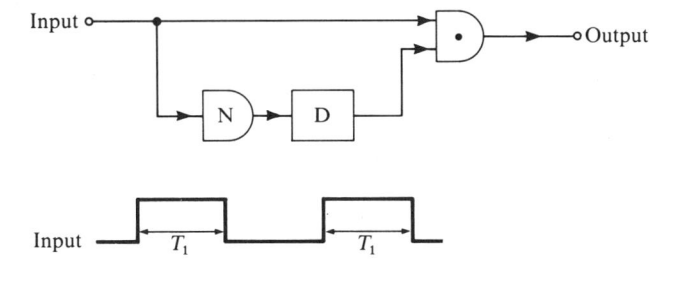

Fig P11-22

11-23 Show the circuit implementation of a DCTL NAND gate with three inputs. If $V_{BE} = 0.7$ V and $V_{CE} = 0.2$ V, what input voltages are required to turn the transistors on? What are the values of the logic levels **1** and **0** (use positive logic)? Comment on the noise immunity.

11-24 (a) Write the truth table for the one-bit digital comparator (equality detector) circuit shown in fig P11-24 and explain its function.
(b) If $A = $ **01010** and $B = $ **00111**, determine C.
(c) Show how two comparator circuits and other gates can be interconnected for a two-bit equality detection.

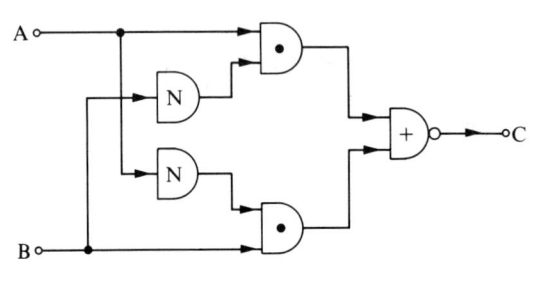

Fig P11-24

11-25 (a) Show that the following expression provides the exclusive-OR function

$$C = (\bar{A} + \bar{B}) \cdot (A + B) = A \oplus B$$

(b) Show at least one logic block realization using the minimum number of AND, OR, and NOT gates.
(c) If $A = $ **01010** and $B = $ **00111**, what is C?

11-26 (a) A half adder (H-A), shown in fig P11-26a, is a logic circuit for the binary-arithmetic addition of two single digits. (It also provides the basic output needed for the arithmetic addition of binary numbers.) Write the Boolean functions S and C in terms of A and B, and show

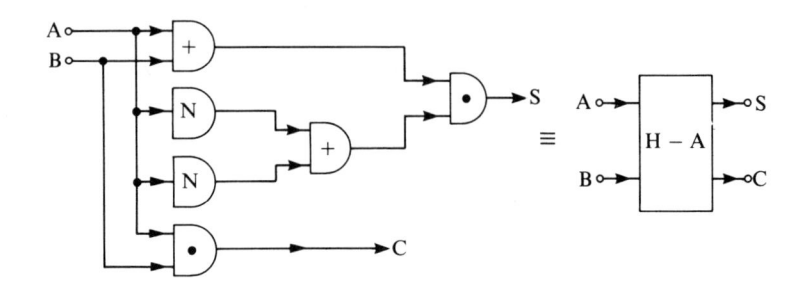

Fig P11-26(a)

that the outputs S and C are indeed the Sum and Carry of the basic addition for binary addition.

(b) A single-bit full adder (F-A), shown in fig P11-26b, has three inputs A, B, and C which are the augend bit, the addend bit, and the carry bit. The outputs S_o and C_o are the sum and output carry bits. Derive the expression for S_o and C_o for the full adder in terms of A, B, and C. Verify your result by using the truth table.

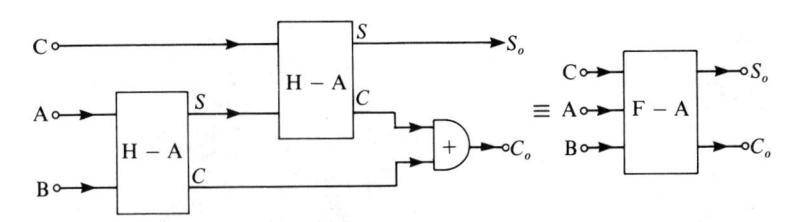

Fig P11-26(b)

11-27 If three NOR gates, each having a propagation delay, t_d, are cascaded in a closed loop, what would be the frequency of oscillation? If four stages were used instead of three, would you still get a square wave oscillator? Derive an expression for the period of the oscillator when N gates are cascaded in a feedback loop. Is there any restriction on N?

APPENDIX A
Summary of Semiconductor and p-n Junction Properties

The *p-n* junctions are the most important and essential part of semiconductor electronic devices. They are to be found in diodes, bipolar transistors, field-effect transistors, and integrated circuits. In order to understand and appreciate the underlying principles of operation of these devices, we summarize the most important properties of semiconductors and *p-n* junctions in this appendix. The discussion is in the form of a review and the reader is referred to the references listed at the end of this appendix for a more complete and detailed discussion of these topics.

A-1 Semiconductors: Intrinsic and Extrinsic

Single-crystal materials may be classified in terms of the energy levels of the outer electrons which form energy bands. In some materials, the lower energy band (valence band) is completely filled and the next higher energy band (conduction band) is completely empty. These bands are separated by a *forbidden energy gap* W_g of several electron volts (eV). If an electric field is applied, current cannot flow. Such materials are called *insulators*. In other materials, the valence band can partially overlap the conduction band with no forbidden energy gap between. If an electric field is applied, current flows readily. These types of materials are called *conductors*. In still other materials, the width of the forbidden gap is less than 1.5 eV, and the material behaves as an insulator at very low temperatures. At very high temperatures, they behave as conductors. This type of material, with a conductivity level between that of insulators and that of conductors, is called a *semiconductor*. The most commonly used semiconductors in electronic devices are silicon (Si) and germanium (Ge). Their energy gaps W_g are 0.72 and 1.12 eV, respectively.

Silicon and germanium are from group IV of the periodic table; therefore, each atom has 4 valence electrons. They are diamond-like crystals and the atoms are bound together by *covalent bonds*, i.e., each atom shares its pairs of electrons. If the temperature of such a crystal is raised, there is an increased probability that an electron is broken loose, or freed, from a covalent bond and a hole is left behind in the bond. The free electron can participate in conduction. A hole is a missing electron in the covalent bond. Through the transfer of electrons from adjacent bonds, the hole is free to participate in conduction, and behaves very much like a positive charge $+q$ in the conduction mechanism. The process of the creation of a free electron and a hole is called *ionization*.

A pure semiconductor crystal which consists entirely of the same type of atoms is called an *intrinsic* semiconductor. The concentrations of atoms in silicon and germanium crystals are approximately 5×10^{22} atom/cm^3 and 4.4×10^{22} atom/cm^3, respectively. A semiconductor whose conductivity is chiefly due to the presence of impurities is called an *extrinsic* semiconductor. As we shall soon see, the addition of minute impurities can cause significant effects in the conductivity of the material. In an intrinsic semiconductor, the concentrations of holes and electrons are equal, i.e.,

$$n = p = n_i$$

where the subscript i denotes intrinsic and n and p designate the concentration of electrons and holes, respectively. In the references, it is shown that the equilibrium concentration, which is strongly dependent on the temperature, is given by

$$np = n_i^2 = KT^3 e^{-(qW_g)/kT} \qquad (A\text{-}1)$$

where

K is a proportionality constant
T is the absolute temperature (°K)
k is Boltzmann's constant (1.38×10^{-23} joules/°K = 8.62×10^{-5} eV/°K)
q is the magnitude of electronic charge (1.6×10^{-19} coul)
W_g is the band-gap energy, i.e., the energy required to break a covalent bond (eV)

For Si and Ge, the equilibrium intrinsic concentrations are

$$n_i^2(T) \simeq 1.5 \times 10^{33} \, T^3 \, e^{-(14,030/T)} \text{ cm}^{-6} \quad \text{for Si} \qquad (A\text{-}2a)$$

$$n_i^2(T) \simeq 3.1 \times 10^{32} \, T^3 \, e^{-9100/T} \text{ cm}^{-6} \quad \text{for Ge} \qquad (A\text{-}2b)$$

At room temperature (≈ 300 K), the concentrations are

$$n_i = p_i \simeq 1.5 \times 10^{10} \text{ holes or electrons/cm}^3 \quad \text{Si} \qquad (A\text{-}3a)$$

$$n_i = p_i \simeq 2.5 \times 10^{13} \text{ holes or electrons/cm}^3 \quad \text{Ge} \qquad (A\text{-}3b)$$

Note that since the concentration of atoms in the crystal (for Si) is about 5×10^{22} cm^{-3}, from (A-3a) one finds that roughly one atom in 10^{12} is ionized at room temperature. Note that by either thermal energy, as described earlier, or light photons of sufficient energy, we can increase the ionization in the crystal considerably.

Extrinsic, or doped, semiconductors are materials in which carriers of one kind, i.e., either hole or electron, predominate. The predominant carrier is called the majority carrier. Semiconductors in which the predominant carriers are electrons are called *n*-type material. An *n*-type semiconductor is formed by doping (adding impurities to) the silicon (or Ge) crystal with elements of group V of the periodic table (phosphorus, arsenic, antimony). When a silicon atom is replaced by one of phosphorus, the impurity atom has an extra electron which is donated to the crystal. The impurity atom in this case is called a *donor* atom. The four electrons of phosphorus form covalent bonds with those of Si, and the fifth electron is left very loosely bound. The binding energy of the fifth electron in the silicon host crystal is small (about 0.04 eV) and almost independent of the type of impurities used from group V.

(For a Ge host crystal, the binding energy of the extra electron is about 0.01 eV.) Since the thermal energy of the crystal is of the same order of magnitude, all impurities are ionized at room temperature and above, and the liberated donor electrons can move freely and contribute to the conduction mechanism. The concentration of impurity atoms is usually in the range 10^{15} to 10^{19} atoms/cm^3, with 5×10^{17} a typical value. Thus, the corresponding fraction of impurities is $\approx (5 \times 10^{17})/(5 \times 10^{22}) = 10^{-5}$, i.e., one part in 10^5. From a mechanical and metallurgical point of view, the crystal is unchanged, but its electrical properties are greatly changed, as we shall presently show. A p-type semiconductor is formed by doping the silicon (or Ge) crystal with elements of group III of the periodic table (boron, aluminum, gallium, indium). These impurity atoms have only 3 valence electrons, and when complete covalent bonds are formed, the impurity atom accepts an electron. Thus, the impurity atom is called an *acceptor* atom. The electron accepted by the acceptor comes from a normal covalent-bonded semiconductor atom, thus producing a hole free to participate in conduction. The binding energy in this case is again small and thus the holes can be produced readily. The conduction in the p-type crystal is then carried mainly by holes, and holes are the majority carriers. In a p-type semiconductor, the free electrons are called the minority carriers. In the n-type semiconductor, the electrons are the majority carriers and the holes are the minority carriers.

In an n-type semiconductor, the donor concentration $N_d \gg n_i$, where N_d is typically 10^{17} cm^{-3} (while $n_i = 1.5 \times 10^{10}$ cm^{-3} in Si). Thus, the majority and minority concentrations are given by

$$n_n \simeq N_d \qquad p_n \simeq \frac{n_i^2}{N_d} \qquad (n\text{-type}) \qquad (A\text{-}4a)$$

Similarly in a p-type the acceptor concentration $N_a \gg n_i$, thus the majority and minority concentrations are given by

$$p_p \simeq N_a \qquad n_p \simeq \frac{n_i^2}{N_a} \qquad (p\text{-type}) \qquad (A\text{-}4b)$$

A-2 Drift and Diffusion Currents

There are two mechanisms by which holes and electrons move through a crystal. One of these is the *drift* current, which is caused by the application of an electric field. The other is the *diffusion* current, which is caused by the net flow of carriers from a region of high concentration to a region of lower concentration.

If an electric field $\mathscr{E}$ is applied to a semiconductor with a hole concentration p, and an electron concentration n, the holes and electrons will attain an average velocity v_{dp} and v_{dn} in the direction of the field proportional to $\mathscr{E}$,

$$v_{dp} = \mu_p \mathscr{E} \tag{A-5a}$$

$$v_{dn} = -\mu_n \mathscr{E} \tag{A-5b}$$

where μ_p and μ_n are the hole and electron mobilities (cm² volt⁻¹ sec⁻¹). The hole current density is then given by

$$J_p = qpv_d = qp\mu_p\mathscr{E} = \sigma_p\mathscr{E} \quad (\text{amp/cm}^2) \tag{A-6}$$

where

$$\sigma_p = qp\mu_p \frac{\text{mhos}}{\text{cm}} \quad (\text{hole conductivity}) \tag{A-7}$$

Similarly for the electron in the p-type material we have

$$J_n = qn\mu_n\mathscr{E} = \sigma_n\mathscr{E} \tag{A-8}$$

where μ_n is the electron mobility and $\sigma_n = qn\mu_n$. Thus the total drift current density is given by

$$J = J_n + J_p = q(n\mu_n + p\mu_p)\mathscr{E} \tag{A-9}$$

Note that the electrons move in the opposite direction from the field but the current is in the same direction as the field, so that the total currents of holes and electrons are in the same direction.

From Ohm's law, the total conductivity is

$$\sigma = q(n\mu_n + p\mu_p) \quad (\text{ohm-cm})^{-1} \tag{A-10}$$

For an intrinsic crystal $n = p = n_i$, thus we have

$$\sigma = q(\mu_n + \mu_p)n_i \quad (\text{for intrinsic}) \tag{A-11}$$

A-2 Drift and Diffusion Currents

The mobilities of Si and Ge at room temperature are

$$\mu_p = 500 \text{ cm}^2/\text{volt-sec} \qquad \mu_n = 1300 \text{ cm}^2/\text{volt-sec} \quad (Si) \qquad \text{(A-12a)}$$
$$\mu_p = 1900 \text{ cm}^2/\text{volt-sec} \qquad \mu_n = 3800 \text{ cm}^2/\text{volt-sec} \quad (Ge) \qquad \text{(A-12b)}$$

Thus, the intrinsic conductivities of Si and Ge are

$$\sigma_i \simeq 4.2 \times 10^{-6} (\text{ohm-cm})^{-1} \quad Si \qquad \text{(A-13a)}$$
$$\sigma_i \approx 2.2 \times 10^{-2} (\text{ohm-cm})^{-1} \quad Ge \qquad \text{(A-13b)}$$

We shall illustrate some of the above relations by an example.

EXAMPLE

A bar of Ge with cross-sectional area $A = 0.01 \text{ cm}^2$ and length $L = 0.2$ cm is doped with donor impurity. When a dc voltage of 10 mV is applied across the bar (at room temperature), the drift current flowing through it is found to be 1 mA. Determine the majority and the minority concentrations of the doped semiconductor.

Solution: The resistance of the bar, from measurement, is $R = V/I = 10 \, \Omega$. But

$$R = \frac{L}{\sigma A}, \quad \text{hence} \quad \sigma = \frac{L}{AR} = \frac{(0.2)}{(0.01)(10)} = 2 (\text{ohm-cm})^{-1}$$

From (A-10) we have

$$q(n\mu_n + p\mu_p) = \sigma = 2$$

For *n*-type, the majority carrier concentration is

$$qn\mu_n \approx \sigma \quad \text{or} \quad n \approx \frac{\sigma}{q\mu_n} = \frac{2}{(1.6 \times 10^{-19})(3.8 \times 10^3)} = 3.29 \times 10^{15} (\text{cm}^{-3})$$

The minority carrier concentration is found from (A-1)

$$p = \frac{n_i^2}{n} = \frac{(2.5 \times 10^{13})^2}{3.29 \times 10^{15}} = 1.90 \times 10^{11} (\text{cm}^{-3})$$

Note that the approximation used above, $n\mu_n \gg p\mu_p$, is clearly justified.

When the carrier concentration is not uniform within the crystal, there will be a net flow of current from the high- to the low-concentration region proportional to the gradient of the carrier concentration. Let us assume that the hole concentration varies along the x direction and is constant in the y and z directions. Then the current density J_p in the x direction is proportional to the slope of the curve $p(x)$, and is given by

$$J_p = -qD_p \frac{dp}{dx} \text{ amp/cm}^2 \qquad \text{(A-14)}$$

Note that the direction of hole current is from the high to the low concentration; hence the negative sign, since the slope of $p(x)$ is negative.

Similarly, for electrons, we have

$$J_n = qD_n \frac{dn}{dx} \text{ amp/cm}^2 \qquad \text{(A-15)}$$

The proportionality constants D_p and D_n are called the diffusion constants and have the following values for Si and Ge:

$$D_p = 13 \frac{\text{cm}^2}{\text{sec}} \qquad D_n = 35 \frac{\text{cm}^2}{\text{sec}} \quad \text{(Si)} \qquad \text{(A-16a)}$$

$$D_p = 50 \frac{\text{cm}^2}{\text{sec}} \qquad D_n = 100 \frac{\text{cm}^2}{\text{sec}} \quad \text{(Ge)} \qquad \text{(A-16b)}$$

Note that in each case the current is given by the current density multiplied by the cross-sectional area perpendicular to the flow of current.

The mobility and diffusion constants are related by *Einstein's relations*, namely,

$$u_p/D_p = q/kT \qquad \text{(A-17a)}$$

$$u_n/D_n = q/kT \qquad \text{(A-17b)}$$

A-3 The *p-n* Junction and its Properties

A *p-n* junction can be formed by diffusing into an *n*-type semiconductor a *p*-type impurity, or vice versa. The transition from a *p*-type to an *n*-type doping may be abrupt, so that it may be considered a step junction, or it may be gradual, which is called the graded junction. Typical concentration profile of a p-n junction is shown in fig A-1a.

The *p*-type semiconductor has many free holes contributed by the acceptor impurity atoms, while the *n*-type semiconductor has many free electrons contributed by the donor impurity atoms. Because of the nonuniform concentrations, there is a strong tendency for holes to move by diffusion from the *p*-type to the *n*-type and, similarly, for electrons to move from the *n*-type to the *p*-type. But when holes cross the junction, they find immediately abundant free electrons and a very large number disappear by *recombination* very quickly. The same situation occurs for electrons passing from the *n*-type region to the *p*-type region across the junction. If such electron and hole migration took place with no restraint, then a current would flow across the junction with no external voltage applied. Since no current can flow in the *p-n* junction without an external voltage, the current must be zero. To accomplish this a so-called *depletion region*, free from mobile charge or, equivalently, a *space-charge* region, is created. This space-charge region arises from the fact that the region near the junction plane within the *n*-type semiconductor is left with a net positive charge, while the region near the junction plane within the *p*-type semiconductor is left with a net negative charge. As a consequence of this dipole layer at the junction, an electric field is established across the junction which results in a potential difference across the junction, as shown in fig A-1b. Since, at thermal equilibrium, no net current can flow, this potential barrier, V_d, prevents the flow of majority carriers to the other side. The potential barrier, which is also called the *contact potential*, is determined from the carrier concentrations by the following simple relations:

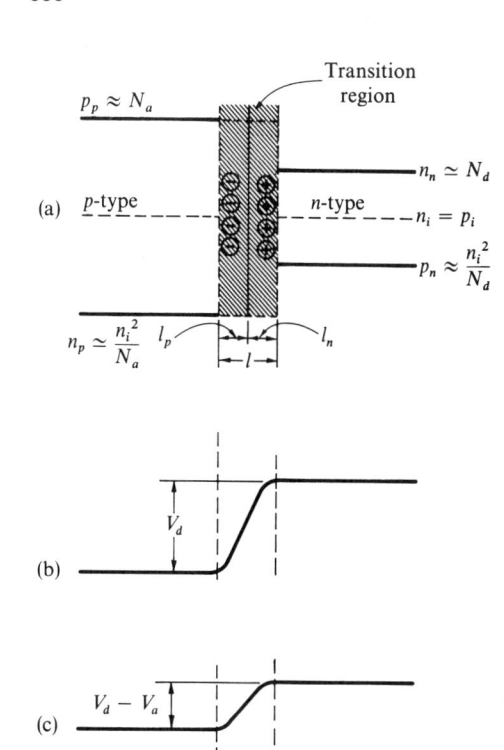

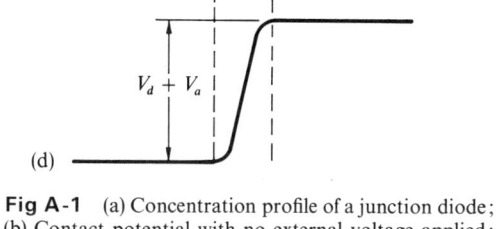

Fig A-1 (a) Concentration profile of a junction diode; (b) Contact potential with no external voltage applied; (c) Potential across the junction for a forward-biased applied voltage; (d) Potential across the junction for a reverse-biased applied voltage

$$\frac{n_n}{n_p} = e^{(q/kT)(V_d)} = \frac{p_p}{p_n} \tag{A-18a}$$

or

$$V_d = \frac{kT}{q} \ln \frac{n_n}{n_p} = \frac{kT}{q} \ln \frac{p_p}{p_n} \tag{A-18b}$$

Equation (A-18b) may also be written as

$$V_d = \frac{kT}{q} \ln \frac{N_a N_d}{n_i^2} = \frac{kT}{q} \ln \frac{\sigma_n \sigma_p}{\mu_n \mu_p q^2 n_i^2} \tag{A-18c}$$

For typical junctions $V_d \simeq 0.2$ to 0.9 volt.

Note that when an external voltage V_a is applied, the potential barrier will naturally be affected. In other words, the carrier concentrations near the junction in such a case is given by

$$\frac{n_n}{n_p} = e^{(q/kT)(V_d \pm V_a)} \tag{A-19a}$$

$$\frac{p_p}{p_n} = e^{(q/kT)(V_d \pm V_a)} \tag{A-19b}$$

where the minus sign is used for forward-biased conditions (in the forward-biased case, the plus polarity of the battery is connected to the p-type) and the plus sign is used for the reverse-biased case. These are shown in figs A-1 (c) and (d), respectively. Observe that a forward-biased case reduces the potential hill, and thus enables many majority carriers to cross the junction by diffusion, while the reverse-biased case increases the potential hill, thus further preventing the flow of majority carriers.

A number of interesting properties of the p-n junction such as the depletion or junction capacitance and the depletion width can be determined by solving Poisson's equation in the depletion region. These are derived in Ref 3. The results are

$$C_j = \frac{\varepsilon A}{l} = \frac{\varepsilon A}{|l_p| + |l_n|} \tag{A-20}$$

where ε is the relative dielectric constant ($\varepsilon = 11.7\,\varepsilon_0$ for Si, $15.8\,\varepsilon_0$ for Ge, and $\varepsilon_0 = 8.85 \times 10^{-12}$ F/m), A is the cross-sectional area, l_p and l_n are the transition widths of the p and n regions. For abrupt junctions the transition widths are given by

$$\left.\begin{array}{l} l_p = \left[\dfrac{2\varepsilon N_d}{qN_a(N_d + N_a)}\right]^{1/2} (V_d - V_a)^{1/2} \qquad\text{(A-21a)} \\[2ex] l_n = \left[\dfrac{2\varepsilon N_a}{qN_d(N_d + N_a)}\right]^{1/2} (V_d - V_a)^{1/2} \qquad\text{(A-21b)} \end{array}\right\} \text{abrupt junction}$$

so that

$$l = |l_p| + |l_n| = \left[\frac{2\varepsilon}{q}\left(\frac{1}{N_d} + \frac{1}{N_a}\right)\right]^{1/2} (V_d - V_a)^{1/2} \tag{A-22}$$

A-3 The p-n Junction and its Properties

For a graded junction

$$l_p = l_n = \left(\frac{3\varepsilon}{2qa}\right)^{1/3}(V_d - V_a)^{1/3} \quad \text{graded junction} \quad \text{(A-23)}$$

where a is the slope of the graded junction impurity profile. In general, we may express the depletion capacitance of a p-n junction by

$$C_j = K(V_d - V_a)^{-1/n} \quad \text{(A-24)}$$

where

K = constant depending on the impurity concentration profile and the area.

n = constant depending on the distribution of impurity near the junction, as given above.

The value of n in a practical case may range from a high of 3 to a low of $\frac{1}{3}$ in various types of p-n junctions.

Note that in (A-22), (A-23), and (A-24), V_a is a positive for forward and negative for a reverse-biased case. The nonlinear relation $C = f(V)$ of a p-n junction enables one to use the p-n junction to obtain amplification. These amplifiers are used for very high frequencies (on the order of 10^9 Hz) and are known as parametric amplifiers. The p-n junctions used for this purpose are specially designed and are known as *varactors*, a shortened term for *variable capacitors* (see Sec 9-14). The fabrication of p-n junctions in integrated circuits is discussed in Appendix B.

An example will illustrate the use of the above relations.

EXAMPLE

Consider a p-n junction as shown schematically in fig A-2. The junction is assumed to be abrupt, and the conductivities of the p-type and n-type Si crystals at room temperature are: $\sigma_p = 100\,(\text{ohm-cm})^{-1}$ and $\sigma_n = 1\,(\text{ohm-cm})^{-1}$. We are to calculate V_d, l, and C_j at zero

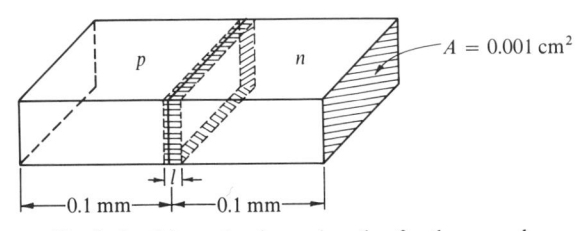

Fig A-2 Schematic of a p-n junction for the example

applied voltage and, also, the values of l and C_j for a reverse-applied potential of -5 volts, and a forward bias potential of $+0.5$ volt.

The donor and acceptor concentrations are found from

$$N_d \simeq n_n \simeq \frac{\sigma_n}{q\mu_n} = 4.80 \times 10^{15} \text{ cm}^{-3}$$

$$N_a \simeq p_p \simeq \frac{\sigma_p}{q\mu_p} = 1.25 \times 10^{18} \text{ cm}^{-3}$$

From (A-18), at room temperature we determine

$$V_d = \frac{kT}{q} \ln \frac{\sigma_n \sigma_p}{\mu_n \mu_p q^2 n_i^2} = (0.025)(30.9) = 0.77 \text{ volt}$$

at no applied voltage, $V_a = 0$ and from (A-22) we have

$$l = \left[\frac{2\varepsilon}{q}\left(\frac{1}{N_d} + \frac{1}{N_a}\right)V_d\right]^{1/2} \simeq \left[\frac{2\varepsilon}{q}\left(\frac{1}{N_d}\right)V_d\right]^{1/2} \approx l_n$$

$$= 4.6 \times 10^{-4} \text{ mm}$$

Note that

$$l_n = \frac{N_a}{N_a + N_d} l \quad \text{and} \quad l_p = \frac{N_d}{N_a + N_d} l$$

From (A-20) and above we obtain

$$C_j(0) = \frac{\varepsilon A}{l} = 22.7 \text{ pF}$$

At a reverse-biased applied voltage $V_a = -5$ volts, the corresponding values of l and C_j are

$$l = 1.2 \times 10^{-3} \text{ mm}, \quad C_j(-5 \text{ volts}) = 8.3 \text{ pF}$$

Similarly, at a forward-biased voltage $V_a = +0.5$ volt, these are

$$l = 2.7 \times 10^{-4} \text{ mm}, \quad C_j(+5 \text{ volts}) = 38.3 \text{ pF}$$

A-4 V-I Characteristic of p-n Junctions

The dc voltage-current relationships of actual *p-n* junction Si and Ge diodes at room temperature are shown in fig A-3. The *V-I* characteristics, exclusive of the breakdown region, are given by the relation

$$I = I_s(e^{\lambda(qV/kT)} - 1) = I_s(e^{\lambda \Lambda V} - 1) \qquad (A\text{-}25)$$

where $\Lambda \equiv q/kT \simeq 40$ volt^{-1} at room temperature
I_s is the reverse saturation current (A)
λ is an empirical constant which lies between 0.5 and 1.0.

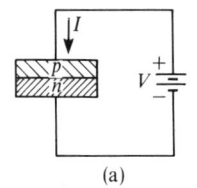

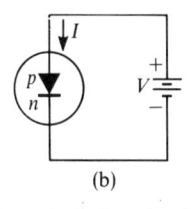

Fig A-3 (a) Schematic of a junction diode with forward bias applied; (b) Graphic symbol for (a)

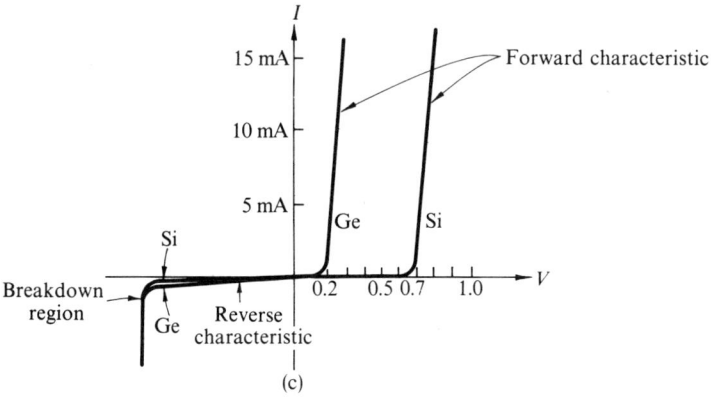

Fig A-3 (c) Typical *V-I* characteristics of Si and Ge diodes (reverse region not ro scale)

We shall now show how (A-25) may be obtained for a *p-n* junction material and also show the strong dependence of I_s on temperature.

The current-voltage relationship of a *p-n* junction diode is obtained from the *continuity equation* and the appropriate boundary conditions.

The continuity equation for holes, assuming one-dimensional diffusion flow, is given by (Refs 2 and 3)

$$\underbrace{\frac{\partial p(x,t)}{\partial t}}_{\text{Time rate of increase in holes in volume}} = \underbrace{-\frac{p - p_n}{\tau_p}}_{\substack{\text{Holes generated} \\ \text{minus holes re-} \\ \text{combined per} \\ \text{unit time per unit} \\ \text{volume}}} + \underbrace{D_p \frac{\partial^2 p(x,t)}{\partial x^2}}_{\substack{\text{Difference between} \\ \text{hole currents} \\ \text{into and leaving} \\ \text{volume}}} \qquad (A\text{-}26)$$

For dc conditions, there is no variation with time, hence $\partial p/\partial t = 0$. Equation (A-26) can be written as

$$D_p \frac{d^2 p}{dx^2} = \frac{p - p_n}{\tau_p} \qquad (A\text{-}27)$$

Equation (A-27), which is referred to as the hole *diffusion equation*, may be written as

$$\frac{d^2(p - p_n)}{dx^2} = \frac{p - p_n}{L_p^2} \qquad (A\text{-}28)$$

where $L_p = \sqrt{D_p \tau_p}$ is the hole diffusion length; p_n, which is constant, is the equilibrium density of holes in the n-region far away from the junction, and τ_p is the hole lifetime in the n-region. Solution of (A-28), subject to the following boundary conditions, is given by (A-29):

(at $x = l_n$ $\quad p = p_n e^{\Lambda V_a}$, and at $x = \infty$ $\quad p = p_n$)

$$p - p_n = p_n(e^{\Lambda V_a} - 1)e^{(l_n - x)/L_p} \qquad (A\text{-}29)$$

Thus, the hole-current density from (A-29) and (A-14) is

$$J_p(x) = q \frac{D_p p_n}{L_p}(e^{\Lambda V_a} - 1)e^{(l_n - x)/L_p} \qquad x \geq l_n \qquad (A\text{-}30)$$

Similarly, the diffusion equation for the electrons in the p-region is

$$D_n \frac{d^2(n - n_p)}{dx^2} = \frac{n - n_p}{\tau_n} \qquad (A\text{-}31)$$

The boundary conditions are

at $x = -l_p$ $\quad n = n_p e^{\Lambda V_a}$ and at $x = -\infty$ $\quad n = n_p$

The electron current density, similar to the above, is

$$J_n(x) = q \frac{D_n n_p}{L_n}(e^{\Lambda V_a} - 1)e^{(x + l_p)/L_n} \qquad x < -l_p \qquad (A\text{-}32)$$

Since the transition width, l, is almost always much smaller than the diffusion length, the recombination in the depletion region can be neglected and we can write

$$J_n(-l_p) = J_n(l_n) \quad \text{and} \quad J_p(l_n) = J_p(-l_p) \tag{A-33}$$

The total diode current, which is constant throughout the crystal, is

$$J = J_n(l_n) + J_p(l_n) = J_n(-l_p) + J_p(l_n) \tag{A-34}$$

or

$$J = q\left(\frac{D_p p_n}{L_p} + \frac{D_n n_p}{L_n}\right)(e^{\Lambda V_a} - 1) \tag{A-35}$$

The total dc current, with the cross-sectional area A, is

$$I = JA = qA\left(\frac{D_p p_n}{L_p} + \frac{D_n n_p}{L_n}\right)(e^{\Lambda V_a} - 1) \tag{A-36}$$

We may rewrite (A-36) as

$$I = I_s(e^{\Lambda V_a} - 1) \tag{A-37}$$

where

$$I_s = qA\left(\frac{D_p p_n}{L_p} + \frac{D_n n_p}{L_n}\right) = qA\left(\frac{D_p}{L_p N_d} + \frac{D_n}{L_n N_a}\right)n_i^2 \tag{A-38}$$

Note that n_i^2 is strongly dependent on temperature, as is evident by (A-1). Hence, I_s is strongly dependent on temperature. For a real diode, an empirical factor λ, i.e., $\Lambda' = \lambda\Lambda$, where λ ranges between 0.5 and 1, is used, so as to fit the V-I characteristics with the measured data. Some authors use m instead of λ, i.e., $\Lambda' = \Lambda/m$ where m ranges between 1 and 2.

The concentration profile and the current density of the silicon example of this section for a forward-bias potential of 0.5 volt are shown in fig A-4. Note that the boundary condition concentration, the hole current, and the electron current are all shown for illustrative purposes. The saturation current of the silicon used for the example is $I_s = 2.7 \times 10^{-15}$ ampere. The actual value of the saturation current in a real silicon function at room temperature, however, is much larger than this value. A typical value is 10^{-9} ampere. This discrepancy is due largely to leakage current and other surface effects which are not included in the analysis. The discrepancy may be resolved if a proportionality constant is introduced with I_s so as to confirm measurements made on the real diode.

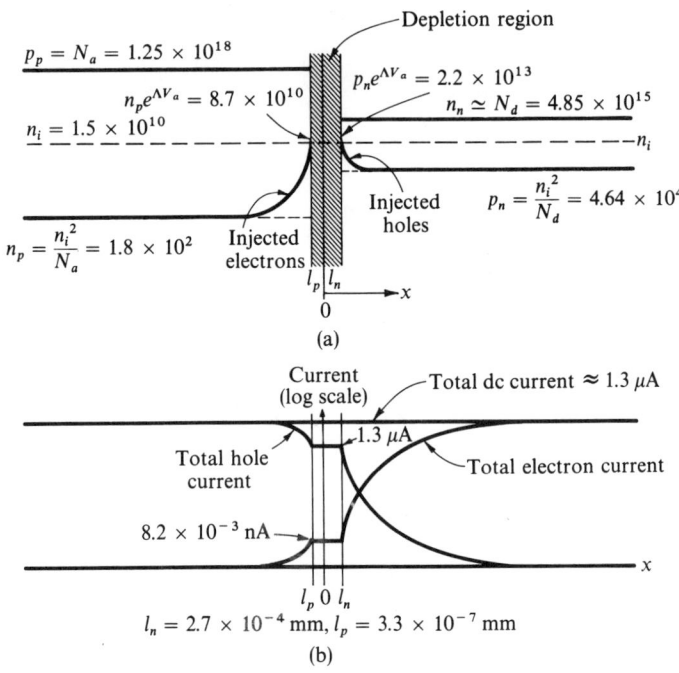

Fig A-4 Silicon crystal of the example with a forward bias of 0.5 volt: (a) Concentration profile; (b) Dc currents (not to scale)

REFERENCES AND SUGGESTED READING

1. Gibbons, J. F., *Semiconductor Electronics*, New York: McGraw-Hill, 1966.
2. Gray, P. E., *Introduction to Electronics*, New York: Wiley, 1967, Chapters 2 and 3.
3. Ghausi, M. S., *Principles and Design of Linear Active Circuits*, New York: McGraw-Hill, 1965, Chapters 6–8.
4. Adler, R., et al., *Introduction to Semiconductor Physics*, SEEC Series, Vol. 1, New York: Wiley, 1964.
5. Gray, P. E., et al., *Physical Electronics and Circuit Models of Transistors*, SEEC series, Vol. 2, New York: Wiley, 1964, Chapters 1–5.

APPENDIX B
Integrated Circuits

In this appendix we will discuss briefly the technique of fabricating circuits which are one or more orders of magnitude smaller, more reliable, and less costly than discrete circuits. *Integrated circuit* (IC), or *microelectronics* technology has made it possible to realize large systems and complex building blocks which were previously considered uneconomical and impracticable. For example, the impact of IC technology in building faster, larger, and lighter digital computers cannot be questioned. The low cost of active elements in IC's has changed the basic ground rules of electronic-circuit design and has led to abundant usage of transistors.

An IC consists of a single-crystal chip of semiconductor material, usually silicon, typically 50 to 100 mils (1.25 to 2.5 mm) on a side, containing both active and passive elements. There are two types of IC's, namely, the monolithic and the hybrid forms. The monolithic is completely integrated, i.e., the entire circuit is fabricated on a single semiconductor substrate. The monolithic form is small and inexpensive in large quantity production, since all components and interconnections are made by a succession of processing steps at the same time. The disadvantage of this form is the restriction imposed on the range of element values, and the tolerance variations of passive components in absolute values, which are typically in the range of 10 to 30 per cent. Ratios of passive component values can be held as close as 3 per cent.

In the hybrid IC, various components, on separate chips, are mounted on an insulating substrate and interconnected. This form of IC is more flexible in that thin-film and diffused components can be used. In small quantity production, a hybrid IC is economical. For large quantity production, the cost of manual assembly and interconnection exceeds that of masks for monolithic IC's and is, therefore, more expensive. In a hybrid IC, however, the parasitic effects can be considerably reduced.

As an example the photomicrograph of a monolithic IC emitter-coupled transistor logic gate (ECL) having three input terminals is

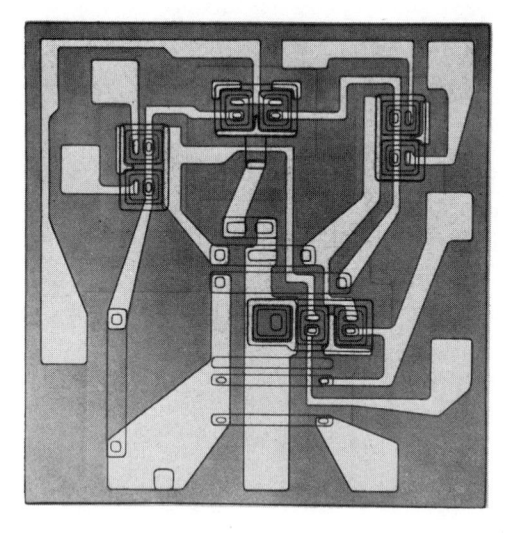

Fig B-1 Photomicrograph of *ECL* three-input *IC* gate (Courtesy Motorola Semiconductor Products)

shown in fig B-1. The die for this circuit is about 25 by 25 mils (1 mil = 0.001 inch). Similarly, IC operational amplifiers using many transistors, diodes, and resistors are sold commercially in a TO-5 package, which is the same size as a single discrete transistor. Other forms of packaging, such as ceramic flatpacks and low-cost plastic dual in-line packages, are also available. Usually the cost of assembly and packaging is a high percentage of the total cost of a chip.

We shall discuss briefly some of the fabrication techniques used in an IC in order to appreciate the constraints and new problems imposed on the circuit designer.

B-1 IC Transistor Fabrication

We shall first show the basic steps in the fabrication of an IC transistor. The fabrication process involves epitaxial growth, oxidation, photomasking, impurity diffusion, and metallization.

A p-type silicon-crystal wafer with a typical resistivity of $10\ \Omega/\text{cm}$ is first cut and polished to a mirror finish and then used as the *substrate* material. An oxide layer is then grown, as shown in fig B-2a. The wafer surface is then coated with a photosensitive emulsion, as shown in fig B-2b. To expose the emulsion selectively, a prescribed mask for the buried-layer diffusion is placed in contact with the substrate and the masked wafer is then exposed to ultraviolet light (this step is not necessary but is almost always used to get an improved transistor as will be described shortly). After exposure, the emulsion is developed and the unexposed emulsion is removed from the buried-layer regions. The wafer is then ready for the first diffusion, as shown in fig B-2c. Arsenic or antimony is used for the buried-layer diffusion, and a heavily doped n-type layer is formed, as shown by the $n+$ layer in fig B-2d. After the buried layer is formed, all oxide is removed, and an n-type layer is grown epitaxially on the surface to serve as the collector region of the transistor, as shown in fig B-2e. At this point several oxidation, photoresist, masking, and diffusion steps take place in the following sequence: An oxide layer is grown over the surface of the n-layer; photoresist is then applied. The isolation pattern is then applied in a manner similar to the buried-layer pattern previously described. The isolation diffusion is then completed through the n-layer by doping the isolation region with acceptors to make it $p+$, in order to provide an isolation region between this transistor and the adjacent n-region on the chip, as shown in fig B-2f. Next we follow the same steps of oxidation, base photoresist, and diffusion in forming the base and resistor regions, as shown in fig B-2g. Once more these steps are repeated for the emitter region, and a heavily doped n^+ region is formed which serves as the emitter of the transistor and also a topside contact to the n-collector region, as shown in fig B-2h. Finally, another mask is used for ohmic contacts, to which aluminum metallic bonds are made. The whole chip is then covered by a silicon dioxide for passivation

Appendix B/Integrated Circuits

(a) Substrate with silicon dioxide layer

(b) Photosensitive emulsion applied

(c) Mask is applied on (b) and then exposed to light

(d) n^+ buried layer diffusion

(e) Oxide removed from (d) and epitaxial n-layer grown

(f) After isolation masking and diffusion

(g) After base masking and diffusion

(h) After emitter masking and diffusion

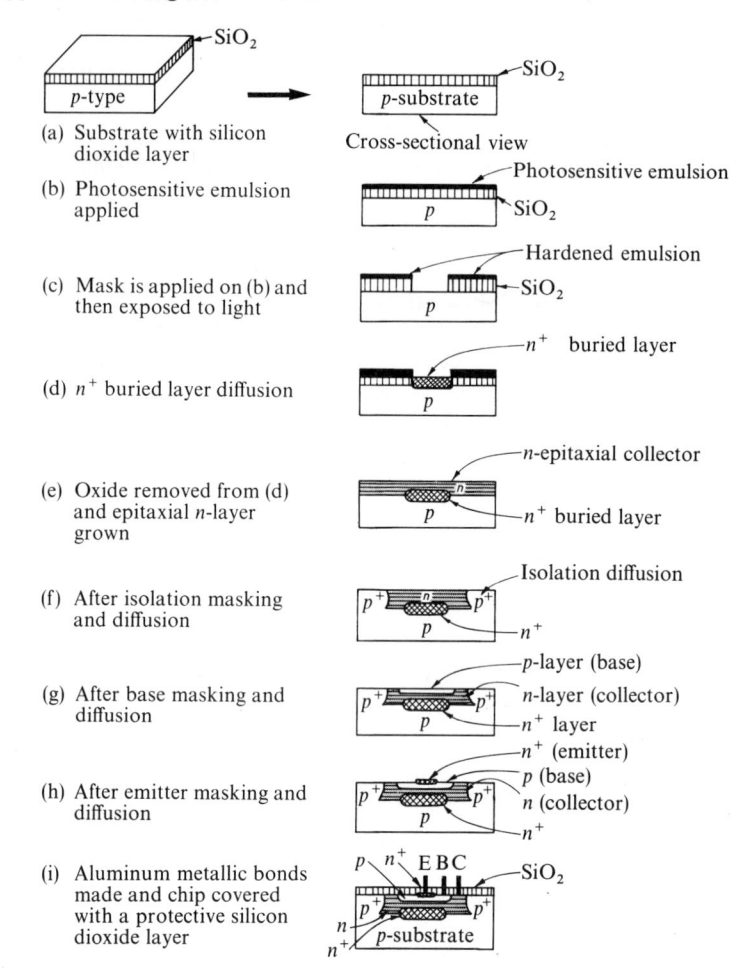

(i) Aluminum metallic bonds made and chip covered with a protective silicon dioxide layer

Fig B-2 Basic steps in the fabrication of an integrated-circuit bipolar transistor

purposes, as shown in fig B-2i. The interconnection of the terminals determines whether the element becomes a transistor, diode, resistor, or capacitor, as will be described in the following. A single transistor on an IC is typically 3 by 5 mils (75 by 125 μm). In monolithic IC fabrication, the whole circuit (a large number of identical ones) is fabricated as an array of dice processed simultaneously on all chips of a wafer. A wafer is generally a circular plate about two inches in diameter and 10 mils thick. Each chip or die (representing a circuit) is a square or rectangle of about 50 to 100 mils on a side. After the fabrication of the circuit is completed, the wafer is scribed and broken into individual dice, each of which is a complete circuit. Figure B-3 shows various stages in the processing of an integrated circuit. At the top is a silicon

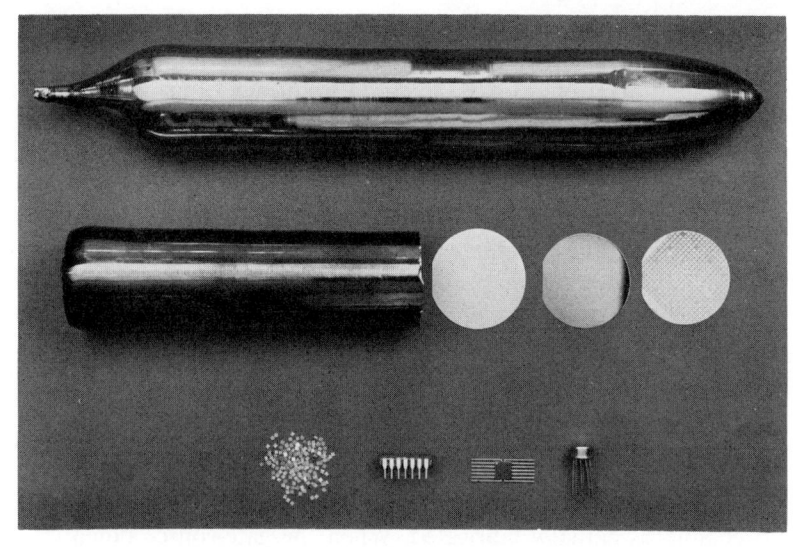

Fig B-3 From grown crystal to integrated circuit: various stages in processing (Courtesy Fairchild Semiconductor)

crystal that has been doped to provide specific electronic characteristics. The middle row shows this crystal, alongside a wafer cut from it, a wafer after polishing, and a wafer that has undergone fabrication into hundreds of identical IC's. The bottom row shows a collection of chips separated from the wafer, followed by three common packages (a dual in-line package, a flatpack, and a TO-5 can).

In order to appreciate the scale and the order of magnitude in the dimensions of the layers, a cross-section of an IC transistor drawn to

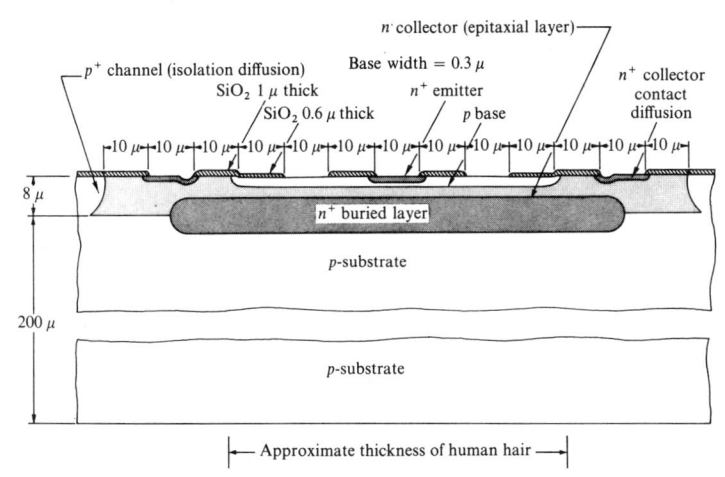

Fig B-4 Cross-section of an integrated-circuit transistor shown to correct scale (after Lynn, et al.; Ref 2)

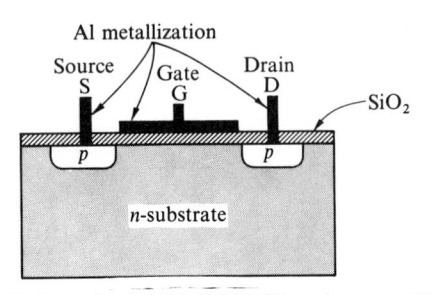

Fig B-5 p-channel *MOSFET* of the enhancement type

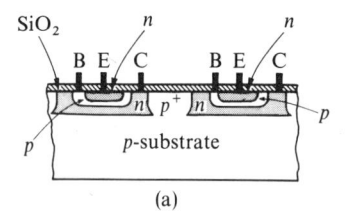

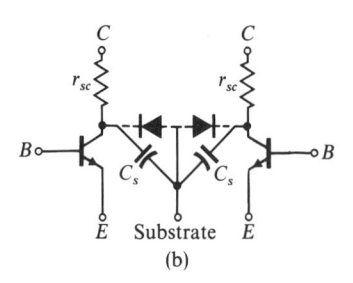

Fig B-6 (a) Two *IC* transistors (with no buried layers); (b) Approximate equivalent circuit of (a)

correct scale is shown in fig B-4. Note that the base width is of the order of 0.3 μm; thus, from (3-47), the cutoff frequency of an IC transistor can reach the GHz region. At this time, high-quality fast *pnp* transistors cannot be fabricated on the same chip with *npn* devices. Typical values of f_T for *pnp* devices are under 20 MHz, while the f_T of *npn* transistors is typically several hundred MHz. With lateral and substrate *pnp*'s, β's of 100, i.e., compatible with those of *npn*'s, are now achieved. Hence, *pnp*'s can be used where ultimate speed is not necessary. For fast circuits, however, all IC transistors are invariably *npn*.

The basic fabrication process of a MOSFET is very simple in IC technology and requires only the diffusion of two *p* regions on an *n*-substrate (or a *p*-substrate with two *n*-regions), as shown in fig B-5. The heavily doped regions form the source and drain contacts. The insulating layer, e.g., SiO_2, Si_3N_4, or Al_2O_3, is then grown and aluminum metallization completes the fabrication. In this type of FET, no channel is actually fabricated. When the gate-source potential is negative, the so-called enhancement mode, a *p*-channel, is induced between the source and the drain.

The MOSFET, in addition to requiring a smaller number of fabrication steps, also requires less area on the chip than the conventional IC bipolar transistor and is, therefore, attractive in this respect. Typical chip area used by a MOSFET is less than 5 square mils. In other words, MOS technology can create circuits with 5 to 10 times as many transistors per unit area as the bipolar transistors.

In IC analysis and design, one has to be careful with parasitic diodes, transistors, resistances, and capacitances. For example, two transistors fabricated in one chip, as shown in fig B-6a, approximate the circuit model shown in fig B-6b. Note that, although the *p*-substrate isolates the two transistors, the collectors of the two transistors form two back-to-back diodes with the *p*-substrate. Note also that the substrate must always be connected to the most negative potential in the circuit to ensure that the isolation diodes are always back-biased, thus yielding very high impedance isolation between components to maintain correct circuit operation. Note also that if the buried n^+ layer is not present, the series collector resistance (r_{sc}), which is typically 10 to 30 Ω or even higher, must be included in the circuit model. The presence of the low resistivity n^+ buried layer in fig B-3d shunts the collector material and reduces r_{sc} to typically 10 Ω or less, and thus r_{sc} can, in most cases, be ignored if a buried layer is present. Because the small value of r_{sc} reduces the transistor's apparent saturation voltage, the buried layer is almost always fabricated in an IC transistor. Finally, note the presence of the parasitic capacitances and transistors which may have to be included in a complete and detailed analysis.

B-2 IC Diodes

IC diodes are usually constructed by proper interconnection of the IC transistor terminals. There are five combinations of diodes possible from a transistor, as shown in fig B-7. Among these, the configurations in (a) and (b) are commonly used. The reverse-breakdown voltage of these diodes is low—approximately 5 to 7 volts. If this presents a problem, the configuration in (c) may be used, since the rating of the collector-base junction is usually high (≥ 15 volts). It is to be noted that a collector-base diode with no emitter diffusion is also used in practice. This type of diode has a high breakdown, smaller area, and lower capacitance. It should also be noted that different storage-time characteristics are available from the different diode configurations.

The circuit model of an IC diode, which includes the capacitance of the reverse-biased parasitic diodes, is shown in fig B-8.

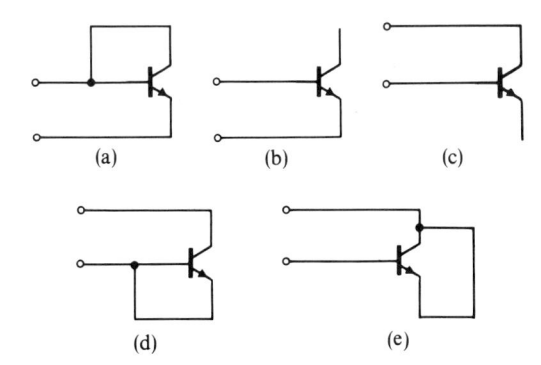

Fig B-7 Various forms of *IC* diodes as obtained from *IC* transistors

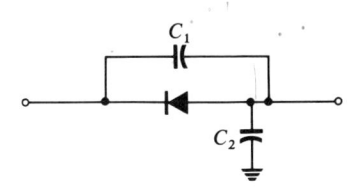

Fig B-8 Circuit model of an *IC* diode

B-3 IC Capacitors

There are many types of capacitors used in IC technology. Among these are the transition capacitors of *p-n* junctions, and thin-film capacitors.

The junction capacitors are the easiest to form in integrated circuits. These polarized capacitors are voltage-dependent and are used primarily as decoupling and bypass capacitors. This type of capacitor has a series resistance of the bulk semiconductor resistivity associated with it which may have to be included in a careful circuit analysis. Typical values of a junction capacitor are about 300 pF/mm², with a tolerance of 30 per cent. Thus, if an accurate capacitor of large value is required, it is made in thin-film form.

The thin-film, specifically MOS capacitors are formed as parallel-plate capacitors by the n^+ region of the emitter and a metal film separated by the silicon dioxide dielectric, as shown in fig B-9. Because of the *m*etal *o*xide and *s*emiconductor layers, it is called an MOS capacitor. The capacitance of this device is typically between 400 to 800 pF/mm². The MOS capacitors are voltage-independent and non-polarized. The series resistance of the MOS capacitor is lower as compared with that of the junction capacitor, because of the low resistivity of the n^+ emitter region forming the lower plate. Typical values of the series resistance of MOS capacitors are 1–10 Ω while 10–50 Ω is typical for junction capacitors.

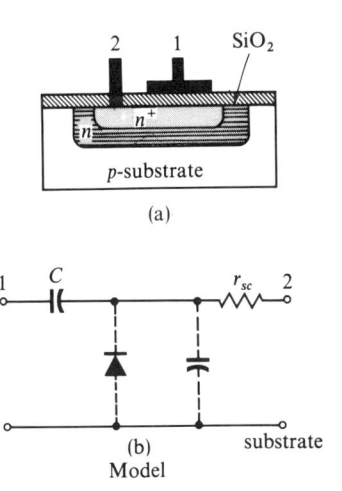

Fig B-9 (a) Schematic of an *MOS* capacitor; (b) Circuit model of (a)

Larger values of capacitors can be obtained by sandwiching the silicon dioxide with layers of aluminum to form the top and the bottom plates. These types of thin-film capacitors require additional steps in the fabrication process, but the capacitance can be high as 3000 pF/mm² with the lowest loss among those mentioned in above.

Note that, in all cases, the maximum size of the capacitors is quite limited. The maximum value of an IC capacitor is under 500 pF. This is because of the area constraints in IC and the breakdown limitations. This is why direct-coupled amplifiers are used in the design of amplifiers for ac applications, since bypass and coupling capacitors of the order of μF are entirely impractical in integrated circuits.

B-4   IC Resistors

Similar to IC capacitors, there are many types of resistors in IC technology. Among these are diffused resistors and thin-film resistors.

The diffused resistors are formed during the p-type base diffusion, as shown in fig B-10. The approximate expression for the resistance, ignoring fringing effects, is

$$R = \frac{\rho l}{dw} \qquad \text{(B-1a)}$$

The resistance of a thin sheet of material is often specified by its sheet resistance (R_o) expressed in ohms-per-square. Thus, if $l = w$, then

$$R_o = \frac{\rho}{d} \text{ ohms/square} \qquad \text{(B-1b)}$$

hence

$$R = R_o \frac{l}{w} \text{ ohms} \qquad \text{(B-2)}$$

Practical values of diffused resistors range from approximately 50 Ω to 20 kΩ. For the lower range, i.e., small values of resistance, the emitter diffusion ($R_o \simeq 2\,\Omega$/square) instead of the base diffusion ($R_o \approx 200\,\Omega$/square) is used. Manufacturing tolerances of diffused resistors are about 20 per cent in absolute value whereas the ratios of resistors can be held to under 5 per cent. This is why one should strive to design the IC circuit such that the performance depends on resistor ratios rather than absolute values. Again notice the presence of the parasitic transistor formed by the pnp layers and the junction capacitors in fig B-10.

Thin-film resistors can be fabricated in the form shown in fig B-11, where the resistive material may be Nichrome, tin oxide, tantalum

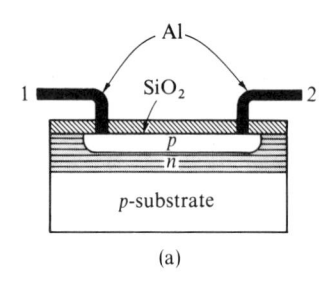

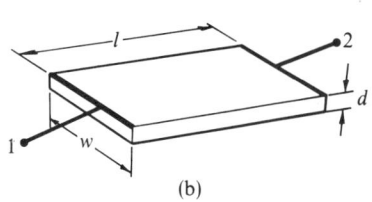

Fig B-10   (a) Schematic of an IC resistor; (b) Simplified geometry of the p-material

nitride, or other resistive film. Large and small values of resistance are obtained by using tin oxide and tantalum nitride, respectively. The sheet resistance of the various materials ranges from 10 to 20,000 ohms/square. Note that for a given resistive material, the large value resistor from (B-2) requires a large number of squares, hence a large ratio of l/w. This can be achieved by reducing w and/or increasing l. The smallest value of w is limited by photographic resolution or power dissipation. w_{min} is about 0.2 mil (0.0002 inch); the largest value of l is limited by the area of the chip. To get the largest l, one usually designs a zigzag pattern in order to get a large number of squares, i.e., a large l/w ratio.

Note that a diffused IC resistor is really a distributed RC network, as shown in fig B-12, where the distributed capacitance is formed by the diffused junction capacitance or the sandwiched dielectric such as a slab of SiO_2 or other dielectric material. The properties and applications of the distributed RC network are discussed in Section 8-9.

If trimming is used after fabrication, tantalum resistors can be manufactured with an accuracy of better than 0.1 per cent and in the range of $5\,\Omega$ to $500\,k\Omega$. Resistor ratios can be held to within 0.01 per cent.

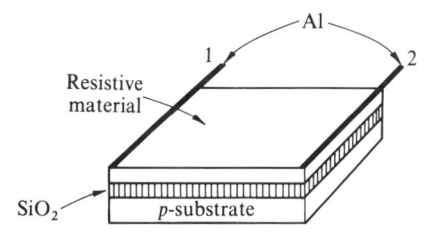

Fig B-11 Schematic of a thin-film resistor

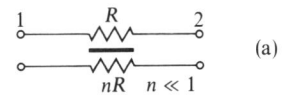

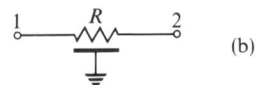

Fig B-12 Circuit model of fig B-11. Note that this is a distributed RC network ($\overline{RC}$)

B-5 IC Inductors

No practical range of inductor values can be fabricated in IC at the present time. In integrated circuit design, one usually tries to avoid using inductors. Apparent inductors can be obtained by active RC circuits such as using negative impedance converters with RC elements or capacitively terminated gyrators (see Chapter 8) or the input of a short-circuited common-base circuit at moderate frequencies (fig P3-46). If an inductor must be used, a discrete component is sometimes connected externally to the hybrid integrated circuit.

B-6 IC Design Guidelines

In integrated circuits, the principal cost factors are the fabrication steps, including the masks, etching, diffusion, etc., and the real estate, i.e., the area occupied by a component on a chip and the assembly and packaging. Since passive components, mainly capacitors, usually require more area than active elements, the active devices usually cost less than the passive elements. Hence the design philosophy should be not to minimize the number of active devices, but to minimize the total area and fabrication steps compatible with other requirements of reliability, heat dissipation, packaging, etc. Furthermore in order to reduce overall costs, which include those of assembly and packaging, the trend is toward several chips in a single package.

In monolithic IC's, the designer should strive to design the circuits such that the circuit performance depends on the resistor ratio rather than absolute values, since the latter cannot be produced to close tolerances.

The standard IC processes provide high quality *npn* transistors. In general, avoid using *pnp* transistors in high-speed circuitry, since a good quality (high f_T) *pnp* transistor cannot be fabricated on the same chip compatibly with *npn* devices.

The substrate should be connected to the most negative potential in the circuit so as to back-bias all the parasitic diodes and ensure proper electrical isolation.

If the parasitic elements are to be drastically reduced, and/or the circuit requires a wide range of component values with close tolerances, thin-film components should be used.

For situations where only a few circuits are to be produced, use a hybrid IC and design as many components on a single chip as practical before interconnecting the chips. The layout should be arranged so as to avoid crossing of bonding wires if possible. Reference 2 gives guidelines for the monolithic layout design and will not be discussed

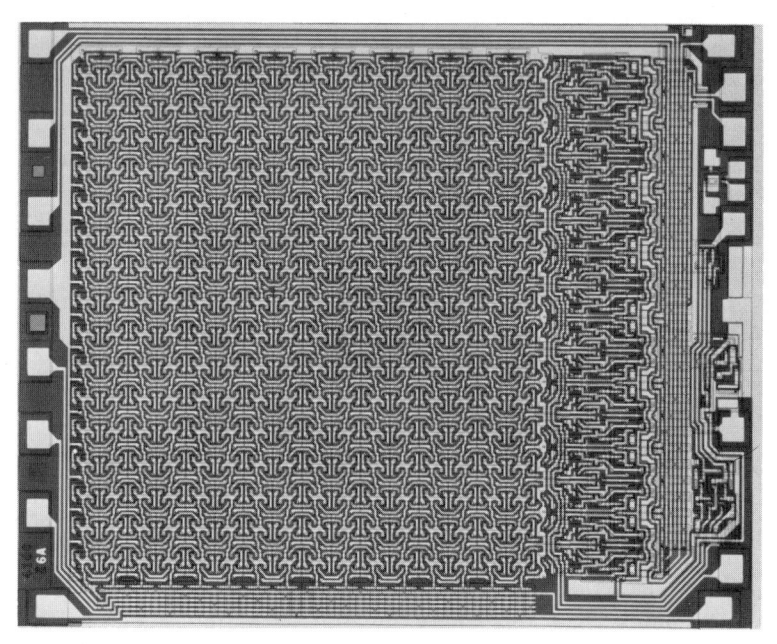

Fig B-13 Photomicrograph of computer memory *IC* having a device density of 160,000/in^2 (Courtesy Fairchild Semiconductor)

here. Digital computers are coming to play an important role in generating the layout and the diffusion masks. Many IC manufacturers use computers to assist designers in the layout of the interconnections of complex digital arrays.

B-7 Large Scale Integration (LSI)

Large scale integration represents the process of fabricating a large number of active and passive components on a silicon chip. These components are interconnected in such a way as to perform a multitude of circuit functions and thus represent a complete subsystem. The component density, i.e., the number of active and passive components per square inch on the silicon chip is very high in LSI. For example, an interconnected functional block consisting of more than several hundred transistors, resistors, and diodes in an area of 0.01 in^2 can be fabricated on a silicon chip. Figure B-13 is a photomicrograph of a computer memory circuit manufactured by Fairchild Semiconductor. It has a density of about 160,000 devices per square inch. Specifically, the chip measures 0.11 by 0.14 inch and contains 1244 transistors, 1170 resistors, and 71 diodes. Figure B-14 shows an LSI memory building block manufactured by Motorola Semiconductor Products. The chip contains 1568 *p*-channel enhancement-mode transistors.

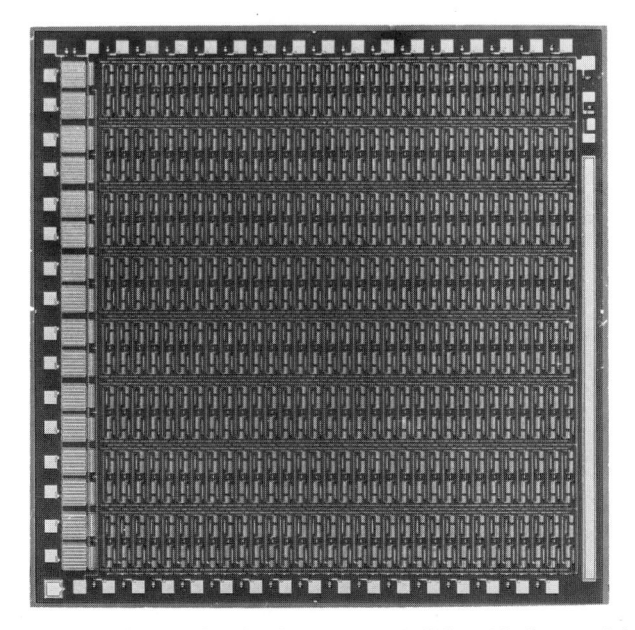

Fig B-14 Photomicrograph of LSI memory building block containing 1568 MOSFET's (Courtesy Motorola Semiconductor)

No other components are used. The component density is 80,500 components per square inch.

Solid-state technology is improving rapidly and very complex subsystems are being made on chips with a good yield. Usually the larger the component density the lower the yield. For this reason the maximum possible component density is intentionally not used on the chip. The so-called MSI, an abbreviation for *medium scale integration*, is now prevalent in industry. At the time of this writing, more than 16,000 components per cm^2 with a reasonable yield have been achieved. Even higher density and yield in LSI production are expected in the future. The problems of analysis, design, and testing in LSI are formidable. LSI has further blurred the lines between components and circuits. The impact of LSI on the development of large-size computers is profound, and the use of computers in the analysis and design of LSI is a necessity.

REFERENCES AND SUGGESTED READING

1. Motorola, Inc. Engineering Staff (Warner and Fordemwalt, editors), *Integrated Circuits*, New York: McGraw-Hill, 1965.
2. Motorola, Inc. Engineering Staff (Lynn, Meyer, and Hamilton, editors), *Analysis and Design of Integrated Circuits*, New York: McGraw-Hill, 1967.
3. Giles, J. N., *Fairchild Semiconductor Linear Integrated Circuits Application Handbook*, Mountain View, Cal: Fairchild Semiconductor, 1967.
4. Radio Corporation of America, *RCA Linear Integrated Circuit Fundamentals*, Harrison, N.J., RCA, 1967.
5. Camenzind, H., *Circuit Design for Integrated Electronics*, Reading, Mass, Addison-Wesley, 1968.
6. Khambata, A. J., *Introduction to Large-Scale Integration*, New York: Wiley, 1969.

APPENDIX C
Introduction to Signal Flow Graphs

The analysis and design of linear electronic circuits frequently consist of determining the relationship between the input and output variables of a system. Such an input–output relationship may be conveniently represented in *block diagram* form, where the various input–output relationships of the significant portions of the system are represented by labeled blocks and interconnecting lines. An alternate equivalent form of this representation is known as the *signal-flow graph*, in which the system transmittances are identified graphically as directed lines with known values, and which offers a simple procedure for relating the desired variables. One such representation is exemplified by a single-loop feedback system, shown in figs C-1a and C-1b, where the input and output variables are denoted by X_1 and X_2, respectively, which are the transforms of voltage and/or current variables.

The signal-flow graph can be used to represent any set of simultaneous linear relations, such as a set of linear equations. A signal-flow graph for purposes of defining various terminologies is shown in fig C-2. In any signal-flow graph we have *branches* and *nodes*. In the

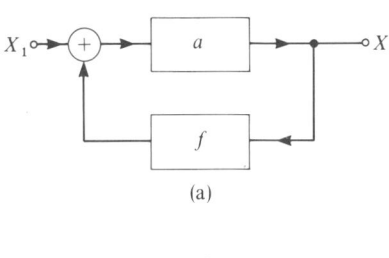

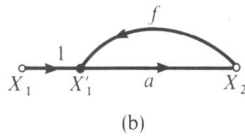

Fig C-1 (a) Block diagram of a single-loop feedback amplifier; (b) Signal-flow graph representation of (a)

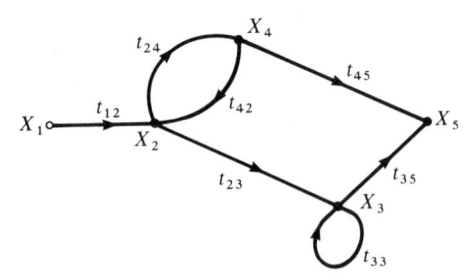

Fig C-2 Signal-flow graph

context of signal-flow graph theory, each *node* represents a variable, e.g., $X_1, X_2, \ldots, X_5$ in fig C-2 could represent transforms of currents and/or voltages. The directed line segments between two nodes (or the same node) are called *branches*. Each branch is assigned a value which represents the functional relation between the variables and is referred to as the *transmittance* of the branch, or the *branch gain*. The transmittances are denoted in fig C-2 by t_{ij}. In other words, a signal X_i traversing a branch in the direction of the arrowhead is multiplied by t_{ij} the transmittance of that branch, and the product is a signal transmission $X_j = t_{ij}X_i$. A signal-flow graph with n nodes has $(n-1)$ dependent variables. The first node, denoted by X_1, is the input or *source*, and is considered to be known. The output node is usually referred to as the *sink*. We may have more than one input and/or output, in which case matrix formulation is used. The equations represented by a signal-flow graph can be written in accordance with the following rule: The value of a variable at any node is equal to the sum of all the signal transmissions *entering* the node. Note that the variables are the Laplace transforms of currents and/or voltages. For example, in fig C-2, the analytic relations are

$$X_2 = t_{12}X_1 + t_{42}X_4 \tag{C-1a}$$

$$X_4 = t_{24}X_2 \tag{C-1b}$$

$$X_3 = t_{23}X_2 + t_{33}X_3 \tag{C-1c}$$

$$X_5 = t_{45}X_4 + t_{35}X_3 \tag{C-1d}$$

From (C-1) observe that a transmittance which begins and terminates at the same node (a self-transmittance) is perfectly acceptable.

Consider now the reverse procedure, where a set of simultaneous linear equations is given and a signal-flow graph which represents the set of equations is to be found. For example, consider three variables of interest related by the following equations:

$$X_2 = a_1X_1 + a_2X_2 + a_3X_3 \tag{C-2a}$$

$$X_1 = b_1X_2 + b_2X_3 \tag{C-2b}$$

To obtain the signal-flow graph representation, the nodes denoting all the variables of interest (X_1, X_2, X_3) are indicated by points. The sources are identified and each equation must be solved for a new dependent variable. The associated transmittances are then drawn according to the equations. The signal-flow graph for (C-2) is shown in fig C-3. Note that for the signal-flow graph representation, the

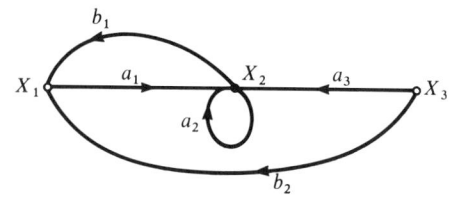

Fig C-3 Signal-flow graph for equation (C-2)

dependent variable is written alone in the left side. For example, in (C-2), X_1 and X_2 are the dependent variables. If X_1 or X_2 is the independent variable which corresponds to a different problem, naturally a different signal-flow graph would result. For example, if X_1 is the independent variable and the variables are related by (C-2), the equations are rewritten as

$$X_2 = a_1 X_1 + a_2 X_2 + a_3 X_3 \tag{C-3a}$$

$$X_3 = \frac{1}{b_2} X_1 - \frac{b_1}{b_2} X_2 \tag{C-3b}$$

The flow graph corresponding to (C-3) is shown in fig C-4.

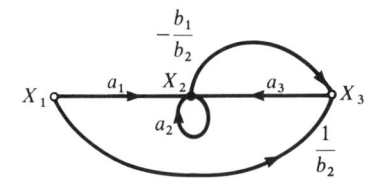

Fig C-4 Signal-flow graph for equation (C-3)

C-1 Simplification of a Signal-Flow Graph

We consider now some techniques for simplifying or reducing a signal-flow graph. The manipulations are straightforward and can be readily verified by writing the pertinent equations from the graph.

Elementary Equivalence

The equations of addition and multiplication, i.e., combination of tandem branches and elimination of an internal node, are illustrated in fig C-5a through d, respectively. The equivalence is demonstrated by the pertinent equations under the corresponding figures.

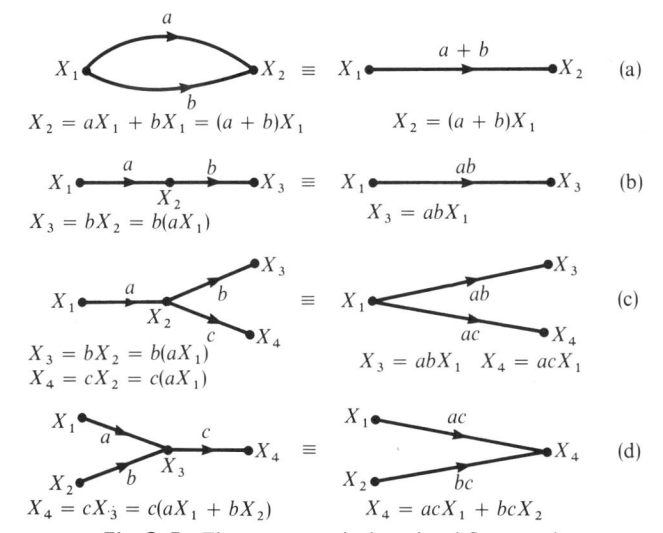

(a) $X_2 = aX_1 + bX_1 = (a+b)X_1 \equiv X_2 = (a+b)X_1$

(b) $X_3 = bX_2 = b(aX_1) \equiv X_3 = abX_1$

(c) $X_3 = bX_2 = b(aX_1)$, $X_4 = cX_2 = c(aX_1) \equiv X_3 = abX_1$, $X_4 = acX_1$

(d) $X_4 = cX_3 = c(aX_1 + bX_2) \equiv X_4 = acX_1 + bcX_2$

Fig C-5 Elementary equivalent signal-flow graphs

Elimination of a Self Loop

A branch or path that starts and ends on the same node, without encountering another node, is called a *self loop*. The elimination of a self loop is illustrated in fig C-6. In other words, a self loop at node 2,

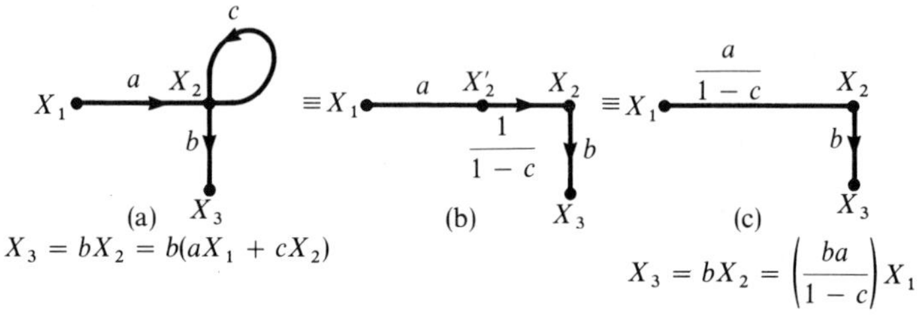

Fig C-6 Signal-flow graph reduction: elimination of a self loop

having a loop gain c, is eliminated by *multiplying only the values of the incident branches by* $[1/(1-c)]$. This is readily seen by writing the corresponding equations:

$$X_3 = bX_2 = b(aX_1 + cX_2) \tag{C-4a}$$

From (C-4a)

$$X_2 = \left(\frac{1}{1-c}\right)aX_1$$

or

$$X_3 = \frac{ba}{1-c}X_1 \tag{C-4b}$$

which is the equation corresponding directly to fig C-6c. A *feedback loop* is a closed path along which no node is encountered more than once (per loop). A feedback loop may be reduced to a self loop, as shown in

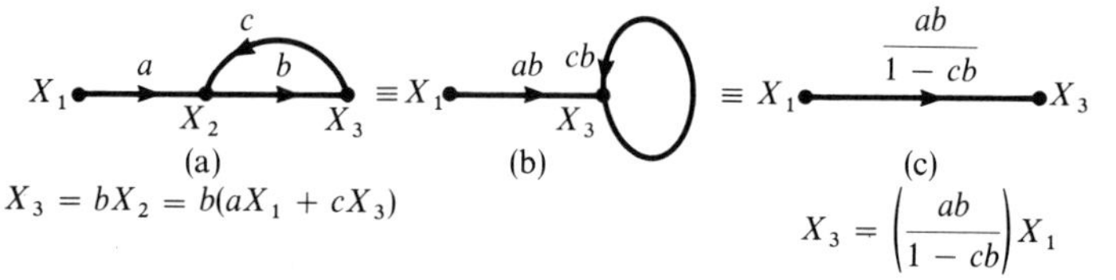

Fig C-7 Signal-flow graph reduction: reduction of a feedback loop

fig C-7b, and then the self loop eliminated as before. This is shown in fig C-7c.

Shift of a Node Entrance or Exit

Sometimes it may be desirable to shift the entrance of a branch from one node to another. This is illustrated in fig C-8, where the entrance to node 2 of branch has been shifted from node 2 to node 3. This step follows directly from fig C-5d. Similarly, the exit of a branch from one node can be shifted to another, as shown in fig C-9. In this case, the exit end of branch d has been shifted from node 3 to node 2.

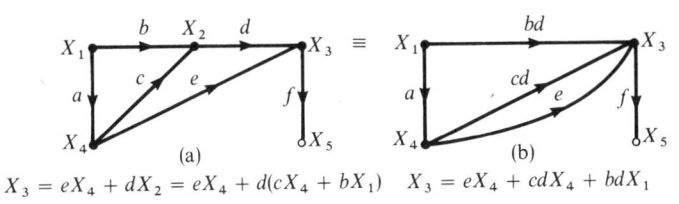

$X_3 = eX_4 + dX_2 = eX_4 + d(cX_4 + bX_1) \quad X_3 = eX_4 + cdX_4 + bdX_1$

Fig C-8 Shift of a node entrance in a signal-flow graph

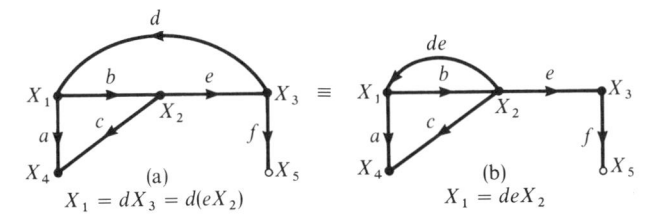

$X_1 = dX_3 = d(eX_2)$ $X_1 = deX_2$

Fig C-9 Shift of a node exit in a signal-flow graph

Node Splitting

It may sometimes be convenient to split a node into two nodes where one node represents the source, i.e., accommodating all entering transmittances, and another the sink, i.e., the group of transmittances leaving the node. This can readily be done, as shown in figs C-10a and b, where node i is split into i and i', joined by a unity transmittance.

These simplifying rules are sufficient for reducing a complicated signal-flow graph to a simple one.

EXAMPLE

As an example, consider the signal-flow graph shown in fig C-11a. The graph can be reduced to a simple form by using the reduction rules as follows: Shift the exit of branch f from node 4 to 3, as in fig C-11b.

Appendix C/Introduction to Signal Flow Graphs

Fig C-10 Node splitting in a signal-flow graph

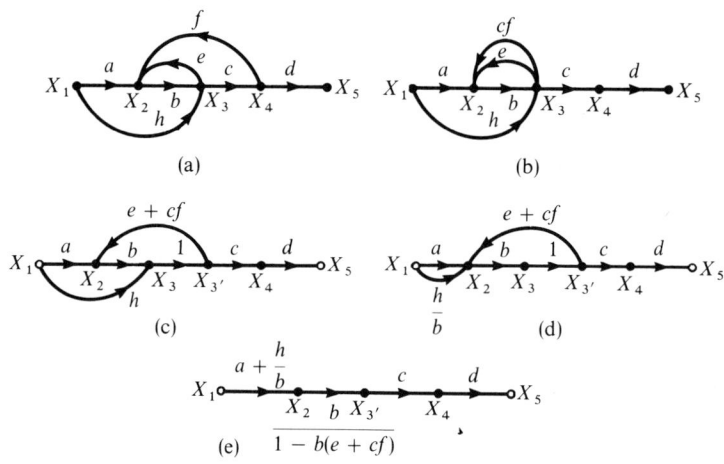

Fig C-11 Signal-flow graph for the example and the equivalent signal-flow graphs using various reduction steps

Now the branches e and cf, which are both parallel paths, can be added and considered as one branch $(e + cf)$. We can next split node 3, as shown in fig C-11c. The entry of branch h is next shifted from node 3 to 2, as shown in fig C-11d. Finally, add parallel forward branches a and h/b, convert the feedback loop according to fig C-7, and the simplified signal-flow graph is shown in fig C-11e. The transfer function can now be written by inspection:

$$\frac{X_5}{X_1} = \left\{\left(a + \frac{h}{b}\right)\left[\frac{b}{1 - b(e + cf)}\right]cd\right\} = \frac{cd(ab + h)}{1 - b(e + cf)} \quad \text{(C-5)}$$

C-2 Mason's Gain Formula

The input–output relationships of a signal-flow graph can be written directly by the application of *Mason's gain formula* and it is not necessary to reduce a graph. Mason's formula, stated without proof (Ref 1), is

$$T(s) = \frac{1}{\Delta_s} \sum_{i=1}^{n} T_i \Delta_i \qquad (C\text{-}6)$$

where the various terms in (C-6) are defined as follows:

$T(s)$ = the desired transfer function.
T_i = the transmittance of the ith forward (open loop) path from input to output
n = the total number of such forward paths
Δ_s = the signal-flow graph determinant which is defined, in turn, as follows:

$$\Delta_s = 1 - \sum_i T_{li} + \sum_{i,j} T_{li} T_{lj} - \sum_{i,j,k} T_{li} T_{lj} T_{lk} + \cdots \qquad (C\text{-}7)$$

where the T_{li} are the feedback loop gains that are singled out before using (C-7). Note that the products of *only nontouching loops* (i.e., loops with no common node) are included in (C-7). If there are only two nontouching loops, the third term in (C-7) contributes only $T_{l1} T_{l2}$, and the rest of the terms beyond the third term are zero.

Δ_i = The value of Δ_s for the part of the graph not touching the ith forward path.

The application of the formula (C-6) is illustrated in the following examples.

C-3 Examples

EXAMPLE 1

Consider the signal-flow graph of a multiloop feedback system shown in fig C-12. To illustrate the various constituents of the graph, we first draw all the forward paths from the input to the output. There are two such paths, as shown in fig C-13a. Thus we have $n = 2$ and

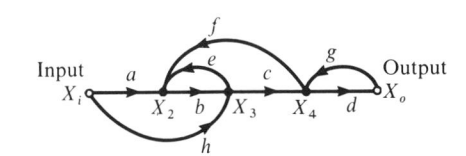

$$T_1 = abcd \qquad T_2 = hcd \qquad (C\text{-}8)$$

Fig C-12 Signal-flow graph for example 1

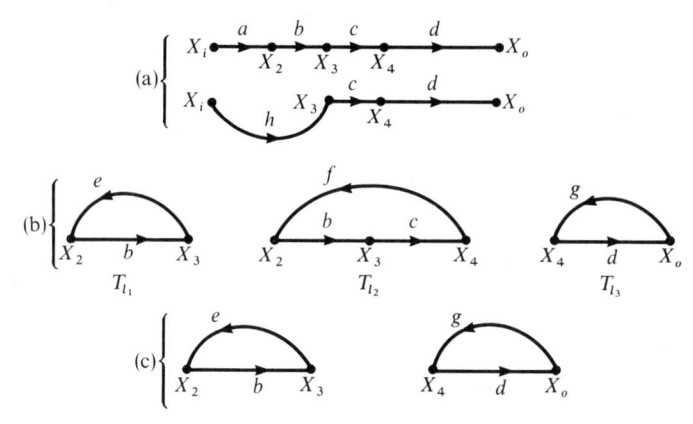

Fig C-13 (a) Forward paths of fig C-12; (b) Feedback loops of fig C-12; (c) Nontouching loops of fig C-12

Since there is no nontouching feedback loop associated with the forward transmittances T_1 and T_2 (see fig C-13), from (C-6), T_{li}, etc., are zero and

$$\Delta_1 = 1 \quad \text{and} \quad \Delta_2 = 1 \tag{C-9}$$

Next, we single out all the feedback loops, as shown in fig C-13b. There are three such loops, hence

$$T_{l1} = eb, \quad T_{l2} = bcf, \quad T_{l3} = dg \tag{C-10}$$

Finally, we pick up the nontouching loops, which are shown in fig C-13c. Since there are only two nontouching loops, the first and only possible combination of two nontouching loops is

$$T_{l1} T_{l3} = (eb)(dg) \tag{C-11}$$

Substitution of (C-11), (C-10), (C-8), and (C-7) into (C-5), and simplification yields

$$T(s) = \frac{X_o(s)}{X_i(s)} = \frac{abcd + hcd}{1 - (eb + bcf + dg) + ebdg} \tag{C-12}$$

Note that for $g = 0$, (C-12) reduces to (C-5), as it should.

EXAMPLE 2

Consider the signal-flow graph of fig C-14. The various components of the graph are identified as follows:

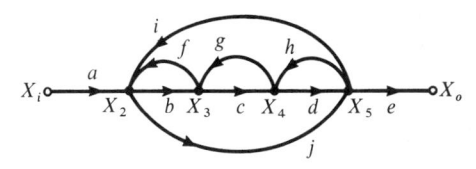

Fig C-14 Signal-flow graph for example 2

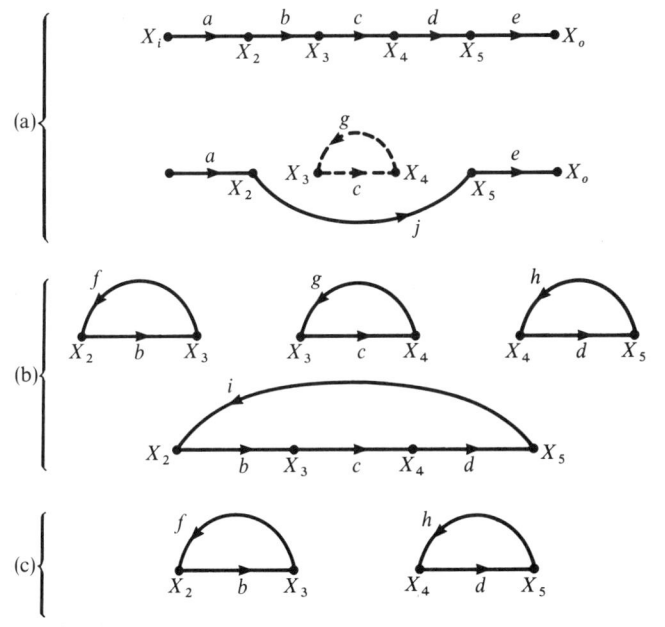

Fig C-15 (a) Forward loops of fig C-14; (b) Feedback loop of fig C-14; (c) Nontouching loops of fig C-14

There are two forward paths, as shown in fig C-15a. The overall transmittances of these forward gains are:

$$T_1 = abcde, \quad T_2 = aje \quad \text{(C-13)}$$

There is no nontouching feedback loop associated with the forward path T_1, hence $\Delta_1 = 1$. The forward path with transmittance T_2, however, does have one nontouching feedback loop, as shown in fig C-15a. Hence, from (C-7)

$$\Delta_2 = 1 - cg \quad \text{(C-14)}$$

The various feedback loops of the graph are shown in fig C-15b. For these we have

$$T_{l1} = bf, \quad T_{l2} = cg \quad \text{(C-15)}$$
$$T_{l3} = dh, \quad T_{l4} = bcdi$$

The nontouching feedback loops are shown in fig C-15c. For these two, we have

$$T_{l1}T_{l3} = bfdh \qquad (C-16)$$

Substitution of (C-16), (C-15), (C-13), and (C-12) into Mason's gain formula, and simplification yields

$$T(s) = \frac{X_o(s)}{X_i(s)} = \frac{abcde + aje(1-cg)}{1 - (bf + cg + dh + bcdi) + bfdh} \qquad (C-17)$$

In a first exposure, Mason's gain formula might seem confusing to the student; however, the simplified version of (C-6) may be used if there are no nontouching loops and paths. Thus *if all the loops and paths are touching*, the simplified version is

$$T(s) = \frac{X_o}{X_i} = \left(\sum_{i=1}^{n} T_i\right) \bigg/ \left(1 - \sum_{i=1}^{m} T_{li}\right) \qquad (C-18)$$

where n is the number of forward paths as defined, and m is the number of feedback loops. In order to use the simplified gain formula above, the nontouching loops and paths are first eliminated by the reduction method, and then (C-18) is applied.

For example, consider fig C-12 by this approach. In fig C-12, the nontouching loops are the loops be and dg. We eliminate either the loop dg or the loop be by the reduction method. These are shown in figs C-16a and C-16b, respectively.

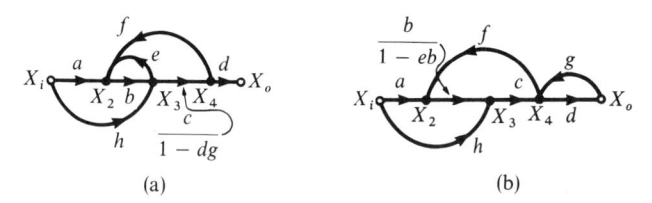

Fig C-16 Signal-flow graph of fig C-12 with nontouching loop be eliminated

From fig C-16a, using (C-18), we have

$$T(s) = \left(\frac{abcd}{1-dg} + \frac{hcd}{1-dg}\right) \bigg/ \left(1 - eb - \frac{bcf}{1-dg}\right)$$
$$= \frac{abcd + hcd}{1 - (eb + dg + bcf) + ebdg} \qquad (C-19)$$

From fig C-16b, using (C-18) we have

$$T(s) = \left(\frac{abcd}{1-eb} + \frac{hcd}{1-eb}\right) \bigg/ \left(1 - \frac{bcf}{1-eb} - dg\right) \qquad \text{(C-20)}$$

$$= \frac{abcd + hcd}{1 - (eb + dg + bcf) + ebdg}$$

which is the same result as before (of course!). The reader might try this approach as an exercise to obtain the transfer function for fig C-14.

REFERENCES AND SUGGESTED READING

1. Mason, S., and Zimmerman, H., *Electronic Circuits, Signals and Systems*, New York: Wiley, 1960. Chapters 4 and 5.
2. Truxal, J. G., *Automatic Feedback Control System Synthesis*, New York: McGraw-Hill, 1955, pp. 88–113.
3. Kuo, B. C., *Automatic Control Systems*, Englewood Cliffs: Prentice-Hall, 1962, Chapter 2.
4. Ruston, H., and Bordogna, J., *Electric Networks: Functions, Filters, Analysis*, New York: McGraw-Hill, 1966. Chapter 6.

The purpose of this appendix is to present data for a few representative discrete devices and IC functional blocks, as supplied by the manufacturers. These data sheets are intended to be used as references accompanying the text in order to familiarize the reader with the types of presentation, and may be used with some of the problem assignments requiring manufacturers' data sheets. The appendix includes the following data sheets:

(1) *pnp* epitaxial planar silicon transistor   (Types 2N3250, 2N-3251, TI)
(2) *npn* diffused silicon planar epitaxial transistor   (Type 2N3011, Fairchild)
(3) *n*-channel epitaxial planar silicon FET   (Type 2N3823, TI)
(4) Silicon MOSFET   (Type 40461, RCA)
(5) Silicon *npn* power transistor   (Types 40546, 40547, RCA)
(6) Operational amplifier   (Type μA702C, Fairchild)
(7) Operational amplifier   (Type MC1520, Motorola)
(8) Integrated circuit logic   (Type MC300 series, Motorola)
(9) MOS integrated circuit   (Type 3102, Fairchild)

Because of the numerous IC transistor digital functional blocks made by many manufacturers, it is impractical and of doubtful value to include the various logic data sheets. Only the data for one type of transistor logic, the ECL gate, because of its high-speed performance, is included for convenience. For complete information on a variety of discrete and IC devices and components, the reader is referred to the following representative manufacturers' data books:

I. Fairchild Semiconductor Data Catalog
II. Motorola Semiconductor Products Catalog
III. RCA Data Book: Semiconductor Devices and Integrated Circuits
IV. Texas Instruments Perpetual Data Book: Semiconductors and Components

D-1 DISCRETE BIPOLAR TRANSISTOR DATA SHEETS

Types 2N3250, 2N3251 P-N-P Epitaxial Planar Silicon Transistors

*absolute maximum ratings at 25°C free-air temperature (unless otherwise noted)

Collector-Base Voltage	−50 v
Collector-Emitter Voltage (See Note 1)	−40 v
Emitter-Base Voltage	−5 v
Collector Current	−200 ma
Continuous Device Dissipation at (or below) 25°C Free-Air Temperature (See Note 2)	0.36 w
Continuous Device Dissipation at (or below) 25°C Case Temperature (See Note 3)	1.2 w
Storage Temperature Range	−65°C to +200°C
Lead Temperature 1/16 Inch from Case for 60 Seconds	300°C

*electrical characteristics at 25°C free-air temperature (unless otherwise noted)

PARAMETER		TEST CONDITIONS	2N3250 MIN	2N3250 MAX	2N3251 MIN	2N3251 MAX	UNIT
$V_{(BR)CBO}$	Collector-Base Breakdown Voltage	$I_C = -10\ \mu a,\ I_E = 0$	−50		−50		v
$V_{(BR)CEO}$	Collector-Emitter Breakdown Voltage	$I_C = -10\ ma,\ I_B = 0$, See Note 4	−40		−40		v
$V_{(BR)EBO}$	Emitter-Base Breakdown Voltage	$I_E = -10\ \mu a,\ I_C = 0$	−5		−5		v
I_{CEV}	Collector Cutoff Current	$V_{CE} = -40\ v,\ V_{BE} = 3\ v$		−20		−20	na
I_{BEV}	Base Cutoff Current	$V_{CE} = -40\ v,\ V_{BE} = 3\ v$		50		50	na
h_{FE}	Static Forward Current Transfer Ratio	$V_{CE} = -1\ v,\ I_C = -0.1\ ma$ See Note 4	40		80		
		$V_{CE} = -1\ v,\ I_C = -1\ ma$	45		90		
		$V_{CE} = -1\ v,\ I_C = -10\ ma$	50	150	100	300	
		$V_{CE} = -1\ v,\ I_C = -50\ ma$	15		30		
V_{BE}	Base-Emitter Voltage	$I_B = -1\ ma,\ I_C = -10\ ma$ See Note 4	−0.6	−0.9	−0.6	−0.9	v
		$I_B = -5\ ma,\ I_C = -50\ ma$		−1.2		−1.2	v
$V_{CE(sat)}$	Collector-Emitter Saturation Voltage	$I_B = -1\ ma,\ I_C = -10\ ma$		−0.25		−0.25	v
		$I_B = -5\ ma,\ I_C = -50\ ma$		−0.5		−0.5	v
h_{ie}	Small-Signal Common-Emitter Input Impedance	$V_{CE} = -10\ v$, $I_C = -1\ ma$, $f = 1\ kc$	1	6	2	12	kΩ
h_{fe}	Small-Signal Common-Emitter Forward Current Transfer Ratio		50	200	100	400	
h_{re}	Small-Signal Common-Emitter Reverse Voltage Transfer Ratio			10×10⁻⁴		20×10⁻⁴	
h_{oe}	Small-Signal Common-Emitter Output Admittance		4	40	10	60	μmho

*electrical characteristics at 25°C free-air temperature

PARAMETER		TEST CONDITIONS	2N3250 MIN	2N3250 MAX	2N3251 MIN	2N3251 MAX	UNIT
$\|h_{fe}\|$	Small-Signal Common-Emitter Forward Current Transfer Ratio	$V_{CE} = -20\ v,\ I_C = -10\ ma,\ f = 100\ Mc$	2.5		3		
f_T	Transition Frequency	$V_{CE} = -20\ v,\ I_C = -10\ ma$, See Note 5	250		300		Mc
C_{obo}	Common-Base Open-Circuit Output Capacitance	$V_{CB} = -10\ v,\ I_E = 0,\ f = 100\ kc$		6		6	pf
C_{ibo}	Common-Base Open-Circuit Input Capacitance	$V_{EB} = -1\ v,\ I_C = 0,\ f = 100\ kc$		8		8	pf
$r_b'C_c$	Collector-Base Time Constant	$V_{CE} = -20\ v,\ I_C = -10\ ma,\ f = 31.8\ Mc$		250		250	psec

NOTE 5: To obtain f_T, the $|h_{fe}|$ response with frequency is extrapolated at the rate of −6 db per octave from $f = 100$ Mc to the frequency at which $|h_{fe}| = 1$.

*operating characteristics at 25°C free-air temperature

PARAMETER		TEST CONDITIONS	2N3250 MAX	2N3251 MAX	UNIT
NF	Spot Noise Figure	$I_C = -100\ \mu a,\ V_{CE} = -5\ v,\ R_G = 1\ k\Omega,\ f = 100\ cps$	6	6	db

*switching characteristics at 25°C free-air temperature

PARAMETER		TEST CONDITIONS§	2N3250 MAX	2N3251 MAX	UNIT
t_d	Delay Time	$I_C = -10\ ma,\ I_{B(1)} = -1\ ma,\ V_{BE(off)} = 0.5\ v$, $R_L = 275\ \Omega$, See Figure 1	35	35	nsec
t_r	Rise Time		35	35	nsec
t_s	Storage Time	$I_C = -10\ ma,\ I_{B(1)} = -1\ ma,\ I_{B(2)} = 1\ ma$, $R_L = 275\ \Omega$, See Figure 2	175	200	nsec
t_f	Fall Time		50	50	nsec

§Voltage and current values shown are nominal; exact values vary slightly with transistor parameters. Nominal base current for delay and rise times is calculated using the minimum value of V_{BE}. Nominal base currents for storage and fall times are calculated using the maximum value of V_{BE}.

NOTES:
1. This value applies between 0 and 200 ma collector current when the base-emitter diode is open circuited.
2. Derate linearly to 200°C free-air temperature at the rate of 2.06 mw/C°.
3. Derate linearly to 200°C case temperature at the rate of 6.9 mw/C°.
4. These parameters must be measured using pulse techniques. PW = 300 μsec, duty cycle ≤ 2%.

687

*PARAMETER MEASUREMENT INFORMATION

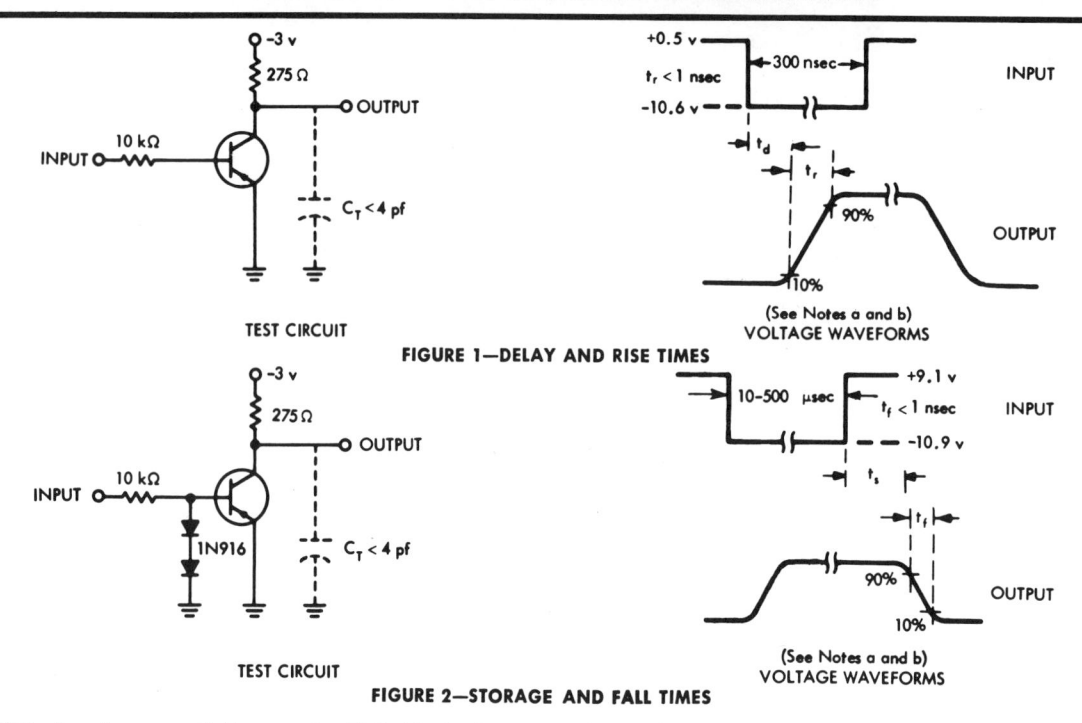

FIGURE 1—DELAY AND RISE TIMES

TEST CIRCUIT

VOLTAGE WAVEFORMS (See Notes a and b)

FIGURE 2—STORAGE AND FALL TIMES

TEST CIRCUIT

VOLTAGE WAVEFORMS (See Notes a and b)

NOTES: a. The input waveforms are supplied by a generator with the following characteristics: $Z_{out} = 50\,\Omega$, duty cycle $= 2\%$.
 b. Waveforms are monitored on an oscilloscope with the following characteristics: $t_r \leq 1$ nsec, $R_{in} \geq 100\,k\Omega$.
*Indicates JEDEC registered data.

NOTE: 2N3250 and 2N3251 are designed for low-power saturated-switching and amplifier applications.

(Courtesy of Texas Instruments Inc.)

2N3011 N-P-N Diffused Silicon Planar Epitaxial Transistor

ABSOLUTE MAXIMUM RATINGS [Note 1]

Maximum Temperatures
- Storage Temperature — $-65°C$ to $+200°C$
- Operating Junction Temperature — $200°C$ Maximum
- Lead Temperature (Soldering, 60 sec time limit) — $300°C$ Maximum

Maximum Power Dissipation
- Total Dissipation at $25°C$ Case Temperature [Notes 2 and 3] — 1.2 Watts
- at $100°C$ Case Temperature [Notes 2 and 3] — 0.68 Watt
- at $25°C$ Ambient Temperature [Notes 2 and 3] — 0.36 Watt

Maximum Voltages
- V_{CBO} Collector to Base Voltage — 30 Volts
- V_{CES} Collector to Emitter Voltage — 30 Volts
- V_{CEO} Collector to Emitter Voltage [Note 4] — 12 Volts
- V_{EBO} Emitter to Base Voltage — 5.0 Volts

ELECTRICAL CHARACTERISTICS ($25°C$ free air temperature unless otherwise noted)

SYMBOL	CHARACTERISTIC	MIN.	TYP.	MAX.	UNITS	TEST CONDITIONS	
h_{FE}	DC Pulse Current Gain [Note 5]	30	70	120		$I_C = 10$ mA	$V_{CE} = 0.35$ V
h_{FE}	DC Pulse Current Gain [Note 5]	25	75			$I_C = 30$ mA	$V_{CE} = 0.4$ V
h_{FE}	DC Pulse Current Gain [Note 5]	12	50			$I_C = 100$ mA	$V_{CE} = 1.0$ V
V_{CE} (sat)	Collector Saturation Voltage		0.17	0.2	Volts	$I_C = 10$ mA	$I_B = 1.0$ mA
V_{CE} (sat)	Collector Saturation Voltage		0.18	0.25	Volts	$I_C = 30$ mA	$I_B = 3.0$ mA
V_{CE} (sat)	Collector Saturation Voltage ($+85°C$)		0.15	0.3	Volts	$I_C = 10$ mA	$I_B = 1.0$ mA
V_{CE} (sat)	Collector Saturation Voltage		0.3	0.5	Volts	$I_C = 100$ mA	$I_B = 10$ mA
V_{BE} (sat)	Base Saturation Voltage	0.72	0.8	0.87	Volts	$I_C = 10$ mA	$I_B = 1.0$ mA
V_{BE} (sat)	Base Saturation Voltage		0.9	1.15	Volts	$I_C = 30$ mA	$I_B = 3.0$ mA
V_{BE} (sat)	Base Saturation Voltage		1.1	1.6	Volts	$I_C = 100$ mA	$I_B = 10$ mA
h_{fe}	High Frequency Current Gain (f = 100 mc)	4.0	6.5			$I_C = 20$ mA	$V_{CE} = 10$ V
C_{ob}	Output Capacitance		2.3	4.0	pf	$I_E = 0$	$V_{CB} = 5.0$ V
I_{CES}	Collector Reverse Current		0.05	0.4	μA	$V_{CE} = 20$ V	$V_{BE} = 0$
I_{CES} ($85°C$)	Collector Reverse Current		1.0	10	μA	$V_{CE} = 20$ V	$V_{BE} = 0$
BV_{CBO}	Collector to Base Breakdown Voltage	30			Volts	$I_C = 10$ μA	$I_E = 0$
BV_{CES}	Collector to Emitter Breakdown Voltage	30			Volts	$I_C = 10$ μA	$V_{EB} = 0$
V_{CEO} (sust)	Collector to Emitter Sustaining Voltage	12			Volts	$I_C = 10$ mA (pulsed)	$I_B = 0$
BV_{EBO}	Emitter to Base Breakdown Voltage	5.0			Volts	$I_E = 100$ μA	$I_C = 0$
τ_s	Charge Storage Time Constant [Note 6]			13	nsec	$I_C = I_{B1} \approx 10$ mA,	$I_{B2} = -10$ mA
t_{on}	Turn On Time [Note 6]			15	nsec	$I_C \approx 30$ mA	$I_{B1} \approx 3.0$ mA
t_{off}	Turn Off Time [Note 6]			20	nsec	$I_C \approx 30$ mA, $I_{B1} \approx 3.0$ mA,	$I_{B2} \approx -3.0$ mA

NOTES:
(1) These ratings are limiting values above which the serviceability of any individual semiconductor device may be impaired.
(2) These are steady state limits. The factory should be consulted on applications involving pulsed or low duty cycle operations.
(3) These ratings give a maximum junction temperature of $200°C$ and junction-to-case thermal resistance of $146°C$/watt (derating factor of 6.85 mW/$°C$). Junction-to-ambient thermal resistance of $486°C$/watt (derating factor of 2.06 mW/$°C$).
(4) Rating refers to a high-current point where collector-to-emitter voltage is lowest. For more information send for Fairchild Publication APP-4.
(5) Pulse Conditions: length = 300 μsec; duty cycle = 1%.
(6) See switching circuits for exact values of I_C, I_{B1}, and I_{B2}.

TYPICAL ELECTRICAL CHARACTERISTICS

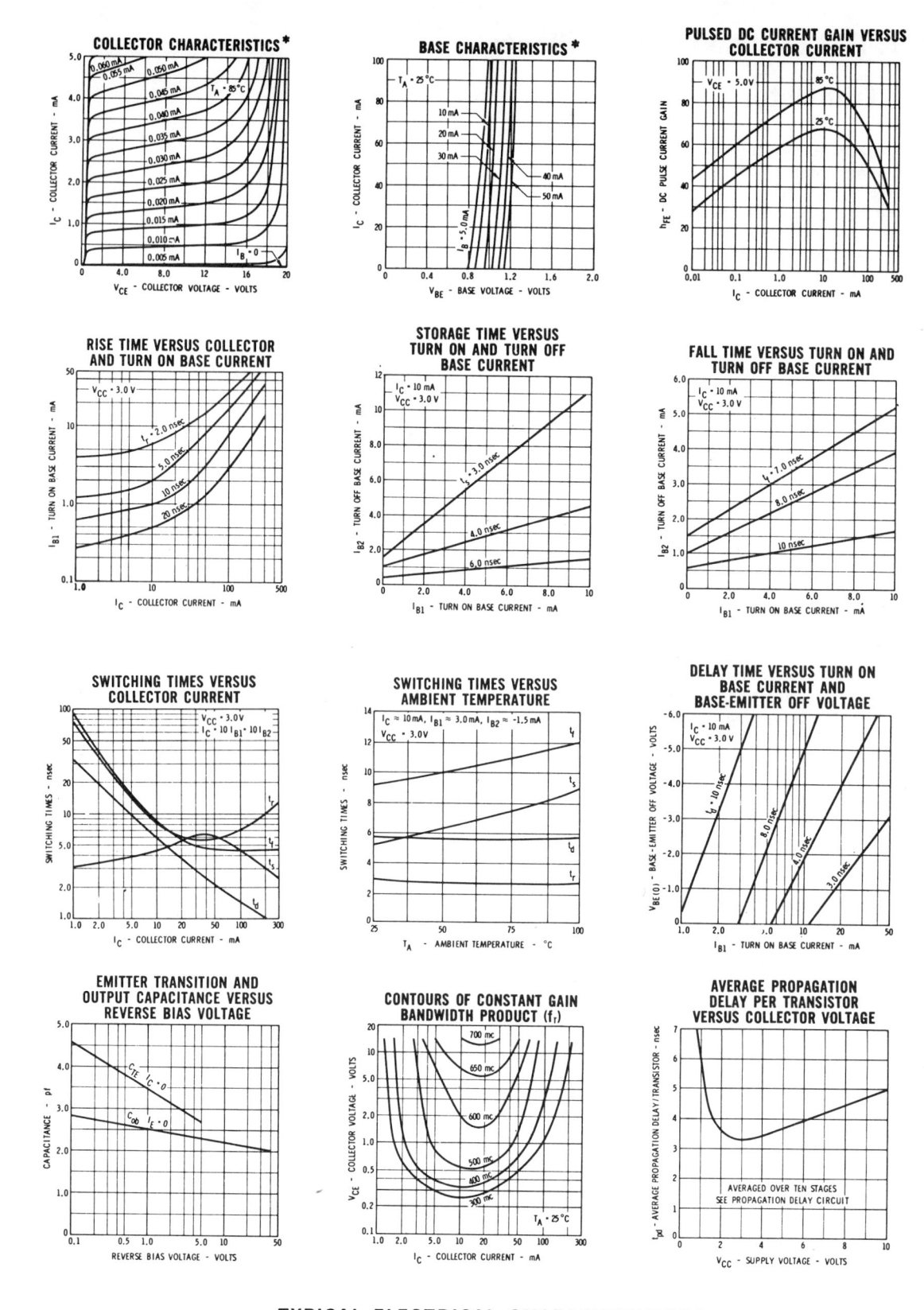

CHARGE STORAGE TIME — CONSTANT TEST CIRCUIT

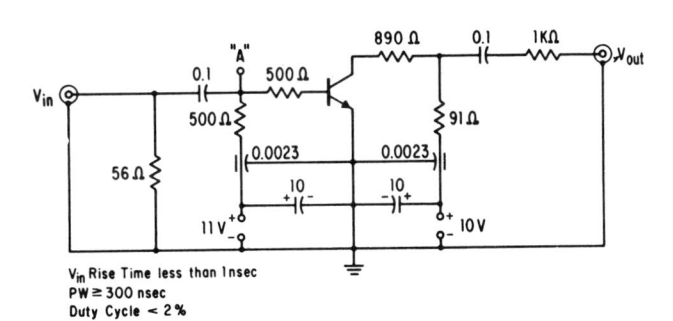

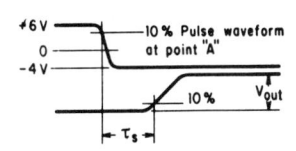

t_{on}-t_{off} MEASUREMENT CIRCUIT

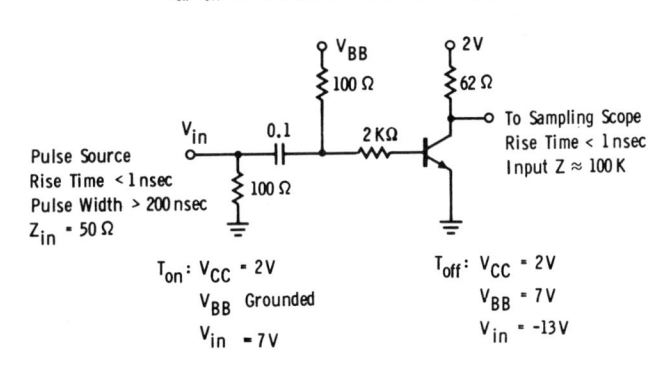

T_{on}: V_{CC} = 2V
V_{BB} Grounded
V_{in} = 7V

T_{off}: V_{CC} = 2V
V_{BB} = 7V
V_{in} = -13V

CIRCUIT FOR MEASUREMENT OF PROPAGATION DELAY

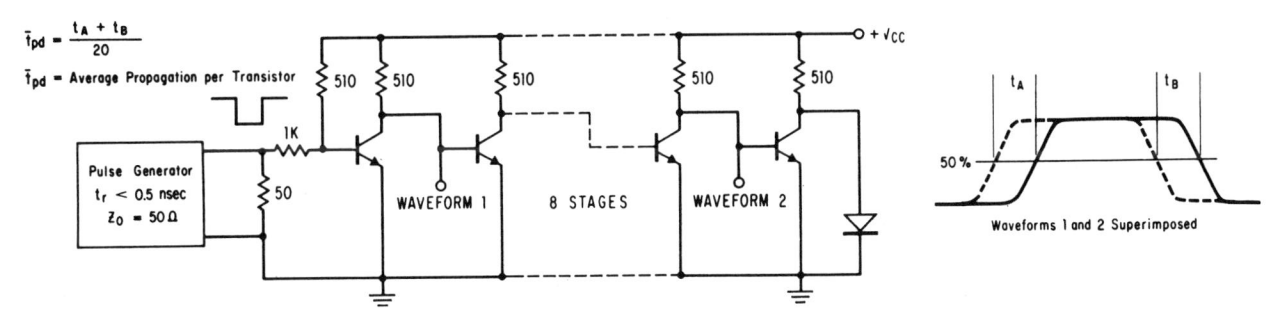

$\bar{t}_{pd} = \dfrac{t_A + t_B}{20}$

$\bar{t}_{pd}$ = Average Propagation per Transistor

Waveforms 1 and 2 Superimposed

NOTE: The 2N3011 is an NPN silicon PLANAR epitaxial transistor designed specifically for high-speed saturated switching applications in the 50-100 mc range at current levels from 100 microamperes to 100 milliamperes. It is suitable for most small-signal, RF, and digital type circuits.

(Courtesy of Fairchild Semiconductor)

D-2 DISCRETE FET DATA SHEET

Type 2N3823 N-Channel Epitaxial Planar Silicon Field-Effect Transistor

* **absolute maximum ratings at 25°C free-air temperature (unless otherwise noted)**

Drain-Gate Voltage	30 v
Drain-Source Voltage	30 v
Reverse Gate-Source Voltage	−30 v
Gate Current	10 ma
Continuous Device Dissipation at (or below) 25°C Free-Air Temperature (See Note 1)	300 mw
Storage Temperature Range	−65°C to +200°C
Lead Temperature 1/16 Inch from Case for 10 Seconds	300°C

***electrical characteristics at 25°C free-air temperature (unless otherwise noted)**

	PARAMETER	TEST CONDITIONS‡	MIN	MAX	UNIT
$V_{(BR)GSS}$	Gate-Source Breakdown Voltage	$I_G = -1\ \mu a$, $V_{DS} = 0$	−30		v
I_{GSS}	Gate Cutoff Current	$V_{GS} = -20\ v$, $V_{DS} = 0$		−0.5	na
		$V_{GS} = -20\ v$, $V_{DS} = 0$, $T_A = 150°C$		−0.5	μa
I_{DSS}	Zero-Gate-Voltage Drain Current	$V_{DS} = 15\ v$, $V_{GS} = 0$, See Note 2	4	20	ma
V_{GS}	Gate-Source Voltage	$V_{DS} = 15\ v$, $I_D = 400\ \mu a$	−1	−7.5	v
$V_{GS(off)}$	Gate-Source Cutoff Voltage	$V_{DS} = 15\ v$, $I_D = 0.5\ na$		−8	v
$|y_{fs}|$	Small-Signal Common-Source Forward Transfer Admittance	$V_{DS} = 15\ v$, $V_{GS} = 0$, $f = 1\ kc$, See Note 2	3500	6500	μmho
$|y_{os}|$	Small-Signal Common-Source Output Admittance	$V_{DS} = 15\ v$, $V_{GS} = 0$, $f = 1\ kc$, See Note 2		35	μmho
C_{iss}	Common-Source Short-Circuit Input Capacitance	$V_{DS} = 15\ v$, $V_{GS} = 0$, $f = 1\ Mc$		6	pf
C_{rss}	Common-Source Short-Circuit Reverse Transfer Capacitance			2	pf
$|y_{fs}|$	Small-Signal Common-Source Forward Transfer Admittance	$V_{DS} = 15\ v$,		3200	μmho
$Re(y_{is})$	Small-Signal Common-Source Input Conductance	$V_{GS} = 0$,		800	μmho
$Re(y_{os})$	Small-Signal Common-Source Output Conductance	$f = 200\ Mc$		200	μmho

NOTES: 1. Derate linearly to 175°C free-air temperature at the rate of 2 mw/C°.
2. These parameters must be measured using pulse techniques. PW = 100 msec, Duty Cycle ≤ 10%.

*Indicates JEDEC registered data.

‡The fourth lead (case) is connected to the source for all measurements.

* **operating characteristics at 25°C free-air temperature**

	PARAMETER	TEST CONDITIONS‡	MAX	UNIT
NF	Common-Source Spot Noise Figure	$V_{DS} = 15\ v$, $V_{GS} = 0$, $f = 100\ Mc$, $R_G = 1\ k\Omega$	2.5	db

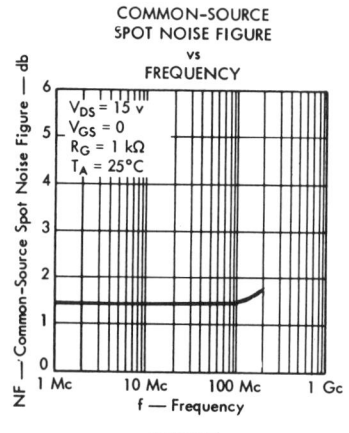

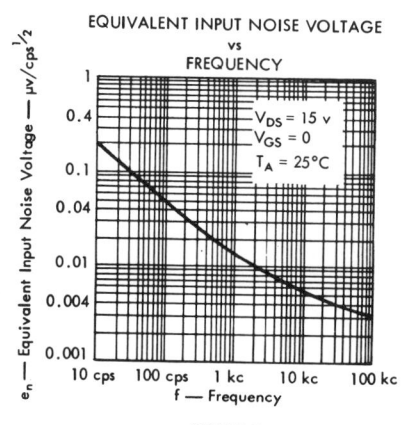

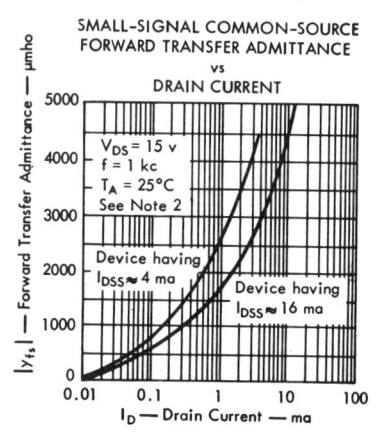

FIGURE 1

FIGURE 2

FIGURE 3

SMALL-SIGNAL COMMON-SOURCE FORWARD TRANSFER ADMITTANCE
vs
GATE-SOURCE VOLTAGE

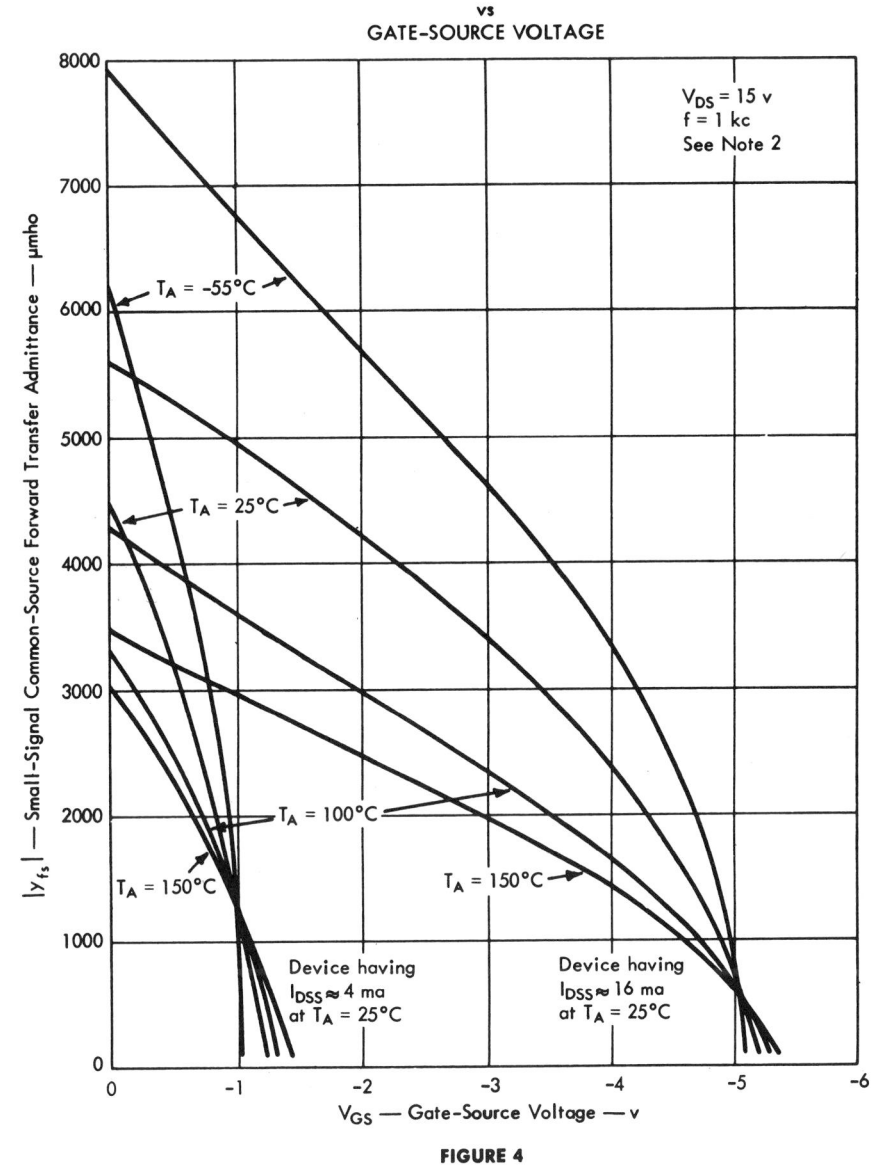

FIGURE 4

TYPICAL CHARACTERISTICS

TYPICAL CHARACTERISTICS

FIGURE 5 — GATE CUTOFF CURRENT vs FREE-AIR TEMPERATURE

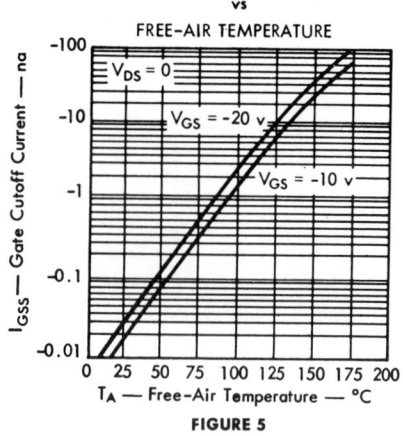

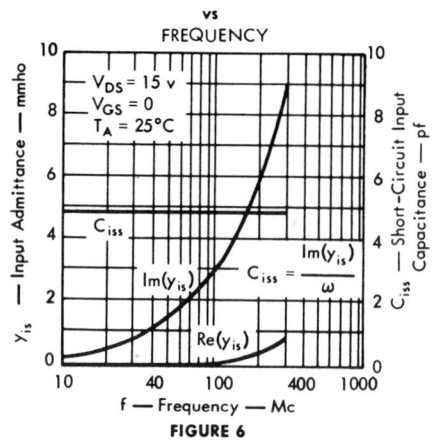

FIGURE 6 — SMALL-SIGNAL COMMON-SOURCE INPUT ADMITTANCE vs FREQUENCY

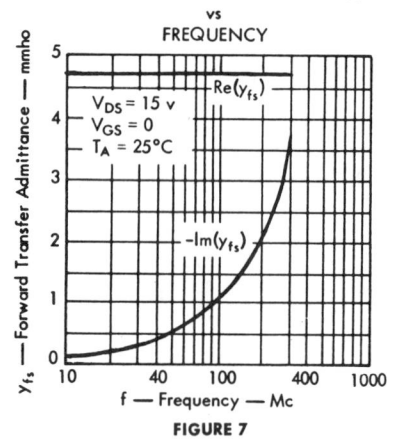

FIGURE 7 — SMALL-SIGNAL COMMON SOURCE FORWARD TRANSFER ADMITTANCE vs FREQUENCY

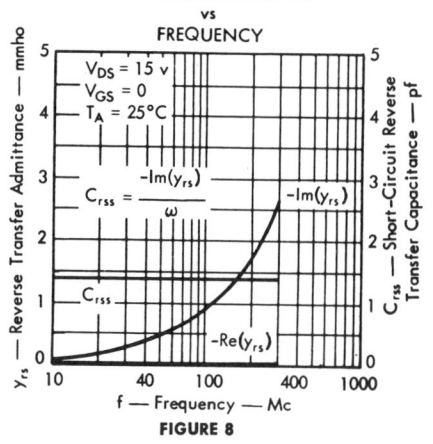

FIGURE 8 — SMALL-SIGNAL COMMON-SOURCE REVERSE TRANSFER ADMITTANCE vs FREQUENCY

FIGURE 9 — SMALL-SIGNAL COMMON-SOURCE OUTPUT ADMITTANCE vs FREQUENCY

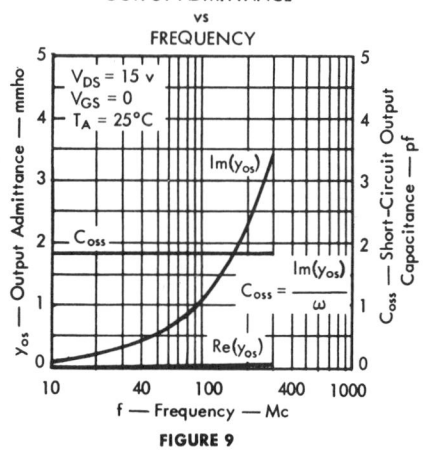

FIGURE 10 — COMMON-SOURCE SHORT-CIRCUIT INPUT AND REVERSE-TRANSFER CAPACITANCES vs GATE-SOURCE VOLTAGE

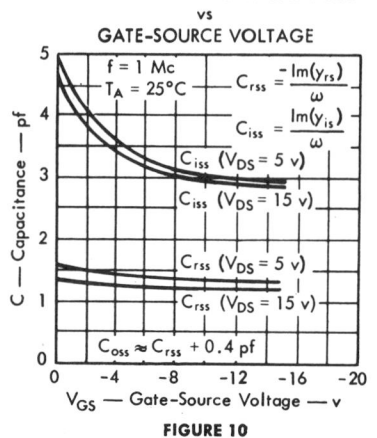

NOTE: 2N3823 is a symmetrical n-channel FET for RF Amplifier and Mixer Applications. It has low noise figure and square-law transfer characteristic.

(Courtesy of Texas Instruments, Inc.)

RCA 40461 Silicon "MOS" (Insulated-Gate Field-Effect) Transistors

Maximum Ratings, *Absolute-Maximum Values:*

(Substrate connected to source unless otherwise specified)

DRAIN-TO-SOURCE VOLTAGE, V_{DS}	25 max.	V
DRAIN-TO-SUBSTRATE VOLTAGE, V_{DB}	$^{+25}_{-0.3}$ max.	V
SOURCE-TO-SUBSTRATE VOLTAGE, V_{SB}	$^{+20}_{-0.3}$ max.	V
DC GATE-TO-SOURCE VOLTAGE, V_{GS}	±10 max.	V
PEAK GATE-TO-SOURCE VOLTAGE, v_{GS}	±25 max.	V
DRAIN CURRENT, I_D	Limited by Dissipation	
TRANSISTOR DISSIPATION, P_T:		
At ambient temperatures from -65 to +125 °C	150 max.	mW
AMBIENT TEMPERATURE RANGE:		
Storage	-65 to +150	°C
Operating	-65 to +125	°C
LEAD TEMPERATURE (During Soldering):		
At distances $\geq$ 1/32 inch from seating surface for 10 seconds max.	265 max.	°C

ELECTRICAL CHARACTERISTICS: At T_A = 25°C Unless Otherwise Specified. Substrate Connected to Source

CHARACTERISTICS	SYMBOLS	TEST CONDITIONS	Min.	Typ.	Max.	Units
Forward Transconductance	g_{fs}	V_{DS} = 12 V, V_{GS} = 0 V_{DS} = 12 V, I_D = 4 mA, f = 1 kHz	- 1600	3500 2500	- -	µmho µmho
Drain Current	I_{DSS}	V_{DS} = 12 V, V_{GS} = 0	4	9	14	mA
Gate Leakage Current	I_{GSS}	V_{DS} = 0, V_{GS} = ±10 V V_{DS} = 0, V_{GS} = ±10 V, T_A = 125°C	- -	0.1 20	10 200	pA pA
Gate-to-Source Cutoff Voltage	$V_{GS(off)}$	V_{DS} = 12 V, I_D = 50 µA	-	-4.5	-6	V
Small-Signal, Short-Circuit, Input Capacitance	C_{iss}	V_{DS} = 12 V, V_{GS} = 0, f = 0.1-1 MHz	-	4	5	pF
Small-Signal, Short-Circuit, Reverse Transfer Capacitance	C_{rss}	V_{DS} = 12 V, V_{GS} = 0, f = 0.1-1 MHz	-	0.9	1.2	pF
Output Resistance	r_d	V_{DS} = 12 V, I_D = 4 mA, f = 1 kHz	9000	13,000	-	Ω
Power Gain	G_{ps}	V_{DS} = 12 V, I_D = 4 mA, f = 60 MHz BW = 1.5 MHz	-	14	-	dB
Noise Figure	NF	V_{DS} = 12 V, I_D = 4 mA, f = 60 MHz BW = 1.5 MHz	-	5.9	-	dB
Equivalent Input Noise Voltage	E_N	V_{DS} = 12 V, I_D = 4 mA, R_g = 0, f = 1 kHz	-	0.16	0.25	µV/$\sqrt{f(Hz)}$
Noise Figure*	NF	V_{DS} = 12 V, I_D = 4 mA, R_g = 1 MΩ, f = 1 kHz	-	4*	-	dB

* Noise Figure = $10 \log_{10} \left[1 + \frac{E_N}{4 \, KTBW R_g} \right]$ where $K = 1.38 \times 10^{-23}$; T = Temperature in °Kelvin; BW = Bandwidth in Hz; R_g = Generator resistance.

TYPICAL CHARACTERISTICS FOR RCA-40461

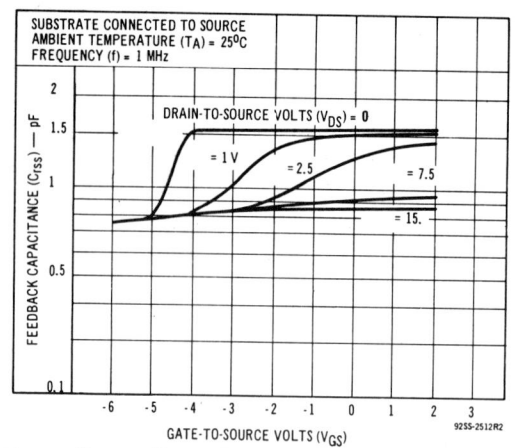

Fig.1 - Gate-to-Drain Capacitance vs. Gate-to-Source Voltage

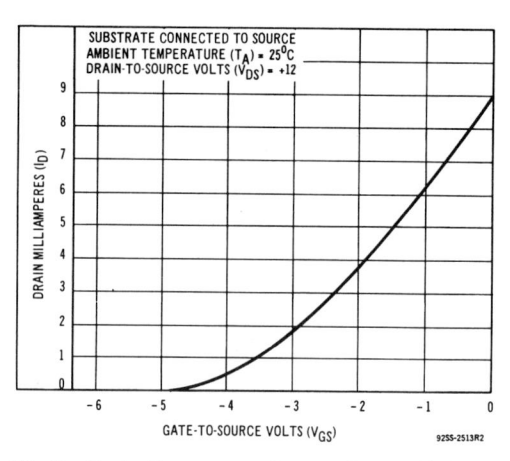

Fig.2 - Drain Current vs. Gate-to-Source Voltage

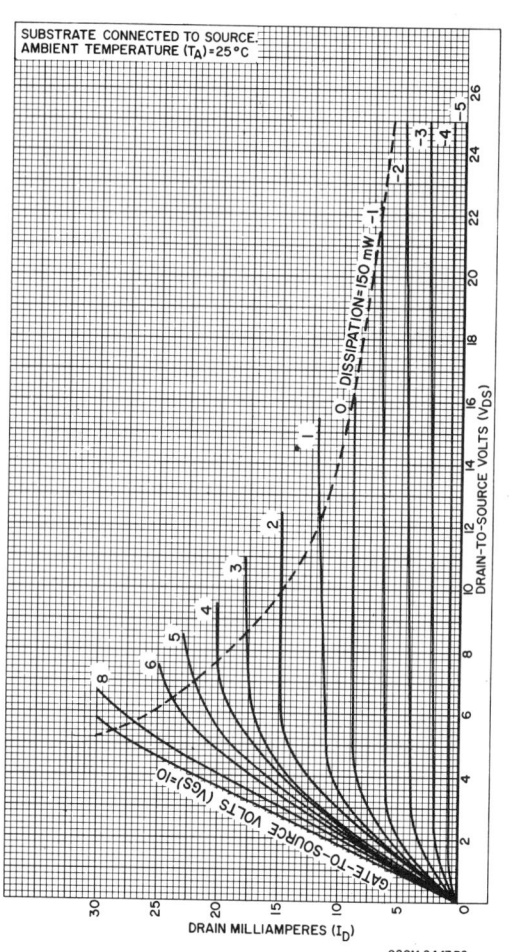

Fig.3 - Drain Current vs. Drain-to-Source Voltage

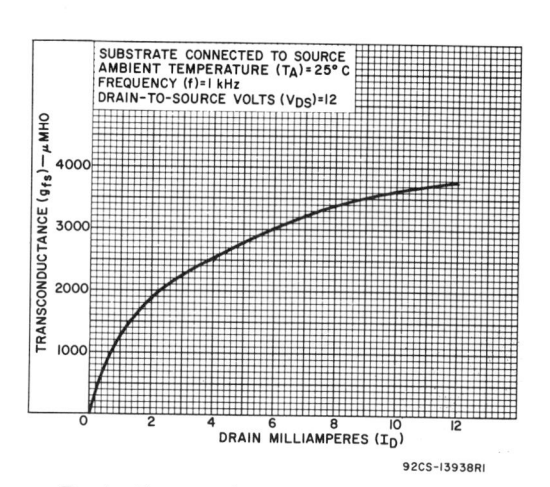

Fig.4 - Transconductance vs. Drain Current Connected to Source

NOTE: For audio, wide-band and tuned amplifier applications up to 60 MHz.

(Courtesy of RCA)

D-3 POWER TRANSISTOR DATA SHEET

RCA 40546, 40547 High-Voltage Silicon N-P-N Power Transistors

MAXIMUM RATINGS

Absolute Maximum Values: 40546 / 40547

COLLECTOR-TO-EMITTER SUSTAINING VOLTAGE:
- For I_C = 5 mA and I_B = 5 μa .... V_{CEX}(sus)   250   V

EMITTER-TO-BASE VOLTAGE .... V_{EBO} 2   V

COLLECTOR CURRENT .... I_C 150   mA

BASE CURRENT .... I_B 150   mA

TRANSISTOR DISSIPATION: .... P_T
- At case temperatures up to 70° C .... 8 W
- At temperatures above 70° C .... Derate linearly to 0 W at 150° C.

TEMPERATURE RANGE
- Storage & Operating (Junction) .... −65 to 150 °C

PIN TEMPERATURE (During Soldering):
- At distances $\geq$ 1/32 in. from seating plane for 10 s max. .... 255 °C

ELECTRICAL CHARACTERISTICS

Case Temperature (T_C) = 25° C

Characteristic	Symbol	DC Voltage (V) V_{CE}	DC Voltage (V) V_{CB}	DC Current (mA) I_C	DC Current (mA) I_E	DC Current (mA) I_B	Type 40546 Min.	Type 40546 Max.	Type 40547 Min.	Type 40547 Max.	Units
Collector-Cutoff Current	I_{CEX}	250				5 μA	—	5	—	5	mA
Collector-Cutoff Current	I_{CBO}		250		0		—	100	—	100	μA
DC Forward-Current Transfer Ratio	h_{FE}	10		50			50	150	20	250	
Emitter-to-Base Breakdown Voltage	V_{EBO}			0	0.1		2	—	2	—	V
Open-Circuit, Common-Base Output Capacitance	C_{ob}		50		0		—	5	—	5	pF
Gain-Bandwidth Product	f_T	50		20			25 (Typ.)		25 (Typ.)		MHz
Thermal Resistance (Junction-to-Case)	θ_{J-C}						—	10	—	10	°C/W

TYPICAL DC BETA FOR TYPES 40546 & 40547

Fig. 1

TYPICAL INPUT CHARACTERISTIC FOR TYPES 40546 & 40547

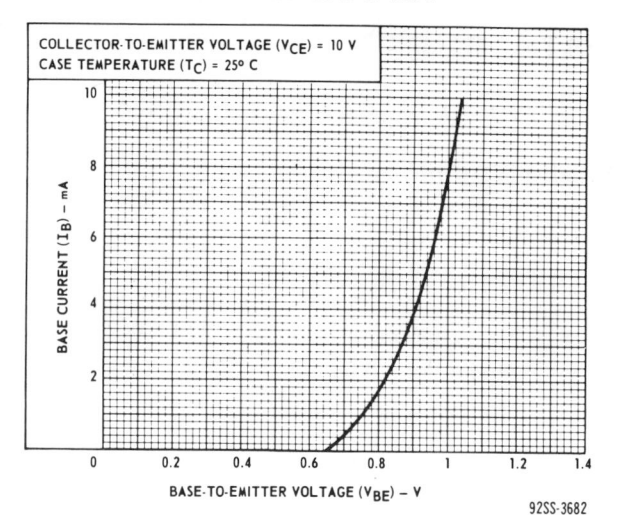

Fig. 2

TYPICAL OUTPUT CHARACTERISTICS FOR TYPES 40546 & 40547

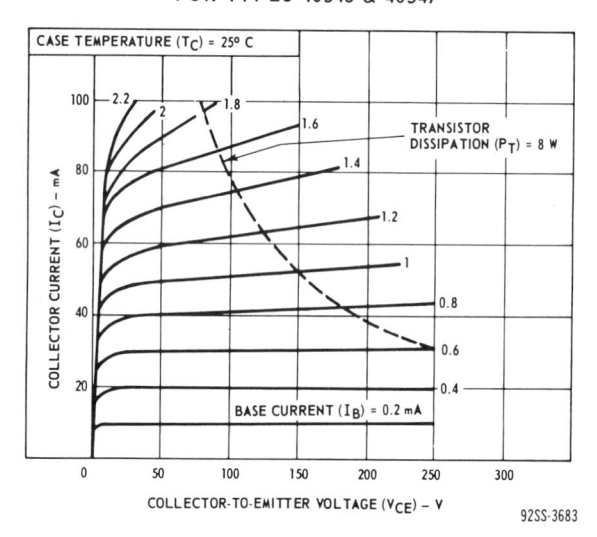

Fig. 3

TYPICAL TRANSFER CHARACTERISTIC FOR TYPES 40546 & 40547

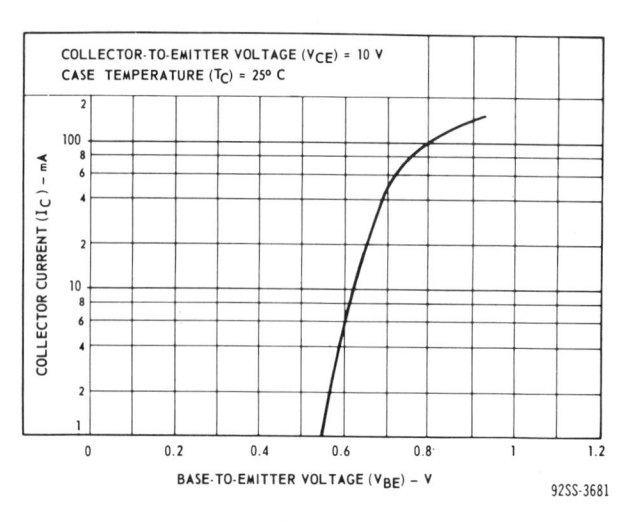

Fig. 4

RCA-40546 and –40547 are triple-diffused, silicon n-p-n power transistors. These types employ the popular TO-66 package; they differ in the limits for DC beta.

The 40546 and 40547 were designed specifically for use in push-pull, class-A, low-cost audio amplifiers.

(Courtesy of RCA)

D-4 OPERATIONAL AMPLIFIER
DATA SHEET

µA702C High Gain, Wideband DC Amplifier

(Fairchild Linear Integrated Circuits)

GENERAL DESCRIPTION — The µA702C is a complete DC amplifier constructed on a single silicon chip, using the Fairchild Planar* epitaxial process. It is intended for use as an operational amplifier in miniaturized analog computers, as a precision instrumentation amplifier, or in other applications requiring a feedback amplifier useful from DC to 30 MHz.
For full temperature range operation (−55°C to +125°C) see µA702A data sheet.

ABSOLUTE MAXIMUM RATINGS

Voltage Between V+ and V− Terminals	21 V
Peak Output Current	50 mA
Differential Input Voltage	±5.0 V
Input Voltage	+1.5 V to −6.0 V
Internal Power Dissipation [Note 1]	
TO-99	300 mW
Flat Package	200 mW
Operating Temperature Range	0°C to +70°C
Storage Temperature Range	−65°C to +150°C
Lead Temperature (Soldering, 60 sec.)	300°C

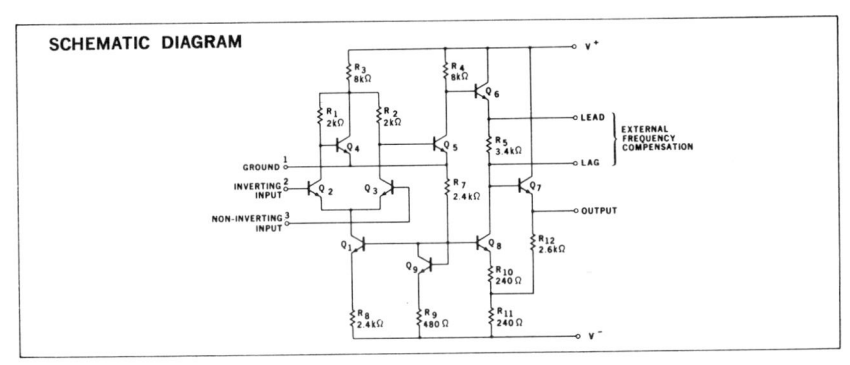

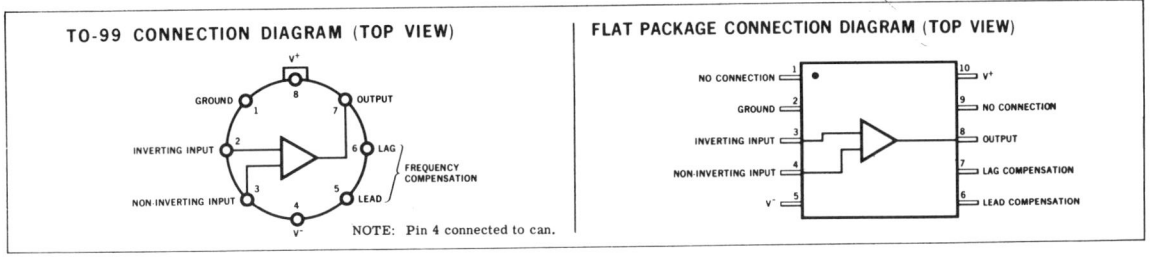

SCHEMATIC DIAGRAM

TO-99 CONNECTION DIAGRAM (TOP VIEW)

NOTE: Pin 4 connected to can.

FLAT PACKAGE CONNECTION DIAGRAM (TOP VIEW)

NOTE 1: Rating applies for ambient temperatures to +70°C.

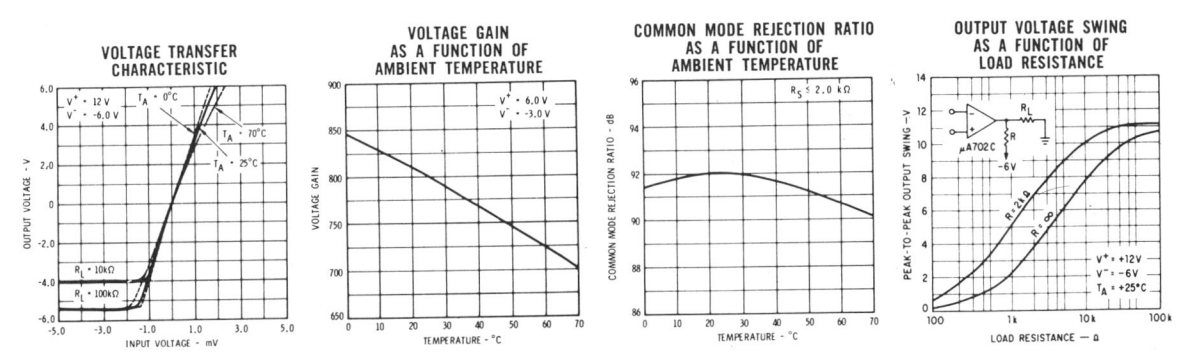

VOLTAGE TRANSFER CHARACTERISTIC

VOLTAGE GAIN AS A FUNCTION OF AMBIENT TEMPERATURE

COMMON MODE REJECTION RATIO AS A FUNCTION OF AMBIENT TEMPERATURE

OUTPUT VOLTAGE SWING AS A FUNCTION OF LOAD RESISTANCE

ELECTRICAL CHARACTERISTICS ($T_A = 25°C$ unless otherwise specified)

PARAMETER (see definitions)	CONDITIONS	$V^+ = 12.0$ V, $V^- = -6.0$ V MIN.	TYP.	MAX.	$V^+ = 6.0$ V, $V^- = -3.0$ V MIN.	TYP.	MAX.	UNITS
Input Offset Voltage	$R_s \leq 2$ kΩ		1.5	5.0		1.7	6.0	mV
Input Offset Current			0.5	2.0		0.3	2.0	μA
Input Bias Current			2.5	7.5		1.5	5.0	μA
Input Resistance		10	32		16	55		kΩ
Input Voltage Range		-4.0		$+0.5$	-1.5		$+0.5$	V
Common Mode Rejection Ratio	$R_s \leq 2$ kΩ, $f \leq 1$ kHz	70	92		70	92		dB
Large-Signal Voltage Gain	$R_L \geq 100$ kΩ, $V_{out} = \pm 5.0$ V	2000	3400	6000				
	$R_L \geq 100$ kΩ, $V_{out} = \pm 2.5$ V				500	800	1500	
Output Resistance			200	600		300	800	Ω
Supply Current	$V_{out} = 0$		5.0	6.7		2.1	3.3	mA
Power Consumption	$V_{out} = 0$		90	120		19	30	mW
Transient Response (unity gain)	$C_i = 0.01$ μF, $R_i = 20\Omega$ $R_L \leq 100$ kΩ, $V_{in} = 10$ mV							
Risetime			25	120				ns
Overshoot	$C_L \leq 100$ pF		10	50				%
Transient Response ($\times 100$ gain)	$C_3 = 50$ pF, $R_L \geq 100$ kΩ, $V_{in} = 1$ mV							
Risetime			10	30				ns
Overshoot			20	40				%

The following specifications apply for $0°C \leq T_A \leq +70°C$:

PARAMETER	CONDITIONS	MIN.	TYP.	MAX.	MIN.	TYP.	MAX.	UNITS
Input Offset Voltage	$R_s \leq 2$ kΩ			6.5			7.5	mV
Average Temperature Coefficient of Input Offset Voltage	$R_s = 50$ Ω, $T_A = +70°C$ to $T_A = 0°C$		5.0	20		7.5	25	μV/°C
Input Offset Current				2.5			2.5	μA
Average Temperature Coefficient of Input Offset Current	$T_A = 25°C$ to $T_A = +70°C$ $T_A = 25°C$ to $T_A = 0°C$		4.0 6.0	10 20		3.0 5.5	8.0 18	nA/°C nA/°C
Input Bias Current	$T_A = 0°C$		4.0	12		2.7	8	μA
Input Resistance		6.0	18		9.0	27		kΩ
Common Mode Rejection Ratio	$R_s \leq 2$ kΩ, $f \leq 1$ kHz	65	86		65	86		dB
Supply Voltage Rejection Ratio	$V^+ = 12$ V, $V^- = 6$ V to $V^+ = 6$ V, $V^- = 3$ V $R_s \leq 2$ kΩ		90	300		90	300	μV/V
Large-Signal Voltage Gain	$R_L \geq 100$ kΩ, $V_{out} = \pm 5.0$ V	1500	7000					
	$R_L \geq 100$ kΩ, $V_{out} = \pm 2.5$ V				400	1750		
Output Voltage Swing	$R_L \geq 100$ kΩ	± 5.0	± 5.3		± 2.5	± 2.7		V
	$R_L \geq 10$ kΩ	± 3.5	± 4.0		± 1.5	± 2.0		V
Supply Current	$V_{out} = 0$		5.0	7.0		2.1	3.9	mA
Power Consumption	$V_{out} = 0$		90	125		19	35	mW

DEFINITION OF TERMS

INPUT OFFSET VOLTAGE — That voltage which must be applied between the input terminals to obtain zero output voltage. The input offset voltage may also be defined for the case where two equal resistances are inserted in series with the input leads.

INPUT OFFSET CURRENT — The difference in the currents into the two input terminals with the output at zero volts.

INPUT RESISTANCE — The resistance looking into either input terminal with the other grounded.

INPUT BIAS CURRENT — The average of the two input currents.

INPUT VOLTAGE RANGE — The range of voltage which, if exceeded on either input terminal, could cause the amplifier to cease functioning properly.

INPUT COMMON MODE REJECTION RATIO — The ratio of the input voltage range to the maximum change in input offset voltage over this range.

SUPPLY VOLTAGE REJECTION RATIO — The ratio of the change in input offset voltage to the change in supply voltage producing it.

LARGE-SIGNAL VOLTAGE GAIN — The ratio of the maximum output voltage swing with load to the change in input voltage required to drive the output from zero to this voltage.

OUTPUT VOLTAGE SWING — The peak output swing, referred to zero, that can be obtained without clipping.

OUTPUT RESISTANCE — The resistance seen looking into the output terminal with the output at null. This parameter is defined only under small signal conditions at frequencies above a few hundred cycles to eliminate the influence of drift and thermal feedback.

POWER CONSUMPTION — The DC power required to operate the amplifier with the output at zero and with no load current.

TRANSIENT RESPONSE — The closed-loop step-function response of the amplifier under small-signal conditions.

PEAK OUTPUT CURRENT — The maximum current that may flow in the output load without causing damage to the unit.

TYPICAL PERFORMANCE CURVES

FREQUENCY RESPONSE FOR VARIOUS CLOSED-LOOP GAINS (LAG COMPENSATION)

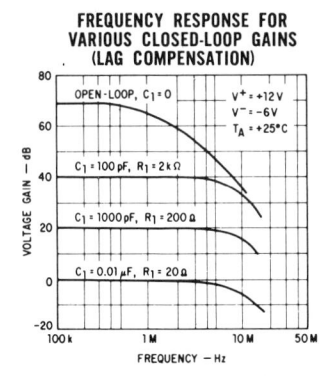

FREQUENCY RESPONSE WITH CONSERVATIVE COMPENSATION NETWORK

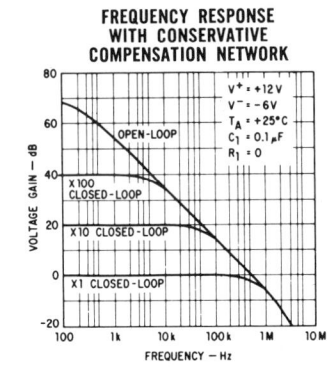

FREQUENCY RESPONSE FOR VARIOUS CLOSED-LOOP GAINS (LEAD-LAG COMPENSATION)

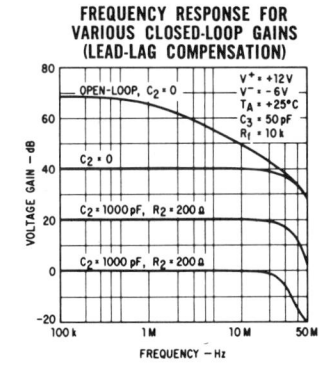

OUTPUT VOLTAGE SWING AS A FUNCTION OF FREQUENCY FOR VARIOUS LAG COMPENSATION NETWORKS

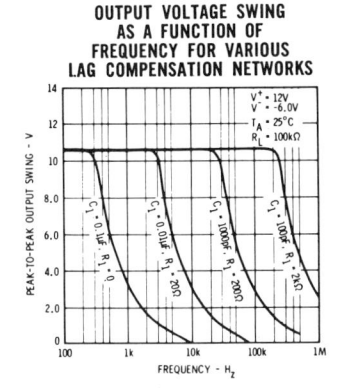

FREQUENCY COMPENSATION CIRCUITS
(Refer to Fairchild APP-117 for further details)

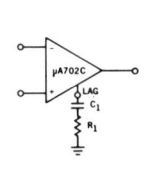

LAG COMPENSATION

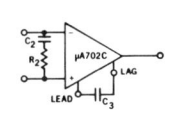

LEAD-LAG COMPENSATION

OUTPUT VOLTAGE SWING AS A FUNCTION OF FREQUENCY WITH LEAD-LAG COMPENSATION

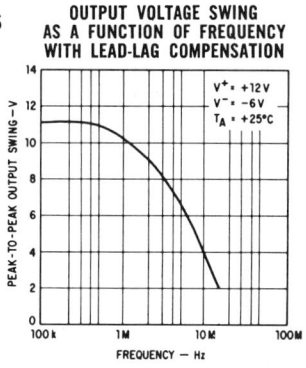

TRANSIENT RESPONSE TEST CIRCUITS

UNITY-GAIN AMPLIFIER (LAG COMPENSATION)

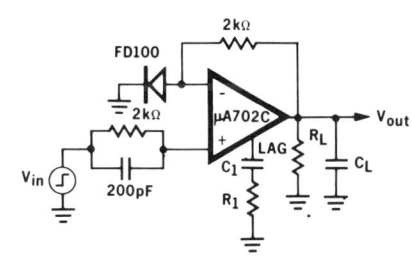

X100 AMPLIFIER (LEAD COMPENSATION)

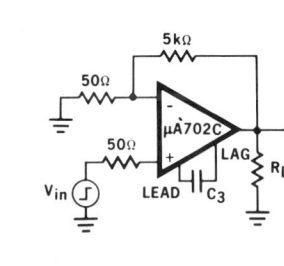

TRANSIENT RESPONSE

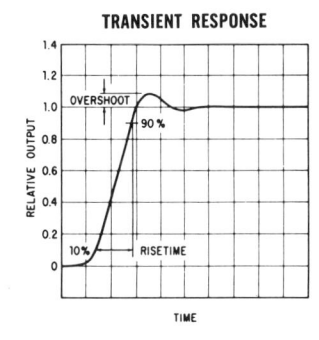

VOLTAGE GAIN AS A FUNCTION OF SUPPLY VOLTAGES

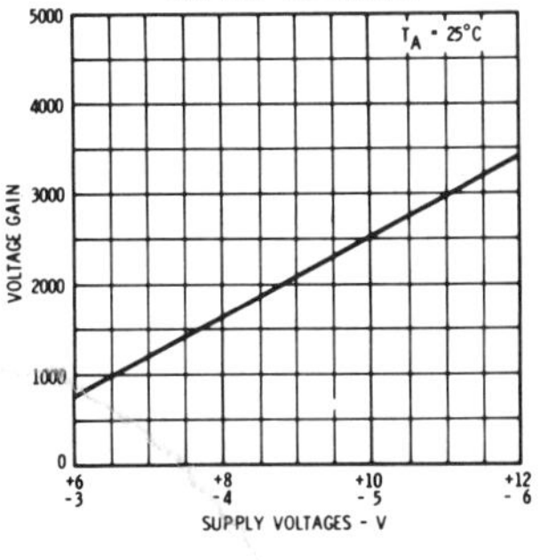

INPUT VOLTAGE RANGE AS A FUNCTION OF SUPPLY VOLTAGES

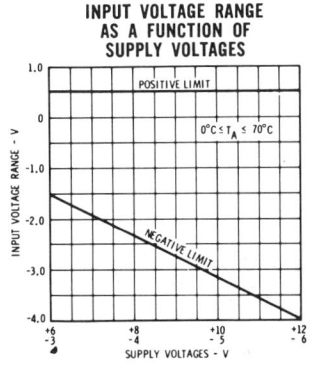

OUTPUT VOLTAGE SWING AS A FUNCTION OF SUPPLY VOLTAGES

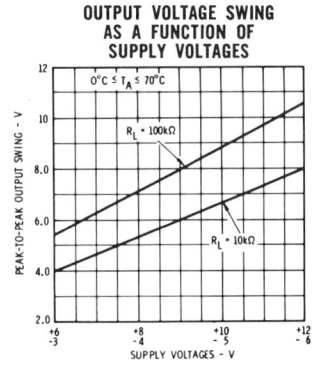

COMMON-MODE REJECTION RATIO AS A FUNCTION OF FREQUENCY

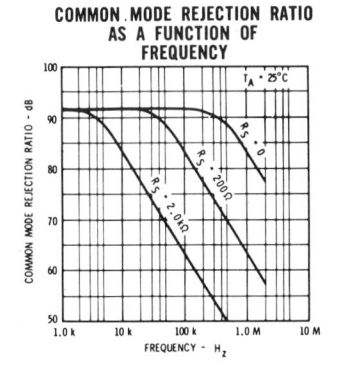

MC1520 Differential Output Monolithic Operational Amplifier

... designed for use in general-purpose or wide band differential amplifier applications, especially those requiring differential outputs.

Typical Characteristics
- Differential Input and Differential Output
- Wide Closed-Loop Bandwidth; 10 MHz
- Differential Gain; 70 dB
- High Input Impedance; 2.0 megohms:
- Low Output Impedance; 50 ohms

MAXIMUM RATINGS (T_A = 25°C unless otherwise noted)

Rating	Symbol	Value	Unit
Power Supply Voltage	V^+ V^-	+8.0 -8.0	Vdc
Differential Input Signal	V_{in}	±8.0	Vdc
Load Current	I_{L1}, I_{L2}	15	mA

Power Dissipation (Package Limitation)
 Metal Can
   Derate above 25°C
 Flat Package
   Derate above 25°C
	P_D	680
		 4.6
		 500
		 3.3
			mW
			 mW/°C
			 mW
			 mW/°C

Operating Temperature Range	T_A	-55 to +125	°C
Storage Temperature Range	T_{stg}	-65 to +150	°C

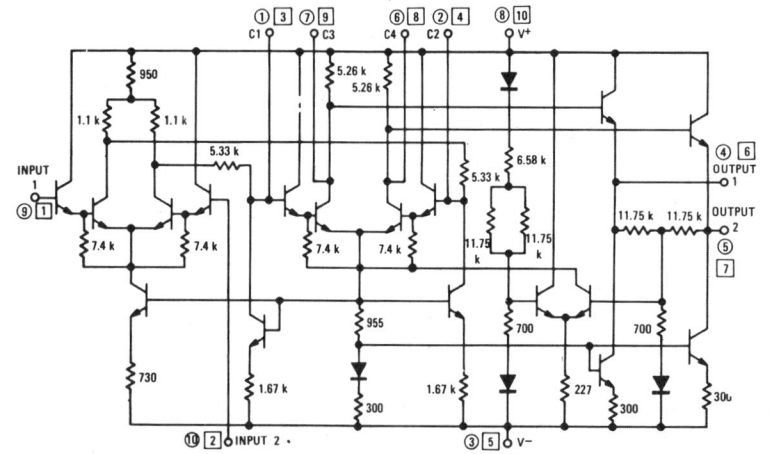

FIGURE 1 - CIRCUIT SCHEMATIC

FIGURE 2 - EQUIVALENT CIRCUIT

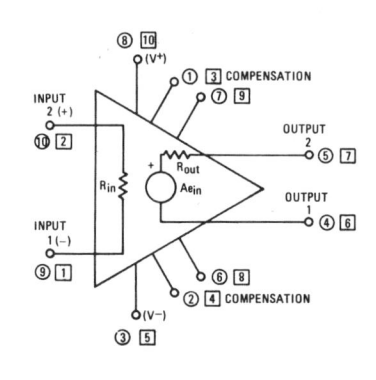

○ contains pin number for metal can package
☐ contains pin number for flat package

SINGLE-ENDED ELECTRICAL CHARACTERISTICS

(V^+ = +6.0 Vdc, V^- = −6.0 Vdc, T_A = 25°C unless otherwise noted)

Characteristic	Symbol	Min	Typ	Max	Unit
Open Loop Voltage Gain ($-55°C \le T_A \le +125°C$)	A_{VOL}	1000 60	1500 64	2500 68	V/V dB
Output Impedance (f = 20 Hz)	Z_{out}	–	50	100	ohms
Input Impedance (f = 20 Hz)	Z_{in}	0.5	2.0	–	megohms
Output Voltage Swing R_L = 7.0 kΩ (Figure 8)	V_{out}	±3.5	±4.0	–	V_{peak}
Input Common Mode Voltage Swing	CMV_{in}	±2.0	±3.0	–	V_{peak}
Common Mode Rejection Ratio	CM_{rej}	75	90	–	dB
Input Bias Current ($I_b = \frac{I_1 + I_2}{2}$), T_A = +25°C	I_b	–	0.8	2.0	µA
Input Offset Current ($I_{io} = I_1 - I_2$) ($I_{io} = I_1 - I_2$, T_A = −55°C) ($I_{io} = I_1 - I_2$, T_A = +125°C)	I_{io}	– – –	30 – –	100 200 200	nA
Input Offset Voltage T_A = 25°C	V_{io}	–	5.0	10	mV
Step Response Gain = 1.0, 10% Overshoot R_1 = 10 kΩ R_2 = 10 kΩ R_3 = 5.0 kΩ C_S = 39 pF	t_f t_{pd} dV_{out}/dt ②	–	80 70 5.0	–	ns ns V/µs
Gain = 10, 10% Overshoot R_1 = 10 kΩ R_2 = 100 kΩ R_3 = 10 kΩ C_S = 10 pF	t_f t_{pd} dV_{out}/dt ②	–	80 70 15	–	ns ns V/µs
Gain = 100, No Overshoot R_1 = 1.0 kΩ R_2 = 100 kΩ R_3 = 1.0 kΩ C_S = 1.0 pF	t_f t_{pd} dV_{out}/dt ②	–	80 70 30	–	ns ns V/µs
Open Loop, No Overshoot R_1 = 50 Ω R_2 = ∞ R_3 = 50 Ω C_S = 0	t_f t_{pd} dV_{out}/dt ②	–	180 70 35	–	ns ns V/µs
Bandwidth: Open Loop (Figure 4) Closed Loop (Unity Gain) (Figure 5)	–	– –	2.0 10	– –	MHz
Input Noise Voltage (Open Loop) (5.0 Hz - 5.0 MHz)	$V_{n(in)}$	–	11	15	µV(rms)
Average Temperature Coefficient of Input Offset Voltage (R_S = 50 Ω, T_A = −55°C to +125°C)	TCV_{io}	–	2.0	–	µV/°C
DC Power Dissipation V_{out} = 0	P_D	–	120	240	mW
Power Supply Sensitivity ($V^\pm$ Constant)	$S^\pm$	–	250	450	µV/V

① All definitions imply linear operation.   ② dV_{out}/dt = Slew Rate

DIFFERENTIAL ELECTRICAL CHARACTERISTICS

(V^+ = + 6.0 Vdc, V^- = –6.0 Vdc, T_A = 25°C unless otherwise noted)

Characteristic Definitions	Characteristic	Symbol	Min	Typ	Max	Unit
e_{in}, Z_{in} — △ — e_{out}, Z_{out}	Gain (Open Loop)	A_{VOL}	2000 66	3000 70	5000 74	
	Input Impedance (f = 20 Hz)	Z_{in}	0.5	2.0	–	megohms
	Output Impedance (f = 20 Hz)	Z_{out}	–	100	200	ohms
	Common Mode Output Voltage	$V_{out(CM)}$	–0.5	0	+0.5	Vdc
	Output Voltage Swing R_L = 7.0 kΩ	V_{out}	±7.0	±8.0	–	V_{peak}

TYPICAL OUTPUT CHARACTERISTICS (V^+ = +6.0 Vdc, V^- = –6.0 Vdc, unless otherwise noted)

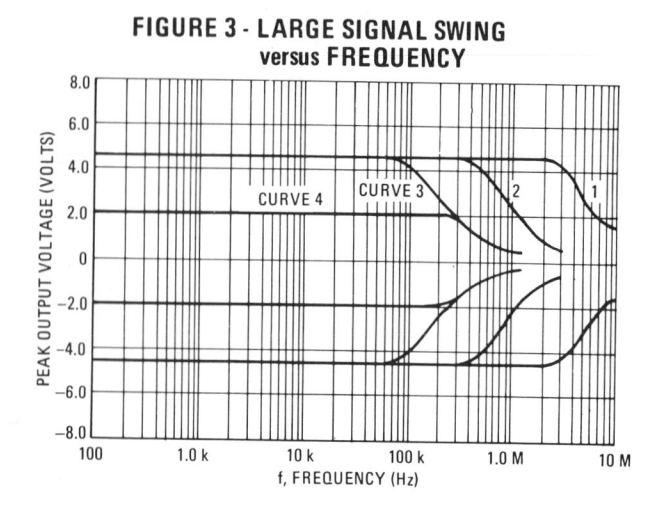

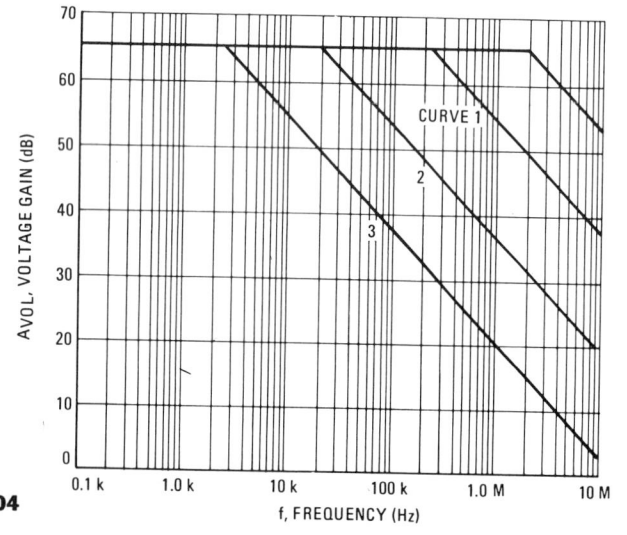

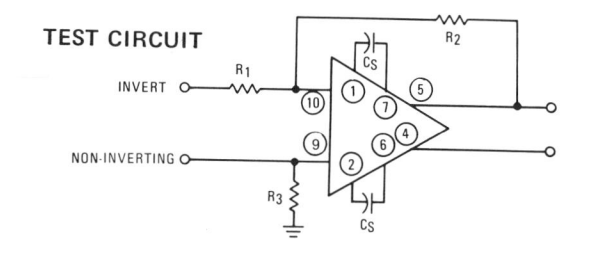

FIGURE 3 - LARGE SIGNAL SWING versus FREQUENCY

TEST CIRCUIT

FIGURE NO.	CURVE NO.	MODE	VOLTAGE GAIN	R_1 (Ω)	R_2 (Ω)	R_3 (Ω)	C_S (pF)	NOISE OUTPUT mV, rms
3	1	INVERTING	100	1.0 k	100 k	1.0 k	1.0	2.0
	2	INVERTING	10	10 k	100 k	10 k	10	0.55
	3	INVERTING	1.0	10 k	10 k	5.0 k	39	0.17
	4	NON INVERTING	1.0	∞	10 k	10 k	39	0.17
4	1	NON-INVERTING	A_{VOL}	0	∞	50	1.0	1.0
	2	NON-INVERTING	A_{VOL}	0	∞	50	10	2.0
	3	NON-INVERTING	A_{VOL}	0	∞	50	39	5.2
5	1	NON-INVERTING	100	100	10 k	100	1.0	2.0
	2	NON-INVERTING	10	1.0 k	9.1 k	910	10	0.55
	3	NON-INVERTING	1.0	∞	10 k	10 k	39	0.17

FIGURE 4 - OPEN LOOP VOLTAGE GAIN

FIGURE 5 - CLOSED LOOP VOLTAGE GAIN versus FREQUENCY

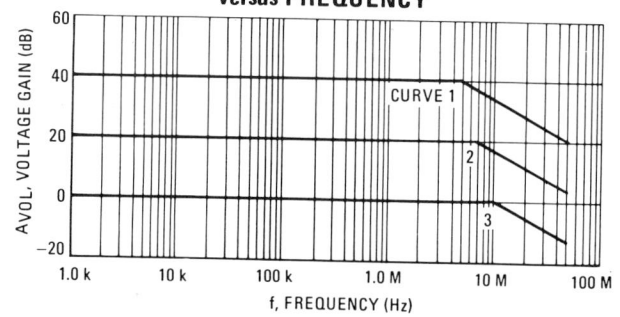

TYPICAL CHARACTERISTICS

FIGURE 6 - POWER DISSIPATION versus POWER SUPPLY VOLTAGE

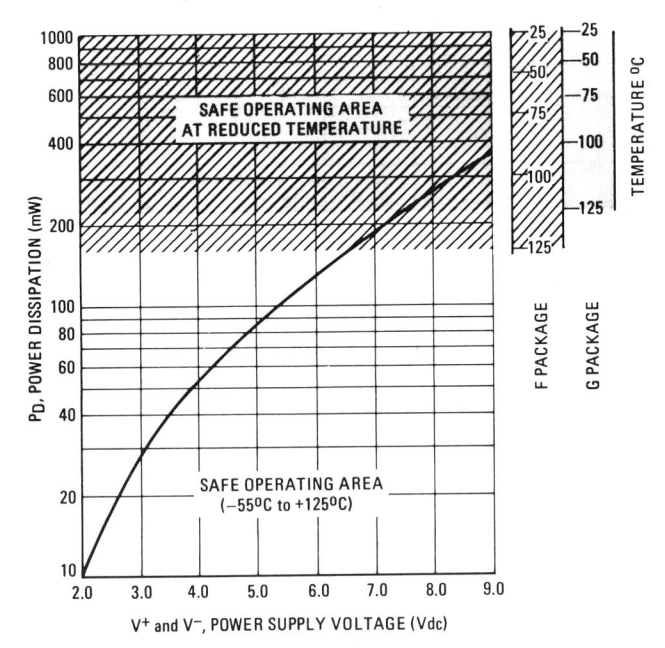

FIGURE 7 - OPEN LOOP VOLTAGE GAIN versus SUPPLY VOLTAGE

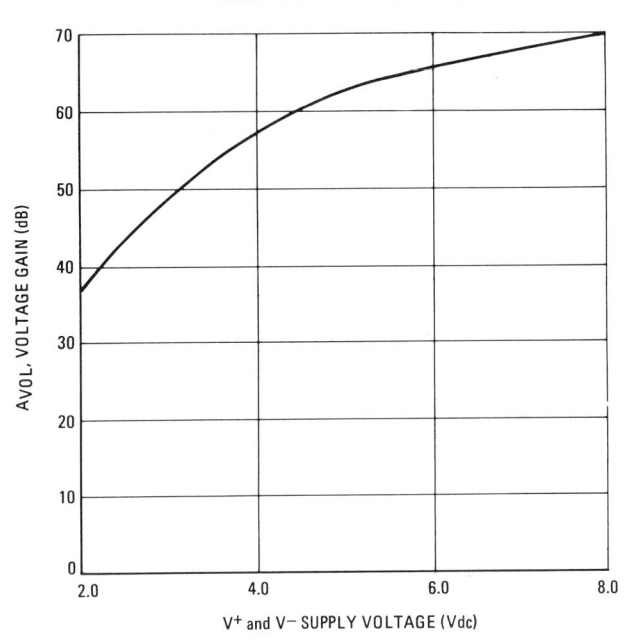

FIGURE 8 - SINGLE ENDED OUTPUT VOLTAGE versus LOAD RESISTANCE

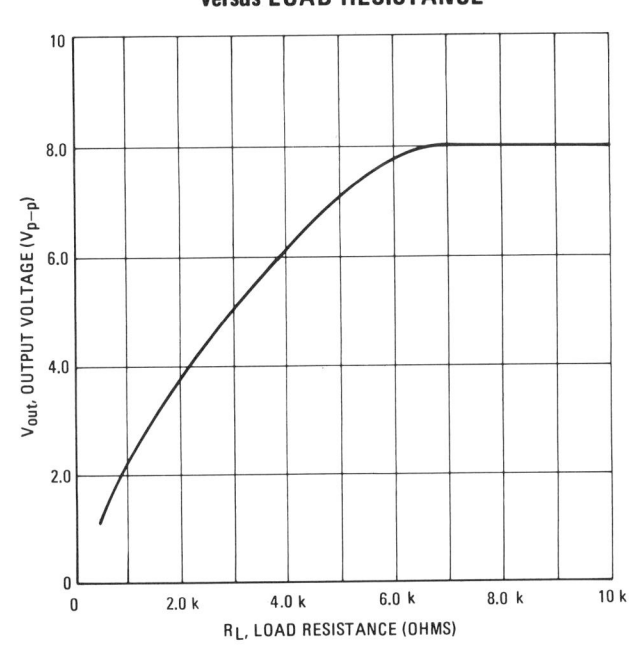

FIGURE 9 - OUTPUT NOISE VOLTAGE versus SOURCE RESISTANCE

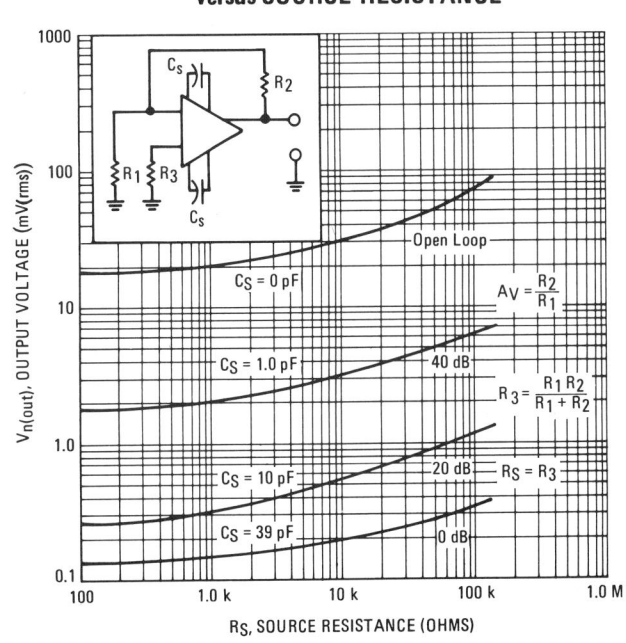

Courtesy of Motorola Semiconductor Products Inc.

705

D-5 IC LOGIC GATES

MC300 Series Integrated Circuits

CIRCUIT DESCRIPTION

BASIC MECL GATE CIRCUIT

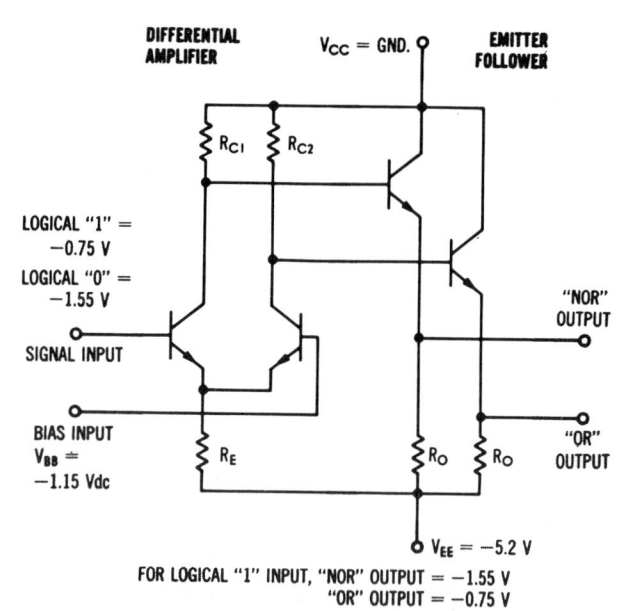

DIFFERENTIAL AMPLIFIER V_{CC} = GND.   EMITTER FOLLOWER

LOGICAL "1" = −0.75 V
LOGICAL "0" = −1.55 V
SIGNAL INPUT

"NOR" OUTPUT
"OR" OUTPUT

BIAS INPUT
V_{BB} = −1.15 Vdc

V_{EE} = −5.2 V

FOR LOGICAL "1" INPUT, "NOR" OUTPUT = −1.55 V
                     "OR" OUTPUT = −0.75 V

FOR LOGICAL "0" INPUT, "NOR" OUTPUT = −0.75 V
                     "OR" OUTPUT = −1.55 V

The MECL line of monolithic integrated logic circuits was designed as a non-saturating form of logic which eliminates transistor storage time as a speed limiting characteristic, and permits extremely high-speed operation.

The typical MECL circuit comprises a differential-amplifier input, with emitter-follower output to restore dc levels. High fan-out operation is possible because of the high input impedance of the differential amplifier and the low output impedance of the emitter followers. Power-supply noise is virtually eliminated by the nearly constant current drain of the differential amplifier, even during the transition period. Basic gate design provides for simultaneous output of both the function and its complement.

POWER-SUPPLY CONNECTIONS

Any one of the power supply nodes, V_{BB}, V_{CC}, or V_{EE} may be used as ground; however, the manufacturer has found it most convenient to ground the V_{CC} node. In such a case: V_{CC} = 0, V_{BB} = −1.15 V, V_{EE} = −5.2 V, as shown in the schematic diagram above.

SYSTEM LOGIC SPECIFICATIONS

The output logic swing of 0.8 V then varies from a low state of V_L = −1.55 V to a high state of V_H = −0.75 V with respect to ground.

Positive logic is used when reference is made to logical "0's" or "1's". Then
"0" = −1.55 V
"1" = −0.75 V } typical

Dynamic logic refers to a change of logic states. Dynamic "0" is a negative going voltage excursion and a dynamic "1" is a positive going voltage excursion.

CIRCUIT OPERATION

A bias of −1.15 volts is applied to the "bias input" of the differential amplifier and the logic signals are applied to the "signal input". If a logical "0" is applied, the current through R_E is supplied by the fixed-biased transistor. A drop of 800 mV occurs across R_{C2}. The OR output then is −1.55 V, or one V_{BE}-drop below 800 mV. Since no current flows in the "signal input" transistor, the NOR output is a V_{BE}-drop below ground, or −0.75 volts. When a logical "1" level is applied to the "signal input", the current through R_{C2} is switched to the "signal input" transistor and a drop of 800 mV occurs across R_{C1}. The OR output then goes to −0.75 volts and the NOR output goes to −1.55 volts.

Note: Any unused input should be connected to V_{EE}.

BIAS VOLTAGE SOURCE

The bias voltage applied to the bias input is obtained from a regulated, temperature-compensated bias driver, type MC304. The temperature characteristics of the bias driver compensate for any variations in circuit operating point over the temperature range or supply voltage changes, to insure that the threshold point is always in the center of the transition region. The bias driver can be used to drive up to 25 logic elements and should be employed for all elements except those with built-in bias networks.

DEFINITIONS

e_{in} AC signal applied to the input
e_{out} AC signal at the output
I_C Amount of current drawn from the positive power supply by the test unit
I_{CEX} Total collector leakage current exhibited by the gate expander when all inputs are at the negative supply potential
I_E Amount of current drawn from the test unit by the negative power supply
I_{in} Current drawn by the input of the test unit when a logical "1" (V_H) is applied to the input
I_L Current drawn from a node when that node is at ground potential
t_{d1} Time required for the output pulse to reach the 50% point of its leading edge when referenced to the 50% point of the input pulse leading edge
t_{d2} Time required for the output pulse to reach the 50% point of its trailing edge when referenced to the 50% point of the input pulse trailing edge
t_{df} Time required for a flip-flop output to reach the 50% point of its negative going edge when referenced to the 50% point of the input pulse leading edge
t_{dr} Time required for a flip-flop output to reach the 50% point of its positive going edge when referenced to the 50% point of the input pulse leading edge
t_f Time required for the output pulse to go more negative from its 90% point to its 10% point
t_r Time required for the output pulse to go more positive from its 10% point to its 90% point

V_1 "NOR" output voltage — logical "1" level output voltage when a logical "0" level (V_L) is applied to the input
V_2 "OR" output voltage — logical "0" level output voltage when a logical "0" level (V_L) is applied to the input
V_3 Saturation breakpoint voltage which corresponds to the "NOR" output characteristic where the rate of change in the output voltage to the rate of change in input voltage is zero
V_4 "NOR" output voltage — logical "0" level output voltage when a logical "1" level ($V_{1\ max}$) level is applied to the input
V_5 "OR" output voltage — logical "1" level output voltage when a logical "1" level ($V_{1\ max}$) level is applied to the input
V_6 Output latch voltage — input voltage to a flip-flop which causes the output voltage to change from a logical "1" level to a logical "0" level and corresponds to the point where the rate of change in the output voltage to the rate of change of the input voltage approaches infinity
V_H Logical "1" input voltage
V_L Logical "0" input voltage
V_{OH} High-level output voltage when the saturated logic circuit output is in an "off" condition
V_{OL} Low-level output voltage when the saturated logic output circuit is in an "on" condition
ΔV_1
ΔV_5 } Change in the "1" level output voltage as the load is varied from no load to full load

WORST-CASE TRANSFER CHARACTERISTICS

The following graphs show minimum and maximum limits of major parameters associated with the transfer characteristics of the MECL line. Min-Max limits, given at three different temperatures can be interpreted for design purposes as 10% to 90% spreads at all points on the curve except for guaranteed points in the Electrical Characteristics tables.

DEFINITIONS

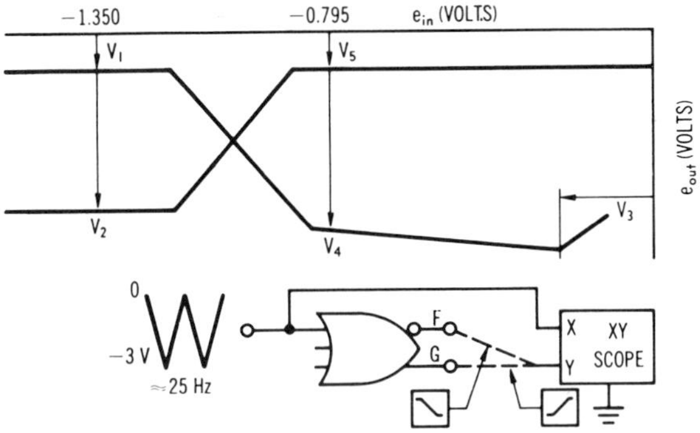

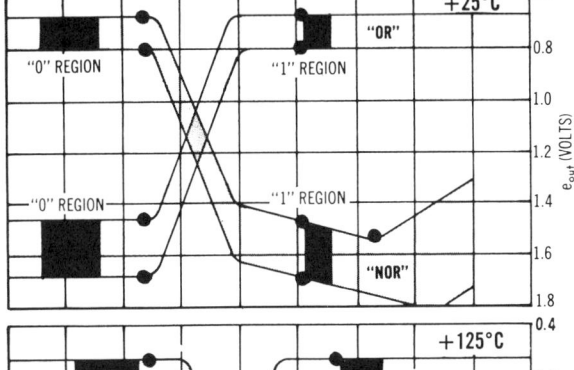

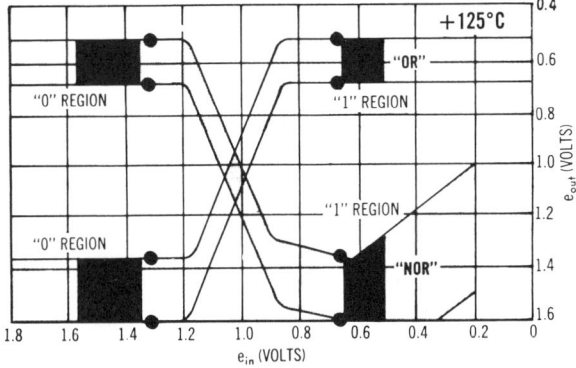

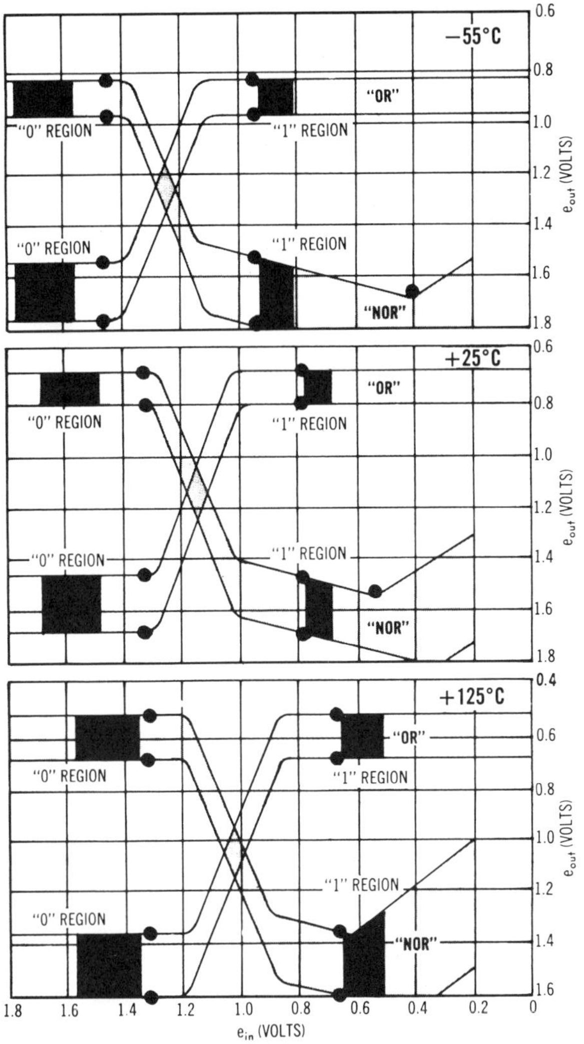

MAXIMUM RATINGS

Characteristic	Symbol	Rating	Unit
Ratings above which device life may be impaired:			
Power Supply Voltage ($V_{CC} = 0$ Vdc)	V_{EE}	−10	Vdc
Bias Input Voltage ($V_{CC} = 0$ Vdc)	V_{in}	0 Vdc to V_{EE}	Vdc
Output Source Current	I_O	20	mAdc
Storage Temperature Range	T_{stg}	−65 to +150	°C
Recommended maximum ratings above which performance may be degraded:			
Operating Temperature Range	T_A	−55 to +125	°C
AC Fan-In (Expandable Gates)	m	18	—
AC Fan-Out* (Gates and Flip-Flops)	n	15	—

*Although a minimum dc fan-out of 25 is guaranteed in each electrical specification, it is recommended that the maximum ac fan-out of 15 be used for high-speed operation.

NOISE MARGINS (90 PERCENTILE)

The following graphs show worst-case Noise Margins as a function of temperature and fan-out. Top graph illustrates the advantage gained through use of MC304 bias driver, as compared with non-compensated fixed bias source, bottom.

Note: Any unused input should be connected to V_{EE}.

707

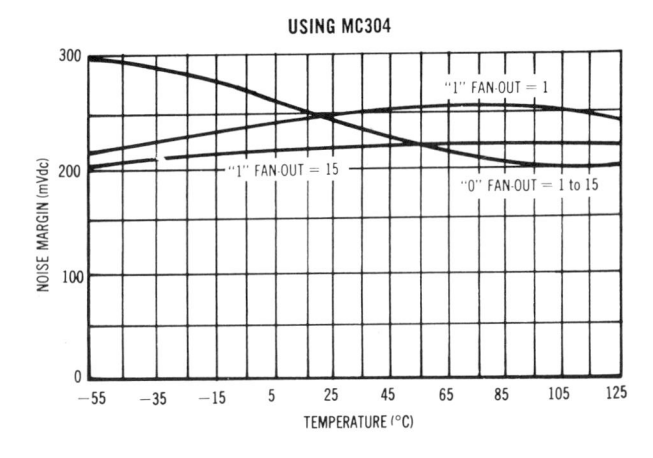

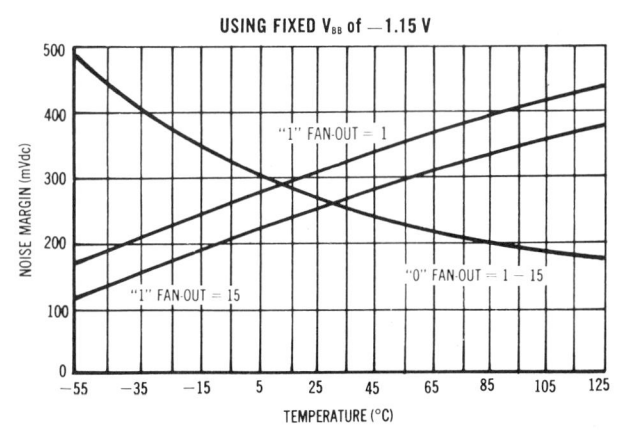

BIAS DRIVER

MC304

Bias driver that compensates for changes in circuit parameters with temperature.

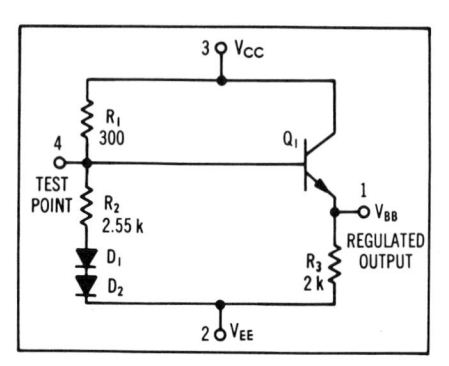

ELECTRICAL CHARACTERISTICS

Characteristic	Test Conditions Vdc ±1% @ Test Temperature	V_{EE} Pin No	I_L Pin No	Ground Pin No	Symbol Pin No in ( )	Test Limits −55°C Min	Max	+25°C Min	Max	+125°C Min	Max	Unit
	−55°C: −5.20 +25°C: −5.20 +125°C: −5.20											
Power Supply Drain Current		2	—	3	I_E(2)	—	4.4	—	4.4	—	4.0	mAdc
Output Voltage		2	1①	3	V_{BB}	−1.19	−1.32	−1.09	−1.22	−0.95	−1.08	Vdc

Pins not listed are left open.

① Current test conditions: no load = 0; full load = −2.5 mAdc ±5%.

CIRCUIT DESCRIPTION

Circuit Operation:

The divider network R_1, R_2, D_1, D_2 compensates for temperature variations of the base-emitter voltages of Q_1, and of the driven gates, producing a bias voltage for the MECL logic circuits that maintains a constant set of dc operating conditions over the temperature range of −55 to +125°C. In addition, compensation for power supply variations is achieved, since the bias output voltage is derived from the system supply.

Either of the supply voltage nodes may be used as ground, however the ground potential of the bias driver must coincide with that of the logic system. Thus, if V_{CC} is grounded in the logic system, then —

$V_{CC} = 0;$ $V_{EE} = -5.2$ V;
$V_{BB} = -1.15$ nominal output voltage at 25°C

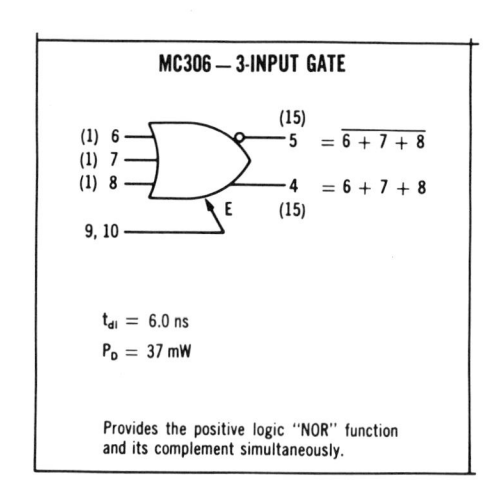

MC306 — 3-INPUT GATE

$t_{dl} = 6.0$ ns
$P_D = 37$ mW

Provides the positive logic "NOR" function and its complement simultaneously.

MC306, 307 3-Input Gates

Expandable 3-input gates that provide the positive logic "NOR" function and its complement simultaneously. MC307 omits output pull-down resistors, permitting reduction of power dissipation.

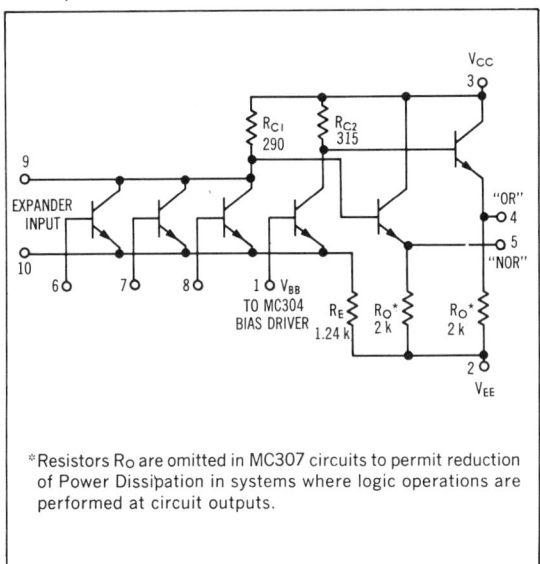

*Resistors R_O are omitted in MC307 circuits to permit reduction of Power Dissipation in systems where logic operations are performed at circuit outputs.

SWITCHING TIME TEST CIRCUIT

INPUT PULSE t_r AND $t_f = 6 \pm 2$ ns

Stray capacitance introduced by the test jig:
$C_S = (n + 12)$ pF where n = number of fan-outs.

PROPAGATION DELAY

RISE AND FALL TIME

Fan-in obtained with MC305 input expanders; all but driven input connected to −5.2 V.

TYPICAL INPUT CHARACTERISTICS

TYPICAL OUTPUT CHARACTERISTICS

TYPICAL SWITCHING TIME VARIATIONS
MC306

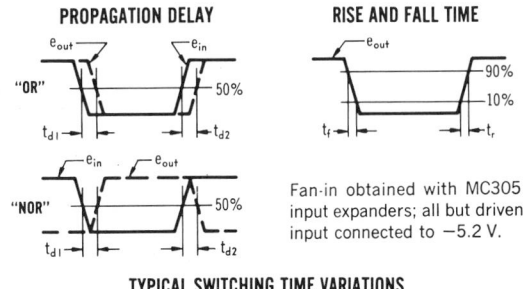

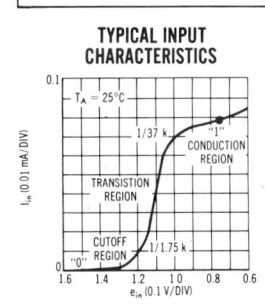

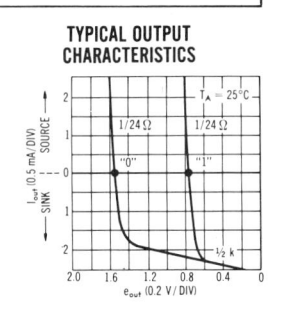

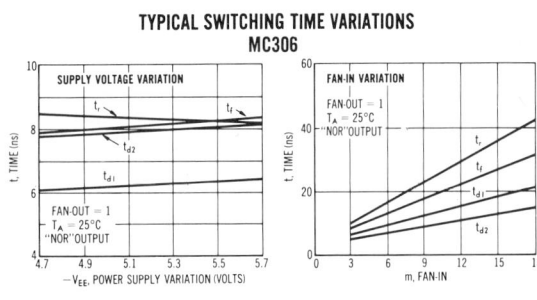

SWITCHING CHARACTERISTICS (10% to 90% distribution)

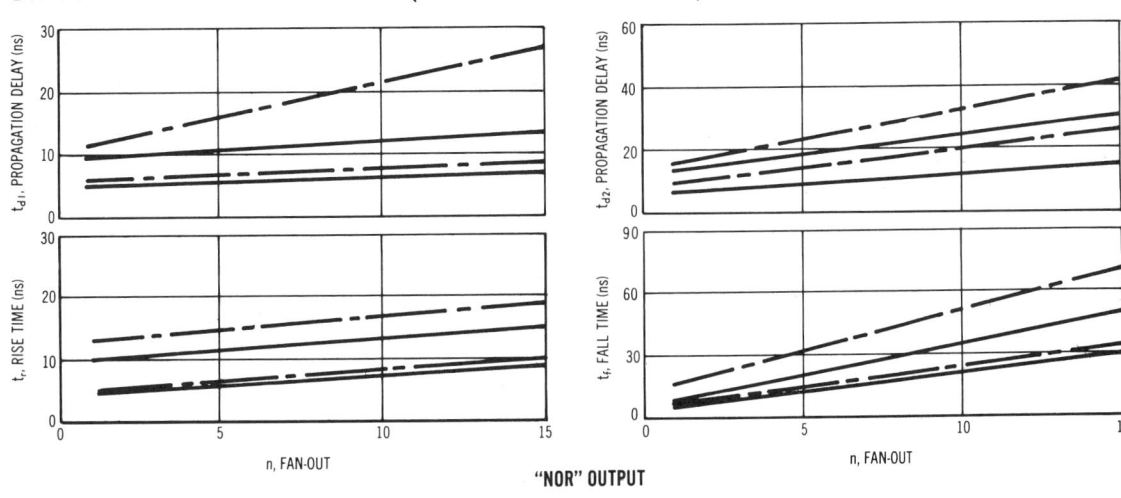

"NOR" OUTPUT

—— −55°C and +25°C
--- +125°C

709

ELECTRICAL CHARACTERISTICS

		Test Conditions Vdc ±1%													
@ Test Temperature	−55°C	—	−0.945	−1.450	−5.20	−1.25									
	+25°C	−0.690	−0.795	−1.350	−5.20	−1.15									
	+125°C	—	−0.655	−1.300	−5.20	−1.00									

Characteristic	V_H Pin No	$V_{I\,max}$ Pin No	V_L Pin No	V_{EE} Pin No	V_{BB} Pin No	dV_{in} Pin No	I_L Pin No	Ground Pin No	Symbol Pin No in ( )	Test Limits −55°C Min	Max	+25°C Min	Max	+125°C Min	Max	Unit
Power Supply MC306	—	—	—	2,6,7,8	1	—	—	3	I_E (2)	—	8.85	—	8.85	—	8.15	mAdc
Drain Current MC307	—	—	—	2,6,7,8	1	—	—	3	I_E (2)	—	3.6	—	3.6	—	3.3	mAdc
Input Current	6	—	—	2,7,8	1	—	—	3	I_{in} (6)	—	—	—	100	—	—	μAdc
	7	—	—	2,6,8	1	—	—	3	I_{in} (7)	—	—	—	↓	—	—	↓
	8	—	—	2,6,7	1	—	—	3	I_{in} (8)	—	—	—	↓	—	—	
"NOR" Logical "1" Output Voltage	—	—	6	2,7,8	1	—	—	3	V_1 (5)	−0.825	−0.945	−0.690	−0.795	−0.525	−0.655	Vdc
	—	—	7	2,6,8	1	—	—	3	V_1 (5)	↓	↓	↓	↓	↓	↓	↓
	—	—	8	2,6,7	1	—	—	3	V_1 (5)							
"NOR" Logical "0" Output Voltage	—	6	—	2,7,8	1	—	—	3	V_4 (5)	−1.560	−1.850	−1.465	−1.750	−1.340	−1.675	Vdc
	—	7	—	2,6,8	1	—	—	3	V_4 (5)	↓	↓	↓	↓	↓	↓	↓
	—	8	—	2,6,7	1	—	—	3	V_4 (5)							
"OR" Logical "1" Output Voltage	—	6	—	2,7,8	1	—	—	3	V_5 (4)	−0.825	−0.945	−0.690	−0.795	−0.525	−0.655	Vdc
	—	7	—	2,6,8	1	—	—	3	V_5 (4)	↓	↓	↓	↓	↓	↓	↓
	—	8	—	2,6,7	1	—	—	3	V_5 (4)							
"OR" Logical "0" Output Voltage	—	—	6	2,7,8	1	—	—	3	V_2 (4)	−1.560	−1.850	−1.465	−1.750	−1.340	−1.675	Vdc
	—	—	7	2,6,8	1	—	—	3	V_2 (4)	↓	↓	↓	↓	↓	↓	↓
	—	—	8	2,6,7	1	—	—	3	V_2 (4)							
"NOR" Output Voltage Change (No load to full load)	—	—	6	2,7,8	1	—	5③	3	ΔV_1 (5)	—	−0.055	—	−0.055	—	−0.060	Volts
"OR" Output Voltage Change (No load to full load)	—	6	—	2,7,8	1	—	4③	3	ΔV_5 (4)	—	−0.055	—	−0.055	—	−0.060	Volts
"NOR" Saturation Breakpoint Voltage	—	—	—	2,7,8	1	6①	—	3	V_3 (5)	—	−0.40	—	−0.55	—	−0.68	Vdc
	—	—	—	2,6,8	1	7①	—	3	V_3 (5)	—	↓	—	↓	—	↓	
	—	—	—	2,6,7	1	8①	—	3	V_3 (5)	—		—		—		

Switching Times	Pulse In	Pulse Out								Typ	Max	Typ	Max	Typ	Max	
Propagation Delay Time	6	4	—	2,7,8	1	—	—	3	t_{d1} (4)	7.0	11.0	7.0	11.5	9.5	14.5	ns
	6	5	—	2,7,8	1	—	—	3	t_{d1} (5)	5.5	10.0	5.5	10.5	7.0	12.5	
	6	4	—	2,7,8	1	—	—	3	t_{d2} (4)	5.5	10.0	5.5	11.0	7.0	12.5	
	6	5	—	2,7,8	1	—	—	3	t_{d2} (5)	7.0	10.5	7.0	11.0	9.5	14.5	
Rise Time	6	4	—	2,7,8	1	—	—	3	t_r (4)	6.0	8.5	6.0	10.0	8.0	13.0	
	6	5	—	2,7,8	1	—	—	3	t_r (5)	7.5	11.5	7.5	12.5	9.5	15.0	
Fall Time	6	4	—	2,7,8	1	—	—	3	t_f (4)	6.5	10.5	6.5	12.0	9.0	15.0	
	6	5	—	2,7,8	1	—	—	3	t_f (5)	6.5	12.0	6.5	12.5	9.0	15.0	

Pins not listed are left open. ① Input voltage is adjusted to obtain dV "NOR/dV_{in} = 0. ③ Current test conditions: no load = 0; full load = −2.5 mAdc ±5%.

SWITCHING CHARACTERISTICS (10% to 90% distribution)

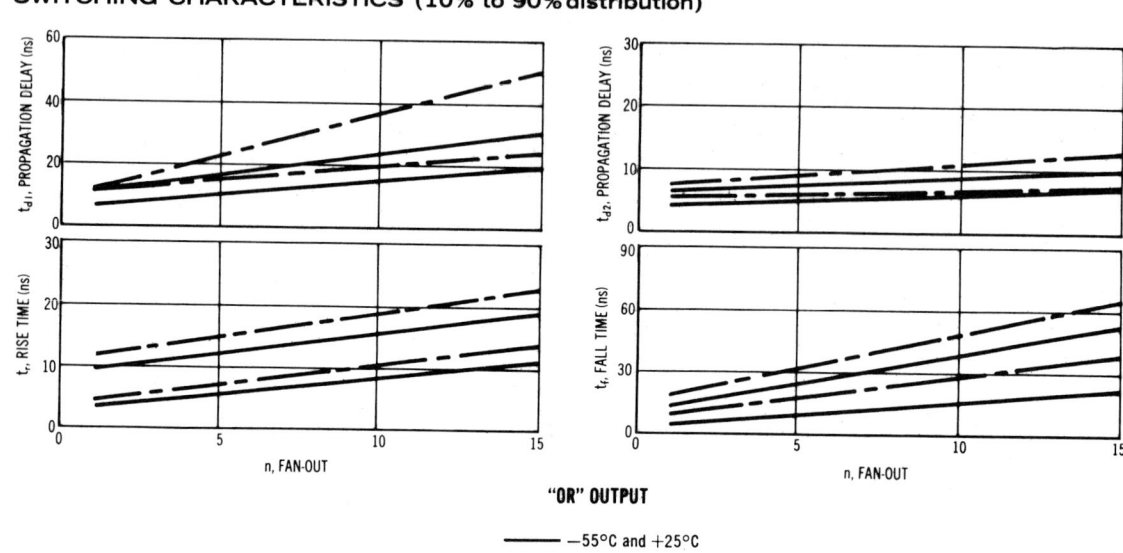

"OR" OUTPUT

——— −55°C and +25°C
− − − +125°C

(Courtesy of Motorola Semiconductor Products, Inc.)

3102 3 Input Gate (MOS Integrated Circuit)

GENERAL DESCRIPTION — The 3102 is a MOS Monolithic 3-Input Gate Integrated Circuit. This device can be utilized as a vehicle in gaining familiarity with MOS integrated circuit logic versatility, as a building block in an all MOS system, or as a breadboarding gate in designing complex custom integrated circuits. Input protection is provided on all gate inputs.

ABSOLUTE MAXIMUM RATINGS (Note 1)

Storage Temperature	−65°C to +150°C
Operating Temperature	−55°C to +85°C
Positive Voltage on any Pin ($V_{Body} = 0$)	+0.3 Volt
Negative Voltage on any Pin ($V_{Body} = 0$)	−35 Volts
Power Dissipation at $T_A = 25°C$	$\leq$ 200 mW

ELECTRICAL CHARACTERISTICS ($V_{Body} = 0$; $T_A = 25°C$)

CHARACTERISTICS	CONDITIONS	MIN.	TYP.	MAX.	UNITS
R_{Load}	$V_{DD} = -27 V \pm 2.0 V$ V_{D1} = Gnd		140		kΩ
R_{ON} (See Figure 1)	$V_{IN} = -20 V$		500		Ω
Input Leakage Current	$V_{IN} = -25 V$			1.0	μA
Threshold Voltage (V_{TH})	$V_D = V_G$, $I_D = -10$ μA	−2.0		−5.5	V
Input Capacitance			4.0		pF
Power Consumption	$V_{DD} = -30 V$, $V_{IN} = -10 V$		6.0		mW

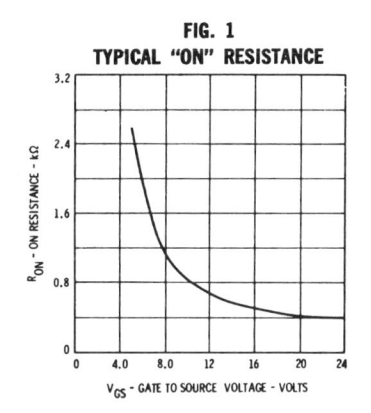

FIG. 1
TYPICAL "ON" RESISTANCE

RESISTANCE CHARACTERISTIC

Parameters	Conditions	TYP.
R_{Load}	$V_{DD} = -25 V$ $V_{D1} = 0 V$	140 kΩ
R_{ON}	$V_{GS} = -20 V$ $V_{DS} \leq -1.0 V$ $V_{D2} = 0 V$	500 Ω
R_{ON}	$V_{GS} = -9.0 V$ $V_{DS} \leq -1.0 V$ $V_{D2} = 0 V$	800 Ω

SCHEMATIC DIAGRAM

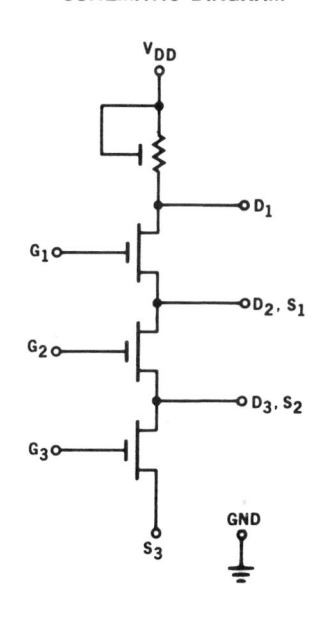

711

FIG. 2 TYPICAL DRAIN TO SOURCE CHARACTERISTICS

FIG. 3 TYPICAL PULL-UP RESISTANCE

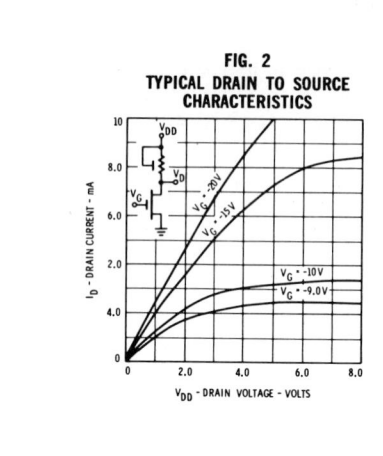

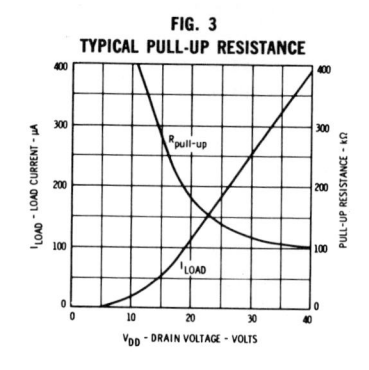

APPLICATIONS — MOS logic will provide the versatility to build many different functions. The following circuits show how to build the functions using one or two 3102 packages.

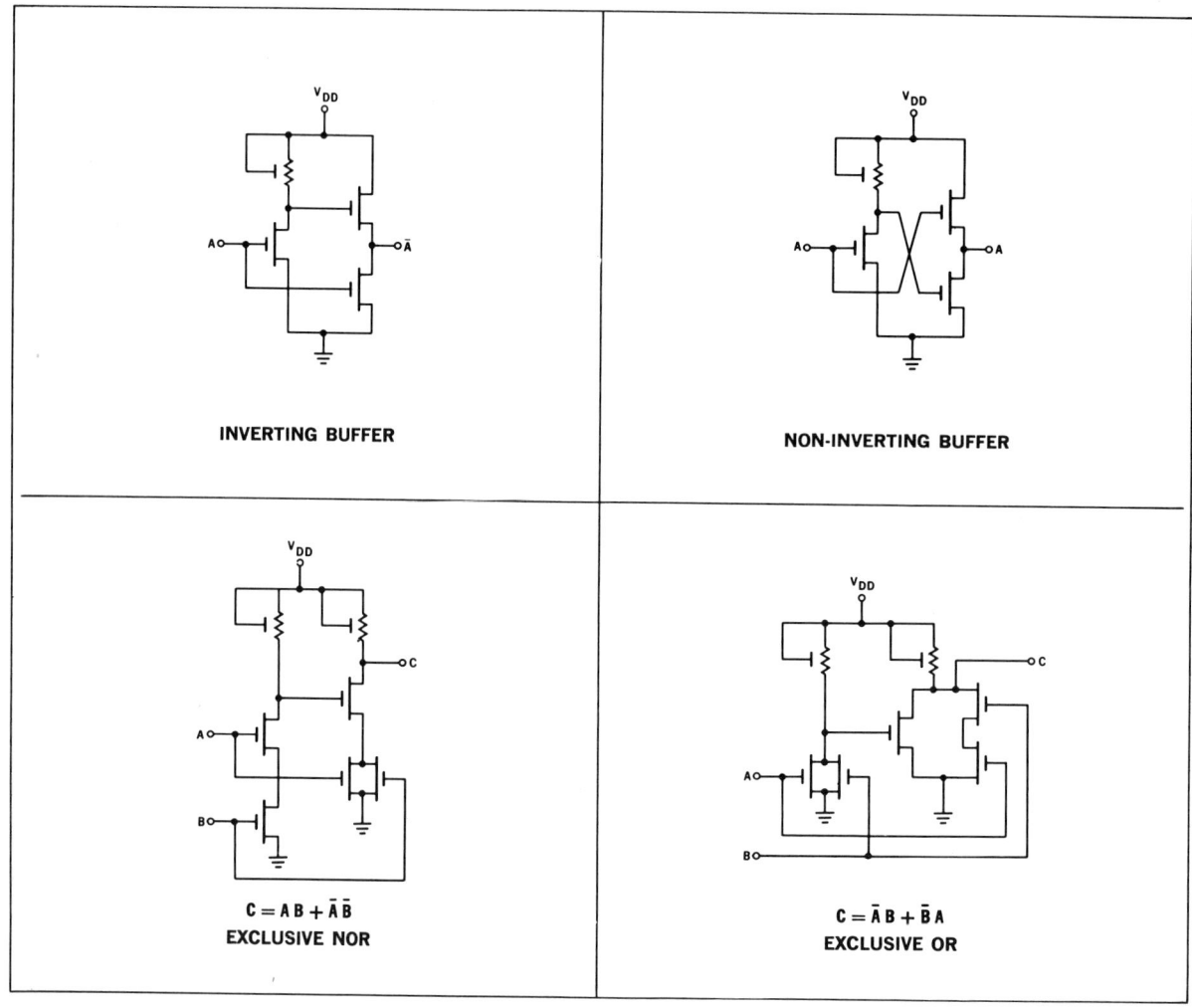

INVERTING BUFFER

NON-INVERTING BUFFER

$C = AB + \bar{A}\bar{B}$
EXCLUSIVE NOR

$C = \bar{A}B + \bar{B}A$
EXCLUSIVE OR

NOR GATE CONFIGURATION

Pin 8 = Pin 10 = V_{OUT}
Pin 1 = Pin 9 = GND
$V_{DD} = -27\text{ V} \pm 2.0\text{ V}$

NAND GATE CONFIGURATION

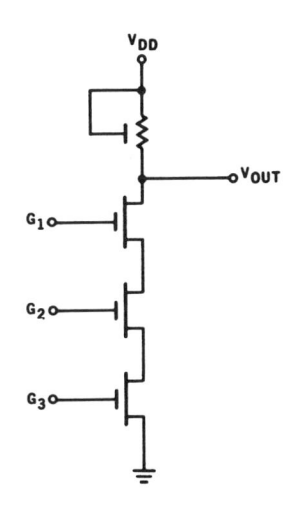

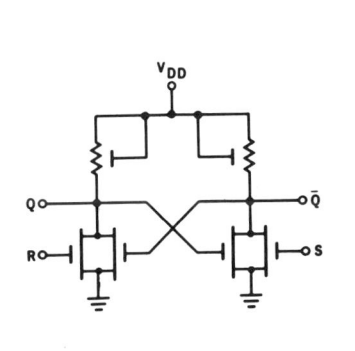

Pin 5 = GND
Pin 8 = OUTPUT
$V_{DD} = -27\text{ V} \pm 2.0\text{ V}$

LOGIC LEVEL	LOGIC LEVEL	VOLTAGE LEVEL MIN.	VOLTAGE LEVEL MAX.
V_{IH}	1	−9.0 V	
V_{IL}	0		−2.0 V
V_{OH}	1	−10 V	
V_{OL}	0		−1.0 V

RS FLIP-FLOP

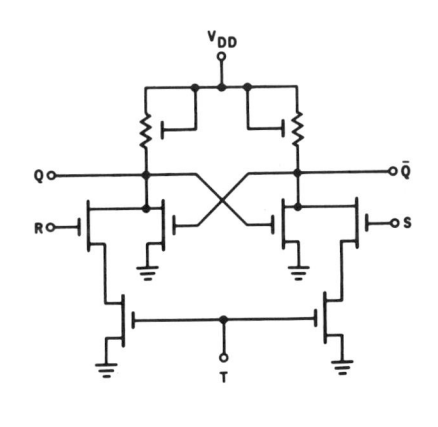

RST FLIP-FLOP

(Courtesy of Fairchild Instrument)

Answers to Selected Problems

(1-2) $x(t) = [(1/2)t^2 + t]e^{-t}u(t)$

(1-5) (a) $(t^2/2)u(t)$ (b) $(e^{-t} - e^{-2t})u(t)$

(1-11) (a)
$$[A] = \begin{bmatrix} -\dfrac{1}{10} & -\dfrac{1}{111} \\ -\dfrac{1}{10} & -\dfrac{101}{111} \end{bmatrix}$$

(b) $\lambda_1 = -0.91, \lambda_2 = -0.10$

(1-13) (a) $x_1(t) = 2e^{-t}, x_2(t) = 2e^{-2t}$

(b) Exact trajectory $x_2(t) = (1/2)x_1^2(t)$

(1-17) $\dfrac{V_2}{V_2} = \dfrac{K/R_1R_2C_1C_2}{s^2 + s\left[\dfrac{1}{C_1}\left(\dfrac{1}{R_1} + \dfrac{1}{R_2} + \dfrac{1}{R_3}\right) + \dfrac{1}{R_2C_2}\right] + \dfrac{1}{R_2C_1C_2}\left(\dfrac{1}{R_1} + \dfrac{1}{R_3} - \dfrac{K}{R_3}\right)}$

Answers to Selected Problems

(1-21) $s_1 = \sqrt{3} - 1, s_2 = -(\sqrt{3} + 1)$

(1-24) Open circuit natural frequencies: $s_1 = -2, s_2, s_2^* = \pm j2$
Short circuit natural frequencies: $s_1 = -1, s_2 = 1$

(1-28) $V_{eq} = [(\alpha R_3 + R_2)/(R_1 + R_2)]V_S$
$R_{eq} = [R_1 R_3 + R_1 R_2 + R_3 R_2(1 - \alpha)]/(R_1 + R_2)$

(1-30) $i \simeq 54$ mA

(2-2) $z_{11} = z_{22} = 0, z_{12} = -z_{21} = \alpha$

(2-5) $h_{11} = 0, h_{12} = -h_{21} = n, h_{22} = [(n-1)^2 R_1 + n^2 R_2]/R_1 R_2$

(2-10) No. $G_T = 4.90$

(2-12) (a) $(Y_s)_{opt} = y_{11}^*, (Y_L)_{opt} = y_{22}^*$
(b) $G_U = |y_{21}|^2/4\text{Re } y_{11} \text{Re } y_{22}$

(2-15) No. $C_L > 0, G_L > 2.49 \times 10^{-3}$ mho for stability

(2-19) (a) unit step response: $[3/2 - e^{-t} - (1/2)e^{-2t}]u(t)$
(b) $y(t) = [e^{-t}(t+1) - e^{-2t}]u(t)$

(2-23) $h(t)|_{\tau=0} = \left(\dfrac{1}{2t+1}\right)^{1/2}$.

Unit step response (for $\tau = 0$) $= 1 - \dfrac{1}{\sqrt{1+2t}}$.

No. Only for time-invariant case it is.

(2-26) $a^3 - 3ab = (a/z)^3$

(2-29) (a) $a_1 = 125/12, a_0 = 625/72$
(b) $a_1 = 25/2, a_0 = 125/8$

(2-32) $z_{11} = 1.80$ kΩ, $z_{12} = z_{21} = 360$ ohms, $z_{22} = 1.08$ kΩ

(2-34) $R = 13.6$ kΩ, $L = 0.617$ mH, $A_v(0) = -13.6$

(3-1) (a) $(R_{ab})_{Ge} = .44$ kΩ (b) $(R_{ab})_{Si} = 2.32$ MΩ,
(c) $(R_{ab})_{Ge} = .55$ kΩ, $(R_{ab})_{Si} = 2.57$ MΩ

(3-4) $N_a = 1.63 \times 10^{18}$ cm^{-3}

(3-6) $V = 0.92$ V, $I = 10.66$ mA

(3-14) (a) $|R| > R_g$ (b) $L_s < C(R_g + R_s)|R|, |R| > R_s + R_g$

(3-16) $V_{BB} \geq 2.98$ V

(3-19) $y_{11} = 5.0$ mhos, $y_{12} = 2$ mhos, $y_{21} = 1.37$ mhos, $y_{22} = 1$ mho

(3-21) (a) $\alpha_0 = 0.961, \omega_\alpha = 1.3 \times 10^8$ rad/sec
(b) $\alpha_0 = 0.990, \omega_\alpha = 5.0 \times 10^8$ rad/sec

(3-24) $r_e = g_m^{-1} = 2.5 \, \Omega, 100 < \beta_0 < 300, \beta_0 \approx 200$
$R_i = 500 \, \Omega, C_c \leq 6$ pF, $f_T \geq 300$ MHz
$r_b \geq 41 \, \Omega, C_\pi \geq 205$ pF, 16 k$\Omega \leq r_0 < 100$ kΩ

(3-27) $h_{11} = 29 \, \Omega, h_{12} = 2 \times 10^{-4}, h_{21} = -0.98, h_{22} = 10^{-6}$ mho

(3-30) $A_0 > (R_0 + R_F + R_i)/R_i$

Answers to Selected Problems

(3-33) (a) $I_C = 3.48$ mA, $V_{CE} = 6.5$ V,
(b) $I_C = 3.86$ mA, $V_{CE} = 5.95$ V

(3-35) $R_s = 500\,\Omega$, $R_{g1} = 100$ kΩ, $g_m = 4 \times 10^{-3}$ mho, $C_{gs} = 10$ pF, $C_{gd} = 2$ pF, $A_v = -2.0$

(3-40) (a) $R_s = 1.25$ kΩ, $R_D = 4.75$ kΩ,
(b) $A_v = -7.8$

(3-43) $R_K = 600\,\Omega$, $R_P = 36.8$ kΩ, $R_S = 197$ kΩ

(3-45) (a) $R_b = 200$ kΩ, $R_C = 1.5$ kΩ
(b) $r_e = g_m^{-1} = 2.5\,\Omega$, $R_i = 250\,\Omega$, $r_b = 50\,\Omega$, $r_0 = 20$ kΩ, $C_c = 5$ pF, $C_\pi = 95$ pF
(c) $V_0/V_{in} = -350$

(4-6) $h_{fe} = -h_{fe}/(1 + h_{fb})$, $h_{oe} = h_{ob}/(1 + h_{fb})$,
$h_{ie} = h_{ib}/(1 + h_{fb})$, $h_{re} = [h_{ib}h_{ob}/(1 + h_{fb})] - h_{rb}$

(4-10) (a) $(G_A)_{CC} = 20 = 13$ dB
$(G_A)_{CE} = 1.6 \times 10^4 = 42$ dB
$(G_A)_{CB} = 760 = 28.8$ dB
(b) CC: inherently stable $(G_A)_{max} = 49 = 16.9$ dB
CE: inherently stable $(G_A)_{max} = 1.7 \times 10^4 = 42.3$ dB
CB: inherently stable $(G_A)_{max} = 2.3 \times 10^3 = 33.6$ dB

(4-15) (a) $C_1 = 0.15\,\mu$F, (b) $C_1 = 0.12\,\mu$F

(4-23) $\omega_{3\,dB} = 7.7 \times 10^5$ rad/sec

(4-25) $\omega_{3\,dB} = 0.51\,p_o$

(4-28) $r_e = g_m^{-1} = 10\,\Omega$, $r_b = 100\,\Omega$, $\beta_0 = 100$, $f_T = 10^8$ Hz, $f_\beta = 10^6$ Hz, $C_c = 6.4$ pF, $C_\pi = 153.6$ pF

(4-29) (a) $R_I = 122\,\Omega$, (b) $R_I = 455\,\Omega$, $A_i(0) = -29.4$

(4-32) $R_1 = 140\,\Omega$, $L = 5.8\,\mu$H, $A_i(0) = -14.3$, $f_{3\,dB} = 3.9$ MHz

(4-35) (b) $R_1 = 28\,\Omega$, $L_s = 21\,\mu$H, $A_i(0) = -3.7$

(4-43) $A_v = 100$, $Z_{in} = 1$ kΩ, $Z_0 = 10$ ohms

(5-1) (a) $I_{C1} = I_{C2} \approx 2$ ma; $V_{CE1} = 5.4$ V, $V_{CE2} = 5.3$ V,
(b) $A_v(0) = -35.8$

(5-3) (a) $A_m = 169$, (b) $\omega_h = 8.35 \times 10^6$ rad/sec, $\omega_l = 9.4 \times 10^3$ rad/sec

(5-6) $A_m = 500$, $\omega_h = 0.71 \times 10^6$ rad/sec, $\omega_l = 158$ rad/sec

(5-9) (a) $n = 3$ stages, (b) $n = 4$ stages

(5-11) $A_v(0) = 37$, $f_{3\,dB} = 95$ MHz

(5-16) $A_v(0) = -19.6$, $f_{3\,dB} = 11.1$ MHz

(5-20) $R_1 = 10$ kΩ, $R_L = 10$ kΩ, $L = 0.5$ mH, $A_v(0) = 2500$

(5-22) (a) $S = 0.804$, (b) $(\omega_{3\,dB})_i \simeq 2\pi(1.25$ MHz$)$

(5-29) $v_N = 1.9\,\mu$V, $F_n = 1.76$ dB

(5-35) $(V_{CE})_{max} = 24$ V, $P_{CM} \geq 7.5$ watts.

Answers to Selected Problems

Turns ratios of the Transformers: Output $n = 1.55$; input $n = 3.17$

(5-40) $R_{b1} = 13.5 \text{ k}\Omega$, $R_{b2} = 125 \text{ k}\Omega$

(6-1) $R_f = 1.1 \text{ k}\Omega$, $\omega_{3\text{ dB}} = 8.7 \times 10^6$ rad/sec
(6-3) $R_f = 90 \text{ }\Omega$, $\omega_{3\text{ dB}} = 2.16 \times 10^7$ rad/sec
(6-6) Pole locations: $p_1 = -1.12$, $p_2, p_2^* = 0.490 \pm j0.245$
$a_i(0) = -1950$, $A_i(0) = -1300$, $\omega_{3\text{ dB}} = 3.4 \times 10^7$ rad/sec
(6-12) (a) $f_0 = 11.7 \times 10^{-4}$, (b) for MFM: $f_0 = 2.6 \times 10^{-4}$, $\Phi_M = 92°$, $G_M = \infty$
(6-15) $T_0 = 23.7$, $A(0) = -405$
(6-19) (a) $0 < K < 5$, (b) $a_0, a_1, a_2 > 0$, $a_2 a_1 > a_0$
(6-23) $A_m = 167$, $\omega_h = 1.32 \times 10^6$ rad/sec; $\omega_l = 1.5$ rad/sec
(6-24) $a_0 = 3.1 \times 10^3$, $f_0 = (1/3)10^{-3}$, $A(0) = 1.5 \times 10^3$, $(R_{\text{in}})_f = 235 \text{ }\Omega$

(6-29) System unstable for: $K \geq \left(\sec \dfrac{\pi}{n}\right)^n$ for $K > 0$, $|K| \geq 1$ for $K < 0$

(6-33) $A_v(s)$

$$= -\dfrac{R_c/R_d R_1 R_2 C_1 C_2}{s^2 + s\left[\left(\dfrac{1}{R_1 C_1}\right)\left(1 + \dfrac{R_1}{R_2} + \dfrac{R_1}{R_3}\right) + \dfrac{1}{R_2 C_2}\left(1 + \dfrac{R_2}{R_d}\right)\right] + \left\{1 + \dfrac{R_1}{R_3}\left(1 + \dfrac{R_c}{R_d} + \dfrac{R_2}{R_d}\right) + \dfrac{R_1}{R_d} + \dfrac{R_2}{R_d}\right\}/R_1 R_2 C_1 C_2}$$

$R_1 = R_2 = 870 \text{ }\Omega$
$C_1 = 1.15 \text{ }\mu\text{F}$, $C_2 = 0.103 \text{ }\mu\text{F}$

(6-39) $\omega_0 = 1/\sqrt{C(L_2 + L_1 + 2M)}$, $\beta_0 = (L_1 + M)/(L_2 + M)$
(6-40) $\omega_0 = 1/RC$, $g_m^2 R_{L1} R_{L2} = 3$

(6-43) (a) $\omega_0 = \sqrt{\dfrac{|R_n| - R_1}{|R_n|LC}}$ $|R_n| > R_1$; $R_1 C = \dfrac{L}{|R_n|}$

(b) $\omega_0 = \sqrt{1/LC}$, $R_2 = |R_n|$

(7-1) $\omega_{3\text{ dB}} = 2.17 \times 10^6$ rad/sec, $\omega_0 = 1.41 \times 10^7$ rad/sec, $A_v(j\omega_0) = -3.70$
(7-4) $R_1 = 500 \text{ }\Omega$, $C_a = 0.039 \text{ }\mu\text{F}$, $L = 2.60 \text{ }\mu\text{H}$
(7-6) $\omega_{3\text{ dB}} = 7.9 \times 10^3$ rad/sec
(7-9) (a) $A_v(j\omega_0) = 25$, $\omega_0 = 3.95 \times 10^8$ rad/sec, $\omega_{3\text{ dB}} = 10^7$ rad/sec
(b) $G_f = 10^{-3}$ mho and 5×10^{-4} mho
(7-11) $C_1 = C_2 = 50.7$ pF, $n_1 = 6.97$, $n_2 = 6.36$

(7-13) $R_L = 633\ \Omega$, $C_a = 1210$ pF, $C_2 = 161$ pF, $L_1 = 0.197\ \mu$H
$L_2 = 1.58\ \mu$H, $R \gg h_{ib}$ (arbitrary)

(7-17) $C_{eq} = C_1$, $R_{eq} = \left(\dfrac{C_2}{C_1}\right)^2 R_L + \dfrac{1}{\omega_0^2 C_1^2 R_L}$,

$C_2 = [C_1^2 R_{eq}/R_L - 1/\omega_0^2 R_L^2]^{1/2}$, $C_1 = 5.07$ pF,
$C_2 = 49$ pF, $|A_v(j\omega_0)| = 71.59$

(7-21) $C_1 = C_2 = 0.226\ \mu$F, $K = 4.717$,
$K = 5$ (Unstable for $K \geq 5$)

(7-25) $S_G^Q = -1$, $S_L^Q = -1/2$, $S_C^Q = 1/2$

(7-33) $n = 3.52$

(8-1) (a) Yes, (b) No, (c) No, (d) No

(8-10) $K = 200$, $R_2 = 1$ kΩ, $C_1 = 0.16\ \mu$F, $C_2 = 0.08\ \mu$F

(8-17) $C_1 = C_2 = 1.59\ \mu$F

(8-24) $(RC)_1 = 2.46\ \mu$sec, $(RC)_2 = 13.16\ \mu$sec, $A_2 = 0.759$

(8-26) $(RC)_1 = 2.25\ \mu$sec, $A_1 = 0.807$, $(RC)_2 = 1.61\ \mu$sec,
$A_2 = 0.544$

(8-28) $R_{eq} = 1$ kΩ, $L_{eq} = 99$ mH, $Q = 9.9$

(8-32) (a) $K_1 = 1.996 \times 10^{-3}$, $K_2 = 0.998$, $K_3 = 1$, $K_4 = 1$
(b) $\omega_0 = 6.289 \times 10^4$ rad/sec, $Q = 500.5$
(c) $\hat{\omega}_0 = 6.280 \times 10^4$ rad/sec, $\hat{Q} = 285.7$
For $\hat{Q} = 500$, $R_s = 149\ \Omega$

(9-18) (a) astable, (b) monostable, (c) bistable

(9-31) $\dot{x}_1 = x_2$, $\dot{x}_2 = -(x_1^2 - 2)x_2^2 + x_1 x_2 + x_1^2$

(9-33) $C_1 = C_o/(V_o + V_d)^{1/n}$, $\delta = V_1/n(V_o + V_d)$

(10-1) $t_d = 2.03$ nsec, $t_r = 1.02$ nsec, $t_{on} = 3.05$ nsec,
$t_s = 14.1$ nsec, $t_f = 2.02$ nsec, $t_{off} = 16.12$ nsec

(10-3) $\beta_0 > \dfrac{R_C + R}{R_C}$ for $R_C \gg r_e$, $R_b \gg R_i$

(10-11) $R_b C = 7.25 \times 10^{-7}$ sec, e.g., $R_b = 5$ kΩ, $C = 145$ pF

(10-13) For Q_1 in saturation: $I_{c1} = 3.63$ mA, $I_{b1} = 0.058$ mA,
$V_E = 2$ V, $V_{c1} = 2.2$ V, $V_{b1} = 2.7$ V, $V_{b2} = 0.73$ V,
$V_{c2} = 9.34$ V, and $\beta_{0\,min} = 66$

(10-16) $T_p = (\beta_0 - n)nL_m/R_B$

(11-1) (a) (10100001110) (b) (101110111)
(c) (10011110.101000111) (d) (11001100.01110)

(11-11) (a) $\bar{A} \cdot B + A \cdot C$, (b) $A \cdot (\bar{C} + B)$, (c) $A \cdot (B + \bar{C})$

(11-17) $v_o(1) = +12$ V, $v_o(0) = 7.3$ V

(11-23) $V_A = 0.7$ V, $V_B = 0.9$ V, $V_C = 1.1$ V
output: $v_o(1) = V_{CC}$, $v_o(0) = 0.6$ V
input: $v_{in}(1) = 1.8$ V, $v_{in}(0) = 0$ V

(11-26) $C = A \cdot B$, $S = \bar{A} \cdot B + A \cdot \bar{B} = A \oplus B$
$C_o = A \cdot B + C(\bar{A} \cdot B + A \cdot \bar{B})$
$S_o = \bar{A} \cdot \bar{B} \cdot C + A \cdot B \cdot C + \bar{A} \cdot B \cdot \bar{C} + A \cdot \bar{B} \cdot \bar{C}$

(11-27) $f = 1/6t_d$, if $N = 4$ the system will not oscillate, $T = 2Nt_d$ for N odd, N must be odd

INDEX

a circuit, 333, 335, 339
A matrix, 23
$ABCD$ parameter, 67
Abrupt junction, 656
Ac load line, 167
Acceptor, 651
Active, 6, 79
Active crystal filter, 444
Active-distributed RC, 499
   design, 502
   phase error, 506
   transient, 511
Active RC synthesis, 461–513
Active RC tuned-amplifier, 444
Active region of operation, 128
Active region model, 130
Admittance parameters, 61
Alignability, 431
Alpha cutoff frequency, ω_α, 139
α definition, 122
   single pole approximation, 139
   transcendental function, 139
Amplification factor, 150, 159
Amplifiers, 183, 249
   audio, 303
   bandpass, 399, 446
   cascaded, 251
   cascode, 271, 423
   emitter-coupled, 320, 457
   high-frequency response of, 206
   feedback, 327
   low frequency response of, 200
   multistage, 249
   narrow-band, 412, 437
   operational, 376, 487
   power, 302
   single stage, 182
   tuned (*see* Bandpass above)
   video, 273, 276
Amplitude modulated, 400
Analog computation, 380
Analysis
   loop, 19
   nodal, 16
   state-variable, 22
AND gate, 617
Anode, 155
Approximation of magnitude, 98
   asymptotic, 83
   equal-ripple, 103
   maximally flat magnitude, 98
Approximation of phase, linear phase, 94
Astable circuits, 539, 586
Asymptotic plot (*see* Bode plot)
Autotransformer, 408
Available power gain, 78

Band, 649
   conduction, 649
   valence, 649
Bandgap, energy, 649
Bandpass amplifier, 399
   active RC, 477
   double-tuned, 416
   mismatch, 429
   narrow-band approximation, 412
   neutralization, 427
   single-tuned, 401
Bandwidth, 183, 258, 262

Bandwidth, degradation factor, D, 214
Bandwidth shrinkage, 259
Barkhausen condition, 385
Base compensation network, 246, 575
Base-emitter voltage, 163
Base region, 122
Base spreading resistor, 136, 138
Basic feedback equation, 329
Beta cutoff frequency, 141
Biasing
  bipolar transistor, 160
  field effect transistor, 172
  vacuum tubes, 173
Binary conversion, 614
Binary variable, 613
Bipolar transistor, 121
  large signal model, 128
  small signal models, 136, 138, 140
Bisection theorem, 288
Bistable circuit, 539, 579
Blocking oscillator, 588
  astable, 592
  monostable, 589
Bode plot, 85
Boltzmann's constant, 650
Boolean relations, 616
Bootstrap circuit, 599
Breakdown diode (*see* Zener dode)
Breakdown voltage, 145
Break frequency, 88
Breakpoint in piecewise-linear analysis, 521
Bridge rectifier, 177
Broadbanding techniques, 224, 276
Buffer, 191
Butterworth filter, 100
Bypass capacitor, 195

Capacitance, high frequency, 140, 151, 159, 160
  p–n junction, 656
  Miller effect, 209
Capacitor, 3

bypass, 194
speed up, 575, 582
Carrier concentration, 649
  majority, 651
  minority, 651
Carrier signal, 400
Cascaded amplifiers, 251
Cascode amplifier, 271
Cascode tuned-amplifier, 423
Catenary filters, 108
Cathode, 157
Cauer forms, piecewise linear network, 525
Cauer forms, *RC* networks, 465
Center frequency (*see* Resonant frequency)
Channel region in FET, 145, 147
Characteristic equation, 330
Charge control, 567
Chebyshev approximation, 103
Chopper circuit, 607
Circuit elements, 2, 495
Circuit models for bipolar transistor, 128, 132, 140
  for diodes, 115, 117
  for JFET, 150, 151
  for MOSFET, 150, 151
  for vacuum tubes, 159, 160
Clipping and clamping circuits, 602
Closed loop gain, 329
Cofactor, 18
Coincidence, 619
Collector, 122
Collector-base leakage current I_{co}, 122
Collector junction capacitance, 138
Common-base configuration, 187
Common-collector configuration, 191
Common-emitter configuration, 189
  π model, 140
  T model, 138
  U model, 213
Common mode, 286
  rejection ratio, 287
Common-source configuration, 193

Index

Colpitts oscillator, 388
Complementary circuit, 316
Complementary-symmetry MOSFET gate, 634
Complex poles, 92
Computer-aided analysis, 44
Conditional stability, 375
Conductance, 2
Conduction band, 649
Conductivity, 2
Conductor, 649
Constant resistance input for shunt-peaked stage, 233
Contact potential, 655
Continuity equation, 659
Controlled-source, 6
  realizations, 468
CORNAP, 49
Coupling capacitors, 194
Counter, 639
Covalent bond, 649
Cramer's rule, 18
Critically coupled doubled-tuned stage, 418
Crossover distortion, 314
Crystal bandpass filter, 437
Crystal model, 389, 437
Crystal oscillator, 389
Current amplification factor
  common base, 122
  common emitter, 123
Current clamping circuit, 611
Current controlled
  current source, 6, 68
  voltage source, 6, 68
Current feedback, 343
Current flow in a p–n junction, 122
Current gain, 74
Current hogging effect, 626
Current-mode logic (CML), 629
Cutoff frequency, 139, 141, 183, 402
Cutoff region, 128
Cutoff region models, 130

D-factor, 214

Dc load line, 113, 167
Dc and differential amplifiers, 285
Darlington configurations, 269
Data sheets, 686–714
DCTL, 625
Degradation factor D, 214
Delay, propagation, 633
Delay function, 97
Delay time, 91, 567
De Morgan theorem, 618
Dependent sources (*see* Controlled sources)
Depletion mode FET, 148
Depletion region, 655
Depletion width, 656
Design of RC-coupled amplifier, 273
Determinant, 18, 21
Difference amplifier (*see* Differential amplifier)
Differential amplifier, 286
Differential mode, 286
Diffused resistor, 670
Diffusion, 651
Diffusion constants, 654
Diffusion current, 654
Diffusion equation, 660
Diffusion length, 660
Digital circuits, 613
Digital comparator, 646
Diode, breakdown, 119
  four layer, 156
  p–n junction, 111
  theory, 659
Diode analysis, 116, 531
Diode equation, 112, 661
Diode logic (DL), 621
Diode model, 115, 117
Diode-transistor logic (DTL), 626
Direct-coupled transistor logic (DCTL), 625
Discrimination factor (*see* Common mode rejection ratio)
Distortion characterization, 309
Distortion, crossover, 314

Distributed RC circuit, 494, 671
  applications, 498
Dominant pole-zero approximation, 216
Dominant time constant, 262
Donor, 650
Doping, 650
Double-tuned interstage, 416
Drain characteristics of FET, 145, 149
Drift current, 651
Drift velocity, 652
Driving-point functions, 38
  piecewise linear, 525
  RC, 463
DTL, 626
Dynamic models of bipolar transistor, 138, 140
  of FET, 151
  of pentodes, 160
  of triodes, 159
Dynamic plate resistance, r_p, 159

Ebers-Moll equations, 123
Ebers-Moll models, 128, 129
ECAP, 45
ECL, 629
Eigenvalues, 25
Einstein's relations, 654
Elastance, 3
Electric field, built in, 655
Electronic charge, 112
Emitter, 122
Emitter bypass capacitor, 194
Emitter-coupled clipper, 576
Emitter-coupled amplifier, low pass, 320
  bandpass, 457
Emitter-coupled logic (ECL), 629
Emitter-coupled multivibrator, 582, 585
Emitter-follower, 191
Emitter-junction, 122
Emitter resistance, 132
Energy gap, 649

Enhancement mode MOS transistor, 148
Epitaxial, 665
Equal ripple approximation, 103
Equivalent circuits, small signal, bipolar transistor, 138, 140
  diode, 178
  field effect transistors, 151
  vacuum tubes, 159, 160
Equivalent noise resistor, 298, 300
Equivalent noise voltage, 300
Exclusive OR, 619
Extrinsic elements in bipolar transistor, 132

f circuit, 335, 339, 351
f_T, 141
Fabrication of integrated circuits, 663
  transistor, 665
Fall time, 567, 572
Fan in, 624
Fan out, 624
Feedback amplifiers, 327-385
  basic equation, 329
  classification, 330
  closed loop gain, 329
  feedback loop, 328
  negative feedback, 330
  positive feedback, 330
  stability tests, 370
Feedback input impedance, 333, 339, 343
  output impedance, 333, 339, 343
Feedback loops, 682
Feedback oscillators, 384
  crystal, 389
  LC, 388
  RC, 385
FET multivibrators, 607
FET parameters, 151
Field effect transistors, 143
Figure of merit of a gate, 632
Filters, 98, 461
Final value theorem, 11
Flip-flop, 634

Index

JK, 636
RS, 636
T, 636
Forbidden band, 649
Forward active region, 128
Foster forms, piecewise linear networks, 525
Foster forms, RC networks, 464
Four layer diode, 156
Fourier-series, 310
Free running multivibrator, 586
Frequency center, 402
    α cutoff, 139
    β cutoff, 141
    oscillation max, 422
Frequency response, 83, 183
Full adder, 647
Full wave rectifier, 177

g-parameters, 66
Gain, closed loop, 329
    current, 73, 74
    power, 76
    voltage, 73, 74
Gain bandwidth product,
    bandpass amplifier, 418
    low-pass amplifier, 222
Gain crossover, 375
Gain margin, 375
Gate, 144
    AND, 617
    NAND, 617, 628, 633
    NOR, 617, 623, 629, 633
    NOT, 617
    OR, 617
Gate to source capacitance, 151
Generalized two-port parameters, 72
Graded junctions, 657
Graphical analysis, 303
Grid, 157, 160
Gyrator, definition, 5, 485
    realization, 486
    synthesis, 516

h-parameters, 65

Half adder, 646
Half-power frequency (*see* Bandwidth), 183
Half-wave rectifier, 116
Hartley oscillator, 388
High frequency model,
    bipolar transistor, 138, 140
    field effect transistor, 151
    tunnel diode, 121
    vacuum tubes, 159, 160
High frequency parameters,
    bipolar transistors, 143
    field effect transistors, 151
High frequency response,
    of bipolar transistors, 206
    of field effect transistors, 220
High-pass realization, 477
Holes, 649
Hybrid integrated circuits, 664
Hybrid parameters, 65
Hybrid π model, 140

Ideal diodes, 114, 522
Ideal filter characteristics, 461
Ideal gyrator, 5
Ideal transformer, 4
Idealized piecewise-linear building blocks, 323
Idealized p–n junction diode equation, 112, 659
IGFET, 147
Immittance, 469
Impedance parameters, 63
Impedance transformation, 408
Impulse response, 91
Impurities, 651
Incremental models,
    bipolar transistor, 138, 140
    diode, 117
    field effect transistor, 151
    pentode, 160
    triode, 159
Independent sources, 5
Inductor, 3
Inductor simulation, 485, 516, 181
Inherently stable, 79

Index

Inhibitor, 618
Initial conditions, 20
Input admittance, 72
Input impedance, 73
Insertion voltage gain, 75
Instability, feedback amplifiers, 370
   tuned amplifiers, 420
Insulator, 649
Integrated circuits, capacitors, 669
   design guidelines, 671
   diodes, 669
   isolation, 666
   fabrication, 663
   transistor, 665
Interelectrode capacitors, 138, 151, 159
Intrinsic semiconductor, 648
Inverse active region, 128
Inverter, 620
Isocline method, 541
I-V characteristics of bipolar transistors, 125, 127
   diodes, 113
   field effect transistor, 145
   pentode, 160
   silicon controlled rectifier, 155
   triode, 158
   tunnel diode, 120
   unijunction transistor, 153
   Zener diode, 119

JFET, 144
JK flip-flop, 636
$j\omega$-axis, 10, 39
Junction capacitance, 656
Junction diodes, 111, 659
Junction field effect transistors, 143
Junction, p–n, 654
Junction, transistors, 121

Laplace transform, 9
   partial fraction expansion, 10
Large scale integration (LSI), 673
Large signal operation, 302, 566
Lattice network, 104
Level shifting circuit, 323

Lienard's method, 547
Life time, 660
Limit cycle, 549
Linear phase response, 94
Linear system, 8
Linearity, 8
Link, 15
Load line, ac, 167
   dc, 113
Logic circuits, 612
   current-mode, 629
   diode-transistor, 626
   emitter-coupled, 629
   resistor transistor, 623
   transistor-transistor, 627
Logic-level shift, 630
Loop, 15
Loop transmission, 329
Low-frequency, design, 199
   response of bipolar transistor, 194
   response of field effect transistor, 200
Low-pass realization, 445
Low-pass response, 184
Low-pass to bandpass transformation, 412
Low-sensitiviey, high Q circuits, 491, 492

Magnetizing inductance, 4
Magnitude and phase, response, 84
   approximation of α, 219
Main amplifier, 251
Majority carriers, 651
Margin, gain, 375
   phase, 375
Mason's gain formula, 681
Matrix, characteristic equation of, 25
Matrix, conversion table, 69
Maximally flat, delay, 94
   magnitude, 98
Maximum frequency of oscillation, 421
Mean gain-bandwidth product, 282

MECA, 532, 564
Medium scale integration (MSI), 674
Memory element, 634
Metal oxide semiconductor, MOS, 669
Metallization, 666
Microelectronics, 663
Midband gain, 184
Midband properties,
  of bipolar transistors, 187, 191
  of field effect transistors, 192
Miller capacitance, 266
Miller effect, 209
Miller sweep circuit, 597
Miller theorem, 59
Minority carriers, 651
Mismatch design, 429
Mobility, 654
Models (see Equivalent circuits)
Monolithic integrated circuits, 663
Monostable circuits, 539, 584, 589
MOS capacitors, 669
MOS gate circuits, 633
MOSFET, 147
  depletion type, 148
  enhancement type, 148
Multiloop feedback, 680, 683
Multiple emitter, 628
Multistage amplifiers, 249
Multivibrators, 577
  astable, 586
  bistable, 579
  monostable, 584
Mutual inductance, 5

NAND, gate, 617, 627, 633
n-type semiconductor, 651
Narrow-band amplifier, 412
Narrow-band crystal filter, 437
Narrow-band approximation, 412
Natural frequencies, 37, 40
  open circuit, 40
  short circuit, 40
Negative feedback (degenerative), 330
  advantages, 344
  disadvantages, 347
Negative impedance converter, NIC, 7, 478
  realization of INIC, 480
  realization of VNIC, 480
  synthesis, 482
Negative resistance, characteristics, 537
  oscillator, 390
Network, biasing, 166
Network function, 37
Network theorems, 30
  reciprocity, 34
  source transformation, 34
  superposition, 30
  Thevenin and Norton, 30
Network topology, 14
  branch, 14
  graph, 15
Neutralization, 427
Nodal analysis, 16
Noise immunity, 620
Noise, models, 294
  figure, 297
  sources, 292
Nonlinear analysis, 520, 540
Nonlinear capacitance, 557
NOR gate, 617
NOT gate, 617
Normal active region, 128
Normalization, 42
Norton theorem, 30
Numerical method of solution for nonlinear network, 553
Nyquist criterion, 374

Octal number system, 614
Offset voltage, 114
Off state, 130, 566
Ohm's law, 652
One port RC synthesis, 463
On state, 131, 566
Open circuit impedance parameters, 63
Open circuit stable, 488

Index

Open circuit voltage ratio, 66
Operating-point, 538
   stable and unstable, 538
Operational amplifiers, 376
   differential input, 488
   ideal, 487
   models 382
   synthesis, 475, 489
Optimum terminations, in two-port networks, 80
OR gate, 617
Oscillators, linear, 384
   crystal, 389
   LC, 388
   RC, 385
Oscillators, nonlinear, 551
Oscillation in tuned-amplifiers, 420
Output, admittance, 72
   impedance, 73
Overshoot, 91
Oxide layer, 666

p-type semiconductor, 651
Packaged computer program, 44
Packages, integrated circuits, 667
Parameter determination, 143, 151
Parameters, admittance, 61
   conversion chart, 69
   hybrid, 65
   impedance, 63
Parameters, of bipolar transistors, 143
   of field effect transistors, 151
   of pentodes, 160
   of triodes, 159
Parametric amplifiers, 557
Paraphase circuit, 247
Parasitic elements in integrated circuits, 668, 669
Passivation in integrated circuits, 670
Passive components, 2, 5
Pentodes, 159
Period of multivibrator, 587
Phase crossover, 375
Phase margin, 376

Phase-plane, 543
Phase response, 85
Phase-shift oscillator, 385
Photomicrograph of integrated circuits, 664
Piecewise-linear analysis, 531
   assumed state method, 532
   breakpoint method, 533
Piecewise-linear approximation, 521
   building block, 522
Piecewise-linear models, 529
Pinchoff voltage, 145
Plate resistance, 159
p–n junctions, 654
Poles, 39
   dominant, 216
   separation, 198
Polynomial decomposition, 469
Port, 60
Positive (regenerative) feedback, 330
Potential barrier, 655
Potential instability, 79
Power amplifiers, 302
   class A, 303
   class B, 312
   class C, 314
   figure of merit, 306
   push-pull, 312
   transformer coupled, 307
Power gain, definitions, 76
Propagation delay, 633
Proper rational function, 10
Push-pull amplifier, 312

Q of a single-tuned circuit, 403
Quasi-stable state, 584
Quiescent point, Q, 113, 161

Radix, 614
Random thermal motion, 293
RC active tuned amplifier, 444, 492
RC active synthesis, 461, 494
RC-coupled amplifiers, bipolar transistors, 273
   field effect transistors, 252

Index

RC, distributed RC, 494
RC-driving-point synthesis, 463
RC-phase shift oscillator, 385
RC waveform shaping, 600
Real pole shrinkage factor, 259
Reciprocal two-port network, 62, 63
Reciprocity, 655
Recombination, 122
Rectifier circuit, 116, 117
Regenerative circuits, 565
Regions of operation, bipolar transistor, 128
   field effect transistor, 146
Register circuit, 637
Regulation with Zener diode, 119
Relaxation oscillators, 539
Resistance, 2
Resistors, integrated circuits, 670
Resistor-transistor logic (RTL), 622
Resonance curve, 404
Resonator circuit, 402
Response, 8
   zero input, 9
   zero state, 9
Response, frequency, 83
   impulse, 91
   step, 91
   transient, 90
Return difference, 330
Return ratio, 330
Reverse active region, 128
Right half-plane poles, 371
Rise time, 91
$RLCM$ circuits, 5
Root-locus, 361
   rules, 367
Routh-Hurwitz test, 371
RTL, 622
Run-away, thermal, 304

s plane, 10, 39
Sag, 205
   compensation, 244
Saturation current, 566

Saturation region, 128
Schmitt trigger, 562
SCR, 155
Semiconductor, 648
   extrinsic, 647
   intrinsic, 647
   n-type, 651
   p-type, 651
Semilattice, 438
Sensitivity, 345, 449
   center frequency, 452
   Q, 451
   root, 451
Series-shunt feedback pair, 356
Sheet resistance, 670
Shift register, 638
Short-circuit, admittance parameters, 62
   current ratio, 65
   stable, 488, 537
Shot noise, 294
Shrinkage factor, identical real poles, 259
   2-pole Butterworth filters, 285
Shunt-peaked amplifier, 229
Shunt-series feedback pair, 349
Shunt-shunt feedback triple, 358
Signal flowgraph, 675
Silicon controlled rectifiers, 154
Silicon dioxide layer, 666
Single stage amplifier, 238
Single-tuned stage, 401
Sources, controlled, 6
   independent, 5
Source transformation, 35
Space charge layer, 655
Speed up capacitor, 575, 620
Square law behavior of FET, 147
Square wave, 587
Stability, of feedback amplifiers, 370
   of operating point, 165
Stagger-tuned design, 425
Stagger-tuning, 283
State, cutoff, 128, 566
State-transition matrix, 25

Index

State-variable, 22
Static characteristics (*see* V–I characteristics)
Static circuit models, of bipolar transistors, 132, 136
  of field effect transistors, 150
  of vacuum tubes, 159
Step response, 91
  of high-pass circuit, 203
  of low-pass circuit, 235
Storage time, 571
Substrate, 665
Superposition principle, 8, 30
Switching circuits, 565
Switching speed, 566
Symbols of, bipolar transistors, 125
  depletion type MOSFET, 144
  diodes, 111
  enhancement type MOSFET, 148
  four layer diode, 156
  JFET, 144
  pentodes, 160
  silicon controlled rectifier, 155
  triode, 159
  tunnel diode, 120
  unijunction transistor, 153
  Zener diode, 119
Synchronously-tuned designed, 425
Synchronous-tuning, 284, 414
Synthesis of active *RC* functions, 468–494
Synthesis of piecewise-linear functions, 524
  driving-point, 525
  transfer function, 526

Table of parameter conversion, 69
Tapped transformer, 409
Taylor series, 133
Tee-equivalent circuit of bipolar transistors, 132, 138
Temperature dependence, 163
Terminated two-port networks, 71
Thermal motion, 293
Thermal runaway, 304

Thermal voltage, 112
Thevenin theorem, 30
Thomson filters, 97
Three-halves power law, 158
T^2L, 627
Time-base generators (*see* Sweep circuits)
  free running, 595
  triggered, 597
Time constants, 569, 571
Time delay, 567
  fall, 572
  off, 567
  on, 566
  rise, 91, 568
  storage, 571
Time-domain response, 90
Time-invariant system, 8
Time sharing system, 52
Toggle, 636
Topology, network, 14
Totem-pole output, 629
Trajectory, 545
Transconductance, 150, 158
Transducer power gain, 77
Transfer characteristics, 145, 149, 158
Transfer function, 38
  synthesis of linear active RC, 468–494
  synthesis of passive RC, 464
  synthesis of piecewise linear, 524
Transformation, of low-pass to bandpass, 412
  of low-pass to band elimination, 412
  of low-pass to high-pass, 412
Transformer, 4
Transient response, 260
Transient, turn off time, 566
  turn on time, 566
Transistor bipolar, 121
  configurations, 125, 189
Transistor, field effect, 143
  configurations, 193

Index

Transistor, fabrication, 665
Transistor models, bipolars, 128, 138, 140
   field effect transistors, 347
Transistor switching times, 566
   off-time, 566
   on-time, 566
Transistor-transistor logic (T^{2}L), 627
Transmittance, 676
Tree, 16
   normal, 22
Triggering, 583
Triode, 157
Truth table, 618
TTL, 627
Tuned-amplifier, 399–437
Tuned-circuit, single, 401
   double, 416
Tuned-oscillator, 388
Tunnel diode, 120
Turn off overdrive, 572
Turn on overdrive, 570
Twin-tee circuit, 445
Two-port conversion chart, 69
Two-port parameters, 60
   ABCD, $\mathscr{ABCD}$, 67
   g,h, 65
   y,z, 63
Two-port parameters of ideal controlled sources, 68
   ideal gyrator, 70
   ideal negative impedance converter, 70

Uniform distributed RC network, 494, 671
Unijunction transistor, 153
Unilateral approximation, 209
Unilateral model, 213
Unilateral power gain, 105
Unipolar transistor, 143
Unity gain block, 447
Unstable operating point, 538

Vacuum tubes, 157
   diodes, 158
   pentodes, 159
   triodes, 157
Valence band, 649
Van der Pols' equation, 551
Varactor, 657
Variable, binary, 613
   state, 22
Venn diagram, 619
Video amplifier, 273, 277, 349
Voltage controlled, current source, 6
   voltage source, 6
Voltage feedback, 343
Voltage gain, 74
Voltage isolation, 242
Voltage multiplier, 611
Voltage transfer ratio, 38, 468

Worst case design, 47, 170
Wien-bridge oscillator, 387

y-parameters, 62

z-parameters, 63
Zener diode, 118